W9-BJG-142

Doing Bayesian Data Analysis

Doing Bayesian Data Analysis

A Tutorial with R, JAGS, and Stan

EDITION 2

JOHN K. KRUSCHKE
Dept. of Psychological and Brain Sciences
Indiana University, Bloomington

AMSTERDAM • BOSTON • HEIDELBERG • LONDON
NEW YORK • OXFORD • PARIS • SAN DIEGO
SAN FRANCISCO • SINGAPORE • SYDNEY • TOKYO
Academic Press is an imprint of Elsevier

Academic Press is an imprint of Elsevier
32 Jamestown Road, London NWI 7BY, UK
525 B Street, Suite 1800, San Diego, CA 92101-4495, USA
225 Wyman Street, Waltham, MA 02451, USA
The Boulevard, Langford Lane, Kidlington, Oxford OX5 1GB, UK

ISBN: 978-0-12-405888-0

Library of Congress Cataloging-in-Publication Data
Kruschke, John K.
Doing Bayesian data analysis : a tutorial with R, JAGS, and Stan / John K. Kruschke. – 2E [edition].
pages cm
Includes bibliographical references.
ISBN 978-0-12-405888-0
1. Bayesian statistical decision theory. 2. R (Computer program language) I. Title.
QA279.5.K79 2014
519.5′42–dc23

2014011293

British Library Cataloguing in Publication Data
A catalogue record for this book is available from the British Library

For information on all Academic Press publications
visit our website at store.elsevier.com

Printed in the United States of America

Last digit is the print number: 9 8 7 6 5

Dedicated to my mother, Marilyn A. Kruschke,
and to the memory of my father, Earl R. Kruschke,
both of whom brilliantly exemplified and taught sound reasoning.
And, in honor of my father,
who dedicated his first book to his children,
I also dedicate this book to mine:
Claire A. Kruschke and Loren D. Kruschke.

CONTENTS

CHAPTER 1

What's in This Book (Read This First!)

Contents

Oh honey I'm searching for love that is true,
But driving through fog is so dang hard to do.
Please paint me a line on the road to your heart,
I'll rev up my pick up and get a clean start.[1]

1.1. REAL PEOPLE CAN READ THIS BOOK

This book explains how to actually *do* Bayesian data analysis, by real people (like you), for realistic data (like yours). The book starts at the basics, with elementary notions of probability and programming. You do not need to already know statistics and programming. The book progresses to advanced hierarchical models that are used in realistic data analysis. This book is speaking to a person such as a first-year graduate student or advanced undergraduate in the social or biological sciences: Someone who grew up in Lake Wobegon,[2] but who is not the mythical being that has the previous training of a nuclear physicist and then decided to learn about Bayesian statistics. (After the publication of the first edition, one of those mythical beings contacted me about

[1] This chapter provides a road map to the book, which hopes to have you fall in love with Bayesian analysis even if you previously had unhappy relationships with statistics. The poem plays with those ideas.

[2] A popular weekly radio show on National Public Radio, called *A Prairie Home Companion*, features fictional anecdotes about a small town named Lake Wobegon. The stories, written and orated by Garrison Keillor, always end with the phrase, "And that's the news from Lake Wobegon, where all the women are strong, all the men are good looking, and all the children are above average." So, if you grew up there, …

Doing Bayesian Data Analysis, 2nd Edition
http://dx.doi.org/10.1016/B978-0-12-405888-0.00001-5

the book! So, even if you *do* have the previous training of a nuclear physicist, I hope the book speaks to you too.)

Details of prerequisites and the contents of the book are presented below. But first things first: As you may have noticed from the beginning of this chapter, the chapters commence with a stanza of elegant and insightful verse composed by a famous poet. The quatrains[3] are formed of dactylic[4] tetrameter,[5] or, colloquially speaking, "country waltz" meter. The poems regard conceptual themes of the chapter via allusion from immortal human motifs in waltz timing.

> If you do not find them to be all that funny,
> If they leave you wanting back all of your money,
> Well, honey, some waltzing's a small price to pay, for
> All the good learning you'll get if you stay.

1.1.1. Prerequisites

There is no avoiding mathematics when doing data analysis. On the other hand, this book is definitely not a mathematical statistics textbook, in that it does *not* emphasize theorem proving or formal analyses.[6] But I *do* expect that you are coming to this book with a dim knowledge of basic calculus. For example, if you understand expressions like $\int x\,dx = \frac{1}{2}x^2$, you're probably good to go. Notice the previous sentence said "understand" the statement of the integral, not "generate" the statement on your own. When mathematical derivations are helpful for understanding, they will usually be presented with a thorough succession of intermediate steps, so you can actually come away feeling secure and familiar with the trip and destination, rather than just feeling car sick after being thrown blindfolded into the back seat and driven around curves at high speed.

The beginnings of your journey will go more smoothly if you have had some basic experience programming a computer, but previous programming experience is not crucial. A computer program is just a list of commands that the computer can execute.

[3] *quatrain* [noun]: Four lines of verse. Unless it's written "*qua* train," in which case it's a philosopher comparing something to a locomotive.

[4] *dactylic* [adjective]: A metrical foot in poetry comprising one stressed and two unstressed syllables. Not to be confused with a pterodactyl, which was a flying dinosaur, and which probably sounded nothing like a dactyl unless it fell from the sky and bounced twice: THUMP-bump-bump.

[5] *tetrameter* [noun]: A line of verse containing four metrical feet. Not to be confused with a quadraped, which has four feet, but is averse to lines.

[6] The first edition mentioned at this point that "any mathematical statistician would be totally bummed at the informality, dude." The statement was meant to be funny, with the slang self-referentially instantiating informality. Even the oracle of truth, Wikipedia, says that "'Dude' is generally used informally to address someone" (http://en.wikipedia.org/wiki/Dude, retrieved February 02, 2014) and "[slang] lowers, if temporarily, 'the dignity of formal or serious speech or writing'" (http://en.wikipedia.org/wiki/Slang, retrieved February 02, 2014). But some readers were offended by such undignified writing, and therefore the joke is now only available to people who read footnotes.

For example, if you've ever typed an equal sign in an Excel spreadsheet cell, you've written a programming command. If you've ever written a list of commands in Java, C, Python, Basic or any other computer programming language, then you're set. We will be using programming languages called R, JAGS, and Stan, which are free and thoroughly explained in this book.

1.2. WHAT'S IN THIS BOOK

This book has three major parts. The first part covers foundations: The basic ideas of Bayesian reasoning, models, probabilities, and programming in R.

The second part covers all the crucial ideas of modern Bayesian data analysis while using the simplest possible type of data, namely dichotomous data such as agree/disagree, remember/forget, male/female, etc. Because the data are so simplistic, the focus can be on Bayesian techniques. In particular, the modern techniques of "Markov chain Monte Carlo" (MCMC) are explained thoroughly and intuitively. Because the data are kept simple in this part of the book, intuitions about the meaning of hierarchical models can be developed in glorious graphic detail. This second part of the book also explores methods for planning how much data will be needed to achieve a desired degree of precision in the conclusions, broadly known as "power analysis."

The third part of the book applies the Bayesian methods to realistic data. The applications are organized around the type of data being analyzed, and the type of measurements that are used to explain or predict the data. Different types of measurement scales require different types of mathematical models, but otherwise the underlying concepts are always the same. More details of coverage are provided below.

The chapters of the book are designed to be read in order, for a "grand tour" of basic applied Bayesian analysis. Especially through parts one and two, the chapters probably make the most sense if read in order. But shorter routes are possible, as described next.

1.2.1. You're busy. What's the least you can read?

Here is a minimalist excursion through the book:

- Chapter 2: The idea of Bayesian inference and model parameters. This chapter introduces important concepts; don't skip it.
- Chapter 3: The R programming language. Read the sections about installing the software, including the extensive set of programs that accompany this book. The rest can be skimmed and returned to later when needed.
- Chapter 4: Basic ideas of probability. Merely skim this chapter if you have a high probability of already knowing its content.
- Chapter 5: Bayes rule!
- Chapter 6: The simplest formal application of Bayes rule, referenced throughout the remainder of the book.

- Chapter 7: MCMC methods. This chapter explains the computing method that makes contemporary Bayesian applications possible. You don't need to study all the mathematical details, but you should be sure to get the gist of the pictures.
- Chapter 8: The JAGS programming language for implementing MCMC.
- Chapter 16: Bayesian estimation of two groups. All of the foundational concepts from the aforementioned chapters, applied to the case of comparing two groups.

1.2.2. You're *really* busy! Isn't there even less you can read?

If all you want is a conceptual foundation and the fastest possible hands-on experience, and if you have some previous knowledge of classical statistics such as a t test, then I recommend the following. First, read Chapter 2 of this book for the conceptual foundation. Then read the article by Kruschke (2013a), which describes Bayesian estimation of two groups (analogous to a traditional t test). Essentially, you've just leapfrogged to Chapter 16 of this book. For your hands-on experience, the article has accompanying software, and there is a version that has been implemented in JavaScript for use in your web browser without need to install other software. For details, see the Web site http://www.indiana.edu/~kruschke/BEST/.

1.2.3. You want to enjoy the view a little longer. But not *too* much longer

After the minimalist excursion suggested above, if you want to delve into further specific applications, you will need to read these sections:

- Chapter 9: Hierarchical models. Many realistic applications involve hierarchical, or "multilevel," structure. One of the things that makes Bayesian methods so exciting is their seamless applicability to hierarchical models.
- Chapter 15: Overview of the generalized linear model. To know what type of model might apply to your data, you need to know the canonical catalog of conventional models, many of which fall under the umbrella of the generalized linear model.
- Individual chapters from Chapters 16–24. Go to the chapter relevant to the data structure you're interested in (which you'll understand because you previously read Chapter 15).
- Chapter 13: Statistical power analysis and planning of research, from a Bayesian perspective. This chapter is not essential on a first reading, but it's important not to skip forever. After all, failing to plan is planning to fail.
- Section 25.1, which has recommendations for how to report a Bayesian analysis. If you want your research to influence other people, you've got to be able to tell them about it. (Well, I suppose there are other means of persuasion, but you'll have to learn those from other sources.)

1.2.4. If you just gotta reject a null hypothesis...

Traditional statistical methods are often focused on rejecting a null hypothesis, as opposed to estimating magnitudes and their uncertainty. For a Bayesian perspective on null hypotheses, read these chapters:

- Chapter 11: The perils of *p* values in traditional null-hypothesis significance testing.
- Chapter 12: Bayesian approaches to null value assessment.

1.2.5. Where's the equivalent of traditional test X in this book?

Because many readers will be coming to this book after having already been exposed to traditional statistics that emphasize null hypothesis significance testing (NHST), this book provides Bayesian approaches to many of the usual topics in NHST textbooks. Table 1.1 lists various tests covered by standard introductory statistics textbooks, along with the location of their Bayesian analogues in this book.

The array of tests mentioned in Table 1.1 are all cases of what is called the "generalized linear model." For those of you already familiar with that term, you can glance ahead to Table 15.3, p. 444, to see which chapters cover which cases. For those of you not yet familiar with that term, do not worry, because all of Chapter 15 is devoted to introducing and explaining the ideas.

A superficial conclusion from Table 1.1 might be, "Gee, the table shows that traditional statistical tests do something analogous to Bayesian analysis in every case, therefore it's pointless to bother with Bayesian analysis." Such a conclusion would be wrong. First, traditional NHST has deep problems, some of which are discussed in Chapter 11. Second, Bayesian analysis yields richer and more informative inferences than NHST, as will be shown in numerous examples throughout the book.

Table 1.1 Bayesian analogues of null hypothesis significance tests.

Traditional analysis name	Bayesian analogue
Binomial test	Chapters 6–9 and 21
t test	Chapter 16
Simple linear regression	Chapter 17
Multiple linear regression	Chapter 18
One-way ANOVA	Chapter 19
Multifactor ANOVA	Chapter 20
Logistic regression	Chapter 21
Multinomial logistic regression	Chapter 22
Ordinal regression	Chapter 23
Chi-square test (contingency table)	Chapter 24
Power analysis (sample size planning)	Chapter 13

1.3. WHAT'S NEW IN THE SECOND EDITION?

The basic progression of topics remains the same as the first edition, but all the details have been changed, from cover to cover. The book and its programs have been completely rewritten. Here are just a few highlights of the changes:

- There are all new programs in JAGS and Stan. The new programs are designed to be much easier to use than the scripts in the first edition. In particular, there are now compact high-level scripts that make it easy to run the programs on your own data. This new programming was a major undertaking by itself.
- The introductory Chapter 2, regarding the basic ideas of how Bayesian inference re-allocates credibility across possibilities, is completely rewritten and greatly expanded.
- There are completely new chapters on the programming languages R (Chapter 3), JAGS (Chapter 8), and Stan (Chapter 14). The lengthy new chapter on R includes explanations of data files and structures such as lists and data frames, along with several utility functions. (It also has a new poem that I am particularly pleased with.) The new chapter on JAGS includes explanation of the RunJAGS package which executes JAGS on parallel computer cores. The new chapter on Stan provides a novel explanation of the concepts of Hamiltonian Monte Carlo. The chapter on Stan also explains conceptual differences in program flow between Stan and JAGS.
- Chapter 5 on Bayes' rule is greatly revised, with a new emphasis on how Bayes' rule re-allocates credibility across parameter values from prior to posterior. The material on model comparison has been removed from all the early chapters and integrated into a compact presentation in Chapter 10.
- What were two separate chapters on the Metropolis algorithm and Gibbs sampling have been consolidated into a single chapter on MCMC methods (as Chapter 7).
- There is extensive new material on MCMC convergence diagnostics in Chapters 7 and 8. There are explanations of autocorrelation and effective sample size. There is also exploration of the stability of the estimates of the highest density interval (HDI) limits. New computer programs display the diagnostics, as well.
- Chapter 9 on hierarchical models includes extensive new and unique material on the crucial concept of shrinkage, along with new examples.
- All the material on model comparison, which was spread across various chapters in the first edition, in now consolidated into a single focused chapter (Chapter 10) that emphasizes its conceptualization as a case of hierarchical modeling.
- Chapter 11 on null hypothesis significance testing is extensively revised. It has new material for introducing the concept of sampling distribution. It has new illustrations of sampling distributions for various stopping rules, and for multiple tests.
- Chapter 12, regarding Bayesian approaches to null value assessment, has new material about the region of practical equivalence (ROPE), new examples of accepting the

null value by Bayes factors, and new explanation of the Bayes factor in terms of the Savage-Dickey method.

- Chapter 13, regarding statistical power and sample size, has an extensive new section on sequential testing, and recommends making the research goal be precision of estimation instead of rejecting or accepting a particular value.
- Chapter 15, which introduces the generalized linear model, is fully revised, with more complete tables showing combinations of predicted and predictor variable types.
- Chapter 16, regarding estimation of means, now includes extensive discussion of comparing two groups, along with explicit estimation of effect size.
- Chapter 17, regarding regression on a single metric predictor, now includes extensive examples of robust regression in JAGS and Stan. New examples of hierarchical regression, including quadratic trend, graphically illustrate shrinkage in estimates of individual slopes and curvatures. The use of weighted data is also illustrated.
- Chapter 18, on multiple linear regression, includes a new section on Bayesian variable selection, in which various candidate predictors are probabilistically included in the regression model.
- Chapter 19, on one-factor ANOVA-like analysis, has all new examples, including a completely worked out example analogous to analysis of covariance (ANCOVA), and a new example involving heterogeneous variances.
- Chapter 20, on multi-factor ANOVA-like analysis, has all new examples, including a completely worked out example of a split-plot design that involves a combination of a within-subjects factor and a between-subjects factor.
- Chapter 21, on logistic regression, is expanded to include examples of robust logistic regression, and examples with nominal predictors.
- There is a completely new chapter (Chapter 22) on multinomial logistic regression. This chapter fills in a case of the generalized linear model (namely, a nominal predicted variable) that was missing from the first edition.
- Chapter 23, regarding ordinal data, is greatly expanded. New examples illustrate single-group and two-group analyses, and demonstrate how interpretations differ from treating ordinal data as if they were metric.
- There is a new section (25.4) that explains how to model censored data in JAGS.
- Many exercises are new or revised.

Oh, and did I mention that the cover is different? The correspondence of the doggies to Bayes' rule is now made explicit: The folded ears of the posterior doggie are a compromise between the perky ears and floppy ears of the likelihood and prior doggies. The marginal likelihood is not usually computed in MCMC methods, so the doggie in the denominator gets sleepy with nothing much to do. I hope that what's between the covers of this book is as friendly and engaging as the doggies on the cover.

1.4. GIMME FEEDBACK (BE POLITE)

I have worked thousands of hours on this book, and I want to make it better. If you have suggestions regarding any aspect of this book, please do email me: johnkruschke@gmail.com. Let me know if you've spotted egregious errors or innocuous infelicities, typo's or thoughto's. Let me know if you have a suggestion for how to clarify something. Especially let me know if you have a good example that would make things more interesting or relevant. I'm interested in complete raw data from research that is interesting to a broad audience, and which can be used with acknowledgement but without fee. Let me know also if you have more elegant programming code than what I've cobbled together. The outside margins of these pages are intentionally made wide so that you have room to scribble your ridicule and epithets before re-phrasing them into kindly stated suggestions in your email to me. Rhyming couplets are especially appreciated.

Since the publication of the first edition I have received hundreds of email messages from readers. I have replied to many, but I am truly embarrassed by the fact that I have not been able to reply to all of them, and some have gone unacknowledged. If I don't respond to your email in a timely manner, it is likely that your message got buried under the avalanche of subsequent emails. You are welcome to send a follow-up email to try to provoke a response from me. In any case, if your email is lengthy or has attachments or asks about a problem in some complicated analysis you are trying, chances are that I'll react by saying to myself, "That's very interesting, but I'll have to think about it when I have more time and reply later." Then, no matter what time it is, later never comes. Whether I am able to reply or not, I appreciate all your messages.

1.5. THANK YOU!

I would like to thank the many readers who posted reviews and recommendations of the first edition on sites such as Amazon.com, Goodreads.com, blogs, and social networking sites. At the time of this writing, there were 46 reviews on Amazon.com, 6 at Amazon.co.uk (United Kingdom), 2 at Amazon.ca (Canada), and 1 at Amazon.cn (China). There were 4 reviews at Goodreads.com plus many ratings. Numerous people have reviewed or recommended the book on their blogs. Many people have given a "shout out" to the book on social networking sites. I am very grateful to all of you for taking the time to write reviews, and pleased that your reviews have generally been very positive! I hope that this second edition elicits continued impulse to post new reviews about the second edition, as the revisions have been a huge effort to make the book even better.

I also thank the authors of professional reviews of first edition, including Andrews (2011), Barry (2011), Colvin (2013), Ding (2011), Fellingham (2012), Goldstein (2011), Smithson (2011), and Vanpaemel and Tuerlinckx (2012). My apologies to any reviewers whom I have missed; please let me know. I think it is valuable to raise the visibility of Bayesian methods in professional circles, and I am very grateful to all these authors for taking the time and effort to compose reviews.

Several people have written computer programs that extended or improved programs related to the first edition of the book. In particular, the programs for Bayesian estimation of two groups ("BEST"; Kruschke, 2013a, cf. Chapter 16 of this book) were re-packaged in R by Mike Meredith, in JavaScript by Rasmus Bååth, and in Python by Andrew Straw. For links to their work, see http://www.indiana.edu/~kruschke/BEST/. Systems for creating hierarchical diagrams, akin to Figure 9.13 (p. 252), were created by Rasmus Bååth for LibreOffice and R, and by Tinu Schneider for LaTeX and TikZ. For links to their work, see http://doingbayesiandataanalysis.blogspot.com/2013/10/diagrams-for-hierarchical-models-new.html. Thank you all for your extensive efforts and contributions to making Bayesian methods accessible to a wider audience.

Many colleagues have organized workshops or courses for *Doing Bayesian Data Analysis*. A list of workshops appears at https://sites.google.com/site/doingbayesiandataanalysis/. At each of those workshops many more people were involved than I can possibly mention by name here, but they include William Jacoby and Dieter Burrell at the Interuniversity Consortium for Political and Social Research Summer Program at the University of Michigan; William Pridemore, James Russell, James Walker, and Jeff DeWitt at the Indiana University Social Science Research Commons; Hans-Joachim Knopf at the University of St. Gallen Summer School in Empirical Research Methods, Switzerland; Hans Olav Melberg at the University of Oslo, Norway; Mark Nawrot and colleagues at North Dakota State University; Ulf Ahlstrom and colleagues at the Federal Aviation Administration Human Factors Lab, New Jersey; Tim Pleskac and colleagues at Michigan State University; John Curtin at the University of Wisconsin, Madison; Michael Roberts and colleagues including Chester Fornari, Bryan Hanson, and Humberto Barreto at DePauw University; Jan Andrews, Ming-Wen An, and colleagues at Vassar College; Kelly Goedert and colleagues at Seton Hall University; Gregory Francis and Zygmunt Pizlo at Purdue University; Boicho Kokinov and colleagues at the New Bulgarian University, Sofia; Nathalie Rothert and the workshop program committees at the Association for Psychological Science; Duncan Brumby and the tutorials program committees of the Cognitive Science Society; Andrew Delamater at the Eastern Psychological Association (and thanks to James McClelland for introducing my talk there); William Merriman at the Midwestern Psychological Association; Xiangen Hu and Michael Jones at the Society for Computers in Psychology. My apologies to those whom I have inadvertently left off this list. I thank you all for your time and effort. I hope that the workshops and courses genuinely facilitate doing Bayesian data analysis by the attendees.

The book has been used by a number of instructors in their courses, and a few of them have sent me notes of their experiences. In particular, Jeffrey Witmer at Oberlin

College sent extensive comments. Adrian Brasoveanu at the University of California, Santa Cruz, and John Miyamoto at the University of Washington, also conveyed or posted information about their courses. The review by Vanpaemel and Tuerlinckx (2012) reported experiences from classroom use. Thanks to all the instructors who have boldly tried the first edition in their courses. I hope that the second edition is even more useful for the classroom and self study.

Over recent years there have been many students in my classes who have made insightful comments and suggestions. Some of these include Young Ahn, Gregory Cox, Junyi Dai, Josh de Lueew, Andrew Jahn, Arash Khodadadi, and Torrin Liddell. Thanks to Torrin Liddell also for help with checking the proofs of the book. I thank Anne Standish for researching and recommending a discussion forum for the book's blog. I am grateful to the many people who have left thoughtful comments at the blog and discussion forum. Thanks also to several careful readers who reported errors in the first edition. And thanks to Jacob Hodes for being so interested in Bayesian analysis that he would travel to a conference to talk with me about it.

The first edition of this book was 6 years in the making, and many colleagues and students provided helpful comments during that period. The most extensive comments came from Luiz Pessoa, Michael Kalish, Jerome Busemeyer, and Adam Krawitz; thank you all! Particular sections were insightfully improved by helpful comments from Michael Erickson, Robert Nosofsky, Geoff Iverson, and James L. (Jay) McClelland. Various parts of the book benefited indirectly from communications with Woojae Kim, Charles Liu, Eric-Jan Wagenmakers, and Jeffrey Rouder. Pointers to data sets were offered by Teresa Treat and Michael Trosset, among others. Very welcome supportive feedback was provided by Michael D. Lee, and also by Adele Diederich. A Bayesian-supportive working environment was provided by many colleagues including Richard Shiffrin, Jerome Busemeyer, Peter Todd, James Townsend, Robert Nosofsky, and Luiz Pessoa. Other department colleagues have been very supportive of integrating Bayesian statistics into the curriculum, including Linda Smith and Amy Holtzworth-Munroe. Various teaching assistants have provided helpful comments; in particular I especially thank Noah Silbert and Thomas Wisdom for their excellent assistance. As this book has evolved over the years, suggestions have been contributed by numerous students, including Aaron Albin, Thomas Smith, Sean Matthews, Thomas Parr, Kenji Yoshida, Bryan Bergert, and perhaps dozens of others who have contributed insightful questions or comments that helped me tune the presentation in the book.

Gratefully acknowledged are the creators of the software R (R Core Team, 2013), RStudio (RStudio, 2013), JAGS (Plummer, 2003, 2012), RunJAGS (Denwood, 2013), BUGS (Lunn, Thomas, Best, & Spiegelhalter, 2000; A. Thomas, O'Hara, Ligges, & Sturtz, 2006), and Stan (Stan Development Team, 2012). Also gratefully acknowledged are the creators of the typesetting software LaTeX (http://www.latex-project.org/) and MikTeX (http://miktex.org/), the editor TeXmaker (http://www.xm1math.net/

texmaker/), and the drawing application of LibreOffice (http://www.libreoffice.org/), in which this book was composed by the author.

To all the people who have made contributions but whom I have inadvertently forgotten to mention, I extend my apologies and genuine appreciation.

Finally, my deepest gratitude goes to Dr. Rima Hanania, who has been my constant and most esteemed companion throughout all the years of writing this book.

PART I

The Basics: Models, Probability, Bayes' Rule, and R

In this part of the book, the basic ideas of Bayesian analysis are presented with intuitive and simple examples. Ideas of probability are defined, and Bayes' rule is extensively introduced. A chapter on the computer language R explains many of its core functions.

CHAPTER 2

Introduction: Credibility, Models, and Parameters

Contents

I just want someone who I can believe in,
Someone at home who will not leave me grievin'.
Show me a sign that you'll always be true,
and I'll be your model of faith and virtue.[1]

The goal of this chapter is to introduce the conceptual framework of Bayesian data analysis. Bayesian data analysis has two foundational ideas. The first idea is that Bayesian inference is reallocation of credibility across possibilities. The second foundational idea is that the possibilities, over which we allocate credibility, are parameter values in meaningful mathematical models. These two fundamental ideas form the conceptual foundation for every analysis in this book. Simple examples of these ideas are presented in this chapter. The rest of the book *merely* fills in the mathematical and computational details for specific applications of these two ideas. This chapter also explains the basic procedural steps shared by every Bayesian analysis.

[1] This chapter introduces ideas of mathematical models, credibility of parameter values, and the semantics of models. The poem plays with the words "model," "believe," and "true" in an everyday context, and hints that Bayesian methods (personified) may be someone to believe in. (And yes, grammatically, the first line should be "in whom I can believe," but the poem is supposed to be colloquial speech. Besides, the grammatically correct version is iambic not dactylic!)

Doing Bayesian Data Analysis, 2nd Edition
http://dx.doi.org/10.1016/B978-0-12-405888-0.00002-7

2.1. BAYESIAN INFERENCE IS REALLOCATION OF CREDIBILITY ACROSS POSSIBILITIES

Suppose we step outside one morning and notice that the sidewalk is wet, and wonder why. We consider all possible causes of the wetness, including possibilities such as recent rain, recent garden irrigation, a newly erupted underground spring, a broken sewage pipe, a passerby who spilled a drink, and so on. If all we know until this point is that some part of the sidewalk is wet, then all those possibilities will have some prior credibility based on previous knowledge. For example, recent rain may have greater prior probability than a spilled drink from a passerby. Continuing on our outside journey, we look around and collect new observations. If we observe that the sidewalk is wet for as far as we can see, as are the trees and parked cars, then we re-allocate credibility to the hypothetical cause of recent rain. The other possible causes, such as a passerby spilling a drink, would not account for the new observations. On the other hand, if instead we observed that the wetness was localized to a small area, and there was an empty drink cup a few feet away, then we would re-allocate credibility to the spilled-drink hypothesis, even though it had relatively low prior probability. This sort of reallocation of credibility across possibilities is the essence of Bayesian inference.

Another example of Bayesian inference has been immortalized in the words of the fictional detective Sherlock Holmes, who often said to his sidekick, Doctor Watson: "How often have I said to you that when you have eliminated the impossible, whatever remains, however improbable, must be the truth?" (Doyle, 1890, chap. 6) Although this reasoning was not described by Holmes or Watson or Doyle as Bayesian inference, it is. Holmes conceived of a set of possible causes for a crime. Some of the possibilities may have seemed very improbable, *a priori*. Holmes systematically gathered evidence that ruled out a number of the possible causes. If all possible causes but one were eliminated, then (Bayesian) reasoning forced him to conclude that the remaining possible cause was fully credible, even if it seemed improbable at the start.

Figure 2.1 illustrates Holmes' reasoning. For the purposes of illustration, we suppose that there are just four possible causes of the outcome to be explained. We label the causes A, B, C, and D. The heights of the bars in the graphs indicate the credibility of the candidate causes. ("Credibility" is synonymous with "probability"; here I use the everyday term "credibility" but later in the book, when mathematical formalisms are introduced, I will also use the term "probability.") Credibility can range from zero to one. If the credibility of a candidate cause is zero, then the cause is definitely not responsible. If the credibility of a candidate cause is one, then the cause definitely *is* responsible. Because we assume that the candidate causes are mutually exclusive and exhaust all possible causes, the total credibility across causes sums to one.

The upper-left panel of Figure 2.1 shows that the prior credibilities of the four candidate causes are equal, all at 0.25. Unlike the case of the wet sidewalk, in which

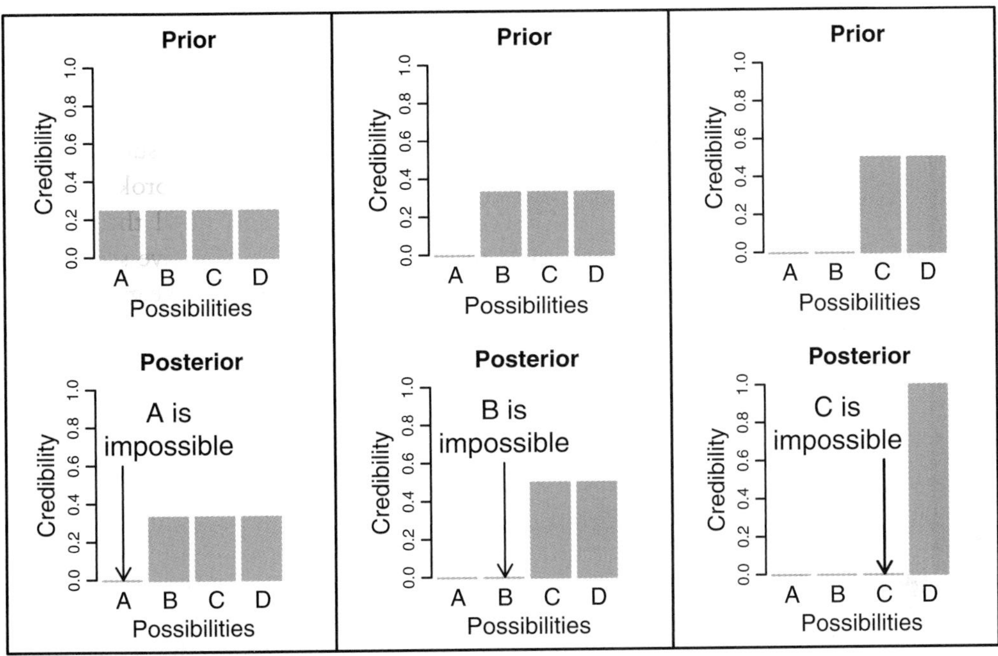

Figure 2.1 The upper-left graph shows the credibilities of the four possible causes for an outcome. The causes, labeled A, B, C, and D, are mutually exclusive and exhaust all possibilities. The causes happen to be equally credible at the outset; hence all have prior credibility of 0.25. The lower-left graph shows the credibilities when one cause is learned to be impossible. The resulting posterior distribution is used as the prior distribution in the middle column, where another cause is learned to be impossible. The posterior distribution from the middle column is used as the prior distribution for the right column. The remaining possible cause is fully implicated by Bayesian reallocation of credibility.

prior knowledge suggested that rain may be a more likely cause than a newly erupted underground spring, the present illustration assumes equal prior credibilities of the candidate causes. Suppose we make new observations that rule out candidate cause A. For example, if A is a suspect in a crime, we may learn that A was far from the crime scene at the time. Therefore, we must re-allocate credibility to the remaining candidate causes, B through D, as shown in the lower-left panel of Figure 2.1. The re-allocated distribution of credibility is called the *posterior distribution* because it is what we believe after taking into account the new observations. The posterior distribution gives zero credibility to cause A, and allocates credibilities of 0.33 (i.e., 1/3) to candidate causes B, C, and D.

The posterior distribution then becomes the prior beliefs for subsequent observations. Thus, the prior distribution in the upper-middle of Figure 2.1 is the posterior distribution from the lower left. Suppose now that additional new evidence rules out candidate cause B. We now must re-allocate credibility to the remaining candidate causes, C and D, as shown in the lower-middle panel of Figure 2.1. This posterior distribution becomes the prior distribution for subsequent data collection, as shown in the upper-right panel of Figure 2.1. Finally, if new data rule out candidate cause C, then all credibility must fall on the remaining cause, D, as shown in the lower-right panel of Figure 2.1, just as Holmes declared. This reallocation of credibility is not only intuitive, it is also what the exact mathematics of Bayesian inference prescribe, as will be explained later in the book.

The complementary form of reasoning is also Bayesian, and can be called judicial *exoneration*. Suppose there are several possible culprits for a crime, and that these suspects are mutually unaffiliated and exhaust all possibilities. If evidence accrues that one suspect is definitely culpable, then the other suspects are exonerated.

This form of exoneration is illustrated in Figure 2.2. The upper panel assumes that there are four possible causes for an outcome, labeled A, B, C, and D. We assume that the causes are mutually exclusive and exhaust all possibilities. In the context of suspects for a crime, the credibility of the hypothesis that suspect A committed the crime is the

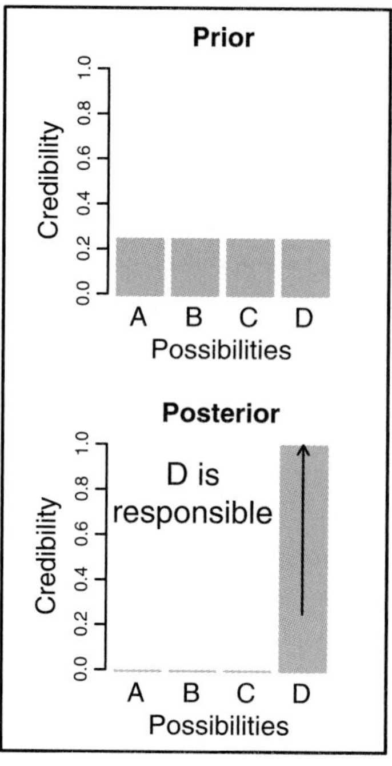

Figure 2.2 The upper graph shows the credibilities of the four possible causes for an outcome. The causes, labeled A, B, C and D, are mutually exclusive and exhaust all possibilities. The causes happen to be equally credible at the outset, hence all have prior credibility of 0.25. The lower graph shows the credibilities when one cause is learned to be responsible. The nonresponsible causes are "exonerated" (i.e., have zero credibility as causes) by Bayesian reallocation of credibility.

culpability of the suspect. So it might be easier in this context to think of culpability instead of credibility. The prior culpabilities of the four suspects are, for this illustration, set to be equal, so the four bars in the upper panel of Figure 2.2 are all of height 0.25. Suppose that new evidence firmly implicates suspect D as the culprit. Because the other suspects are known to be unaffiliated, they are exonerated, as shown in the lower panel of Figure 2.2. As in the situation of Holmesian deduction, this exoneration is not only intuitive, it is also what the exact mathematics of Bayesian inference prescribe, as will be explained later in the book.

2.1.1. Data are noisy and inferences are probabilistic

The cases of Figures 2.1 and 2.2 assumed that observed data had definitive, deterministic relations to the candidate causes. For example, the fictional Sherlock Holmes may have found a footprint at the scene of the crime and identified the size and type of shoe with complete certainty, thereby completely ruling out or implicating a particular candidate suspect. In reality, of course, data have only probabilistic relations to their underlying causes. A real detective might carefully measure the footprint and the details of its tread, but these measurements would only probabilistically narrow down the range of possible shoes that might have produced the print. The measurements are not perfect, and the footprint is only an imperfect representation of the shoe that produced it. The relation between the cause (i.e., the shoe) and the measured effect (i.e., the footprint) is full of random variation.

In scientific research, measurements are replete with randomness. Extraneous influences contaminate the measurements despite tremendous efforts to limit their intrusion. For example, suppose we are interested in testing whether a new drug reduces blood pressure in humans. We randomly assign some people to a test group that takes the drug, and we randomly assign some other people to a control group that takes a placebo. The procedure is "double blind" so that neither the participants nor the administrators know which person received the drug or the placebo (because that information is indicated by a randomly assigned code that is decrypted after the data are collected). We measure the participants' blood pressures at set times each day for several days. As you can imagine, blood pressures for any single person can vary wildly depending on many influences, such as exercise, stress, recently eaten foods, etc. The measurement of blood pressure is itself an uncertain process, as it depends on detecting the sound of blood flow under a pressurized sleeve. Blood pressures are also very different from one person to the next. The resulting data, therefore, are extremely messy, with tremendous variability within each group, and tremendous overlap across groups. Thus, there will be many measured blood pressures in the drug group that are higher than blood pressures in the placebo group, and vice versa. From these two dispersed and overlapping heaps of numbers, we want to infer how big a difference there is between the groups, and how certain we can be about that difference. The problem is that for any particular real difference between the drug and the placebo, its measurable effect is only a random impression.

All scientific data have some degree of "noise" in their values. The techniques of data analysis are designed to infer underlying trends from noisy data. Unlike Sherlock Holmes, who could make an observation and completely rule out some possible causes, we can collect data and only incrementally adjust the credibility of some possible trends. We will see many realistic examples later in the book. The beauty of Bayesian analysis is that the mathematics reveal exactly how much to re-allocate credibility in realistic probabilistic situations.

Here is a simplified illustration of Bayesian inference when data are noisy. Suppose there is a manufacturer of inflated bouncy balls, and the balls are produced in four discrete sizes, namely diameters of 1.0, 2.0, 3.0, and 4.0 (on some scale of distance such as decimeters). The manufacturing process is quite variable, however, because of randomness in degrees of inflation even for a single size ball. Thus, balls of manufactured size 3 might have diameters of 1.8 or 4.2, even though their average diameter is 3.0. Suppose we submit an order to the factory for three balls of size 2. We receive three balls and measure their diameters as best we can, and find that the three balls have diameters of 1.77, 2.23, and 2.70. From those measurements, can we conclude that the factory correctly sent us three balls of size 2, or did the factory send size 3 or size 1 by mistake, or even size 4?

Figure 2.3 shows a Bayesian answer to this question. The upper graph shows the four possible sizes, with blue bars at positions 1, 2, 3, and 4. The prior credibilities of the four sizes are set equal, at heights of 0.25, representing the idea that the factory received the order for three balls, but may have completely lost track of which size was ordered, hence any size is equally possible to have been sent.

At this point, we must specify the form of random variability in ball diameters. For purposes of illustration, we will suppose that ball diameters are centered on their manufactured size, but could be bigger or smaller depending on the amount of inflation. The bell-shaped curves in Figure 2.3 indicate the probability of diameters produced by each size. Thus, the bell-shaped curve centered on size 2 indicates that size-2 balls are usually about 2.0 units in diameter, but could be much bigger or smaller because of randomness in inflation. The horizontal axis in Figure 2.3 is playing double duty as a scale for the ball sizes (i.e., blue bars) and for the measured diameters (suggested by the bell-shaped distributions).

The lower panel of Figure 2.3 shows the three measured diameters plotted as circles on the horizontal axis. You can see that the measured diameters are closest to sizes 2 or 3, but the bell-shaped distributions reveal that even size 1 could sometimes produce balls of those diameters. Intuitively, therefore, we would say that size 2 is most credible, given the data, but size 3 is also somewhat possible, and size 1 is remotely possible, but size 4 is rather unlikely. These intuitions are precisely reflected by Bayesian analysis,

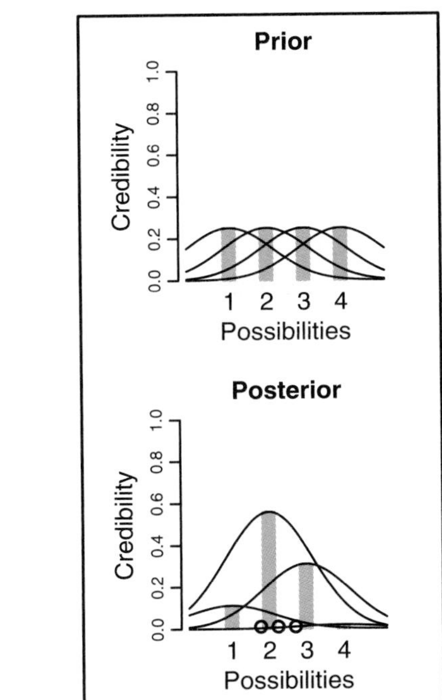

Figure 2.3 The upper graph shows the prior credibilities of the four candidate means in normal distributions, located at values of 1, 2, 3, and 4. Superimposed on the means are the corresponding normal distributions. The horizontal axis is playing double duty as a scale for the means (marked by the blue bars) and for the data (suggested by the normal distributions). The three observed data values are plotted as circles on the floor of the lower panel. Bayesian reallocation of credibility across the four candidate means indicates that the mean at 2 is most credible given the data, the mean at 3 is somewhat credible, and so on.

which is shown in the lower panel of Figure 2.3. The heights of the blue bars show the exact reallocation of credibility across the four candidate sizes. Given the data, there is 56% probability that the balls are size 2, 31% probability that the balls are size 3, 11% probability that the balls are size 1, and only 2% probability that the balls are size 4.

Inferring the underlying manufactured size of the balls from their "noisy" individual diameters is analogous to data analysis in real-world scientific research and applications. The data are noisy indicators of the underlying generator. We hypothesize a range of possible underlying generators, and from the data we infer their relative credibilities.

As another example, consider testing people for illicit drug use. A person is taken at random from a population and given a blood test for an illegal drug. From the result of the test, we infer whether or not the person has used the drug. But, crucially, the test is not perfect, it is noisy. The test has a non-trivial probability of producing false positives and false negatives. And we must also take into account our prior knowledge that the drug is used by only a small proportion of the population. Thus, the set of possibilities has two values: The person uses the drug or does not. The two possibilities have prior credibilities based on previous knowledge of how prevalent drug use is in the population. The noisy datum is the result of the drug test. We then use Bayesian inference to re-allocate credibility across the possibilities. As we will see quantitatively later in the book, the posterior probability of drug use is often surprisingly small even when the test result is positive, because the prior probability of drug use is small and the test is noisy. This is true not only for tests of drug use, but also for tests of diseases such as cancer. A related real-world application of Bayesian inference is detection of spam in email. Automated spam filters often use Bayesian inference to compute a posterior probability that an incoming message is spam.

In summary, the essence of Bayesian inference is reallocation of credibility across possibilities. The distribution of credibility initially reflects prior knowledge about the possibilities, which can be quite vague. Then new data are observed, and the credibility is re-allocated. Possibilities that are consistent with the data garner more credibility, while possibilities that are not consistent with the data lose credibility. Bayesian analysis is the mathematics of re-allocating credibility in a logically coherent and precise way.

2.2. POSSIBILITIES ARE PARAMETER VALUES IN DESCRIPTIVE MODELS

A key step in Bayesian analysis is defining the set of possibilities over which credibility is allocated. This is not a trivial step, because there might always be possibilities beyond the ones we include in the initial set. (For example, the wetness on the sidewalk might have been caused by space aliens who were crying big tears.) But we get the process going by choosing a set of possibilities that covers a range in which we are interested. After the analysis, we can examine whether the data are well described by the most credible possibilities in the considered set. If the data seem not to be well described, then we can consider expanding the set of possibilities. This process is called a posterior predictive check and will be explained later in the book.

Consider again the example of the blood-pressure drug, in which blood pressures are measured in one group that took the drug and in another group that took a placebo. We want to know how much difference there is in the tendencies of the two groups: How big is the difference between the typical blood pressure in one group versus the typical blood pressure in the other group, and how certain can we be of the difference? The magnitude of difference *describes* the data, and *our goal is to assess which possible descriptions are more or less credible.*

In general, data analysis begins with a family of candidate descriptions for the data. The descriptions are mathematical formulas that characterize the trends and spreads in the data. The formulas themselves have numbers, called parameter values, that determine the exact shape of mathematical forms. You can think of *parameters as control knobs on mathematical devices that simulate data generation*. If you change the value of a parameter, it changes a trend in the simulated data, just like if you change the volume control on a music player, it changes the intensity of the sound coming out of the player.

In previous studies of statistics or mathematics, you may have encountered the so-called normal distribution, which is a bell-shaped distribution often used to describe data. It was alluded to above in the example of the inflated bouncy balls (see Figure 2.3). The normal distribution has two parameters, called the mean and standard deviation. The mean is a control knob in the mathematical formula for the normal distribution that controls the location of the distribution's central tendency. The mean is sometimes called a *location parameter*. The standard deviation is another control knob in the mathematical formula for the normal distribution that controls the width or dispersion of the distribution. The standard deviation is sometimes called a *scale parameter*. The mathematical formula for the normal distribution converts the parameter values to a particular bell-like shape for the probabilities of data values.

Figure 2.4 shows some data with candidate normal distributions superimposed. The data are shown as a histogram, which plots vertical bars that have heights indicating

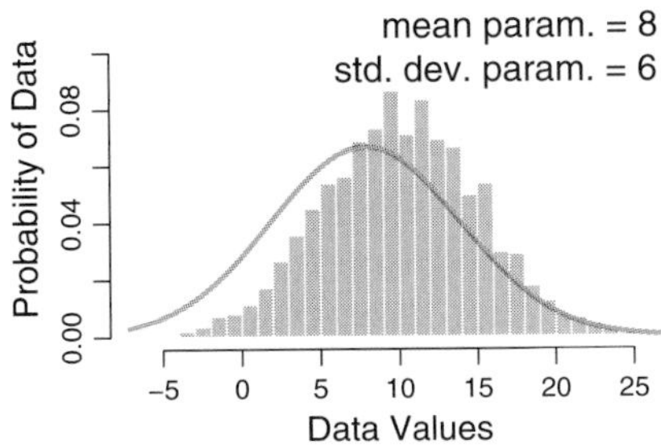

Figure 2.4 The two graphs show the same data histogram but with two different candidate descriptions by normal distributions. Bayesian analysis computes the relative credibilities of candidate parameter values.

how much of the data falls within the small range spanned by the bar. The histogram appears to be roughly unimodal and left-right symmetric. The upper panel superimposes a candidate description of the data in the form of a normal distribution that has a mean of 10 and a standard deviation of 5. This choice of parameter values appears to describe the data well. The lower panel shows another choice of parameter values, with a mean of 8 and a standard deviation of 6. While this candidate description appears to be plausible, it is not as good as the upper panel. The role of Bayesian inference is to compute the exact relative credibilities of candidate parameter values, while also taking into account their prior probabilities.

In realistic applications, the candidate parameter values can form an infinite continuum, not only a few discrete options. The location parameter of the normal distribution can take on any value from negative to positive infinity. Bayesian inference operates without trouble on infinite continuums.

There are two main desiderata for a mathematical description of data. First, the mathematical form should be comprehensible with meaningful parameters. Just as it would be fruitless to describe the data in a language that we do not know, it would be fruitless to describe the data with a mathematical form that we do not understand, with parameters that we cannot interpret. In the case of a normal distribution, for example, the mean parameter and standard-deviation parameter are directly meaningful as the location and scale of the distribution. Throughout this book, we will use mathematical descriptions that have meaningful parameters. Thus, Bayesian analysis reallocates credibility among parameter values within a meaningful space of possibilities defined by the chosen model.

The second desideratum for a mathematical description is that it should be descriptively adequate, which means, loosely, that the mathematical form should "look like" the data. There should not be any important systematic discrepancies between trends in the data and the form of the model. Deciding whether or not an apparent discrepancy is important or systematic is not a definite process. In early stages of research, we might be satisfied with a rough, "good enough" description of data, because it captures meaningful trends that are interesting and novel relative to previous knowledge. As the field of research matures, we might demand more and more accurate descriptions of data. Bayesian analysis is very useful for assessing the relative credibility of different candidate descriptions of data.

It is also important to understand that mathematical descriptions of data are not necessarily causal explanations of data. To say that the data in Figure 2.4 are well described by a normal distribution with mean of 10 and standard deviation of 5 does not explain what process caused the data to have that form. The parameters are "meaningful" only in the context of the familiar mathematical form defined by the normal distribution; the parameter values have no necessary meaning with respect to causes in the world. In some applications, the mathematical model might be motivated as a description of a natural process that generated the data, and thereby the parameters and mathematical form can refer to posited states and processes in the world. For example, in the case of the inflated bouncy balls (Figure 2.3), the candidate parameter values were interpreted as "sizes" at the manufacturer, and the underlying size, combined with random inflation, caused the observed data value. But reference to physical states or processes is not necessary for merely describing the trends in a sample of data. In this book, we will be focusing on generic data description using intuitively accessible model forms that are broadly applicable across many domains.

2.3. THE STEPS OF BAYESIAN DATA ANALYSIS

In general, Bayesian analysis of data follows these steps:

1. Identify the data relevant to the research questions. What are the measurement scales of the data? Which data variables are to be predicted, and which data variables are supposed to act as predictors?
2. Define a descriptive model for the relevant data. The mathematical form and its parameters should be meaningful and appropriate to the theoretical purposes of the analysis.
3. Specify a prior distribution on the parameters. The prior must pass muster with the audience of the analysis, such as skeptical scientists.
4. Use Bayesian inference to re-allocate credibility across parameter values. Interpret the posterior distribution with respect to theoretically meaningful issues (assuming that the model is a reasonable description of the data; see next step).
5. Check that the posterior predictions mimic the data with reasonable accuracy (i.e., conduct a "posterior predictive check"). If not, then consider a different descriptive model.

Perhaps the best way to explain these steps is with a realistic example of Bayesian data analysis. The discussion that follows is abbreviated for purposes of this introductory chapter, with many technical details suppressed. For this example, suppose we are interested in the relationship between weight and height of people. We suspect from everyday experience that taller people tend to weigh more than shorter people, but we would like to know by how much people's weights tend to increase when height increases, and how certain we can be about the magnitude of the increase. In particular, we might be interested in predicting a person's weight based on their height.

The first step is identifying the relevant data. Suppose we have been able to collect heights and weights from 57 mature adults sampled at random from a population of interest. Heights are measured on the continuous scale of inches, and weights are measured on the continuous scale of pounds. We wish to predict weight from height. A scatter plot of the data is shown in Figure 2.5.

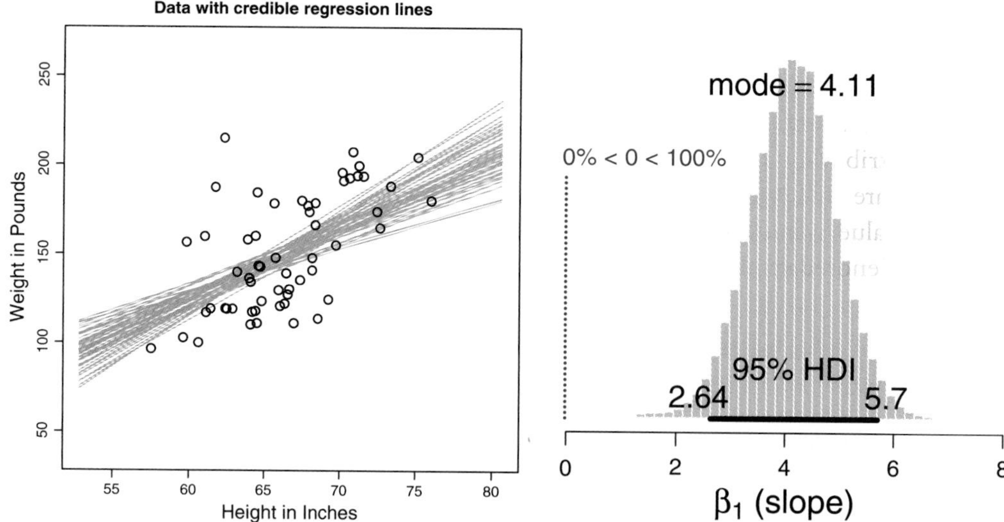

Figure 2.5 Data are plotted as circles in the scatter plot of the left panel. The left panel also shows a smattering of credible regression lines from the posterior distribution superimposed on the data. The right panel shows the posterior distribution of the slope parameter (i.e., β_1 in Equation 2.1).

The second step is to define a descriptive model of the data that is meaningful for our research interest. At this point, we are interested merely in identifying a basic trend between weight and height, and it is not absurd to think that weight might be proportional to height, at least as an approximation over the range of adult weights and heights. Therefore, we will describe predicted weight as a multiplier times height plus a baseline. We will denote the predicted weight as $\hat{y}$ (spoken "y hat"), and we will denote the height as x. Then the idea that predicted weight is a multiple of height plus a baseline can be denoted mathematically as follows:

$$\hat{y} = \beta_1 x + \beta_0 \tag{2.1}$$

The coefficient, β_1 (Greek letter "beta"), indicates how much the predicted weight increases when the height goes up by one inch.[2] The baseline is denoted β_0 in Equation 2.1, and its value represents the weight of a person who is zero inches tall. You might suppose that the baseline value should be zero, *a priori*, but this need not be the case for describing the relation between weight and height of mature adults, who have a limited range of height values far above zero. Equation 2.1 is the form of a line,

[2] Here is a proof that β_1 indicates how much that $\hat{y}$ goes up when x increases by 1 unit. First, at height x, the predicted weight is $\hat{y}_x = \beta_1 x + \beta_0$. Second, at height $x + 1$, the predicted weight is $\hat{y}_{x+1} = \beta_1(x + 1) + \beta_0 = \beta_1 x + \beta_1 + \beta_0$. Therefore, the change in predicted weight is $\hat{y}_{x+1} - \hat{y}_x = \beta_1$.

in which β_1 is the slope and β_0 is the intercept, and this model of trend is often called linear regression.

The model is not complete yet, because we have to describe the random variation of actual weights around the predicted weight. For simplicity, we will use the conventional normal distribution (explained in detail in Section 4.3.2.2), and assume that actual weights y are distributed randomly according to a normal distribution around the predicted value $\hat{y}$ and with standard deviation denoted σ (Greek letter "sigma"). This relation is denoted symbolically as

$$y \sim \text{normal}(\hat{y}, \sigma) \tag{2.2}$$

where the symbol "$\sim$" means "is distributed as." Equation 2.2 is saying that y values near $\hat{y}$ are most probable, and y values higher or lower than $\hat{y}$ are less probable. The decrease in probability around $\hat{y}$ is governed by the shape of the normal distribution with width specified by the standard deviation σ.

The full model, combining Equations 2.1 and 2.2, has three parameters altogether: the slope, β_1, the intercept, β_0, and the standard deviation of the "noise," σ. Note that the three parameters are meaningful. In particular, the slope parameter tells us how much the weight tends to increase when height increases by an inch, and the standard deviation parameter tells us how much variability in weight there is around the predicted value. This sort of model, called linear regression, is explained at length in Chapters 15, 17, and 18.

The third step in the analysis is specifying a prior distribution on the parameters. We might be able to inform the prior with previously conducted, and publicly verifiable, research on weights and heights of the target population. Or we might be able to argue for a modestly informed prior based on consensual experience of social interactions. But for purposes of this example, I will use a noncommittal and vague prior that places virtually equal prior credibility across a vast range of possible values for the slope and intercept, both centered at zero. I will also place a vague and noncommittal prior on the noise (standard deviation) parameter, specifically a uniform distribution that extends from zero to a huge value. This choice of prior distribution implies that it has virtually no biasing influence on the resulting posterior distribution.

The fourth step is interpreting the posterior distribution. Bayesian inference has re-allocated credibility across parameter values, from the vague prior distribution, to values that are consistent with the data. The posterior distribution indicates combinations of β_0, β_1, and σ that together are credible, given the data. The right panel of Figure 2.5 shows the posterior distribution on the slope parameter, β_1 (collapsing across the other two parameters). It is important to understand that Figure 2.5 shows a distribution of parameter values, not a distribution of data. The blue bars of Figure 2.5 indicate the credibility across the *continuum* of candidate slope values, analogous to the blue bars in the examples of Sherlock Holmes, exoneration, and discrete candidate means (in Figures 2.1–2.3). The posterior distribution in Figure 2.5 indicates that the most credible value of the slope is about 4.1, which means that weight increases about 4.1 pounds for every 1-inch increase in height. The posterior distribution also shows the uncertainty in that estimated slope, because the distribution shows the relative credibility of values across the continuum. One way to summarize the uncertainty is by marking the span of values that are most credible and cover 95% of the distribution. This is called the *highest density interval* (HDI) and is marked by the black bar on the floor of the distribution in Figure 2.5. Values within the 95% HDI are more credible (i.e., have higher probability "density") than values outside the HDI, and the values inside the HDI have a total probability of 95%. Given the 57 data points, the 95% HDI goes from a slope of about 2.6 pounds per inch to a slope of about 5.7 pounds per inch. With more data, the estimate of the slope would be more precise, meaning that the HDI would be narrower.

Figure 2.5 also shows where a slope of zero falls relative to the posterior distribution. In this case, zero falls far outside any credible value for the slope, and therefore we could decide to "reject" zero slope as a plausible description of the relation between height and weight. But this discrete decision about the status of zero is separate from the Bayesian analysis *per se*, which provides the complete posterior distribution.

Many readers may have previously learned about null hypothesis significance testing (NHST) which involves *sampling distributions* of summary statistics such as t, from which are computed p values. (If you do not know these terms, do not worry. NHST will be discussed in Chapter 11.) It is important to understand that the posterior distribution in Figure 2.5 is *not* a sampling distribution and has nothing to do with p values.

Another useful way of understanding the posterior distribution is by plotting examples of credible regression lines through the scatter plot of the data. The left panel of Figure 2.5 shows a random smattering of credible regression lines from the posterior distribution. Each line plots $\hat{y} = \beta_1 x + \beta_0$ for credible combinations of β_1 and β_0. The bundle of lines shows a range of credible possibilities, given the data, instead of plotting only a single "best" line.

The fifth step is to check that the model, with its most credible parameter values, actually mimics the data reasonably well. This is called a "posterior predictive check." There is no single, unique way to ascertain whether the model predictions systematically and meaningfully deviate from the data, because there are innumerable ways in which to define systematic deviation. One approach is to plot a summary of predicted data from the model against the actual data. We take credible values of the parameters, β_1, β_0, and σ, plug them into the model Equations 2.1 and 2.2, and randomly generate simulated y values (weights) at selected x values (heights). We do that for many, many

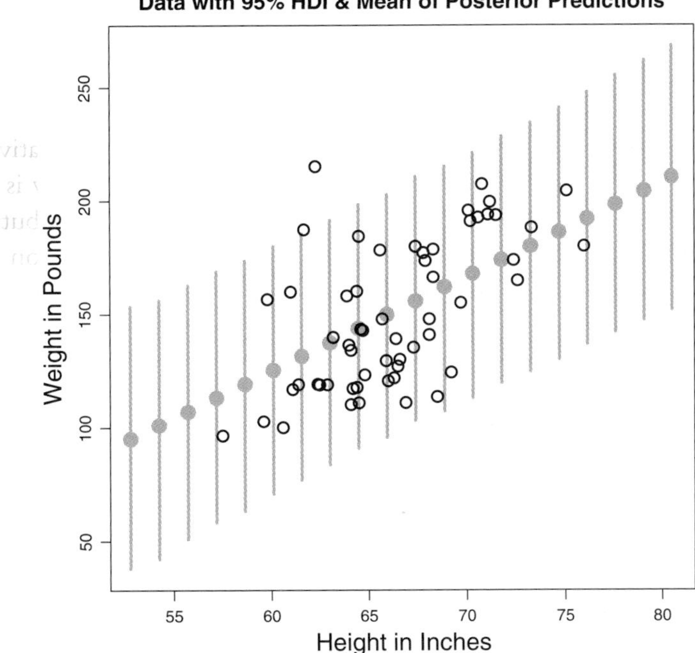

Figure 2.6 The data of Figure 2.5 are shown with posterior predicted weight values superimposed at selected height values. Each vertical bar shows the range of the 95% most credible predicted weight values, and the dot at the middle of each bar shows the mean predicted weight value.

credible parameter values to create representative distributions of what data would look like according to the model. The results of this simulation are shown in Figure 2.6. The predicted weight values are summarized by vertical bars that show the range of the 95% most credible predicted weight values. The dot at the middle of each bar shows the mean of the predicted weight values. By visual inspection of the graph, we can see that the actual data appear to be well described by the predicted data. The actual data do not appear to deviate systematically from the trend or band predicted from the model.

If the actual data did appear to deviate systematically from the predicted form, then we could contemplate alternative descriptive models. For example, the actual data might appear to have a nonlinear trend. In that case, we could expand the model to include nonlinear trends. It is straightforward to do this in Bayesian software, and easy to estimate the parameters that describe nonlinear trends. We could also examine the distributional properties of the data. For example, if the data appear to have outliers relative to what is predicted by a normal distribution, we could change the model to use a heavy-tailed distribution, which again is straightforward in Bayesian software.

We have seen the five steps of Bayesian analysis in a fairly realistic example. This book explains how to do this sort of analysis for many different applications and types of descriptive models. For a shorter but detailed introduction to Bayesian analysis for comparing two groups, with explanation of the perils of the classical *t* test, see the article by Kruschke (2013a). For an introduction to Bayesian analysis applied to multiple linear regression, see the article by Kruschke, Aguinis, and Joo (2012). For a perspective on posterior predictive checks, see the article by Kruschke (2013b) and Section 17.5.1 (among others) of this book.

2.3.1. Data analysis without parametric models?

As outlined above, Bayesian data analysis is based on meaningfully parameterized descriptive models. Are there ever situations in which such models cannot be used or are not wanted?

One situation in which it might appear that parameterized models are not used is with so-called *non*parametric models. But these models are confusingly named because they actually do have parameters; in fact they have a potentially infinite number of parameters. As a simple example, suppose we want to describe the weights of dogs. We measure the weights of many different dogs sampled at random from the entire spectrum of dog breeds. The weights are probably not distributed unimodally, instead there are probably subclusters of weights for different breeds of dogs. But some different breeds might have nearly identical distributions of weights, and there are many dogs that cannot be identified as a particular breed, and, as we gather data from more and more dogs, we might encounter members of new subclusters that had not yet been included in the previously collected data. Thus, it is not clear how many clusters we should include in the descriptive model. Instead, we infer, from the data, the relative credibilities of different clusterings. Because each cluster has its own parameters (such as location and scale parameters), the number of parameters in the model is inferred, and can grow to infinity with infinite data. There are many other kinds of infinitely parameterized models. For a tutorial on Bayesian nonparametric models, see Gershman and Blei (2012); for a recent review, see Müller and Mitra (2013); and for textbook applications, see Gelman et al. (2013). We will not be considering Bayesian nonparametric models in this book.

There are a variety of situations in which it might seem at first that no parameterized model would apply, such as figuring out the probability that a person has some rare disease if a diagnostic test for the disease is positive. But Bayesian analysis does apply even here, although the parameters refer to discrete states instead of continuous distributions. In the case of disease diagnosis, the parameter is the underlying health status of the individual, and the parameter can have one of two values, either "has disease" or "does not have disease." Bayesian analysis re-allocates credibility over those two parameter values based on the observed test result. This is exactly analogous to the discrete possibilities considered by Sherlock Holmes in Figure 2.1, except that the test results yield probabilistic information instead of perfectly conclusive information. We will do exact Bayesian computations for this sort of situation in Chapter 5 (see specifically Table 5.4).

Finally, there might be some situations in which the analyst is loathe to commit to any parameterized model of the data, even tremendously flexible infinitely parameterized models. If this is the case, then Bayesian methods cannot apply. These situations are rare, however, because mathematical models are enormously useful tools. One case of trying to make inferences from data without using a model is a method from NHST called *resampling* or *bootstrapping*. These methods compute p values to make decisions, and p values have fundamental logical problems that will be discussed in Chapter 11. These methods also have very limited ability to express degrees of certainty about characteristics of the data, whereas Bayesian methods put expression of uncertainty front and center.

2.4. EXERCISES

Look for more exercises at https://sites.google.com/site/doingbayesiandataanalysis/

Exercise 2.1. [Purpose: To get you actively manipulating mathematical models of probabilities.] Suppose we have a four-sided die from a board game. On a tetrahedral die, each face is an equilateral triangle. When you roll the die, it lands with one face down and the other three faces visible as a three-sided pyramid. The faces are numbered 1-4, with the value of the bottom face printed (as clustered dots) at the bottom edges of all three visible faces. Denote the value of the bottom face as x. Consider the following three mathematical descriptions of the probabilities of x. Model A: $p(x) = 1/4$. Model B: $p(x) = x/10$. Model C: $p(x) = 12/(25x)$. For each model, determine the value of $p(x)$ for each value of x. Describe in words what kind of bias (or lack of bias) is expressed by each model.

Exercise 2.2. [Purpose: To get you actively thinking about how data cause credibilities to shift.] Suppose we have the tetrahedral die introduced in the previous exercise, along with the three candidate models of the die's probabilities. Suppose that initially, we are not sure what to believe about the die. On the one hand, the die might be fair, with each face landing with the same probability. On the other hand, the die might be biased, with the faces that have more dots landing down more often (because the dots are created by embedding heavy jewels in the die, so that the sides with more dots are more likely to land on the bottom). On yet another hand, the die might be biased such that more dots on a face make it less likely to land down (because maybe the dots are bouncy rubber or protrude from the surface). So, initially, our beliefs about the three models can be described as $p(A) = p(B) = p(C) = 1/3$. Now we roll the die 100 times and find these results: $\#1\text{'}s = 25$, $\#2\text{'}s = 25$, $\#3\text{'}s = 25$, $\#4\text{'}s = 25$. Do these data change our beliefs about the models? Which model now seems most likely? Suppose when we rolled the die 100 times we found these results: $\#1\text{'}s = 48$, $\#2\text{'}s = 24$, $\#3\text{'}s = 16$, $\#4\text{'}s = 12$. Now which model seems most likely?

CHAPTER 3

The R Programming Language

Contents

Doing Bayesian Data Analysis, 2nd Edition
http://dx.doi.org/10.1016/B978-0-12-405888-0.00003-9

You said, dear Descartes, that "je pense, donc je suis,"
Deriving existence from uncertainty.
Now, you are gone, and we say, "au revoir,"
Doubtless we think, René, therefore we R.[1]

In this book, you will learn how to actually *do* Bayesian data analysis. For any but the simplest models, that means using a computer. Because the computer results are so central to doing real Bayesian data analysis, examples of using the R computer programming language will be integrated into the simplest "toy" problems, so that R will not be an extra hurdle later.

The material in this chapter is rather dull reading because it basically amounts to a list (although a carefully scaffolded list) of basic commands in R along with illustrative examples. After reading the first few pages and nodding off, you may be tempted to skip ahead, and I wouldn't blame you. But much of the material in this chapter is crucial, and all of it will eventually be useful, so you should at least skim it all so you know where to return when the topics arise later.

Here are a few of the essential points of this chapter:

- **Software installation.** Section 3.1 describes how to install the R programming language. Section 3.1.1 describes how to install the R editor called RStudio. Section 3.2.1 describes how to install the programs written for this book.
- **Data formats.** Section 3.5.1 describes how to read data files into R. To understand the resulting format, you'll need to understand the `data.frame` structure from Section 3.4.4, which in turn requires understanding the `list` and `matrix` and `factor` structures (which is why those structures are explained before getting to the `data.frame` structure).
- **Running programs.** Section 3.7.2 explains how to run programs in R, and points out that it is important to set R's working directory to the folder in which the programs reside.

The R programming language is great at doing Bayesian statistics for a number of reasons. First, it's free! You can get it via the web and easily install it on your computer. Second, it's already a popular language for doing Bayesian statistics, so there are lots of resources available. Third, it is a powerful and easy, general-purpose computing language, so you can use it for many other applications too. The Bayesian MCMC packages that we will rely on later can be accessed from other programming environments but we'll use R.

[1] This chapter introduces the programming language R. The poem provides motivation for using R, primarily in the form of an extended setup for the final pun on the word "are." Further background: The French philosopher and mathematician, René Descartes (1596–1650), wondered how he could be certain of anything. The only thing he could be certain of was his own thoughts of uncertainty, and therefore he, as thinker, must exist. In English, the idea is captured by the phrase, "I think therefore I am." Changed to plural, the phrase becomes "we think therefore we are."

If you would like to learn more about the history of R, see its web page at http://www.r-project.org/ which also has extensive documentation about the R language.

3.1. GET THE SOFTWARE

It's easy to get and install R, but there are a lot of optional details in the process, and the hardest part of installation is figuring out which little details do *not* apply to you!

Basic installation of R is easy. Go to http://cran.r-project.org/. (Web site addresses occasionally change. If the address stated here does not work, please search the web for "R language." Be sure the site is legitimate before downloading anything to your computer.) At the top of that web page is a section headed "Download and Install R" followed by three links: Linux, MacOS, and Windows. These three links refer to the type of operating system used on your computer. Get the version that matches your computer's operating system. For example, if you click Windows, the next page that appears has a variety of links. The only one you need to click is "base," which will get you to the most recent version of R for downloading. There are a few subsequent details specific to each operating system that you have to navigate on your own, but remember that centuries ago lots of people crossed various oceans in tiny wooden boats without any electronics, so you can navigate the small perils of R installation.[2]

3.1.1. A look at RStudio

The R programming language comes with its own basic user interface that is adequate for modest applications. But larger applications become unwieldy in the basic R user interface, and therefore it helps to install a more sophisticated R-friendly editor. There are a number of useful editors available, many of which are free, and they are constantly evolving. At the time of this writing, I recommend RStudio, which can be obtained from http://www.rstudio.com/. (Web site addresses occasionally change. If the address stated here does not work, please search the web for "RStudio." Be sure the site is legitimate before downloading anything to your computer.) Just go to the web site, find the download link, and get the version appropriate to your computer operating system.

There are various ways to invoke R. One way is by starting R directly. For example, in Windows, you could just double-click the R icon, or find R in the list of programs from the Start menu. A second way to invoke R is via your favorite R editor. For example, invoke RStudio, which then automatically and transparently communicates with R. A third way to invoke R, and *the best way for our purposes*, is to associate files that have a ".R" extension on their filename with RStudio. Then, when the file is opened, your computer knows to invoke RStudio and open the file in RStudio. An example of doing this will be provided soon.

[2] Of course, lots of people failed to cross the ocean, but that's different.

3.2. A SIMPLE EXAMPLE OF R IN ACTION

Whether you are using the basic R interface or an editor such as RStudio, the primary window is a command-line interface. This window is constantly attentive to your every whim (well, every whim you can express in R). All you have to do is type in your wish and R will execute it as a command. The command line in R is marked by a prompt in the form of a greater-than sign: ">." For example, if we want to know what $2 + 3$ is, we can type it into R and R will provide the answer as text on a subsequent displayed line, like this:

```
> 2+3
[1] 5
```

Again, the ">" symbol above indicates the R command prompt, at which we typed "`2+3`." The next line above, "`[1] 5`," shows R's reply. The answer, of course, is `5`, but the line begins with a bracketed "`[1]`" to indicate that the first component of the answer is `5`. As we will see later, variables in R are often multi-component structures, and therefore it can be informative to the user to display which component is being shown.

Throughout this book, programming commands are typeset in a distinctive font, `like this`, to distinguish them from English prose and to help demarcate the scope of the programming command when it is embedded in an English sentence.

A *program*, also known as a *script*, is just a list of commands that R executes. For example, you could first type in `x=2` and then, as a second command, type in `x+x`, to which R will reply `4`. This is because R assumes that when you type in `x+x`, you are really asking for the value of the sum of the value of `x` with the value of `x`, and you are not asking for an algebraic reformulation such as `2*x`. We could save the list of commands as a program, or script, in a simple text file. Then we could run the program and R would step through the list of commands, in order. In R, running a script is called `source`-ing the script, because you are redirecting the source of commands from the command line to the script. To run a program stored in a file named `MyProgram.R`, we could type "`source("MyProgram.R")`" at R's command line, or click the "source" button in the R or RStudio menu. The example in the next section should help clarify.

As a simple example of what R can do, let's plot a quadratic function: $y = x^2$. What looks like a smooth curve on a graph is actually a set of points connected by straight lines, but the lines are so small that the graph looks like a smooth curve. We must tell R where all those densely packed points should be positioned.

Every point is specified by its x and y coordinates, so we have to provide R with a list of x values and a list of corresponding y values. Let's arbitrarily select x values from -2 to $+2$, separated by intervals of 0.1. We have R set up the list of x values by using the built-in *sequence* function: `x = seq( from = -2, to = 2, by = 0.1 )`. Inside R, the variable `x` now refers to a list of 31 values: -2.0, -1.9,

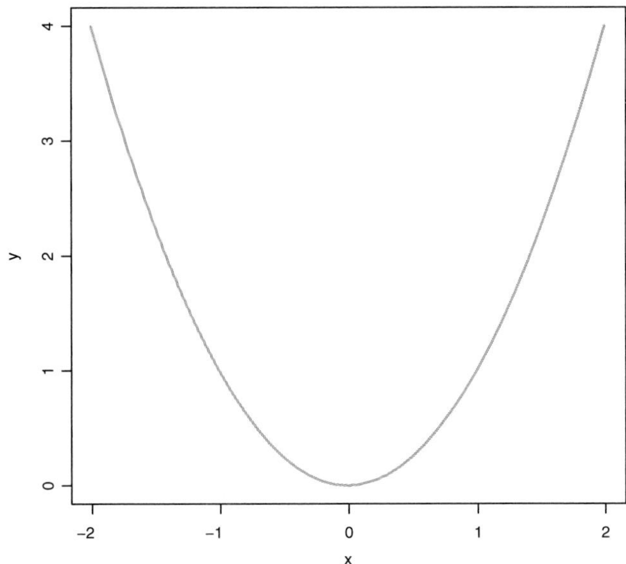

Figure 3.1 A simple graph drawn by R.

−1.8, …, +2.0. This sort of ordered list of numerical values is called a "vector" in R. We'll discuss vectors and other structures in more detail later.

Next we tell R to create the corresponding y values. We type in `y = x^2`. R interprets "`^`" to mean raising values to a power. Inside R, the variable `y` now refers to a vector of 41 values: 4.0, 3.61, 3.24, …, 4.0.

All that remains is telling R to make a plot of the x and y points, connected by lines. Conveniently, R has a built-in function called `plot`, which we call by entering `plot (x , y , type="l")`. The segment of code, `type="l"` (that's a letter "`l`" not a numeral "1") tells R to plot connecting lines between the points, with no distinctive symbols marking the points. If we omitted that part of the command, then R would plot only points by default, not the connecting lines. The resulting plot is shown in Figure 3.1, and the complete R code that generated the graph is shown below:

```
x = seq( from = -2 , to = 2 , by = 0.1 )   # Specify vector of x values.
y = x^2                                     # Specify corresponding y values.
plot( x , y , col="skyblue" , type="l" )   # Plot the x,y points as a blue line.
```

The command lines above include comments on each line to explain to the human reader what the line is intended to do. A comment in R begins with the "#" symbol, called a number sign or a pound sign. Anything on a line after the comment symbol is ignored by R.

3.2.1. Get the programs used with this book

The program above is stored as a file named SimpleGraph.R in the folder of programs that accompany the book. For the latest updates about programs used with this book, see the book's web site at https://sites.google.com/site/doingbayesiandataanalysis/, and the book's blog at http://doingbayesiandataanalysis.blogspot.com/. (Web site addresses occasionally change. If either of the addresses stated here does not work, please search the web for "Doing Bayesian Data Analysis" in quotes. Be sure the site is legitimate before downloading anything to your computer.) Specifically, find the link to the zip file that contains all the programs used with this book. Save the zip file at a convenient location on your computer where you normally save your data files and documents. (Do *not* save it in a protected area such as Programs in the Windows operating system.) Be sure to "unzip" the file or "extract all" of the programs from the zipped folder.

The simplest way to open the program is to associate the file with RStudio and then open the file. For example, in the Windows operating system, find the file SimpleGraph.R. Right-click it and select "Open with..." and then browse to RStudio. Be sure to select (i.e., check) the box that says to always use RStudio to open files of this type. Then, the file will open in the RStudio editor, ready for running and additional editing as desired.

One of the benefits of opening a program directly and letting it automatically invoke RStudio, instead of opening RStudio and subsequently loading the program, is that RStudio sets the *working directory* to the folder in which the program resides. The working directory is where R first looks for auxiliary programs and where R saves output files. If you invoke RStudio from scratch, instead of via a program file, then the working directory is a default generic folder and consequently R will not know where to find auxiliary programs used with this book. More information about setting the working directory will appear later.

One of the programs accompanying this book has filename `ExamplesOfR.R`. That file contains most of the examples in this chapter (other than `SimpleGraph.R`). I recommend that you open it in R and try the examples as you read about them. Or just type in the examples at the command line.

3.3. BASIC COMMANDS AND OPERATORS IN R

The rest of this chapter describes some of the important features of the R programming language. There are many other sources available online, including the introductory manual, the latest version of which should be available at this address: http://cran.r-project.org/doc/manuals/r-release/R-intro.pdf. If that link no longer works by the time you read this, do a little searching of documentation on the R home page or on the web.

3.3.1. Getting help in R

The `plot` function has many optional details that you can specify, such as the axis limits and labels, font sizes, etc. You can learn more about those details by getting help from R. Type the command `?"plot"` and you can read all about it. In particular, the information directs you to another command, `par`, that controls all the plot parameters. To learn about it, type `?"par"`. In general, it actually *is* helpful to use R's built-in help. To get a list of all sorts of online documentation, much of it written in readable prose instead of telegraphic lists, type `help.start()` including the empty parentheses. In particular, the list of manuals displayed by `help.start()` includes *An Introduction to R*, which is highly recommended. Another useful way to find help with R is through web search. In your favorite web searcher, type in the R terms you want help with.

Also very useful is the double question mark, which is an abbreviation for the `help.search` function, and which searches the entire help database for entries that contain the specified word or phrase. While `?"plot"` returns the single help page that explains the `plot` function, `??"plot"` returns a list of dozens of help pages that contain the word "plot." (For more details, you can type `?"?"` and `?"??"` at R's command prompt.)

A highly recommended resource is a summary of basic R commands that can be found on a compact list available at this address:
http://cran.r-project.org/doc/contrib/Short-refcard.pdf
Other versions of reference cards can be found by searching the web with the phrase "R reference card."

Much of the time you'll learn about features of R on an as-needed basis, and usually that means you'll look for examples of the sort of thing you want to do and then imitate the example. (Or, the example might at least provoke you into realizing that there is a better way to do it than the method in the example!) Therefore, most of the examples in this book have their full R code included. Hopefully, it will help you to study those examples as needed.

If you are already familiar with the programming languages Matlab or Python, you can find thesauruses of synonymous commands in R at this Web site: http://mathesaurus.sourceforge.net/.

3.3.2. Arithmetic and logical operators

The four arithmetic operators are typed as `+`, `-`, `*`, and `/`. Power is typed as `^`. When operators are specified successively, they are processed left to right, *except* that power has precedence over multiplication and division, which have precedence over addition and subtraction. Make sure you understand the output returned by each of the following examples:

```
> 1+2*3^2        # power first, then multiplication, then addition
[1] 19
> (1+2)*3^2      # parentheses force addition before multiplication
[1] 27
> (1+2*3)^2      # operations inside parentheses done before power
[1] 49
> ((1+2)*3)^2    # nested parentheses
[1] 81
```

In general, it's a good idea to include explicit parentheses to be sure that the operations are carried out in the order that you intended. Details about operator precedence can be found in R by typing `?"Syntax"`.

R also deals with the logical values `TRUE` and `FALSE`, along with logical operations of negation, `!`, conjunction, `&`, and disjunction, `|`. Negation has precedence, followed by conjunction then disjunction. For example,

```
> !TRUE  # negation
[1] FALSE
> TRUE & FALSE  # conjunction
[1]  FALSE
> TRUE | FALSE  # disjunction
[1]  TRUE
> TRUE | TRUE & FALSE  # conjunction has precedence over disjunction
[1] TRUE
> ( TRUE | TRUE ) & FALSE  # parentheses force disjunction first
[1] FALSE
```

3.3.3. Assignment, relational operators, and tests of equality

We can assign values to named variables. For example, `x = 1` commands R to assign the value 1 to a variable named x. There is a synonymous syntax for the assignment operator that looks like an arrow: `x <- 1`. Originally, R only used the arrow-like symbol to indicate assignment, but in recent years, due to popular demand from users of other programming languages, R also allows the equal sign. This book will usually use the equal sign for assignment.

Purists avoid using the single equal sign for assignment, because it can be confused with the meaning(s) of the mathematical equal sign. In particular, it is perfectly valid in R to state `x = x + 1`, even though it does not make much sense in ordinary mathematical notation to say $x = x + 1$ (because that implies $0 = 1$, which is false in ordinary arithmetic). In R, "`x = x + 1`" means to add 1 to the current value of `x` and assign the result to the variable named `x`. The new value of `x` replaces the old value of `x`. For example, consider the following sequence of R commands:

```
> x = 2       # assign the value 2 to x
> x = x + 1   # add 1 to the value of x and assign the result to x
> x           # show the value of x
[1] 3
```

In practice, this possible misinterpretation of an R command as a mathematical statement of equality is rare or nonexistent. Besides, R uses the single equal sign for assignment-like operations in arguments of functions and in definitions of list structures (as will be explained subsequently). Therefore, I am not deterred from using the single equal sign for assignment.

Beware, however, not be confuse the assignment operator = with a test for equality, which is denoted in R by a *double* equal sign. The double equal sign tests for equality and returns a value of TRUE or FALSE. For example:

```
> x = 2    # assign the value 2 to the variable named x
> x        # show the value of x
[1] 2
> x == 2   # check whether the value of x is equal to 2
[1] TRUE
```

There are also relational operators for inequalities:

```
> x != 3   # check whether the value of x is NOT equal to 3
[1] TRUE
> x < 3    # check whether the value of x is less than 3
[1] TRUE
> x > 3    # check whether the value of x is greater than 3
[1] FALSE
```

If you use <- for the assignment operator instead of =, be certain not to accidentally insert a space: The expression x <- 3 assigns the value 3 to x, but the expression x < - 3 tests whether x is less than the value -3 and returns TRUE or FALSE (or an error message if x does not exist).

The double equal sign for testing equality, used above, is brittle and can give unintended results due to the limited precision of representing numbers in the computer's memory. For example, in most computers, the value of $0.5 - 0.3$ does not equal the value of $0.3 - 0.1$, even though mathematically they are equivalent. Therefore, R has another function, all.equal, for testing equality up to the degree of precision for the computer being used. For example:

```
> x = 0.5 - 0.3
> y = 0.3 - 0.1
> x == y   # although mathematically TRUE, it's FALSE for limited precision
[1] FALSE
> all.equal(x,y)  # equal up to precision of computer
[1] TRUE
```

Therefore, it's safe practice to use `all.equal` instead of == when testing equality, especially for nonlogical or noninteger values.

3.4. VARIABLE TYPES

All objects in R have a *class* which can indicate the structure and type of content in the object, and which can be recognized by some functions so that the functions treat different classes of objects different ways. For example, objects can be of class `matrix`, `array`, `list`, `factor`, `data.frame`, etc. The `summary` function, described later, detects the class of its argument and will summarize a numeric vector differently than a factor. In this section, a few important classes of objects are described.

3.4.1. Vector

A vector is simply an ordered list of elements of the same type. A vector is not technically a class in R, but a vector is the fundamental data structure. For example, a vector could be an ordered list of numbers like this, $\langle 2.718, 3.14, 1.414 \rangle$, or an ordered list of strings like this, ⟨"now", "is", "the", "time"⟩, or an ordered list of logical values like this, ⟨TRUE, FALSE, FALSE, TRUE⟩. By "ordered" I mean that R knows the elements by their position in the list, not that the elements themselves somehow go in increasing or decreasing value. For example, R knows that the third element of the vector $\langle 2.718, 3.14, 1.414 \rangle$ is 1.414 because the elements are ordered.

3.4.1.1 The combine function

The *combine* function, `c` (yes, the single letter "c"), makes vectors. Actually, the combine function can combine different types of data structures, but at this point we consider only its ability to create vectors by combining elements. For example, `c( 2.718 , 3.14 , 1.414 )` combines the three numbers into a vector. We could then assign that vector to a variable named "x" by the command `x = c( 2.718 , 3.14 , 1.414 )`.

3.4.1.2 Component-by-component vector operations

R defaults to component-by-component vector operations. In particular, if `x` and `y` are two vectors of the same length, then `x*y` is the vector consisting of the products of corresponding components. For example:

```
> c(1,2,3) * c(7,6,5)
[1]  7 12 15
```

R automatically applies a scalar operation to all elements of a vector. For example:

```
> 2 * c(1,2,3)
[1] 2 4 6
> 2 + c(1,2,3)
[1] 3 4 5
```

3.4.1.3 The colon operator and sequence function

The colon operator, :, makes sequences of integers. For example, 4:7 creates the vector $\langle 4, 5, 6, 7 \rangle$. The combine function and the colon operator are used very often in R programming. The colon operator has precedence over basic arithmetical operators, but not over the power operator. Consider carefully the output of these examples:

```
> 2+3:6        # colon operator has precedence over addition
[1] 5 6 7 8
> (2+3):6      # parentheses override default precedence
[1] 5 6
> 1:3^2        # power operator has precedence over colon operator
[1] 1 2 3 4 5 6 7 8 9
> (1:3)^2      # parentheses override default precedence
[1] 1 4 9
```

In general, to be sure that the computations are executed in the order that you intend, include explicit parentheses.

The sequence function, seq, is very handy for creating vectors that consist of regular sequences of numbers. In its basic form, the user specifies the starting value of the sequence, the ending value, and the increment between successive values, like this:

```
> seq( from=0 , to=3 , by=0.5 )          # length not specified
[1] 0.0 0.5 1.0 1.5 2.0 2.5 3.0
```

If the increment is not an exact divisor of the distance between the starting and ending values, the sequence will not exceed the ending value. For example:

```
> seq( from=0 , to=3 , by=0.5001 )       # will not exceed end value
[1] 0.0000 0.5001 1.0002 1.5003 2.0004 2.5005
```

The seq function is clever and will infer whatever value is omitted by the user. This capability can be very useful, for example if we want a sequence of a certain length and do not care about the exact end point or the increment. Consider these examples:

```
> seq( from=0 , by=0.5 , length.out=7 )  # end not specified
[1] 0.0 0.5 1.0 1.5 2.0 2.5 3.0
> seq( from=0 , to=3 , length.out=7 )    # increment not specified
[1] 0.0 0.5 1.0 1.5 2.0 2.5 3.0
> seq( to=3 , by=0.5 , length.out=7 )    # start not specified
[1] 0.0 0.5 1.0 1.5 2.0 2.5 3.0
```

3.4.1.4 The replicate function

The *replicate* function, rep, is also very useful for creating vectors. Here are some examples:

```
> ABC = c("A","B","C")  # define a vector for replication
> rep( ABC, 2 )
[1] "A" "B" "C" "A" "B" "C"
> rep( ABC, times=2 )
[1] "A" "B" "C" "A" "B" "C"
> rep( ABC, times=c(4,2,1) )
[1] "A" "A" "A" "A" "B" "B" "C"
> rep( ABC, each=2 )
[1] "A" "A" "B" "B" "C" "C"
> rep( ABC, each=2, length=10)
 [1] "A" "A" "B" "B" "C" "C" "A" "A" "B" "B"
> rep( ABC, each=2, times=3)
 [1] "A" "A" "B" "B" "C" "C" "A" "A" "B" "B" "C" "C" "A" "A" "B" "B" "C" "C"
```

Notice above that rep has three optional arguments after the first argument (which is the structure to be replicated). When only one integer is entered, as in the second line above, then rep assumes it is the number of times that the entire structure should be replicated. According to the R help file, "Normally just one of the additional arguments is specified, but if each is specified with either of the other two, its replication is performed first, and then that implied by times or length.out." If you intend that each component should be replicated first, be sure to use the each argument explicitly, or use the times argument in its vector form.

Here is an example to test your understanding of the rep function:

```
> rep( ABC, each=2, times=c(1,2,3,1,2,3) )
[1] "A" "A" "A" "B" "B" "B" "B" "C" "C" "C" "C" "C"
```

What happened in that example is that the each=2 argument was applied first, thereby creating (internally) the vector c("A","A","B","B","C","C"), which has six components. After that, the times=c(1,2,3,1,2,3) argument was applied, which created one copy of the first "A," followed by two copies of the second "A," followed by three copies of the first "B," and so forth. In other words, the same result could be created by the following nested application of rep:

```
> rep( rep( ABC, each=2 ) , times=c(1,2,3,1,2,3) )
 [1] "A" "A" "A" "B" "B" "B" "B" "C" "C" "C" "C" "C"
```

3.4.1.5 Getting at elements of a vector

Yoohoo, time to wake up! The methods of this section will be used often. In R, we can get at elements of a vector by referencing their position in the vector. There are three ways to reference elements: by numerical position, by logical inclusion, and by name. All three ways refer to positions by using square brackets after the vector name, like this: `x[...]`. What differs between the methods is what goes between the square brackets.

We can refer to elements by their numerical position. For example, suppose we define a vector `x=c( 2.718 , 3.14 , 1.414 , 47405 )`. We can get the elements of the vector by referring to the rank positions inside square brackets after the vector name. Thus, `x[c(2,4)]` returns the second and fourth elements, as the vector $\langle 3.14, 47405 \rangle$. Another way to get at elements is by telling R which position *not* to include, by using a negative sign. For example, if you want all the elements *except* the first and third, you can type `x[c(-1,-3)]`.

Another way to access elements of a vector is by a sequence of logical true and false values that specify whether or not to return the corresponding element. For example, `x[c(FALSE,TRUE,FALSE,TRUE)]` also returns the vector $\langle 3.14, 47405 \rangle$.

The elements of a vector can also, optionally, have *names*. For example, we can name the elements of the vector `x` with the command, `names(x)=c( "e" , "pi" , "sqrt2" , "zipcode" )`. Notice that the names are in quotes. Then we get at the components of the vector by specifying their names in square brackets. For example, `x[c("pi","zipcode")]` returns the vector $\langle 3.14, 47405 \rangle$ along with the names of those components.

To recapitulate and summarize, here are some different ways to get at the elements of a vector:

```
> x = c( 2.718 , 3.14 , 1.414 , 47405 )          # define the vector
> names(x) = c( "e" , "pi" , "sqrt2" , "zipcode" )  # name the components
> x[c(2,4)]                     # which indices to include
     pi  zipcode
   3.14 47405.00
> x[c(-1,-3)]                   # which indices to exclude
     pi  zipcode
   3.14 47405.00
> x[c(FALSE,TRUE,FALSE,TRUE)]   # for each position, include it?
     pi  zipcode
   3.14 47405.00
```

```
> x[c("pi","zipcode")]          # names of indices to include
      pi  zipcode
    3.14 47405.00
```

3.4.2. Factor

Factors are a type of vector in R for which the elements are *categorical* values that could also be ordered. The values are stored internally as integers with labeled levels. The best way to explain factors is by example.

Suppose we have data regarding the socio-economic status of five people, coded as one of "low," "medium," or "high." Here are the data, in the vector named `x`:

```
> x = c( "high" , "medium" , "low" , "high" , "medium" )
```

The terms "high," "medium," and "low" get a bit unwieldy in large data sets, and therefore it can be useful to recode them into categorical indices, such as "1" for "low," "2" for "medium," and "3" for "high." The data become a vector of indices along with a legend from translating from the indices to names for the indices. The resulting structure is called a "factor." In R, the `factor` function converts the vector `x` to a factor:

```
> xf = factor( x )
```

To see what the factor `xf` looks like, we can type "`xf`" at R's command line and see what R returns:

```
> xf
[1] high   medium low    high   medium
Levels: high low medium
```

Notice that R has extracted "levels" from the vector and listed them after the contents of the factor. The `factor` function read the contents of vector x and kept track of all the distinct elements, calling them the *levels* of the factor. By default, it ordered the levels alphabetically. It then *recoded the contents of the vector in terms of the integer indices of the alphabetized levels*. Thus, the original element "high" becomes integer "1" because "high" is alphabetically first among the three levels. You can think of the real contents of the factor as being the integer indices of the levels, along with a legend that decodes each integer into a level name. To see the integer indices of the levels explicitly, we can ask R to show us the factor as a numeric vector:

```
> as.numeric(xf)
[1] 1 3 2 1 3
```

The `factor` command extracts the levels *and orders them alphabetically by default*. Unfortunately, alphabetical order depends on local language customs (e.g., uppercase letters might go before lowercase letters or after), and alphabetical order might not be the most meaningful order for a given application. If we want to specify a particular order for the levels, we can do so with additional arguments in the `factor` function, like this:

```
> xfo = factor( x , levels=c("low","medium","high") , ordered=TRUE )
> xfo
[1] high   medium low    high   medium
Levels: low < medium < high
```

Notice now that the levels are displayed with less-than signs to indicate that they are ordered. And, when we examine the integer indices internal to the factor, they are coded according to the desired order, such that the first element, "high," is coded as integer 3, not as 1:

```
> as.numeric(xfo)
[1] 3 2 1 3 2
```

It is good practice to explicitly specify the levels and their ordering when making a factor.

Sometimes we will want to reorder the levels of a factor that has already been created. This can happen, for example, when reading in a data file using default settings, but subsequently reordering to make the levels more meaningful. To reorder, just use the `factor` function again, explicitly indicating the desired levels and order. For example, we create factor `xf` as before:

```
> x = c( "high" , "medium" , "low" , "high" , "medium" )
> xf = factor( x )  # results in default ordering
> xf
[1] high   medium low    high   medium
Levels: high low medium
> as.numeric(xf)
[1] 1 3 2 1 3
```

But now we reorder the levels by applying the factor function to the existing factor `xf`:

```
> xf = factor( xf , levels=c("low","medium","high") , ordered=TRUE )
> xf
[1] high   medium low    high   medium
Levels: low < medium < high
> as.numeric(xf)
[1] 3 2 1 3 2
```

Notice that explicitly reordering the levels has also changed the underlying integer coding of the elements in the vector, as it should. It is good practice to specify the ordering of the levels explicitly.

We might want to relabel the levels of a factor. For example, the vector x used elements named low, medium, and high. The levels argument of the factor function must refer to those exact terms for reading in the data. But for the resulting output, the levels can be relabeled to whatever might be more meaningful for the application, such as "Bottom SES," "Middle SES," and "Top SES." We accomplish the relabeling in R with the labels argument:

```
> xfol = factor( x , levels=c("low","medium","high") , ordered=TRUE ,
                 labels=c("Bottom SES","Middle SES","Top SES") )
> xfol
[1] Top SES    Middle SES Bottom SES Top SES    Middle SES
Levels: Bottom SES < Middle SES < Top SES
```

Notice that the original data terms (low, medium, high) are no longer accessible in the factor. *It is important to realize that relabeling the levels does not change their order or their integer coding.* For example, if we specify labels=c("Right","Left", "UpsideDown"), the integers will be unchanged, but their labeling will no longer be meaningful.

Factors will appear frequently when R reads data from files. Factors can be very useful for some functions in R. But factors can be confusing if you are not aware that a variable is being stored by R as a factor and not as a vector. Factors can also be confusing if the default ordering of their levels is not intuitive.

3.4.3. Matrix and array

A matrix is simply a *two*-dimensional array of values of the same type. A matrix can be created in R using the matrix command. The first argument specifies the contents of the matrix, in order. Other arguments specify the size of the matrix, the order in which the matrix should be filled, and, optionally, the names of the dimensions and rows and columns. (The names are specified as a list, which is a structure that is described explicitly very soon, in Section 3.4.4.) Please note the comments in the following examples:

```
> matrix( 1:6 , ncol=3 )  # contents are 1:6, filled by column
     [,1] [,2] [,3]
[1,]    1    3    5
[2,]    2    4    6
> matrix( 1:6 , nrow=2 )  # or you can specify number of rows
     [,1] [,2] [,3]
[1,]    1    3    5
[2,]    2    4    6
```

```
> matrix( 1:6 , nrow=2 , byrow=TRUE )  # filled by row instead of by column
     [,1] [,2] [,3]
[1,]    1    2    3
[2,]    4    5    6
> matrix( 1:6 , nrow=2 ,         # with names of dimensions and rows and columns
+         dimnames=list( TheRowDimName=c("Row1Name","Row2Name") ,
+                        TheColDimName=c("Col1Name","Col2Name", "Col3Name") ) )
             TheColDimName
TheRowDimName Col1Name Col2Name Col3Name
     Row1Name        1        3        5
     Row2Name        2        4        6
```

In the final example above, the "+" symbols at the beginning of some command lines are merely R's way of displaying that those lines are continuations of the same command, instead of new commands. You do *not* type in the plus symbols when you enter the command, just as you do *not* type in the greater-than symbol at the beginning of a command.

Just as with vectors, the components of matrices can be accessed via their indices or their names. The following shows two ways to access the element in the second row and third column of the matrix x:

```
> x = matrix( 1:6 , nrow=2 ,
+             dimnames=list( TheRowDimName=c("Row1Name","Row2Name") ,
+                            TheColDimName=c("Col1Name","Col2Name", "Col3Name") ) )
> x[2,3]  # use numerical indices
[1] 6
> x["Row2Name","Col3Name"]  # use row, column names
[1] 6
```

As you may have inferred by now, the indices are ordered such that the first index refers to the row and the second index refers to the column.

An entire row or column of a matrix can be accessed by specifying its entire range or by leaving its range unspecified. For example:

```
> x[2,1:3]  # specify range of columns for inclusion
Col1Name Col2Name Col3Name
       2        4        6
> x[2,]  # leave range of columns blank to include all columns
Col1Name Col2Name Col3Name
       2        4        6
> x[,3]  # all rows from column 3, returned as a vector
Row1Name Row2Name
       5        6
```

Be very careful to include the comma when specifying a row or column of a matrix. If you use a numerical index for a row or column and accidentally leave out the comma, R will not complain, but will instead return the element at that position in the ordered contents. Consider these examples:

```
> x[2,] # 2nd row (returned as vector)
Col1Name Col2Name Col3Name
       2        4        6
> x[,2] # 2nd column (returned as vector)
Row1Name Row2Name
       3        4
> x[2] # no comma; returns 2nd element.
[1] 2
```

Notice in the final example above, when there was no comma, R returned the second element in the contents of the matrix. Although it is perfectly valid to refer to the elements of a matrix this way, without using row and column referents, it should usually be avoided unless there is some specific application that is made more efficient.

An array is a generalization of a matrix to multiple dimensions. There is really no need for a separate `matrix` function because it is merely the two-dimensional case of the `array` function. In the `array` function, the first argument specifies the ordered contents, the second argument specifies the size of each dimension, and an optional third argument specifies the names of the dimensions and levels within dimensions. (The names are specified as a `list`, which is a structure that is described explicitly very soon, in Section 3.4.4.) It is important to understand that the array function fills the array by incrementing the first index (row) first, then incrementing the second index (column) next, then incrementing the third index (layer) next, and so forth. Unlike the `matrix` function, there is no built-in way to load the contents into the array in a different ordering of dimensions.

Here is an example of a three-dimensional array. I have referred to the third dimension as a "layer." Notice that the contents are the integers 1-24, and they are filled into the array by first incrementing the row, then the column, and then the layer.

```
> a = array( 1:24 , dim=c(3,4,2) , # 3 rows, 4 columns, 2 layers
+              dimnames = list( RowDimName = c("R1","R2","R3") ,
+                               ColDimName = c("C1","C2","C3","C4") ,
+                               LayDimName = c("L1","L2") ) )
> a
, , LayDimName = L1
          ColDimName
RowDimName C1 C2 C3 C4
        R1  1  4  7 10
        R2  2  5  8 11
        R3  3  6  9 12
, , LayDimName = L2
          ColDimName
RowDimName C1 C2 C3 C4
        R1 13 16 19 22
        R2 14 17 20 23
        R3 15 18 21 24
```

```
> a["R3",,"L2"]  # returns all columns of R3 and L2, as a vector
C1 C2 C3 C4
15 18 21 24
> a["R3","C4",]  # returns all layers of R3 and C4, as a vector
L1 L2
12 24
```

3.4.4. List and data frame

The `list` structure is a generic vector in which components can be of different types, and named. The list structure was used in previous examples to specify dimension names in the `matrix` and `array` functions. Below is an example of a list in which the first element is a vector of integers named "a," the second element is a matrix of integers named "b," and the third element is a string named "c."

```
> MyList = list( "a"=1:3 , "b"=matrix(1:6,nrow=2) , "c"="Hello, world." )
> MyList
$a
[1] 1 2 3

$b
     [,1] [,2] [,3]
[1,]    1    3    5
[2,]    2    4    6

$c
[1] "Hello, world."
```

The named components of a list can be referred to by appending the list name with a "$" and the component name. Consider these examples:

```
> MyList$a         # the contents of the list item named "a"
[1] 1 2 3

> MyList$a[2]      # the second element of the list item named "a"
[1] 2
```

The contents of the list can also be accessed with (numerical) indices inside square brackets. There is an additional layer of nuance for referring to named components, however. We can get element `i` from a list, *including its name*, by putting `i` inside *single* square brackets. We can get the *contents* of element `i` by putting `i` inside *double* square brackets. Consider these examples:

```
> MyList[[1]]      # the contents of the first list item
[1] 1 2 3
```

```
> MyList[[1]][2]  # the second element of the first list item
[1] 2
> MyList[1]       # the first list item, including its name
$a
[1] 1 2 3
> MyList[1][2]    # does not make sense in this case
$<NA>
NULL
```

A *data frame* is much like a matrix, insofar as it has several columns of equal length. But each column can be of a different type, and, in particular, columns can be factors. A data frame is really a type of `list` in which each component is thought of as a named column of a matrix, with different columns possibly of different types. The elements of a data frame can be accessed as if it were a list or as if it were a matrix. Consider this example:

```
> d = data.frame( Integers=1:3 , NumberNames=c("one","two","three") )
> d
  Integers NumberNames
1        1         one
2        2         two
3        3       three
```

In the display of data frame `d`, above, the column of numbers on the far left side shows the row names supplied, by default, by the `data.frame` function. Do not confuse those row names with the contents of the columns.

The elements of the data frame can be accessed as for a list, by using names or single brackets or double brackets:

```
> d$NumberNames   # notice this is a factor
[1] one   two   three
Levels: one three two

> d[[2]]          # the second element contents
[1] one   two   three
Levels: one three two

> d[2]            # the second element with its name
  NumberNames
1         one
2         two
3       three
```

The elements of the data frame can also be accessed as for a matrix, using row and column indices:

```
> d[,2]           # elements can be accessed as if it's a matrix
[1] one   two   three
Levels: one three two
```

```
> d[2,]          # elements can be accessed as if it's a matrix
  Integers NumberNames
2        2         two
```

Data frames are important because they are the default format for data loaded into R by the often-used function `read.table`, as we will explore next.

3.5. LOADING AND SAVING DATA

This book uses a typical convention for arranging data in which every row of a data file contains one measurement instance, which might be one person in a survey, or one trial in a response-time experiment. Each row contains the value of the key measurement to be explained or predicted, and each row also contains values that indicate predictors or explanatory variables for that instance. Examples, shown below, will clarify.

3.5.1. The `read.csv` and `read.table` functions

Consider a small sample of people, from whom we record their gender, hair color, and ask a random number between 1 and 10. We also record each person's first name, and which of two groups they will be assigned to. It is typical to save this sort of data as a computer file in a format called *comma separated values*, or *CSV* format. In CSV format, there is a column for each type of measurement, and a row for each person (or item) measured. The first row specifies the column names, and the information in each column is separated by commas. Here is an example:

```
Hair,Gender,Number,Name,Group
black,M,2,Alex,1
brown,F,4,Betty,1
blond,F,3,Carla,1
black,F,7,Diane,2
black,M,1,Edward,2
red,M,7,Frank,2
brown,F,10,Gabrielle,2
```

CSV files are especially useful because they are generic text files that virtually any computer system can read. A disadvantage of CSV files is that they can take up a lot of memory space, but these days computer memory is cheap and plentiful.

CSV files are easily loaded into R's memory using the `read.csv` function. Suppose that the data above are saved in a file called `HGN.csv`. Then the data can be loaded into a variable I've named `HGNdf` as follows:

```
> HGNdf = read.csv( "HGN.csv" )
```

The resulting variable, `HGNdf`, is a *data frame* in R. Thus, the columns of `HGNdf` are vectors or factors, named according to the words in the first row of the CSV file, and all of length equal to the number of data rows in the CSV file.

It is important to note that columns with any character (non-numeric) entries are turned into *factors* (recall that Section 3.4.2 described factors). For example, the Hair column is a factor:

```
> HGNdf$Hair
[1] black brown blond black black red brown
Levels: black blond brown red

> as.numeric(HGNdf$Hair)
[1] 1 3 2 1 1 4 3
```

The levels of the factor are alphabetical by default. We might want to reorder the levels to be more meaningful. For example, in this case we might want to reorder the hair colors from lightest to darkest. We can do that after loading the data, like this:

```
> HGNdf$Hair = factor( HGNdf$Hair , levels=c("red","blond","brown", "black"))
>
> HGNdf$Hair
[1] black brown blond black black red brown
Levels: red blond brown black
>
> as.numeric(HGNdf$Hair)
[1] 4 3 2 4 4 1 3
```

There might be times when we do not want a column with character entries to be treated as a factor. For example, the Name column is treated by `read.csv` as a factor because it has character entries:

```
> HGNdf$Name
[1] Alex  Betty  Carla  Diane  Edward  Frank  Gabrielle
Levels: Alex Betty Carla Diane Edward Frank Gabrielle
```

Because the names are never repeated, there are as many factor levels as entries in the column, and there might be little use in structuring it as a factor. To convert a factor to an ordinary vector, use the function `as.vector`, like this:

```
> HGNdf$Name = as.vector( HGNdf$Name )
> HGNdf$Name
[1] "Alex"  "Betty"  "Carla"  "Diane"  "Edward"  "Frank"  "Gabrielle"
```

There might be times when a column of integers is read by `read.csv` as a numeric vector, when you intended it to be treated as indexical levels for grouping the data. In other words, you would like the column to be treated as a factor, not as a numeric vector. This happened in the present example with the Group column:

```
> HGNdf$Group
[1] 1 1 1 2 2 2 2
```

Notice above that no levels are associated with the Group column. It is easy to convert the column to a factor:

```
> HGNdf$Group = factor( HGNdf$Group )
> HGNdf$Group
[1] 1 1 1 2 2 2 2
Levels: 1 2
```

The `read.csv` function is a special case of the more general `read.table` function. You can learn more about it by typing `?"read.table"` at R's command prompt. For example, you can turn off the default action of making character columns into factors.

3.5.2. Saving data from R

In most real research situations, data are obtained from some measurement device or survey outside of R. The collected data are then loaded into R using a function such as `read.csv`, as explained above. Sometimes, however, we want to save reformatted or transformed data from R.

One way to save a data table is with the `write.csv` function. For example, the command

```
write.csv( HGNdf , file="HGN.csv" , row.names=FALSE , quote=FALSE )
```

saves the data frame `HGNdf` as the file named HGN.csv, without including the row names and without putting all character elements inside double quotes (which `write.csv` would otherwise do by default). It is important to understand that the resulting file loses all information about levels in factors, because it is only a raw text file with no summary information.

If you want to save data frames with all the factor information intact, then you can use the `save` command. The resulting file is in a special R format, not generic text, and the standard filename extension for this type of file is ".Rdata." For example, the command

```
save( HGNdf , file="HGN.Rdata" )
```

saves the data frame HGNdf as the file named HGN.Rdata with all its factor information, and with the name of data frame (i.e., "HGNdf"). If desired, several different variables can all be saved in a single Rdata file, simply by specifying them as the initial comma-separated arguments in the save command. All the variables are saved along with their names.

To retrieve an Rdata file, use the load function:

```
> load( "HGN.Rdata" )
```

The load function executes without explicitly showing the user any changes in R's internal state. But R has, in fact, loaded the variables into its working memory. You can see the objects that R has in its active memory by typing

```
> objects()
[1] "HGNdf"
```

The output from the objects function shows that R has HGNdf in its memory, in this case because it loaded that variable from the file HGN.Rdata. In the editor RStudio, a convenient way to see what objects are in R's memory is by looking at the Workspace window (usually in the upper right or RStudio's display). The Workspace window shows all the objects, organized by type.

3.6. SOME UTILITY FUNCTIONS

A function is a process that takes some input, called the *arguments*, and produces some output, called the *value*. This section reviews some useful utility functions in R.

The summary function detects the type, or "class," of the argument provided to it, and returns a summary appropriate for that class of object. For example, if we construct a vector consisting of numerical values, summary provides the minimum value in the vector, median value, etc., as shown by this example:

```
> x = c( rep(1,100) , rep(2,200) , rep(3,300) )  # 100 1's, 200 2's, 300 3's
> summary(x)
   Min. 1st Qu.  Median    Mean 3rd Qu.    Max.
  1.000   2.000   2.500   2.333   3.000   3.000
```

However, if we convert the vector to a factor, then the summary function provides a table with the frequency of each level:

```
> xf = factor(x)
> summary(xf)
  1   2   3
100 200 300
```

If you put a data frame into the `summary` function, it will provide a summary of each column appropriate to the class of information in each column.

Other useful functions for examining data objects are `head`, `tail`, and `str`. The `head` function returns the first few components of the variable put in its argument. The `str` function returns a compact display of the structure of its argument. Type `?"head"` (etc.) in R to learn more.

The `aggregate` function is very useful for summarizing data according to factor characteristics. To illustrate, recall the hair-gender-number data from Section 3.5.1, in which each person provided a self-generated number between 1 and 10. The data were read into a data frame named `HGNdf`, with the `Group` column converted to a factor. Suppose we want to know the median value of the numbers selected by each gender within each hair color. The `aggregate` function provides the answer. The first argument, specified as "`x=`," is the data we want summarized. The "`by=`" argument is a list of factors for grouping the data. The "`FUN=`" argument is the function that should be applied to the groups of data. For example:

```
> aggregate( x=HGNdf$Number , by=list(HGNdf$Gender,HGNdf$Hair) , FUN=median )
  Group.1 Group.2   x
1       M     red 7.0
2       F   blond 3.0
3       F   brown 7.0
4       F   black 7.0
5       M   black 1.5
```

Notice that the names of the columns in the output were given defaults that are unrelated to the original variable names. If we want more meaningful output, we have to explicitly name the variables:

```
> aggregate( x=list(Number=HGNdf$Number) ,
+            by=list(Gender=HGNdf$Gender,Hair=HGNdf$Hair) , FUN=median )
  Gender  Hair Number
1      M   red    7.0
2      F blond    3.0
3      F brown    7.0
4      F black    7.0
5      M black    1.5
```

Alternatively, we can use the *formula* format for the arguments. In formula format, the first argument is a formula that uses the column names from the data frame to express what we want aggregated. For example, to specify that we want the Number aggregated by Group and Gender, we would use the formula `Number ~ Group + Gender`. But we also have to tell the function what data frame we are referring to, and this is done with the `data` argument. Thus,

```
> aggregate( Number ~ Gender + Hair , data=HGNdf , FUN=median )
  Gender  Hair Number
1      M   red    7.0
2      F blond    3.0
3      F brown    7.0
4      F black    7.0
5      M black    1.5
```

The formula format is used by many functions in R that take data frames as arguments. The `aggregate` function has several other arguments, and it can also be used to summarize time series. Of course, you can type `?"aggregate"` for more information.

The `aggregate` function is also useful for counting how many times various levels of a factor occur, or how many times combinations of factor levels occur. For example, suppose we would like to know how many times each hair color occurs for each gender. We use the `aggregate` function by applying the `sum` function, aggregating on the gender and hair-color factors. But what value do we sum? We want to sum the number of rows of each combination, so we create a new column that contains all 1s, to explicitly indicate that each row contains a count of 1. Thus,

```
> aggregate( x=list(Count=rep(1,NROW(HGNdf))) , # column of 1's
+            by=list(Gender=HGNdf$Gender,Hair=HGNdf$Hair) , FUN=sum )
  Gender  Hair Count
1      M   red     1
2      F blond     1
3      F brown     2
4      F black     1
5      M black     2
```

The result, above, shows that there was 1 male with red hair, 1 female with blond hair, 2 females with brown hair, and so on. Notice that combinations with zero counts, such as females with red hair, are not displayed.

Another way to generate a table of counts is with the `table` function. Here is an example of its use and its output:

```
> table(list(Gender=HGNdf$Gender,Hair=HGNdf$Hair))
      Hair
Gender red blond brown black
     F   0     1     2     1
     M   1     0     0     2
```

The result, above, displays a table of all levels of gender crossed with all levels of hair color, explicitly marking combinations that do not occur with counts of zero. This form of output can be useful for human comprehension. But because there are several data values per row, it does not comply with the usual format for data files that are used in this book, which assume a single measurement "instance" per row. The format produced by the `aggregate` function, in the previous paragraph, is appropriate for the usual data files.

The `apply` function is handy for collapsing arrays across specified dimensions, and applying a function to the data within the collapsed dimensions. For example, recall from Section 3.4.3 the three-dimensional array named `a`:

```
> a
, , LayDimName = L1

          ColDimName
RowDimName C1 C2 C3 C4
        R1  1  4  7 10
        R2  2  5  8 11
        R3  3  6  9 12

, , LayDimName = L2

          ColDimName
RowDimName C1 C2 C3 C4
        R1 13 16 19 22
        R2 14 17 20 23
        R3 15 18 21 24
```

Suppose we want to know the sums of the values within columns and layers, collapsed across rows. This means we want to retain the second and third dimensions, while collapsing across the other dimension. The appropriate command is:

```
> apply( a , MARGIN=c(2,3) , FUN=sum )

          LayDimName
ColDimName L1 L2
        C1  6 42
        C2 15 51
        C3 24 60
        C4 33 69
```

Notice that the `MARGIN=` argument specifies which dimensions to retain, uncollapsed. If the dimensions are named, you can specify the retained dimensions by their names instead of by their indexical number.

The `melt` command, from the `reshape2` package, is very useful for rearranging data so that there is one datum per row. To use the package, you must first install it on your computer:

```
install.packages("reshape2")
```

You only need to install the package once, then never again. For any new session of R, however, you must load the package into R's working memory, like this:

```
library(reshape2)
```

At this point, R knows the various commands defined in the `reshape2` package. To illustrate the `melt` command, first recall the array `a` defined earlier:

```
> a
, , LayDimName = L1

          ColDimName
RowDimName C1 C2 C3 C4
        R1  1  4  7 10
        R2  2  5  8 11
        R3  3  6  9 12

, , LayDimName = L2

          ColDimName
RowDimName C1 C2 C3 C4
        R1 13 16 19 22
        R2 14 17 20 23
        R3 15 18 21 24
```

Notice that the array is three-dimensional, with a datum in every one of the 24 cells. *We would like to rearrange the data so that there is one datum per row, with each row also specifying the levels of the array from which the datum came.* This is done by the `melt` command:

```
> am = melt(a)
> am
   RowDimName ColDimName LayDimName value
1          R1         C1         L1     1
2          R2         C1         L1     2
3          R3         C1         L1     3
4          R1         C2         L1     4
5          R2         C2         L1     5
6          R3         C2         L1     6
7          R1         C3         L1     7
8          R2         C3         L1     8
9          R3         C3         L1     9
10         R1         C4         L1    10
11         R2         C4         L1    11
12         R3         C4         L1    12
13         R1         C1         L2    13
14         R2         C1         L2    14
15         R3         C1         L2    15
16         R1         C2         L2    16
17         R2         C2         L2    17
18         R3         C2         L2    18
19         R1         C3         L2    19
20         R2         C3         L2    20
21         R3         C3         L2    21
22         R1         C4         L2    22
23         R2         C4         L2    23
24         R3         C4         L2    24
```

Notice that the values that were in the array `a`, namely the numbers 1-24, are now arranged one per row, in the column named "`value`." The result of `melt` is a data frame, with the level identifiers being factors. You can read more about the many other capabilities of the `reshape2` package in the article by Wickham (2007), and in the reference manual available at http://cran.r-project.org/web/packages/reshape2/.

3.7. PROGRAMMING IN R

Instead of typing all the commands one at a time at the command line, you can type them into a text document and then have R execute the document. The document is called a program or script. We saw an example in the program SimpleGraph.R, back in Section 3.2.

RStudio is a nice environment for developing programs in R. If you have associated filenames that end in ".R" with RStudio, then opening the file will open it in RStudio's editing window. For starting a new program, in RStudio use the pull-down menu items File → New → R Script. A blank editing window will open, where you can type your program.

Some important points for newbie programmers:

- Be sure you save your program in a folder where you can find it again, with a filename that is easy to recognize weeks later.
- Be sure to save the program every time you make a small *working* change.
- If you are about to make a big change, save the current working version and start the modified version with a new filename. This way, when your modified version doesn't work, you still have the old working version to fall back on.
- Put lots of explanatory comments in your code, so that you can understand what you were doing when you come back to the program months later. To include a comment in a program, simply type a "#" character, and everything after that character, on the same line, will be ignored by the R interpreter. There is a peril inherent in comments, however: They easily become obsolete and confusing if you change the programming code without also changing the corresponding comment. This is a problem I frequently encounter in my own programming, and I hope that the programs I've created for this book do not have too many obsolete comments.

3.7.1. Variable names in R

You should name variables meaningfully, so that the programming commands are easy for a reader to understand. If you name your variables cryptically, you will curse your poor judgment when you return to the program weeks later and you have no idea what your program does.

You can use fairly long, descriptive names. If the names get too long, however, then the program becomes unwieldy to type and read. For example, suppose you want to name the crucial final output of a program. You could name it `tempfoo`, but that's not very meaningful, and might even lead you to think that the variable is unimportant. Instead, you could name it `theCrucialFinalOutputThatWillChangeTheWorldForever`, but that would be burdensome to type and read as it gets reused in the program. So, you might best name it something like `finalOutput`, which is meaningful but not too long.

Computer programmers typically use a naming convention called *camelBack notation*. This is a way of connecting several words into a contiguous variable name without using spaces between words. For example, suppose you want to name a variable "final output." You are not allowed to name a variable with a space in it because R (and most other languages) interprets spaces as separators of variables. One way to avoid using spaces is to connect the words with explicit connectors such as an underscore or a dot, like this: `final_output` or `final.output`. Many programmers *do* use and recommend those naming conventions. But the underscore notation can be difficult to read in some displays, and the dot notation is interpreted by some programming languages (including R in some contexts) as referring to subcomponents or classes of structured variables, which can confuse people (or computers) who are familiar with that meaning of a dot. Therefore, the spaces are simply dropped, with successive words capitalized: `finalOutput`. The initial word is typically not capitalized, but some people have different uses for initial-capitalized variable names. R is case sensitive: the variable `myVar` is different than the variable `myvar`!

I will try to use camelBack notation in all the programs in this book. I may occasionally lapse from bactrian beauty, instead slithering into snakeback notation (`finaloutput`) or gooseneck notation (`final_output`) or ant notation (`final.output`). If you see these lower forms, quietly shun them, knowing that when you create your own programs, you will use the more highly evolved dromedary design.

3.7.2. Running a program

Running a program is easy, but exactly how to do it depends on how you are interacting with R.

Important: First set the working directory. Many programs read data from another computer file, or save data to another file. The programs need to know where to find or put those files. The default location might have no relevance to your work and may cause problems. Therefore, you must be sure that R knows what directory (also known as a folder) to use for getting and saving data. This location is called the *working directory*. Typically you want the working directory to be the folder in which the current program resides. In R's basic command window, set the working directory by selecting menu items File → Change dir. In RStudio, set the working directory by selecting menu items Session → Set Working Directory. (The menu structures occasionally change with updated versions of R and RStudio.) In RStudio, to be sure that the working directory really is set properly, look in the lower-right window and click the Files tab. Then click the circular arrow to the right of the files tab to refresh the file listing. Check that it displays the items in the intended working directory. If it does not show the intended working directory contents, then use the folder navigation links at the top of the list of files to find the desired folder, and click the tab marked "More" to set the working directory.

As has been described previously, the easiest way to have R's working directory set to the folder of your program is by invoking RStudio via the program, instead of by opening RStudio before opening the program. With R and RStudio closed, open the desired .R program file, for example in Windows by double clicking the file icon. If files of type .R are not yet associated with an application, your computer will ask you what application to use to open the file. Select the application RStudio, and select the option that tells the computer always to use RStudio for files of that type. Then RStudio will open with the .R program open in its editing window, and, importantly, with that program's folder as R's working directory.

To run an entire program, we tell R to get its commands from a source other than the interactive command line. We give it the `source` command at the interactive command prompt, to tell it to accept all the commands from the alternative source. For example, to run the `SimpleGraph.R` program from Section 3.2, we could just type `source("SimpleGraph.R")`. This command will work only if the file `SimpleGraph.R` is in R's current working directory.

Typically you will be working interactively with a program that is open in an editing window. If you have already associated filenames that have .R extensions with RStudio, then when you open the file it will automatically open in RStudio's editing window. Alternatively, you can navigate to a program from RStudio's menu, using File → Open File. Be sure to set the working directory appropriately (see above).

Once the program is open in an editing window, it is easy to run all or only part of it via menu buttons on the editor. For example, in RStudio, the top right of the editing window has buttons marked "Run" and "Source." The Run button will run the line on which the cursor is presently positioned, or multiple lines if they are presently selected, and will echo the lines in the command window. The Source button will run the entire program without echoing the lines in the command window.

Notice that the `source` function is different than the `load` function. The `source` function reads the referred-to text file as if it were a script of commands for R to execute.

The commands might create new variables, but they are read as commands. The `load` function, on the other hand, expects to read a file that is in compressed Rdata format (not a text file) and specifies variables and their values.

3.7.3. Programming a function

A function takes input values, called "arguments," and does something with them. In general, a function in R in defined by code of the form:

```
functionName = function( arguments ) { commands }
```

The commands inside the curly braces can extend over many lines of code. When the function is called, it takes the values of the arguments in parentheses and uses them in the commands in the braces. You invoke the function by commanding R thus:

```
functionName( arguments=argumentValues )
```

As a simple example, consider this definition:

```
asqplusb = function( a , b=1 ) {
  c = a^2 + b
  return( c )
}
```

The function is named `asqplusb` and it takes the value `a`, squares it, and adds the value `b`. The function then returns the result as output. Here is an example:

```
> asqplusb( a=3 , b=2 )
[1] 11
```

In a function call, explicitly labeled arguments may go in any order:

```
> asqplusb( b=2 , a=3 )
[1] 11
```

However, arguments without explicit labels must be in the order used in the definition of the function. Notice that these two function calls give different results:

```
> asqplusb( 3 , 2 )
[1] 11
> asqplusb( 2 , 3 )
[1] 7
```

The function definition gave argument `b` a default value of 1 by specifying `b=1` in the list of arguments. This means that if the function is called without an explicit value for `b` provided, the default value will be used. The argument `a` was not given a default value, however, and must always be specified in a call to the function. Consider these examples:

```
> asqplusb( a=2 )  # b gets default value
[1] 5
> asqplusb( b=1 )  # error: a has no default value
Error in a^2 : 'a' is missing
> asqplusb( 2 )    # argument value is assigned to first argument
[1] 5
```

In the last example above, there was only one unlabeled argument provided in the function call. R assigns that value to the first argument in the definition, regardless of whether the argument was defined with a default value.

3.7.4. Conditions and loops

There are many situations in which we want a command to be executed only under certain conditions. A special case of that is executing a command a certain number of times. The R language can be programmed with conditions and loops to satisfy these needs.

The basic conditional statement uses an if-else structure. This might be best explained by example. Suppose we want to enter a value, x, and have R reply with "small" if $x \leq 3$ and with "big" if $x > 3$. We can type

```
if ( x <= 3 ) {   # if x is less than or equal to 3
  show("small")   # display the word "small"
} else {          # otherwise
  show("big")     # display the word "big"
}                 # end of 'else' clause
```

If we tell R that x=5 *before* executing the lines above, then the word "big" appears in the command window as a reply from R.

Notice the arrangement of curly braces across lines in the if-else structure above. In particular, the line containing "else" begins with a closing curly brace, which tells R that the "else" clause is continuing the preceding "if." A line that begins with "else" causes an error:

```
> if ( x <= 3 ) { show("small") }
> else { show("big") }
Error: unexpected 'else' in "else"
```

There are a variety of ways to have R execute a set of commands repeatedly. The most basic is the *for* loop. We specify a vector of values that R steps through, executing a set of commands for each value of the vector. Here is an example:

```
> for ( countDown in 5:1 ) {
+    show(countDown)
+ }

[1] 5
[1] 4
[1] 3
[1] 2
[1] 1
```

Notice that the index `countDown` took on each of the values in the vector `5:1`, and the command `show(countDown)` was executed. The entries in the vector do not need to be numeric. For example, they can be strings:

```
> for ( note in c("do","re","mi") ) {
+    show(note)
+ }

[1] "do"
[1] "re"
[1] "mi"
```

There will be many examples of conditions and loops in the programs used later in the book. R has other ways for constructing loops and conditions (such as `while`), and ways for breaking out of loops (such as `break`). Type `?"Control"` to read about them.

3.7.5. Measuring processing time

It can be useful to know how long a process is taking in R, either to anticipate when a very long process will be completed, or to assess where a program is being inefficient. A simple way to measure processing time is with the `proc.time` function, which returns the current computer-system time. To measure the duration of a process, use `proc.time` at the beginning and end of the process, and compute the difference between the times. Here is an example:

```
startTime = proc.time()
y = vector(mode="numeric",length=1.0E6)
for ( i in 1:1.0E6 ) { y[i] = log(i) }
stopTime = proc.time()
elapsedTime = stopTime - startTime
show(elapsedTime)
   user  system elapsed
   3.17    0.02    3.20
```

In other words, it took R 3.17 seconds (on my computer) to compute the logarithm of every integer from 1 to 1,000,000 and store the results in a vector. Instead of using a loop, let's do the same computation in vectorized form and see how long it takes:

```
startTime = proc.time()
y = log(1:1.0E6)
stopTime = proc.time()
elapsedTime = stopTime - startTime
show(elapsedTime)
   user  system elapsed
   0.09    0.01    0.11
```

You can see that the vector operation took only about 3% of the time needed for the loop! In general, `for` loops are slow relative to vectorized operations.

3.7.6. Debugging

This book includes dozens of programs that can be used without modification for real research; all you need to do is load your data. An important goal of the book is also for you to be able to modify the programs for application to different data structures and models. For this purpose, you will need to be able to debug programs as you modify them. To "debug" means to find, diagnose, and correct errors in programs.

To avoid producing errors in the first place, here are a few hints:

- When creating a new program, always start with a new file name. Do not overwrite existing, working programs with a new program of the same name.
- When creating `for` loops, beware of naming the for-loop index as merely a single letter such as "`i`" or "`j`". Such single letters are extremely easy to confuse and mistype and misperceive! Instead, name the indices meaningfully and distinctly for each index. For example, name a row index something like `rowIdx` instead of just `i`, and name a column index something like `colIdx` instead of just `j`.
- Use explicit parentheses to be sure that operators are applied in the intended order. In particular, beware of operator precedence when using the colon operator to define index limits in `for` loops. For example, suppose there are N components of vector `theVector` and you want to display all but the last two components. This command, `for ( vIdx in 1:N-2 ) { show(theVector[vIdx]) }`, will not do what you want! There should be parentheses around `N-2`.
- Use visual white space to make your code easily readable by real human beings. Some R programmers like to suppress spaces because it makes the code more compact. But in my experience, space-free code can be extremely difficult to read and can be much more prone to errors.
- Use indenting to meaningfully group sections of the program. In particular, use RStudio's automatic indenting and parenthesis matching to help keep track of embedded loops and functions. In RStudio, select a section of a program or the entire program and then press ctrl-I (or menu Code → Reindent Lines) to see embedded sections properly indented.

A list of common mistakes in R could go on and on. For example, it is easy to mistakenly use a single equal sign, =, when intending to test for equality, which should use a double equal sign, == (or the `all.equal` function instead). If you want to check for the maximum or minimum across components of a set of vectors, it is easy to mistakenly use the `min` function instead of the `pmin` function. The `pmin` function has not been discussed above, but it will be described in Section 5.5. If you use the `attach` function successively on two different data frames that happen to have some shared column names, then the later `attach`ed variables will mask the earlier variables. (The `attach` function has not been previously discussed; you know how to get help.) It is natural to encounter many mistakes as you advance through programming in any language. But do not be deterred because, as an anonymous sage once remarked, "If we learn from our mistakes, shouldn't we try to make as many mistakes as possible?"

When you encounter an error, here are some hints regarding how to diagnose and repair the problem.

- The error messages displayed by R can sometimes be cryptic but are usually very helpful to isolate where the error occurred and what caused it. When an error message appears, don't just ignore what it says. Be sure to actually read the error message and see if it makes sense.
- Isolate the first point in the code that causes an error. Run the code sequentially a line at a time, or in small blocks, until the first error is encountered. Fix this error, and it might cure subsequent problems too.
- When you have found that a particular line of code causes an error, and that line involves a complex nesting of embedded functions, check the nested functions from the inside out. Do this by selecting, with the cursor, the inner-most variable or function and running it. Then work outward. An example of doing this is presented in Section 4.5.
- If you have defined a new function that had been working but mysteriously has stopped working, be sure that it is not relying on a variable that is defined outside the function. For examples, suppose that you specify `N=30` at the command line without putting it into the program. Then, in the program, you define a function, `addN = function( x ) { x+N }`. The function will work without complaint until you start a new R session that does not have `N` defined. In general, be wary of using variables inside a function without explicitly putting them as arguments in the function definition.

When you are dealing with complex programs that involve functions calling functions calling functions, the simple suggestions above can be inadequate. R has several advanced facilities devoted to interactive debugging. These include the functions `debug`, `browser`, and `traceback`. A thorough explanation of these functions could easily fill a large chapter by itself, and the required examples would be more complex than we can tackle after only this short introduction to R. Therefore I invite you to investigate the functions in more detail when you discover need for them. RStudio also has editing facilities for advanced debugging. You can investigate them at the RStudio web page, http://www.rstudio.com/ide/docs/debugging/overview.

3.8. GRAPHICAL PLOTS: OPENING AND SAVING

Creating and saving graphs in R is easy. Unfortunately, because graphical devices are different on different computer operating systems, the R commands for opening display windows and saving graphs can be different on different systems. I have created simple functions for opening graphical windows and saving their contents that can be used identically on Windows, Macintosh, and Linux operating systems. The functions are `openGraph` and `saveGraph`, and they are defined in the file `DBDA2E-utilities.R`. Here is an example of their use:

```
source("DBDA2E-utilities.R")                 # read defn. of openGraph, saveGraph
openGraph( width=3 , height=4 )              # open a graphics window
plot( x=1:4 , y=c(1,3,2,4) , type="o" )  # make a plot in the screen window
saveGraph( file="temp" , type="pdf" )        # save the graph as "temp.pdf"
```

Notice that the first line above `source`d the program `DBDA2E-utilities.R`, which holds the definitions of the functions. If that program is not in the current working directory, then R will not know where to find the program and R will return an error. Therefore, be sure to have the program in your current working directory. The `saveGraph` command saves the graph in the current working directory, so make sure that the working directory is set appropriately.

The `openGraph` command takes `width` and `height` arguments, as shown in the example above. It defaults to a 7 × 7 window if no width and height are specified. The `saveGraph` command takes `file` and `type` arguments. The `file` should specify the root part of the filename, because the `saveGraph` function will append the format `type` as an extension to the filename. The `saveGraph` function allows the following formats: pdf, eps (encapsulated postscript), jpg, jpeg (same as jpg), and png. You can use multiple `saveGraph` commands successively to save the same graph in different formats.

Chapter 4.5 (p. 93) provides an example of using various other arguments in the `plot` command. The example in that section also shows some ways to superimpose lines and annotations on a plot.

3.9. CONCLUSION

This chapter has surveyed only the most basic ideas and capabilities of R. It is worth repeating here that an excellent summary of basic R functions is available at http://cran.r-project.org/doc/contrib/Short-refcard.pdf. R has many elaborate packages that have been developed for different applications. I would list various web resources here, but aside from the main sites for R and RStudio, other sites are continuously evolving and changing. Therefore, search the web for the latest packages and documentation. There are also numerous books about R and specific applications of R.

3.10. EXERCISES

Look for more exercises at https://sites.google.com/site/doingbayesiandataanalysis/

Exercise 3.1. [Purpose: Actually doing Bayesian statistics, eventually, and the next exercises, immediately.] Install R on your computer. (And if that's not exercise, I don't know what is.)

Exercise 3.2. [Purpose: Being able to record and communicate the results of your analyses.] Open the program `ExampleOfR.R`. At the end, notice the section that produces a simple graph. Your job is to save the graph so you can incorporate it into documents in the future, as you would for reporting actual data analyses. Save the graph in a format that is compatible with your word processing software. Import the saved file into your document and explain, in text, what you did. (Notice that for some word processing systems you could merely copy and paste directly from R's graphic window to the document window. But the problem with this approach is that you have no other record of the graph produced by the analysis. We want the graph to be saved separately so that it can be incorporated into various reports at a future time.)

Exercise 3.3. [Purpose: Getting experience with the details of the command syntax within R.] Adapt the program `SimpleGraph.R` so that it plots a cubic function ($y = x^3$) over the interval $x \in [-3, +3]$. Save the graph in a file format of your choice. Include a properly commented listing of your code, along with the resulting graph.

CHAPTER 4

What Is This Stuff Called Probability?

Contents

Oh darlin' you change from one day to the next,
I'm feelin' deranged and just plain ol' perplexed.
I've learned to put up with your raves and your rants:
The mean I can handle but not variance.[1]

Inferential statistical techniques assign precise measures to our uncertainty about possibilities. Uncertainty is measured in terms of *probability*, and therefore we must establish the properties of probability before we can make inferences about it. This chapter introduces the basic ideas of probability. If this chapter seems too abbreviated for you, an excellent beginner's introduction to the topics of this chapter has been written by Albert and Rossman (2001, pp. 227-320).

[1] This chapter discusses ideas of probability distributions. Among those ideas are the technical definitions of the *mean* and *variance* of a distribution. The poem plays with colloquial meanings of those words.

Doing Bayesian Data Analysis, 2nd Edition
http://dx.doi.org/10.1016/B978-0-12-405888-0.00004-0

4.1. THE SET OF ALL POSSIBLE EVENTS

Suppose I have a coin that I am going to flip. How likely is it to come up a head? How likely is it to come up a tail?[2] How likely is it to come up a torso? Notice that when we contemplate the likelihood of each outcome, we have in mind a set of all possible outcomes. Torso is not one of the possible outcomes. Notice also that a single flip of a coin can result in only one outcome; it cannot be both heads and tails in a single flip. The outcomes are mutually exclusive.

Whenever we ask about how likely an outcome is, we always ask with a set of possible outcomes in mind. This set exhausts all possible outcomes, and the outcomes are all mutually exclusive. This set is called the *sample space*. The sample space is determined by the measurement operation we use to make an observation of the world. In all of our applications throughout the book, we take it for granted that there is a well-defined operation for making a measurement. For example, in flipping a coin, we take it for granted that there is a well-defined way to launch the coin and catch it, so that we can decide exactly when the coin has stopped its motion and is stable enough to be declared one outcome or the other.[3] As another example, in measuring the height of a person, we take it for granted that there is a well-defined way to pose a person against a ruler and decide exactly when we have a steady enough reading of the scale to declare a particular value for the person's height. The mechanical operationalization, mathematical formalization, and philosophical investigation of measurement could each have entire books devoted to them. We will have to settle for this single paragraph.

Consider the probability that a coin comes up heads when it is flipped. If the coin is fair, it should come up heads in about 50% of the flips. If the coin (or its flipping mechanism) is biased, then it will tend to come up heads more than or less than 50% of the flips. The probability of coming up heads can be denoted with parameter label θ (Greek letter theta); for example, a coin is fair when $\theta = 0.5$ (spoken "theta equals point five").

We can also consider our degree of belief that the coin is fair. We might know that the coin was manufactured by a government mint, and therefore we have a high degree of belief that the coin is fair. Alternatively, we might know that the coin was manufactured by Acme Magic and Novelty Company, and therefore we have a high degree of belief that the coin is biased. The degree of belief about a parameter can be denoted $p(\theta)$. If the coin was minted by the federal government, we might have a strong belief that the coin

[2] Many coins minted by governments have the picture of an important person's head on one side. This side is called "heads" or, technically, the "obverse." The reverse side is colloquially called "tails" as the natural opposite of "heads" even though there is rarely if ever a picture of a tail on the other side!

[3] Actually, it has been argued that *flipped* coins always have a 50% probability of coming up heads, and only *spun* coins can exhibit unequal head-tail probabilities (Gelman & Nolan, 2002). If this flip-spin distinction is important to you, please mentally substitute "spin" for "flip" whenever the text mentions flipping a coin. For empirical and theoretical studies of coin-flip probabilities, see, e.g., Diaconis, Holmes, and Montgomery (2007).

is fair; for example we might believe that $p(\theta = 0.5) = 0.99$, spoken "the probability that theta equals 0.5 is 99 percent." If the coin was minted by the novelty company, we might have a strong belief that the coin is biased; for example we might believe that $p(\theta = 0.5) = 0.01$ and that $p(\theta = 0.9) = 0.99$.

Both "probability" of head or tail outcome and "degree of belief" in biases refer to sample spaces. The sample space for flips of a coin consists of two possible outcomes: head and tail. The sample space for coin bias consists of a continuum of possible values: $\theta = 0.0$, $\theta = 0.01$, $\theta = 0.02$, $\theta = 0.03$, and all values in between, up to $\theta = 1.0$. When we flip a given coin, we are sampling from the space of head or tail. When we grab a coin at random from a sack of coins, in which each coin may have a different bias, we are sampling from the space of possible biases.

4.1.1. Coin flips: Why you should care

The fairness of a coin might be hugely consequential for high stakes games, but it isn't often in life that we flip coins and care about the outcome. So why bother studying the statistics of coin flips?

Because coin flips are a surrogate for myriad other real-life events that we do care about. For a given type of heart surgery, we may classify the patient outcome as survived more than a year or not, and we may want to know what is the probability that patients survive more than one year. For a given type of drug, we may classify the outcome as having a headache or not, and we may want to know the probability of headache. For a survey question, the outcome might be agree or disagree, and we want to know the probability of each response. In a two-candidate election, the two outcomes are candidate A and candidate B, and before the election itself we want to estimate, from a poll, the probability that candidate A will win. Or perhaps you are studying arithmetic ability by measuring accuracy on a multi-item exam, for which the item outcomes are correct or wrong. Or perhaps you are researching brain lateralization of a particular cognitive process in different subpopulations, in which case the outcomes are right-lateralized or left-lateralized, and you are estimating the probability of being left-lateralized in the subpopulation.

Whenever we are discussing coin flips, which might not be inherently fascinating to you, keep in mind that we could be talking about some domain in which you are actually interested! The coins are merely a generic representative of a universe of analogous applications.

4.2. PROBABILITY: OUTSIDE OR INSIDE THE HEAD

Sometimes we talk about probabilities of outcomes that are "out there" in the world. The face of a flipped coin is such an outcome: We can observe the flip, and the probability of coming up heads can be estimated by observing several flips.

But sometimes we talk about probabilities of things that are not so clearly "out there," and instead are just possible beliefs "inside the head." Our belief about the fairness of a coin is an example of something inside the head. The coin may have an intrinsic physical bias, but now I am referring to our *belief* about the bias. Our beliefs refer to a space of mutually exclusive and exhaustive possibilities. It might be strange to say that we randomly sample from our beliefs, like we randomly sample from a sack of coins. Nevertheless, the mathematical properties of probabilities outside the head and beliefs inside the head are the same in their essentials, as we will see.

4.2.1. Outside the head: Long-run relative frequency

For events outside the head, it's intuitive to think of probability as being the long-run relative frequency of each possible outcome. For example, if I say that for a fair coin the probability of heads is 0.5, what I mean is that if we flipped the coin many times, about 50% of the flips would come up heads. In the long run, after flipping the coin many, many times, the relative frequency of heads would be very nearly 0.5.

We can determine the long-run relative frequency by two different ways. One way is to approximate it by actually sampling from the space many times and tallying the number of times each event happens. A second way is by deriving it mathematically. These two methods are now explored in turn.

4.2.1.1 Simulating a long-run relative frequency

Suppose we want to know the long-run relative frequency of getting heads from a fair coin. It might seem blatantly obvious that we should get about 50% heads in any long sequence of flips. But let's pretend that it's not so obvious: All we know is that there's some underlying process that generates an "H" or a "T" when we sample from it. The process has a parameter called θ, whose value is $\theta = 0.5$. If that's all we know, then we can approximate the long-run probability of getting an "H" by simply repeatedly sampling from the process. We sample from the process N times, tally the number of times an "H" appeared, and estimate the probability of H by the relative frequency.

It gets tedious and time-consuming to sample a process manually, such as flipping a coin. Instead, we can let the computer do the repeated sampling much faster (and hopefully the computer feels less tedium than we would). Figure 4.1 shows the results of a computer simulating many flips of a fair coin. The R programming language has pseudo-random number generators built into it, which we will use often.[4] On the first flip, the computer randomly generates a head or a tail. It then computes the proportion

[4] Pseudo-random number generators (PRNGs) are not actually random; they are in fact deterministic. But the properties of the sequences they generate mimic the properties of random processes. The methods used in this book rely heavily on the quality of PRNGs, which is an active area of intensive research (e.g., Deng & Lin, 2000; Gentle, 2003).

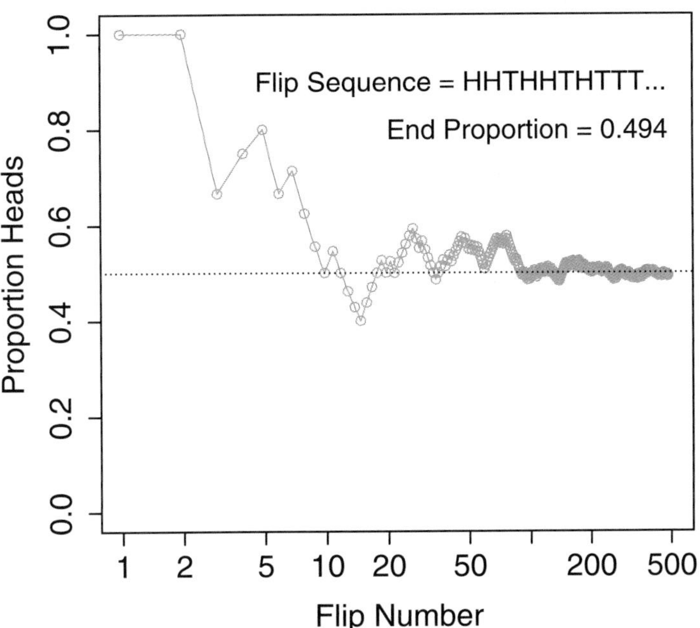

Figure 4.1 Running proportion of heads when flipping a coin. The x-axis is plotted on a logarithmic scale so that you can see the details of the first few flips but also the long-run trend after many flips. R code for producing this figure is discussed in Section 4.5.

of heads obtained so far. If the first flip was a head, then the proportion of heads is $1/1 = 1.0$. If the first flip was a tail, then the proportion of heads is $0/1 = 0.0$. Then the computer randomly generates a second head or tail, and computes the proportion of heads obtained so far. If the sequence so far is HH, then the proportion of heads is $2/2 = 1.0$. If the sequence so far is HT or TH, then the proportion of heads is $1/2 = 0.5$. If the sequence so far is TT, then the proportion of heads is $0/2 = 0.0$. Then the computer generates a third head or tail, and computes the proportion of heads so far, and so on for many flips. Figure 4.1 shows the running proportion of heads as the sequence continues.

Notice in Figure 4.1 that at the end of the long sequence, the proportion of heads is *near* 0.5 but not necessarily exactly equal to 0.5. This discrepancy reminds us that even this long run is still just a finite random sample, and there is no guarantee that the relative frequency of an event will match the true underlying probability of the event. That's why we say we are *approximating* the probability by the long-run relative frequency.

4.2.1.2 Deriving a long-run relative frequency

Sometimes, when the situation is simple enough mathematically, we can derive the exact long-run relative frequency. The case of the fair coin is one such simple situation. The sample space of the coin consists of two possible outcomes, head and tail. By the assumption of fairness, we know that each outcome is equally likely. Therefore, the long-run relative frequency of heads should be exactly one out of two, i.e., 1/2, and the long-run relative frequency of tails should also be exactly 1/2.

This technique is easily extended to other simple situations. Consider, for example, a standard six-sided die. It has six possible outcomes, namely 1 dot, 2 dots, ..., 6 dots. If we assume that the die is fair, then the long-run relative frequency of each outcome should be exactly 1/6.

Suppose that we put different dots on the faces of the six-side die. In particular, suppose that we put 1 dot on one face, 2 dots on two faces, and 3 dots on the remaining three faces. We still assume that each of the six faces is equally likely. Then the long-run relative frequency of 1 dot is exactly 1/6, and the long-run relative frequency of 2 dots is exactly 2/6, and the long-run relative frequency of 3 dots is exactly 3/6.

4.2.2. Inside the head: Subjective belief

How strongly do you believe that a coin minted by the US government is fair? If you believe that the coin could be slightly different than exactly fair, then how strongly do you believe that the probability of heads is $\theta = 0.51$? Or $\theta = 0.49$? If instead you are considering a coin that is ancient, asymmetric, and lopsided, do you believe that it inherently has $\theta = 0.50$? How about a coin purchased at a magic shop? We are not talking here about the true, inherent probability that the coin will come up heads. We are talking about our degree of belief in each possible probability.

To specify our subjective beliefs, we have to specify how likely we think each possible outcome is. It can be hard to pin down mushy intuitive beliefs. In the next section, we explore one way to "calibrate" subjective beliefs, and in the subsequent section we discuss ways to mathematically describe degrees of belief.

4.2.2.1 Calibrating a subjective belief by preferences

Consider a simple question that might affect travelers: How strongly do you believe that there will be a snowstorm that closes the interstate highways near Indianapolis next New Year's Day? Your job in answering that question is to provide a number between 0 and 1 that accurately reflects your belief probability. One way to come up with such a number is to calibrate your beliefs relative to other events with clear probabilities.

As a comparison event, consider a marbles-in-sack experiment. In a sack we put 10 marbles: 5 red, and 5 white. We shake the sack and then draw a marble at random. The probability of getting a red marble is, of course, $5/10 = 0.5$. We will use this sack of marbles as a comparison for considering snow in Indianapolis on New Year's Day.

Consider the following two gambles that you can choose from:

- Gamble A: You get $100 if there is a traffic stopping snowstorm in Indianapolis next New Year's Day.
- Gamble B: You get $100 if you draw a red marble from a sack of marbles with 5 red and 5 white marbles.

Which gamble would you prefer? If you prefer Gamble B, that means you think there is less than a 50-50 chance of a traffic-stopping snowstorm in Indy. So at least you now know that your subjective belief about the probability of traffic-stopping snowstorm is less than 0.5.

We can narrow down the degree of belief by considering other comparison gambles. Consider these two gambles:

- Gamble A: You get $100 if there is a traffic stopping snowstorm in Indianapolis next New Year's Day.
- Gamble C: You get $100 if you draw a red marble from a sack of marbles with 1 red and 9 white marbles.

Which gamble would you prefer? If you now prefer Gamble A, that means you think there is more than a 10% chance of traffic-stopping snowstorm in Indy on New Year's Day. Taken together, the two comparison gambles have told you that your subjective probability lies somewhere between 0.1 and 0.5. We could continue to consider preferences against other candidate gambles to calibrate your subjective belief more accurately.

4.2.2.2 Describing a subjective belief mathematically

When there are several possible outcomes in a sample space, it might be too much effort to try to calibrate your subjective belief about every possible outcome. Instead, you can use a mathematical function to summarize your beliefs.

For example, you might believe that the average American woman is 5′4″ tall, but be open to the possibility that the average might be somewhat above or below that value. It is too tedious and may be impossible to specify your degree of belief that the average height is 4′1″, or 4′2″, or 4′3″, and so on up through 6′1″, 6′2″, and 6′3″ etc. So you might instead describe your degree of belief by a bell-shaped curve that is highest at 5′4″ and drops off symmetrically above and below that most-likely height. You can change the width and center of the curve until it seems to best capture your subjective belief. Later in the book, we will talk about exact mathematical formulas for functions like these, but the point now is merely to understand the idea that mathematical functions can define curves that can be used to describe degrees of belief.

4.2.3. Probabilities assign numbers to possibilities

In general, a probability, whether it's outside the head or inside the head, is just a way of assigning numbers to a set of mutually exclusive possibilities. The numbers, called "probabilities," merely need to satisfy three properties (Kolmogorov, 1956):

1. A probability value must be nonnegative (i.e., zero or positive).
2. The sum of the probabilities across all events in the entire sample space must be 1.0 (i.e., one of the events in the space must happen, otherwise the space does not exhaust all possibilities).
3. For any two mutually exclusive events, the probability that one *or* the other occurs is the *sum* of their individual probabilities. For example, the probability that a fair six-sided die comes up 3-dots *or* 4-dots is $1/6 + 1/6 = 2/6$.

Any assignment of numbers to events that respects those three properties will also have all the properties of probabilities that we will discuss below. So whether a probability is thought of as a long-run relative frequency of outcomes in the world, or as a magnitude of a subjective belief, it behaves the same way mathematically.

4.3. PROBABILITY DISTRIBUTIONS

A probability *distribution* is simply a list of all possible outcomes and their corresponding probabilities. For a coin, the probability distribution is trivial: We list two outcomes (head and tail) and their two corresponding probabilities (θ and $1 - \theta$). For other sets of outcomes, however, the distribution can be more complex. For example, consider the height of a randomly selected person. There is some probability that the height will be 60.2″, some probability that the height will be 68.9″, and so forth, for every possible exact height. When the outcomes are continuous, like heights, then the notion of probability takes on some subtleties, as we will see.

4.3.1. Discrete distributions: Probability mass

When the sample space consists of discrete outcomes, then we can talk about the probability of each distinct outcome. For example, the sample space of a flipped coin has two discrete outcomes, and we talk about the probability of head or tail. The sample space of a six-sided die has six discrete outcomes, and we talk about the probability of 1 dot, 2 dots, and so forth.

For continuous outcome spaces, we can *discretize* the space into a finite set of mutually exclusive and exhaustive "bins." For example, although heights of people are a continuous scale, we can divide the scale into a finite number of intervals, such as $< 51''$, $51''$ to $53''$, $53''$ to $55''$, $55''$ to $57''$, ..., $> 83''$. Then we can talk about the probability that a randomly selected person falls into any of those intervals. Suppose that we randomly sample 10,000 people and measure the heights very accurately. The top panel of Figure 4.2 shows a scatter plot of the 10,000 measurements, with vertical dashed lines marking the intervals. In particular, the number of measurements that fall within the interval $63''$ to $65''$ is 1,473, which means that the (estimated) probability of falling in that interval is $1,473/10,000 = 0.1473$.

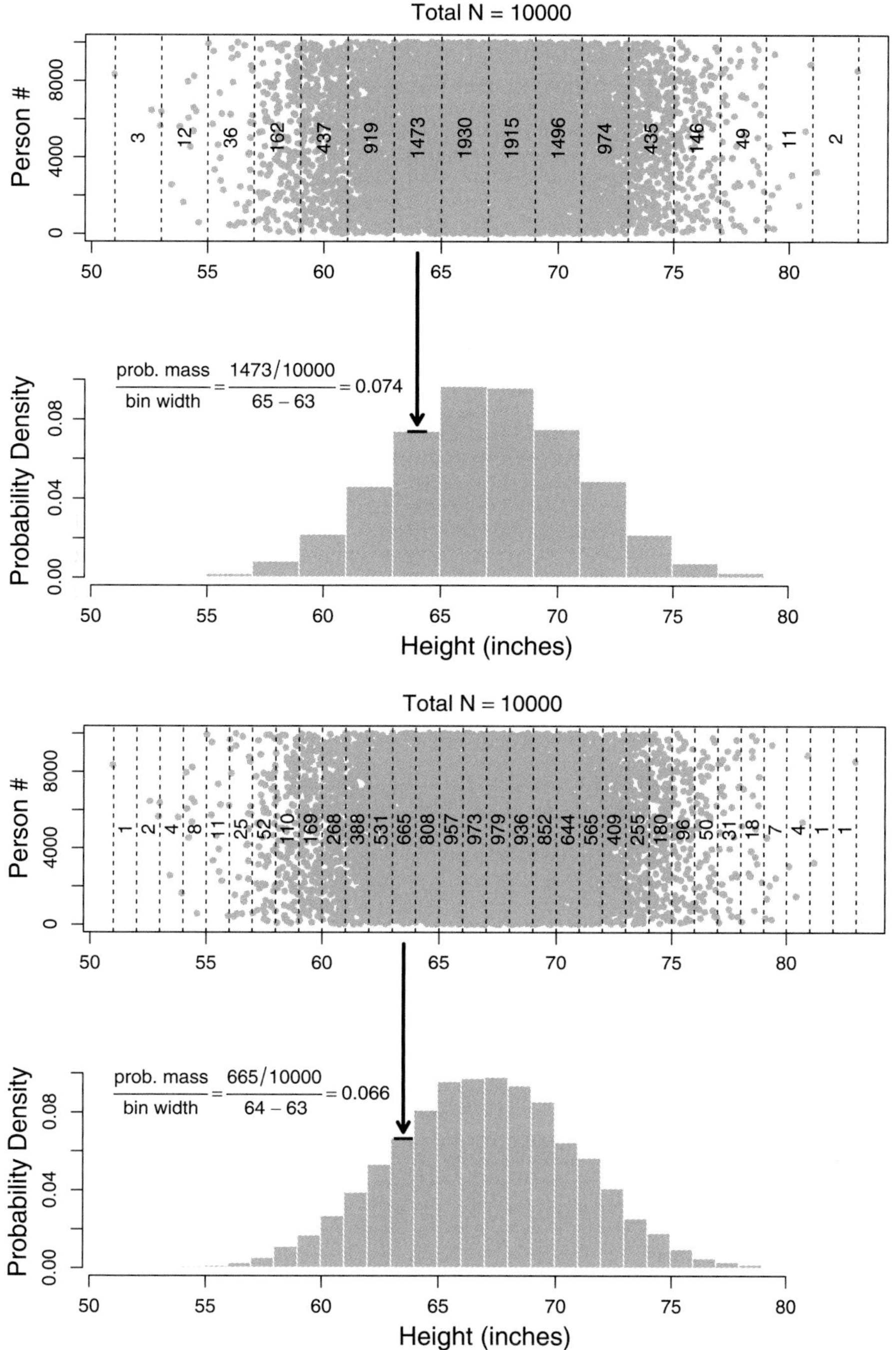

Figure 4.2 Examples of computing probability density. Within each main panel, the upper plot shows a scatter of 10,000 heights of randomly selected people, and the lower plot converts into probability density for the particular selection of bins depicted.

The probability of a discrete outcome, such as the probability of falling into an interval on a continuous scale, is referred to as a probability *mass*. Loosely speaking, the term "mass" refers the amount of stuff in an object. When the stuff is probability and the object is an interval of a scale, then the mass is the proportion of the outcomes in the interval. Notice that the sum of the probability masses across the intervals must be 1.

4.3.2. Continuous distributions: Rendezvous with density[5]

If you think carefully about a continuous outcome space, you realize that it becomes problematic to talk about the probability of a specific value on the continuum, as opposed to an interval on the continuum. For example, the probability that a randomly selected person has height (in inches) of exactly 67.21413908 ... is essentially nil, and that is true for *any* exact value you care to think of. We can, however, talk about the probability mass of intervals, as we did in the example above. The problem with using intervals, however, is that their widths and edges are arbitrary, and wide intervals are not very precise. Therefore, what we will do is make the intervals infinitesimally narrow, and instead of talking about the infinitesimal probability mass of each infinitesimal interval, we will talk about the ratio of the probability mass to the interval width. That ratio is called the probability *density*.

Loosely speaking, density is the amount of stuff per unit of space it takes up. Because we are measuring amount of stuff by its mass, then density is the mass divided by the amount space it occupies. Notice that a small mass can have a high density: A milligram of the metal lead has a density of more than 11 grams per cubic centimeter, because the milligram takes up only 0.000088 cubic centimeters of space. Importantly, we can conceive of density *at a point* in space, as the ratio of mass to space when the considered space shrinks to an infinitesimal region around the point.

Figure 4.2 shows examples of this idea. As previously mentioned, the upper panel shows a scatter plot of heights (in inches) of 10,000 randomly selected people, with intervals of width 2.0. To compute the average probability density in the interval $63''$ to $65''$, we divide the interval's probability mass by the interval's width. The probability mass is (estimated as) $1,473/10,000 = 0.1473$, and the interval width is 2.0 units (i.e., $65 - 63$), hence the average probability density in the interval is 0.074 (rounded). This is the average probability density over the interval. For a more precise density over a narrower interval, consider the lower panel of Figure 4.2. The interval $63''$ to $64''$ has (estimated) mass of $665/10,000$, and hence the average probability density in the interval is $(665/10,000)/(64 - 63) = 0.066$ (rounded). We can continue narrowing the intervals and computing density.

[5] "There is a mysterious cycle in human events. To some generations much is given. Of other generations much is expected. This generation of Americans has a rendezvous with destiny." Franklin Delano Roosevelt, 1936.

The example in Figure 4.2 illustrates the estimation of density from a finite sample across noninfinitesimal intervals. But to compute density for an infinitesimal interval, we must conceive of an infinite population continuously spread across the scale. Then, even an infinitesimal interval may contain some nonzero (though infinitesimal) amount of probability mass, and we can refer to probability density at a point on the scale. We will soon see mathematical examples of this idea.

Figure 4.3 shows another example, to emphasize that probability densities can be larger than 1, even though probability mass cannot exceed 1. The upper panel of Figure 4.3 shows heights in inches of 10,000 randomly selected *doors* that are manufactured to be 7 feet (84 inches) tall. Because of the regularity of the manufacturing process, there is only a little random variation among the heights of the doors, as can be seen in the figure by the fact that the range of the scale is small, going only from 83.6″ to 84.4″. Thus, all the probability mass is concentrated over a small range of the scale.

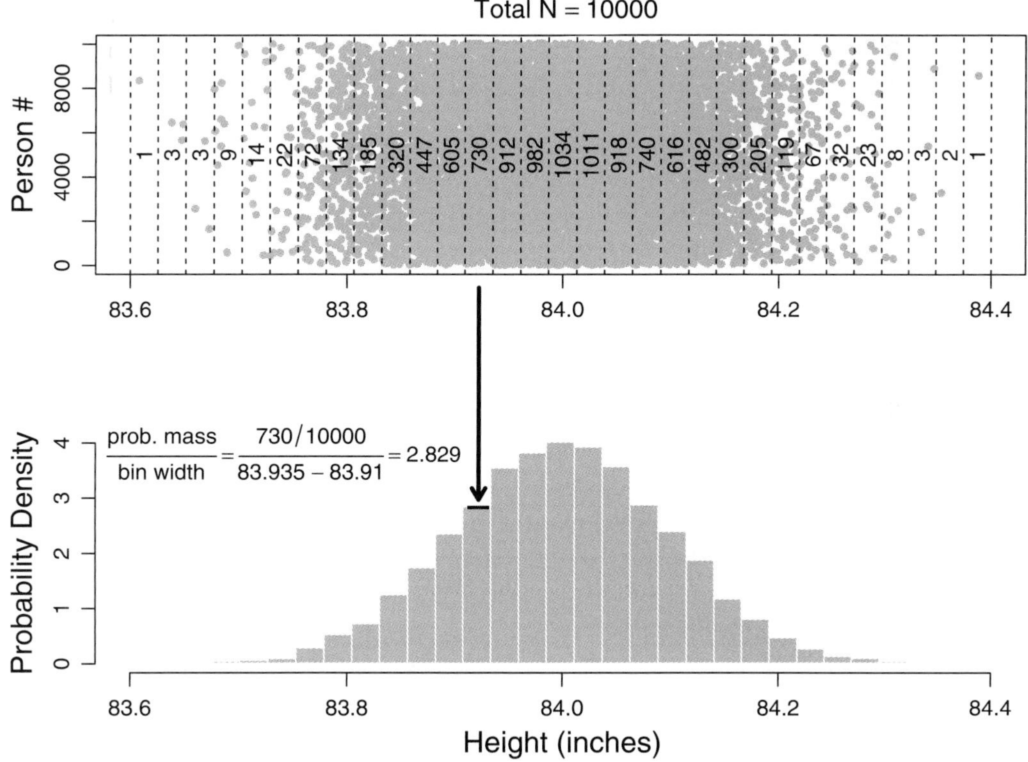

Figure 4.3 Example of probability density greater than 1.0. Here, all the probability mass is concentrated into a small region of the scale, and therefore the density can be high at some values of the scale. The annotated calculation of density uses rounded interval limits for display. (For this example, we can imagine that the points refer to manufactured doors instead of people, and therefore the y axis of the top panel should be labeled "Door" instead of "Person.")

Consequently, the probability density near values of 84 inches exceeds 1.0. For example, in the interval 83.9097″ to 83.9355″, there is a probability mass of $730/10,000 = 0.073$. But this mass is concentrated over a bin width of only $83.9355 - 83.9097 = 0.0258$, hence the average density within the interval is $0.073/0.0258 = 2.829$. There is nothing mysterious about probability densities larger than 1.0; it means merely that there is a high concentration of probability mass relative to the scale.

4.3.2.1 Properties of probability density functions

In general, for any continuous value that is split up into intervals, the sum of the probability masses of the intervals must be 1, because, by definition of making a measurement, some value of the measurement scale must occur. We can write that fact as an equation, but we need to define some notation first. Let the continuous variable be denoted x. The width of an interval on x is denoted Δx (the symbol "Δ" is the Greek letter, capital delta). Let i be an index for the intervals, and let $[x_i, x_i{+}\Delta x]$ denote the interval between x_i and $x_i{+}\Delta x$. The probability *mass* of the ith interval is denoted $p([x_i, x_i{+}\Delta x])$. Then the sum of those probability masses must be 1, which is denoted as follows:

$$\sum_i p\left([x_i, x_i{+}\Delta x]\right) = 1 \tag{4.1}$$

Recall now the definition of probability density: It is the ratio of probability mass over interval width. We can rewrite Equation 4.1 in terms of the density of each interval, by dividing and multiplying by Δx, as follows:

$$\sum_i \Delta x \frac{p\left([x_i, x_i{+}\Delta x]\right)}{\Delta x} = 1 \tag{4.2}$$

In the limit, as the interval width becomes infinitesimal, we denote the width of the interval around x as $\mathrm{d}x$ instead of Δx, and we denote the probability *density* in the infinitesimal interval around x simply as $p(x)$. The probability density $p(x)$ is not to be confused with $p([x_i, x_i{+}\Delta x])$, which was the probability mass in an interval. Then the summation in Equation 4.2 becomes an integral:

$$\underbrace{\sum_i}_{\int} \underbrace{\Delta x}_{\mathrm{d}x} \underbrace{\frac{p([x_i, x_i{+}\Delta x])}{\Delta x}}_{p(x)} = 1 \quad \text{that is,} \quad \int \mathrm{d}x\, p(x) = 1 \tag{4.3}$$

In this book, integrals will be written with the $\mathrm{d}x$ term next to the integral sign, as in Equation 4.3, instead of at the far right end of the expression. Although this placement is not the most conventional notation, it is neither wrong nor unique to this book. The placement of $\mathrm{d}x$ next to the integral sign makes it easy to see what variable is being integrated over, without have to put subscripts on the integral sign. This usage can be especially helpful if we encounter integrals of functions that involve multiple variables. The placement of dx next to the integral sign also maintains grouping of terms when rewriting discrete sums and integrals, such that $\sum_x$ becomes $\int \mathrm{d}x$ without having to move the dx to the end of the expression.

To reiterate, in Equation 4.3, $p(x)$ is the probability density in the infinitesimal interval around x. Typically, we let context tell us whether we are referring to a probability mass or a probability density, and use the same notation, $p(x)$, for both. For example, if x is the value of the face of a six-sided die, then $p(x)$ is a probability mass. If x is the exact point-value of height, then $p(x)$ is a probability density. There can be "slippage" in the usage, however. For example, if x refers to height, but the scale is discretized into intervals, then $p(x)$ is really referring to the probability mass of the interval in which x falls. Ultimately, you'll have to be attentive to context and tolerant of ambiguity.

4.3.2.2 The normal probability density function

Any function that has only nonnegative values and integrates to 1 (i.e., satisfies Equation 4.3) can be construed as a probability density function. Perhaps the most famous probability density function is the *normal* distribution, also known as the Gaussian distribution. A graph of the normal curve is a well-known bell shape; an example is shown in Figure 4.4.

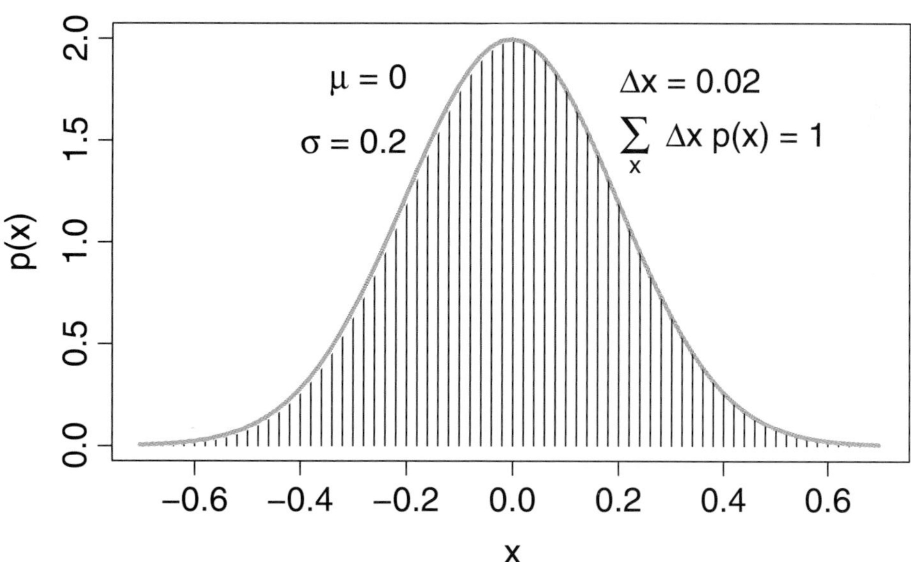

Figure 4.4 A normal probability density function, shown with a comb of narrow intervals. The integral is approximated by summing the width times height of each interval.

The mathematical formula for the normal probability density has two parameters: μ (Greek mu) is called the *mean* of the distribution and σ (Greek sigma) is called the *standard deviation*. The value of μ governs where the middle of the bell shape falls on the x-axis, so it is called a location parameter, and the value of σ governs how wide the bell is, so it is called a scale parameter. As discussed in Section 2.2, you can think of the parameters as control knobs with which to manipulate the location and scale of the distribution. The mathematical formula for the normal probability density is

$$p(x) = \frac{1}{\sigma\sqrt{2\pi}} \exp\left(-\frac{1}{2}\left[\frac{x-\mu}{\sigma}\right]^2\right). \tag{4.4}$$

Figure 4.4 shows an example of the normal distribution for specific values of μ and σ as indicated. Notice that the peak probability density can be greater than 1.0 when the standard deviation, σ, is small. In other words, when the standard deviation is small, a lot of probability mass is squeezed into a small interval, and consequently the probability density in that interval is high.

Figure 4.4 also illustrates that the area under the normal curve is, in fact, 1. The x axis is divided into a dense comb of small intervals, with width denoted Δx. The integral of the normal density is approximated by summing the masses of all the tiny intervals as in Equation 4.2. As can be seen in the text within the graph, the sum of the interval areas is essentially 1.0. Only rounding error, and the fact that the extreme tails of the distribution are not included in the sum, prevent the sum from being exactly 1.

4.3.3. Mean and variance of a distribution

When we have a numerical (not just categorical) value x that is generated with probability $p(x)$, we can wonder what would be its average value in the long run, if we repeatedly sampled values of x. For example, if we have a fair six-sided die, then each of its six values should come up 1/6th of the time in the long run, and so the long-run average value of the die is $(1/6)1 + (1/6)2 + (1/6)3 + (1/6)4 + (1/6)5 + (1/6)6 = 3.5$. As another example, if we play a slot machine for which we win \$100 with probability 0.001, we win \$5 with probability 0.14, and otherwise we lose \$1, then in the long run our payoff is $(0.001)(\$100) + (0.14)(\$5) + (0.859)(-\$1) = -\$0.059$. In other words, in the long run we lose about 6 cents per pull of the bandit's arm. Notice what we did in those calculations: We weighted each possible outcome by the probability that it happens. This procedure defines the *mean* of a probability distribution, which is also called the *expected value*, and which is denoted $E[x]$:

$$E[x] = \sum_x p(x)\, x \tag{4.5}$$

Equation 4.5 applies when the values of x are discrete, and so $p(x)$ denotes a probability mass. When the values of x are continuous, then $p(x)$ denotes a probability density and the sum becomes an integral over infinitesimal intervals:

$$E[x] = \int dx\, p(x)\, x \tag{4.6}$$

The conceptual meaning is the same whether x is discrete or continuous: $E[x]$ is the long-run average of the values.

The mean value of a distribution typically lies near the distribution's middle, intuitively speaking. For example, the mean of a normal distribution turns out to be the value of its parameter μ. In other words, it turns out to be the case that $E[x] = \mu$. A specific example of that fact is illustrated in Figure 4.4, where it can be seen that the bulk of the distribution is centered over $x = \mu$; see the text in the figure for the exact value of μ.

Here's an example of computing the mean of a continuous distribution, using Equation 4.6. Consider the probability density function $p(x) = 6x(1 - x)$ defined over the interval $x \in [0, 1]$. This really is a probability density function: It's an upside down parabola starting at $x = 0$, peaking over $x = 0.5$, and dropping down to baseline again at $x = 1$. Because it is a symmetric distribution, intuition tells us that the mean should be at its midpoint, $x = 0.5$. Let's check that it really is:

$$\begin{aligned} E[x] &= \int dx\, p(x)\, x \\ &= \int_0^1 dx\ 6x(1 - x)\, x \\ &= 6 \int_0^1 dx\ \left(x^2 - x^3\right) \\ &= 6 \left[\frac{1}{3}x^3 - \frac{1}{4}x^4 \right]_0^1 \\ &= 6 \left[\left(\frac{1}{3}1^3 - \frac{1}{4}1^4 \right) - \left(\frac{1}{3}0^3 - \frac{1}{4}0^4 \right) \right] \\ &= 0.5 \end{aligned} \tag{4.7}$$

We will be doing relatively little calculus in this book, and Equation 4.7 is about as advanced as we'll get. If your knowledge of calculus is rusty, don't worry, just keep reading for conceptual understanding.

The *variance* of a probability distribution is a number that represents the dispersion of the distribution away from its mean. There are many conceivable definitions of how far the values of x are dispersed from their mean, but the definition used for the specific term "variance" is based on the squared difference between x and the mean. The definition of variance is simply the mean squared deviation (MSD) of the x values from their mean:

$$\text{var}_x = \int \mathrm{d}x\, p(x)\ (x - E[x])^2 \tag{4.8}$$

Notice that Equation 4.8 is just like the formula for the mean (Equation 4.6) except that instead of integrating x weighted by x's probability, we're integrating $(x - E[x])^2$ weighted by x's probability. In other words, the variance is just the average value of $(x - E[x])^2$. For a discrete distribution, the integral in Equation 4.8 becomes a sum, analogous to the relationship between Equations 4.5 and 4.6. The square root of the variance, sometimes referred to as root mean squared deviation (RMSD), is called the *standard deviation* of the distribution.

The variance of the normal distribution turns out to be the value of its parameter σ squared. Thus, for the normal distribution, $\text{var}_x = \sigma^2$. In other words, the standard deviation of the normal distribution is the value of the parameter σ. In a normal distribution, about 34% of the distribution lies between μ and $\mu + \sigma$ (see Exercise 4.5). Take a look at Figure 4.4 and visually identify where μ and $\mu + \sigma$ lie on the x axis (the values of μ and σ are indicated in the text within the figure) to get a visual impression of how far one standard deviation lies from the mean. Be careful, however, not to overgeneralize to distributions with other shapes: Non-normal distributions can have very different areas between their mean and first standard deviation.

A probability distribution can refer to probability of measurement values or of parameter values. The probability can be interpreted either as how much a value could be sampled from a generative process, or as how much credibility the value has relative to other values. When $p(\theta)$ represents credibility values of θ, instead of the probability of sampling θ, then the mean of $p(\theta)$ can be thought of as a value of θ that represents a typical credible value. The standard deviation of θ, which measures how wide the distribution is, can be thought of as a measure of uncertainty across candidate values. If the standard deviation is small, then we believe strongly in values of θ near the mean. If the standard deviation is large, then we are not very certain about what value of θ to believe in. This notion of standard deviation as representing uncertainty will reappear often. A related measure of the width of a distribution is the highest density interval, described below.

4.3.3.1 Mean as minimized variance

An alternative conceptual emphasis starts with the definition of variance and derives a definition of mean, instead of starting with the mean and working to a definition of variance. Under this alternative conception, the goal is to define a value for the *central tendency* of a probability distribution. A value represents the central tendency of the distribution if the value is close to the highly probable values of the distribution.

Therefore, we define the central tendency of a distribution as whatever value M minimizes the long-run expected distance between it and all the other values of x. But how should we define "distance" between values? One way to define distance is as squared difference: The distance between x and M is $(x - M)^2$. One virtue of this definition is that the distance from x to M is the same as the distance from M to x, because $(x - M)^2 = (M - x)^2$. But the primary virtue of this definition is that it makes a lot of subsequent algebra tractable (which will not be rehearsed here). The central tendency is, therefore, the value M that minimizes the expected value of $(x - M)^2$. Thus, we want the value M that minimizes $\int dx\, p(x)\, (x - M)^2$. Does that look familiar? It's essentially the formula for the variance of the distribution, in Equation 4.8, but here thought of as a function of M. Here's the punch line: It turns out that the value of M that minimizes $\int dx\, p(x)\, (x - M)^2$ is $E[x]$. In other words, the mean of the distribution is the value that minimizes the expected squared deviation. In this way, the mean is a central tendency of the distribution.

As an aside, if the distance between M and x is defined instead as $|x - M|$, then the value that minimizes the expected distance is called the *median* of the distribution. An analogous statement applies to the *modes* of a distribution, with distance defined as zero for any exact match, and one for any mismatch.

4.3.4. Highest density interval (HDI)

Another way of summarizing a distribution, which we will use often, is the highest density interval, abbreviated HDI.[6] The HDI indicates which points of a distribution are most credible, and which cover most of the distribution. Thus, the HDI summarizes the distribution by specifying an interval that spans most of the distribution, say 95% of it, such that every point inside the interval has higher credibility than any point outside the interval.

Figure 4.5 shows examples of HDIs. The upper panel shows a normal distribution with mean of zero and standard deviation of one. Because this normal distribution is symmetric around zero, the 95% HDI extends from -1.96 to $+1.96$. The area under the curve between these limits, and shaded in grey in Figure 4.5, has area of 0.95. Moreover, the probability density of any x within those limits has higher probability density than any x outside those limits.

[6] Some authors refer to the HDI as the HDR, which stands for highest density *region*, because a region can refer to multiple dimensions, but an interval refers to a single dimension. Because we will almost always consider the HDI of one parameter at a time, I will use "HDI" in an effort to reduce confusion. Some authors refer to the HDI as the HPD, to stand for highest probability density, but which I prefer not to use because it takes more space to write "HPD interval" than "HDI." Some authors refer to the HDI as the HPD, to stand for highest *posterior* density, but which I prefer not to use because *prior* distributions also have HDIs.

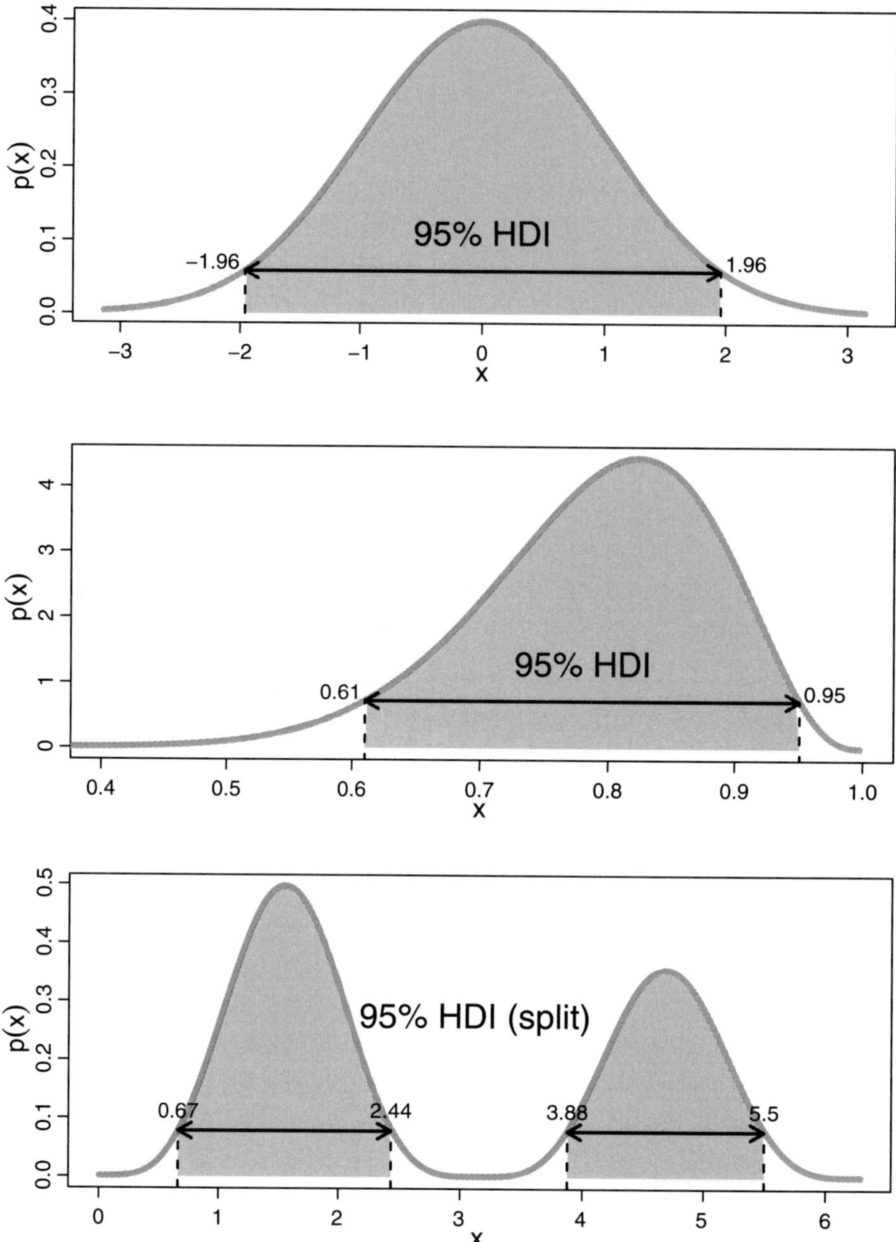

Figure 4.5 Examples of 95% highest density intervals (HDIs). For each example, all the x values inside the interval have higher density than any x value outside the interval, and the total mass of the points inside the interval is 95%. The 95% area is shaded, and it includes the zone below the horizontal arrow. The horizontal arrow indicates the width of the 95% HDI, with its ends annotated by (rounded) x values. The height of the horizontal arrow marks the minimal density exceeded by all x values inside the 95% HDI.

The middle panel of Figure 4.5 shows a 95% HDI for a skewed distribution. By definition, the area under the curve between the 95% HDI limits, shaded in grey in the figure, has area of 0.95, and the probability density of any x within those limits is higher than any x outside those limits. Importantly, notice that the area in the left tail, less than the left HDI limit, is larger than the area in right tail, greater than the right HDI limit. In other words, the HDI does not necessarily produce equal-area tails outside the HDI. (For those of you who have previously encountered the idea of equal-tailed credible intervals, you can look ahead to Figure 12.2, p. 342, for an explanation of how HDIs differ from equal-tailed intervals.)

The lower panel of Figure 4.5 shows a fanciful bimodal probability density function. In many realistic applications, multimodal distributions such as this do not arise, but this example is useful for clarifying the definition of an HDI. In this case, the HDI is split into two subintervals, one for each mode of the distribution. However, the defining characteristics are the same as before: The region under the curve within the 95% HDI limits, shaded in grey in the figure, has total area of 0.95, and any x within those limits has higher probability density than any x outside those limits.

The formal definition of an HDI is just a mathematical expression of the two essential characteristics. The 95% HDI includes all those values of x for which the density is at least as big as some value W, such that the integral over all those x values is 95%. Formally, the values of x in the 95% HDI are those such that $p(x) > W$ where W satisfies $\int_{x\,:\,p(x)>W} \mathrm{d}x\; p(x) = 0.95$.

When the distribution refers to credibility of values, then the width of the HDI is another way of measuring uncertainty of beliefs. If the HDI is wide, then beliefs are uncertain. If the HDI is narrow, then beliefs are relatively certain. As will be discussed at length in Chapter 13, sometimes the goal of research is to obtain data that achieve a reasonably high degree of certainty about a particular parameter value. The desired degree of certainty can be measured as the width of the 95% HDI. For example, if μ is a measure of how much a drug decreases blood pressure, the researcher may want to have an estimate with a 95% HDI width no larger than 5 units on the blood pressure scale. As another example, if θ is a measure of a population's preference for candidate A over candidate B, the researcher may want to have an estimate with a 95% HDI width no larger than 10 percentage points.

4.4. TWO-WAY DISTRIBUTIONS

There are many situations in which we are interested in the conjunction of two outcomes. What is the probability of being dealt a card that is both a queen *and* a heart? What is the probability of meeting a person with both red hair *and* green eyes? When playing a board game involving a die and a spinner, we have degrees of belief about both the die *and* the spinner being fair.

Table 4.1 Proportions of combinations of hair color and eye color

	Hair color				
Eye color	**Black**	**Brunette**	**Red**	**Blond**	**Marginal (eye color)**
Brown	0.11	0.20	0.04	0.01	0.37
Blue	0.03	0.14	0.03	0.16	0.36
Hazel	0.03	0.09	0.02	0.02	0.16
Green	0.01	0.05	0.02	0.03	0.11
Marginal (hair color)	0.18	0.48	0.12	0.21	1.0

Some rows or columns may not sum exactly to their displayed marginals because of rounding error from the original data. Data adapted from Snee (1974).

As a specific example for developing these ideas, consider Table 4.1, which shows the probabilities of various combinations of people's eye color and hair color. The data come from a particular convenience sample (Snee, 1974), and are not meant to be representative of any larger population. Table 4.1 considers four possible eye colors, listed in its rows, and four possible hair colors, listed across its columns. In each of its main cells, the table indicates the *joint probability* of particular combinations of eye color and hair color. For example, the top-left cell indicates that the joint probability of brown eyes and black hair is 0.11 (i.e., 11%). Notice that not all combinations of eye color and hair color are equally likely. For example, the joint probability of blue eyes and black hair is only 0.03 (i.e., 3%). We denote the joint probability of eye color e and hair color h as $p(e, h)$. The notation for joint probabilities is symmetric: $p(e, h) = p(h, e)$.

We may be interested in the probabilities of the eye colors overall, collapsed across hair colors. These probabilities are indicated in the right margin of the table, and they are therefore called *marginal* probabilities. They are computed simply by summing the joint probabilities in each row, to produce the row sums. For example, the marginal probability of green eyes, irrespective of hair color, is 0.11. The joint values indicated in the table do not all sum exactly to the displayed marginal values because of rounding error from the original data. The marginal probability of eye color e is denoted $p(e)$, and it is computed by summing the joint probabilities across the hair colors: $p(e) = \sum_h p(e, h)$.

Of course, we can also consider the marginal probabilities of the various hair colors. The marginal probabilities of the hair colors are indicated on the lower margin of Table 4.1. For example, the probability of black hair, irrespective of eye color, is 0.18. The marginal probabilities are computed by summing the joint probabilities within each column. Thus, $p(h) = \sum_e p(e, h)$.

In general, consider a row variable r and a column variable c. When the row variables are continuous instead of discrete, then $p(r, c)$ is a probability density, and the summation for computing the marginal probability becomes an integral, $p(r) = \int \mathrm{d}c\, p(r, c)$, where the resulting marginal distribution, $p(r)$, is also a probability density. This summation process is called *marginalizing over c* or *integrating out* the variable c. Of course, we can also determine the probability $p(c)$ by marginalizing over r: $p(c) = \int \mathrm{d}r\, p(r, c)$.

4.4.1. Conditional probability

We often want to know the probability of one outcome, given that we know another outcome is true. For example, suppose I sample a person at random from the population referred to in Table 4.1. Suppose I tell you that this person has blue eyes. Conditional on that information, what is the probability that the person has blond hair (or any other particular hair color)? It is intuitively clear how to compute the answer: We see from the blue-eye row of Table 4.1 that the total (i.e., marginal) amount of blue-eyed people is 0.36, and that 0.16 of the population has blue eyes and blond hair. Therefore, of the 0.36 with blue eyes, the fraction 0.16/0.36 has blond hair. In other words, of the blue-eyed people, 45% have blond hair. We also note that of the blue-eyed people, 0.03/0.36 = 8% have black hair. Table 4.2 shows this calculation for each of the hair colors.

The probabilities of the hair colors represent the credibilities of each possible hair color. For this group of people, the general probability of having blond hair is 0.21, as can be seen from the marginal distribution of Table 4.1. But when we learn that a person from this group has blue eyes, then the credibility of that person having blond hair increases to 0.45, as can be seen from Table 4.2. This reallocation of credibility across the possible hair colors *is* Bayesian inference! But we are getting ahead of ourselves; the next chapter will explain the basic mathematics of Bayesian inference in detail.

The intuitive computations for conditional probability can be denoted by simple formal expressions. We denote the conditional probability of hair color given eye color as $p(h|e)$, which is spoken "the probability of *h given e*." The intuitive calculations above are then written $p(h|e) = p(e, h)/p(e)$. This equation is taken as the *definition* of conditional probability. Recall that the marginal probability is merely the sum of the cell probabilities, and therefore the definition can be written $p(h|e) = p(e, h)/p(e) = p(e, h)/\sum_h p(e, h)$. That equation can be confusing because the h in the numerator is a specific value of hair color, but the h in the denominator is a variable that takes on all possible values of hair color. To disambiguate the two meanings of h, the equation can be written $p(h|e) = p(e, h)/p(e) = p(e, h)/\sum_{h^*} p(e, h^*)$, where h^* indicates possible values of hair color.

Table 4.2 Example of conditional probability

	Hair color				
Eye color	**Black**	**Brunette**	**Red**	**Blond**	**Marginal (eye color)**
Blue	0.03/0.36 = 0.08	0.14/0.36 = 0.39	0.03/0.36 = 0.08	0.16/0.36 = 0.45	0.36/0.36 = 1.0

Of the blue-eyed people in Table 4.1, what proportion have hair color h? Each cell shows $p(h|\text{blue}) = p(\text{blue}, h)/p(\text{blue})$ rounded to two decimal points.

The definition of conditional probability can be written using more general variable names, with *r* referring to an arbitrary row attribute and *c* referring to an arbitrary column attribute. Then, for attributes with discrete values, conditional probability is defined as

$$p(c|r) = \frac{p(r,c)}{\sum_{c^*} p(r,c^*)} = \frac{p(r,c)}{p(r)} \tag{4.9}$$

When the column attribute is continuous, the sum becomes an integral:

$$p(c|r) = \frac{p(r,c)}{\int \mathrm{d}c\, p(r,c)} = \frac{p(r,c)}{p(r)} \tag{4.10}$$

Of course, we can conditionalize on the other variable, instead. That is, we can consider $p(r|c)$ instead of $p(c|r)$. It is important to recognize that, in general, $p(r|c)$ is *not* equal to $p(c|r)$. For example, the probability that the ground is wet, given that it's raining, is different than the probability that it's raining, given that the ground is wet. The next chapter provides an extended discussion of the relationship between $p(r|c)$ and $p(c|r)$.

It is also important to recognize that there is no temporal order in conditional probabilities. When we say "the probability of x given y" we do *not* mean that y has already happened and x has yet to happen. All we mean is that we are restricting our calculations of probability to a particular subset of possible outcomes. A better gloss of $p(x|y)$ is, "among all joint outcomes with value y, this proportion of them also has value x." So, for example, we can talk about the conditional probability that it rained the previous night given that there are clouds the next morning. This is simply referring to the proportion of all cloudy mornings that had rain the night before.

4.4.2. Independence of attributes

Suppose I have a six-sided die and a coin. Suppose they are fair. I flip the coin and it comes up heads. Given this result from the coin, what is the probability that the rolled die will come up 3? In answering this question, you probably thought, "the coin has no influence on the die, so the probability of the die coming up 3 is 1/6 regardless of the result from the coin." If that's what you thought, you were assuming that the spinner and the coin are *independent.*

In general, when the value of y has no influence on the value of x, we know that the probability of x given y simply is the probability of x in general. Written formally, that idea is expressed as $p(x|y) = p(x)$ for all values of x and y. Let's think a moment about what that implies. We know from the definition of conditional probability, in Equations 4.9 or 4.10, that $p(x|y) = p(x,y)/p(y)$. Combining those equations implies that $p(x) = p(x,y)/p(y)$ for all values of x and y. After multiplying both sides by $p(y)$, we get the implication that $p(x,y) = p(x)p(y)$ for all values of x and y. The implication goes the other way, too: When $p(x,y) = p(x)p(y)$ for all values of x and y, then $p(x|y) =$

$p(x)$ for all values of x and y. Therefore, either of these conditions is our mathematical definition of independence of attributes. To reiterate, to say that attributes x and y are independent means that $p(x|y) = p(x)$ for all values of x and y, which is mathematically equivalent to saying that $p(x, y) = p(x)p(y)$ for all values of x and y.

Consider the example of eye color and hair color back in Table 4.1 (page 90). Are the attributes independent? Intuitively from everyday experience, we know that the answer should be no, but we can show it mathematically. All we need, to disprove independence, is some eye color e and some hair color h for which $p(h|e) \neq p(h)$, or equivalently for which $p(e, h) \neq p(e)p(h)$. We already dealt with such a case, namely blue eyes and blond hair. Table 4.1 shows that the marginal probability of blond hair is $p(\text{blond}) = 0.21$, while Table 4.2 shows that the conditional probability of blond hair given blue eyes is $p(\text{blond}|\text{blue}) = 0.45$. Thus, $p(\text{blond}|\text{blue}) \neq p(\text{blond})$. We can instead disprove independence by cross-multiplying the marginal probabilities and showing that they do not equal the joint probability: $p(\text{blue}) \cdot p(\text{blond}) = 0.36 \cdot 0.21 = 0.08 \neq 0.16 = p(\text{blue}, \text{blond})$.

As a simple example of two attributes that *are* independent, consider the suit and value of playing cards in a standard deck. There are four suits (diamonds, hearts, clubs, and spades), and thirteen values (ace, 2, ..., 9, jack, queen, king) of each suit, making 52 cards altogether. Consider a randomly dealt card. What is the probability that it is a heart? (Answer: 13/52 = 1/4.) Suppose I look at the card without letting you see it, and I tell you that it is a queen. Now what is the probability that it is a heart? (Answer: 1/4.) In general, telling you the card's value does not change the probabilities of the suits, therefore value and suit are independent. We can verify this in terms of cross multiplying marginal probabilities, too: Each combination of value and suit has a 1/52 chance of being dealt (in a fairly shuffled deck). Notice that 1/52 is exactly the marginal probability of any one suit (1/4) times the marginal probability of any one value (1/13).

Among other contexts, independence will come up when we are constructing mathematical descriptions of our beliefs about more than one parameter. We will create a mathematical description of our beliefs about one parameter, and another mathematical description of our beliefs about the other parameter. Then, to describe what we believe about combinations of parameters, we will often assume independence, and simply multiply the separate credibilities to specify the joint credibilities.

4.5. APPENDIX: R CODE FOR FIGURE 4.1

Figure 4.1 was produced by the script `RunningProportion.R`. To run it, simply type `source("RunningProportion.R")` at R's command line (assuming that your working directory contains the file!). Each time you run it, you will get a different result because of the pseudo-random number generator. If you want to set the pseudo-random number generator to a specific starting state, use the `set.seed` command. The example in Figure 4.1 was created by typing `set.seed(47405)` and then `source("RunningProportion.R")`.

If you want to control the window size created for the graph and subsequently save the figure, you can load the graphics functions defined in the utility programs that accompany this book. Here is a sequence of commands that open and save a separate graphics window:

```
source("DBDA2E-utilities.R") # Definitions of openGraph, saveGraph, etc.
set.seed(47405)              # Optional, for a specific pseudo-random sequence.
openGraph(width=6,height=6)
source("RunningProportion.R")
saveGraph(file="RunningProportionExample",type="jpg")
```

The previous paragraphs explain how to use the script `RunningProportion.R`, but if you want to explore its internal mechanics, you can open it in RStudio's editing window. You will see a script of commands like this:

```
N = 500        # Specify the total number of flips, denoted N.
pHeads = 0.5   # Specify underlying probability of heads.
# Generate a random sample of N flips (heads=1, tails=0):
flipSequence = sample( x=c(0,1), prob=c(1-pHeads,pHeads), size=N, replace=TRUE)
# Compute the running proportion of heads:
r = cumsum( flipSequence ) # Cumulative sum: Number of heads at each step.
n = 1:N                    # Number of flips at each step.
runProp = r / n            # Component by component division.
# Graph the running proportion:
plot( n , runProp , type="o" , log="x" , col="skyblue" ,
      xlim=c(1,N) , ylim=c(0.0,1.0) , cex.axis=1.5 ,
      xlab="Flip Number" , ylab="Proportion Heads" , cex.lab=1.5 ,
      main="Running Proportion of Heads" , cex.main=1.5 )
# Plot a dotted horizontal reference line:
abline( h=pHeads , lty="dotted" )
# Display the beginning of the flip sequence:
flipLetters = paste( c("T","H")[flipSequence[1:10]+1] , collapse="" )
displayString = paste0( "Flip Sequence = " , flipLetters , "..." )
text( N , .9 , displayString , adj=c(1,0.5) , cex=1.3 )
# Display the relative frequency at the end of the sequence.
text( N , .8 , paste("End Proportion =",runProp[N]) , adj=c(1,0.5) , cex=1.3 )
```

The first two commands, above, merely specify the number of flips and the underlying probability of heads for the simulated coin. The fourth line introduces a new function that is predefined in standard distributions of R, namely the `sample` function. It generates pseudo-random samples from the set of elements defined by the user-supplied argument `x`, which in this case is a vector containing a 0 (zero) and a 1 (one). The argument `prob` specifies the probability with which each component of `x` should be sampled. You can read more about the `sample` function by typing `?sample` at R's command line.

The sixth line, above, uses the cumulative sum function, `cumsum`, which is predefined in R. You can read about the `cumsum` function by typing `?cumsum` at R's command line. The function computes, at each component of a vector, the cumulative sum up to that component. For example, `cumsum( c(1,1,0,0,1) )` produces the vector `1 2 2 2 3`.

As you read through the remaining lines of R code, above, it helps to run each line individually in R to see what it does. For example, to see the contents of the variable `runProp`, just type `runProp` and R's command line. To learn about the various arguments of the `plot` function, type `?plot` at R's command line.

The line that defines the variable `flipLetters` may seem a bit mysterious. Whenever you encounter a complex command like this in R, it can be a good strategy to unpack the commands from the inside out (as was mentioned in Section 3.7.6 in relation to diagnosing programming errors). Try entering these commands in R (or just select the text in RStudio and click Run):

```
                 flipSequence[1:10]
                 flipSequence[1:10]+1
      c("T","H")[flipSequence[1:10]+1]
paste( c("T","H")[flipSequence[1:10]+1] , collapse="" )
```

4.6. EXERCISES

Look for more exercises at https://sites.google.com/site/doingbayesiandataanalysis/

Exercise 4.1. [Purpose: To gain experience with the `apply` function in R, while dealing with a concrete example of computing conditional probabilities.] The eye-color hair-color data from Table 4.1 are built into R as the array named HairEyeColor. The array is frequencies of eye and hair color for males and females. Run the following code in R:

```
show( HairEyeColor ) # Show data
EyeHairFreq = apply( HairEyeColor, c("Eye","Hair"), sum ) # Sum across sex
EyeHairProp = EyeHairFreq / sum( EyeHairFreq )  # joint proportions, Table 4.1
show( round( EyeHairProp , 2 ) )
HairFreq = apply( HairEyeColor , c("Hair") , sum ) # Sum across sex and eye
HairProp = HairFreq / sum( HairFreq ) # marginal proportions, Table 4.1
show( round( HairProp , 2 ) )
EyeFreq = apply( HairEyeColor , c("Eye") , sum ) # Sum across sex and eye
EyeProp = EyeFreq / sum( EyeFreq ) # marginal proportions, Table 4.1
show( round( EyeProp , 2 ) )
EyeHairProp["Blue",] / EyeProp["Blue"] # conditional prob, Table 4.2
```

In your write-up, include each line above and its results. *Explain* what each line does (in a bit more detail than the inline comments). Extend the above commands by also computing the probabilities of the hair colors given Brown eyes, and the probabilities of the eye colors given Brown hair.

Exercise 4.2. [Purpose: To give you some experience with random number generation in R.] Modify the coin flipping program in Section 4.5 `RunningProportion.R` to simulate a biased coin that has $p(H) = 0.8$. Change the height of the reference line in the plot to match $p(H)$. Comment your code. *Hint:* Read the help for the `sample` command.

Exercise 4.3. [Purpose: To have you work through an example of the logic presented in Section 4.2.1.2.] Determine the exact probability of drawing a 10 from a shuffled pinochle deck. (In a pinochle deck, there are 48 cards. There are six values: 9, 10, Jack, Queen, King, Ace. There are two copies of each value in each of the standard four suits: hearts, diamonds, clubs, spades.)

(A) What is the probability of getting a 10?

(B) What is the probability of getting a *10 or Jack*?

Exercise 4.4. [Purpose: To give you hands-on experience with a simple probability density function, in R and in calculus, and to reemphasize that density functions can have values larger than 1.] Consider a spinner with a [0,1] scale on its circumference. Suppose that the spinner is slanted or magnetized or bent in some way such that it is biased, and its probability density function is $p(x) = 6x(1-x)$ over the interval $x \in [0, 1]$.

(A) Adapt the program `IntegralOfDensity.R` to plot this density function and approximate its integral. Comment your code. Be careful to consider values of x only in the interval $[0, 1]$. *Hint:* You can omit the first couple of lines regarding `meanval` and `sdval`, because those parameter values pertain only to the normal distribution. Then set `xlow=0` and `xhigh=1`, and set `dx` to some small value.

(B) Derive the exact integral using calculus. *Hint:* See the example, Equation 4.7.

(C) Does this function satisfy Equation 4.3?

(D) From inspecting the graph, what is the maximal value of $p(x)$?

Exercise 4.5. [Purpose: To have you use a normal curve to describe beliefs. It's also handy to know the area under the normal curve between μ and σ.]

(A) Adapt the code from `IntegralOfDensity.R` to determine (approximately) the probability mass under the normal curve from $x = \mu - \sigma$ to $x = \mu + \sigma$. Comment your code. *Hint:* Just change `xlow` and `xhigh` appropriately, and change the `text` location so that the `area` still appears within the plot.

(B) Now use the normal curve to describe the following belief. Suppose you believe that women's heights follow a bell-shaped distribution, centered at 162 cm with about two-thirds of all women having heights between 147 and 177 cm. What should be the μ and σ parameter values?

Exercise 4.6. [Purpose: Recognize and work with the fact that Equation 4.9 can be solved for the joint probability, which will be crucial for developing Bayes' theorem.] School children were surveyed regarding their favorite foods. Of the total sample, 20% were 1st graders, 20% were 6th graders, and 60% were 11th graders. For each grade, the following table shows the proportion of respondents that chose each of three foods as their favorite:

	Ice cream	Fruit	French fries
1st graders	0.3	0.6	0.1
6th graders	0.6	0.3	0.1
11th graders	0.3	0.1	0.6

From that information, construct a table of joint probabilities of grade and favorite food. Also, say whether grade and favorite food are independent or not, and how you ascertained the answer. *Hint:* You are given p(grade) and p(food|grade). You need to determine p(grade,food).

CHAPTER 5

Bayes' Rule

Contents

I'll love you forever in every respect
(I'll marginalize all your glaring defects)
But if you could change some to be more like me
I'd love you today unconditionally.[1]

On a typical day at your location, what is the probability that it is cloudy? Suppose you are told it is raining, now what is the probability that it is cloudy? Notice that those two probabilities are not equal, because we can be pretty sure that $p(\text{cloudy}) < p(\text{cloudy}|\text{raining})$. Suppose instead you are told that everyone outside is wearing sunglasses. Most likely, it is true that $p(\text{cloudy}) > p(\text{cloudy}|\text{sunglasses})$. Notice how we have reasoned in this meteorological example. We started with prior credibility allocated over two possible states of the sky: cloudy or sunny. Then we took into account some other data, namely, that it is raining or that people are wearing sunglasses. Conditional on the new data, we re-allocated credibility across the possible states of the sky. When the data indicated rain, then cloudy was more credible than when we started. When the data instead indicated sunglasses, then cloudy was less credible than when we started. Bayes'

[1] This chapter is about Bayes' rule, which shows how *marginal* probabilities relate to *conditional* probabilities when taking data into account. The terms "marginal" and (un-)"conditional" are used in the poem with their colloquial meanings. The poem also plays with the reversal of meaning between conditional and unconditional: The poem says that the conditional love, $p(\text{love}|\text{change})$, is greater than the marginal love, $p(\text{love})$, but ironically says that satisfying the condition would bring unconditional love.

Doing Bayesian Data Analysis, 2nd Edition
http://dx.doi.org/10.1016/B978-0-12-405888-0.00005-2

rule is merely the mathematical relation between the prior allocation of credibility and the posterior reallocation of credibility conditional on data.

5.1. BAYES' RULE

Thomas Bayes (1702-1761) was a mathematician and Presbyterian minister in England. His famous theorem was published posthumously in 1763, thanks to the extensive editorial efforts of his friend, Richard Price (Bayes & Price, 1763). The simple rule has vast ramifications for statistical inference, and therefore as long as his name is attached to the rule, we'll continue to see his name in textbooks.[2] But Bayes himself probably was not fully aware of these ramifications, and many historians argue that it is Bayes' successor, Pierre-Simon Laplace (1749-1827), whose name should really label this type of analysis, because it was Laplace who independently rediscovered and extensively developed the methods (e.g., Dale, 1999; McGrayne, 2011).

There is another branch of statistics, called *frequentist*, which does not use Bayes' rule for inference and decisions. Chapter 11 describes aspects of the frequentist approach and its perils. This approach is often identified with another towering figure from England who lived about 200 years later than Bayes, named Ronald Fisher (1890-1962). His name, or at least the first letter of his last name, is immortalized in one of the most common measures used in frequentist analysis, the F-ratio.[3] It is curious and re-assuring that the overwhelmingly dominant Fisherian approach of the 20th century is giving way in the 21st century to a Bayesian approach that had its genesis in the 18th century (e.g., Lindley, 1975; McGrayne, 2011).[4]

5.1.1. Derived from definitions of conditional probability

Recall from the intuitive definition of conditional probability, back in Equation 4.9 on p. 92, that

$$p(c|r) = \frac{p(r, c)}{p(r)} \tag{5.1}$$

In words, the definition simply says that the probability of c given r is the probability that they happen together relative to the probability that r happens at all.

Now we do some very simple algebraic manipulations. First, multiply both sides of Equation 5.1 by $p(r)$ to get

$$p(c|r)\, p(r) = p(r, c) \tag{5.2}$$

[2] The first edition of this book visited Bayes' tomb. You can see photos at the blog, http://doingbayesiandataanalysis.blogspot.com/2012/02/book-visits-bayes.html

[3] But Fisher did not advocate the type of null hypothesis significance testing that contemporary social science performs; see Gigerenzer, Krauss, and Vitouch (2004).

[4] The first edition of this book visited Fisher's remains. You can see photos at the blog, http://doingbayesiandataanalysis.blogspot.com/2012/03/book-visits-fisher.html

Second, notice that we can do the analogous manipulation starting with the definition $p(r|c) = p(r,c)/p(c)$ to get

$$p(r|c)\,p(c) = p(r,c) \tag{5.3}$$

Equations 5.2 and 5.3 are two different expressions equal to $p(r,c)$, so we know those expressions are equal each other:

$$p(c|r)\,p(r) = p(r|c)\,p(c) \tag{5.4}$$

Now divide both sides of that last expression by $p(r)$ to arrive at

$$p(c|r) = \frac{p(r|c)\,p(c)}{p(r)} \tag{5.5}$$

But we are not quite done yet, because we can re-write the denominator in terms of $p(r|c)$, just as we did back in Equation 4.9 on p. 92. Thus,

$$p(c|r) = \frac{p(r|c)\,p(c)}{\sum_{c^*} p(r|c^*)\,p(c^*)} \tag{5.6}$$

In Equation 5.6, the c in the numerator is a specific fixed value, whereas the c^* in the denominator is a variable that takes on all possible values. Equations 5.5 and 5.6 are called *Bayes' rule*. This simple relationship lies at the core of Bayesian inference. It may not seem to be particularly auspicious in the form of Equation 5.6, but we will soon see how it can be applied in powerful ways to data analysis. Meanwhile, we will build up a little more intuition about how Bayes' rule works.

5.1.2. Bayes' rule intuited from a two-way discrete table

Consider Table 5.1, which shows the joint probabilities of a row attribute and a column attribute, along with their marginal probabilities. In each cell, the joint probability $p(r,c)$

Table 5.1 A table for making Bayes' rule not merely special but spatial

	Column			
Row	...	c	...	**Marginal**
⋮		⋮		
r	...	$p(r,c) = p(r|c)\,p(c)$	...	$p(r) = \sum_{c^*} p(r|c^*)\,p(c^*)$
⋮		⋮		
Marginal		$p(c)$		

When conditionalizing on row value r, the conditional probability $p(c|r)$ is simply the cell probability, $p(r,c)$, divided by the marginal probability, $p(r)$. When algebraically re-expressed as shown in the table, this is Bayes' rule. Spatially, Bayes' rule gets us from the lower marginal distribution, $p(c)$, to the conditional distribution $p(c|r)$ when focusing on row value r.

is re-expressed by the equivalent form $p(r|c)\,p(c)$ from the definition of conditional probability in Equation 5.3. The marginal probability $p(r)$ is re-expressed by the equivalent form $\sum_{c^*} p(r|c^*)\,p(c^*)$, as was done in Equations 4.9 and 5.6. Notice that the numerator of Bayes' rule is the joint probability, $p(r, c)$, and the denominator of Bayes' rule is the marginal probability, $p(r)$. Looking at Table 5.1, you can see that Bayes' rule gets us from the lower marginal distribution, $p(c)$, to the conditional distribution $p(c|r)$ when focusing on row value r. In summary, the key idea is that conditionalizing on a known row value is like restricting attention to only the row for which that known value is true, and then normalizing the probabilities in that row by dividing by the row's total probability. This act of spatial attention, when expressed in algebra, yields Bayes' rule.

A concrete example of going from marginal to conditional probabilities was provided in the previous chapter, regarding eye color and hair color. Let's revisit it now. Table 5.2 shows the joint and marginal probabilities of various combinations of eye color and hair color. Without knowing anything about a person's eye color, all we believe about hair colors is expressed by the marginal probabilities of the hair colors, at the bottom of Table 5.2. However, if we are told that a randomly selected person's eyes are blue, then we know that this person comes from the "blue" row of the table, and we can focus our attention on that row. We compute the conditional probabilities of the hair colors, given the eye color, as shown in Table 5.3. Notice that we have gone from the "prior" (marginal) beliefs about hair color before knowing eye color, to the "posterior" (conditional) beliefs about hair color given the observed eye color. For example, without knowing the eye color, the probability of blond hair in this population is 0.21. But with knowing that the eyes are blue, the probability of blond hair is 0.45.

The example involving eye color and hair color illustrates conditional reallocation of credibility across column values (hair colors) when given information about a row value (eye color). But the example uses joint probabilities $p(r, c)$ that are directly provided

Table 5.2 Proportions of combinations of hair color and eye color

	Hair color				
Eye color	**Black**	**Brunette**	**Red**	**Blond**	**Marginal (Eye color)**
Brown	0.11	0.20	0.04	0.01	0.37
Blue	0.03	0.14	0.03	0.16	0.36
Hazel	0.03	0.09	0.02	0.02	0.16
Green	0.01	0.05	0.02	0.03	0.11
Marginal (hair color)	0.18	0.48	0.12	0.21	1.0

Some rows or columns may not sum exactly to their displayed marginals because of rounding error from the original data. Data adapted from Snee (1974). This is a Table 4.1 duplicated here for convenience.

Table 5.3 Example of conditional probability

	Hair color				
Eye color	**Black**	**Brunette**	**Red**	**Blond**	**Marginal (Eye color)**
Blue	$0.03/0.36 = 0.08$	$0.14/0.36 = 0.39$	$0.03/0.36 = 0.08$	$0.16/0.36 = 0.45$	$0.36/0.36 = 1.0$

Of the blue-eyed people in Table 5.2, what proportion has hair color h? Each cell shows $p(h|\text{blue}) = p(\text{blue}, h)/p(\text{blue})$ rounded to two decimal points. This is a Table 4.2 duplicated here for convenience.

as numerical values, whereas Bayes' rule instead involves joint probabilities expressed as $p(r|c)\,p(c)$, as shown in Equation 5.6 and Table 5.1. The next example provides a concrete situation in which it is natural to express joint probabilities as $p(r|c)\,p(c)$.

Consider trying to diagnose a rare disease. Suppose that in the general population, the probability of having the disease is only one in a thousand. We denote the true presence or absence of the disease as the value of a parameter, θ, that can have the value $\theta = ☹$ if disease is present in a person, or the value $\theta = ☺$ if the disease is absent. The base rate of the disease is therefore denoted $p(\theta = ☹) = 0.001$. This is our prior belief that a person selected at random has the disease.

Suppose that there is a test for the disease that has a 99% hit rate, which means that if a person has the disease, then the test result is positive 99% of the time. We denote a positive test result as $T = +$, and a negative test result as $T = -$. The observed test result is the datum that we will use to modify our belief about the value of the underlying disease parameter. The hit rate is expressed formally as $p(T = + \mid \theta = ☹) = 0.99$. Suppose also that the test has a false alarm rate of 5%. This means that 5% of the time when the disease is absent, the test falsely indicates that the disease is present. We denote the false alarm rate as $p(T = + \mid \theta = ☺) = 0.05$.

Suppose we sample a person at random from the population, administer the test, and it comes up positive. What is the posterior probability that the person has the disease? Mathematically expressed, we are asking, what is $p(\theta = ☹ \mid T = +)$? Before determining the answer from Bayes' rule, generate an intuitive answer and see if your intuition matches the Bayesian answer. Most people have an intuition that the probability of having the disease is near the hit rate of the test (which in this case is 99%).

Table 5.4 shows how to conceptualize disease diagnosis as a case of Bayes' rule. The base rate of the disease is shown in the lower marginal of the table. Because the background probability of having the disease is $p(\theta = ☹) = 0.001$, it is the case that the probability of not having the disease is the complement, $p(\theta = ☺) = 1 - 0.001 = 0.999$. Without any information about test results, this lower marginal probability is our prior belief about a person having the disease.

Table 5.4 shows the joint probabilities of test results and disease states in terms of the hit rate, false alarm rate, and base rate. For example, the joint probability of the test

Table 5.4 Joint and marginal probabilities of test results and disease states

	Disease		
Test result	$\theta =$ ☹ **(present)**	$\theta =$ ☺ **(absent)**	**Marginal (test result)**
$T = +$	$p(+\|$☹$)\,p($☹$)$ $= 0.99 \cdot 0.001$	$p(+\|$☺$)\,p($☺$)$ $= 0.05 \cdot (1 - 0.001)$	$\sum_\theta p(+\|\theta)\,p(\theta)$
$T = -$	$p(-\|$☹$)\,p($☹$)$ $= (1 - 0.99) \cdot 0.001$	$p(-\|$☺$)\,p($☺$)$ $= (1 - 0.05) \cdot (1 - 0.001)$	$\sum_\theta p(-\|\theta)\,p(\theta)$
Marginal (disease)	$p($☹$) = 0.001$	$p($☺$) = 1 - 0.001$	1.0

For this example, the base rate of the disease is 0.001, as shown in the lower marginal. The test has a hit rate of 0.99 and a false alarm rate of 0.05, as shown in the row for $T = +$. For an actual test result, we restrict attention to the corresponding row of the table and compute the conditional probabilities of the disease states via Bayes' rule.

being positive and the disease being present is shown in the upper-left cell as $p(T = +, \theta =$ ☹$) = p(T = + | \theta =$ ☹$)\, p(\theta =$ ☹$) = 0.99 \cdot 0.001$. In other words, the joint probability of the test being positive and the disease being present is the hit rate of the test times the base rate of the disease. Thus, it is natural in this application to express the joint probabilities as $p(\text{row}|\text{column})p(\text{column})$.

Suppose we select a person at random and administer the diagnostic test, and the test result is positive. To determine the probability of having the disease, we should restrict attention to the row marked $T = +$ and compute the conditional probabilities of $p(\theta|T{=}+)$ via Bayes' rule. In particular, we find that

$$p(\theta{=}\text{☹}|T{=}+) = \frac{p(T{=}+|\theta{=}\text{☹})\,p(\theta{=}\text{☹})}{\sum_\theta p(T{=}+|\theta)\,p(\theta)}$$

$$= \frac{0.99 \cdot 0.001}{0.99 \cdot 0.001 + 0.05 \cdot (1 - 0.001)}$$

$$= 0.019$$

Yes, that's correct: Even with a positive test result for a test with a 99% hit rate, the posterior probability of having the disease is only 1.9%. This low probability is a consequence of the low-prior probability of the disease and the non-negligible false alarm rate of the test. A caveat regarding interpretation of the results: Remember that here we have assumed that the person was selected at random from the population, and there were no other symptoms that motivated getting the test. If there were other symptoms that indicated the disease, those data would also have to be taken into account.

To summarize the example of disease diagnosis: We started with the prior credibility of the two disease states (present or absent) as indicated by the lower marginal of Table 5.4. We used a diagnostic test that has a known hit rate and false alarm rate, which are the conditional probabilities of a positive test result for each disease state. When an observed test result occurred, we restricted attention to the corresponding row of Table 5.4 and computed the conditional probabilities of the disease states in that row via Bayes' rule. These conditional probabilities are the posterior probabilities of the disease states. The conditional probabilities are the re-allocated credibilities of the disease states, given the data.

5.2. APPLIED TO PARAMETERS AND DATA

The key application that makes Bayes' rule so useful is when the row variable represents data values and the column variable represents parameter values. A model of data specifies the probability of particular data values given the model's structure and parameter values. The model also indicates the probability of the various parameter values. In other words, a model specifies

$$p(\text{data values} \mid \text{parameters values})$$
$$\text{along with the prior, } p(\text{parameters values})$$

and we use Bayes' rule to convert that to what we really want to know, which is how strongly we should believe in the various parameter values, given the data:

$$p(\text{parameters values} \mid \text{data values})$$

Comprehension can be aided by thinking about the application of Bayes' rule to data and parameters in terms of a two-way table, shown in Table 5.5. The columns of Table 5.5 correspond to specific values of the parameter, and the rows of Table 5.5 correspond to specific values of the data. Each cell of the table holds the joint probability of the specific combination of parameter value θ and data value D, denoted $p(D, \theta)$, and which we know can be algebraically re-expressed as $p(D|\theta)\, p(\theta)$. The prior probability

Table 5.5 Applying Bayes' rule to data and parameters

	Model parameter			
Data	...	***θ* value**	...	**Marginal**
⋮		⋮		⋮
***D* value**	...	$p(D, \theta) = p(D\|\theta)\, p(\theta)$	...	$p(D) = \sum_{\theta^*} p(D\|\theta^*)\, p(\theta^*)$
⋮		⋮		⋮
Marginal	...	$p(\theta)$	...	

When conditionalizing on row value D, the conditional probability $p(\theta|D)$ is the cell probability $p(D, \theta)$ divided by the marginal probability $p(D)$. When these probabilities are algebraically re-expressed as shown in the table, this is Bayes' rule. This table is merely Table 5.1 with its rows and columns re-named.

of the parameter values is the marginal distribution, $p(\theta)$, which appears in the lower margin of Table 5.5. Table 5.5 is merely Table 5.1 with its rows and columns re-named.

When we observe a particular data value, D, we are restricting our attention to one specific row of Table 5.5, namely, the row corresponding to the observed value, D. The posterior distribution on θ is obtained by dividing the joint probabilities in that row by the row marginal, $p(D)$. In other words, the posterior distribution on θ is obtained by conditionalizing on the row with the observed data value, and that operation *is* Bayes' rule.

It might help at this point to recapitulate the progression of ideas in the last few sections. In particular, Table 5.1 showed Bayes' rule in general form, using generic rows and columns. That table emphasized how the factors of Bayes' rule are positioned in a joint probability table, and how Bayes' rule amounts to moving attention from a margin of the table to a row of the table. Next, Table 5.4 showed a numerical example in which columns were underlying states of health (i.e., having or nor having a disease) and in which rows were observed data (i.e., testing positive or negative). That table emphasized a concrete case of the general form in Table 5.1. In disease diagnosis, we started with the marginal probabilities of the disease in Table 5.4, then re-allocated credibility across the disease states when the test result moved our attention to one row of the table. Finally, we re-expressed Table 5.1 as Table 5.5 by supposing that the rows are data values and the columns are parameter values. Bayes' rule is shifting attention from the prior, marginal distribution of the parameter values to the posterior, conditional distribution of parameter values in a specific datum's row. It is this form, in Table 5.5, that we will use throughout the remainder of the book. Additional examples will be presented very soon.

The factors of Bayes' rule have specific names that will be used regularly throughout the book, as indicated here:

$$\underbrace{p(\theta|D)}_{\text{posterior}} = \underbrace{p(D|\theta)}_{\text{likelihood}} \; \underbrace{p(\theta)}_{\text{prior}} \; / \; \underbrace{p(D)}_{\text{evidence}} \tag{5.7}$$

The denominator is

$$p(D) = \sum_{\theta^*} p(D|\theta^*)p(\theta^*) \tag{5.8}$$

where the superscript asterisk in θ^* is merely a reminder that the denominator's θ^* is distinct from the specific θ value in the numerator of Equation 5.7. The "prior," $p(\theta)$, is the credibility of the θ values without the data D. The "posterior," $p(\theta|D)$, is the credibility of θ values with the data D taken into account. The "likelihood," $p(D|\theta)$, is the probability that the data could be generated by the model with parameter value θ. The "evidence" for the model, $p(D)$, is the overall probability of the data according to the model, determined by averaging across all possible parameter values weighted by the strength of belief in those parameter values.

The denominator of Bayes' rule, labeled in Equation 5.7 as the *evidence* for the model, is also called the *marginal likelihood*. The term "evidence" is intuitively suggestive and takes fewer syllables and characters than "marginal likelihood," but the term "evidence" is also more ambiguous. The term "marginal likelihood" refers specifically to the operation of taking the average of the likelihood, $p(D|\theta)$, across all values of θ, weighted by the prior probability of θ. In this book, I will use the terms "evidence" and "marginal likelihood" interchangeably.

Up to this point, Bayes' rule has been presented only in the context of discrete-valued variables. It also applies to continuous variables, but probability masses become probability densities and sums become integrals. For continuous variables, the only change in Bayes' rule is that the marginal likelihood changes from the sum in Equation 5.8 to an integral:

$$p(D) = \int \mathrm{d}\theta^* p(D|\theta^*) p(\theta^*) \tag{5.9}$$

where the superscript asterisk in θ^* is merely a reminder that it is distinct from the specific θ value in Equation 5.7. This continuous-variable version of Bayes' rule is what we will deal with most often in real applications later in the book.[5]

5.2.1. Data-order invariance

Bayes' rule in Equation 5.7 gets us from a prior belief, $p(\theta)$, to a posterior belief, $p(\theta|D)$, when we take into account some data D. Now suppose we observe some *more* data, which we'll denote D'. We can then update our beliefs again, from $p(\theta|D)$ to $p(\theta|D', D)$. Here's the question: Does our final belief depend on whether we update with D first and D' second, or update with D' first and D second?

The answer is: It depends! In particular, it depends on the model function that defines the likelihood, $p(D|\theta)$. In many models, the probability of data, $p(D|\theta)$, does not depend in any way on *other* data. That is, the joint probability $p(D, D'|\theta)$ equals $p(D|\theta) \cdot p(D'|\theta)$. In other words, in this sort of model, the data probabilities are *independent* (recall that independence was defined in Section 4.4.2). Under this condition, then the order of updating has no effect of the final posterior.

This invariance to ordering of the data makes sense intuitively: If the likelihood function has no dependence on data ordering, then the posterior shouldn't have any dependence on data ordering. But it's trivial to prove mathematically, too. We simply

[5] Recall from the discussion after Equation 4.3 on p. 82 that integrals in this book are written with the differential, such as $\mathrm{d}\theta$, placed next to the integral sign instead of at the end of the formula. Such placement is neither wrong nor unique to this book and has notational and conceptual advantages discussed on p. 82.

write down Bayes' rule and use the independence assumption that $p(D', D|\theta) = p(D'|\theta)$ $p(D|\theta)$:

$$
\begin{aligned}
p(\theta|D', D) &= \frac{p(D', D|\theta)\, p(\theta)}{\sum_{\theta^*} p(D', D|\theta^*)\, p(\theta^*)} && \text{Bayes' rule} \\
&= \frac{p(D'|\theta)p(D|\theta)\, p(\theta)}{\sum_{\theta^*} p(D'|\theta^*)p(D|\theta^*)\, p(\theta^*)} && \text{by assumption of independence} \\
&= \frac{p(D|\theta)p(D'|\theta)\, p(\theta)}{\sum_{\theta^*} p(D|\theta^*)p(D'|\theta^*)\, p(\theta^*)} && \text{multiplication is commutative} \\
&= p(\theta|D, D') && \text{Bayes' rule}
\end{aligned}
$$

In all of the examples in this book, we use mathematical likelihood functions that generate independent data, and we assume that the observed data came from a procedure that can be usefully described by such likelihood functions. One way of thinking about this assumption is as follows: We assume that every datum is equally representative of the underlying process, regardless of when the datum was observed, and regardless of any data observed before or after.

5.3. COMPLETE EXAMPLES: ESTIMATING BIAS IN A COIN

We will now consider an extensive set of examples that will build your understanding of how prior distributions and data interact to produce posterior distributions. These examples all involve estimating the underlying bias in a coin. Recall from Section 4.1.1 that we don't necessarily care about coins, per se, but coins represent things we do care about. When we observe the number of heads that result from flipping a coin, and we estimate its underlying probability of coming up heads, the exercise represents, for example, observing the number of correct responses on a multiple choice exam and estimating the underlying probability of being correct, or observing the number of headaches cured by a drug and estimating the underlying probability of curing.

A note about terminology: When I refer to the "bias" in a coin, I will sometimes be referring to its underlying probability of coming up heads. Thus, *when a coin is fair, it has a "bias" of 0.50*. Other times, I might use the term "bias" in its colloquial sense of a *departure from fairness*, as in "head biased" or "tail biased." Although I will try to be clear about which meaning is intended, there will be times that you will have to rely on context to determine whether "bias" means the probability of heads or departure from fairness. I hope the ambiguity does not bias you against me.

Recall from Section 2.3 (p. 25) that the first step in Bayesian data analysis is identifying the type of data being described. In this case, the data consist of heads and tails. We will denote the outcome of a flip as y. When the outcome is heads, we say $y = 1$, and when the outcome is tails, we say $y = 0$. Although we are denoting the outcome with numerical values for mathematical convenience later, it should be remembered that heads and tails are merely categorical, with no metric attributes. In particular, heads is not greater than tails, nor are heads and tails separated by a distance of 1 unit, nor is heads something while tails is nothing. The values of 1 and 0 are used merely as labels with useful algebraic properties in subsequent formulas.

The next step in Bayesian data analysis is creating a descriptive model with meaningful parameters. We denote the underlying probability of coming up heads as $p(y = 1)$. We want to describe that underlying probability with some meaningfully parameterized model. We will use a particularly simple model and directly describe the underlying probability of heads as the value of the parameter θ (Greek letter theta). This idea can be written formally as $p(y{=}1|\theta) = \theta$, which could be said aloud this way: "The probability that the outcome is heads, given a value of parameter θ, is the value of θ." Notice that this definition requires that the value of θ is between 0 and 1. A nice quality of this model is that θ is intuitively meaningful: its value is directly interpretable as the probability that the coin comes up heads.[6]

For our formal model, we need a mathematical expression of the likelihood function in Bayes' rule (Equation 5.7). We already have an expression for the probability of heads, namely $p(y = 1|\theta) = \theta$, but what about the probability of tails? A moment's thought suggests that the probability of tails is simply the complement of the probability of heads, because heads and tails are the only possible outcomes. Therefore, $p(y{=}0|\theta) = 1{-}\theta$. We can combine the equations for the probabilities of heads and tails into a single expression:

$$p(y|\theta) = \theta^{y} (1 - \theta)^{(1-y)} \tag{5.10}$$

Notice that when $y = 1$, Equation 5.10 reduces to $p(y{=}1|\theta) = \theta$, and when $y = 0$, Equation 5.10 reduces to $p(y{=}0|\theta) = 1 - \theta$. The probability distribution expressed by Equation 5.10 is called the Bernoulli distribution, named after the mathematician Jacob Bernoulli (1655-1705).[7]

As an aside, from Equation 5.10 we can also figure out the formula for the likelihood of a whole set of outcomes from multiple flips. Denote the outcome of the ith flip as y_i

[6] We could have created a different model of the underlying probability of coming up heads. For example, we could have defined $p(y{=}1|\phi) = \phi/2{+}0.5$, where ϕ (Greek letter phi) can range from -1 to $+1$. This model could be said aloud this way: "The probability that the outcome is heads, given a value of parameter ϕ, is the value of ϕ divided by 2 plus 0.5." In this model, the coin is fair when $\phi = 0$, because then $p(y{=}1|\phi{=}0) = 0.5$. The coin is utterly tail-biased when $\phi = -1$ because then $p(y{=}1|\phi{=}{-1}) = 0$. The coin is utterly head-biased when $\phi = +1$ because then $p(y{=}1|\phi{=}1) = 1$. Notice that the parameter ϕ is meaningful as an expression of bias, but the value of ϕ does not *directly* express the probability of heads.

[7] The first edition of this book visited Jacob Bernoulli's grave. You can see photos at the blog, http://doingbayesiandataanalysis.blogspot.com/2013/06/doing-bayesisan-data-analysis-at-jacob.html

and denote the set of outcomes as $\{y_i\}$. We assume that the outcomes are independent of each other, which means that the probability of the set of outcomes is the multiplicative product of the probabilities of the individual outcomes. Therefore, we derive the following formula for the probability of the set of outcomes:

$$\begin{aligned} p(\{y_i\}|\theta) &= \prod_i p(y_i|\theta) && \text{by assumption of independence} \\ &= \prod_i \theta^{y_i} (1-\theta)^{(1-y_i)} && \text{from Equation 5.10} \\ &= \theta^{\sum_i y_i} (1-\theta)^{\sum_i (1-y_i)} && \text{by algebra} \\ &= \theta^{\#\text{heads}} (1-\theta)^{\#\text{tails}} && (5.11) \end{aligned}$$

The final line of Equation 5.11 is a consequence of the fact that y_i is 1 for heads and y_i is 0 for tails; hence, $\sum_i y_i = \#\text{heads}$ and $\sum_i 1 - y_i = \#\text{tails}$. Later, we will also refer to #heads as z, the number of flips as N, and hence the #tails as $N - z$. Equation 5.11 will be useful later for numerical computations of Bayes' rule.

Now that we have defined the likelihood function in Equation 5.10, the next step (recall Section 2.3, p. 25) is establishing a prior distribution over the parameter values. At this point, we are merely building intuitions about Bayes' rule, and therefore, we will use an unrealistic but illustrative prior distribution. For this purpose, we suppose that there are only a few discrete values of the parameter θ that we will entertain, namely $\theta = 0.0$, $\theta = 0.1$, $\theta = 0.2$, $\theta = 0.3$, and so forth up to $\theta = 1.0$. You can think of this as meaning that there is a factory that manufactures coins, and the factory generates coins of only those 11 types: Some coins have $\theta = 0.0$, some coins have $\theta = 0.1$, some coins have $\theta = 0.2$, and so forth. The prior distribution indicates what we believe about the factory's production of those types. Suppose we believe that the factory tends to produce fair coins, with θ near 0.5, and we assign lower prior credibility to biases far above or below $\theta = 0.5$. One exact expression of this prior distribution is shown in the top panel of Figure 5.1. That panel shows a bar plotted at each candidate value of θ, with the height of the bar indicating the prior probability, $p(\theta)$. You can see that the heights have a triangular shape, with the peak of the triangle at $\theta = 0.5$, and the heights decreasing toward either extreme, such that $p(\theta = 0) = 0$ and $p(\theta = 1) = 0$. Notice that this graph is just like the Sherlock Holmes examples in Figure 2.1 (p. 17), the exoneration example in Figure 2.2 (p. 18), and especially the normal-means example in Figure 2.3 (p. 21), where we considered a discrete set of candidate possibilities.

The next steps of Bayesian data analysis (recall Section 2.3, p. 25) are collecting the data and applying Bayes' rule to re-allocate credibility across the possible parameter values. Suppose that we flip the coin once and observe heads. In this case, the data D consist of a single head in a single flip, which we annotate as $y = 1$ or, equivalently, $z = 1$ with $N = 1$. From the formula for the likelihood function in Equation 5.10,

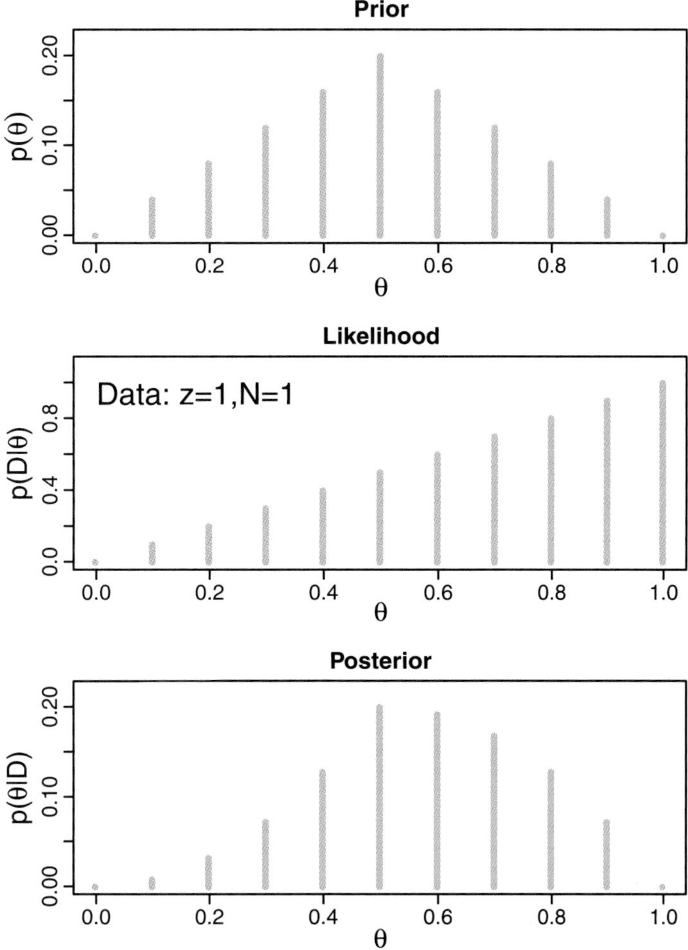

Figure 5.1 Bayes' rule applied to estimating the bias of a coin. There are discrete candidate values of θ. At each value of θ, the posterior is computed as prior times likelihood, normalized. In the data, denoted *D*, the number of heads is *z* and the number of flips is *N*.

we see that for these data the likelihood function becomes $p(D|\theta) = \theta$. The values of this likelihood function are plotted at the candidate values of θ in the middle panel of Figure 5.1. For example, the bar at $\theta = 0.2$ has height $p(D|\theta) = 0.2$, and the bar at $\theta = 0.9$ has height $p(D|\theta) = 0.9$.

The posterior distribution is shown in the lower panel of Figure 5.1. At each candidate value of θ, the posterior probability is computed from Bayes' rule (Equation 5.7) as the likelihood at θ times the prior at θ divided by $p(D)$. For example, consider $\theta = 0.2$, scanning vertically across panels. In the lower panel at $\theta = 0.2$, the posterior probability is the likelihood from the middle panel at $\theta = 0.2$ times the prior in the upper panel at

$\theta = 0.2$, divided by the sum, $p(D) = \sum_{\theta^*} p(D|\theta^*)p(\theta^*)$. This relationship is true at all values of θ.

Notice that the overall contour of the posterior distribution is different from the prior distribution. Because the data showed a head, the credibility of higher θ values has increased. For example, in the prior distribution, $p(\theta = 0.4)$ equals $p(\theta = 0.6)$, but in the posterior distribution, $p(\theta = 0.4|D)$ is less than $p(\theta = 0.6|D)$. Nevertheless, the prior distribution has a notable residual effect on the posterior distribution because it involves only a single flip of the coin. In particular, despite the data showing 100% heads (in a sample consisting of a single flip), the posterior probability of large θ values such as 0.9 is low. This illustrates a general phenomenon in Bayesian inference: The posterior is a compromise between the prior distribution and the likelihood function. Sometimes this is loosely stated as a compromise between the prior and the data. The compromise favors the prior to the extent that the prior distribution is sharply peaked and the data are few. The compromise favors the likelihood function (i.e., the data) to the extent that the prior distribution is flat and the data are many. Additional examples of this compromise are explained next.

5.3.1. Influence of sample size on the posterior

Suppose we fill in the candidate values of θ with 1,001 options, from 0.000, 0.001, 0.002, up to 1.000. Figure 5.1 would then be filled in with a dense forest of vertical bars, side by side with no visible space between them. But Bayes' rule applies the same as before, despite there being 1,001 candidate values of θ instead of only 11. Figure 5.2 shows two cases in which we start with the triangular prior used previously, but now filled in with 1,001 candidate values for θ. In the left side of Figure 5.2, the data consist of 25% heads for a small sample of $N = 4$ flips. In the right side of Figure 5.2, the data again consist of 25% heads, but for a larger sample of $N = 40$ flips. Notice that the likelihood functions in both cases have their modal (i.e., maximal) value at $\theta = 0.25$. This is because the probability of 25% heads is maximized in the likelihood function (Equation 5.11) when $\theta = 0.25$. Now inspect the posterior distributions at the bottom of Figure 5.2. For the small sample size, on the left, the mode (peak) of the posterior distribution is at $\theta = 0.40$, which is closer to the mode of the prior distribution (at 0.5) than the mode of the likelihood function. For the larger sample size, on the right, the mode of the posterior distribution is at $\theta = 0.268$, which is close to the mode of the likelihood function. In both cases, the mode of the posterior distribution is between the mode of the prior distribution and the mode of the likelihood function, but the posterior mode is closer to the likelihood mode for larger sample sizes.

Notice also in Figure 5.2 that the width of the posterior highest density intervals (HDI) is smaller for the larger sample size. Even though both samples of data showed 25% heads, the larger sample implied a smaller range of credible underlying biases in

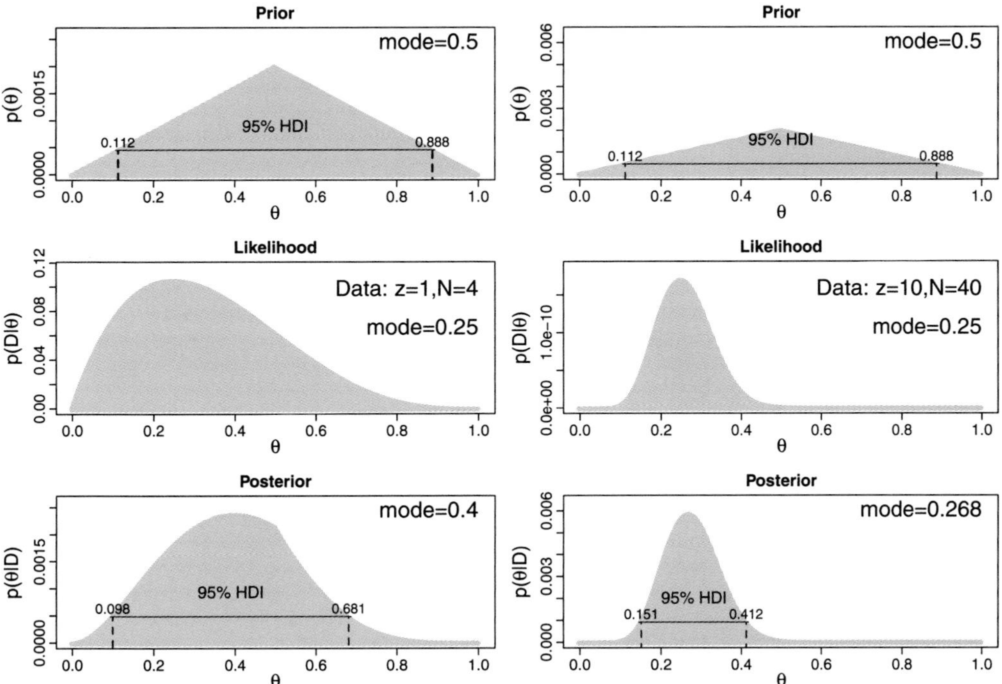

Figure 5.2 The two columns show different sample sizes with the same proportion of heads. The prior is the same in both columns but plotted on a different vertical scale. The influence of the prior is overwhelmed by larger samples, in that the peak of the posterior is closer to the peak of the likelihood function. Notice also that the posterior HDI is narrower for the larger sample.

the coin. In general, the more data we have, the more precise is the estimate of the parameter(s) in the model. Larger sample sizes yield greater precision or certainty of estimation.

5.3.2. Influence of the prior on the posterior

In Figure 5.3, the left side is the same small sample as the left side of Figure 5.2, but the prior in Figure 5.3 is flatter. Notice that the posterior distribution in Figure 5.3 is very close to the likelihood function, despite the small sample size, because the prior is so flat. In general, whenever the prior distribution is relatively broad compared with the likelihood function, the prior has fairly little influence on the posterior. In most of the applications in this book, we will be using broad, relatively flat priors.

The right side of Figure 5.3 is the same larger sample as the right side of Figure 5.2, but the prior in Figure 5.3 is sharper. In this case, despite the fact that the sample has a larger size of $N = 40$, the prior is so sharp that the posterior distribution is noticeably influenced by the prior. In real applications, this is a reasonable and intuitive inference,

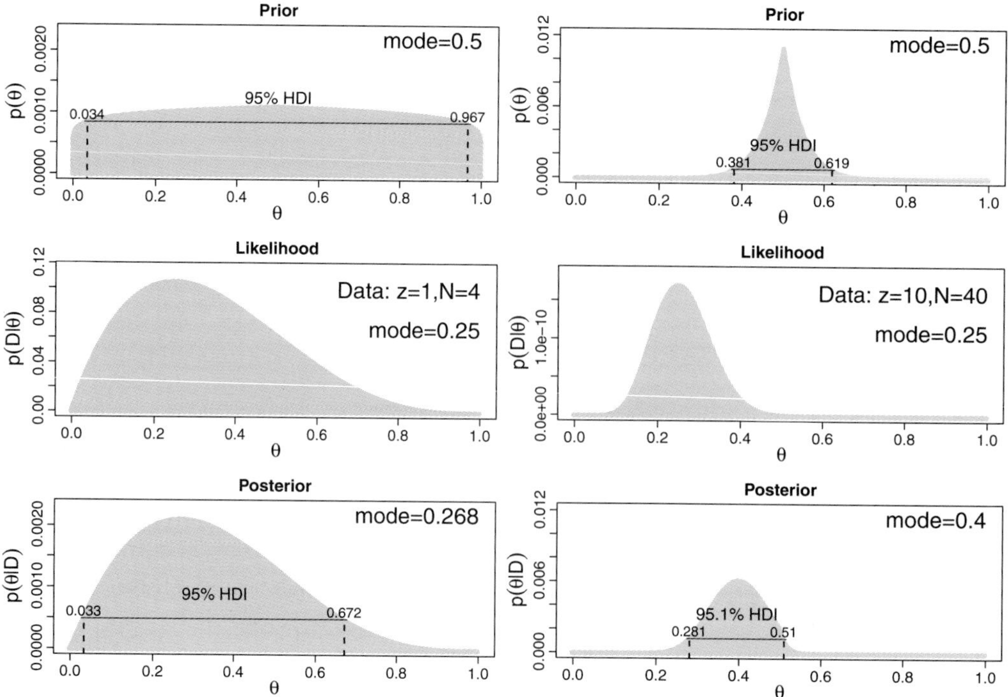

Figure 5.3 The left side is the same small sample as the left side of Figure 5.2 but with a flatter prior. The right side is the same larger sample as the right side of Figure 5.2 but with a sharper prior.

because in real applications a sharp prior has been informed by genuine prior knowledge that we would be reluctant to move away from without a lot of contrary data.

In other words, Bayesian inference is intuitively rational: With a strongly informed prior that uses a lot of previous data to put high credibility over a narrow range of parameter values, it takes a lot of novel contrary data to budge beliefs away from the prior. But with a weakly informed prior that spreads credibility over a wide range of parameter values, it takes relatively little data to shift the peak of the posterior distribution toward the data (although the posterior will be relatively wide and uncertain).

These examples have used arbitrary prior distributions to illustrate the mechanics of Bayesian inference. In real research, the prior distributions are either chosen to be broad and noncommittal on the scale of the data or specifically informed by publicly agreed prior research. As was the case with disease diagnosis in Table 5.4 (p. 104), it can be advantageous and rational to start with an informed prior, and it can be a serious blunder not to use strong prior information when it is available. Prior beliefs *should* influence rational inference from data, because the role of new data is to modify our beliefs from whatever they were without the new data. Prior beliefs are *not* capricious and idiosyncratic and covert, but instead are based on publicly agreed facts or theories. Prior beliefs used in data analysis must be admissible by a skeptical scientific audience. When scientists disagree about prior beliefs, the analysis can be conducted with multiple priors, to assess the robustness of the posterior against changes in the prior. Or, the multiple priors can be mixed together into a joint prior, with the posterior thereby incorporating the uncertainty in the prior. In summary, for most applications, specification of the prior turns out to be technically *un*problematic, although it is conceptually very important to understand the consequences of one's assumptions about the prior.

5.4. WHY BAYESIAN INFERENCE CAN BE DIFFICULT

Determining the posterior distribution directly from Bayes' rule involves computing the evidence (a.k.a. marginal likelihood) in Equations 5.8 and 5.9. In the usual case of continuous parameters, the integral in Equation 5.9 can be impossible to solve analytically. Historically, the difficulty of the integration was addressed by restricting models to relatively simple likelihood functions with corresponding formulas for prior distributions, called *conjugate* priors, that "played nice" with the likelihood function to produce a tractable integral. A couple cases of this conjugate-prior approach will be illustrated later in the book, but this book will instead emphasize the flexibility of modern computer methods. When the conjugate-prior approach does not work, an alternative is to approximate the actual functions with other functions that are easier to work with, and then show that the approximation is reasonably good under typical conditions. This approach goes by the name "variational approximation." This book does not provide any examples of variational approximation, but see Grimmer (2011) for an overview and pointers to additional sources.

Instead of analytical mathematical approaches, another class of methods involves numerical approximation of the integral. When the parameter space is small, one numerical approximation method is to cover the space with a comb or grid of points and compute the integral by exhaustively summing across that grid. This was the approach we used in Figures 5.2 and 5.3, where the domain of the parameter θ was represented by a fine comb of values, and the integral across the continuous parameter θ was approximated by a sum over the many discrete representative values. This method will not work for models with many parameters, however. In many realistic models, there are dozens or even hundreds of parameters. The space of possibilities is the *joint* parameter space involving all *combinations* of parameter values. If we represent each parameter with a comb of, say, 1,000 values, then for P parameters there are $1{,}000^P$ combinations of parameter values. When P is even moderately large, there are far too many combinations for even a modern computer to handle.

Another kind of approximation involves randomly sampling a large number of representative combinations of parameter values from the posterior distribution. In recent decades, many such algorithms have been developed, generally referred to as Markov chain Monte Carlo (MCMC) methods. What makes these methods so useful is that they can generate representative parameter-value combinations from the posterior distribution of complex models *without* computing the integral in Bayes' rule. It is the development of these MCMC methods that has allowed Bayesian statistical methods to gain practical use. This book focuses on MCMC methods for realistic data analysis.

5.5. APPENDIX: R CODE FOR FIGURES 5.1, 5.2, etc.

Figures 5.1, 5.2, etc., were created with the program `BernGrid.R`, which is in the program files for this book. The file name refers to the facts that the program estimates the bias in a coin using the Bernoulli likelihood function and a grid approximation for the continuous parameter. The program `BernGrid.R` defines a function named `BernGrid`. Function definitions and their use were explained in Section 3.7.3 (p. 64). The `BernGrid` function requires three user-specified arguments and has several other arguments with function-supplied default values. The three user-specified arguments are the grid of values for the parameter, `Theta`, the prior probability masses at each of those values, `pTheta`, and the data values, `Data`, which consists of a vector of 0's and 1's. Before calling the `BernGrid` function, its definition must be read into R using the `source` command. The program also uses utility functions defined in `DBDA2E-utilities.R`. Here is an example of how to use BernGrid, assuming that R's working directory contains the relevant programs:

```
source("DBDA2E-utilities.R")  # Load definitions of graphics functions etc.
source("BernGrid.R")          # Load the definition of the BernGrid function

Theta = seq( 0 , 1 , length=1001 )  # Specify fine comb for Theta.
pTheta = pmin( Theta , 1-Theta )    # Triangular shape for pTheta.
pTheta = pTheta/sum(pTheta)         # Make pTheta sum to 1.0
Data = c(rep(0,3),rep(1,1))         # Same as c(0,0,0,1). 25% heads with N=4.

openGraph(width=5,height=7)  # Open a graphics window.
posterior = BernGrid( Theta, pTheta , Data , plotType="Bars" ,
                      showCentTend="Mode" , showHDI=TRUE , showpD=FALSE )
saveGraph(file="BernGridExample",type="jpg")
```

The first two lines above use the `source` function to read in R commands from files. The `source` function was explained in Section 3.7.2. The next line uses the `seq` command to create a fine comb of θ values spanning the range from 0 to 1. The `seq` function was explained in Section 3.4.1.3.

The next line uses the `pmin` command to establish a triangular-shaped prior on the `Theta` vector. The `pmin` function has not been used previously in this book. To learn what it does, you could type ?pmin at the R command line. There you will find that pmin determines the component-wise minimum values of vectors, instead of the single minimum of all components together. For example, pmin(c(0,.25,.5,.75,1), c(1,.75,.5,.25,0)) yields c(0,.25,.5,.25,0). Notice how the values in the result go up then down, to form a triangular trend. This is how the triangular trend on pTheta is created.

The next line uses the sum command. It takes a vector argument and computes the sum of its components. The next line uses the rep function, which was explained in Section 3.4.1.4, to create a vector of zeros and ones that represent the data.

The BernGrid function is called in the penultimate line. Its first three arguments are the previously defined vectors, Theta, pTheta, and Data. The next arguments are optional. The plotType argument can be "Bars" (the default) or "Points". Try running it both ways to see the effect of this argument. The showCentTend argument specifies which measure of central tendency is shown and can be "Mode" or "Mean" or "None" (the default). The showHDI argument can be TRUE or FALSE (the default). The mass of the HDI is specified by the argument HDImass, which has a default value of 0.95. You will notice that for sparse grids on θ, it is usually impossible to span grid values that exactly achieve the desired HDI, and therefore, the bars that minimally exceed the HDI mass are displayed. The showpD argument specifies whether or not to display the value of the evidence from Equation 5.8. We will have no use for this value until we get into the topic of model comparison, in Section 10.1.

The previous paragraphs explained how to use the BernGrid function. If you want to know its internal mechanics, then you can open the function definition, BernGrid.R, in RStudio's editing window. You will discover that the vast majority of the program is devoted to producing the graphics at the end, and the next largest part is devoted to checking the user-supplied arguments for consistency at the beginning. The Bayesian part of the program consists of only a few lines:

```
# Create summary values of Data
z = sum( Data ) # number of 1's in Data
N = length( Data )
# Compute the Bernoulli likelihood at each value of Theta:
pDataGivenTheta = Theta^z * (1-Theta)^(N-z)
# Compute the evidence and the posterior via Bayes' rule:
pData = sum( pDataGivenTheta * pTheta )
pThetaGivenData = pDataGivenTheta * pTheta / pData
```

The Bernoulli likelihood function in the R code is in the form of Equation 5.11. Notice that Bayes' rule appears written in the R code very much like Equations 5.7 and 5.8.

5.6. EXERCISES

Look for more exercises at https://sites.google.com/site/doingbayesiandataanalysis/

Exercise 5.1. [Purpose: Iterative application of Bayes' rule, and seeing how posterior probabilities change with inclusion of more data.] This exercise extends the ideas of Table 5.4, so at this time, please review Table 5.4 and its discussion in the text. Suppose that the same randomly selected person as in Table 5.4 gets re-tested after the first test result was positive, and on the re-test, the result is negative. When taking into account the results of both tests, what is the probability that the person has the disease? *Hint:* For the prior probability of the re-test, use the posterior computed from the Table 5.4. Retain as many decimal places as possible, as rounding can have a surprisingly big effect on the results. One way to avoid unnecessary rounding is to do the calculations in R.

Exercise 5.2. [Purpose: Getting an intuition for the previous results by using "natural frequency" and "Markov" representations]

(A) Suppose that the population consists of 100,000 people. Compute how many people would be expected to fall into each cell of Table 5.4. To compute the expected frequency of people in a cell, just multiply the cell probability by the size of the population. To get you started, a few of the cells of the frequency table are filled in here:

	$\theta = \ddot{\frown}$	$\theta = \ddot{\smile}$	
$D = +$	$\text{freq}(D{=}+, \theta{=}\ddot{\frown})$ $= p(D{=}+, \theta{=}\ddot{\frown})\, N$ $= p(D{=}+ \mid \theta{=}\ddot{\frown})\, p(\theta{=}\ddot{\frown})\, N$ $= 99$	$\text{freq}(D{=}+, \theta{=}\ddot{\smile})$ $= p(D{=}+, \theta{=}\ddot{\smile})\, N$ $= p(D{=}+ \mid \theta{=}\ddot{\smile})\, p(\theta{=}\ddot{\smile})\, N$ $=$	$\text{freq}(D{=}+)$ $= p(D{=}+)\, N$ $=$
$D = -$	$\text{freq}(D{=}-, \theta{=}\ddot{\frown})$ $= p(D{=}-, \theta{=}\ddot{\frown})\, N$ $= p(D{=}- \mid \theta{=}\ddot{\frown})\, p(\theta{=}\ddot{\frown})\, N$ $= 1$	$\text{freq}(D{=}-, \theta{=}\ddot{\smile})$ $= p(D{=}-, \theta{=}\ddot{\smile})\, N$ $= p(D{=}- \mid \theta{=}\ddot{\smile})\, p(\theta{=}\ddot{\smile})\, N$ $=$	$\text{freq}(D{=}-)$ $= p(D = -)\, N$ $=$
	$\text{freq}(\theta = \ddot{\frown})$ $= p(\theta = \ddot{\frown})\, N$ $= 100$	$\text{freq}(\theta = \ddot{\smile})$ $= p(\theta = \ddot{\smile})\, N$ $= 99{,}900$	N $= 100{,}000$

Notice the frequencies on the lower margin of the table. They indicate that out of 100,000 people, only 100 have the disease, while 99,900 do not have the disease. These marginal frequencies instantiate the prior probability that $p(\theta = \ddot{\frown}) = 0.001$. Notice also the cell frequencies in the column $\theta = \ddot{\frown}$, which indicate that of 100 people with the disease, 99 have a positive test result and 1 has a negative test result. These cell frequencies instantiate the hit rate of 0.99. Your job for this part of the exercise is to fill in the frequencies of the remaining cells of the table.

(B) Take a good look at the frequencies in the table you just computed for the previous part. These are the so-called "natural frequencies" of the events, as opposed to the somewhat unintuitive expression in terms of conditional probabilities (Gigerenzer & Hoffrage, 1995). From the cell frequencies alone, determine the proportion of people who have the disease, given that their test result is positive. Before computing the exact answer arithmetically, first give a rough intuitive answer merely by looking at the relative frequencies in the row $D = +$. Does your intuitive answer match the intuitive answer you provided when originally reading about Table 5.4? Probably not. Your intuitive answer here is probably much closer to the correct answer. Now compute the exact answer arithmetically. It should match the result from applying Bayes' rule in Table 5.4.

(C) Now we'll consider a related representation of the probabilities in terms of natural frequencies, which is especially useful when we accumulate more data. This type of representation is called a "Markov" representation by Krauss, Martignon, and Hoffrage (1999). Suppose now we start with a population of $N = 10,000,000$ people. We expect 99.9% of them (i.e., 9,990,000) not to have the disease, and just 0.1% (i.e., 10,000) to have the disease. Now consider how many people we expect to test positive. Of the 10,000 people who have the disease, 99%, (i.e., 9,900) will be expected to test positive. Of the 9,990,000 people who do not have the disease, 5% (i.e., 499,500) will be expected to test positive. Now consider re-testing everyone who has tested positive on the first test. How many of them are expected to show a negative result on the retest? Use this diagram to compute your answer:

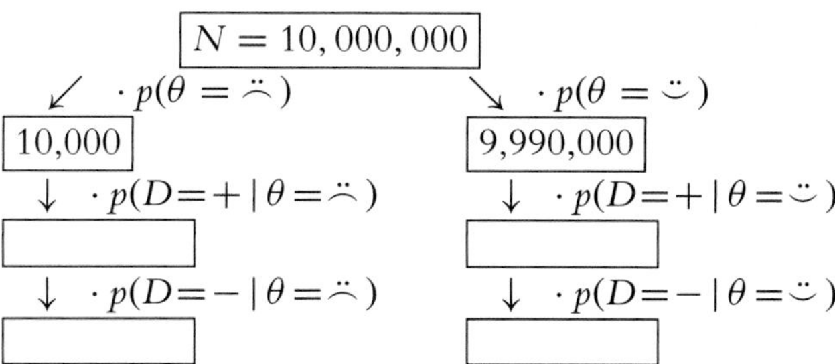

When computing the frequencies for the empty boxes above, be careful to use the proper conditional probabilities!

(D) Use the diagram in the previous part to answer this: What proportion of people, who test positive at first and then negative on retest, actually have the disease? In other words, of the total number of people at the bottom of the diagram in the previous part (those are the people who tested positive then negative), what proportion of them are in the left branch of the tree? *How does the result compare with your answer to Exercise 5.1?*

Exercise 5.3. [Purpose: To see a hands-on example of data-order invariance.] Consider again the disease and diagnostic test of the previous two exercises.

(A) Suppose that a person selected at random from the population gets the test and it comes back negative. Compute the probability that the person has the disease.

(B) The person then gets re-tested, and on the second test the result is positive. Compute the probability that the person has the disease. *How does the result compare with your answer to Exercise 5.1?*

Exercise 5.4. [Purpose: To gain intuition about Bayesian updating by using `BernGrid`.] Open the program `BernGridExample.R`. You will notice there are several examples of using the function `BernGrid`. Run the script. For each example, include the R code and the resulting graphic and explain what idea the example illustrates. Hints: Look back at Figures 5.2 and 5.3, and look ahead to Figure 6.5. Two of the examples involve a single flip, with the only difference between the examples being whether the prior is uniform or contains only two extreme options. The point of those two examples is to show that a single datum implies little when the prior is vague, but a single datum can have strong implications when the prior allows only two very different possibilities.

PART II

All the Fundamentals Applied to Inferring a Binomial Probability

In the next few chapters, we will develop all the foundational concepts and methods of Bayesian data analysis, which are applied to the simplest type of data. Because of the simplicity of the data, we can focus on the Bayesian methods and scaffold the concepts clearly and efficiently. The subsequent part of the book applies the methods developed in this part to more complex data structures.

CHAPTER 6

Inferring a Binomial Probability via Exact Mathematical Analysis

Contents

I built up my courage to ask her to dance
By drinking too much before taking the chance.
I fell on my butt when she said see ya later;
Less priors might make my posterior beta.[1]

This chapter presents an example of how to do Bayesian inference using pure analytical mathematics without any approximations. Ultimately, we will not use the pure analytical approach for complex applications, but this chapter is important for two reasons. First, the relatively simple mathematics in this chapter nicely reveal the underlying concepts of Bayesian inference on a continuous parameter. The simple formulas show how the continuous allocation of credibility changes systematically as data accumulate. The examples provide an important conceptual foundation for subsequent approximation methods, because the examples give you a clear sense of *what* is being approximated. Second, the distributions introduced in this chapter, especially the beta distribution, will be used repeatedly in subsequent chapters.

We continue with situations in which the observed datum has two nominal levels, and we would like to estimate the underlying probability of the two possible outcomes.

[1] This chapter is about using the beta distribution as a prior distribution for the Bernoulli likelihood function, in which case the posterior distribution is also a beta distribution. The poem explains another way to make a posterior beta.

Doing Bayesian Data Analysis, 2nd Edition
http://dx.doi.org/10.1016/B978-0-12-405888-0.00006-4

One stereotypical case is the flip of a coin, in which the observed outcome could be heads or tails, and we want to estimate the underlying probabilities of the two outcomes. As has been emphasized before (e.g., Section 4.1.1, p. 73), the coin merely stands in for a real-world application we care about, such as estimating the success probability of a drug, or the probability correct on an exam, or the probability of a person being left handed, or the probability of successful free throw by a player in basketball, or the probability that a baby is a girl, or the probability that a heart surgery patient will survive more than a year after surgery, or the probability that a person will agree with a statement on a survey, or the probability that a widget on an assembly line is faulty, and so on. While we talk about heads and tails for coins, keep in mind that the methods could be applied to many other interesting real-world situations.

We require in this scenario that the space of possibilities for each datum has just two values that are mutually exclusive. These two values have no ordinal or metric relationship with each other, they are just nominal values. Because there are two nominal values, we refer to this sort of data as "dichotomous," or "nominal with two levels," or "binomial." We also assume that each observed datum is independent of the others. Typically, we will also assume that the underlying probability is stationary through time. Coin flipping is the standard example of this situation: There are two possible outcomes (head or tail), and we assume that the flips are independent of each other and that the probability of getting a head is stationary through time.

In a Bayesian analysis, we begin with some prior allocation of credibility over possible probabilities of the coin coming up heads. Then, we observe some data that consist of a set of results from flipping the coin. Then, we infer the posterior distribution of credibility using Bayes' rule. Bayes' rule requires us to specify the likelihood function and that is the topic of the next section.

6.1. THE LIKELIHOOD FUNCTION: BERNOULLI DISTRIBUTION

The Bernoulli distribution was defined back in Equation 5.10 (p. 109). We repeat it here for convenience. The outcome of a single flip is denoted y and can take on values of 0 or 1. The probability of each outcome is given by a function of parameter θ:

$$p(y|\theta) = \theta^{y}\,(1-\theta)^{(1-y)} \tag{6.1}$$

Notice that Equation 6.1 reduces to $p(y = 1|\theta) = \theta$, so the parameter θ can be interpreted as the underlying probability of heads for the coin flip. The formula in Equation 6.1 expresses the *Bernoulli distribution*, which is a probability distribution over the two discrete values of y, for any fixed value of θ. In particular, the sum of the probabilities is 1, as must be true of a probability distribution: $\sum_y p(y|\theta) = p(y = 1|\theta) + p(y{=}0|\theta) = \theta + (1{-}\theta) = 1$.

Another perspective on Equation 6.1 is to think of the data value y as fixed by an observation, and the value of θ as variable. Equation 6.1 then specifies the probability of the fixed y value as a function of candidate values of θ. Different values of θ yield different probabilities of the datum y. When thought of in this way, Equation 6.1 is the *likelihood function* of θ.

Notice that the likelihood function is a function of a continuous value θ, whereas the Bernoulli distribution is a discrete distribution over the two values of y. The likelihood function, although it specifies a probability at each value of θ, is *not* a probability distribution. In particular, it does not integrate to 1. For example, suppose we have observed that $y = 1$. Then, the integral of the likelihood function is $\int_0^1 \mathrm{d}\theta \, \theta^y \, (1-\theta)^{(1-y)} = \int_0^1 \mathrm{d}\theta \, \theta = \frac{1}{2}$, which does not equal 1.

In Bayesian inference, the function $p(y|\theta)$ is usually thought of with the data, y, known and fixed, and the parameter, θ, uncertain and variable. Therefore, $p(y|\theta)$ is usually called the likelihood function for θ, and Equation 6.2 is called the *Bernoulli likelihood function*. Don't forget, however, that the very same formula is also the probability of the datum, y, and can be called the Bernoulli distribution if θ is considered to be fixed and y is thought of as the variable.

We also previously figured out the formula for the probability of a *set* of outcomes, back in Equation 5.11 (p. 110), which again we repeat here for convenience. Denote the outcome of the ith flip as y_i, and denote the set of outcomes as $\{y_i\}$. We assume that the outcomes are independent of each other, which means that the probability of the set of outcomes is the multiplicative product of the probabilities of the individual outcomes. If we denote the number of heads as $z = \sum_i y_i$ and the number of tails as $N - z = \sum_i (1 - y_i)$, then

$$
\begin{aligned}
p(\{y_i\}|\theta) &= \prod_i p(y_i|\theta) && \text{by assumption of independence} \\
&= \prod_i \theta^{y_i} \, (1-\theta)^{(1-y_i)} && \text{from Equation 6.1} \\
&= \theta^{\sum_i y_i} \, (1-\theta)^{\sum_i (1-y_i)} && \text{by algebra} \\
&= \theta^z \, (1-\theta)^{N-z} && (6.2)
\end{aligned}
$$

This formula is useful for applications of Bayes' rule to large data sets. I will sometimes lapse terminologically sloppy and refer to the formula in Equation 6.2 as the Bernoulli likelihood function for a set of flips, but please remember that the Bernoulli likelihood function really refers to a single flip in Equation 6.1.[2]

6.2. A DESCRIPTION OF CREDIBILITIES: THE BETA DISTRIBUTION

In this chapter, we use purely mathematical analysis, with no numerical approximation, to derive the mathematical form of the posterior credibilities of parameter values. To do this, we need a mathematical description of the prior allocation of credibilities. That is, we need a mathematical formula that describes the prior probability for each value of the parameter θ on the interval $[0, 1]$.

In principle, we could use any probability density function supported on the interval $[0, 1]$. When we intend to apply Bayes' rule (Equation 5.7, p. 106), however, there are two desiderata for mathematical tractability. First, it would be convenient if the product of $p(y|\theta)$ and $p(\theta)$, which is in the numerator of Bayes' rule, results in a function of the same form as $p(\theta)$. When this is the case, the prior and posterior beliefs are described using the same form of function. This quality allows us to include subsequent additional data and derive another posterior distribution, again of the same form as the prior. Therefore, no matter how much data we include, we always get a posterior of the same functional form. Second, we desire the denominator of Bayes' rule (Equation 5.9, p. 107), namely $\int d\theta\, p(y|\theta)p(\theta)$, to be solvable analytically. This quality also depends on how the form of the function $p(\theta)$ relates to the form of the function $p(y|\theta)$. When the forms of $p(y|\theta)$ and $p(\theta)$ combine so that the posterior distribution has the same form as the prior distribution, then $p(\theta)$ is called a *conjugate prior* for $p(y|\theta)$. Notice that the prior is conjugate only with respect to a particular likelihood function.

In the present situation, we are seeking a functional form for a prior density over θ that is conjugate to the Bernoulli likelihood function in Equation 6.1. If you think about it a minute, you'll notice that if the prior is of the form $\theta^a(1-\theta)^b$, then when you multiply the Bernoulli likelihood with the prior, you'll again get a function of the same form, namely $\theta^{(y+a)}(1-\theta)^{(1-y+b)}$. Therefore, to express the prior beliefs over θ, we seek a probability density function involving $\theta^a(1-\theta)^b$.

[2] Some readers might be familiar with the binomial distribution, $p(z|N,\theta) = \binom{N}{z}\theta^z(1-\theta)^{(N-z)}$, and wonder why it is not used here. The reason is that here we are considering each flip of the coin to be a distinct event, whereby each observation has just two possible values, $y \in \{0, 1\}$. Therefore, the likelihood function is the Bernoulli distribution, which has two possible outcome values. The probability of the *set* of events is then the product of the individual event probabilities, as in Equation 6.2. If we instead considered a single "event" to be the flipping of N coins, then an observation of a *single* event could have $N + 1$ possible values, namely $z \in \{0, 1, \ldots, N\}$. Then we would need a likelihood function that provided the probabilities of those $N + 1$ possible outcomes, given the fixed value N that defines a single observational event. In this case, the probabilities of the values would be given by the binomial distribution. The binomial distribution is explained in the section accompanying Equation 11.5 on p. 303.

A probability density of that form is called a *beta distribution*. Formally, a beta distribution has two parameters, called a and b, and the density itself is defined as

$$\begin{aligned} p(\theta|a,b) &= \text{beta}(\theta|a,b) \\ &= \theta^{(a-1)}\,(1-\theta)^{(b-1)}/B(a,b) \end{aligned} \tag{6.3}$$

where $B(a, b)$ is simply a normalizing constant that ensures that the area under the beta density integrates to 1.0, as all probability density functions must. In other words, the normalizer for the beta distribution is the beta *function*

$$B(a,b) = \int_0^1 \mathrm{d}\theta\,\theta^{(a-1)}\,(1-\theta)^{(b-1)} \tag{6.4}$$

Remember that the beta distribution (Equation 6.3) is only defined for values of θ in the interval [0, 1], and the values of a and b must be positive. Notice also that in the definition of the beta distribution (Equation 6.3), the value of θ is raised to the power $a-1$, not the power a, and the value of $(1-\theta)$ is raised to the power $b-1$, not the power b.

Be careful to distinguish the beta *function*, $B(a, b)$ in Equation 6.4, from the beta *distribution*, beta$(\theta|a, b)$ in Equation 6.3. The beta function is not a function of θ because θ has been "integrated out," the function only involves the variables a and b. In the programming language R, beta$(\theta|a, b)$ is `dbeta(θ,a,b)`, and $B(a, b)$ is `beta(a,b)`.[3]

Examples of the beta distribution are displayed in Figure 6.1. Each panel of Figure 6.1 plots $p(\theta|a, b)$ as a function of θ for particular values of a and b, as indicated inside each panel. Notice that as a gets bigger (left to right across columns of Figure 6.1), the bulk of the distribution moves rightward over higher values of θ, but as b gets bigger (top to bottom across rows of Figure 6.1), the bulk of the distribution moves leftward over lower values of θ. Notice that as a and b get bigger together, the beta distribution gets narrower. The variables a and b are called the *shape parameters* of the beta distribution because they determine its shape, as can be seen in Figure 6.1. Although Figure 6.1 features mostly integer values of a and b, the shape parameters can have any positive real value.

6.2.1. Specifying a beta prior

We would like to specify a beta distribution that describes our prior beliefs about θ. You can think of a and b in the prior as if they were previously observed data, in which there

[3] Whereas it is true that $B(a,b) = \int_0^1 \mathrm{d}\theta\,\theta^{(a-1)}\,(1-\theta)^{(b-1)}$, the beta function can also be expressed as $B(a,b) = \Gamma(a)\Gamma(b)/\Gamma(a+b)$, where Γ is the *Gamma function*: $\Gamma(a) = \int_0^\infty \mathrm{d}t\,t^{(a-1)}\exp(-t)$. The Gamma function is a generalization of the factorial function, because, for integer valued a, $\Gamma(a) = (a-1)!$. In R, $\Gamma(a)$ is `gamma(a)`. Many sources define the beta function this way, in terms of the Gamma function. We will not be using the Gamma function.

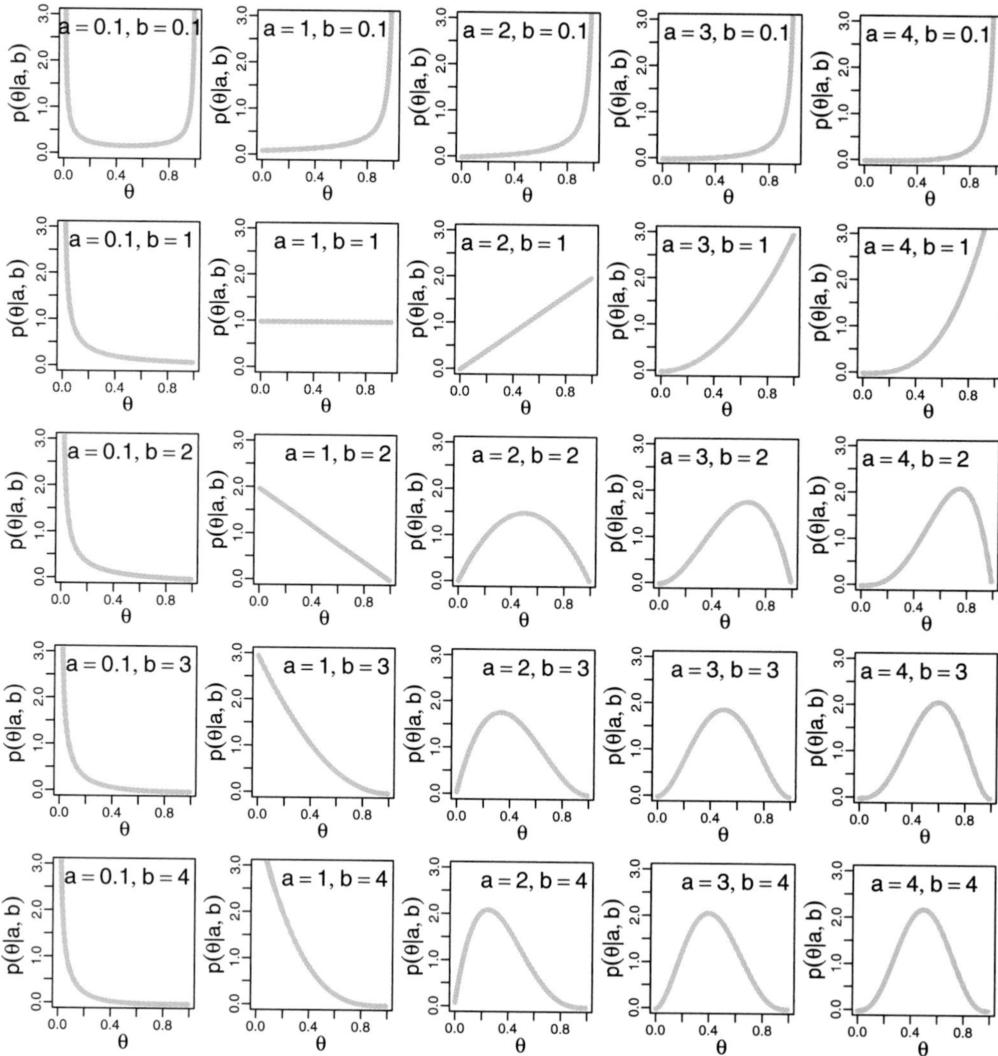

Figure 6.1 Examples of the beta distribution (Equation 6.1). The shape parameter *a* increases from left to right across the columns, while the shape parameter *b* increases from top to bottom across the rows.

were a heads and b tails in a total of $n = a + b$ flips. For example, if we have no prior knowledge other than the knowledge that the coin has a head side and a tail side, that's tantamount to having previously observed one head and one tail, which corresponds to $a = 1$ and $b = 1$. You can see in Figure 6.1 that when $a = 1$ and $b = 1$, the beta distribution is uniform: All values of θ are equally probable. As another example, if we think that the coin is probably fair but we're not very sure, then we can imagine that the previously observed data had, say, $a = 4$ heads and $b = 4$ tails. You can see in Figure 6.1 that when $a = 4$ and $b = 4$, the beta distribution is peaked at $\theta = 0.5$, but higher or lower values of θ are moderately probable too.

Often we think of our prior beliefs in terms of a central tendency and certainty about that central tendency. For example, in thinking about the probability of left handedness in the general population of people, we might think from everyday experience that it's around 10%. But if we are not very certain about that value, we might consider the equivalent previous sample size to be small, say, $n = 10$, which means that of 10 previously observed people, 1 of them was left handed. As another example, in thinking about the probability that a government-minted coin comes up heads, we might believe that it is very nearly 50%, and because we are fairly certain, we could set the equivalent previous sample size to, say, $n = 200$, which means that of 200 previously observed flips, 100 were heads. Our goal is to convert a prior belief expressed in terms of central tendency and sample size into equivalent values of a and b in the beta distribution.

Toward this goal, it is useful to know the central tendency and spread of the beta distribution expressed in terms of a and b. It turns out that the mean of the beta$(\theta|a, b)$ distribution is $\mu = a/(a + b)$ and the mode is $\omega = (a - 1)/(a + b - 2)$ for $a > 1$ and $b > 1$ (μ is Greek letter mu and ω is Greek letter omega). Thus, when $a = b$, the mean and mode are 0.5. When $a > b$, the mean and mode are greater than 0.5, and when $a < b$, the mean and mode are less than 0.5. The spread of the beta distribution is related to the "concentration" $\kappa = a + b$ (κ is Greek letter kappa). You can see from Figure 6.1 that as $\kappa = a + b$ gets larger, the beta distribution gets narrower or more concentrated. Solving those equations for a and b yields the following formulas for a and b in terms of the mean μ, the mode ω, and the concentration κ:

$$a = \mu\kappa \quad \text{and} \quad b = (1 - \mu)\kappa \tag{6.5}$$

$$a = \omega(\kappa - 2) + 1 \quad \text{and} \quad b = (1 - \omega)(\kappa - 2) + 1 \;\; \text{for } \kappa > 2 \tag{6.6}$$

The value we choose for the prior κ can be thought of this way: It is the number of new flips of the coin that we would need to make us teeter between the new data and the prior belief about μ. If we would only need a few new flips to sway our beliefs, then our prior beliefs should be represented by a small κ. If we would need a large number of new flips to sway us away from our prior beliefs about μ, then our prior beliefs are worth a very large κ. For example, suppose that I think the coin is fair, so $\mu = 0.5$, but I'm *not* highly confident about it, so maybe I imagine I've seen only $\kappa = 8$ previous flips. Then, $a = \mu\kappa = 4$ and $b = (1 - \mu)\kappa = 4$, which, as we saw before, is a beta distribution peaked at $\theta = 0.5$ and with higher or lower values less probable.

The mode can be more intuitive than the mean, especially for skewed distributions, because the mode is where the distribution reaches its tallest height, which is easy to visualize. The mean in a skewed distribution is somewhere away from the mode, in the direction of the longer tail. For example, suppose we want to create a beta distribution that has its mode at $\omega = 0.80$, with a concentration corresponding to $\kappa = 12$. Then, using Equation 6.6, we find that the corresponding shape parameters are $a = 9$ and $b = 3$. The lower panels of Figure 6.2 show plots of beta distributions for which the mode falls at $\theta = 0.8$. On the other hand, if we create a beta distribution that has its *mean* at 0.8, we get the distributions shown in the upper panels of Figure 6.2, where it can be seen that the modes are considerably to the right of the mean, not at $\theta = 0.8$.

Yet another way of establishing the shape parameters is by starting with the mean and standard deviation, σ, of the desired beta distribution. You must be careful with this approach, because the standard deviation must make sense in the context of a beta density. In particular, the standard deviation should typically be less than 0.28867, which

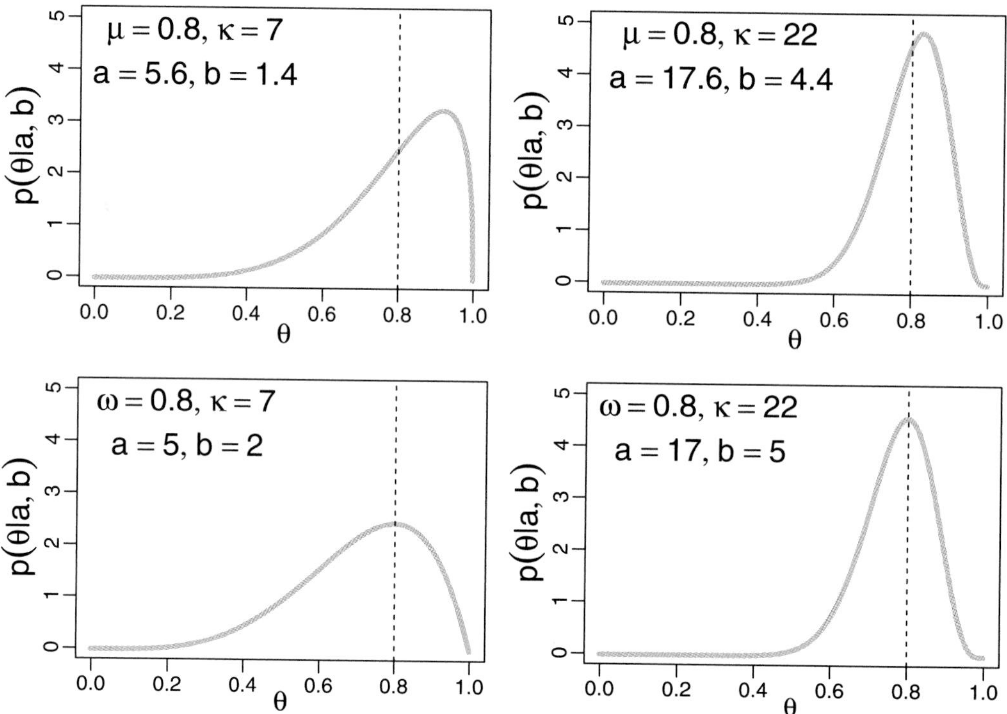

Figure 6.2 Beta distributions with a *mean* of $\mu = 0.8$ in the upper panels and a *mode* of $\omega = 0.8$ in the lower panels. Because the beta distribution is usually skewed, it can be more intuitive to think in terms of its mode instead of its mean. When κ is smaller, as in the left column, the beta distribution is wider than when κ is larger, as in the right column.

is the standard deviation of a uniform distribution.[4] For a beta density with mean μ and standard deviation σ, the shape parameters are

$$a = \mu \left(\frac{\mu(1-\mu)}{\sigma^2} - 1 \right) \quad \text{and} \quad b = (1-\mu) \left(\frac{\mu(1-\mu)}{\sigma^2} - 1 \right) \tag{6.7}$$

For example, if $\mu = 0.5$ and $\sigma = 0.28867$, Equation 6.7 implies that $a = 1$ and $b = 1$. As another example, if $\mu = 0.5$ and $\sigma = 0.1$, then $a = 12$ and $b = 12$, which is to say that a beta$(\theta|12, 12)$ distribution has a standard deviation of 0.1.

I have created convenient utility functions in R that implement the parameter transformations in Equations 6.5–6.7. The functions are loaded into R by typing `source("DBDA2E-utilities.R")`, when that file is in the current working directory. Hopefully, the function names are self-explanatory, and here is an example of their use:

```
> betaABfromMeanKappa( mean=0.25 , kappa=4 )
$a
[1] 1
$b
[1] 3
> betaABfromModeKappa( mode=0.25 , kappa=4 )
$a
[1] 1.5
$b
[1] 2.5
> betaABfromMeanSD( mean=0.5 , sd=0.1 )
$a
[1] 12
$b
[1] 12
```

The functions return a `list` of named components. Therefore, if you assign the result of the function to a variable, you can get at the individual parameters by their component names, as in the following example:

```
> betaParam = betaABfromModeKappa( mode=0.25 , kappa=4 )
> betaParam$a
[1] 1.5
> betaParam$b
[1] 2.5
```

In most applications, we will deal with beta distributions for which $a \geq 1$ and $b \geq 1$, that is, $\kappa > 2$. This reflects prior knowledge that the coin has a head side and a

[4] The standard deviation of the beta distribution is $\sqrt{\mu(1-\mu)/(a+b+1)}$. Notice that the standard deviation gets smaller when the concentration $\kappa = a + b$ gets larger. While this fact is nice to know, we will not have use for it in our applications.

tail side. In these situations when we know $\kappa > 2$, it can be most intuitive to use the parameterization of the beta distribution in terms of the mode in Equation 6.6. There are some situations, however, in which it may be convenient to use beta distributions in which $a < 1$ and/or $b < 1$, or for which we cannot be confident that $\kappa > 2$. For example, we might believe that the coin is a trick coin that nearly always comes up heads or nearly always comes up tails, but we don't know which. In these situations, we cannot use the parameterization in terms of the mode, which requires $\kappa > 2$, and instead we can use the parameterization of the beta distribution in terms of the mean in Equations 6.5.

6.3. THE POSTERIOR BETA

Now that we have determined a convenient prior for the Bernoulli likelihood function, let's figure out exactly what the posterior distribution is when we apply Bayes' rule (Equation 5.7, p. 106). Suppose we have a set of data comprising N flips with z heads. Substituting the Bernoulli likelihood (Equation 6.2) and the beta prior distribution (Equation 6.3) into Bayes' rule yields

$$
\begin{aligned}
p(\theta|z,N) &= p(z,N|\theta)p(\theta)/p(z,N) && \text{Bayes' rule} \\
&= \theta^{z}(1-\theta)^{(N-z)}\,\frac{\theta^{(a-1)}(1-\theta)^{(b-1)}}{B(a,b)}\Big/p(z,N) && \\
& && \text{by definitions of Bernoulli and beta distributions} \\
&= \theta^{z}\,(1-\theta)^{(N-z)}\;\theta^{(a-1)}\,(1-\theta)^{(b-1)}\Big/\big[B(a,b)p(z,N)\big] && \text{by rearranging factors} \\
&= \theta^{((z+a)-1)}\,(1-\theta)^{((N-z+b)-1)}\Big/\big[B(a,b)\,p(z,N)\big] && \text{by collecting powers} \\
&= \theta^{((z+a)-1)}\,(1-\theta)^{((N-z+b)-1)}\Big/B(z+a,N-z+b) && (6.8)
\end{aligned}
$$

The last step in the above derivation, from $B(a,b)p(z,N)$ to $B(z+a,N-z+b)$, was not made via some elaborate covert analysis of integrals. Instead, the transition was made simply by thinking about what the normalizing factor for the numerator must be. The numerator is $\theta^{((z+a)-1)}\,(1-\theta)^{((N-z+b)-1)}$, which is the numerator of a $\text{beta}(\theta|z+a,\ N-z+b)$ distribution. For the function in Equation 6.8 to be a probability distribution, *as it must be*, the denominator *must* be the normalizing factor for the corresponding beta distribution, which is $B(z+a,N-z+b)$ by definition of the beta function.[5]

Equation 6.8 tells us a key point: If the prior distribution is $\text{beta}(\theta|a,b)$, and the data have z heads in N flips, then the posterior distribution is $\text{beta}(\theta|z+a,N-z+b)$.

[5] As an aside, because $B(a,b)\,p(z,N) = B(z+a,N-z+b)$, we re-arrange to find that $p(z,N) = B(z+a,N-z+b)/B(a,b)$, which will be useful in Section 10.2.1.

The simplicity of this updating formula is one of the beauties of the mathematical approach to Bayesian inference. You can think about this updating formula with reference to Figure 6.1. Suppose the prior is beta$(\theta|1, 1)$, as shown in the second row and second column of Figure 6.1. We flip the coin once and observe heads. The posterior distribution is then beta$(\theta|2, 1)$, as shown in the second row and third column of Figure 6.1. Suppose we flip the coin again and observe tails. The posterior distribution is now updated to beta$(\theta|2, 2)$, as shown in the third row and third column of Figure 6.1. This process continues for any amount of data. If the initial prior is a beta distribution, then the posterior distribution is always a beta distribution.

6.3.1. Posterior is compromise of prior and likelihood

The posterior distribution is always a compromise between the prior distribution and the likelihood function. The previous chapter (specifically Section 5.3) gave examples by using grid approximation, but now we can illustrate the compromise with a mathematical formula. For a prior distribution expressed as beta$(\theta|a, b)$, the prior mean of θ is $a/(a + b)$. Suppose we observe z heads in N flips, which is a proportion of z/N heads in the data. The posterior mean is $(z + a)/[(z + a) + (N - z + b)] = (z+a)/(N+a+b)$. It turns out that the posterior mean can be algebraically re-arranged into a weighted average of the prior mean, $a/(a + b)$, and the data proportion, z/N, as follows:

$$\underbrace{\frac{z+a}{N+a+b}}_{\text{posterior}} = \underbrace{\frac{z}{N}}_{\text{data}} \underbrace{\frac{N}{N+a+b}}_{\text{weight}} + \underbrace{\frac{a}{a+b}}_{\text{prior}} \underbrace{\frac{a+b}{N+a+b}}_{\text{weight}}. \tag{6.9}$$

Equation 6.9 indicates that the posterior mean is always somewhere between the prior mean and the proportion in the data. The mixing weight on the data proportion increases as N increases. Thus, the more data we have, the less is the influence of the prior, and the posterior mean gets closer to the proportion in the data. In particular, when $N = a + b$, the mixing weights are 0.5, which indicates that the prior mean and the data proportion have equal influence in the posterior. This result echoes what was said earlier (Equation 6.5) regarding how to set a and b to represent our prior beliefs: The choice of prior n (which equals $a + b$) should represent the size of the new data set that would sway us away from our prior toward the data proportion.

Figure 6.3 illustrates an example of Equation 6.9. The prior has $a = 5$ and $b = 5$, hence a prior mean of $a/(a + b) = 0.5$, and the data show $z = 1$ with $N = 10$, hence a proportion of heads of $z/N = 0.1$. The weight on the prior mean is $(a + b)/(N + a + b) = 0.5$, as is the weight on the data proportion. Hence, the mean of the posterior should be $0.5 \cdot 0.5 + 0.5 \cdot 0.1 = 0.3$, which indeed it is, as shown in Figure 6.3.

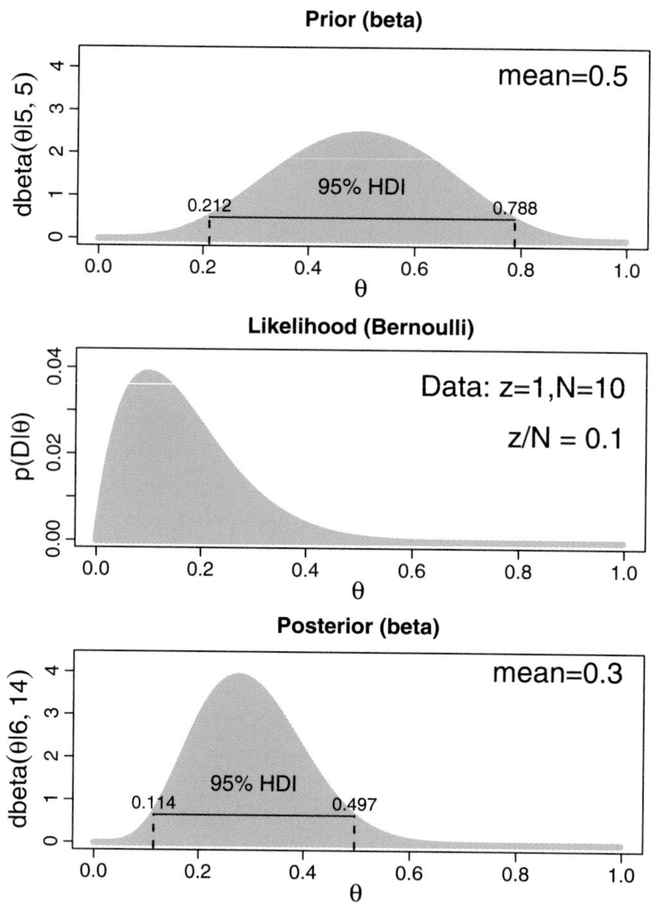

Figure 6.3 An illustration of Equation 6.9, showing that the mean of the posterior is a weighted combination of the mean of the prior and the proportion of heads in the data.

6.4. EXAMPLES

6.4.1. Prior knowledge expressed as a beta distribution

Suppose someone has a coin that we know to be a regular, unaltered coin freshly minted by a federal government. The person flips (or spins) the coin 20 times and it happens to come up heads 17 times, that is, 85% heads. What do you believe to be the underlying probability of heads for the coin? Despite the result from the 20 flips, the strong prior knowledge about the coin suggests that the result was a fluke and that the underlying probability of heads is, nevertheless, around 0.5. The left column of Figure 6.4 illustrates this reasoning. The top-left panel shows a beta prior distribution that expresses strong prior knowledge that the coin is fair. The prior uses a mode of $\omega = 0.5$ and an effective prior sample size of $\kappa = 500$, which translates into beta shape parameters of $a = 250$ and $b = 250$ using Equation 6.6. Scanning down to the lower-left panel, you can see that the posterior beta distribution is still loaded heavily over $\theta = 0.5$. It would take a lot more data to budge us away from the strong prior.

Consider a different situation, in which we are trying to estimate the probability that a particular professional basketball player will successfully make free throws. Suppose that all we know about the player is that he is in the professional league. Suppose we also know that professional players tend to make about 75% of their free throws, with most players making at least 50% but at most about 90%. This prior knowledge is expressed in the upper-middle panel of Figure 6.4, which used a beta distribution with mode $\omega = 0.75$ and equivalent prior sample of $\kappa = 25$, so that the width of the 95% HDI of the prior captures our prior knowledge about the range of abilities in the professional league. Suppose we observe 20 free throws of the player, who successfully makes 17 of the attempts. This 85% success rate in the sample is impressive, but is this our best estimate of the player's ability? If we appropriately take into account the fact that most professional players only make about 75%, our estimate of this particular player's ability is tempered. The posterior distribution in the lower-middle panel of Figure 6.4 shows a mode just under 80%.

Finally, suppose we study a newly discovered substance on a distant planet, using a remotely controlled robot. We notice that the substance can be blue or green, and we would like to estimate the underlying probability of these two forms. The robot takes 20 random samples and finds that 17 are blue. The right column of Figure 6.4 shows

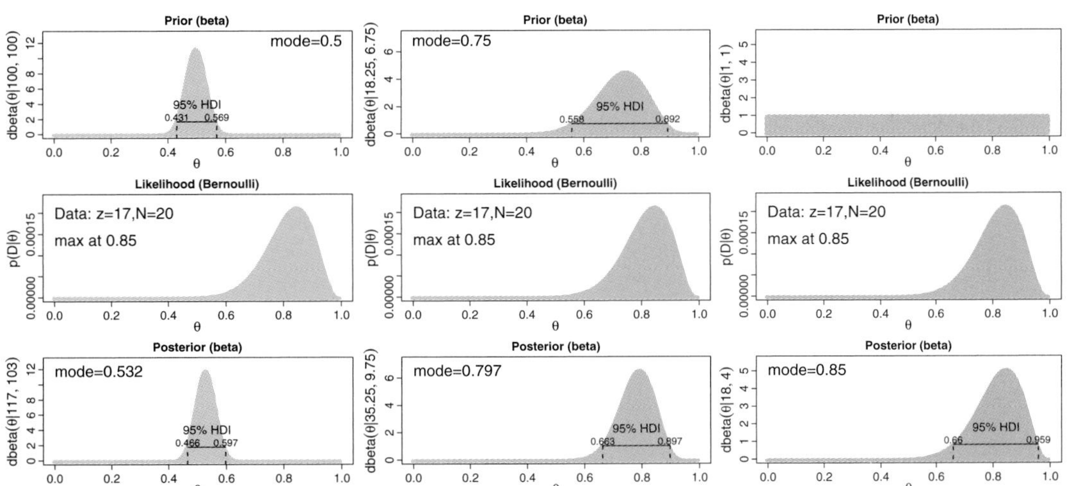

Figure 6.4 Examples of updating a beta prior distribution. The three columns show the same data with different priors. R code for this figure is described in Section 6.6.

the estimate of this probability, starting with a prior that uses only the knowledge that two colors exist. In this case, the mode of the posterior distribution is at 85%.

6.4.2. Prior knowledge that cannot be expressed as a beta distribution

The beauty of using a beta distribution to express prior knowledge is that the posterior distribution is again exactly a beta distribution, and therefore, no matter how much data we include, we always have an exact representation of the posterior distribution and a simple way of computing it. But not all prior knowledge can be expressed by a beta distribution, because the beta distribution can only be in the forms illustrated by Figure 6.1. If the prior knowledge cannot be expressed as a beta distribution, then we must use a different method to derive the posterior. In particular, we might revert to grid approximation as was explained in Section 5.5 (p. 116). I provide an example here to illustrate the limits of using a beta prior.

Suppose that we are estimating the underlying probability that a coin comes up heads, but we know that this coin was manufactured by the Acme Magic and Novelty Company, which produces coins of two types: Some have probability of heads near 25%, and others have probability of heads near 75%. In other words, our prior knowledge indicates a bimodal prior distribution over θ, with peaks over $\theta = 0.25$ and $\theta = 0.75$. Unfortunately, there is no beta distribution that has this form.

We can instead try to express the prior knowledge as a grid of discrete values over θ. In this case, there is no uniquely correct way to do this, because the prior knowledge is not expressed in any specific mathematical form. But we can improvise. For example, we might express the prior as two triangular peaks centered over 0.25 and 0.75, like this:

```
Theta = seq( 0 , 1 , length=1000 )  # Fine teeth for Theta.
# Two triangular peaks on a small non-zero floor:
pTheta = c( rep(1,200),seq(1,100,length=50),seq(100,1,length=50), rep(1,200) ,
            rep(1,200),seq(1,100,length=50),seq(100,1,length=50), rep(1,200) )
pTheta = pTheta/sum(pTheta)      # Make pTheta sum to 1.0
```

The expression above for `pTheta` is specifying the relative height of $p(\theta)$ at each grid point in `Theta`. Because `Theta` has 1,000 components, we need `pTheta` to have 1,000 components. The first part of `pTheta` is `rep(1,200)`, which says to repeat the height 1 a total of 200 times. Thus, the first 200 components of `pTheta` are a flat floor. The next part of `pTheta` is an incline: `seq(1,100,length=50)` goes from a height of 1 to a height of 100 in 50 steps. Then, the next part is a decline in 50 steps: `seq(100,1,length=50)`. This gets us back to the floor, and the process repeats. The resulting prior can be seen in the top panel of Figure 6.5.

Now suppose we flip the coin and observe 13 tails and 14 heads. We can enter the data and compute the posterior distribution using these commands:

```
Data = c(rep(0,13),rep(1,14))
posterior = BernGrid( Theta, pTheta , Data , plotType="Bars" ,
                      showCentTend="None" , showHDI=FALSE , showpD=FALSE )
```

(Don't forget you must source `BernGrid.R` before using it.) The results are shown in Figure 6.5. Notice that the posterior has *three* bumps, which clearly could not be described by a beta distribution. The three bumps are a compromise between the two peaks of the prior and the middle peak of the likelihood. In this case, it seems that the prior and the data conflict: The prior says the coin should be biased away from $\theta = 0.5$, but the data suggest that the coin might be fair. If we collected a lot more data, the

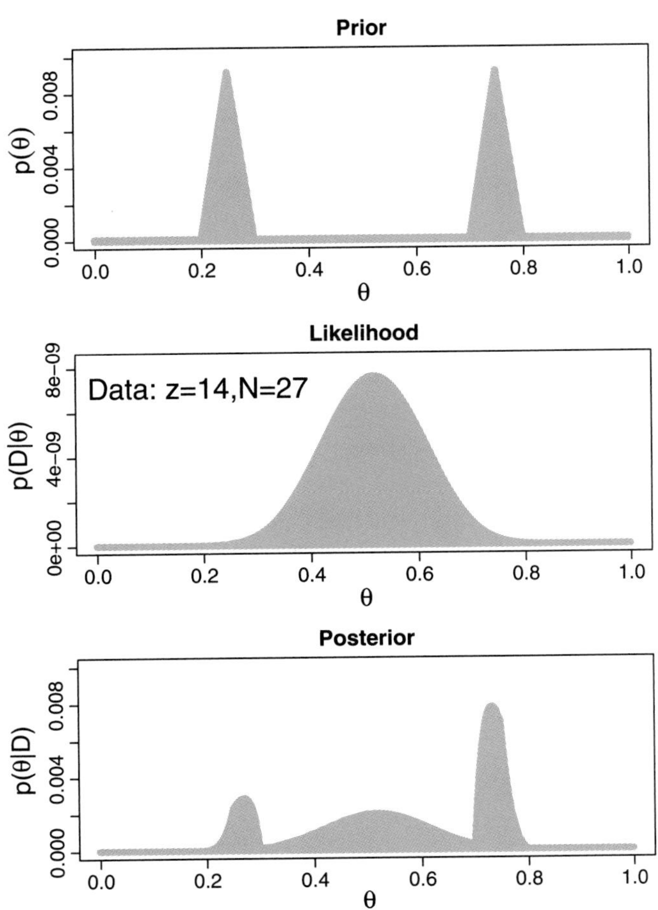

Figure 6.5 An example for which the prior distribution cannot be expressed by a beta distribution.

posterior would eventually overwhelm the prior, regardless of whether or not the prior was consistent with the data.

6.5. SUMMARY

The main point of this chapter was to demonstrate how Bayesian inference works when Bayes' rule can be solved analytically, using mathematics alone, without numerical approximation. More specifically, this chapter illustrated a case in which the likelihood has a conjugate prior distribution, whereby the posterior distribution has the same mathematical form as the prior distribution. This situation is especially convenient because we have a simple and exact mathematical description of the posterior distribution, no matter what data are included.

Unfortunately, there are two severe limitations with this approach. First, only simple likelihood functions have conjugate priors. In realistic applications that we encounter later, the complex models have no conjugate priors, and the posterior distributions have no simple form. Second, even if a conjugate prior exists, not all prior knowledge can be expressed in the mathematical form of the conjugate prior. Thus, although it is interesting and educational to see how Bayes' rule can be solved analytically, we will have to abandon exact mathematical solutions when doing complex applications. We will instead use Markov chain Monte Carlo (MCMC) methods.

This chapter also introduced you to the beta distribution, which we *will* continue to use frequently throughout the remainder of the book. Thus, despite the fact that we will not be using analytical mathematics to solve Bayes' rule, we will be using the beta distribution to express prior knowledge in complex models.

6.6. APPENDIX: R CODE FOR FIGURE 6.4

The program `BernBeta.R` defines a function `BernBeta` for producing graphs like those in Figure 6.4. The function behaves much like the function `BernGrid` that was explained in Section 5.5 (p. 116). Here is an example of how to use `BernBetaExample.R`. You must have R's working directory set to the folder that contains the files `BernBeta.R` and `DBDA2E-utilities.R`.

```
source("DBDA2E-utilities.R")  # Load definitions of graphics functions etc.
source("BernBeta.R")          # Load the definition of the BernBeta function
# Specify the prior:
t = 0.75              # Specify the prior mode.
n = 25                # Specify the effective prior sample size.
a = t*(n-2) + 1       # Convert to beta shape parameter a.
b = (1-t)*(n-2) + 1   # Convert to beta shape parameter b.
Prior = c(a,b)        # Specify Prior as vector with the two shape parameters.
```

```
# Specify the data:
N = 20                        # The total number of flips.
z = 17                        # The number of heads.
Data = c(rep(0,N-z),rep(1,z))  # Convert N and z into vector of 0's and 1's.
openGraph(width=5,height=7)
posterior = BernBeta( priorBetaAB=Prior, Data=Data , plotType="Bars" ,
                      showCentTend="Mode" , showHDI=TRUE , showpD=FALSE )
saveGraph(file="BernBetaExample",type="jpg")
```

The first two lines above use the `source` function to read in R commands from files. The `source` function was explained in Section 3.7.2.

The next section of code, above, specifies the prior. The function `BernBeta` assumes that the prior is a beta distribution, and the function requires the user to provide the beta shape parameters, a and b, as a vector for the argument `priorBetaAB`, in a form like this: `BernBeta( priorBetaAB=c(a,b) , ...)`. Because it can sometimes be unintuitive to think directly in terms of a and b, the code above instead starts by specifying the mode t and effective prior sample size n, and then converts to the equivalent a and b shape parameters by implementing Equation 6.6.

The next section of code, above, specifies the data. The function `BernBeta` needs the data to be specified as a vector of 0's and 1's. The example instead begins by specifying the number of flips and the number of heads, and then uses the `rep` function to create a corresponding data vector.

The `BernBeta` function itself is called near the end of the script above. The user must specify the arguments `priorBetaAB` and `Data`, but there are also several optional arguments. They behave much like the corresponding arguments in function `BernGrid` explained in Section 5.5 (p. 116). In particular, you might want to try `showCentTend="Mean"`, which displays the means of the distributions instead of the modes.

If you want to specify the prior in terms of its mean instead of its mode, then you must implement Equation 6.5:

```
# Specify the prior:
m = 0.75        # Specify the prior mean.
n = 25          # Specify the effective prior sample size.
a = m*n         # Convert to beta shape parameter a.
b = (1-m)*n     # Convert to beta shape parameter b.
Prior = c(a,b)  # Specify Prior as vector with the two shape parameters.
```

The output of `BernBeta`, other than a graphical display, is the vector of a and b shape parameters for the posterior.

6.7. EXERCISES

Look for more exercises at https://sites.google.com/site/doingbayesiandataanalysis/

Exercise 6.1. [Purpose: For you to see the influence of the prior in each successive flip, and for you to see another demonstration that the posterior

is invariant under re-orderings of the data.] For this exercise, use the R function explained in Section 6.6 (`BernBeta.R`). (Don't forget to `source` the function before calling it.) Notice that the function returns the posterior beta values each time it is called, so you can use the returned values as the prior values for the next function call.

(A) Start with a prior distribution that expresses some uncertainty that a coin is fair: beta$(\theta|4, 4)$. Flip the coin once; suppose we get a head. What is the posterior distribution?

(B) Use the posterior from the previous flip as the prior for the next flip. Suppose we flip again and get a head. Now what is the new posterior? (*Hint:* If you type `post = BernBeta( c(4,4) , c(1) )` for the first part, then you can type `post = BernBeta( post , c(1) )` for the next part.)

(C) Using that posterior as the prior for the next flip, flip a third time and get a tail. Now what is the new posterior? (*Hint:* Type `post = BernBeta( post , c(0) )`.)

(D) Do the same three updates but in the order T, H, H instead of H, H, T. Is the final posterior distribution the same for both orderings of the flip results?

Exercise 6.2. [Purpose: Connecting HDIs to the real world, with iterative data collection.] Suppose an election is approaching, and you are interested in knowing whether the general population prefers candidate A or candidate B. There is a just-published poll in the newspaper, which states that of 100 randomly sampled people, 58 preferred candidate A and the remainder preferred candidate B.

(A) Suppose that before the newspaper poll, your prior belief was a uniform distribution. What is the 95% HDI on your beliefs after learning of the newspaper poll results?

(B) You want to conduct a follow-up poll to narrow down your estimate of the population's preference. In your follow-up poll, you randomly sample 100 other people and find that 57 prefer candidate A and the remainder prefer candidate B. Assuming that peoples' opinions have not changed between polls, what is the 95% HDI on the posterior?

Exercise 6.3. [Purpose: Apply the Bayesian method to real data analysis. These data are representative of real data (Kruschke, 2009).] Suppose you train people in a simple learning experiment, as follows. When people see the two words, "radio" and "ocean," on the computer screen, they should press the F key on the computer keyboard. They see several repetitions and learn the response well. Then you introduce another correspondence for them to learn: Whenever the words "radio" and "mountain" appear, they should press the J key on the computer keyboard. You keep training them until they know both correspondences well. Now you probe what they've learned by asking them about two novel test items. For the first test, you show them the word "radio" by itself and instruct them to make the best response (F or J) based on what they learned before.

For the second test, you show them the two words "ocean" and "mountain" and ask them to make the best response. You do this procedure with 50 people. Your data show that for "radio" by itself, 40 people chose F and 10 chose J. For the word combination "ocean" and "mountain," 15 chose F and 35 chose J. Are people biased toward F or toward J for either of the two probe types? To answer this question, assume a uniform prior, and use a 95% HDI to decide which biases can be declared to be credible. (Consult Chapter 12 for how to declare a parameter value to be not credible.)

Exercise 6.4. [Purpose: To explore an unusual prior and learn about the beta distribution in the process.] Suppose we have a coin that we know comes from a magic-trick store, and therefore we believe that the coin is strongly biased either usually to come up heads or usually to come up tails, but we don't know which. Express this belief as a beta prior. (*Hint:* See Figure 6.1, upper-left panel.) Now we flip the coin 5 times and it comes up heads in 4 of the 5 flips. What is the posterior distribution? (Use the R function of Section 6.6 (`BernBeta.R`) to see graphs of the prior and posterior.)

Exercise 6.5. [Purpose: To get hands on experience with the goal of predicting the next datum, and to see how the prior influences that prediction.]

(A) Suppose you have a coin that you know is minted by the government and has not been tampered with. Therefore you have a strong prior belief that the coin is fair. You flip the coin 10 times and get 9 heads. What is your predicted probability of heads for the 11th flip? Explain your answer carefully; justify your choice of prior.

(B) Now you have a different coin, this one made of some strange material and marked (in fine print) "Patent Pending, International Magic, Inc." You flip the coin 10 times and get 9 heads. What is your predicted probability of heads for the 11th flip? Explain your answer carefully; justify your choice of prior. *Hint:* Use the prior from Exercise 6.4.

CHAPTER 7

Markov Chain Monte Carlo

Contents

You furtive posterior: coy distribution.
Alluring, curvaceous, evading solution.
Although I can see what you hint at is ample,
I'll settle for one representative sample.[1]

[1] This chapter is about methods for approximating a posterior distribution by collecting from it a large representative sample. These methods are important because complex posterior distributions are otherwise very challenging to get a handle on. The poem says merely that complexly shaped posterior distributions are evasive, but instead of demanding a precise solution, we will do practical analysis with a representative sample. Some people have suggested that the poem seems to allude to something else, but I don't know what they could mean.

Doing Bayesian Data Analysis, 2nd Edition
http://dx.doi.org/10.1016/B978-0-12-405888-0.00007-6

This chapter introduces the methods we will use for producing accurate approximations to Bayesian posterior distributions for realistic applications. The class of methods is called Markov chain Monte Carlo (MCMC), for reasons that will be explained later in the chapter. It is MCMC algorithms and software, along with fast computer hardware, that allow us to do Bayesian data analysis for realistic applications that would have been effectively impossible 30 years ago. This chapter explains some of the essential ideas of MCMC methods that are crucial to understand before using the sophisticated software that make MCMC methods easy to apply.

In this chapter, we continue with the goal of inferring the underlying probability θ that a coin comes up heads, given an observed set of flips. In Chapter 6, we considered the scenario when the prior distribution is specified by a function that is conjugate to the likelihood function, and thus yields an analytically solvable posterior distribution. In Chapter 5, we considered the scenario when the prior is specified on a dense grid of points spanning the range of θ values, and thus the posterior is numerically generated by summing across the discrete values.

But there are situations in which neither of those previous methods will work. We already recognized the possibility that our prior beliefs about θ could not be adequately represented by a beta distribution or by any function that yields an analytically solvable posterior function. Grid approximation is one approach to addressing such situations. When we have just one parameter with a finite range, then approximation by a grid is a useful procedure. But what if we have several parameters? Although we have, so far, only been dealing with models involving a single parameter, it is much more typical, as we will see in later chapters, to have models involving several parameters. In these situations, the parameter space cannot be spanned by a grid with a reasonable number of points. Consider, for example, a model with, say, six parameters. The parameter space is, therefore, a six-dimensional space that involves the *joint* distribution of all *combinations* of parameter values. If we set up a comb on each parameter that has 1,000 values, then the six-dimensional parameter space has $1{,}000^6 = 1{,}000{,}000{,}000{,}000{,}000{,}000$ combinations of parameter values, which is too many for any computer to evaluate. In anticipation of those situations when grid approximation will not work, we explore a new method called Markov chain Monte Carlo, or MCMC for short. In this chapter, we apply MCMC to the simple situation of estimating a single parameter. In real research you would probably not want to apply MCMC to such simple one-parameter models, instead going with mathematical analysis or grid approximation. But it is very useful to *learn* about MCMC in the one-parameter context.

The method described in this chapter assumes that the prior distribution is specified by a function that is easily evaluated. This simply means that if you specify a value for θ, then the value of $p(\theta)$ is easily determined, especially by a computer. The method also assumes that the value of the likelihood function, $p(D|\theta)$, can be computed for any specified values of D and θ. (Actually, all that the method really demands is that the prior and likelihood can be easily computed up to a multiplicative constant.) In other words, the method does *not* require evaluating the difficult integral in the denominator of Bayes' rule.

What the method produces for us is an approximation of the posterior distribution, $p(\theta|D)$, in the form of a large number of θ values sampled from that distribution. This heap of representative θ values can be used to estimate the central tendency of the posterior, its highest density interval (HDI), etc. The posterior distribution is estimated by randomly generating a lot of values from it, and therefore, by analogy to the random events at games in a casino, this approach is called a Monte Carlo method.

7.1. APPROXIMATING A DISTRIBUTION WITH A LARGE SAMPLE

The concept of representing a distribution by a large representative sample is foundational for the approach we take to Bayesian analysis of complex models. The idea is applied intuitively and routinely in everyday life and in science. For example, polls and surveys are founded on this concept: By randomly sampling a subset of people from a population, we estimate the underlying tendencies in the entire population. The larger the sample, the better the estimation. What is new in the present application is that the population from which we are sampling is a mathematically defined distribution, such as a posterior probability distribution.

Figure 7.1 shows an example of approximating an exact mathematical distribution by a large random sample of representative values. The upper-left panel of Figure 7.1 plots an exact beta distribution, which could be the posterior distribution for estimating the underlying probability of heads in a coin. The exact mode of the distribution and the 95% HDI are also displayed. The exact values were obtained from the mathematical formula of the beta distribution. We can approximate the exact values by randomly sampling a large number of representative values from the distribution. The upper-right panel of Figure 7.1 shows a histogram of 500 random representative values. With only 500 values, the histogram is choppy, not smooth, and the estimated mode and 95% HDI are in the vicinity of the true values, but unstable in the sense that a different random sample of 500 representative values would yield noticeably different estimates. The lower-left panel shows a larger sample size of 5,000 representative values, and the lower-right panel shows a larger sample yet, with 50,000 representative values. You can see that for larger samples sizes, the histogram becomes smoother, and the estimated values become closer (on average) to the true values. (The computer generates pseudorandom values; see Footnote 4, p. 74.)

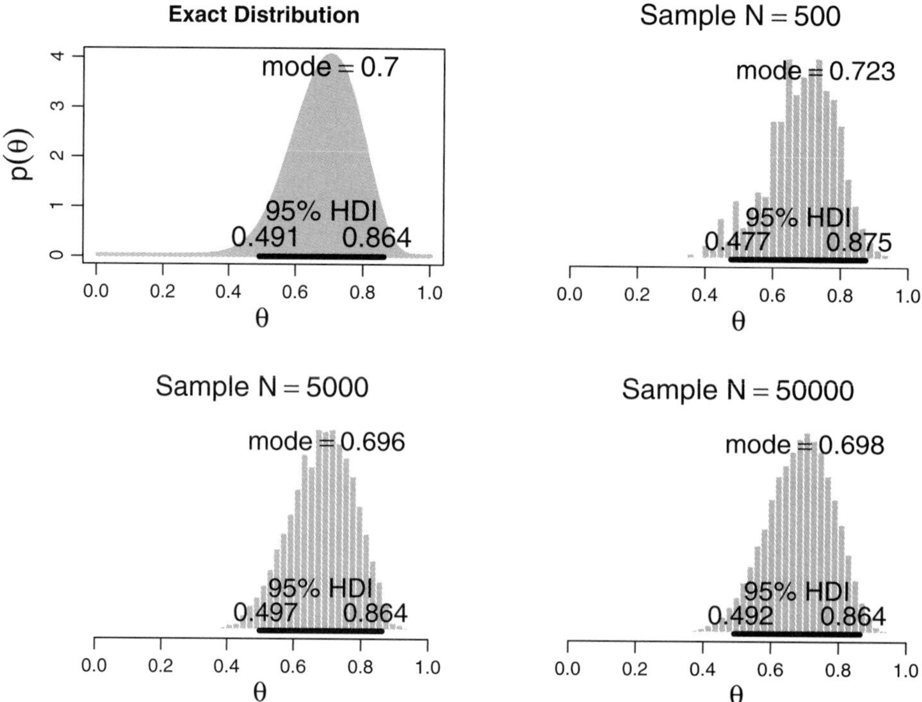

Figure 7.1 Large representative samples approximate the continuous distribution in the upper-left panel. The larger the sample, the more accurate the approximation. (This happens to be a beta(θ|15, 7) distribution.)

7.2. A SIMPLE CASE OF THE METROPOLIS ALGORITHM

Our goal in Bayesian inference is to get an accurate representation of the posterior distribution. One way to do that is to sample a large number of representative points from the posterior. The question then becomes this: How can we sample a large number of representative values from a distribution? For an answer, let's ask a politician.

7.2.1. A politician stumbles upon the Metropolis algorithm

Suppose an elected politician lives on a long chain of islands. He is constantly traveling from island to island, wanting to stay in the public eye. At the end of a grueling day of photo opportunities and fundraising,[2] he has to decide whether to (i) stay on the current island, (ii) move to the adjacent island to the west, or (iii) move to the adjacent island to

[2] Maybe I shouldn't blithely make cynical jokes about politicians, because I believe that most elected representatives really do try to do some good for their constituencies. But saying so isn't as entertaining as the cheap joke.

the east. His goal is to visit all the islands proportionally to their relative population, so that he spends the most time on the most populated islands, and proportionally less time on the less populated islands. Unfortunately, he holds his office despite having no idea what the total population of the island chain is, and he doesn't even know exactly how many islands there are! His entourage of advisers is capable of some minimal information gathering abilities, however. When they are not busy fundraising, they can ask the mayor of the island they are on how many people are on the island. And, when the politician proposes to visit an adjacent island, they can ask the mayor of that adjacent island how many people are on that island.

The politician has a simple heuristic for deciding whether to travel to the proposed island: First, he flips a (fair) coin to decide whether to propose the adjacent island to the east or the adjacent island to the west. If the proposed island has a larger population than the current island, then he definitely goes to the proposed island. On the other hand, if the proposed island has a smaller population than the current island, then he goes to the proposed island only probabilistically, to the extent that the proposed island has a population as big as the current island. If the population of the proposed island is only half as big as the current island, the probability of going there is only 50%.

In more detail, denote the population of the proposed island as P_{proposed}, and the population of the current island as P_{current}. Then he moves to the less populated island with probability $p_{\text{move}}=P_{\text{proposed}}/P_{\text{current}}$. The politician does this by spinning a fair spinner marked on its circumference with uniform values from zero to one. If the pointed-to value is between zero and p_{move}, then he moves.

What's amazing about this heuristic is that it works: In the long run, the probability that the politician is on any one of the islands exactly matches the relative population of the island!

7.2.2. A random walk

Let's consider the island hopping heuristic in a bit more detail. Suppose that there are seven islands in the chain, with relative populations as shown in the bottom panel of Figure 7.2. The islands are indexed by the value θ, whereby the leftmost, western island is $\theta = 1$ and the rightmost, eastern island is $\theta = 7$. The relative populations of the islands increase linearly such that $P(\theta) = \theta$. Notice that uppercase $P()$ refers to the *relative* population of the island, not its absolute population and not its probability mass. To complete your mental picture, you can imagine islands to the left of 1 and to the right of 7 that have populations of zero. The politician can propose to jump to those islands, but the proposal will always be rejected because the population is zero.

The middle panel of Figure 7.2 shows one possible trajectory taken by the politician. Each day corresponds to one-time increment, indicated on the vertical axis. The plot of the trajectory shows that on the first day ($t = 1$), the politician happens to start on the

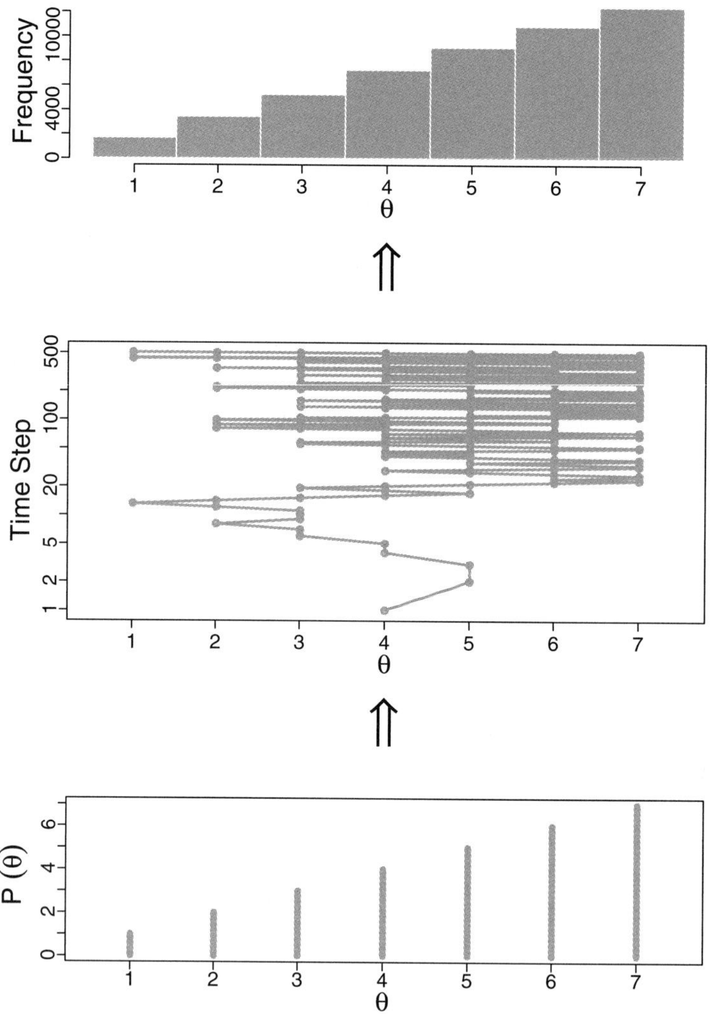

Figure 7.2 Illustration of a simple Metropolis algorithm. The bottom panel shows the values of the target distribution. The middle panel shows one random walk, at each time step proposing to move either one unit right or one unit left, and accepting the proposed move according the heuristic described in the main text. The top panel shows the frequency distribution of the positions in the walk.

middle island in the chain, hence $\theta_{\text{current}}=4$. To decide where to go on the second day, he flips a coin to propose moving either one position left or one position right. In this case, the coin proposed moving right, hence $\theta_{\text{proposed}} = 5$. Because the relative population at the proposed position is greater than the relative population at the current position (i.e., $P(5) > P(4)$), the proposed move is accepted. The trajectory shows this move, because when $t = 2$, then $\theta = 5$.

Consider the next day, when $t=2$ and $\theta = 5$. The coin flip proposes moving to the left. The probability of accepting this proposal is $p_{\text{move}} = P(\theta_{\text{proposed}})/P(\theta_{\text{current}}) = 4/5 = 0.80$. The politician then spins a fair spinner that has a circumference marked from 0 to 1, which happens to come up with a value greater than 0.80. Therefore, the politician rejects the proposed move and stays at the current island. Hence the trajectory shows that θ is still 5 when $t=3$. The middle panel of Figure 7.2 shows the trajectory for the first 500 steps in this random walk across the islands. The scale of the time step is plotted logarithmically so you can see the details of the early steps but also see the trend of the later steps. There are many thousands of steps in the simulation.

The upper panel of Figure 7.2 shows a histogram of the frequencies with which each position is visited during this junket. Notice that the sampled relative frequencies closely mimic the actual relative populations in the bottom panel! In fact, a sequence generated this way will converge, as the sequence gets longer, to an arbitrarily close approximation of the actual relative probabilities.

7.2.3. General properties of a random walk

The trajectory shown in Figure 7.2 is just one possible sequence of positions when the movement heuristic is applied. At each time step, the direction of the proposed move is random, and if the relative probability of the proposed position is less than that of the current position, then acceptance of the proposed move is also random. Because of the randomness, if the process were started over again, then the specific trajectory would almost certainly be different. Regardless of the specific trajectory, in the long run the relative frequency of visits mimics the target distribution.

Figure 7.3 shows the probability of being in each position as a function of time. At time $t=1$, the politician starts at $\theta=4$. This starting position is indicated in the upper-left panel of Figure 7.3, labeled $t=1$, by the fact that there is 100% probability of being at $\theta=4$.

We want to derive the probability of ending up in each position at the next time step. To determine the probabilities of positions for time $t=2$, consider the possibilities from the movement process. The process starts with the flip of a fair coin to decide which direction to propose moving. There is a 50% probability of proposing to move right, to $\theta=5$. By inspecting the target distribution of relative probabilities in the lower-right panel of Figure 7.3, you can see that $P(\theta{=}5) > P(\theta{=}4)$, and therefore, a rightward move is always accepted whenever it is proposed. Thus, at time $t=2$, there is a 50% probability of ending up at $\theta=5$. The panel labeled $t=2$ in Figure 7.3 plots this probability as a bar of height 0.5 at $\theta=5$. The other 50% of the time, the proposed move is to the left, to $\theta=3$. By inspecting the target distribution of relative probabilities in the lower-right panel of Figure 7.3, you can see that $P(\theta{=}3)=3$, whereas $P(\theta{=}4)=4$, and therefore, a leftward move is accepted only 3/4 of the times it is proposed. Hence, at time $t=2$,

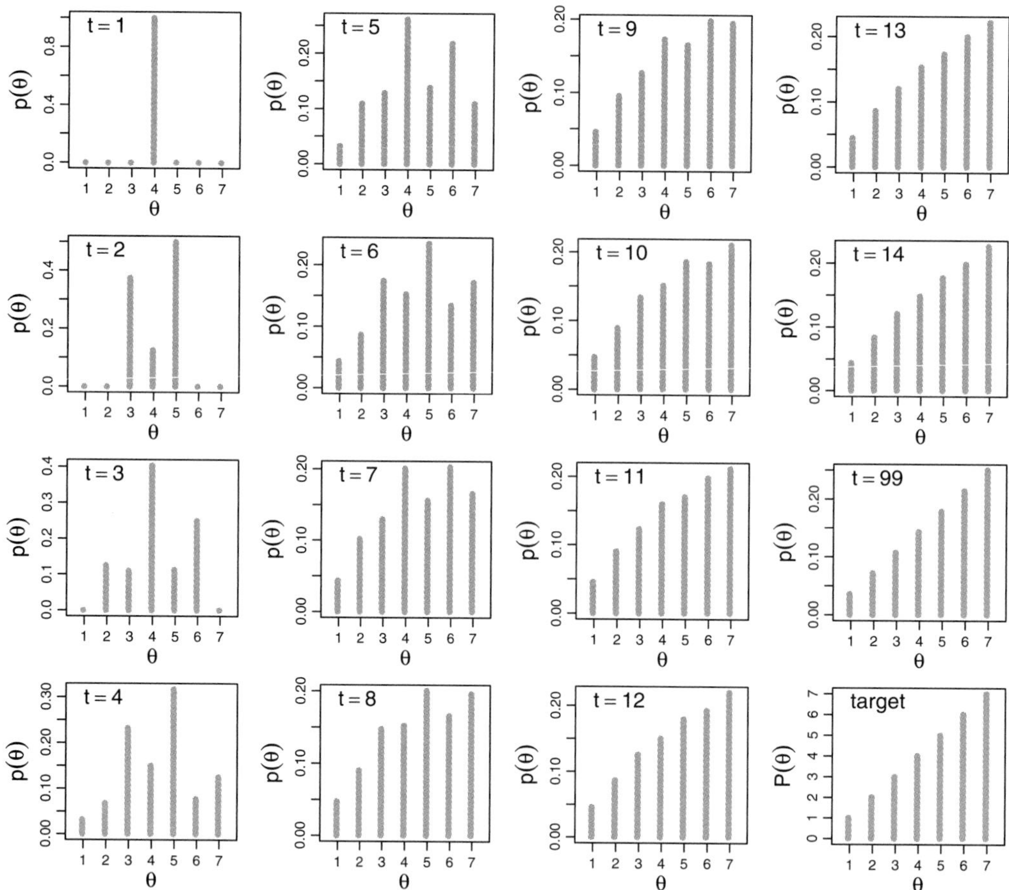

Figure 7.3 The probability of being at position θ, as a function of time t, when a simple Metropolis algorithm is applied to the target distribution in the lower-right panel. The time in each panel corresponds to the step in a random walk, an example of which is shown in Figure 7.2. The target distribution is shown in the lower-right panel.

the probability of ending up at $\theta = 3$ is $50\% \cdot 3/4 = 0.375$. The panel labeled $t = 2$ in Figure 7.3 shows this as a bar of height 0.375 at $\theta = 3$. Finally, if a leftward move is proposed but not accepted, we just stay at $\theta = 5$. The probability of this happening is only $50\% \cdot (1 - 3/4) = 0.125$.

This process repeats for the next time step. I won't go through the arithmetic details for each value of θ. But it is important to notice that after two proposed moves, i.e., when $t = 3$, the politician could be at any of the positions $\theta = 2$ through $\theta = 6$, but not yet at $\theta = 1$ or $\theta = 7$, because he could be at most two positions away from where he started.

The probabilities continue to be computed the same way at every time step. You can see that in the early time steps, the probability distribution is not a straight incline like the target distribution. Instead, the probability distribution has a bulge over the starting position. As you can see in Figure 7.3, by time $t = 99$, the position probability is virtually indistinguishable from the target distribution, at least for this simple distribution. More complex distributions require a longer duration to achieve a good approximation to the target.

The graphs of Figure 7.3 show the *probability* that the moving politician is at each value of θ. But remember, at any given time step, the politician is at only one particular position, as shown in Figure 7.2. To approximate the target distribution, we let the politician meander around for many time steps while we keep track of where he has been. When we have a long record of where the traveler has been, we can approximate the target probability at each value of θ by simply counting the relative number times that the traveler visited that value.

Here is a summary of our algorithm for moving from one position to another. We are currently at position θ_{current}. We then propose to move one position right or one position left. The specific proposal is determined by flipping a coin, which can result in 50% heads (move right) or 50% tails (move left). The range of possible proposed moves, and the probability of proposing each, is called the *proposal distribution*. In the present algorithm, the proposal distribution is very simple: It has only two values with 50-50 probabilities.

Having proposed a move, we then decide whether or not to accept it. The acceptance decision is based on the value of the target distribution at the proposed position, relative to the value of the target distribution at our current position. Specifically, if the target distribution is greater at the proposed position than at our current position, then we definitely accept the proposed move: We always move higher if we can. On the other hand, if the target position is less at the proposed position than at our current position, we accept the move probabilistically: We move to the proposed position with probability $p_{\text{move}} = P(\theta_{\text{proposed}})/P(\theta_{\text{current}})$, where $P(\theta)$ is the value of the target distribution at θ. We can combine these two possibilities, of the target distribution being higher or lower at the proposed position than at our current position, into a single expression for the probability of moving to the proposed position:

$$p_{\text{move}} = \min\left(\frac{P(\theta_{\text{proposed}})}{P(\theta_{\text{current}})}, 1\right) \tag{7.1}$$

Notice that Equation 7.1 says that when $P(\theta_{\text{proposed}}) > P(\theta_{\text{current}})$, then $p_{\text{move}} = 1$. Notice also that the target distribution, $P(\theta)$, does not need to be normalized, which means it does not need to sum to 1 as a probability distribution must. This is because what matters for our choice is the *ratio*, $P(\theta_{\text{proposed}})/P(\theta_{\text{current}})$, not the absolute magnitude of $P(\theta)$. This property was used in the example of the island-hopping politician: The target distribution was the population of each island, not a normalized probability.

Having proposed a move by sampling from the proposal distribution and having then determined the *probability* of accepting the move according to Equation 7.1, we then actually accept or reject the proposed move by sampling a value from a uniform distribution over the interval [0, 1]. If the sampled value is between 0 and p_{move}, then we actually make the move. Otherwise, we reject the move and stay at our current position. The whole process repeats at the next time step.

7.2.4. Why we care

Notice what we must be able to do in the random-walk process:

- We must be able to generate a random value from the proposal distribution, to create θ_{proposed}.
- We must be able to evaluate the target distribution at any proposed position, to compute $P(\theta_{\text{proposed}})/P(\theta_{\text{current}})$.
- We must be able to generate a random value from a uniform distribution, to accept or reject the proposal according to p_{move}.

By being able to do those three things, we are able to do *indirectly* something we could not necessarily do directly: We can generate random samples from the target distribution. Moreover, we can generate those random samples from the target distribution even when the target distribution is not normalized.

This technique is profoundly useful when the target distribution $P(\theta)$ is a posterior proportional to $p(D|\theta)p(\theta)$. Merely by evaluating $p(D|\theta)p(\theta)$, without normalizing it by $p(D)$, we can generate random representative values from the posterior distribution. This result is wonderful because the method obviates direct computation of the evidence $p(D)$, which, as you'll recall, is one of the most difficult aspects of Bayesian inference. By using MCMC techniques, we can do Bayesian inference in rich and complex models. It has only been with the development of MCMC algorithms and software that Bayesian inference is applicable to complex data analysis, and it has only been with the production of fast and cheap computer hardware that Bayesian inference is accessible to a wide audience.

7.2.5. Why it works

In this section, I'll explain a bit of the mathematics behind why the algorithm works. Despite the mathematics presented here, however, *you will not need to use any of this mathematics for applied data analysis later in the book*. The math is presented here only to help you understand why MCMC works. Just as you can drive a car without being able to build an engine from scratch, you can apply MCMC methods without being able to program a Metropolis algorithm from scratch. But it can help you interpret the results of MCMC applications if you know a bit of what is going on "under the hood."

To get an intuition for why this algorithm works, consider two adjacent positions and the probabilities of moving from one to the other. We'll see that the relative transition probabilities, between adjacent positions, exactly match the relative values of the target distribution. Extrapolate that result across all the positions, and you can see that, in the long run, each position will be visited proportionally to its target value. Now the details: Suppose we are at position θ. The probability of moving to $\theta + 1$, denoted $p(\theta \rightarrow \theta + 1)$, is the probability of proposing that move times the probability of accepting it if proposed, which is $p(\theta \rightarrow \theta + 1) = 0.5 \cdot \min(P(\theta + 1)/P(\theta), 1)$. On the other hand, if we are presently at position $\theta + 1$, the probability of moving to θ is the probability of proposing that move times the probability of accepting it if proposed, which is $p(\theta + 1 \rightarrow \theta) = 0.5 \cdot \min(P(\theta)/P(\theta + 1), 1)$. The ratio of the transition probabilities is

$$
\begin{aligned}
\frac{p(\theta \rightarrow \theta+1)}{p(\theta+1 \rightarrow \theta)} &= \frac{0.5 \min(P(\theta+1)/P(\theta), 1)}{0.5 \min(P(\theta)/P(\theta+1), 1)} \\
&= \begin{cases} \frac{1}{P(\theta)/P(\theta+1)} & \text{if } P(\theta + 1) > P(\theta) \\ \frac{P(\theta+1)/P(\theta)}{1} & \text{if } P(\theta + 1) < P(\theta) \end{cases} \\
&= \frac{P(\theta+1)}{P(\theta)}
\end{aligned} \tag{7.2}
$$

Equation 7.2 tells us that during transitions back and forth between adjacent positions, the relative probability of the transitions exactly matches the relative values of the target distribution. That might be enough for you to get the intuition that, in the long run, adjacent positions will be visited proportionally to their relative values in the target distribution. If that's true for adjacent positions, then, by extrapolating from one position to the next, it must be true for the whole range of positions.

To make that intuition more defensible, we have to fill in some more details. To do this, I'll use matrix arithmetic. This is the only place in the book where matrix arithmetic appears, so if the details here are unappealing, feel free to skip ahead to the next section (Section 7.3, p. 156). What you'll miss is an explanation of the mathematics underlying Figure 7.3, which depicts the key idea that the target distribution is *stable*: If the current probability of being in a position matches the target probabilities, then the Metropolis algorithm keeps it that way.

Consider the probability of transitioning from position θ to some other position. The proposal distribution, in the present simple scenario, considers only positions $\theta + 1$ and $\theta - 1$. The probability of moving to position $\theta - 1$ is the probability of proposing that position times the probability of accepting the move if it is proposed: $0.5 \min(P(\theta - 1)/P(\theta), 1)$. The probability of moving to position $\theta + 1$ is the probability of proposing that position times the probability of accepting the move if it is proposed: $0.5 \min(P(\theta+1)/P(\theta), 1)$. The probability of staying at position θ is simply the complement of those two move-away probabilities: $0.5\,[1 - \min(P(\theta-1)/P(\theta), 1)] + 0.5\,[1 - \min(P(\theta+1)/P(\theta), 1)]$.

We can put those transition probabilities into a matrix. Each row of the matrix is a possible current position, and each column of the matrix is a candidate moved-to position. Below is a *sub*matrix from the full transition matrix T, showing rows $\theta - 2$ to $\theta + 2$, and columns $\theta - 2$ to $\theta + 2$:

$$\begin{bmatrix} \ddots & p(\theta-2 \to \theta-1) & 0 & 0 & 0 \\ \ddots & p(\theta-1 \to \theta-1) & p(\theta-1 \to \theta) & 0 & 0 \\ 0 & p(\theta \to \theta-1) & p(\theta \to \theta) & p(\theta \to \theta+1) & 0 \\ 0 & 0 & p(\theta+1 \to \theta) & p(\theta+1 \to \theta+1) & \ddots \\ 0 & 0 & 0 & p(\theta+2 \to \theta+1) & \ddots \end{bmatrix}$$

which equals

$$\begin{bmatrix} \ddots & 0.5\min\left(\frac{P(\theta-1)}{P(\theta-2)}, 1\right) & 0 & 0 & 0 \\ \ddots & \begin{matrix} 0.5\left[1-\min\left(\frac{P(\theta-2)}{P(\theta-1)}, 1\right)\right] \\ +\,0.5\left[1-\min\left(\frac{P(\theta)}{P(\theta-1)}, 1\right)\right] \end{matrix} & 0.5\min\left(\frac{P(\theta)}{P(\theta-1)}, 1\right) & 0 & 0 \\ 0 & 0.5\min\left(\frac{P(\theta-1)}{P(\theta)}, 1\right) & \begin{matrix} 0.5\left[1-\min\left(\frac{P(\theta-1)}{P(\theta)}, 1\right)\right] \\ +\,0.5\left[1-\min\left(\frac{P(\theta+1)}{P(\theta)}, 1\right)\right] \end{matrix} & 0.5\min\left(\frac{P(\theta+1)}{P(\theta)}, 1\right) & 0 \\ 0 & 0 & 0.5\min\left(\frac{P(\theta)}{P(\theta+1)}, 1\right) & \begin{matrix} 0.5\left[1-\min\left(\frac{P(\theta)}{P(\theta+1)}, 1\right)\right] \\ +\,0.5\left[1-\min\left(\frac{P(\theta+2)}{P(\theta+1)}, 1\right)\right] \end{matrix} & \ddots \\ 0 & 0 & 0 & 0.5\min\left(\frac{P(\theta+1)}{P(\theta+2)}, 1\right) & \ddots \end{bmatrix} \tag{7.3}$$

The usefulness of putting the transition probabilities into a matrix is that we can then use matrix multiplication to get from any current location to the probability of the next locations. Here's a reminder of how matrix multiplication operates. Consider a matrix T. The value in its rth row and cth column is denoted T_{rc}. We can multiply the matrix on its *left* side by a *row* vector w, which yields another row vector. The cth component of the product wT is $\sum_r w_r T_{rc}$. In other words, to compute the cth component of the result, take the row vector w and multiply its components by the corresponding components in the cth column of W, and sum up those component products.[3]

To use the transition matrix in Equation 7.3, we put the *current* location probabilities into a row vector, which I will denote w because it indicates *w*here we are. For example, if at the current time, we are definitely in location $\theta = 4$, then w has 1.0 in its $\theta = 4$ component, and zeros everywhere else. To determine the probability of the locations at the next time step, we simply multiply w by T. Here's a key example to think through: When $w = [\ldots, 0, 1, 0, \ldots]$ with a 1 only in the θ position, then wT is simply the row of T corresponding to θ, because the cth component of wT is $\sum_r w_r T_{rc} = T_{\theta c}$, where I'm using the subscript θ to stand for the index that corresponds to value θ.

Matrix multiplication is a very useful procedure for keeping track of position probabilities. At every time step, we just multiply the current position probability vector w by the transition probability matrix T to get the position probabilities for the next time step. We keep multiplying by T, over and over again, to derive the long-run position probabilities. This process is exactly what generated the graphs in Figure 7.3.

Here's the climactic implication: When the vector of position probabilities is the target distribution, it stays that way on the next time step! In other words, the position probabilities are stable at the target distribution. We can actually prove this result without much trouble. Suppose the current position probabilities are the target probabilities, i.e., $w = [\ldots, P(\theta - 1), P(\theta), P(\theta + 1), \ldots]/Z$, where $Z = \sum_\theta P(\theta)$ is the normalizer for the target distribution. Consider the θ component of wT. We will demonstrate that the θ component of wT is the same as the θ component of w, for any component θ. The θ component of wT is $\sum_r w_r T_{r\theta}$. Look back at the transition matrix in Equation 7.3, and you can see then that the θ component of wT is

$$
\begin{aligned}
\sum_r w_r T_{r\theta} &= P(\theta-1)/Z \cdot 0.5\min\left(\frac{P(\theta)}{P(\theta-1)}, 1\right) \\
&\quad + P(\theta)/Z \cdot \left(0.5\left[1 - \min\left(\frac{P(\theta-1)}{P(\theta)}, 1\right)\right] + 0.5\left[1 - \min\left(\frac{P(\theta+1)}{P(\theta)}, 1\right)\right]\right) \\
&\quad + P(\theta+1)/Z \cdot 0.5\min\left(\frac{P(\theta)}{P(\theta+1)}, 1\right)
\end{aligned}
\tag{7.4}
$$

To simplify that equation, we can consider separately the four cases: Case 1: $P(\theta) > P(\theta - 1)$ and $P(\theta) > P(\theta + 1)$; Case 2: $P(\theta) > P(\theta - 1)$ and $P(\theta) < P(\theta + 1)$; Case 3: $P(\theta) < P(\theta - 1)$ and $P(\theta) > P(\theta + 1)$; Case 4: $P(\theta) < P(\theta - 1)$ and $P(\theta) < P(\theta+1)$. In each case, Equation 7.4 simplifies to $P(\theta)/Z$. For example, consider Case 1, when $P(\theta) > P(\theta-1)$ and $P(\theta) > P(\theta+1)$. Equation 7.4 becomes

[3] Although we don't do it here, we can also multiply a matrix on its *right* side by a *column* vector, which yields another column vector. For a column vector v, the rth component of Tv is $\sum_c T_{rc} v_c$.

$$
\begin{aligned}
\sum_r w_r T_{r\theta} &= P(\theta-1)/Z \cdot 0.5 \\
&\quad + P(\theta)/Z \cdot \left(0.5\left[1-\left(\frac{P(\theta-1)}{P(\theta)}\right)\right] + 0.5\left[1-\left(\frac{P(\theta+1)}{P(\theta)}\right)\right]\right) \\
&\quad + P(\theta+1)/Z \cdot 0.5 \\
&= 0.5\,P(\theta-1)/Z \\
&\quad + 0.5P(\theta)/Z - 0.5P(\theta)/Z\frac{P(\theta-1)}{P(\theta)} + 0.5P(\theta)/Z - 0.5P(\theta)/Z\frac{P(\theta+1)}{P(\theta)} \\
&\quad + 0.5\,P(\theta+1)/Z \\
&= P(\theta)/Z
\end{aligned}
$$

If you work through the other cases, you'll find that it always reduces to $P(\theta)/Z$. In conclusion, when the θ component starts at $P(\theta)/Z$, it stays at $P(\theta)/Z$.

We have shown that the target distribution is stable under the Metropolis algorithm, for our special case of island hopping. To prove that the Metropolis algorithm realizes the target distribution, we would need to show that the process actually gets us to the target distribution regardless of where we start. Here, I'll settle for intuition: You can see that no matter where you start, the distribution will naturally diffuse and explore other positions. Examples of this were shown in Figures 7.2 and 7.3. It's reasonable to think that the diffusion will settled into *some* stable state and we've just shown that the target distribution is *a* stable state. To make the argument complete, we'd have to show that there are no other stable states, and that the target distribution is actually an attractor into which other states flow, rather than a state that is stable if it is ever obtained but impossible to actually attain. This complete argument is far beyond what would be useful for the purposes of this book, but if you're interested you could take a look at the book by Robert and Casella (2004).

7.3. THE METROPOLIS ALGORITHM MORE GENERALLY

The procedure described in the previous section was just a special case of a more general procedure known as the Metropolis algorithm, named after the first author of a famous article (Metropolis, Rosenbluth, Rosenbluth, Teller, & Teller, 1953).[4] In the previous section, we considered the simple case of (i) discrete positions, (ii) on one dimension, and (iii) with moves that proposed just one position left or right. That simple situation made it relatively easy (believe it or not) to understand the procedure and how it works. The general algorithm applies to (i) continuous values, (ii) on any number of dimensions, and (iii) with more general proposal distributions.

[4] Nicholas Metropolis (1915–1999) was first author of the article, but history suggests that he had little to do with the invention of the algorithm itself, which might better be attributed to the second and third authors of the article, Marshall and Arianna Rosenbluth (Gubernatis, 2005).

The essentials of the general method are the same as for the simple case. First, we have some target distribution, $P(\theta)$, over a multidimensional continuous parameter space from which we would like to generate representative sample values. We must be able to compute the value of $P(\theta)$ for any candidate value of θ. The distribution, $P(\theta)$, does not have to be normalized, however. It merely needs to be nonnegative. In typical applications, $P(\theta)$ is the unnormalized posterior distribution on θ, which is to say, it is the product of the likelihood and the prior.

Sample values from the target distribution are generated by taking a random walk through the parameter space. The walk starts at some arbitrary point, specified by the user. The starting point should be someplace where $P(\theta)$ is nonzero. The random walk progresses at each time step by proposing a move to a new position in parameter space and then deciding whether or not to accept the proposed move. Proposal distributions can take on many different forms, with the goal being to use a proposal distribution that efficiently explores the regions of the parameter space where $P(\theta)$ has most of its mass. Of course, we must use a proposal distribution for which we have a quick way to generate random values! For our purposes, we will consider the generic case in which the proposal distribution is normal, centered at the current position. (Recall the discussion of the normal distribution back in Section 4.3.2.2, p. 83.) The idea behind using a normal distribution is that the proposed move will typically be near the current position, with the probability of proposing a more distant position dropping off according to the normal curve. Computer languages such as R have built-in functions for generating pseudorandom values from a normal distribution. For example, if we want to generate a proposed jump from a normal distribution that has a mean of zero and a standard deviation (SD) of 0.2, we could command R as follows: `proposedJump=rnorm(1, mean=0, sd=0.2)`, where the first argument, 1, indicates that we want a single random value, not a vector of many random values.

Having generated a proposed new position, the algorithm then decides whether or not to accept the proposal. The decision rule is exactly what was already specified in Equation 7.1. In detail, this is accomplished by computing the ratio $p_{\text{move}} = P(\theta_{\text{proposed}})/P(\theta_{\text{current}})$. Then a random number from the uniform interval $[0, 1]$ is generated; in R, this can be accomplished with the command `runif(1)`. If the random number is between 0 and p_{move}, then the move is accepted. The process repeats and, in the long run, the positions visited by the random walk will closely approximate the target distribution.

7.3.1. Metropolis algorithm applied to Bernoulli likelihood and beta prior

In the scenario of the island-hopping politician, the islands represented candidate parameter values, and the relative populations represented relative posterior probabilities. In the scenario of coin flipping, the parameter θ has values that range on a continuum from zero to one, and the relative posterior probability is computed as likelihood times prior. To apply the Metropolis algorithm, we conceive of the parameter dimension as a dense chain of infinitesimal islands, and we think of the (relative) population of each infinitesimal island as its (relative) posterior probability density. And, instead of the proposed jump being only to immediately adjacent islands, the proposed jump can be to islands farther away from the current island. We need a proposal distribution that will let us visit any parameter value on the continuum. For this purpose, we will use the familiar normal distribution (recall Figure 4.4, p. 83).

We will apply the Metropolis algorithm to the following familiar scenario. We flip a coin N times and observe z heads. We use a Bernoulli likelihood function, $p(z, N|\theta) = \theta^z (1-\theta)^{(N-z)}$. We start with a prior $p(\theta) = \text{beta}(\theta|a, b)$. For the proposed jump in the Metropolis algorithm, we will use a normal distribution centered at zero with standard deviation (SD) denoted as σ. We denote the proposed jump as $\Delta\theta \sim \text{normal}(\mu = 0, \sigma)$, where the symbol "$\sim$" means that the value is randomly sampled from the distribution (cf. Equation 2.2, p. 27). Thus, the proposed jump is usually close to the current position, because the mean jump is zero, but the proposed jump can be positive or negative, with larger magnitudes less likely than smaller magnitudes. Denote the current parameter value as θ_{cur} and the proposed parameter value as $\theta_{\text{pro}} = \theta_{\text{cur}} + \Delta\theta$.

The Metropolis algorithm then proceeds as follows. Start at an arbitrary initial value of θ (in the valid range). This is the current value, denoted θ_{cur}. Then:

1. Randomly generate a proposed jump, $\Delta\theta \sim \text{normal}(\mu = 0, \sigma)$ and denote the proposed value of the parameter as $\theta_{\text{pro}} = \theta_{\text{cur}} + \Delta\theta$.
2. Compute the probability of moving to the proposed value as Equation 7.1, specifically expressed here as

$$
\begin{aligned}
p_{\text{move}} &= \min\left(1, \frac{P(\theta_{\text{pro}})}{P(\theta_{\text{cur}})}\right) && \text{generic Metropolis form} \\
&= \min\left(1, \frac{p(D|\theta_{\text{pro}})p(\theta_{\text{pro}})}{p(D|\theta_{\text{cur}})p(\theta_{\text{cur}})}\right) && P \text{ is likelihood times prior} \\
&= \min\left(1, \frac{\text{Bernoulli}(z, N|\theta_{\text{pro}})\text{beta}(\theta_{\text{pro}}|a, b)}{\text{Bernoulli}(z, N|\theta_{\text{cur}})\text{beta}(\theta_{\text{cur}}|a, b)}\right) \\
& && \text{Bernoulli likelihood and beta prior} \\
&= \min\left(1, \frac{\theta_{\text{pro}}^z (1-\theta_{\text{pro}})^{(N-z)} \theta_{\text{pro}}^{(a-1)} (1-\theta_{\text{pro}})^{(b-1)} / B(a, b)}{\theta_{\text{cur}}^z (1-\theta_{\text{cur}})^{(N-z)} \theta_{\text{cur}}^{(a-1)} (1-\theta_{\text{cur}})^{(b-1)} / B(a, b)}\right) \\
& && \text{by Equations 6.2 and 6.3}
\end{aligned}
$$

   If the proposed value θ_{pro} happens to land outside the permissible bounds of θ, the prior and/or likelihood is set to zero, hence p_{move} is zero.
3. Accept the proposed parameter value if a random value sampled from a [0, 1] uniform distribution is less than p_{move}, otherwise reject the proposed parameter value and tally the current value again.

Repeat the above steps until it is judged that a sufficiently representative sample has been generated. The judgment of "sufficiently representative" is not trivial and is an issue that will be discussed later in this chapter. For now, the goal is to understand the mechanics of the Metropolis algorithm.

Figure 7.4 shows specific examples of the Metropolis algorithm applied to a case with a beta(θ|1, 1) prior, $N = 20$, and $z = 14$. There are three columns in Figure 7.4, for three different runs of the Metropolis algorithm using three different values for σ in the proposal distribution. In all three cases, θ was arbitrarily started at 0.01, merely for purposes of illustration.

The middle column of Figure 7.4 uses a moderately sized SD for the proposal distribution, namely $\sigma = 0.2$, as indicated in the title of the upper-middle panel where it says "Prpsl.SD = 0.2." This means that at any step in the chain, for whatever value θ happens to be, the proposed jump is within ± 0.2 of θ about 68% of the time (because a normal distribution has about 68% of its mass between -1 and $+1$ SDs). In this case, the proposed jumps are accepted roughly half the time, as indicated in the center panel by the annotation $N_{acc}/N_{pro} = 0.495$, which is the number of accepted proposals divided by

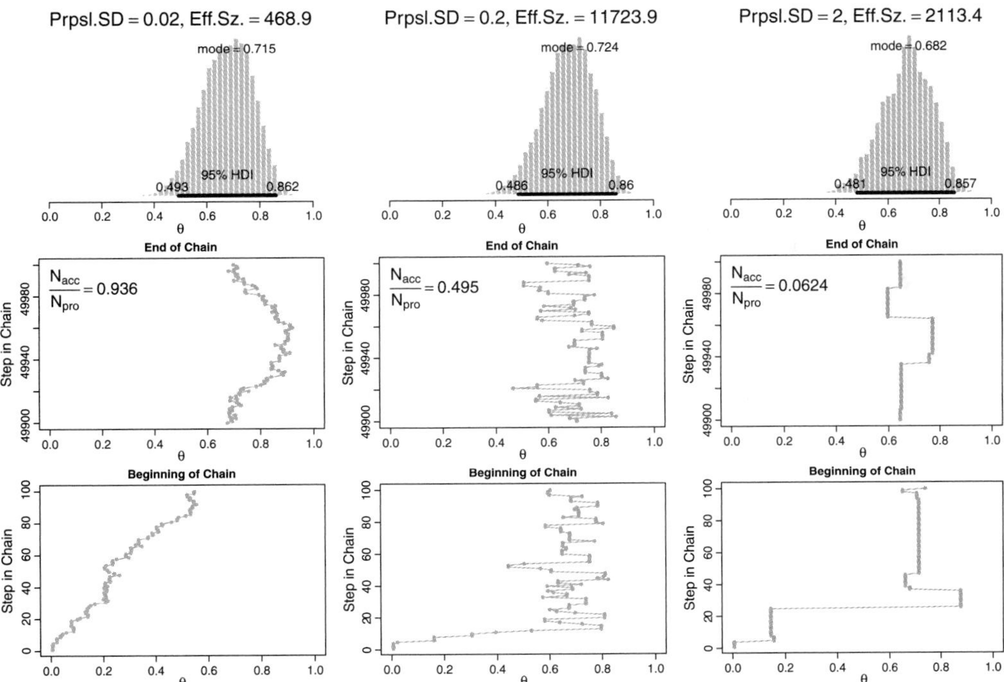

Figure 7.4 Metropolis algorithm applied to Bernoulli likelihood with beta(θ|1, 1) prior and $z = 14$ with $N = 20$. For each of the three columns, there are 50,000 steps in the chain, but for the left column, the proposal standard deviation (SD) is 0.02, for the middle column SD = 0.2, and for the right column SD = 2.0.

the total number of proposals in the chain. This setting of the proposal distribution allows the chain to move around parameter space fairly efficiently. In particular, you can see in the lower-middle panel that the chain moves away quickly from the unrepresentative starting position of $\theta = 0.01$. And, the chain visits a variety of representative values in relatively few steps. That is, the chain is not very clumpy. The upper-middle panel shows a histogram of the chain positions after 50,000 steps. The histogram looks smooth and is an accurate approximation of the true underlying posterior distribution, which we know in this case is a beta$(\theta|15, 7)$ distribution (cf. Figure 7.1, p. 146). Although the chain is not very clumpy and yields a smooth-looking histogram, it does have some clumpiness because each step is linked to the location of the previous step, and about half the steps don't change at all. Thus, there are not 50,000 independent representative values of the posterior distribution in the chain. If we take into account the clumpiness, then the so-called "effective size" of the chain is less, as indicated in the title of upper-middle panel where it says "Eff.Sz. = 11723.9." This is the equivalent number of values if they were sampled independently of each other. The technical meaning of effective size will be discussed later in the chapter.

The left column of Figure 7.4 uses a relatively small proposal SD, namely $\sigma = 0.02$, as indicated in the title of the upper-left panel where it says "Prpsl.SD = 0.02." You can see that successive steps in the chain make small moves because the proposed jumps are small. In particular, in the lower-left panel you can see that it takes many steps for the chain to move away from the unrepresentative starting position of $\theta = 0.01$. The chain explores values only very gradually, producing a snake-like chain that lingers around particular values, thereby producing a form of clumpiness. In the long run, the chain will explore the posterior distribution thoroughly and produce a good representation, but it will require a very long chain. The title of the upper-left panel indicates that the effective size of this 50,000 step chain is only 468.9.

The right column of Figure 7.4 uses a relatively large proposal SD, namely $\sigma = 2$, as indicated in the title of the upper-left panel where it says "Prpsl.SD = 2." The proposed jumps are often far away from the bulk of the posterior distribution, and therefore, the proposals are often rejected and the chain stays at one value for many steps. The process accepts new values only occasionally, producing a very clumpy chain. In the long run, the chain will explore the posterior distribution thoroughly and produce a good representation, but it will require a very long chain. The title of the upper-right panel indicates that the effective size of this 50,000 step chain is only 2113.4.

Regardless of the which proposal distribution in Figure 7.4 is used, the Metropolis algorithm will eventually produce an accurate representation of the posterior distribution, as is suggested by the histograms in the upper row of Figure 7.4. What differs is the efficiency of achieving a good approximation. The moderate proposal distribution will achieve a good approximation in fewer steps than either of the extreme proposal distributions. Later in the chapter, we will discuss criteria for deciding that a chain has produced a sufficiently good approximation, but for now suppose that our goal is to achieve an effective size of 10,000. The proposal distribution in the middle column of Figure 7.4 has achieved this goal. For the proposal distribution in the left column of Figure 7.4, we would need to run the chain more than 20 times longer because the effective size is less than 1/20th of the desired size. Sophisticated implementations of the Metropolis algorithm have an automatic preliminary phase that adjusts the width of the proposal distribution so that the chain moves relatively efficiently. A typical way to do this is to adjust the proposal distribution so that the acceptance ratio is a middling value such as 0.5. The steps in the adaptive phase are not used to represent the posterior distribution.

7.3.2. Summary of Metropolis algorithm

In this section, I recapitulate the main ideas of the Metropolis algorithm and explicitly point out the analogy between the island-hopping politician and the θ-hopping coin bias.

The motivation for methods like the Metropolis algorithm is that they provide a high-resolution picture of the posterior distribution, even though in complex models we cannot explicitly solve the mathematical integral in Bayes' rule. The idea is that we get a handle on the posterior distribution by generating a large sample of representative values. The larger the sample, the more accurate is our approximation. As emphasized previously, this is a sample of representative credible parameter values from the posterior distribution; it is not a resampling of data (there is a fixed data set).

The cleverness of the method is that representative parameter values can be randomly sampled from complicated posterior distributions without solving the integral in Bayes' rule, and by using only simple proposal distributions for which efficient random number generators already exist. All we need to decide whether to accept a proposed parameter value is the mathematical formulas for the likelihood function and prior distribution, and these formulas can be directly evaluated from their definitions.

We have seen two examples of the Metropolis algorithm in action. One was the island-hopping politician in Figure 7.2. The other was the θ-hopping coin bias in Figure 7.4. The example of the island-hopping politician was presented to demonstrate the Metropolis algorithm in its most simple form. The application involved only a finite space of discrete parameter values (i.e., the islands), the simplest possible proposal distribution (i.e., a single step right or left), and a target distribution that was directly evaluated (i.e., the relative population of the island) without any reference to mathematical formulas for likelihood functions and prior distributions. The simple forms of the discrete space and proposal distribution also allowed us to explore some of the basic mathematics of transition probabilities, to get some sense of why the Metropolis algorithm works.

The example of the θ-hopping coin bias in Figure 7.4 used the Metropolis algorithm in a more realistic setting. Instead of a finite space of discrete parameter values, there was a continuum of possible parameter values across the interval from zero to one. This is like an infinite string of adjacent infinitesimal islands. Instead of a proposal distribution that could only go a single step left or right, the normal proposal distribution could jump anywhere on the continuum, but it favored nearby values as governed by the SD of its bell shape. And, instead of a simple "relative population" at each parameter value, the target distribution was the relative density of the posterior distribution, computed by evaluating the likelihood function times the prior density.

7.4. TOWARD GIBBS SAMPLING: ESTIMATING TWO COIN BIASES

The Metropolis method is very useful, but it can be inefficient. Other methods can be more efficient in some situations. In particular, another type of sampling method that can be very efficient is *Gibbs sampling*. Gibbs sampling typically applies to models with multiple parameters, and therefore, we need to introduce an example in which there is more than one parameter. A natural extension of previous examples, which involved estimating the bias of a single coin, is estimating the biases of two coins.

In many real-world situations we observe two proportions, which differ by some amount in the specific random samples, but for which we want to infer what underlying difference is credible for the broader population from which the sample came. After all, the observed flips are just a noisy hint about the actual underlying biases. For example, suppose we have a sample of 97 people suffering from a disease. We give a random subset of them a promising drug, and we give the others a placebo. After 1 week, 12 of the 51 drug-treated people have gotten better, and 5 of the 46 placebo-treated people have gotten better. Did the drug actually work better than the placebo, and by how much? In other words, based on the observed difference in proportions, 12/51 versus 5/46, what underlying differences are actually credible? As another example, suppose you want to find out if mood affects cognitive performance. You manipulate mood by having 83 people sit through mood-inducing movies. A random subset of your participants is shown a bittersweet film about lovers separated by circumstances of war but who never forget each other. The other participants are shown a light comedy about high school pranks. Immediately after seeing the film, all participants are given some cognitive tasks, including an arithmetic problem involving long division. Of the 42 people who saw the war movie, 32 correctly solved the long division problem. Of the 41 people who saw the comedy, 27 correctly solved the long division problem. Did the induced mood actually affect cognitive performance? In other words, based on the observed difference in proportions, 32/42 versus 27/41, what underlying differences are actually credible?

To discuss the problem more generally and with mathematical precision, we need to define some notation. To make the notation generic, we will talk about heads and tails of flips of coins, instead of outcomes of participants in groups. What was called a group is now called a coin, and what was the outcome of a participant is now called the outcome of a flip. We'll use the same sort of notation that we've been using previously, but with subscripts to indicate which of the two coins is being referred to. Thus, the hypothesized bias for heads in coin j ,where $j=1$ or $j=2$, is denoted θ_j. The actual number of heads observed in a sample of N_j flips is z_j, and the ith individual flip in coin j is denoted y_{ji}.

We assume that the data from the two coins are independent: The performance of one coin has no influence on the performance of the other. Typically, we design research to make sure that the assumption of independence holds. In the examples above, we assumed independence in the disease-treatment scenario because we assumed that social interaction among the patients was minimal. We assumed independence in the mood-induction experiment because the experiment was designed to enforce zero social interaction between participants after the movie. The assumption of independence is crucial for all the mathematical analyses we will perform. If you have a situation in which the groups are not independent, the methods of this section do not directly apply. In situations where there are dependencies in the data, a model can try to formally express the dependency, but we will not be pursuing those situations here.

7.4.1. Prior, likelihood and posterior for two biases

We are considering situations in which there are *two* underlying biases, namely θ_1 and θ_2, for the two coins. We are trying to determine what we should believe about these biases after we have observed some data from the two coins. Recall that I am using the term "bias" as the name of the parameter θ, and *not* to indicate that the value of θ deviates from 0.5. The colloquial meaning of bias, as unfair, might also sneak into my writing from time to time. Thus, a colloquially "unbiased" coin technically "has a bias of 0.5."

To estimate the biases of the coins in a Bayesian framework, we must first specify what we believe about the biases without the data. Our prior beliefs are about *combinations* of parameter values. To specify a prior belief, we must specify the credibility, $p(\theta_1, \theta_2)$, for every combination θ_1, θ_2. If we were to make a graph of $p(\theta_1, \theta_2)$, it would be three dimensional, with θ_1 and θ_2 on the two horizontal axes, and $p(\theta_1, \theta_2)$ on the vertical axis. Because the credibilities form a probability density function, their integral across the parameter space must be one: $\iint \mathrm{d}\theta_1 \mathrm{d}\theta_2 p(\theta_1, \theta_2) = 1$.

For simplicity in these examples, we will assume that our beliefs about θ_1 are independent of our beliefs about θ_2. For example, if I have a coin from Canada and a coin from India, my belief about bias in the coin from Canada could be completely separate from my belief about bias in the coin from India. Independence of attributes was discussed in Section 4.4.2, p. 92. Mathematically, independence means that $p(\theta_1, \theta_2) = p(\theta_1)p(\theta_2)$ for every value of θ_1 and θ_2, where $p(\theta_1)$ and $p(\theta_2)$ are the marginal belief distributions. When beliefs about two parameters are independent, the mathematical manipulations can be greatly simplified. Beliefs about the two parameters do not need to be independent, however. For example, perhaps I believe that coins are minted in similar ways across countries, and so if a Canadian coin is biased (i.e., has a θ value different than 0.5), then an Indian coin should be similarly biased. At the extreme, if I believe that θ_1 always exactly equals θ_2, then the two parameter values are completely dependent upon each other. Dependence does not imply direct causal relationship, it merely implies that knowing the value of one parameter constrains beliefs about the value of the other parameter.

Along with the prior beliefs, we have some observed data. We assume that the flips within a coin are independent of each other and that the flips across coins are independent of each other. The veracity of this assumption depends on exactly how the observations are actually made, but, in properly designed experiments, we have reasons to trust this assumption. Notice that we will always assume independence of *data* within and across groups, regardless of whether we assume independence in our *beliefs* about the biases in the groups. Formally, the assumption of independence in the data means the following. Denote the result of a flip of coin 1 as y_1, where the result can be $y_1 = 1$ (heads) or $y_1 = 0$ (tails). Similarly, the result of a flip of coin 2 is denoted y_2. Independence of the data across the two coins means that the data from coin 1 depend only on the bias in coin 1, and the data from coin 2 depend only on the bias in coin 2, which can be expressed formally as $p(y_1|\theta_1, \theta_2) = p(y_1|\theta_1)$ and $p(y_2|\theta_1, \theta_2) = p(y_2|\theta_2)$.

From one coin we observe the data D_1 consisting of z_1 heads out of N_1 flips, and from the other coin we observe the data D_2 consisting of z_2 heads out of N_2 flips. In other words, $z_1 = \sum_{i=1}^{N_1} y_{1i}$, where y_{1i} denotes the ith flip of the first coin. Notice that $z_1 \in \{0, \ldots, N_1\}$ and $z_2 \in \{0, \ldots, N_2\}$. We denote the whole set of data as $D = \{z_1, N_1, z_2, N_2\}$. Because of independence of sampled flips, the probability of D is the product of the Bernoulli distribution functions for the individual flips:

$$
\begin{aligned}
p(D|\theta_1, \theta_2) &= \prod_{y_{1i} \in D_1} p(y_{1i}|\theta_1, \theta_2) \prod_{y_{2j} \in D_2} p(y_{2j}|\theta_1, \theta_2) \\
&= \prod_{y_{1i} \in D_1} \theta_1^{y_{1i}} (1 - \theta_1)^{(1-y_{1i})} \prod_{y_{2j} \in D_2} \theta_2^{y_{2j}} (1 - \theta_2)^{(1-y_{2j})} \\
&= \theta_1^{z_1} (1 - \theta_1)^{(N_1 - z_1)} \, \theta_2^{z_2} (1 - \theta_2)^{(N_2 - z_2)} \qquad (7.5)
\end{aligned}
$$

The posterior distribution of our beliefs about the underlying bias is derived in the usual way by applying Bayes' rule, but now the functions involve two parameters:

$$
\begin{aligned}
p(\theta_1, \theta_2|D) &= p(D|\theta_1, \theta_2) p(\theta_1, \theta_2) \Big/ p(D) \\
&= p(D|\theta_1, \theta_2) p(\theta_1, \theta_2) \Big/ \iint d\theta_1 \, d\theta_2 \, p(D|\theta_1, \theta_2) p(\theta_1, \theta_2) \qquad (7.6)
\end{aligned}
$$

Remember, as always in the expression of Bayes' rule, the θ_j's in left side of the equation and in the numerator of the right side are referring to specific values of θ_j, but the θ_j's in the integral in the denominator range over all possible values of θ_j.

What has just been described in the previous few paragraphs is the general Bayesian framework for making inferences about two biases when the likelihood function consists of independent Bernoulli distributions. What we have to do next is specify a particular mathematical form for the prior distribution. We will work through the mathematics of a particular case for two reasons: First, it will allow us to explore graphical displays of two-dimensional parameter spaces, which will inform our intuitions about Bayes' rule and sampling from the posterior distribution. Second, the mathematics will set the stage for a specific example of Gibbs sampling. Later in the book when we do applied Bayesian analysis, we will *not* be doing any of this sort of mathematics. We are doing the math now, for simple cases, to understand how the methods work so we can properly interpret their outputs in realistically complex cases.

7.4.2. The posterior via exact formal analysis

Suppose we want to pursue a solution to Bayes' rule, in Equation 7.6 above, using formal analysis. What sort of prior probability function would make the analysis especially tractable? Perhaps you can guess the answer by recalling the discussion of Chapter 6. We learned there that the beta distribution is conjugate to the Bernoulli likelihood function. This suggests that a product of beta distributions would be conjugate to a product of Bernoulli functions.

This suggestion turns out to be true. We pursue the same logic as was used for Equation 6.8 (p. 132). First, recall that a beta distribution has the form $\text{beta}(\theta|a,b) = \theta^{(a-1)}(1-\theta)^{(b-1)}/B(a,b)$, where $B(a,b)$ is the beta normalizing function, which by definition is $B(a,b) = \int_0^1 d\theta\, \theta^{(a-1)}(1-\theta)^{(b-1)}$. We assume a $\text{beta}(\theta_1|a_1,b_1)$ prior on θ_1, and an independent $\text{beta}(\theta_2|a_2,b_2)$ prior on θ_2, so that $p(\theta_1,\theta_2) = \text{beta}(\theta_1|a_1,b_1)\cdot\text{beta}(\theta_2|a_2,b_2)$. Then:

$$
\begin{aligned}
p(\theta_1,\theta_2|D) &= p(D|\theta_1,\theta_2)p(\theta_1,\theta_2)/p(D) && \text{general form of Bayes' rule} \\
&= \theta_1^{z_1}(1-\theta_1)^{(N_1-z_1)}\,\theta_2^{z_2}(1-\theta_2)^{(N_2-z_2)}p(\theta_1,\theta_2)/p(D) && \text{Bernoulli likelihood from Equation 7.5} \\
&= \frac{\theta_1^{z_1}(1-\theta_1)^{(N_1-z_1)}\,\theta_2^{z_2}(1-\theta_2)^{(N_2-z_2)}\,\theta_1^{(a_1-1)}(1-\theta_1)^{(b_1-1)}\,\theta_2^{(a_2-1)}(1-\theta_2)^{(b_2-1)}}{p(D)B(a_1,b_1)B(a_2,b_2)} && \text{independent beta prior} \\
&= \frac{\theta_1^{(z_1+a_1-1)}(1-\theta_1)^{(N_1-z_1+b_1-1)}\theta_2^{(z_2+a_2-1)}(1-\theta_2)^{(N_2-z_2+b_2-1)}}{p(D)B(a_1,b_1)B(a_2,b_2)} && (7.7)
\end{aligned}
$$

We know that the left side of Equation 7.7 must be a probability density function, and we see that the numerator of the right side has the form of a product of beta distributions, namely $\text{beta}(\theta_1|z_1+a_1, N_1-z_1+b_1)$ times $\text{beta}(\theta_2|z_2+a_2, N_2-z_2+b_2)$. Therefore, the denominator of Equation 7.7 must be the corresponding normalizer for the product of beta distributions:[5]

$$p(D)B(a_1, b_1)B(a_2, b_2) = B(z_1+a_1, N_1-z_1+b_1)\, B(z_2+a_2, N_2-z_2+b_2) \qquad (7.9)$$

Recapitulation: When the prior is a product of independent beta distributions, the posterior is also a product of independent beta distributions, with each beta obeying the update rule we derived in Chapter 6. Explicitly, if the prior is $\text{beta}(\theta_1|a_1, b_1) \cdot \text{beta}(\theta_2|a_2, b_2)$, and the data are z_1, N_1, z_2, N_2, then the posterior is $\text{beta}(\theta_1|z_1+a_1, N_1-z_1+b_1) \cdot \text{beta}(\theta_2|z_2+a_2, N_2-z_2+b_2)$.

One way of understanding the posterior distribution is to visualize it. We want to plot the probability densities as a function of *two* parameters, θ_1 and θ_2. One way to do this is by placing the two parameters, θ_1 and θ_2, on two horizontal axes, and placing the probability density, $p(\theta_1, \theta_2)$, on a vertical axis. The surface can then be displayed in a picture as if it were a landscape viewed in perspective. Alternatively, we can place the two parameter axes flat on the drawing plane and indicate the probability density with level contours, such that any one contour marks points that have the same specific level. An example of these plots is described next.

Figure 7.5 shows graphs for an example of Bayesian updating from Equation 7.6. In this example, the prior begins with mild beliefs that each bias is about 0.50, using a $\text{beta}(\theta|2, 2)$ distribution for both biases. The upper panels show a perspective plot and a contour plot for the prior distribution. Notice that it is gently peaked at the center of the parameter space, which reflects the prior belief that the two biases are around 0.5, but without much certainty. The perspective plot shows vertical slices of the prior density parallel to θ_1 and θ_2. Consider slices parallel to the θ_1 axis, with different slices at different values of θ_2. The profile of the density on every slice has the same shape, with merely different heights. In particular, at every level of θ_2, the profile of the slice along θ_1 is a $\text{beta}(\theta_1|2, 2)$ shape, with merely an overall height that depends on the level of θ_2. When the shape of the profile on the slices does not change, as exemplified here, then the joint distribution is constructed from the product of independent marginal distributions.

[5] By rearranging terms of Equation 7.9, a convenient consequence is that

$$p(D) = \frac{B(z_1+a_1, N_1-z_1+b_1)\, B(z_2+a_2, N_2-z_2+b_2)}{B(a_1, b_1)B(a_2, b_2)} \qquad (7.8)$$

This is analogous to the result we found previously for one parameter, in Equation 6.8 and Footnote 5 on p. 132.

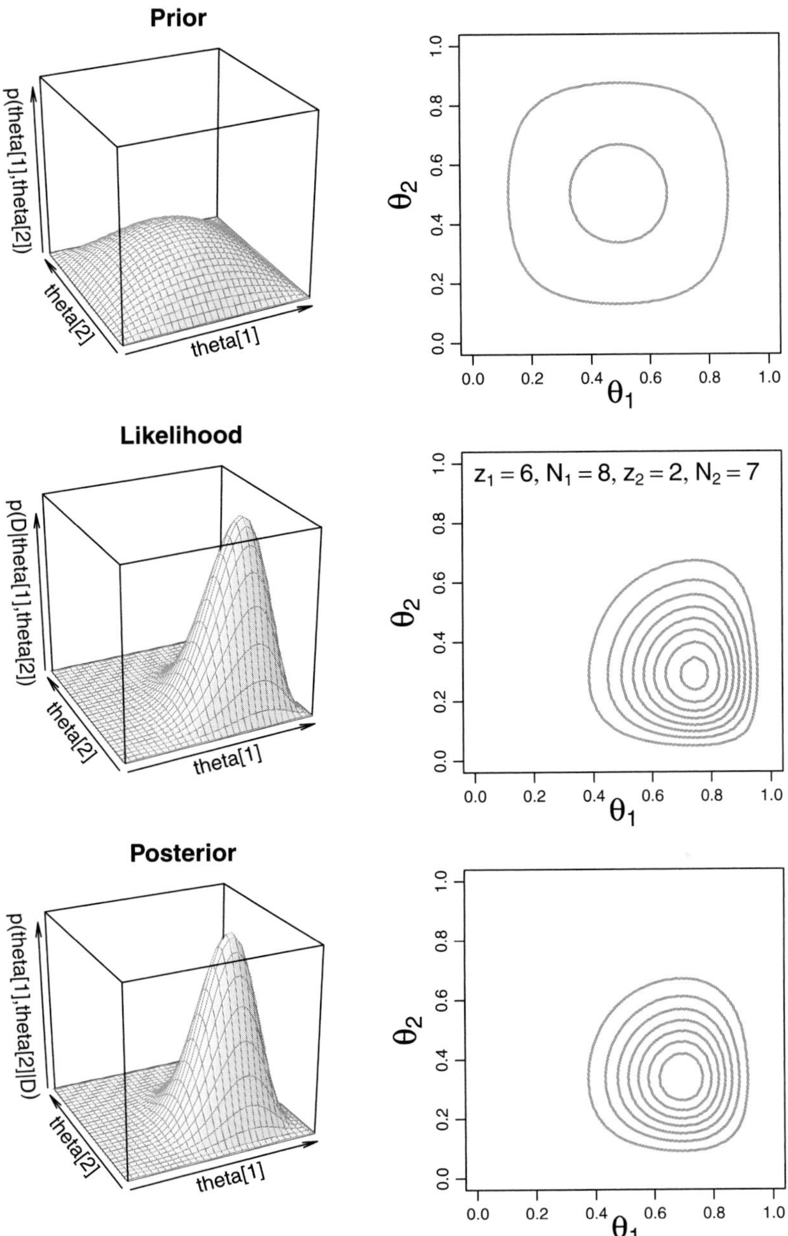

Figure 7.5 Bayesian updating of independent beta(θ|2, 2) priors with the data annotated in the middle-right panel. Left panels show perspective surface plots; right panels show contour plots of the same distributions.

The contour plot in the upper-right panel of Figure 7.5 shows the same distribution as the upper-left panel. Instead of showing vertical slices through the distribution, the contour plot shows horizontal slices. Each contour corresponds to a slice at a particular height. Contour plots can be challenging to interpret because it is not immediately obvious what the heights of the contours are, or even whether two adjacent contours belong to different heights. Contours can be labeled with numerical values to indicate their heights, but then the plot can become very cluttered. Therefore, if the goal is a quick intuition about the general layout of the distribution, then a perspective plot is preferred over a contour plot. If the goal is instead a more detailed visual determination of the parameter values for a particular peak in the distribution, then a contour plot may be preferred.

The middle row of Figure 7.5 shows the likelihood functions for the specific data displayed in the panels. The plots show the likelihood for each possible combination of θ_1 and θ_2. Notice that the likelihood is maximized at θ values that match the proportions of heads in the data.

The resulting posterior distribution is shown in the lowest row of Figure 7.5. At each point in the θ_1, θ_2 parameter space, the posterior is the product of the prior and likelihood values at that point, divided by the normalizer, $p(D)$.

Summary: This section provided a graphical display of a prior, likelihood, and posterior on a two-parameter space, in Figure 7.5. In this section, we emphasized the use of mathematical forms, with priors that are conjugate to the likelihood. The particular mathematical form, involving beta distributions, will be used again in a subsequent section that introduces Gibbs sampling, and that is another motivation for including the mathematical formulation of this section. Before getting to Gibbs sampling, however, we first apply the Metropolis algorithm to this two-parameter situation. We will see later that Gibbs sampling generates a posterior sample more efficiently than the Metropolis algorithm.

7.4.3. The posterior via the Metropolis algorithm

Although we have already solved the present example with exact mathematical analysis, we will apply the Metropolis algorithm to expand our understanding of the algorithm for a two-parameter case and subsequently to compare the Metropolis algorithm with Gibbs sampling. Recall that the Metropolis algorithm is a random walk through the parameter space that starts at some arbitrary point. We propose a jump to a new point in parameter space, with the proposed jump randomly generated from a proposal distribution from which it is easy to generate values. For our present purposes, the proposal distribution is a *bivariate* normal. You can visualize a bivariate normal distribution by imagining a one-dimensional normal (as in Figure 4.4, p. 83), sticking a pin down vertically through its peak, and spinning it around the pin, to form a bell-shaped surface. The use of a bivariate normal proposal distribution implies that the proposed jumps will usually be near the current position. Notice that proposed jumps can be in any direction in parameter space relative to the current position.[6] The proposed jump is definitely accepted if the posterior is taller (more dense) at the proposed position than at the current position, and the proposed jump is probabilistically accepted if the posterior is shorter (less dense) at the proposed position than at the current position. If the proposed jump is rejected, the current position is counted again. Notice that a position in parameter space represents a *combination* of jointly credible parameter values, $\langle \theta_1, \theta_2 \rangle$.

Figure 7.6 shows the Metropolis algorithm applied to the case of Figure 7.5 (p. 167), so that you can directly compare the results of the Metropolis algorithm with the results of formal analysis and grid approximation. By comparing with the contour plot in the lower-right panel of Figure 7.5 (p. 167), you can see that the points do indeed appear to explore the bulk of the posterior distribution. The sampled points in Figure 7.6 are connected by lines so that you can get a sense of the trajectory taken by the random

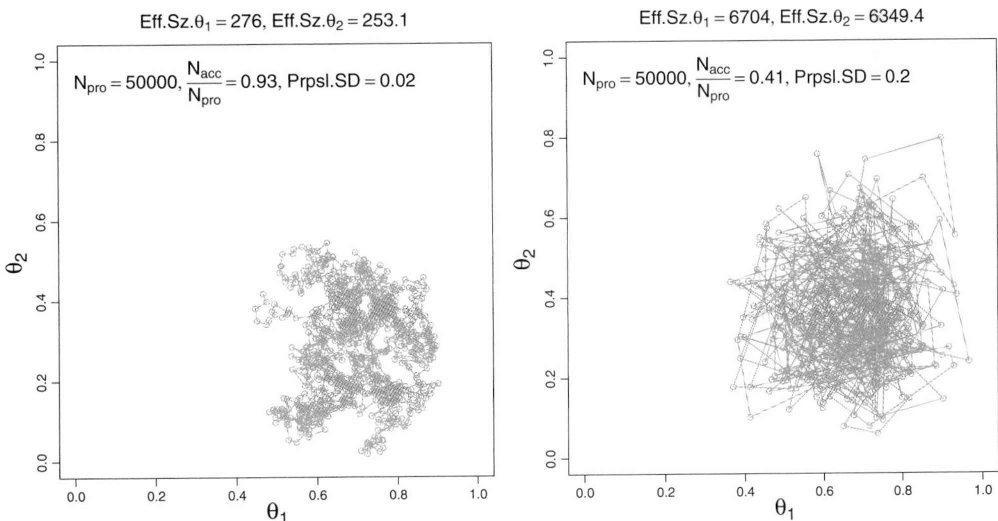

Figure 7.6 Metropolis algorithm applied to the prior and likelihood shown in Figure 7.5, p. 167. The left panel shows a chain with a narrow proposal distribution and the right panel shows a chain with a moderate-width proposal distribution, as indicated by annotation "Prpsl.SD" in each panel. N_{pro} is the number of proposed jumps, and N_{acc} is the number of accepted proposals. *Many of the plotted points have multiple superimposed symbols where the chain lingered during rejected proposals.* Notice that the effective size of the chain, indicated at the top of the plot, is far less than the length of the chain (N_{pro}). Only 1,000 of the N_{pro} steps are displayed here.

[6] The proposal distribution does not have to be a rotationally symmetric bivariate normal. For example, it could be a bivariate normal with nonzero covariances, so that proposals are more likely to be made in some diagonal directions more than others. The proposal distribution could even be nonnormal. It is only for the present illustrative purposes that we assume a simple symmetric proposal distribution.

walk. But the ultimate estimates regarding the posterior distribution do not care about the sequence in which the sampled points appeared, and the trajectory is irrelevant to anything but your understanding of the Metropolis algorithm.

The two panels of Figure 7.6 show results from using two different widths for the proposal distribution. The left panel shows results from a relatively narrow proposal distribution, that had a standard deviation of only 0.02. You can see that there were only tiny changes from one step to the next. Visually, the random walk looks like a clumpy strand that gradually winds through the parameter space. Because the proposed jumps were so small, the proposals were usually accepted, but each jump yielded relatively little new information and consequently the effective sample size (ESS) of the chain is very small, as annotated at the top of the panel.

The right panel of Figure 7.6 shows results from using a moderate width for the proposal distribution, with a SD of 0.2. The jumps from one position to the next are larger than the previous example, and the random walk explores the posterior distribution more efficiently than the previous example. The ESS of the chain is much larger than before. But the ESS of the chain is still far less than the number of proposed jumps, because many of the proposed jumps were rejected, and even when accepted, the jumps tended to be near the previous step in the chain.

In the limit of infinite random walks, the Metropolis algorithm yields arbitrarily accurate representations of the underlying posterior distribution. The left and right panels of Figure 7.6 would eventually converge to an identical and highly accurate approximation to the posterior distribution. But in the real world of finite random walks, we care about how efficiently the algorithm generates an accurate representative sample. We prefer to use the proposal distribution from the right panel of Figure 7.6 because it will, typically, produce a more accurate approximation of the posterior than the proposal distribution from left panel, for the same number of proposed jumps.

7.4.4. Gibbs sampling

The Metropolis algorithm is very general and broadly applicable. One problem with it, however, is that the proposal distribution must be properly tuned to the posterior distribution if the algorithm is to work well. If the proposal distribution is too narrow or too broad, a large proportion of proposed jumps will be rejected and the trajectory will get bogged down in a localized region of the parameter space. Even at its most efficient, the effective size of the chain is far less than the number of proposed jumps. It would be nice, therefore, if we had another method of sample generation that was more efficient. Gibbs sampling is one such method.

Gibbs sampling was introduced by Geman and Geman (1984), who were studying how computer vision systems could infer properties of an image from its pixelated camera input. The method is named after the physicist Josiah Willard Gibbs (1839–1903), who studied statistical mechanics and thermodynamics. Gibbs sampling can be especially useful for hierarchical models, which we explore in Chapter 9. It turns out that Gibbs sampling is a special case of the Metropolis-Hastings algorithm, which is a generalized form of the Metropolis algorithm, as will be briefly discussed later. An influential article by Gelfand and Smith (1990) introduced Gibbs sampling to the statistical community; a brief review is provided by Gelfand (2000), and interesting details of the history of MCMC can be found in the book by McGrayne (2011).

The procedure for Gibbs sampling is a type of random walk through parameter space, like the Metropolis algorithm. The walk starts at some arbitrary point, and at each point in the walk, the next step depends only on the current position, and on no previous positions. What is different about Gibbs sampling, relative to the Metropolis algorithm, is how each step is taken. At each point in the walk, one of the component parameters is selected. The component parameter could be selected at random, but typically the parameters are cycled through, in order: $\theta_1, \theta_2, \theta_3, \ldots, \theta_1, \theta_2, \theta_3, \ldots$. The reason that parameters are cycled rather than selected randomly is that for complex models with many dozens or hundreds of parameters, it would take too many steps to visit every parameter by random chance alone, even though they would be visited about equally often in the long run. Suppose that parameter θ_i has been selected. Gibbs sampling then chooses a new value for that parameter by generating a random value directly from the conditional probability distribution $p(\theta_i|\{\theta_{j\neq i}\}, D)$. The new value for θ_i, combined with the unchanged values of $\theta_{j\neq i}$, constitutes the new position in the random walk. The process then repeats: Select a component parameter and select a new value for that parameter from its conditional posterior distribution.

Figure 7.7 illustrates this process for a two-parameter example. In the first step, we want to select a new value for θ_1. We conditionalize on the values of all the other parameters, $\theta_{j\neq 1}$, from the previous step in the chain. In this example, there is only one other parameter, namely θ_2. The upper panel of Figure 7.7 shows a slice through the joint distribution at the current value of θ_2. The heavy curve is the posterior distribution conditional on this value of θ_2, denoted $p(\theta_1|\{\theta_{j\neq 1}\}, D)$, which is $p(\theta_1|\theta_2, D)$ in this case because there is only one other parameter. If the mathematical form of the conditional distribution is appropriate, a computer can directly generate a random value of θ_1. Having thereby generated a new value for θ_1, we then conditionalize on it and determine the conditional distribution of the next parameter, θ_2, as shown in the lower panel of Figure 7.7. The conditional distribution is denoted formally as $p(\theta_2|\{\theta_{j\neq 2}\}, D)$, which equals $p(\theta_2|\theta_1, D)$ in this case of only two parameters. If the mathematical form is convenient, our computer can directly generate a random value of θ_2 from the conditional distribution. We then conditionalize on the new value θ_2, and the cycle repeats.

Gibbs sampling can be thought of as just a special case of the Metropolis algorithm, in which the proposal distribution depends on the location in parameter space and the

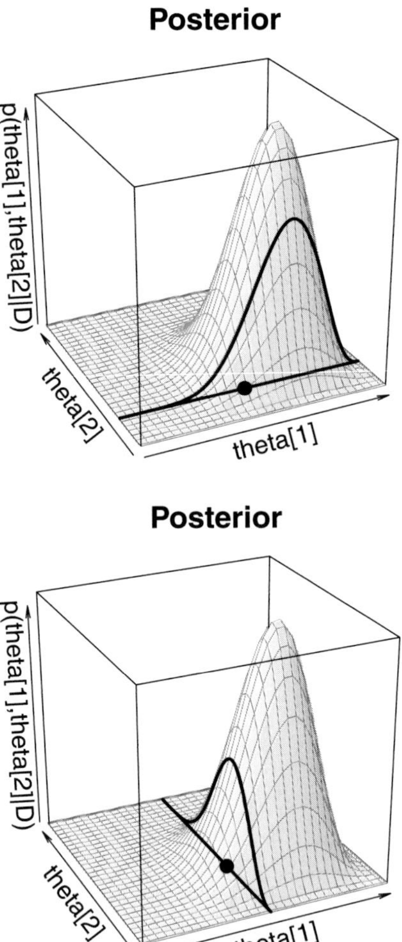

Figure 7.7 Two steps in a Gibbs sampling. In the upper panel, the heavy lines show a slice through the posterior conditionalized on a particular value of θ_2, and the large dot shows a random value of θ_1 sampled from the conditional density. The lower panel shows a random generation of a value for θ_2, conditional on the value for θ_1 determined by the previous step. The heavy lines show a slice through the posterior at the conditional value of θ_1, and the large dot shows the random value of θ_2 sampled from the conditional density.

component parameter selected. At any point, a component parameter is selected, and then the proposal distribution for that parameter's next value *is* the conditional posterior probability of that parameter. *Because the proposal distribution exactly mirrors the posterior probability for that parameter, the proposed move is always accepted.* A rigorous proof of this idea requires development of a generalized form of the Metropolis algorithm, called the Metropolis-Hastings algorithm (Hastings, 1970). A technical overview of the relation is provided by Chib and Greenberg (1995), and an accessible mathematical tutorial is given by Bolstad (2009).

Gibbs sampling is especially useful when the complete joint posterior, $p(\{\theta_i\}|D)$, cannot be analytically determined and cannot be directly sampled, but all the conditional distributions, $p(\theta_i|\{\theta_{j\neq i}\}, D)$, can be determined and directly sampled. We will not encounter such a situation until later in the book, but the process of Gibbs sampling can be illustrated now for a simpler situation.

We continue now with estimating two coin biases, θ_1 and θ_2, when assuming that the prior belief distribution is a product of beta distributions. The posterior distribution is again a product of beta distributions, as was derived in Equation 7.7 (p. 165). This joint posterior is easily dealt with directly, and so there is no real need to apply Gibbs sampling, but we will go ahead and do Gibbs sampling of this posterior distribution for purposes of illustration and comparison with other methods.

To accomplish Gibbs sampling, we must determine the conditional posterior distribution for each parameter. By definition of conditional probability,

$$
\begin{aligned}
p(\theta_1|\theta_2, D) &= p(\theta_1, \theta_2|D)/p(\theta_2|D) && \text{conditional is joint divided by marginal} \\
&= p(\theta_1, \theta_2|D) \Big/ \int d\theta_1\, p(\theta_1, \theta_2|D) && \text{marginal is integral of joint}
\end{aligned}
$$

For our current application, the joint posterior is a product of beta distributions, as was derived in Equation 7.7, p. 165. Therefore, substituting into the equation above, we have

$$
\begin{aligned}
p(\theta_1|\theta_2, D) &= p(\theta_1, \theta_2|D) \Big/ \int d\theta_1\, p(\theta_1, \theta_2|D) \\
&= \frac{\text{beta}(\theta_1|z_1+a_1, N_1-z_1+b_1)\, \text{beta}(\theta_2|z_2+a_2, N_2-z_2+b_2)}{\int d\theta_1\, \text{beta}(\theta_1|z_1+a_1, N_1-z_1+b_1)\, \text{beta}(\theta_2|z_2+a_2, N_2-z_2+b_2)} \\
&= \frac{\text{beta}(\theta_1|z_1+a_1, N_1-z_1+b_1)\, \text{beta}(\theta_2|z_2+a_2, N_2-z_2+b_2)}{\text{beta}(\theta_2|z_2+a_2, N_2-z_2+b_2) \int d\theta_1\, \text{beta}(\theta_1|z_1+a_1, N_1-z_1+b_1)} \\
&\qquad \text{because beta}(\theta_2|\ldots) \text{ is constant w.r.t. } \theta_1 \\
&= \frac{\text{beta}(\theta_1|z_1+a_1, N_1-z_1+b_1)\, \text{beta}(\theta_2|z_2+a_2, N_2-z_2+b_2)}{\text{beta}(\theta_2|z_2+a_2, N_2-z_2+b_2)} \\
&\qquad \text{because integral of prob. distrib. must be 1} \\
&= \text{beta}(\theta_1|z_1+a_1, N_1-z_1+b_1) \qquad (7.10)
\end{aligned}
$$

From these considerations, you can also see that the other conditional posterior probability is $p(\theta_2|\theta_1, D) = \text{beta}(\theta_2|z_2+a_2, N_2-z_2+b_2)$. We have just derived what may have already been intuitively clear: Because the posterior is a product of independent beta distributions, it makes sense that $p(\theta_1|\theta_2, D) = p(\theta_1|D)$. Nevertheless, the derivation illustrates the sort of analytical procedure that is needed in general. In general, the posterior distribution will not be a product of independent marginal distributions as in this simple case. Instead, the formula for $p(\theta_i|\theta_{j\neq i}, D)$ will typically involve the values of $\theta_{j\neq i}$.

The conditional distributions just derived were displayed graphically in Figure 7.7. The upper panel shows a slice conditional on θ_2, and the heavy curve illustrates $p(\theta_1|\theta_2, D)$ which is $\text{beta}(\theta_1|z_1{+}a_1, N_1{-}z_1{+}b_1)$. The lower panel shows a slice conditional on θ_1, and the heavy curve illustrates $p(\theta_2|\theta_1, D)$, which is $\text{beta}(\theta_2|z_2{+}a_2, N_2{-}z_2{+}b_2)$.

Having successfully derived the conditional posterior probability distributions, we now figure out whether we can directly sample from them. In this case, the answer is yes, we can, because the conditional probabilities are beta distributions, and our computer software comes prepackaged with generators of random beta values.

Figure 7.8 shows the result of applying Gibbs sampling to this scenario. The left panel shows every step, changing one parameter at a time. Notice that each step in the random walk is parallel to a parameter axis, because each step changes only one component parameter. You can also see that at each point, the walk direction changes to the other parameter, rather than doubling back on itself or continuing in the same direction. This is because the walk cycled through the component parameters, $\theta_1, \theta_2, \theta_1, \theta_2, \theta_1, \theta_2, \ldots$, rather than selecting them randomly at each step.

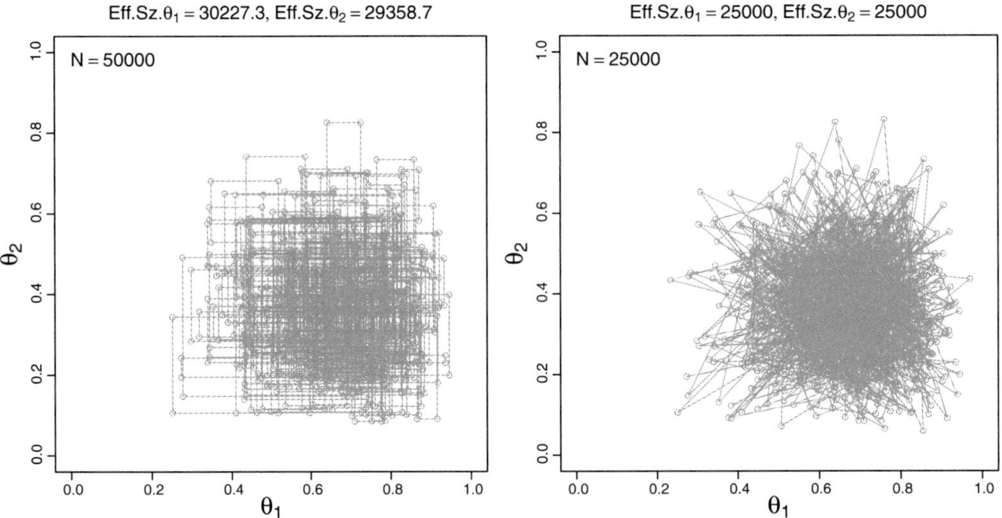

Figure 7.8 Gibbs sampling applied to the posterior shown in Figure 7.5, p. 167. The left panel shows all the intermediate steps of chain, changing one parameter at a time. The right panel shows only the points after complete sweeps through all (two) parameters. Both are valid samples from the posterior distribution. Only 1,000 of the *N* steps are displayed here. Compare with the results of the Metropolis algorithm in Figure 7.6. Notice that the effective size of the Gibbs sample is larger than the effective size of the Metropolis sample for the same length of chain.

The right panel of Figure 7.8 only shows the steps after each complete cycle through the parameters. It is the same chain as in the left panel, only with the intermediate steps not shown. Both the left and the right panels are representative samples from the posterior distribution, and they would converge to the same continuous distribution in the limit of an infinitely long chain. The software we will use later in the book (called JAGS) records only complete cycles, not intermediate single-parameter steps.

If you compare the results of Gibbs sampling in Figure 7.8 with the results of the Metropolis algorithm in Figure 7.6 (p. 169), you will see that the points produced by Gibbs sampling and the Metropolis algorithm look similar in where they fall, if not in the trajectory of the walk that produces them. In fact, in the infinitely long run, the Gibbs and Metropolis methods converge to the same distribution. What differs is the efficiency of getting to any desired degree of approximation accuracy in a finite run. Notice that the effective size of the Gibbs sampler, displayed in the panel annotations of Figure 7.8, is much greater than the effective size of the Metropolis algorithm displayed in Figure 7.6. (The right panel of Figure 7.8 shows that the effective size is the same as the number of complete cycles of the Gibbs sampler in this case, but this is not generally true and happens here because the conditional distributions are independent of each other.)

By helping us visualize *how* Gibbs sampling works, Figure 7.8 also helps us understand better *why* it works. Imagine that instead of changing the component parameter at every step, we linger a while on a component. Suppose, for example, that we have a fixed value of θ_1, and we keep generating new random values of θ_2 for many steps. In terms of Figure 7.8, this amounts to lingering on a vertical slice of the parameter space, lined up over the fixed value of θ_1. As we continue sampling within that slice, we build up a good representation of the posterior distribution for that value of θ_1. Now we jump to a different value of θ_1, and again linger a while at the new value, filling in a new vertical slice of the posterior. If we do that enough, we will have many vertical slices, each representing the posterior distribution along that slice. We can use those vertical slices to represent the posterior, *if* we have also lingered in each slice proportionally to the posterior probability of being in that slice! Not all values of θ_1 are equally likely in the posterior, so we visit vertical slices according to the conditional probability of θ_1. Gibbs sampling carries out this process, but lingers for only one step before jumping to a new slice.

So far, I have emphasized the advantages of Gibbs sampling over the Metropolis algorithm, namely, that there is no need to tune a proposal distribution and no inefficiency of rejected proposals. I also mentioned restrictions: We must be able to derive the conditional probabilities of each parameter on the others and be able to generate random samples from those conditional probability distributions.

But there is one other disadvantage of Gibbs sampling. Because it only changes one parameter value at a time, its progress can be stalled by highly correlated parameters. We will encounter applications later in which credible combinations of parameter values can be very strongly correlated; see, for example, the correlation of slope and intercept parameters in Figure 17.3, p. 481. Here I hope merely to plant the seeds of an intuition for later development. Imagine a posterior distribution over two parameters. Its shape is a narrow ridge along the *diagonal* of the parameter space, and you are *inside, within,* this narrow ridge, like being inside a long-narrow hallway that runs diagonally to the parameter axes. Now imagine doing Gibbs sampling from this posterior. You are in the hallway, and you are contemplating a step along a parameter axis. Because the hallway is narrow and diagonal, a step along a parameter axis quickly encounters a wall, and your step size must be small. This is true no matter which parameter axis you face along. Therefore, you can take only small steps and only very gradually explore the length of the diagonal hallway. On the other hand, if you were stepping according to a Metropolis sampler, whereby your proposal distribution included changes of both parameters at once, then you could jump in the diagonal direction and quickly explore the length of the hallway.

7.4.5. Is there a difference between biases?

Until this point, we have focused on the mechanics of Gibbs sampling (and the Metropolis algorithm) without answering the question, How different are the biases of the two coins? Fortunately, the representative sample from the joint posterior distribution gives us an answer. Notice that every step in the chain (either from Gibbs or Metropolis) is a combination of θ_1 and θ_2 values that are jointly credible. Therefore, at each step in the chain, a credible difference is $\theta_1 - \theta_2$. Across the full chain, we have a large representative sample of credible differences. (We cannot get a sense of the difference of parameters merely by looking at the marginal distributions of those parameters; see Section 12.1.2.1, p. 340.)

Figure 7.9 shows histograms of $\theta_1 - \theta_2$ from the posterior distribution, generated by each of the chains in Figures 7.6 and 7.8. In the limit of an infinite chain, they would all converge to the same true underlying posterior distribution, but any finite chain is an approximation. In principle, on average, the results from a chain with larger effective size should be more accurate than a smaller effective size, although the smaller might get lucky and happen to be more accurate on a given random walk. In this case, therefore, the results from the Gibbs sampling are probably more accurate than the results from the Metropolis algorithm.

Figure 7.9 reveals that a difference of zero is clearly among the 95% most credible differences (i.e., within the 95% HDI), and we would not want to declare that there is a difference. The Bayesian posterior gives complete information about credible differences

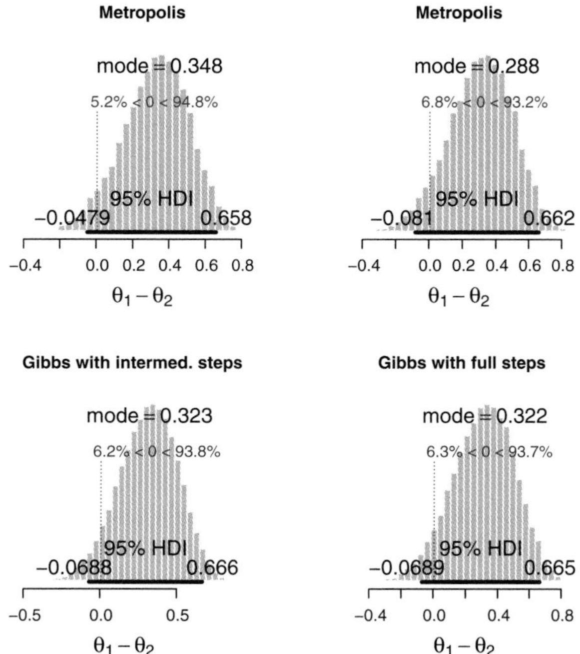

Figure 7.9 Credible posterior differences between biases. Upper panels come from results of Metropolis algorithm in Figure 7.6 using two different proposal standard deviations; lower panels come from results of Gibbs sampling in Figure 7.8. The four distributions are nearly the same, and in the limit for infinitely long chains, should be identical. For these finite chains, the ones with longer effective size (i.e., the Gibbs sampled) are more accurate on average.

and indicates that there is a suggestion of difference, but our uncertainty is large because the amount of data is small. Notice that the decision rule is a separate process from Bayes' rule, and the result of a decision rule leaves behind most of the posterior information. See Section 12.1.1, p. 336, for complete discussion of HDI and decisions.

7.4.6. Terminology: MCMC

Assessing the properties of a target distribution by generating representative random values is a case of a Monte Carlo simulation. Any simulation that samples a lot of random values from a distribution is called a Monte Carlo simulation, named after the dice and spinners and shufflings of the famous casino locale. The appellation "Monte Carlo" is attributed (Eckhardt, 1987) to the mathematicians Stanislaw Ulam (1909–1984) and John von Neumann (1903–1957).

The Metropolis algorithm and Gibbs sampling are specific types of Monte Carlo process. They generate random walks such that each step in the walk is completely independent of the steps before the current position. Any such process in which each step has no memory for states before the current one is called a (first-order) Markov process, and a succession of such steps is a Markov chain (named after the mathematician Andrey Markov, 1856–1922). The Metropolis algorithm and Gibbs sampling are examples of a *Markov chain Monte Carlo (MCMC)* process. There are many others. It is the invention of MCMC algorithms, along with software for automatically creating samplers for complex models (such as BUGS, JAGS, and Stan), and fast cheap computers, that allow us to do Bayesian data analysis for complex realistic data.

7.5. MCMC REPRESENTATIVENESS, ACCURACY, AND EFFICIENCY

We have three main goals in generating an MCMC sample from the posterior distribution:

1. The values in the chain must be *representative* of the posterior distribution. They should not be unduly influenced by the arbitrary initial value of the chain, and they should fully explore the range of the posterior distribution without getting stuck.
2. The chain should be of sufficient size so that estimates are *accurate and stable.* In particular, the estimates of the central tendency (such as median or mode), and the limits of the 95% HDI, should not be much different if the MCMC analysis is run again (using different seed states for the pseudorandom number generators).
3. The chain should be generated *efficiently*, with as few steps as possible, so not to exceed our patience or computing power.

These goals are achieved only more or less, not absolutely, in real applications. In principle, the mathematics of MCMC guarantee that infinitely long chains will achieve a perfect representation of the posterior distribution, but unfortunately we do not have infinite time or computer memory. Therefore, we must check the quality of finite MCMC chains to determine whether there are signs of *un*representativeness or *in*stability. For typical models of moderate complexity such as those in this book, the MCMC chains are usually well behaved, and checking them is routine. As models increase in complexity, their MCMC chains may be more problematic and checking them can be more important and more challenging.

7.5.1. MCMC representativeness

Checks for unrepresentativeness usually look for lingering influence of the initial value and for orphaned chains that have somehow settled into unusual regions of parameter space. There is (at the time of this writing) no single universally best method for conducting these checks. Current practice often focuses on two methods: visual examination of the trajectory, and consideration of a numerical description of convergence.

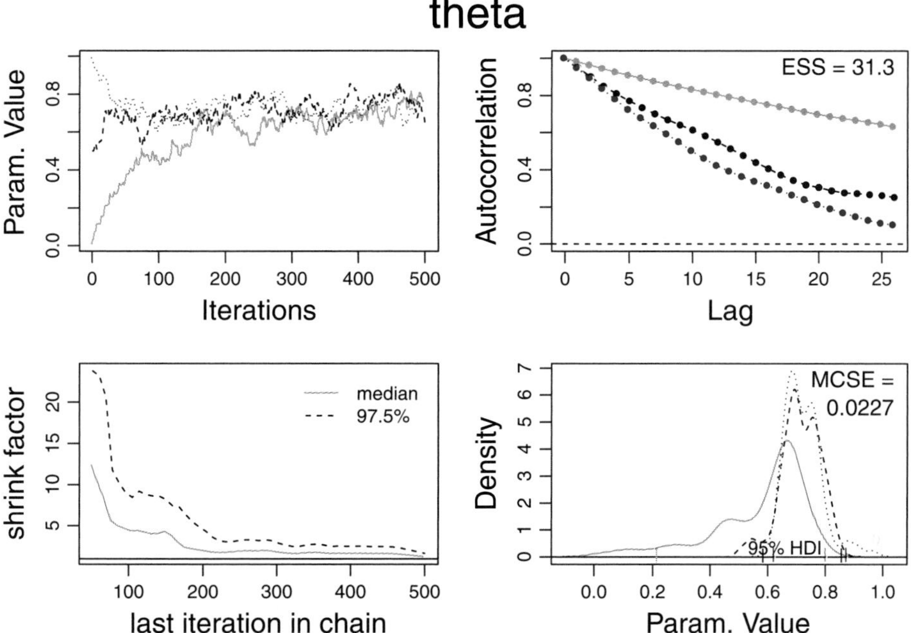

Figure 7.10 Illustration of MCMC diagnostics. Three chains were generated by starting a Metropolis algorithm at different initial values, with proposal SD=0.02 (cf. Figure 7.4) for data $z = 35$, $N = 50$. Only steps 1–500 are shown here. See Figure 7.11 for later steps in the chain.

The first method to detect unrepresentativeness is a visual examination of the chain trajectory. A graph of the sampled parameter values as a function of step in the chain is called a *trace plot*. Examples appeared previously in Figure 7.4, and new examples appear in Figures 7.10 and 7.11. One way to enhance the visibility of unrepresentative parts of the chain is to superimpose two or more chains (that have been sampled with independent pseudo-random numbers). If the chains are all representative of the posterior distribution, they should overlap each other. Figure 7.10 shows the early steps of three MCMC trajectories started at different initial values. The upper-left panel of Figure 7.10 shows the trace plot. The vertical axis is the parameter value, and the horizontal axis is the step in the chain, labeled as "iterations." The trajectories reveal that it takes a few hundred steps for the three chains to converge to the same region of the parameter. This visual assessment suggests that the first several hundred steps of the chain should be excluded from the sample because they are not representative. The preliminary steps, during which the chain moves from its unrepresentative initial value to the modal region of the posterior, is called the *burn-in* period. For realistic applications, it is routine to apply a burn-in period of several hundred to several thousand steps. Later steps in the chains are shown in Figure 7.11. In particular, the upper-left panel of Figure 7.11

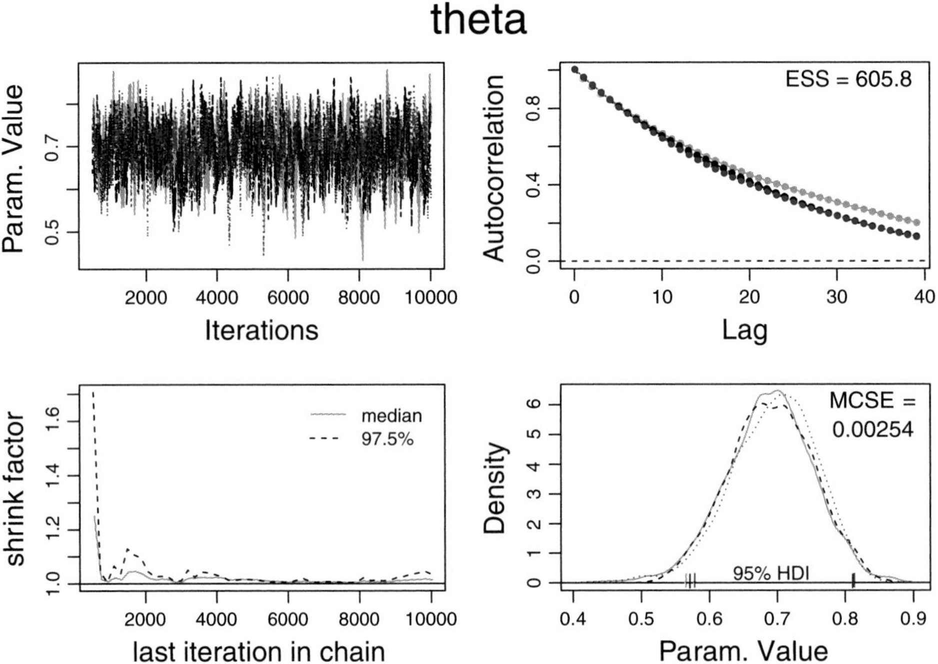

Figure 7.11 Illustration of MCMC diagnostics. Three chains were generated by starting a Metropolis algorithm at different initial values, with proposal SD=0.02 (cf. Figure 7.4) for data $z = 35$, $N = 50$. Steps 500–10,000 are shown here. See Figure 7.10 for earlier steps in the chain.

shows the trace plot, where it can be seen that the three chains meander fairly smoothly and overlap each other. If any chain were isolated from the others, it would be a sign that convergence had not been achieved. If any chain lingered for extended durations at (nearly) the same value, or changed values only very gradually, it might also be a sign of failure to converge. In the present case, fortunately, the chains are thoroughly overlapping, which is a good sign that suggests representativeness. But, the chains do meander relatively slowly, which is a sign of inefficiency (to be discussed more below). It is important to understand that the logic of checking for representativeness goes only one way: If the chains are representative, then they should overlap and mix well. But the converse is not necessarily true. The chains might overlap, but all be stuck in the same unrepresentative part of parameter space. Fortunately, this is rarely a problem for moderately complex models with chains started at reasonable initial values.

Another useful visual representation appears in the lower-right panels of Figures 7.10 and 7.11. These plots show smoothed histograms of the parameter values sampled in each chain. Smoothed histograms are also called *density plots*. Unlike histograms, which show the exact proportion of points in each histogram bin, density plots average across overlapping intervals to produce a smooth representation of probability density. (You can learn about the details by typing `?density` at the command line in R.) The lower-right panel of Figure 7.10 shows that the density plots of the three chains do *not* overlap very well during the burn-in period. This is a clear visual indicator that the chains have not converged. On the other hand, the lower-right panel of Figure 7.11 shows that the density plots of the three chains *do* overlap well after the burn-in period. This suggests, but does not guarantee, that the chains are producing representative values from the posterior distribution. Notice also that the density plots display the estimated 95% HDI for each chain. The HDI limits are slightly different for each chain, of course, because each chain is a finite random sample from the posterior distribution. In the limit, for infinite chains, the HDI limits of different chains will all converge to the same values. For finite chains, the display provides a visual impression of the variability in the estimates due to the finite sample size of the chain. (The meaning of "MCSE," displayed in the density plots, is explained below.)

Besides the visual checks of convergence, there are also numerical checks. One popular numerical check is a measure of how much variance there is between chains relative to how much variance there is within chains. The idea is that if all the chains have settled into a representative sampling, then the average difference between the chains should be the same as the average difference (across steps) within the chains. But, if one or more chains is orphaned or stuck, it will increase the between-chain variance relative to the within-chain variance. The lower-left panels of Figures 7.10 and 7.11 show plots of this measure. You can see that during the burn-in period (Figure 7.10), the measure greatly exceeds 1.0. After the burn-in period (Figure 7.11), the measure quickly gets very close to 1.0.

The specific numerical measure is called the Gelman-Rubin statistic (Gelman & Rubin, 1992), or the Brooks-Gelman-Rubin statistic (Brooks & Gelman, 1998), or the "potential scale reduction factor," or simply the "shrink factor" as plotted in Figures 7.10 and 7.11. Intuitively, its value is 1.0 if the chains are fully converged, but its value is larger than 1.0 if there are orphaned or stuck chains. As a heuristic, if the Gelman-Rubin statistic is greater than 1.1 or so, you should worry that perhaps the chains have not converged adequately. The exact definition of the Gelman-Rubin statistic involves a lot of details that are not essential for the purposes of this book. Therefore, interested readers are encouraged to learn more by reading the original articles or secondary summaries such as Gill (2009, p. 478) or Ntzoufras (2009, p. 143). The plots in the left columns of Figures 7.10 and 7.11 were produced by functions from the `coda` package created by Martyn Plummer et al. Later in the book, you will be introduced to the JAGS system for MCMC sampling, also created by Martyn Plummer, and the `coda` package will come along with the JAGS system. Meanwhile, another way to learn more about the Gelman-Rubin statistic is by installing the `coda` package into R. At R's command line, type `install.packages("coda")`, then load it into memory by typing `library(coda)`, and then find out about the diagnostic measure by typing `?gelman.diag`.

7.5.2. MCMC accuracy

After we have some assurance that the chains are genuinely representative samples from the posterior distribution, the second main goal is to have a large enough sample for stable and accurate numerical estimates of the distribution. The larger the sample, the more stable and accurate (on average) will be the estimates of the central tendency and HDI limits. But, as we saw in Figure 7.4 (p. 159), some chains can be more clumpy than others. Successive steps in a clumpy chain do not provide independent information about the parameter distribution.

What we need, therefore, are measures of chain length and accuracy that take into account the clumpiness of the chain. And for that, we need a measure of clumpiness. We will measure clumpiness as *autocorrelation*, which is simply the correlation of the chain values with the chain values k steps ahead. There is a different autocorrelation for each choice of k.

Figure 7.12 shows an example of computing autocorrelation. The upper panels show an MCMC chain of 70 steps, superimposed with the same chain translated a certain number of steps ahead. The vertical axis plots the parameter value, and the horizontal axis plots the step in the chain, here labeled as the "index." The number of steps between the chain and its superimposed copy is called the *lag*. The three columns show three different lags. The middle row shows the values of the original chain (on the vertical axis) plotted against the lagged values of the chain (on the horizontal axis). The scatter plot makes it easy to visualize the correlation of the original and lagged values. To help understand the correspondence of the upper row and middle row, a point in each scatter plot is marked by a square, and the corresponding values in the upper row are framed by a rectangle. Specifically, the value of the chain at about index 50 is nearly 16. At a lag of 10 (about index 60), the value of the chain is about 13. The square in the middle-right panel surrounds the point plotted at about $\langle 13, 16 \rangle$, and the rectangle in the upper-right panel surrounds those values at about index 50 in the original chain.

The autocorrelation can be computed for any lag of interest. The *autocorrelation function* is the autocorrelation across a spectrum of candidate lags. For our applications, we are typically interested in lags from around 1 to 20 or so. The lower panel of Figure 7.12 shows the autocorrelation function for the chain. Because lags are discrete, the autocorrelations are plotted as bars at each integer lag. The autocorrelation at lag k is denoted ACF(k), and you can see the correspondence of the ACF annotating the three scatterplots and the bar heights. The plot of the ACF reveals that this chain is highly autocorrelated, which means that it changes only gradually from step to step. In other words, this chain is fairly clumpy.

ACFs were also displayed in Figures 7.10 and 7.11, in their upper-right panels. There are multiple chains being plotted, with a distinct ACF for each chain. Instead of using bar graphs to plot the ACFs, the figures use points connected by lines because it is easier to

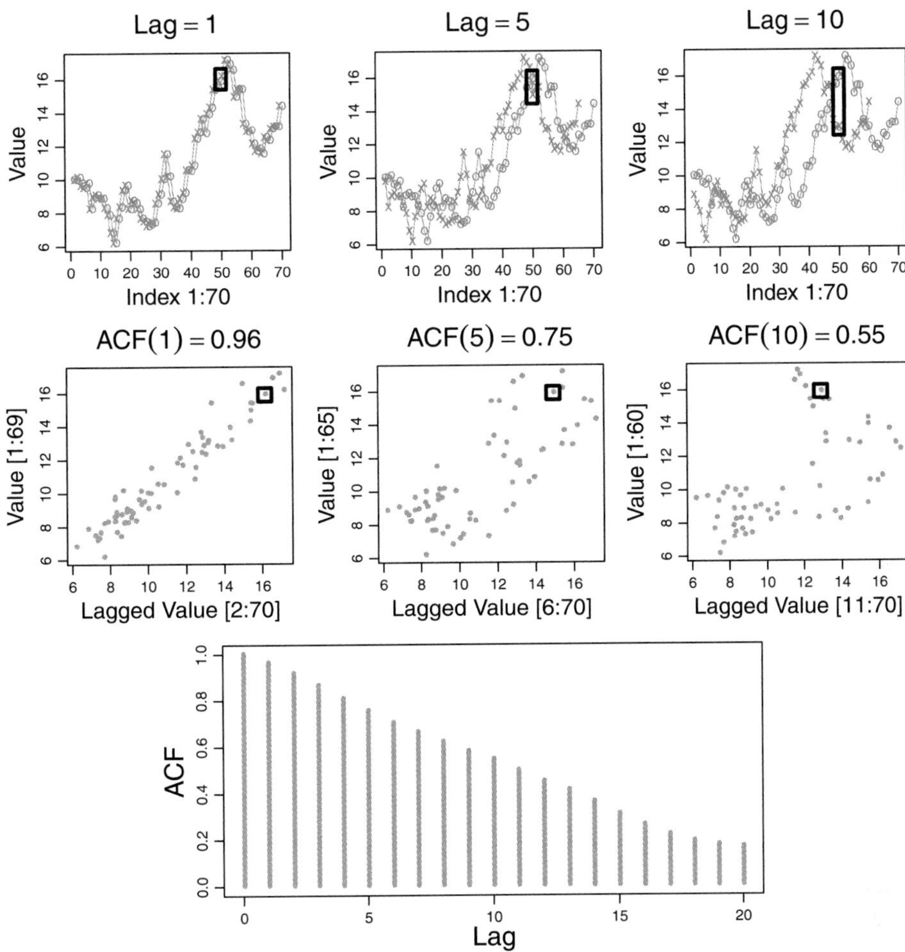

Figure 7.12 Autocorrelation of a chain. Upper panels show examples of lagged chains. Middle panels show scatter plots of chain values against lagged chain values, with their correlation annotated. Lowest panel shows the autocorrelation function (ACF).

see the superimposed ACFs. You can see that the chains are highly autocorrelated, insofar as the autocorrelations remain well above zero for large lags. Therefore, the parameter values at successive steps in the chain are not providing independent information about the posterior distribution, because each successive step is partially redundant with the previous step. The upshot is that it takes a lot of steps of these chains to generate independent information about the posterior distribution.

We would like some measure of how much independent information there is in autocorrelated chains. In particular, we can ask, what would be the sample size of a completely *non*-autocorrelated chain that yielded the same information? An answer to this question is provided by a measure called the *effective sample size* (proposed by Radford Neal in Kass, Carlin, Gelman, & Neal, 1998, p. 99), which divides the actual sample size by the amount of autocorrelation. Formally, denote the actual number of steps in the chain as N. The ESS is

$$\text{ESS} = N \Bigg/ \left(1 + 2\sum_{k=1}^{\infty} \text{ACF}(k)\right) \tag{7.11}$$

where ACF(k) is the autocorrelation of the chain at lag k. For practical computation, the infinite sum in the definition of ESS may be stopped when $\text{ACF}(k) < 0.05$ (because, typically, $\text{ACF}(k+1) < \text{ACF}(k)$). In R, the ESS can be computed by the `effectiveSize` function in the `coda` package (which uses a different algorithm than Equation 7.11).

The upper-right panels of Figures 7.10 and 7.11 annotate the ACFs with the *total* ESS across the multiple chains. For example, in Figure 7.11, there are three chains, each with $N = 9{,}500$ (i.e., each chain goes from "iteration" 500–10,000), yielding 28,500 steps overall. But the autocorrelation is so high that the ESS is only 605.8. The first decimal value is included in the display of ESS merely as a reminder that the ESS is an estimated continuous number, not to be confused with an actual sample.

How big should the ESS be for an accurate and stable picture of the posterior distribution? The answer depends on which detail of the posterior distribution you want a handle on. For aspects of the distribution that are strongly influenced by dense regions, such as the median in unimodal distributions, the ESS does not need to be huge. But for aspects of the distribution that are strongly influenced by sparse regions, such as the limits of the 95% HDI, the ESS needs to be relatively large. The reason is that sparse regions of the distribution are relatively rarely sampled in the chain, and therefore, a long chain is required to generate a high-resolution picture of sparse regions such as 95% HDI limits. One simple guideline is this: For reasonably accurate and stable estimates of the limits of the 95% HDI, an ESS of 10,000 is recommended. This is merely a heuristic based on experience with practical applications, it is not a requirement. If accuracy of the HDI limits is not crucial for your application, then a smaller ESS may be sufficient.

To get an intuition for the (in-)stability of the estimates of the 95% HDI limits, we will repeatedly generate MCMC chains from a known distribution, which has precisely known true 95% HDI limits. In particular, a standardized normal distribution has 95% HDI limits at very nearly -1.96 and $+1.96$. We will repeatedly generate MCMC chains from the normal distribution, each time using an ESS of 10,000. For each MCMC sample, we will estimate the 95% HDI. Then we will look at the estimates and assess how much they depart from the known true values.

Figure 7.13 shows the results of 50,000 repetitions (each using an ESS of 10,000). The upper panels show that the estimate of each HDI limit, where it can be seen that

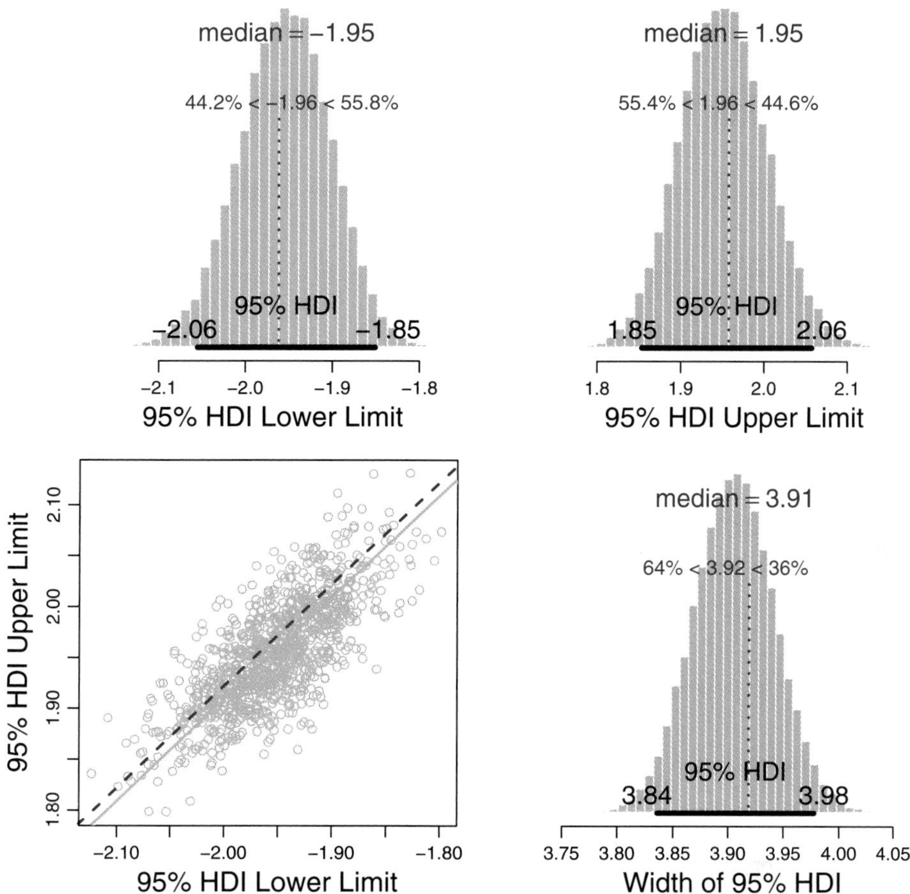

Figure 7.13 Estimated 95% HDI limits for random samples from a standardized normal distribution that have an *ESS* of *10,000*. Repeated runs yield a distribution of estimates as shown here; there were 50,000 repetitions. Upper panels show estimate of HDI limits. Lower panels show estimate of HDI width. True values are indicated by dashed lines.

the median estimate of ±1.95 is a little less extreme than the true value of ±1.96. The distribution of the estimated HDI limit has a SD of about 0.053. The sample of parameter values should have a SD of 1.0 because it comes from a standardized normal. Hence, the SD of the estimate of the 95% HDI limit, when using an ESS of 10,000, is roughly 5% of the SD of the MCMC chain.

The width of the HDI is slightly underestimated, but not by as much as might be suggested by the marginal distributions of the HDI limits. The estimates of the HDI limits are strongly correlated, as shown in the lower-left panel of Figure 7.13. (See Section 12.1.2.1, p. 340, for more discussion.) The distribution of the estimated widths is shown in the lower right panel, where it can be seen that the median estimated width

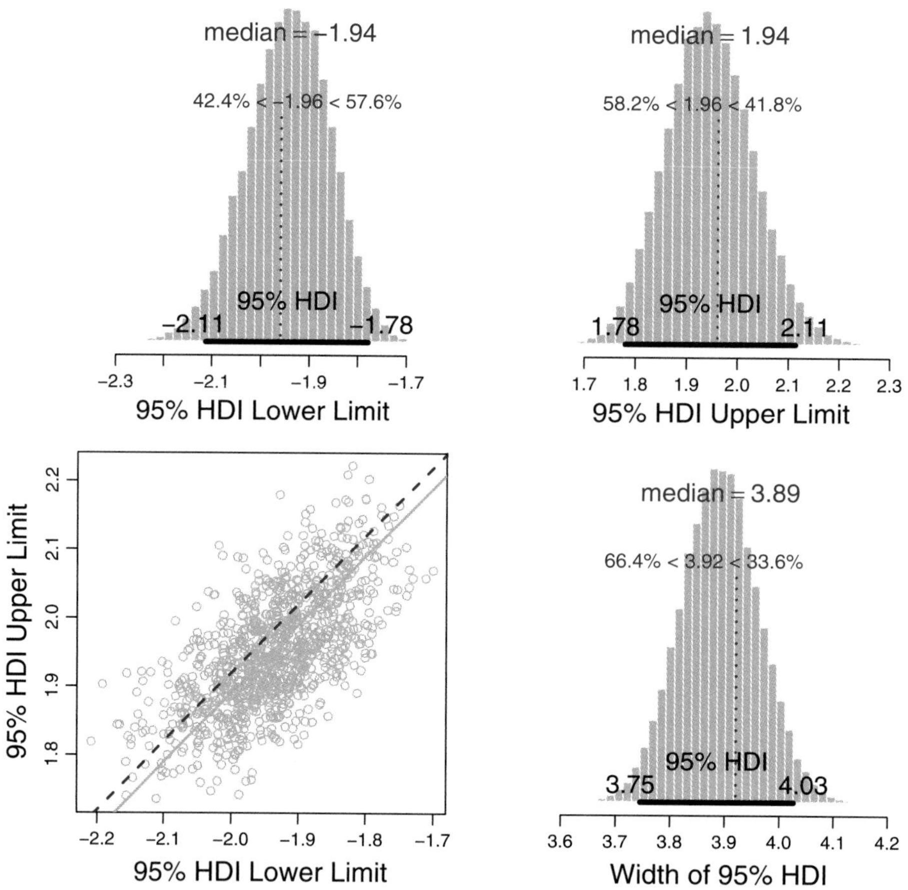

Figure 7.14 Estimated 95% HDI limits for random samples from a standardized normal distribution that have an *ESS of only 2,500*. Repeated runs yield a distribution of estimates as shown here; there were 50,000 repetitions. Upper panels show estimate of HDI limits. Lower panels show estimate of HDI width. True values are indicated by dashed lines.

is only slightly less than the true width. The true width of the 95% HDI is 3.920, and the median of the estimated widths is 3.907, which indicates that the estimated width is about 99.7% of the true width in this case.

The SD of the HDI estimate gets bigger for skewed tails of distributions, and for smaller ESS. For instance, when the ESS is only 2,500 instead of 10,000, the estimated HDI is more unstable across repeated Monte Carlo runs, as shown in Figure 7.14. When compared with Figure 7.13, it can be seen that the smaller ESS produces, on average, slightly worse underestimates of the HDI limits and width, and less stable estimates.

Another useful measure of the effective accuracy of the chain is the Monte Carlo standard error (MCSE). First, some background concepts and terminology. Consider randomly sampling values $x_1, \ldots, x_N$ from a normal distribution, and computing the mean of the sample, $\overline{x} = \frac{1}{N}\sum_{i=1}^{N} x_i$. The sample mean will be a noisy indicator of the underlying mean μ of the normal generating distribution. If we sample N values many times, sometimes the sample mean $\overline{x}$ will be greater than μ and sometimes less than μ. The SD of the sample mean, across repetitions, is called the *standard error* of the sample mean, and its estimated value is simply $\text{SE} = \text{SD}/\sqrt{N}$, where SD is the standard deviation of the sample. Thus, as the sample size increases, the standard error decreases. In other words, the bigger the sample, the less noise there is in the estimate of the underlying mean, and the standard error itself provides a quantitative suggestion of how big the estimation noise is. The notion of standard error for independent values is extended to Markov chains merely by substituting the ESS for the actual sample size. Thus, a simple formulation of the MCSE is

$$\text{MCSE} = \text{SD}/\sqrt{\text{ESS}} \tag{7.12}$$

where SD is the standard deviation of the chain and ESS is as defined in Equation 7.11. This is a simple version of MCSE; for more elaborate considerations see Flegal, Haran, and Jones (2008).

Examples of the MCSE are displayed in the lower-right panels of Figures 7.10 and 7.11. The MCSE indicates the estimated SD of the sample mean in the chain, on the scale of the parameter value. In Figure 7.11, for example, despite the small ESS, the mean of the posterior appears to be estimated very stably.

Here is a recapitulation regarding accuracy and stability of MCMC results. Visual inspection of the trace plots and density plots, and the Gelman-Rubin statistic, can suggest whether the burn-in period has been suitably passed. Second, those indicators can also suggest whether or not the chains are well mixed and representative of the posterior. Remember, the diagnostics logically can only probabilistically indicate violations of representativeness and cannot guarantee representativeness. Next, the measures of ESS and MCSE suggest how stable and accurate the chain is. As a heuristic, if you want reasonable stability in the estimates of the limits of the 95% HDI, an ESS of (at least) 10,000 is desirable. If you want a particular accuracy in the estimate of the posterior mean, consult the MCSE, which is interpreted on the scale of the parameter.

7.5.3. MCMC efficiency

The third of our main goals for MCMC is efficiency, so we do not exceed our patience and computing power. It is often the case in realistic applications that there is strong autocorrelation for some parameters, and therefore, an extremely long chain is required to achieve an adequate ESS or MCSE. There are various ways to (attempt to) improve the efficiency of the MCMC process.

- Run chains on parallel hardware. Most new computers have four or more processors, and each can be recruited to run a chain simultaneously, in parallel. Running parallel chains does not improve the amount of information per step in each chain, but it does improve the amount of information per increment of real time. The package `runjags` is nice for running parallel chains, as we will see in Chapter 8.
- Adjust the sampling method; for example, use a Gibbs sampler instead of a Metropolis sampler, as illustrated in the improvement in ESS from Figures 7.6 to 7.8. This approach requires great knowledge of various samplers and what types of model structures and substructures work best with which samplers. We will not explore these nuances in this book and will instead let sophisticated software tackle this issue for us. One sampling method that can be relatively efficient is Hamiltonian Monte Carlo, which is implemented in the Stan software introduced in Chapter 14.
- Change the parameterization of the model. In some cases, the various parameters of the model can be algebraically re-expressed in equivalent different forms but MCMC operates more efficiently in one form than another. For example, two parameters α and β might be re-expressed as $\mu = (\alpha + \beta)/2$ and $\delta = (\alpha - \beta)/2$, hence $\alpha = \mu + \delta$ and $\beta = \mu - \delta$. (This particular reparameterization might have little effect on MCMC sampling but is mentioned to illustrate the idea.) One application of this idea is mean-centering of data in regression analysis, as we will see in Chapter 17. We will see few other examples of this approach because applications can require a nuanced understanding of technicalities specific to individual models, which goes beyond the intended purpose of this book.

A method for reducing autocorrelation that does not improve efficiency is *thinning* the chain. In thinning, only every kth step in the chain is stored in the computer's memory. For example, when thinning to every 10th step, only steps 1, 11, 21, 31, and so on, are retained. The resulting thinned chain is less autocorrelated than the original complete chain. But the thinned chain also has less information than the original chain, and estimates from a thinned chain are (on average) less stable and accurate than from the original chain (Link & Eaton, 2012). After all, each step in the original chain did provide some new information. And, of course, the thinned chain takes just as many steps as the original chain to generate, so there is no savings in time during the generation of the chain. The only reason to thin a chain is if storing the full original chain would take too much computer memory, or if subsequent processing of the full original chain would take too much time.

7.6. SUMMARY

Let's regain perspective on the forest of Bayesian inference after focusing on the trees of MCMC. Recall that the overarching goal of Bayesian analysis is identifying the credibility of parameter values in a descriptive model of data. Bayes' rule provides an exact mathematical formulation for the posterior distribution on the parameter values. But the exact form requires evaluation of an integral that might be intractable for realistically complex models. Therefore, we approximate the posterior distribution, to arbitrarily high accuracy, using MCMC methods. Because of recent developments in MCMC algorithms, software that cleverly applies them in complex models, and hardware that runs them incredibly quickly, we can now use MCMC methods to analyze realistically complex models that would have been impossible only a few decades ago.

This chapter focused on explaining the concepts of MCMC, applied to simple examples that illustrate the concepts. (Software for realistic applications is introduced in the next chapter.) This chapter introduced a simple case of the Metropolis algorithm in the guise of a politician hopping across adjacent islands. Some basic mathematical intuitions were established in that case of discrete parameter values. A more general form of the Metropolis algorithm was then applied to estimating a continuous parameter, namely, the bias in a coin. After that, Gibbs sampling was introduced and applied to the estimation of the difference of biases from two coins. The Metropolis and Gibbs methods are two types of MCMC samplers. There are many others that are not discussed in this chapter. All the MCMC methods converge to an accurate representation of the posterior distribution in the infinitely long run, but they differ in how efficiently they generate representative samples for different models in finite runs.

The chapter concluded, in Section 7.5, with a discussion of diagnostics and heuristics for assessing whether an MCMC chain is representative of the posterior distribution and is sufficiently accurate. Figures 7.10 and 7.11 showed visual and numerical representations of MCMC chains, including trace plot, density plot, autocorrelation plot, and plot of the Gelman-Rubin statistic, along with two numerical indicators, the ESS and MCSE.

In subsequent chapters, we will use software that *automatically* sets up MCMC samplers for complex models. We will not need to manually tune proposal distributions in Metropolis samplers. We will not need to derive conditional distributions for Gibbs samplers. We will not need to figure out which of various sampling algorithms should be applied. But we will need to evaluate the output of the MCMC samplers and decide whether it is sufficiently representative and accurate. Thus, it is crucial to retain from this chapter the concepts of what MCMC does, and the details of how to assess it.

7.7. EXERCISES

Look for more exercises at https://sites.google.com/site/doingbayesiandataanalysis/

Exercise 7.1. [Purpose: Experiment with the Metropolis algorithm as displayed in Figure 7.4.] Open the program named `BernMetrop.R` from the files that accompany this book. The script implements a Metropolis algorithm for Figure 7.4 Midway through the script, you will find a line that specifies the SD of the proposal distribution:

```
proposalSD = c(0.02,0.2,2.0)[2]
```

The line may look strange but it's merely a vector of constants with an index at the end to specify which component should be used. Thus, it's a simple way of specifying three options and then selecting one option. Run the script three times, once with each option (i.e., once with `[1]`, once with `[2]`, and once with `[3]`). *There is also a line that specifies the seed for the random number generator; comment it out so that you get a different trajectory than the ones shown in Figure 7.4.* Notice at the end of the script that you can specify the format of the graphic file for saving the resulting graph. Include the graphs in your write-up and describe whether they show similar behavior as the corresponding trajectories in Figure 7.4. Be sure to discuss the ESS.

Exercise 7.2. [Purpose: To explore the autocorrelation function in Figure 7.12.] At the end of the script `BernMetrop.R`, add these lines:

```
openGraph(height=7,width=3.5)
layout(matrix(1:2,nrow=2))
acf( acceptedTraj , lag.max=30 , col="skyblue" , lwd=3 )
Len = length( acceptedTraj )
Lag = 10
trajHead = acceptedTraj[ 1        : (Len-Lag) ]
trajTail = acceptedTraj[ (1+Lag) :  Len       ]
plot( trajHead , trajTail , pch="." , col="skyblue" ,
      main=bquote( list( "Prpsl.SD" == .(proposalSD) ,
                        lag == .(Lag) ,
                        cor == .(round(cor(trajHead,trajTail),3)))) )
```

(A) Before each line, add a comment that explains what the line does. Include the commented code in your write-up.

(B) Repeat the previous exercise, with the lines above appended to the script. Include the resulting new graphs in your write-up. For each run, verify that the height of the ACF bar at the specified lag matches the correlation in the scatterplot.

(C) When the proposal distribution has SD=2, why does the scatter plot have a dense line of points on the diagonal? (*Hint:* Look at the trajectory.)

Exercise 7.3. [Purpose: Using a multimodal prior with the Metropolis algorithm, and seeing how chains can transition across modes or get stuck within them.] In this exercise, you will see that the Metropolis algorithm operates with multimodal distributions.

(A) Consider a prior distribution on coin bias that puts most credibility at 0.0, 0.5, and 1.0, which we will formulate as $p(\theta) = (\cos(4\pi\theta) + 1)^2/1.5$.

(B) Make a plot of the prior. *Hint:* `theta = seq(0,1,length=501); plot (theta , (cos(4*pi*theta)+1)^2/1.5 )`

(C) In the script `BernMetrop.R`, find the function definition that specifies the prior distribution. Inside that function definition, comment out the line that assigns a beta density to `pTheta`, and instead put in a trimodal prior like this:

```
#pTheta = dbeta( theta , 1 , 1 )
pTheta = (cos(4*pi*theta)+1)^2/1.5
```

To have the Metropolis algorithm explore the prior, we give it empty data. Find the line in the script that specifies the data and set `myData = c()`. Run the script, using a proposal SD=0.2. Include the graphical output in your write-up. Does the histogram of the trajectory look like the graph of the previous part of the exercise?

(D) Repeat the previous part but now with `myData = c(0,1,1)`. Include the graphical output in your write-up. Does the posterior distribution make sense? Explain why.

(E) Repeat the previous part but now with proposal SD=0.02. Include the graphical output in your write-up. Does the posterior distribution make sense? Explain why *not*; what has gone wrong? If we did not know from the previous part that this output was unrepresentative of the true posterior, how could we try to check? *Hint:* See next part.

(F) Repeat the previous part but now with the initial position at 0.99: `trajectory[1] = 0.99`. In conjunction with the previous part, what does this result tell us?

CHAPTER 8

JAGS

Contents

I'm hurtin' from all these rejected proposals;
My feelings, like peelings, down garbage disposals.
S'pose you should go your way and I should go mine,
We'll both be accepted somewhere down the line.[1]

8.1. JAGS AND ITS RELATION TO R

JAGS is a system that automatically builds Markov chain Monte Carto (MCMC) samplers for complex hierarchical models (Plummer, 2003, 2012). JAGS stands for just another Gibbs sampler, because it succeeded the pioneering system named BUGS, which stands for Bayesian inference using Gibbs sampling (Lunn, Jackson, Best, Thomas, & Spiegelhalter, 2013; Lunn, Thomas, Best, & Spiegelhalter, 2000). In 1997, BUGS had a Windows-only version called WinBUGS, and later it was reimplemented in OpenBUGS which also runs best on Windows operating systems. JAGS (Plummer, 2003, 2012)

[1] This chapter is about the software package "JAGS," which stands for Just Another Gibbs Sampler. In Gibbs sampling, unlike Metropolis sampling, all proposed jumps are accepted, but all jumps are along a line parallel to a parameter axis. The quatrain personifies two different parameters in Gibbs sampling: they go orthogonal directions but both are accepted somewhere down the line.

Doing Bayesian Data Analysis, 2nd Edition
http://dx.doi.org/10.1016/B978-0-12-405888-0.00008-8

retained many of the design features of BUGS, but with different samplers under the hood and better usability across different computer-operating systems (Windows, MacOS, and other forms of Linux/Unix).

JAGS takes a user's description of a hierarchical model for data, and returns an MCMC sample of the posterior distribution. Conveniently, there are packages of commands that let users shuttle information in and out of JAGS from R. Figure 8.1 shows the packages that we will be using for this book. In particular, in this chapter you will learn how to use JAGS from R via the rjags and runjags packages. Later, in Chapter 14, we will take a look at the Stan system.

To install JAGS, see http://mcmc-jags.sourceforge.net/. (Web site addresses occasionally change. If the address stated here does not work, please search the web for "Just Another Gibbs Sampler." Be sure the site is legitimate before downloading anything to your computer.) On the JAGS web page, follow the links under "Downloads." As you navigate through the links, you want to be sure to follow the ones for JAGS and for Manuals. Because installation details can change through time, and because the procedures are different for different computer platforms, I will not recite detailed installation instructions here. If you encounter problems, please check that you have the latest versions of R and RStudio installed before installing JAGS, and then try installing JAGS again. Note that merely saving the JAGS installation executable on your computer does not install it; you must run the executable file to install JAGS, analogous to installing R and RStudio.

Once JAGS is installed on your computer, invoke R and, at its command line, type `install.packages("rjags")` and `install.packages("runjags")`. As suggested by Figure 8.1, these packages let you communicate with JAGS via commands in R.

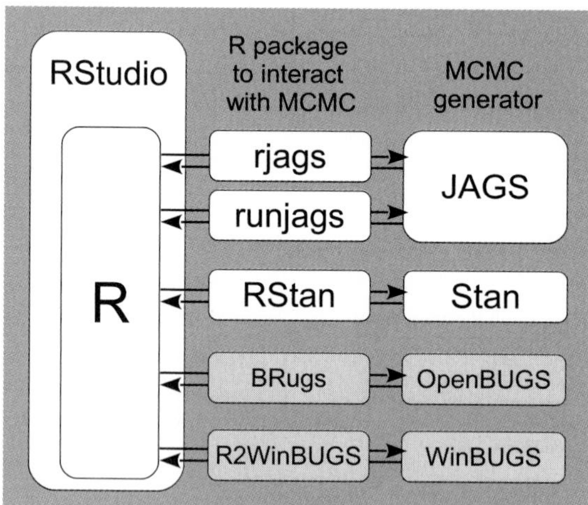

Figure 8.1 Relation of R programming language to other software tools. On the left, RStudio is an editor for interacting with R. The items on the right are various programs for generating MCMC samples of posterior distributions. The items in the middle are packages in R that interact with the MCMC generators.

8.2. A COMPLETE EXAMPLE

Because of its simplicity and clarity, we have repeatedly considered the Bayesian estimation of the bias of a coin. The idea was introduced in Section 5.3, where the problem was addressed via grid approximation. Then, Chapter 6 showed how Bayes' rule could be solved analytically when the prior distribution is a beta density. Next, Section 7.3.1 showed how the posterior distribution could be approximated with a Metropolis sampler. In the present section, we will again approximate the posterior distribution with an MCMC sample, but we will let the JAGS system figure out what samplers to use. All we have to do is tell JAGS the data and the model.

A parameterized model consists of a likelihood function, which specifies the probability of data given the parameter values, and a prior distribution, which specifies the probability (i.e., credibility) of candidate parameter values without taking into account the data. In the case of estimating the bias of a coin, the likelihood function is the Bernoulli distribution (recall Equation 6.2, p. 125). For MCMC, we must specify a prior distribution that can be computed quickly for any candidate value of the parameter, and we will use the beta density (recall Equation 6.3, p. 127). Here, we introduce some new notation for these familiar likelihood and prior distributions. To indicate that the data, y_i, come from a Bernoulli distribution that has parameter θ, we write:

$$y_i \sim \text{dbern}(\theta) \tag{8.1}$$

Spoken aloud, Equation 8.1 would be stated, "y_i is distributed as a Bernoulli distribution with parameter θ." We should also say that Equation 8.1 applies for all i. Equation 8.1 is merely a mnemonic way to express the relation between the data and the model parameters; what it really means mathematically is the formula in Equation 6.2 (p. 125). The mnemonic forms will be used extensively throughout the remainder of the book. To indicate that the prior distribution is a beta density with shape constants A and B, we write:

$$\theta \sim \text{dbeta}(A, B) \tag{8.2}$$

Spoken aloud, Equation 8.2 would be stated, "θ is distributed as a beta density with shape constants A and B." What it really means mathematically is the formula in Equation 6.3 (p. 127), but this mnemonic form will be very useful.

Figure 8.2 shows a diagram that illustrates Equations 8.1 and 8.2. Each equation corresponds to an arrow in the diagram. Starting at the bottom, we see that the data value y_i is pointed at by an arrow coming from an icon of a Bernoulli distribution. The horizontal axis is not marked, to reduce clutter, but implicitly the two vertical bars are located at $y = 0$ and $y = 1$ on the horizontal axis. The heights of the two bars indicate the probability of $y = 0$ and $y = 1$, given the parameter value θ. The height of the bar at $y = 1$ is the value of the parameter θ, because $p(y = 1|\theta) = \theta$. The arrow pointing to y_i is labeled with the "is distributed as" symbol, "$\sim$," to indicate the relation between y_i and the distribution. The arrow is also labeled with "i" to indicate that this relation holds for all instances of y_i. Continuing up the diagram, we see that

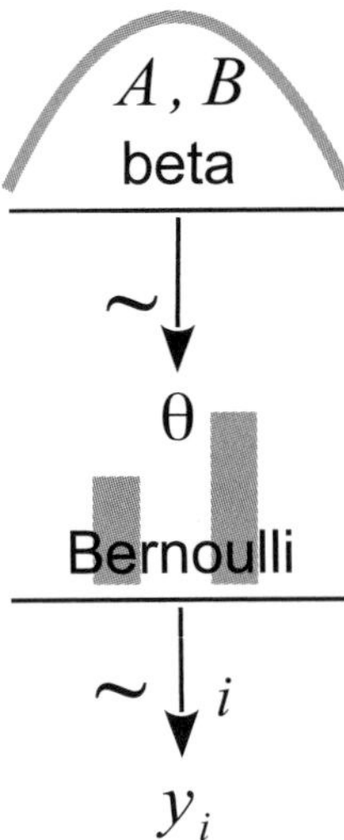

Figure 8.2 Diagram of model with Bernoulli likelihood and beta prior. The pictures of the distributions are intended as stereotypical icons, and are not meant to indicate the exact forms of the distributions. Diagrams like this should be scanned from the bottom up, starting with the data y_i and working upward through the likelihood function and prior distribution. Every arrow in the diagram has a corresponding line of code in a JAGS model specification.

the parameter θ is pointed at by an arrow coming from an icon of a beta distribution. The horizontal axis is not marked, to reduce clutter, but implicitly the icon suggests that values of θ on the horizontal axis are restricted to a range from 0 to 1. The specific shape of the beta distribution is determined by the values of the shape parameters, A and B. The icon for the beta distribution is based roughly on a dbeta(2, 2) distribution, but this shape is merely suggestive to visually distinguish it from other distributions such as uniform, normal, and gamma.

Diagrams like Figure 8.2 should be scanned from the bottom up. This is because models of data always start with the data, then conceive of a likelihood function that describes the data values in terms of meaningful parameters, and finally determine a prior distribution over the parameters. Scanning the diagram from bottom to top may seem awkward at first, but it is necessitated by the conventions for plotting distributions: the dependent value is displayed on the bottom axis and the parameters of the function are best displayed above the bottom axis. If this book's written language were also to flow from bottom to top, instead of from top to bottom, then we would not feel any inconsistency.[2] To reiterate: when describing a model, the description logically

[2] There is another directional inconsistency between written English and mathematical notation that you are probably so familiar with that you forgot it exists. When writing an integer, ascending powers of 10 are written right-to-left instead of left-to-right. Therefore, during reading of English, when you encounter the leftmost digit of an integer you cannot determine what power of 10 it indicates until you find the rightmost digit and work your way backwards. If you can get used to scanning numbers right-to-left, you can get used to scanning hierarchical diagrams bottom-to-top.

flows from data to likelihood (with its parameters) to prior. If the representational medium is graphs of distributions, then the logical flow goes from bottom to top. If the representational medium is text such as English or computer code, then the logical flow goes from top to bottom.

Diagrams like the one in Figure 8.2 are very useful for two important reasons:

1. They help with conceptual clarity, especially as we deal with more complex models. The diagrams capture the conceptual dependencies among data and parameters without getting bogged down with details of the mathematical formulas for the distributions. Although Equations 8.1 and 8.2 fully specify relations of the variables in the model, the diagram explicitly spatially connects the corresponding variables so that it is easy to visually comprehend the chain of dependencies among variables. Whenever I am creating a model, I draw its diagram to be sure that I really have all its details thought out coherently.
2. The diagrams are very useful for implementing the model in JAGS, because every arrow in the diagram has a corresponding line of code in the JAGS model specification. We will soon see an example of specifying a model in JAGS.

We will now go through a complete example of the sequence of commands in R that generate an MCMC sample from JAGS for this model. The script that contains the commands is named `Jags-ExampleScript.R`, in the bundle of programs that accompany this book. You are encouraged to open that script in RStudio and run the corresponding commands as you read the following sections.

8.2.1. Load data

Logically, models of data start with the data. We must know their basic scale and structure to conceive of a descriptive model. Thus, the first part of a program for analyzing data is loading the data into R.

In real research, data are stored in a computer data file created by the research. A generic format is text with comma separated values (CSV) on each row. Files of this format usually have a file name with a `.csv` extension. They can be read into R with the `read.csv` function which was introduced in Section 3.5.1 (p. 53). The usual format for data files is one y value per row, with other fields in each row indicating identifying information for the y value, or covariates and predictors of the y value. In the present application, the data values are 1's and 0's from flips of a single coin. Therefore, each row of the data file has a single 1 or 0, with no other fields on each row because there is only one coin.

To bundle the data for JAGS, we put it into a `list` structure, which was introduced in Section 3.4.4 (p. 51). JAGS will also have to know how many data values there are altogether, so we also compute that total and put it into the `list`,[3] as follows:

[3] It is possible to compute the number of data values inside a JAGS `data` block instead of computing the value in R and delivering it to JAGS, but this aspect of JAGS model specification will not be introduced until later in the book.

```
myData = read.csv("z15N50.csv") # Read the data file; result is a data.frame.
y = myData$y          # The y values are the component named y.
Ntotal = length(y)  # Compute the total number of flips.
dataList = list(    # Put the information into a list.
  y = y ,
  Ntotal = Ntotal
)
```

The syntax in the `list` may seem strange, because it seems to make tautological statements: `y = y` and `Ntotal = Ntotal`. But the two sides of the equal sign are referring to different things. On the left side of the equal sign is the name of that component of the list. On the right side of the equal sign is the value of that component of the list. Thus, the component of the list that says `y = y` means that this component is named `y` (i.e., the left side of equal sign) and its value is the value of the variable `y` that currently exists in R, which is a vector of 0's and 1's. The component of the list that says `Ntotal = Ntotal` means that this component is named `Ntotal` (i.e., the left side of equal sign) and its value is the value of the variable `Ntotal` that currently exists in R, which is an integer.

The `dataList` thereby created in R will subsequently be sent to JAGS. The list tells JAGS the names and values of variables that define the data. Importantly, the names of the components must match variable names in the JAGS model specification, which is explained in the next section.

8.2.2. Specify model

The diagram in Figure 8.2 is very useful for implementing the model in JAGS, because every arrow in the diagram has a corresponding line of code in the JAGS model specification. The model specification begins with the key word `model`, and whatever follows, inside curly braces, is a textual description of the dependencies between the variables. Here is the diagram of Figure 8.2 expressed in JAGS:

```
model {
  for ( i in 1:Ntotal ) {
    y[i] ~ dbern( theta )  # likelihood
  }
  theta ~ dbeta( 1 , 1 )   # prior
}
```

The arrow at the bottom of Figure 8.2, pointing to y_i, is expressed in JAGS as `y[i] ~ dbern(theta)`. Because that relation is true for all instances of y_i, the statement is put inside a `for` loop. The `for` loop is merely a shortcut for telling JAGS to copy the statement for every value of i from 1 to `Ntotal`. Notice that the data-specifying variable names, `y` and `Ntotal`, match the names used in the `dataList` defined in the previous section. This correspondence of names is crucial for JAGS to know the values of the data in the model specification.

The next line in the model statement progresses logically to the specification of the prior, that is, the next arrow up the diagram in Figure 8.2. This particular version uses shape constants of $A = 1$ and $B = 1$ in the prior, thus: `theta ~ dbeta(1,1)`.

The model statement progresses logically from likelihood to prior only for the benefit of the human reader. JAGS itself does not care about the ordering of the statements, because JAGS does not execute the model statement as an ordered procedure. Instead, JAGS merely reads in the declarations of the dependencies and then compiles them together to see if they form a coherent structure. The model specification is really just a way of saying, "the ankle bone's connected to the shin bone" and "the shin bone's connected to the knee bone."[4] Both statements are true regardless of which one you say first. Thus, the following model specification, which declares the prior first, is equivalent to the model specification above:

```
model {
  theta ~ dbeta( 1 , 1 )    # prior
  for ( i in 1:Ntotal ) {
    y[i] ~ dbern( theta )   # likelihood
  }
}
```

While stating the prior first is not a logical problem for JAGS, stating the prior first can be a conceptual problem for humans, for whom it may be difficult to understand the role of `theta` before knowing it is a parameter in a Bernoulli function, and before knowing that the Bernoulli function is being used to describe dichotomous data.

To get the model specification into JAGS, we create the specification as a character string in R, then save the string to a temporary text file, and subsequently send the text file to JAGS. Here is the R code for setting the model specification as a string and writing it to a file arbitrarily named `TEMPmodel.txt`:

```
modelString = " # open quote for modelString
  model {
    for ( i in 1:Ntotal ) {
      y[i] ~ dbern( theta )
    }
    theta ~ dbeta( 1 , 1 )
  }
" # close quote for modelString
writeLines( modelString , con="TEMPmodel.txt" ) # write to file
```

[4] The bone linkages allude to the well-known spiritual song, Dem Bones (a.k.a., Dry Bones or Dem Dry Bones), composed by James Weldon Johnson (1871–1938). The lyrics of the song are not exactly as phrased here. The original song goes through the linkages from foot to head, using ascending pitches, and subsequently from head to foot, using descending pitches.

The bundling of the model specification into a string, and saving it to a file, is the same for every model we will ever write for JAGS. Only the model specification changes across applications.

You may have noticed that there was a comment inside modelString, above. JAGS employs comment syntax just like R. Therefore, we can embed comments in JAGS model specifications, which can be very useful in complicated models. It is worth repeating that the model specification is not interpreted by R, it is merely a character string in R. The string is sent to JAGS, and JAGS processes and interprets the model specification.

8.2.3. Initialize chains

This section is about how to specify the starting values of the MCMC chains. JAGS can do this automatically using its own defaults. Therefore, you do not have to fully comprehend this section on a first reading, but you should definitely be aware of it for subsequent review. Although JAGS can automatically start the MCMC chains at default values, the efficiency of the MCMC process can sometimes be improved if we intelligently provide reasonable starting values to JAGS. To do this, we have to figure out values for the parameters in the model that are a reasonable description of the data, and might be in the midst of the posterior distribution.

In general, a useful choice for initial values of the parameters is their maximum likelihood estimate (MLE). The MLE is the value of the parameter that maximizes the likelihood function, which is to say, the value of the parameter that maximizes the probability of the data. The MLE is a reasonable choice because the posterior distribution is usually not radically far from the likelihood function if the prior is noncommittal. For example, review Figure 7.5, p. 167, which showed a case in which the peak of the likelihood function (i.e., the MLE) was not far from the peak of the posterior distribution.

It turns out that for the Bernoulli likelihood function, the MLE is $\theta = z/N$. In other words, the value of θ that maximizes $\theta^z(1-\theta)^{(N-z)}$ is $\theta = z/N$. An example of this fact can be seen in the likelihood function plotted in Figure 6.3, p. 134. For our current application, the data are 0's and 1's in the vector y, and therefore z is sum(y) and N is length(y). We will name the ratio thetaInit, and then bundle the initial value of the parameter in a list that will subsequently get sent to JAGS:

```
thetaInit = sum(y)/length(y)
initsList = list( theta=thetaInit )
```

In initsList, each component must be named as a parameter in the JAGS model specification. Thus, in the list statement above, the component name, theta, refers to theta in the JAGS model specification. The initsList is subsequently sent to JAGS.

When there are multiple chains, we can specify different or identical initial values for each chain. The rjags package provides three ways of initializing chains. One way is to specify a single initial point for the parameters, as a single named `list` as in the example above, and having all chains start there. A second way is to specify a `list` of `lists`, with as many sub-`lists` as chains, and specifying specific initial values in each sub-`list`. A third way is to define a function that returns initial values when called. (Defining of functions in R was described in Section 3.7.3, p. 64.) The rjags package calls the function as many times as there are chains. An example of this third method is provided below.

Starting the chains together at the MLE is not universally recommended. Some practitioners recommend starting different chains at greatly dispersed points in parameter space, so that the chains' eventual convergence may be taken as a sign that burn-in has been properly achieved and that the full parameter space has been explored. In particular, this procedure is recommended for interpreting the Gelman-Rubin convergence statistic. In my experience, however, for the relatively simple models used in this book, appropriate burn-in and convergence is rarely a problem. And, for some complex models, if the chains are initialized at randomly dispersed arbitrary points, they might never converge during a reasonable finite run, with some chains being orphaned indefinitely (although orphaned chains are usually less of a problem in Stan than in JAGS).

A compromise approach is to start the chains at random points near the MLE. One way to do this is, for each chain, to resample from the data and compute the MLE for the resampled data. Resampled data will tend to have approximately the same proportion of heads, but sometimes more and sometimes less. Here is an example of a function that returns a named `list` with a different value for `theta` every time it is called:

```
initsList = function() {
  resampledY = sample( y , replace=TRUE )        # resample values from y
  thetaInit = sum(resampledY)/length(resampledY) # compute proportion (MLE)
  thetaInit = 0.001+0.998*thetaInit              # keep away from 0,1
  return( list( theta=thetaInit ) )              # return as a named list
}
```

The third line in the function, above, is a protective measure that keeps the initial value of `theta` in a valid range for JAGS. For example, a prior of beta($\theta|2,2$) has zero density at $\theta = 0$ and $\theta = 1$, and JAGS balks at zero prior probabilities, and therefore the chain cannot be started at those extremes.

To demonstrate how the `initsList` function works, suppose the data `y` consist of 75% 1's:

```
> y = c(rep(0,25),rep(1,75))
```

When the `initsList` function is called, it returns a list with a component named `theta` that has a value near 75%, as you can see in the following sequence of repeated calls:

```
> initsList()
$theta
[1] 0.70958
> initsList()
$theta
[1] 0.77944
> initsList()
$theta
[1] 0.72954
```

8.2.4. Generate chains

Now that we have assembled the data, composed the model specification, and established the initial values of the chains, we are ready to have JAGS actually generate the MCMC sample from the posterior distribution. We do this in three steps. The first step gets all the information into JAGS and lets JAGS figure out appropriate samplers for the model. The next step runs the chains for a burn-in period. Finally, the third step runs and records the MCMC sample that we will subsequently examine.

The first step is accomplished using the `jags.model` function from the rjags package. This function sends the model specification, data list, and initial-value list to JAGS and tells JAGS to figure out appropriate MCMC samplers. Here is an example of the command:

```
jagsModel = jags.model( file="TEMPmodel.txt" , data=dataList , inits=initsList ,
                        n.chains=3 , n.adapt=500 )
```

The first argument is the name of the file in which the model specification is stored, that was created at the end of Section 8.2.2. The `data` argument specifies the list of data that was created in Section 8.2.1. The `inits` argument specifies the list of initial values that was created in Section 8.2.3. If you want JAGS to create its own initial values for the chains, simply omit the `inits` argument entirely. The next two arguments specify the number of chains and the number of steps to take for adapting (or tuning) the samplers. These arguments default to 1 chain and 1000 adaptation steps if they are not specified by the user. The `jags.model` function returns an object that here has been named `jagsModel`. The object is a list of functions that encapsulate, in JAGS terms, the model and samplers. We will not delve into the `jagsModel` object. Instead, we merely pass it along to subsequent functions that generate the MCMC sample.

When executing the `jags.model` command, JAGS checks the model specification, data list, and initial values, for coherence and consistency. If anything is awry, JAGS will issue an error statement. Sometimes the error statements can be cryptic, but they usually contain enough information that you can track down the problem. Sometimes the problem is as simple as a misplaced parenthesis or brace. Sometimes a variable name will be misspelled. Or sometimes you might have a logical problem in the model, or impossible initial values.

After JAGS has created its model, we tell it to run the chains some number of steps to accomplish burn in. We do this by using the `update` command from package rjags:

```
update( jagsModel , n.iter=500 )
```

The first argument is the JAGS object previously created by `jags.model`. The argument `n.iter` is the number of iterations, or steps, to take in the MCMC chains. For the present simple application, only a short burn-in is specified. The `update` function returns no values, it merely changes the internal state of the `jagsModel` object. It does *not* record the sampled parameter values during the updating.

After burn-in, we tell JAGS to generate MCMC samples that we will actually use to represent the posterior distribution. The chains of the parameter values are arranged in a specialized format so that various functions from the `coda` package can be used to examine the chains. Therefore, the function that generates the MCMC samples is called `coda.samples` in the rjags package. Here is an example of its use:

```
codaSamples = coda.samples( jagsModel , variable.names=c("theta") ,
                            n.iter=3334 )
```

As in the `update` function, the first argument is the JAGS object previously created by `jags.model`. The argument `n.iter` is the number of iterations, or steps, taken by each chain. Here, `n.iter` is set to 3334, which will yield a total of 10,002 steps because there were three chains specified in the `jags.model` function. Crucially, the `variable.names` argument specifies which parameters will have their values recorded during the MCMC walk. JAGS will only record the trajectories of parameters that you explicitly tell it to! The `variable.names` argument must be a vector of character strings. In the present application, there is only one parameter, so the vector has only one element.

The result of the `coda.samples` function is a `coda`-formatted object, here named `codaSamples`, that contains all the sampled parameter values in all the chains. It is a list of matrices. Each component of the list corresponds to a chain, hence in the current example, which has three chains, the list has three components. In each component is a matrix, which has rows indexed by the step in the chain and has columns corresponding to each parameter.

8.2.5. Examine chains

When examining the MCMC samples, the first task is to check that the chains appear to be well mixed and suitably representative of the posterior distribution. Figure 8.3 shows diagnostic information for the θ parameter in our ongoing example. The trace plot in the

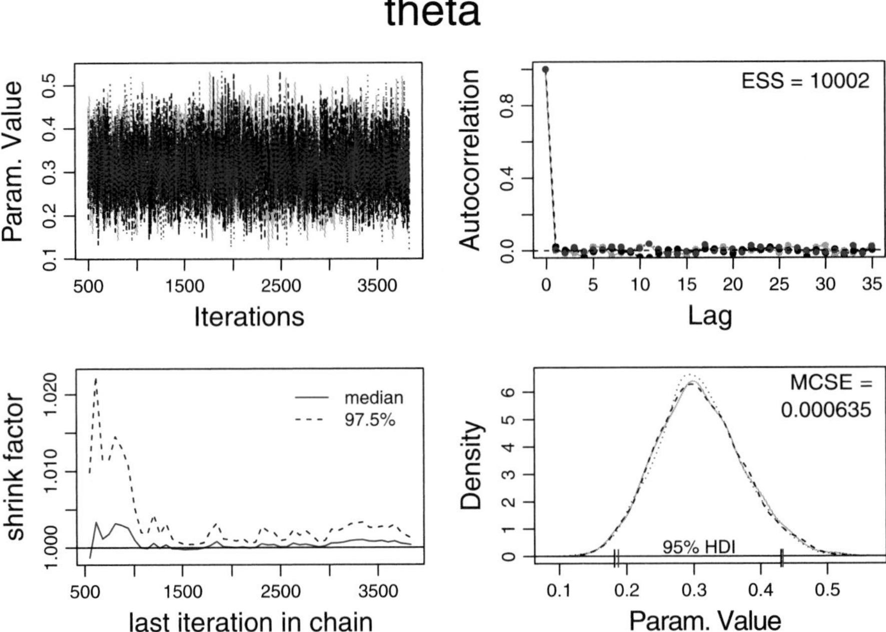

Figure 8.3 Convergence diagnostics for output of JAGS.

upper-left panel shows no signs of orphaned chains. The density plot in the lower-right shows that the three subchains are well super-imposed, which is echoed by the Gelman-Rubin statistic in the lower-left panel being very close to 1.0. The autocorrelation plot in the upper-right panel shows essentially zero autocorrelation for this simple model, and the effective sample size is essentially as large as the full length of the chain.

Figure 8.3 was made by a function I created, which, in turn, calls some functions from the `coda` package. The function I created is named `diagMCMC`, which stands for diagnostics of MCMC. It is defined in the script called `DBDA2E-utilities.R`. To use the function, its definition must first be loaded into R's working memory:

```
source("DBDA2E-utilities.R")
```

The `source` command will only work if the file, `DBDA2E-utilities.R`, is in R's current working directory. Otherwise, you must specify the directory path to the folder in which the file resides, such as, `source("C:/[yourpath]/DBDA2E- utilities.R")`. Then the diagnostic-plotting function itself is called:

```
diagMCMC( codaObject=codaSamples , parName="theta" )
```

The `diagMCMC` function has two arguments. The first specifies the MCMC coda object that was created by JAGS. The second argument specifies which parameter to examine.

In models with many parameters, the function is called repeatedly, once for each parameter of interest.

8.2.5.1 *The* `plotPost` *function*

Figure 8.4 shows the MCMC sample from the posterior distribution. It is much like the density plot of Figure 8.3, but plotted as a histogram instead of as a smoothed-density curve, and it is annotated with some different information. Figure 8.4 was created by a function I created called `plotPost`, which is defined in `DBDA2E-utilities.R`. The left panel of Figure 8.4 was created using this command:

```
plotPost( codaSamples[,"theta"] , main="theta" , xlab=bquote(theta) )
```

The first argument specified what values to plot; in this case the column named theta from the `codaSamples`. The `plotPost` function can take either `coda` objects or regular vectors for the values to plot. The next arguments merely specify the main title for the plot and the x-axis label. You can see from the plot in the left panel of Figure 8.4 that the annotations default to displaying the estimated mode of the distribution and the estimated 95% HDI. The mode indicates the value of the parameter that is most credible, given the data, but unfortunately the estimate of the mode from an MCMC sample can be rather unstable because the estimation is based on a smoothing algorithm that can be sensitive to random bumps and ripples in the MCMC sample.

The right panel of Figure 8.4 was created using this command:

```
plotPost( codaSamples[,"theta"] , main="theta" , xlab=bquote(theta) ,
          cenTend="median" , compVal=0.5 , ROPE=c(0.45,0.55) , credMass=0.90 )
```

Notice that there were four additional arguments specified. The `cenTend` argument specifies which measure of central tendency will be displayed; the options are `"mode"`,

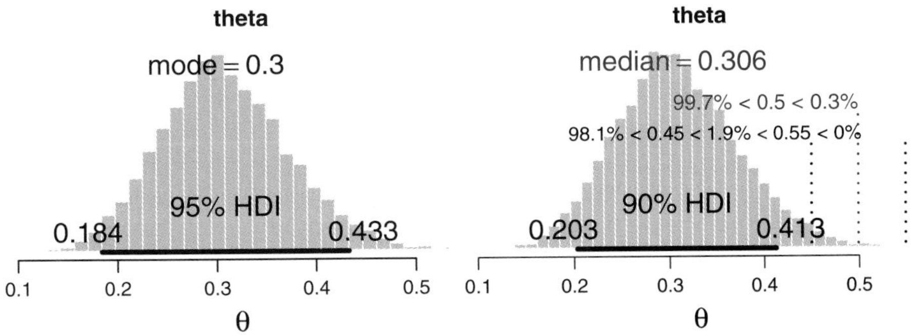

Figure 8.4 Posterior distribution based on output from JAGS, plotted twice using different options in the `plotPost` function.

"median", and "mean". The mode is most meaningful but also unstable in its MCMC approximation. The median is typically the most stable. The compVal argument specifies a "comparison value." For example, we might want to know whether the estimated bias differs from the "fair" value of $\theta = 0.5$. The plot shows the comparison value and annotates it with the percentage of the MCMC sample below or above the comparison value. The next argument specifies the limits of the region of practical equivalence, which will be discussed in depth in Section 12.1.1. For now, we merely note that it is a buffer interval around the comparison value, and the plot annotates its limits with the percentage of the distribution below, within, and above the region of practical equivalence. Finally, the credMass argument specifies the mass of the HDI to display. It defaults to 0.95, but here I have specified 0.90 instead merely to illustrate its use.

The plotPost function has other optional arguments. The argument HDItextPlace specifies where the numerical labels for the HDI limits will be placed. If HDItextPlace=0, the numerical labels are placed completely inside the HDI. If HDItextPlace=1, the numerical labels are placed completely outside the HDI. The default is HDItextPlace=0.7, so that the labels fall 70% outside the HDI. The argument showCurve is a logical value which defaults to FALSE. If showCurve=TRUE, then a smoothed kernel density curve is displayed instead of a histogram.

The plotPost function also takes optional graphical parameters that get passed to R's plotting commands. For example, when plotting histograms, plotPost defaults to putting white borders on each bar. If the bars happen to be very thin, the white borders can obscure the thin bars. Therefore, the borders can be plotted in a color that matches the bars by specifying border="skyblue". As another example, sometimes the default limits of the x-axis can be too extreme because the MCMC chain happens to have a few outlying values. You can specify the limits of the x-axis using the xlim argument, such as xlim=c(0,1).

8.3. SIMPLIFIED SCRIPTS FOR FREQUENTLY USED ANALYSES

As was mentioned at the beginning of this example, several pages back, the complete script for the preceding analysis is available in the file Jags-ExampleScript.R. It is valuable for you to understand all the steps involved in the script so you know how to interpret the output and so you can create scripts for your own specialized applications in the future. But dealing with all the details can be unwieldy and unnecessary for typical, frequently used analyses. Therefore, I have created a catalog of simplified, high-level scripts.

The scripts in the catalog have elaborate names, but the naming structure allows an extensive array of applications to be addressed systematically. The file naming system works as follows. First, all scripts that call JAGS begin with "Jags-." Then the nature of the modeled data is indicated. The modeled data are generically called y in mathematical expressions, so the file names use the letter "Y" to indicate this variable. In the present application, the data are dichotomous (i.e., 0 and 1), and therefore the file name begins with "Jags-Ydich-." The next part of the file name indicates the nature of the predictor variables, sometimes also called covariates or regressors, and generically called x in mathematical expressions. In our present application, there is only a single subject. Because the subject identifier is a nominal (not metric) variable, the file name begins with "Jags-Ydich-Xnom1subj-." Finally, the nature of the model is indicated. In the present application, we are using a Bernoulli likelihood with beta prior, so we complete the file name as "Jags-Ydich-Xnom1subj-MbernBeta.R." Other scripts in the catalog use different descriptors after Y, X, and M in their file names. The file Jags-Ydich-Xnom1subj-MbernBeta.R defines the functions that will be used, and the file Jags-Ydich-Xnom1subj-MbernBeta-Example.R contains the script that calls the functions.

What follows below are condensed highlights from the script Jags-Ydich-Xnom1subj-MbernBeta-Example.R. Please read through it, paying attention to the comments.

```
# Load the data
myData = read.csv("z15N50.csv") # must have a component named y
# Load the functions genMCMC, smryMCMC, and plotMCMC:
source("Jags-Ydich-Xnom1subj-MbernBeta.R")
# Generate the MCMC chain:
mcmcCoda = genMCMC( data=myData , numSavedSteps=10000 )
# Display diagnostics of chain, for specified parameter:
diagMCMC( mcmcCoda , parName="theta" )
# Display numerical summary statistics of chain:
smryMCMC( mcmcCoda )
# Display graphical posterior information:
plotMCMC( mcmcCoda , data=y )
```

Notice that the JAGS process—that was previously an extensive script that assembled the data list, the model specification, the initial values, and the running of the chains—has here been collapsed into a single function called genMCMC. Hooray! Thus, by using this script, you can analyze any data that have this structure merely by reading in the appropriate data file at the beginning, and without changing anything else in the script.

The high-level scripts should operate without changes for any data file structured in the way expected by the function genMCMC. If you want to change the functions in some way, then you will probably need to do some debugging in the process. Please review the section on debugging functions, back in Section 3.7.6 (p. 67).

8.4. EXAMPLE: DIFFERENCE OF BIASES

In the previous chapter, we also estimated the difference of biases between coins (or subjects). It is easy to implement that situation in JAGS. This section will first illustrate the model structure and its expression in JAGS, and then show the high-level script that makes it easy to run the analysis for any data structured this way.

As usual, we start our analysis with the structure of the data. In this situation, each subject (or coin) has several trials (or flips) that have dichotomous (0/1) outcomes. The data file therefore has one row per measurement instance, with a column for the outcome y and a column for the subject identifier s. For example, the data file might look like this:

```
y,s
1,Reginald
0,Reginald
1,Reginald
1,Reginald
1,Reginald
1,Reginald
1,Reginald
0,Reginald
0,Tony
0,Tony
1,Tony
0,Tony
0,Tony
1,Tony
0,Tony
```

Notice that the first row contains the column names (`y` and `s`), and the columns are comma-separated. The values under `y` are all 0's or 1's. The values under `s` are unique identifiers for the subjects. The column names must come first, but after that, the rows can be in any order.

The program reads the CSV file and converts the subject identifiers to consecutive integers so that they can be used as indices inside JAGS:

```
myData = read.csv("z6N8z2N7.csv")
y = myData$y
s = as.numeric(myData$s) # converts character to consecutive integer levels
```

The result of the above is a vector `y` of 0's and 1's, and a vector `s` of 1's for Reginald followed by 2's for Tony. (The conversion of factor levels to integers was discussed in Section 3.4.2, following p. 46.) Next, the data are bundled into a `list` for subsequent shipment to JAGS:

```
Ntotal = length(y)
Nsubj = length(unique(s))
dataList = list(
  y = y ,
  s = s ,
  Ntotal = Ntotal ,
  Nsubj = Nsubj
)
```

To construct a model specification for JAGS, we first make a diagram of the relations between variables. The data consist of observed outcomes $y_{i|s}$, where the subscript refers to instances i within subjects s. For each subject, we want to estimate their bias, that is, their underlying probability of generating an outcome 1, which we will denote θ_s. We model the data as coming from a Bernoulli distribution:

$$y_{i|s} \sim \text{dbern}(\theta_s)$$

and this corresponds with the lower arrow in Figure 8.5. Each subject's bias, θ_s, has an independent beta-distributed prior:

$$\theta_s \sim \text{dbeta}(A, B)$$

and this corresponds with the upper arrow in Figure 8.5. To keep the example consistent with the earlier demonstrations, we set $A = 2$ and $B = 2$ (recall the use of a beta$(\theta|2, 2)$ prior in Figure 7.5, p. 167, Figure 7.6, p. 169, and Figure 7.8, p. 174).

Having created a diagram for the model, we can then express it in JAGS:

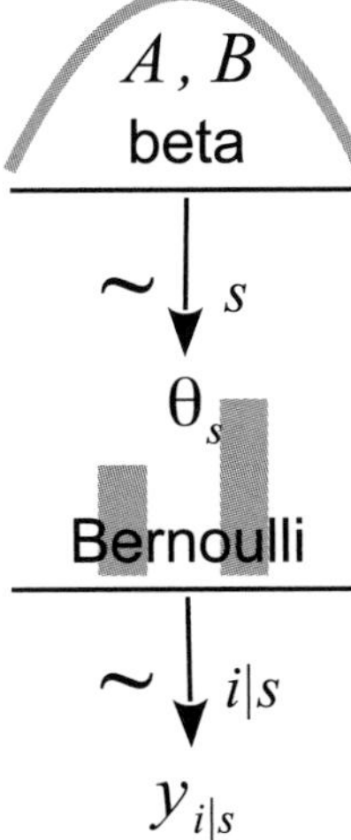

Figure 8.5 Diagram of model with Bernoulli likelihood and beta prior for multiple subjects, *s*. Notice the indices on the variables and arrows. Every arrow in the diagram has a corresponding line of code in JAGS model specification.

```
model {
  for ( i in 1:Ntotal ) {
    y[i] ~ dbern( theta[s[i]] )  # notice nested indexing
  }
  for ( s in 1:Nsubj ) {
    theta[s] ~ dbeta(2,2)
  }
}
```

In the above code, notice that the index `i` goes through the total number of data values, which is the total number of rows in the data file. So you can also think of `i` as the row number. The important novelty of this model specification is the use of "nested indexing" for `theta[s[i]]` in the `dbern` distribution. For example, consider when the `for` loop gets to `i` = 12. Notice from the data file that `s[12]` is 2. Therefore, the statement `y[i]` ~ `dbern(theta[s[i]])` becomes `y[12]` ~ `dbern(theta[s[12]])` which becomes `y[12]` ~ `dbern(theta[2])`. Thus, the data values for subject s are modeled by the appropriate corresponding θ_s.

The details of the JAGS model specification were presented to show how easy it is to express the model in JAGS. JAGS then does the MCMC sampling on its own! There is no need for us to decide whether to apply a Metropolis sampler or a Gibbs sampler, and no need for us to build a Metropolis proposal distribution or derive conditional distributions for a Gibbs sampler. JAGS figures that out by itself.

The JAGS program for this data structure has been assembled into a set of functions that can be called from a high-level script, so you do not have to mess with the detailed model specification if you do not want to. The high-level script is analogous in structure to the example in the previous section. The only thing that changes is the script file name, which now specifies `Ssubj` instead of `1subj`, and, of course, the data file name. What follows below are condensed highlights from the script `Jags-Ydich-XnomSsubj-MbernBeta-Example.R`. Please read through it, paying attention to the comments.

```
# Load the data
myData = read.csv("z6N8z2N7.csv") # myData is a data frame.
# Load the functions genMCMC, smryMCMC, and plotMCMC:
source("Jags-Ydich-XnomSsubj-MbernBeta.R")
# Generate the MCMC chain:
mcmcCoda = genMCMC( data=myData , numSavedSteps=10000 )
# Display diagnostics of chain, for specified parameter:
diagMCMC( mcmcCoda , parName="theta[1]" )
# Display numerical summary statistics of chain:
smryMCMC( mcmcCoda , compVal=NULL , compValDiff=0.0 )
# Display graphical posterior information:
plotMCMC( mcmcCoda , data=myData , compVal=NULL , compValDiff=0.0 )
```

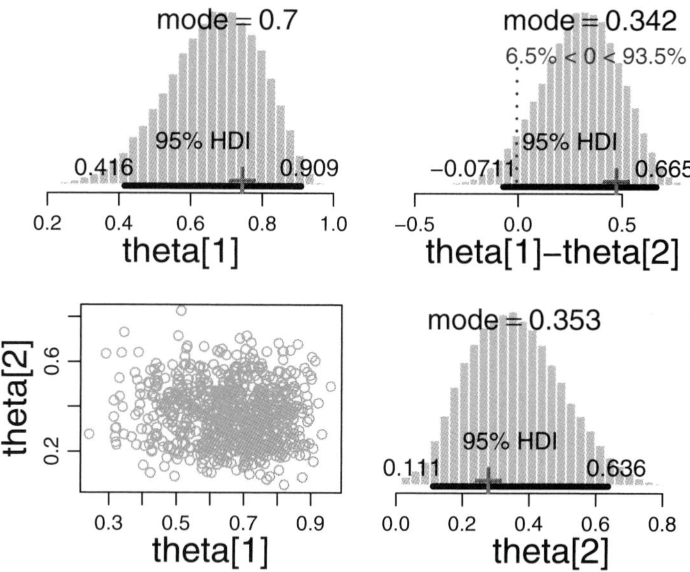

Figure 8.6 Posterior distribution created by JAGS. Compare the upper-right panel with Figure 7.9.

The overall structure of the script is the same as the previous high-level script for the single-subject data. In particular, the functions `genMCMC`, `smryMCMC`, and `plotMCMC` have the same names as the previous script, but they are defined differently to be specific to this data structure. Therefore, it is crucial to source the appropriate function definitions (as is done above) before calling the functions.

Figure 8.6 shows the result of the `plotMCMC` function. It displays the marginal distribution of each individual parameter and the difference of the parameters. It also annotates the distributions with information about the data. Specifically, the value of z_s/N_s is plotted as a large "+" symbol on the horizontal axis of each marginal distribution, and the value $z_1/N_1 - z_2/N_2$ is plotted on the horizontal axis of the difference distribution. The difference distribution also plots a reference line at zero, so we can easily see the relation of zero to the HDI.

If you compare the upper-right panel of Figure 8.6 with the results from Figure 7.9 (p. 177), you can see that JAGS produces the same answer as our "home grown" Metropolis and Gibbs samplers.

8.5. SAMPLING FROM THE PRIOR DISTRIBUTION IN JAGS

There are various reasons that we may want to examine the prior distribution in JAGS:

- We might want to check that the implemented prior actually matches what we intended:

- For example, perhaps what we imagine a beta($\theta|A, B$) prior looks like is not really what it looks like.
- Or, perhaps we simply want to check that we did not make any programming errors.
- Or, we might want to check that JAGS is functioning correctly. (JAGS is well developed, and we can have high confidence that it is functioning correctly. Yet, as the proverb says, "trust but verify."[5])

- We might want to examine the implied prior on parameter combinations that we did not explicitly specify. For example, in the case of estimating the difference of two biases, we put independent beta($\theta|A, B$) priors on each bias separately, but we never considered what shape of prior that implies for the difference of biases, $\theta_1 - \theta_2$. We will explore this case below.
- We might want to examine the implied prior on mid-level parameters in hierarchical models, when the explicit prior was specified on higher-level parameters. We will use hierarchical models extensively in later chapters and will see an example of this in Section 9.5.1.

It is straight forward to have JAGS generate an MCMC sample from the prior: We simply run the program with no data included. This directly implements the meaning of the prior, which is the allocation of credibility across parameter values without the data. Intuitively, this merely means that when there are no new data, the posterior is the prior, by definition. Another way to conceptualize it is by thinking of the likelihood function for null data as yielding a constant value. That is, $p(\varnothing|\theta) = k$, a constant (greater than zero). You can think of this as merely an algebraic convenience, or as a genuine probability that the data-sampling process happens to collect no data. In either case, notice that it asserts that the probability of collecting no data does not depend on the value of the parameter. When a constant likelihood is plugged into Bayes' rule (Equation 5.7, p. 106), the likelihood cancels out (in numerator and denominator) and the posterior equals the prior.

To run JAGS without the data included, we must omit the y values, but we must retain all constants that define the *structure* of the model, such as the number of (absent) data values and the number of (absent) subjects and so forth. Thus, at the point in the program that defines the `list` of data shipped to JAGS, we comment out only the line that specifies `y`. For example, in `Jags-Ydich-XnomSsubj-MbernBeta.R`, comment out the data, `y`, but retain all the structural constants, like this:

[5] According to a web page by author Suzanne Massie (http://www.suzannemassie.com/reaganYears.html, accessed 15 September 2013), she taught the Russian proverb, "trust but verify," to the American President, Ronald Reagan. The phrase has been used often in the high-stakes world of international politics, and therefore its use here is intended to be humorous. Trust me.

```
dataList = list(
# y = y ,
  s = s ,
  Ntotal = Ntotal ,
  Nsubj = Nsubj
)
```

Then just run the program with the usual high-level script. Be sure you have saved the program so that the high-level script calls the modified version. Also be sure that if you save the output you use file names that indicate they reflect the prior, not the posterior.

Figure 8.7 shows the prior sampled by JAGS. Notice that the distributions on the individual biases (θ_1 and θ_2) are, as intended, shaped like beta($\theta|2, 2$) distributions (cf. Figure 6.1, p. 128). Most importantly, notice the implied prior on the difference of biases, shown in the upper-right panel of Figure 8.7. The prior does not extend uniformly across the range from -1 to $+1$, and instead shows that extreme differences are suppressed by the prior. This might or might not reflect the intended prior on the difference. Exercise 8.4 explores the implied prior on $\theta_1 - \theta_2$ for different choices of priors on θ_1 and θ_2 individually.

8.6. PROBABILITY DISTRIBUTIONS AVAILABLE IN JAGS

JAGS has a large collection of frequently used probability distributions that are built-in. These distributions include the beta, gamma, normal, Bernoulli, and binomial along

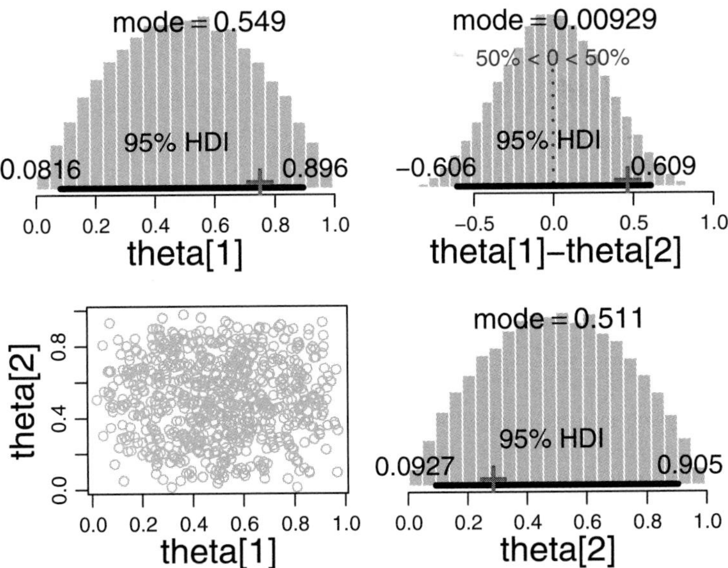

Figure 8.7 The prior distribution for difference of biases. Compare with Figure 8.6.

with many others. A complete list of distributions, and their JAGS names, can be found in the JAGS user manual.

8.6.1. Defining new likelihood functions

Despite the breadth of offerings, there may be occasions when you would like to use a probability distribution that is not built into JAGS. In particular, how can we get JAGS to use a likelihood function that is not prepackaged? One way is to use the so-called "Bernoulli ones trick" or the analogous "Poisson zeros trick" (e.g., Lunn et al., 2013).

Suppose we have some likelihood function, $p(y|\text{parameters})$, for which we have a mathematical formula. For example, the normal probability distribution has a mathematical formula specified in Equation 4.4, p. 84. Suppose, however, that JAGS has no built-in probability density function (pdf) of that form. Therefore, unfortunately, we cannot specify in JAGS something like `y[i] ~ pdf(parameters)`, where "`pdf`" stands for the name of the probability density function we would like.

To understand what we will do instead, it is important to realize that what we would get from `y[i] ~ pdf(parameters)` is the value of the pdf when `y[i]` has its particular value and when the parameters have their particular randomly generated MCMC values. *All we need to do is specify any valid JAGS code that yields that result.*

Here is the "Bernoulli ones trick." First, notice that $\text{dbern}(1|\theta) = \theta$. In other words, if we write JAGS code `1 ~ dbern( theta )`, it yields a likelihood that is the value of theta. If we define, in another line of JAGS code, the value of theta as the desired likelihood at `y[i]`, then the result is the likelihood at `y[i]`. Thus, the JAGS model statements

```
spy[i] <- pdf( y[i] , parameters ) / C # where pdf is a formula
1 ~ dbern( spy[i] )
```

together yield the same thing as `y[i] ~ pdf ( parameters )`. The variable name `spy[i]` stands for scaled probability of y_i.

Important caveat: the value of the parameter in dbern must be between zero and one, and therefore the value of `spy[i]` must be scaled so that, no matter what the random parameter values happen to be, `spy[i]` is always less than 1. We achieve this by dividing by a large positive constant C, which must be determined separately for each specific application. It is okay to use a scaled likelihood instead of the correct-magnitude likelihood because MCMC sampling uses only relative posterior densities of current and proposed positions, not absolute posterior densities.

In JAGS, the vector of 1's must be defined outside the model definition. One way to do that is with a `data` block that precedes the `model` block in JAGS:

```
data {
  C <- 10000 # JAGS does not warn if too small!
  for (i in 1:N) {
    ones[i] <- 1
  }
}
model {
  for (i in 1:N) {
    spy[i] <-  pdf( y[i] , parameters )/C # where pdf is a formula
    ones[i] ~ dbern( spy[i] )
  }
  parameters ~ dprior...
}
```

As a specific example, suppose that the values of y are continuous metric values, and we want to describe the data as a normal distribution. Suppose that JAGS had no `dnorm` built into it. We could use the method above, along with the mathematical formula for the normal density from Equation 4.4 (p. 84). Therefore, we would specify

```
spy[i] <- (exp(-0.5*((y[i]-mu)/sigma)^2)/(sigma*(2*3.1415926)^0.5))/C
```

Alternatively, there is the "Poisson zeros trick" which uses the same approach but with the fact that $\text{dpois}(0, \theta) = \exp(-\theta)$. Therefore, the JAGS statement `0` $\sim$ `dpois( -log(theta[i]) )` yields the value `theta[i]`. So we can define theta[i] = pdf(y[i],parameters) and put -log(theta[i]) into dpois to get the desired likelihood value in JAGS. But notice that `-log(theta[i])` must be a positive value (to be valid for the Poisson distribution), therefore the likelihood must have a constant added to it to be sure it is positive no matter what values of `y[i]` and `parameters` gets put into it.

Finally, with a bit of programming effort, JAGS can be genuinely extended with new distributions, completely bypassing the tricks above. The `runjags` package implements a few additional probability densities and explains a way to extend JAGS (see Denwood, 2013). Another approach to extending JAGS is explained by Wabersich and Vandekerckhove (2014).

8.7. FASTER SAMPLING WITH PARALLEL PROCESSING IN RUNJAGS

Even when using multiple chains in JAGS, each chain is computed sequentially on a single central processing unit, or "core," in the computer. At least, that is the usual default procedure for JAGS called from rjags. But most modern computers have multiple cores. Instead of using only one core while the others sit idle, we can run chains in parallel on multiple cores, thereby decreasing the total time needed. For example, suppose we desire 40,000 total steps in the chain. If we run four chains of 10,000 steps simultaneously on four cores, it takes only about 1/4th the time as running them successively on a single core. (It is a bit more than the reciprocal of the number of chains because all the chains must be burned in for the full number of steps needed to reach convergence.) For the simple examples we have explored so far, the time to run the MCMC chains is trivially brief. But for models involving large data sets and many parameters, this can be a huge time savings.

The package `runjags` (Denwood, 2013) makes it easy to run parallel chains in JAGS. The package provides many modes for running JAGS, and a variety of other facilities, but here I will focus only on its ability to run chains in parallel on multiple processors. For more information about its other abilities, see the article by Denwood (2013) and the runjags manual available at http://cran.r-project.org/web/packages/runjags/.

To install the runjags package into R, follow the usual procedure for installing packages. At R's command line, type

```
install.packages("runjags")
```

You only need to do that once. Then, for any session in which you want to use the runjags functions, you must load the functions into R's memory by typing

```
library("runjags")
```

With runjags loaded into R's working memory, you can get more information about it by typing `?runjags` at the command line.

The essential difference between running JAGS via `rjags` and `runjags` is merely the functions used to invoke JAGS. As an example, suppose our script has already defined the model as the file `model.txt`, the data list as `dataList`, and the initial values as a list or function `initsList`. Then, suppose the script specifies the details of the chains as follows:

```
nChains=3
nAdaptSteps=1000
nBurninSteps=500
nUseSteps=10000  # total number of used steps
nThinSteps=2
```

If we were calling JAGS via `rjags`, we could use this familiar sequence of commands:

```
library(rjags)
jagsModel = jags.model(      file="model.txt" ,
                             data=dataList ,
                             inits=initsList ,
                             n.chains=nChains ,
                             n.adapt=nAdaptSteps )
update(                      jagsModel ,
                             n.iter=nBurninSteps )
codaSamples = coda.samples( jagsModel ,
                             variable.names=c("theta") ,
                             n.iter=ceiling(nUseSteps*nThinSteps/nChains) ,
                             thin=nThinSteps )
```

Across the sequence of three commands above, I aligned the arguments vertically, with one argument per line, so that you could easily see them all. To call JAGS via `runjags`, we use the same argument values like this:

```
library(runjags)
runJagsOut <- run.jags( method="parallel" ,
                        model="model.txt" ,
                        monitor=c("theta") ,
                        data=dataList ,
                        inits=initsList ,
                        n.chains=nChains ,
                        adapt=nAdaptSteps ,
                        burnin=nBurninSteps ,
                        sample=ceiling(nUseSteps/nChains) ,
                        thin=nThinSteps ,
                        summarise=FALSE ,
                        plots=FALSE )
codaSamples = as.mcmc.list( runJagsOut )
```

A key advantage of runjags is specified by the first argument in the `run.jags` command above, named `method`. This is where we can tell runjags to run parallel chains on multiple cores. The `method` argument has several options, which you can read about in the help file. For our purposes, the two options of primary interest are `"rjags,"` which runs in ordinary rjags mode, and `"parallel,"` which runs the chains in parallel on different computer cores.

Some of the arguments are slightly renamed in runjags relative to rjags, but the correspondence is easy to understand, and there is detailed help available by typing `?run.jags` at R's command line (and in the runjags manual, of course). For example, the argument for specifying the number of steps to take after burnin is `n.iter` in rjag's `coda.samples` command, and corresponds to `sample` in runjag's `run.jags` command, but because of the way thinning is implemented, the values must be adjusted as shown. (This is for rjags version 3-12 and runjags version 1.2.0-7; beware of the possibility of changes in subsequent versions.)

There are two new arguments in the `run.jags` command: `summarise` and `plots`. The `summarise` argument tells runjags whether or not to compute diagnostic statistics for every parameter. The diagnostics can take a long time to compute, especially for models with many parameters and long chains. The `summarise` argument has a default value of `TRUE`. Be careful, however, to diagnose convergence in other ways when runjags is told not to.

After `run.jags` returned its output, the final line of code, above, converted the runjags object to coda format. This conversion allows `coda`-oriented graphical functions to be used later in the script.

Some important caveats: if your computer has K cores, you will usually want to run only $K-1$ chains in parallel, because while JAGS is running you will want to reserve one core for other uses, such as word processing, or answering email, or watching cute puppy videos to improve your productivity (Nittono, Fukushima, Yano, & Moriya, 2012). If you use all K cores for running JAGS, your computer may be completely swamped by those demands and effectively lock you out until it is done.[6] Another important consideration is that JAGS has only four distinct (pseudo-)random number generators (RNGs). If you run up to four parallel chains, runjags (and rjags too) will use four different RNGs, which gives some assurance that the four chains will be independent. If you run more than four parallel chains, however, you will want to explicitly initialize the RNGs to be sure that you are generating different chains. The runjags manual suggests how to do this.

"Okay," you may be saying, "that's all terrific, but how do I know how many cores my computer has?" One way is by using a command in R from the `parallel` package, which gets installed along with runjags. At the R command prompt, type `library("parallel")` to load the package, then type `detectCores()`, with the empty parentheses.

8.8. TIPS FOR EXPANDING JAGS MODELS

Often, the process of programming a model is done is stages, starting with a simple model and then incrementally incorporating complexifications. At each step, the model is checked for accuracy and efficiency. This procedure of incremental building is useful for creating a desired complex model from scratch, for expanding a previously created model for a new application, and for expanding a model that has been found to be inadequate in a posterior predictive check. The best way to explain how to expand a JAGS model is by example. Section 17.5.2 (p. 507) summarizes the necessary steps, exemplified by expanding a linear regression model to a quadratic trend model.

8.9. EXERCISES

Look for more exercises at https://sites.google.com/site/doingbayesiandataanalysis/

Exercise 8.1. [Purpose: Run the high level scripts with other data to see how easy they are.] Consider the high-level script, Jags-Ydich-XnomSsubj-MbernBeta-Example.R. For this exercise, you will use that script with a new data file, and notice that you only need to change a single line, namely the one that loads the data file.

[6] If you do lock up your computer by using all of its processors, and therefore cannot watch cute puppy videos, you can still improve your productivity by gazing contentedly at the cute doggies on the cover of this book.

In RStudio, open a new blank file by selecting the menu items File → New → Text file. Manually type in new fictional data in the same format as the data shown in Section 8.4, p. 208, but with three subjects instead of two. Use whatever names you fancy, and as many trials for each subject as you fancy. Perhaps put a preponderance of 0's for one subject and a preponderance of 1's for another subject. Use a lot of trials of one subject, and relatively few trials for another. Save the file with a filename that ends with ".csv." Then, in the script, use that file name in the `read.csv` command. In your report, include the data file and the graphical output of the analysis. Are the estimates reasonable? What is the effect of different sample sizes for the estimates of different subjects?

Exercise 8.2. [Purpose: Pay attention to the output of `smryMCMC`.] The graphical plots from `plotMCMC` are useful for understanding, but they lack some numerical details. Run the high-level script, Jags-Ydich-XnomSsubj-MbernBeta-Example.R, and explain the details in the output from `smryMCMC`. Explain what the `rope` and `ropeDiff` arguments do.

Exercise 8.3. [Purpose: Notice what gets saved by the high-level scripts.] Run the high-level script, Jags-Ydich-XnomSsubj-MbernBeta-Example.R, and notice what files are created (i.e., saved) in your computer's working directory. Explain what these files are, and why they might be useful for future reference. Hint: In the file Jags-Ydich-XnomSsubj-MbernBeta.R, search for the word "save."

Exercise 8.4. [Purpose: Explore the prior on a difference of parameters implied from the priors on the individual parameters.]

(A) Reproduce Figure 8.7 in Section 8.5. Explain how you did it.

(B) Change the priors on the individual θ's to $\text{beta}(\theta|1, 1)$ and produce the figure anew. Describe its panels and explain.

(C) Change the priors on the individual θ's to $\text{beta}(\theta|0.5, 0.5)$ and produce the figure again. Describe its panels and explain.

CHAPTER 9

Hierarchical Models

Contents

Oh darlin', for love, it's on you I depend.
Well, I s'pose Jack Daniels is also my friend.
But you keep him locked up in the ladies' loo,
S'pose that means his spirits depend on you too.[1]

Hierarchical models are mathematical descriptions involving multiple parameters such that the credible values of some parameters meaningfully depend on the values of other parameters. For example, consider several coins minted from the same factory. A head-biased factory will tend to produce head-biased coins. If we denote the bias of the factory by ω (Greek letter omega), and we denote the biases of the coins as θ_s (where the subscript indexes different coins), then the credible values of θ_s depend on the values of ω. The estimate of the bias of any one coin depends on the estimate of factory bias, which is influenced by the data from all the coins. There are many realistic situations that involve meaningful hierarchical structure. Bayesian modeling software makes it straightforward to specify and analyze complex hierarchical models. This chapter begins with the simplest

[1] This chapter is about hierarchical models. In hierarchical models, the joint prior distribution can be factored into terms that have some parameters dependent on other parameters. In other words, there are chains of dependencies among parameters. The poem expresses other situations in which there are dependencies and codependencies. Wordplay: Jack Daniels is a brand of whiskey, here being referred to as a person. The word "spirits" is playing double duty to refer to the alcoholic beverage and the person's mood.

Doing Bayesian Data Analysis, 2nd Edition
http://dx.doi.org/10.1016/B978-0-12-405888-0.00009-X

possible case of a hierarchical model and then incrementally scaffolds to more complex hierarchical structures.

There are many examples of data that naturally suggest hierarchical descriptions. For example, consider the batting ability of baseball players who have different primary fielding positions. The data consist of (i) the number of hits each player attains, (ii) the number of opportunities at bat, and (iii) the primary fielding position of the player. Each player's ability is described by a parameter θ_s, which depends on a parameter ω that describes the typical ability of players in that fielding position. As another example, consider the probability that a child bought lunch from the school cafeteria, in different schools in different districts. The data consist of (i) the number of days each child bought lunch, (ii) the number of days the child was at school, and (iii) the school and district of the child. Each child's probability is described by a parameter θ_s, which depends on a parameter ω that describes the typical buying probability of children in the school. In this example, the hierarchy could be extended to a higher level, in which additional parameters describe the buying probabilities of districts. As yet another example, consider the probability that patients recover from heart surgery performed by different surgical teams within different hospitals in different cities. In this case, we can think of each surgical team as a coin, with patient outcomes being flips of the surgical-team coin. The data consist of (i) the number of recoveries, (ii) the number of surgeries, and (iii) the hospital and city of the surgery. Each team's probability is described by a parameter θ_s that depends on a parameter ω that describes the typical recovery probability for the hospital. We could expand the hierarchy to estimate typical recovery probabilities for cities. In all these examples, what makes hierarchical modeling tremendously effective is that the estimate of each individual parameter is simultaneously informed by data from all the other individuals, because all the individuals inform the higher-level parameters, which in turn constrain all the individual parameters. These structural dependencies across parameters provide better informed estimates of all the parameters.

The parameters at different levels in a hierarchical model are all merely parameters that coexist in a joint parameter space. We simply apply Bayes' rule to the joint parameter space, as we did for example when estimating two coin biases back in Figure 7.5, p. 167. To say it a little more formally with our parameters θ and ω, Bayes' rule applies to the joint parameter space: $p(\theta, \omega|D) \propto p(D|\theta, \omega)\, p(\theta, \omega)$. What is special to hierarchical models is that the terms on the right-hand side can be factored into a chain of dependencies, like this:

$$\begin{aligned} p(\theta, \omega|D) &\propto p(D|\theta, \omega)\, p(\theta, \omega) \\ &= p(D|\theta)\, p(\theta|\omega)\, p(\omega) \end{aligned} \tag{9.1}$$

The refactoring in the second line means that the data depend only on the value of θ, in the sense that when the value θ is set then the data are independent of all other parameter values. Moreover, the value of θ depends on the value of ω and the value of

θ is conditionally independent of all other parameters. Any model that can be factored into a chain of dependencies like Equation 9.1 is a hierarchical model.

Many multiparameter models can be reparameterized, which means that the same mathematical structure can be re-expressed by recombining the parameters. For example, a rectangle can be described by its length and width, or by its area and aspect ratio (where area is length times width and aspect ratio is length divided by width). A model that can be factored as a chain of dependencies under one parameterization might not have a simple factorization as a chain of dependencies under another parameterization. Thus, a model is hierarchical with respect to a particular parameterization. We choose a parameterization to be meaningful and useful for the application at hand.

The dependencies among parameters become useful in several respects. First, the dependencies are meaningful for the given application. Second, because of the dependencies across parameters, all the data can jointly inform all the parameter estimates. Third, the dependencies can facilitate efficient Monte Carlo sampling from the posterior distribution, because clever algorithms can take advantage of conditional distributions (recall, for example, that Gibbs sampling uses conditional distributions).

As usual, we consider the scenario of flipping coins because the mathematics are relatively simple and the concepts can be illustrated in detail. And, as usual, keep in mind that the coin flip is just a surrogate for real-world data involving dichotomous outcomes, such as recovery or nonrecovery from a disease after treatment, recalling or forgetting a studied item, choosing candidate A or candidate B in an election, etc. Later in the chapter we will see applications to real data involving therapeutic touch, extrasensory perception, and baseball batting ability.

9.1. A SINGLE COIN FROM A SINGLE MINT

We begin with a review of the likelihood and prior distribution for our single-coin scenario. Recall from Equation 8.1 that the likelihood function is the Bernoulli distribution, expressed as

$$y_i \sim \text{dbern}(\theta) \tag{9.2}$$

and the prior distribution is a beta density, expressed (recall Equation 8.2) as

$$\theta \sim \text{dbeta}(a, b) \tag{9.3}$$

Recall also, from Equation 6.6 (p. 129), that the shape parameters of the beta density, a and b, can be re-expressed in terms of the mode ω and concentration κ of the beta distribution: $a = \omega(\kappa - 2) + 1$ and $b = (1 - \omega)(\kappa - 2) + 1$, where $\omega = (a - 1)/(a + b - 2)$ and $\kappa = a + b$. Therefore Equation 9.3 can be re-expressed as

$$\theta \sim \text{dbeta}\,(\,\omega(\kappa - 2) + 1\,,\ (1 - \omega)(\kappa - 2) + 1\,) \tag{9.4}$$

Think for a moment about the meaning of Equation 9.4. It says that the value of θ depends on the value of ω. When ω is, say, 0.25, then θ will be near 0.25. The value of κ governs how near θ is to ω, with larger values of κ generating values of θ more concentrated near ω. Thus, the magnitude of κ is an expression of our prior certainty regarding the dependence of θ on ω. For the purposes of the present example, we will treat κ as a constant, and denote it as K.

Now we make the essential expansion of our scenario into the realm of hierarchical models. Instead of thinking of ω as fixed by prior knowledge, we think of it as another parameter to be estimated. Because of the manufacturing process, the coins from the mint tend to have a bias near ω. Some coins randomly have a value of θ greater than ω, and other coins randomly have a value of θ less than ω. The larger K is, the more consistently the mint makes coins with θ close to ω. Now suppose we flip a coin from the mint and observe a certain number of heads and tails. From the data, we can estimate the bias θ of the coin, and we can also estimate the tendency ω of the mint. The estimated credible values of θ will (typically) be near the proportion of heads observed. The estimated credible values of ω will be similar to θ if K is large, but will be much less certain if K is small.

To infer a posterior distribution over ω, we must supply a prior distribution. Our prior distribution over ω expresses what we believe about how mints are constructed. For the sake of simplicity, we suppose that the prior distribution on ω is a beta distribution,

$$p(\omega) = \text{beta}(\omega|A_\omega, B_\omega) \tag{9.5}$$

where A_ω and B_ω are constants. In this case, we believe that ω is typically near $(A_\omega - 1)/(A_\omega + B_\omega - 2)$, because that is the mode of the beta distribution.

The scenario is summarized in Figure 9.1. The figure shows the dependency structure among the variables. Downward pointing arrows denote how higher-level variables generate lower-level variables. For example, because the coin bias θ governs the result of a coin flip, we draw an arrow from the θ-parameterized Bernoulli distribution to the datum y_i. As was emphasized in Section 8.2 in the discussion of Figure 8.2 (p. 196), these sorts of diagrams are useful for two reasons. First, they make the structure of the model visually explicit, so it is easy to mentally navigate the meaning of the model. Second, they facilitate expressing the model in JAGS (or Stan) because every arrow in the diagram has a corresponding line of code in the JAGS model specification.

Let's now consider how Bayes' rule applies to this situation. If we treat this situation as simply a case of two parameters, then Bayes' rule is merely $p(\theta, \omega|y) = p(y|\theta, \omega)p(\theta, \omega)/p(y)$. There are two aspects that are special about our present situation. First, the likelihood function does not involve ω, so $p(y|\theta, \omega)$ can be rewritten as $p(y|\theta)$. Second, because by definition $p(\theta|\omega) = p(\theta, \omega)/p(\omega)$, the prior on the joint parameter space can be factored thus: $p(\theta, \omega) = p(\theta|\omega)\, p(\omega)$. Therefore, Bayes' rule for our current hierarchical model has the form

$$p(\theta, \omega|y) = \frac{p(y|\theta, \omega)\, p(\theta, \omega)}{p(y)}$$
$$= \frac{p(y|\theta)\, p(\theta|\omega)\, p(\omega)}{p(y)} \tag{9.6}$$

Notice in the second line that the three terms of the numerator are given specific expression by our particular example. The likelihood function, $p(y|\theta)$, is a Bernoulli distribution in Equation 9.2. The dependence of θ on ω, $p(\theta|\omega)$, is specified to be a beta density in Equation 9.4. And the prior distribution on ω, $p(\omega)$, is defined to be a beta density in Equation 9.5. These specific forms are summarized in Figure 9.1. Thus, the generic expression in the first line of Equation 9.6 gains specific hierarchical meaning because of its factorization in the second line.

The concept expressed by Equation 9.6 is important. In general, we are merely doing Bayesian inference on a joint parameter space. But sometimes, as in this case, the likelihood and prior can be re-expressed as a hierarchical chain of dependencies among parameters. There are two reasons this is important. First, the semantics of the model matter for human interpretation, because the hierarchy is meaningful. Second, hierarchical models can sometimes be parsed for MCMC samplers very effectively. The algorithms in JAGS, for example, look for particular model substructures to which specific samplers can be applied.

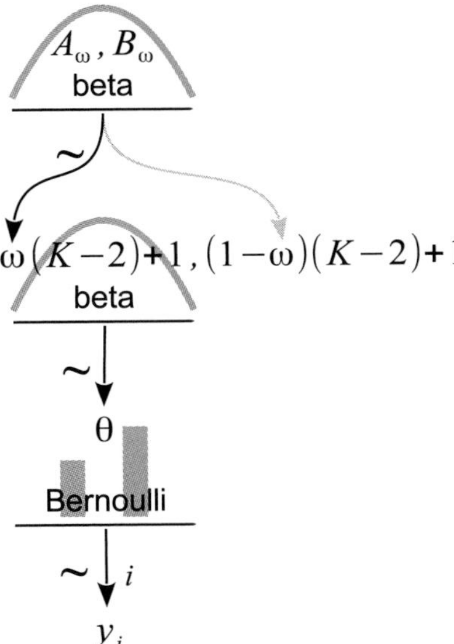

Figure 9.1 A model of hierarchical dependencies for data from a single coin. The chain of arrows illustrates the chain of dependencies in Equations 9.2, 9.4, and 9.5. (At the top of the diagram, the second instantiation of the arrow to ω is shown in grey instead of black merely to suggest that it is the same dependence already indicated by the first arrow in black.)

Whether or not a model structure can be factored into a chain of hierarchical dependencies depends on the model's parameterization. Under some parameterizations there may be a natural factorization of dependencies, but under another parameterization there might not be. In typical applications of hierarchical models, the meaningful chain of dependencies is primary in the conceptual genesis of the model, and thinking of the model as a joint distribution on a joint parameter space is secondary.

9.1.1. Posterior via grid approximation

It turns out that direct formal analysis of Equation 9.6 does not yield a simple formula for the posterior distribution. We can, nevertheless, use grid approximation to get a thorough picture of what is going on. When the parameters extend over a finite domain, and there are not too many of them, then we can approximate the posterior via grid approximation. In our present situation, we have the parameters θ and ω that both have finite domains, namely the interval $[0, 1]$. Therefore, a grid approximation is tractable and the distributions can be readily graphed.

Figure 9.2 shows an example in which the prior distribution has $A_\omega = 2$, $B_\omega = 2$, and $K = 100$. The left and middle panels of the top row of Figure 9.2 show the *joint* prior distribution: $p(\theta, \omega) = p(\theta|\omega)p(\omega) = \text{beta}(\theta|\omega(100 - 2) + 1, (1 - \omega)(100 - 2) + 1)\, \text{beta}(\omega|2, 2)$. The contour plot in the middle panel shows a top-down view of the perspective plot in the left panel. As this is a grid approximation, the joint prior $p(\theta, \omega)$ was computed by first multiplying $p(\theta|\omega)$ and $p(\omega)$ at every grid point and then dividing by their sum across the entire grid. The normalized probability masses were then converted to estimates of probability density at each grid point by dividing each probability mass by the area of a grid cell.

The right panel of the top row (Figure 9.2), showing $p(\omega)$ tipped on its side, is a "marginal" distribution of the prior: if you imagine collapsing (i.e., summing) the joint prior across θ, the resulting pressed flower[2] is the graph of $p(\omega)$. The scale on the $p(\omega)$ axis is density, not mass, because the mass at each interval of the comb on ω was divided by the width of the interval to approximate density.

The implied marginal prior on θ is shown in the second row of Figure 9.2. It was computed numerically by collapsing the joint prior across ω. But notice, we did not explicitly formulate the marginal prior on θ. Instead, we formulated the dependence of θ on ω. The dependence is shown in the right panel of the second row, which plots $p(\theta|\omega)$ for two values of ω. Each plot is a "slice" through the joint prior at a particular value of ω. The slices are renormalized, however, so that they are individually proper probability densities that sum to 1.0 over θ. One slice is $p(\theta|\omega = 0.75)$, and the other

[2] A popular craft is preservation of flowers by pressing them flat between boards and letting them dry. The compression of a three-dimensional object into two dimensions is used here as an analogy for marginalizing a probability distribution.

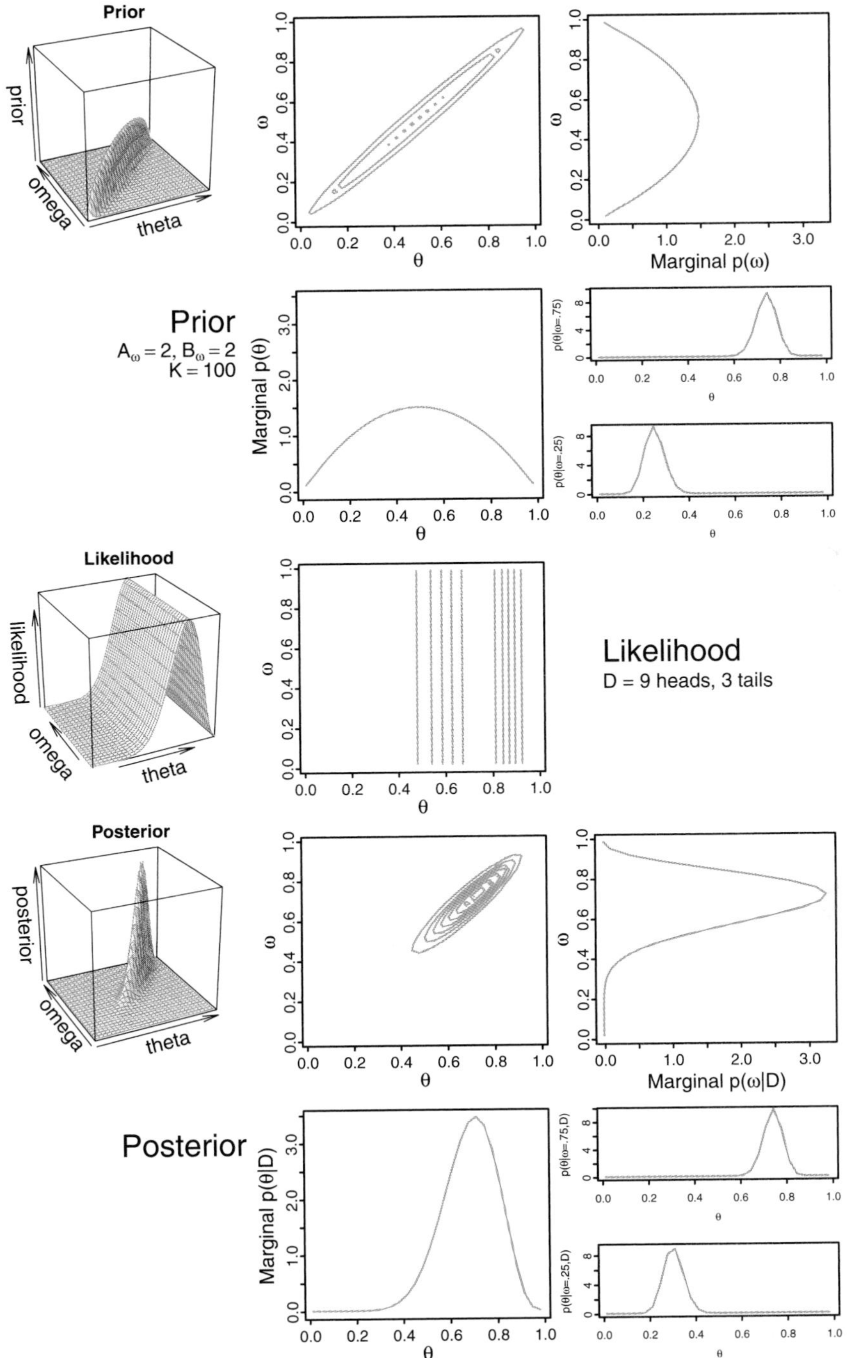

Figure 9.2 The prior has low certainty regarding ω, but high certainty regarding the dependence of θ on ω. The posterior shows that the distribution of ω has been altered noticeably by the data (see sideways plots of marginal $p(\omega)$), but the dependence of θ on ω has not been altered much (see small plots of $p(\theta|\omega)$). Compare with Figure 9.3, which uses the same data but a different prior.

slice is $p(\theta|\omega=0.25)$. You can see that the prior puts most credibility on θ values near ω. In fact, the slices are beta densities: $\text{beta}(\theta|(0.75)(100-2)+1, (1-0.75)(100-2)+1) = \text{beta}(\theta|74.5, 25.5)$ and $\text{beta}(\theta|(0.25)(100-2)+1, (1-0.25)(100-2)+1) = \text{beta}(\theta|25.5, 74.5)$.

The likelihood function is shown in the middle row of Figure 9.2. The data comprise nine heads and three tails. The likelihood distribution is a product of Bernoulli distributions, $p(\{y_i\}|\theta) = \theta^9(1-\theta)^3$. Notice in the graph that all the contour lines are parallel to the ω axis, and orthogonal to the θ axis. These parallel contours are the graphical signature of the fact that the likelihood function depends only on θ and not on ω.

The posterior distribution in the fourth row of Figure 9.2 is determined by multiplying, at each point of the grid on θ, ω space, the joint prior and the likelihood. The point-wise products are normalized by dividing the sum of those values across the parameter space.

When we take a slice through the joint posterior at a particular value of ω, and renormalize by dividing the sum of discrete probability masses in that slice, we get the conditional distribution $p(\theta|\omega, D)$. The bottom right panel of Figure 9.2 shows the conditional for two values of ω. Notice in Figure 9.2 that there is not much difference in the graphs of the prior $p(\theta|\omega)$ and the posterior $p(\theta|\omega, \{y_i\})$. This is because the prior beliefs regarding the dependency of θ on ω had little uncertainty.

The distribution of $p(\omega|\{y_i\})$, shown in the right panel of the fourth row of Figure 9.2, is determined by summing the joint posterior across all values of θ. This marginal distribution can be imagined as collapsing the joint posterior along the θ axis, with the resulting pressed flower being silhouetted in the third panel of the bottom row. Notice that the graphs of the prior $p(\omega)$ and the posterior $p(\omega|\{y_i\})$ are rather different. The data have had a noticeable impact on beliefs about how ω is distributed because the prior was very uncertain and was therefore easily influenced by the data.

As a contrasting case, consider instead what happens when there is high prior certainty regarding ω, but low prior certainty regarding the dependence of θ on ω. Figure 9.3 illustrates such a case, where $A_\omega = 20$, $B_\omega = 20$, and $K = 6$. The joint prior distribution is shown in the top row of Figure 9.3, left and middle panels. The marginal prior on ω is shown in the top row, right panel, where it can be seen that $p(\omega)$ is fairly sharply peaked over $\omega = .5$. However, the conditional distributions $p(\theta|\omega)$ are very broad, as can be seen in right panel of the second row.

The same data as for Figure 9.2 are used here, so the likelihood graphs look the same in the two figures. Notice again that the contour lines of the likelihood function are parallel to the ω axis, indicating that ω has no influence on the likelihood.

The posterior distribution is shown in the lower two rows of Figure 9.3. Consider the marginal posterior distribution on ω. Notice it has changed fairly little from the prior, because the distribution began with high certainty. Consider now the conditional posterior distributions in the lower right panels. They are very different from their priors,

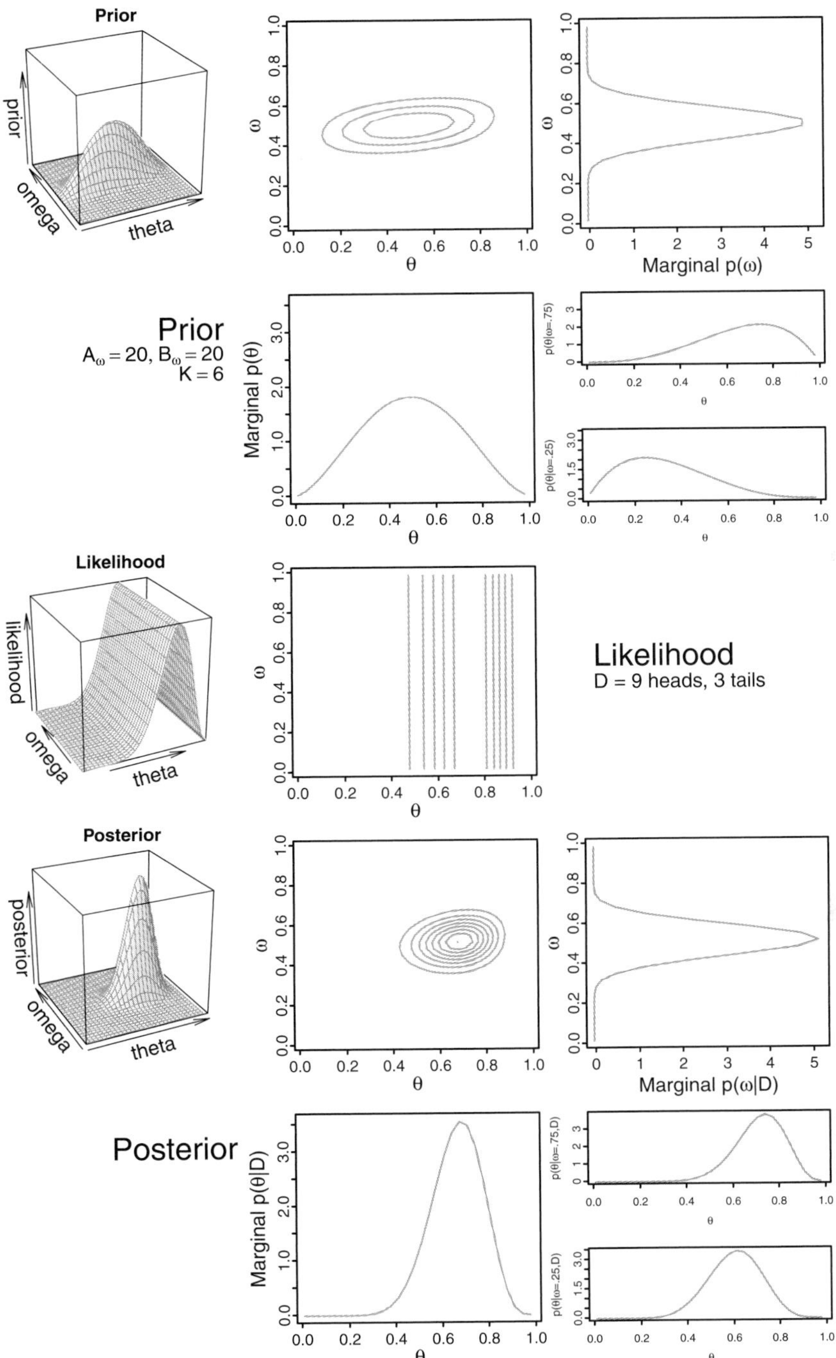

Figure 9.3 The prior has high certainty regarding ω, but low certainty regarding the dependence of θ on ω. The posterior shows that the distribution of ω has not been altered much by the data (see sideways plots of marginal $p(\omega)$), but the dependence of θ on ω has been altered noticeably (see small plots of $p(\theta|\omega)$). Compare with Figure 9.2, which uses the same data but a different prior.

because they began with low certainty, and therefore the data can have a big impact on these distributions. In this case, the data suggest that θ depends on ω in a rather different way than we initially suspected. For example, even when $\omega = 0.25$, the conditional posterior puts the most credible value of θ up near 0.6, not near 0.25.

In summary, Bayesian inference in a hierarchical model is merely Bayesian inference on a joint parameter space, but we look at the joint distribution (e.g., $p(\theta, \omega)$) in terms of its marginal on a subset of parameters (e.g., $p(\omega)$) and its conditional distribution for other parameters (e.g., $p(\theta|\omega)$). We do this primarily because it is meaningful in the context of particular models. The examples in Figures 9.2 and 9.3 graphically illustrated the joint, marginal, and conditional distributions. Those examples also illustrated how Bayesian inference has its greatest impact on aspects of the prior distribution that are least certain. In Figure 9.2, there was low prior certainty on ω (because A_ω and B_ω were small) but high prior certainty about the dependence of θ on ω (because K was large). The posterior distribution showed a big change in beliefs about ω, but not much change in beliefs about the dependence of θ on ω. Figure 9.3 showed the complementary situation, with high prior certainty on ω and low prior certainty about the dependence of θ on ω.

The aim of this section has been to illustrate the essential ideas of hierarchical models in a minimalist, two-parameter model, so that all the ideas could be graphically illustrated. While useful for gaining an understanding of the ideas, this simple, contrived example is also unrealistic. In subsequent sections, we progressively build more complex and realistic models.

9.2. MULTIPLE COINS FROM A SINGLE MINT

The previous section considered a scenario in which we flip a single coin and make inferences about the bias θ of the coin and the parameter ω of the mint. Now we consider an interesting extension: what if we collect data from more than one coin created by the mint? If each coin has its own distinct bias θ_s, then we are estimating a distinct parameter value for each coin, and using all the data to estimate ω.

This sort of situation arises in real research. Consider, for example, a drug that is supposed to affect memory. The researcher gives the drug to several people, and gives each person a memory test, in which the person tries to remember a list of words. Each word is remembered or not (i.e., a dichotomous measurement). In this application, each person plays the role of a coin, whose probability of remembering a word is dependent on the tendency induced by the drug. Let's index the subjects in the experiment by the letter s. Then the probability of remembering by subject s is denoted θ_s. We assume that there is some random variation of subjects around the drug-induced tendency denoted as ω. In particular, we model the subjects' variation around ω as $\theta_s \sim \text{dbeta}(\omega(K-2)+1, (1-\omega)(K-2)+1)$, where K is a fixed constant for now. We also assume that the performance of each subject depends only on that subject's individual θ_s, which is

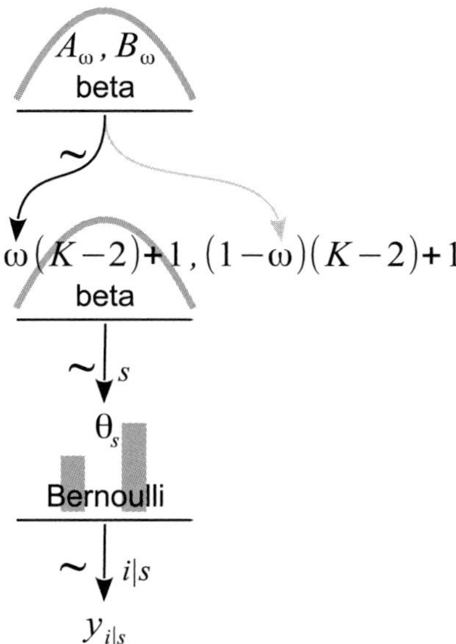

Figure 9.4 A model of hierarchical dependencies for data from several coins created independently from the same mint. A datum $y_{i|s}$, from the ith flip of the sth coin, depends on the value of the bias parameter θ_s for the coin. The values of θ_s depend on the value of the hyperparameter ω for the mint that created the coins. The ω parameter has a prior belief distributed as a beta distribution with shape parameters A_ω and B_ω.

modeled as $y_{i|s} \sim \text{dbern}(\theta_s)$, where the subscript, $i|s$, indicates the ith observation within subject s. The scenario is summarized in Figure 9.4. It is very much like Figure 9.1, but with changes on the subscripts to indicate multiple subjects. Notice that the model involves more than two parameters. If there are S subjects altogether, then there are $S+1$ parameters (namely $\theta_1, \ldots, \theta_S$, and ω) being estimated simultaneously. If our primary research interest is the overall effect of the drug, not the reactions of individual subjects, then we are most interested in the estimate of ω.

9.2.1. Posterior via grid approximation

As a concrete example, suppose we have only two subjects in the same condition (i.e., two coins from the same mint). We want to estimate the biases θ_1 and θ_2 of the two subjects, and simultaneously estimate ω of the drug that influenced them. Figures 9.5 and 9.6 show grid approximations for two different choices of prior distributions.

For the case illustrated in Figure 9.5, the prior on ω is gently peaked over $\omega = 0.5$, in the form of a $\text{beta}(\omega|2, 2)$ distribution; that is, $A_\omega = B_\omega = 2$ in the top-level formula of Figure 9.4. The biases of the coins are only weakly dependent on ω according to the prior $p(\theta_j|\omega) = \text{beta}(\theta_j|\omega(5-2)+1, (1-\omega)(5-2)+1)$; that is, $K = 5$ in the middle-level formula of Figure 9.4.

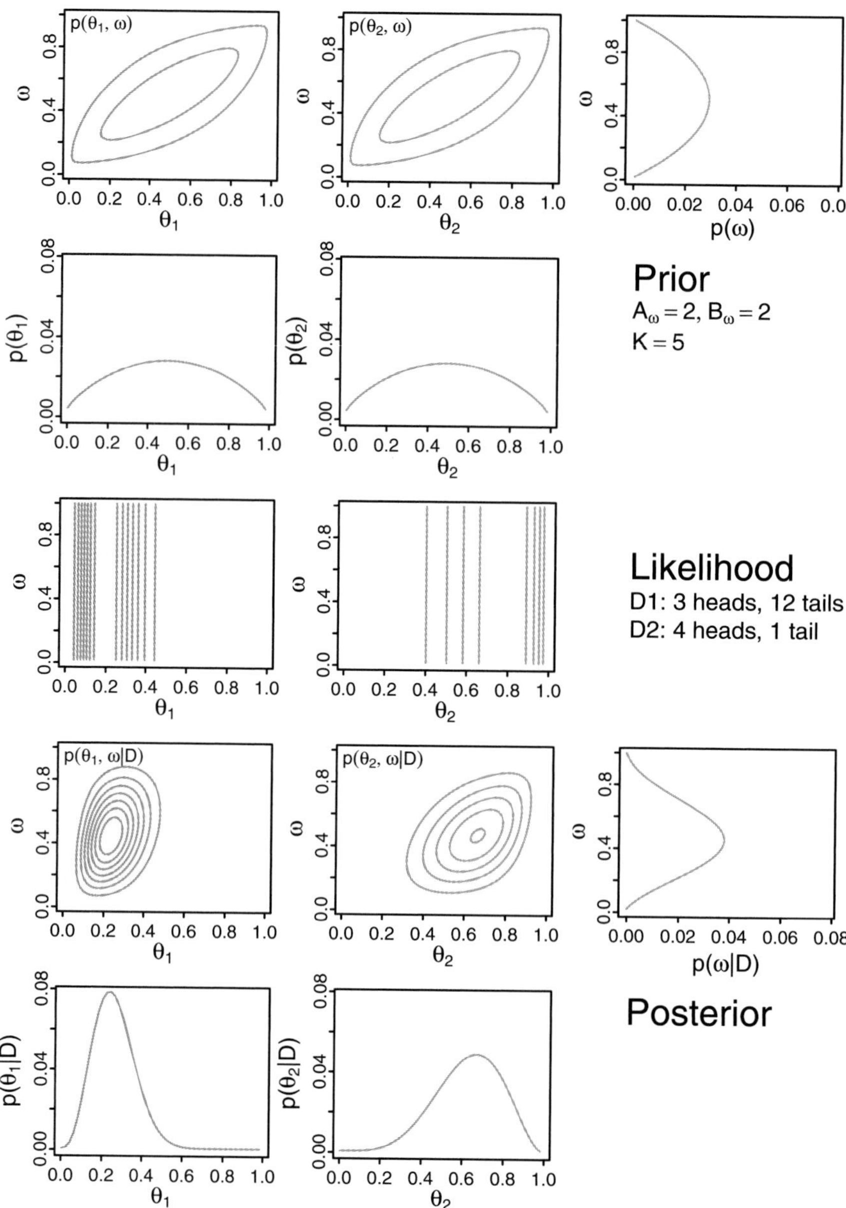

Figure 9.5 The prior imposes only a weak dependence of θ on ω (i.e., K is small), so the posteriors on θ_1 and θ_2 (bottom row) are only weakly influenced by each other's data. Compare with Figure 9.6, which uses the same data but a prior that has a strong dependence.

The full prior distribution is a joint distribution over three parameters: ω, θ_1, and θ_2. In a grid approximation, the prior is specified as a three-dimensional (3D) array that holds the prior probability at various grid points in the 3D space. The prior probability at point $\langle \omega, \theta_1, \theta_2 \rangle$ is $p(\theta_1|\omega) \cdot p(\theta_2|\omega) \cdot p(\omega)$ with exact normalization enforced by summing across the entire grid and dividing by the total.

Because the parameter space is 3D, a distribution on it cannot easily be displayed on a 2D page. Instead, Figure 9.5 shows various marginal distributions. The top row shows two contour plots: one is the marginal distribution $p(\theta_1, \omega)$ which collapses across θ_2, and the other is the marginal distribution $p(\theta_2, \omega)$ which collapses across θ_1. From these contour plots you can see the weak dependence of θ_s on ω. The upper-right panel shows the marginal prior distribution $p(\omega)$, which collapses across θ_1 and θ_2. You can see that the marginal on $p(\omega)$ does indeed have the shape of a beta(2, 2) distribution (but the $p(\omega)$ axis is scaled for probability masses at the arbitrary grid points, not for density).

The middle row of Figure 9.5 shows the likelihood function for the data, which comprise 3 heads out of 15 flips of the first coin, and four heads out of five flips of the second coin. Notice that the contours of the likelihood plot are parallel to the ω axis, indicating that the likelihood does not depend on ω. Notice that the contours are more tightly grouped for the first coin than for the second, which reflects the fact that we have more data from the first coin (i.e., 15 flips v. 5 flips).

The lower two rows of Figure 9.5 show the posterior distribution. Notice that the marginal posterior on θ_1 is centered near the proportion $3/15 = 0.2$ in its subject's data, and the posterior on θ_2 is centered near the proportion $4/5 = 0.8$ in its subject's data. The marginal posterior on θ_1 has less uncertainty than the marginal posterior on θ_2, as indicated by the widths of the distributions. Notice also that the marginal posterior on ω, at the right of the fourth row, remains fairly broad.

That result should be contrasted with the result in Figure 9.6, which uses the same data with a different prior. In Figure 9.6, the prior on ω is the same gentle peak, but the prior dependency of θ_s on ω is much stronger, with $K = 75$ instead of $K = 5$. The dependency can be seen graphically in the top two panels of Figure 9.6, which show contour plots of the marginals $p(\theta_s, \omega)$. The contours reveal that the distribution of θ_s is tightly linked to the value of ω, along a ridge or spindle in the joint parameter space.

The plots of the posterior distribution, in the lower rows of Figure 9.6, reveal some interesting results. Because the biases θ_s are strongly dependent on the parameter ω, the posterior estimates are fairly tightly constrained to be near the estimated value of ω. Essentially, because the prior emphasizes a relatively narrow spindle in the joint parameter space, the posterior is restricted to a zone within that spindle. Not only does this cause the posterior to be relatively peaked on all the parameters, it also pulls all the estimates in toward the focal zone. Notice, in particular, that the posterior on θ_2 is peaked around 0.4, far from the proportion $4/5 = 0.8$ in its coin's data! This shift away from the data proportion of subject 2 is caused by the fact that the other coin had a larger sample size, and

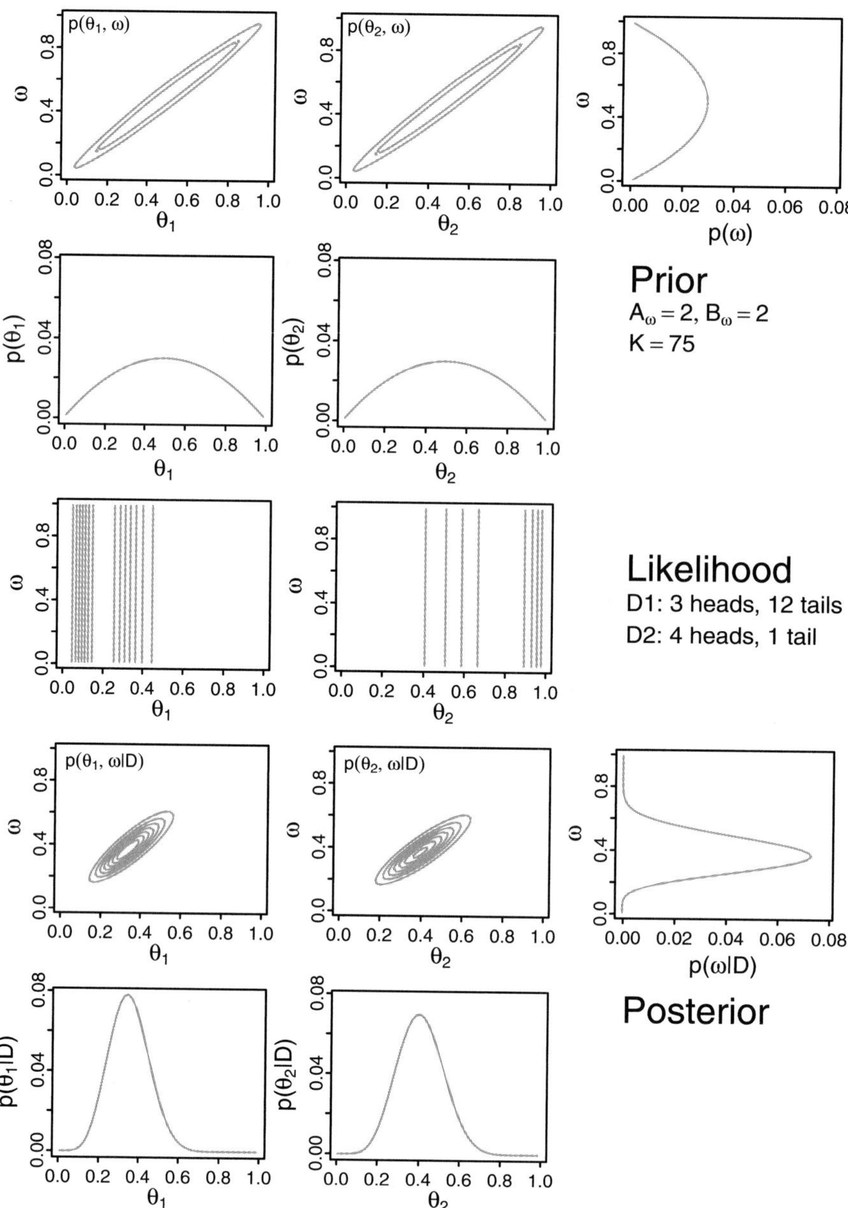

Figure 9.6 The prior imposes a strong dependency of θ on ω (i.e., K is large), so the posteriors on θ_1 and θ_2 (bottom row) are strongly influenced by each other's data, with θ_2 being pulled toward θ_1 because $N_1 > N_2$. Compare with Figure 9.5, which uses the same data but a prior that has a weak dependence.

therefore had more influence on the estimate of ω, which in turn influenced the estimate of θ_2. Of course, the data from subject 2 also influenced ω and θ_1, but with less strength.

One of the desirable aspects of using grid approximation to determine the posterior is that we do not require any analytical derivation of formulas. Instead, our computer simply keeps track of the values of the prior and likelihood at a large number of grid points and sums over them to determine the denominator of Bayes' rule. Grid approximation can use mathematical expressions for the prior as a convenience for determining the prior values at all those thousands of grid points. What's nice is that we can use, for the prior, any (non-negative) mathematical function we want, without knowing how to formally normalize it, because it will be normalized by the grid approximation. My choice of the priors for this example, summarized in Figure 9.4, was motivated merely by the pedagogical goal of using functions that you are familiar with, not by any mathematical restriction.

The grid approximation displayed in Figures 9.5 and 9.6 used combs of only 50 points on each parameter (ω, θ_1, and θ_2). This means that the 3D grid had $50^3 = 125,000$ points, which is a size that can be handled easily on an ordinary desktop computer of the early 21st century. It is interesting to remind ourselves that the grid approximation displayed in Figures 9.5 and 9.6 would have been on the edge of computability 50 years ago, and would have been impossible 100 years ago. The number of points in a grid approximation can get hefty in a hurry. If we were to expand the example by including a third coin, with its parameter θ_3, then the grid would have $50^4 = 6,250,000$ points, which already strains small computers. Include a fourth coin, and the grid contains over 312 million points. Grid approximation is not a viable approach to even modestly large problems, which we encounter next.

9.2.2. A realistic model with MCMC

The previous sections have used simplified hierarchical models for the purpose of being able to graphically display the distributions over parameter space and to gain clear intuitions about how Bayesian inference works. In this section, we will make the model realistic by including another parameter and including more subjects.

The previous examples arbitrarily fixed the degree of dependency of θ on ω. The degree of dependency was specified as the fixed value K of the concentration or consistency parameter, κ. When κ was fixed at a large value, then the individual θ_s values stayed close to ω, but when κ was fixed at a small value, then the individual θ_s values could roam quite far from ω.

In real situations, we do not know the value of κ in advance, and instead we let the data inform us regarding its credible values. Intuitively, when the data from different coins show very similar proportions of heads, we have evidence that κ is high. But when the data from different coins show diverse proportions of heads, then we have evidence that κ is small.

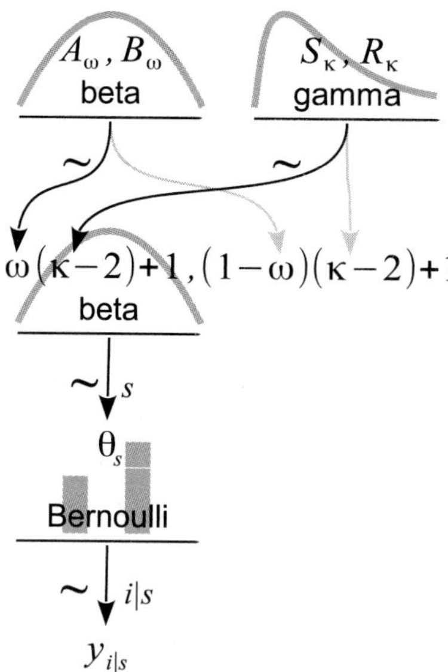

Figure 9.7 A model of hierarchical dependencies for data from several coins created independently from the same mint, with the characteristics of the mint parameterized by its mode ω and concentration κ. The value of $\kappa - 2$ has a prior distributed as a gamma density with shape and rate parameters of S_κ and R_κ.

If we want to estimate a parameter, we must have prior uncertainty about it, otherwise we would not be trying to estimate it. If we want to estimate κ with Bayesian inference, we must formally express our prior uncertainty about it. Therefore, we expand the hierarchical model to include a prior distribution on κ, as shown in Figure 9.7. This model is just like the hierarchy of Figure 9.4, except that what was a constant K is now a parameter κ with its own prior distribution. Instead of specifying a single value of K, we allow a distribution of values κ.

Because the value of $\kappa - 2$ must be non-negative, the prior distribution on $\kappa - 2$ must not allow negative values. There are many distributions with this property, and we must use one that adequately expresses our prior knowledge. For our present application, we want the prior to be vague and noncommittal, allowing a broad range of values for the concentration parameter. One distribution that is convenient for this purpose, because it is familiar in traditional mathematical statistics and available in JAGS (and in Stan and BUGS), is the *gamma distribution*. We will be using gamma distributions in a variety of contexts, just as we have been using beta distributions a lot.

The gamma$(\kappa|s, r)$ distribution is a probability density for $\kappa \geq 0$, with two parameters that determine its exact form, called its shape s and rate r parameters.[3]

[3] The formula for the gamma density is $\text{gamma}(\kappa|s, r) = (r^s/\Gamma(s))\, \kappa^{s-1} e^{-r\kappa}$, where $\Gamma(s)$ is the gamma function: $\Gamma(s) = \int_0^\infty dt\, t^{s-1} e^{-t}$. These mathematical details can be useful for analytic derivations of Bayesian posterior distributions, especially when using conjugate priors, but we will not be pursuing that method here.

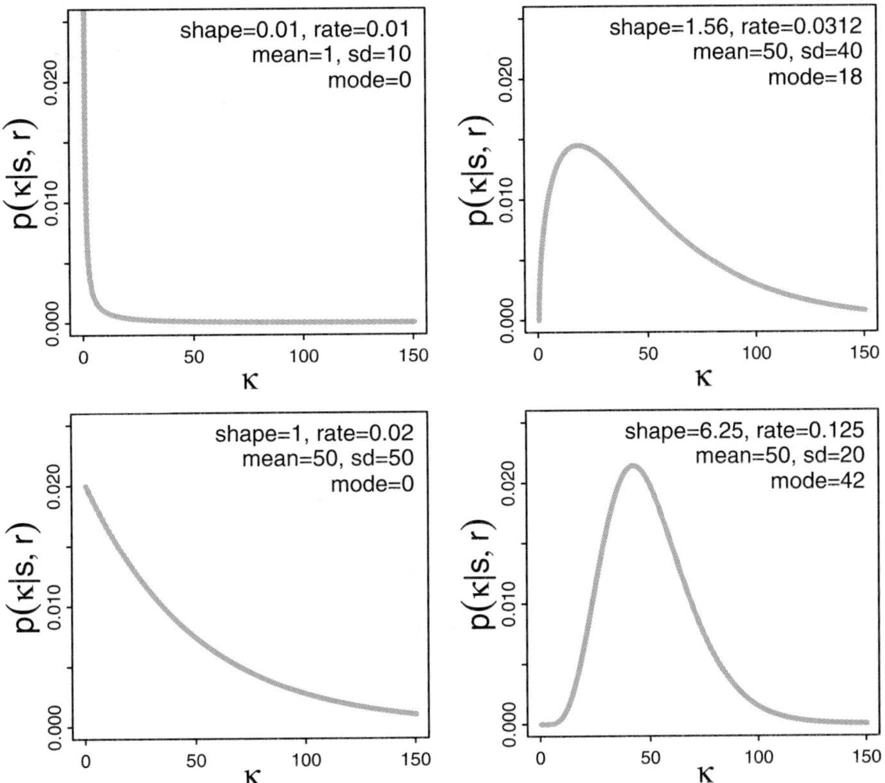

Figure 9.8 Examples of the gamma distribution. The vertical axis is $p(\kappa|s, r)$ where s is the shape and r is the rate, whose values are annotated in each panel.

Figure 9.8 shows examples of the gamma distribution with different values of the shape and rate parameters. The annotations in the panels of Figure 9.8 also show the corresponding mean, standard deviation, and mode of the distribution. The top-left panel shows a frequently used version in which both the shape and rate are set to equivalent small values. This setting results in the mean being 1 and the standard deviation being relatively large, with the mode at zero. The other three panels of Figure 9.8 show cases in which the mean is set to 50, but the standard deviations and modes differ. In the lower left, the standard deviation equals the mean, in which case the gamma distribution has shape $= 1$ (and is, in fact, an exponential distribution). In the right panels, the standard deviation is less than the mean, and you can see how the gamma distribution becomes less skewed as the standard deviation gets smaller, with the mode getting closer to the mean. I will use the generic shape of the upper-right panel of Figure 9.8 as the iconic gamma distribution in model diagrams such as Figure 9.7, because this shape suggests that the distribution has a left boundary of zero but an infinite extent on positive values.

(Other distributions can have shapes that are visually similar, and therefore the icons are labeled with the name of the intended distribution.)

As you can see from Figure 9.8, it is not intuitively obvious how particular values of the shape and rate parameters correspond with particular visual shapes of the gamma distribution. Moreover, prior beliefs are most intuitively expressed in terms of the central tendency and a width of the distribution. Therefore, it would be useful to be able to start with values for central tendency and width, and convert them into corresponding shape and rate values. It turns out that the mean of the gamma distribution is $\mu = s/r$, the mode is $\omega = (s-1)/r$ for $s > 1$, and the standard deviation is $\sigma = \sqrt{s}/r$. Algebraic manipulation then yields the following formulas for the shape and rate values in terms of the mean and standard deviation, or mode and standard deviation:

$$s = \frac{\mu^2}{\sigma^2} \quad \text{and} \quad r = \frac{\mu}{\sigma^2} \quad \text{for mean } \mu > 0 \tag{9.7}$$

$$s = 1 + \omega r \quad \text{where} \quad r = \frac{\omega + \sqrt{\omega^2 + 4\sigma^2}}{2\sigma^2} \quad \text{for mode } \omega > 0 \tag{9.8}$$

For example, suppose we desire a gamma distribution with a mode of $\omega = 42$ and a standard deviation of $\sigma = 20$. We compute the corresponding shape and rate from Equation 9.8, which yields $r = (42 + \sqrt{42^2 + 4 \cdot 20^2})/(2 \cdot 20^2) = 0.125$ and $s = 1 + 42 \cdot 0.125 = 6.25$, as shown in the lower-right panel of Figure 9.8.

I have created convenient utility functions in R that implement the parameter transformations in Equations 9.7 and 9.8. The functions are loaded into R by typing `source("DBDA2E-utilities.R")`, when that file is in the current working directory. Hopefully the function names are self-explanatory, and here is an example of their use:

```
> gammaShRaFromMeanSD( mean=10 , sd=100 )
$shape
[1] 0.01
$rate
[1] 0.001
> gammaShRaFromModeSD( mode=10 , sd=100 )
$shape
[1] 1.105125
$rate
[1] 0.01051249
```

The functions return a `list` of named components. Therefore, if you assign the result of the function to a variable, you can get at the individual parameters by their component names, as in the following example:

```
> gammaParam = gammaShRaFromModeSD( mode=10 , sd=100 )
> gammaParam$shape
[1] 1.105125
> gammaParam$rate
[1] 0.01051249
```

You may find those functions useful when you are setting the constants for prior distributions. If you want to graph a gamma distribution in R, the gamma density is provided by `dgamma(x,shape=s,rate=r)`. As we will see in the next section, JAGS parameterizes the gamma distribution with shape and rate parameters in that order.

9.2.3. Doing it with JAGS

Look again at the hierarchical diagram in Figure 9.7 (p. 236). The arrows in that diagram indicate the dependencies between the variables. The key thing to understand is that every arrow in the hierarchical diagram has a corresponding expression in the JAGS model specification. The JAGS model specification is a verbal code for the graphical diagram. Here is a JAGS model specification that corresponds to Figure 9.7:

```
model {
  for ( i in 1:Ntotal ) {
    y[i] ~ dbern( theta[s[i]] )
  }
  for ( s in 1:Nsubj ) {
    theta[s] ~ dbeta( omega*(kappa-2)+1 , (1-omega)*(kappa-2)+1 )
  }
  omega ~ dbeta( 1 , 1 )
  kappa <- kappaMinusTwo + 2
  kappaMinusTwo ~ dgamma( 0.01 , 0.01 )  # mean=1 , sd=10 (generic vague)
}
```

The beginning of the model specification expresses the arrow for the Bernoulli likelihood function and uses nested indexing in `theta[s[i]]`, as was explained in Section 8.4 (p. 208). Moving up the diagram in Figure 9.7, the arrow expressing the prior on ω is coded in JAGS as `omega` $\sim$ `dbeta(1,1)`, which is a uniform distribution. In general, the prior on ω can be whatever is an appropriate description of real prior knowledge. In the present implementation, a vague and noncommittal prior is used for generic application.

Finally, the prior on $\kappa - 2$ is implemented by two lines of code in the JAGS model specification, repeated here:

```
kappa <- kappaMinusTwo + 2
kappaMinusTwo ~ dgamma( 0.01 , 0.01 )  # mean=1, sd=10 (generic vague)
```

Notice that the `dgamma` distribution generates a variable called `kappaMinusTwo`, and then `kappa` itself is `kappaMinusTwo + 2`. This two-line implementation is needed because the gamma distribution covers zero to infinity, but κ must be no smaller than 2. The choice of shape and rate constants in the `dgamma` distribution makes the prior allow a very broad range of κ values, yielding a sensible prior on θ_s as will be shown later. Exercise 9.1 explores different priors on κ.

These details of the model specification in JAGS have been presented to demonstrate how easy it is to implement a hierarchical model in JAGS. If you can draw a coherent hierarchical diagram, then you can implement it in JAGS. This functionality provides great flexibility for defining meaningful descriptive models of data. I have created scripts for using the model that require no knowledge of the implementation details, as explained in the following example.

9.2.4. Example: Therapeutic touch

Therapeutic touch is a nursing technique in which the practitioner manually manipulates the "energy field" of a patient who is suffering from a disease. The practitioner holds her or his hands near but not actually touching the patient, and repatterns the energy field to relieve congestion and restore balance, allowing the body to heal. Rosa, Rosa, Sarner, and Barrett (1998) reported that therapeutic touch has been widely taught and widely used in nursing colleges and hospitals despite there being little if any evidence of its efficacy.

Rosa et al. (1998) investigated a key claim of practitioners of therapeutic touch, namely, that the practitioners can sense a body's energy field. If this is true, then practitioners should be able to sense which of their hands is near another person's hand, even without being able to see their hands. The practitioner sat with her hands extended through cutouts in a cardboard screen, which prevented the practitioner from seeing the experimenter. On each trial, the experimenter flipped a coin and held her hand a few centimeters above one or the other of the practitioner's hands, as dictated by the flip of the coin. The practitioner then responded with her best guess regarding which of her hand's was being hovered over. Each trial was scored as correct or wrong. The experimenter (and co-author of the article) was 9-years old at the time.

Each practitioner was tested for 10 trials. There were a total of 21 practitioners in the study, seven of whom were tested twice approximately a year apart. The retests were counted by the authors as separate subjects, yielding 28 nominal subjects. The proportions correct for the 28 subjects are shown in Figure 9.9. Chance performance is 0.50. The question is how much the group as a whole differed from chance performance, and how much any individuals differed from chance performance. The hierarchical model of Figure 9.7 is perfect for these data, because it estimates the underlying ability of each subject while simultaneously estimating the modal ability of the group and

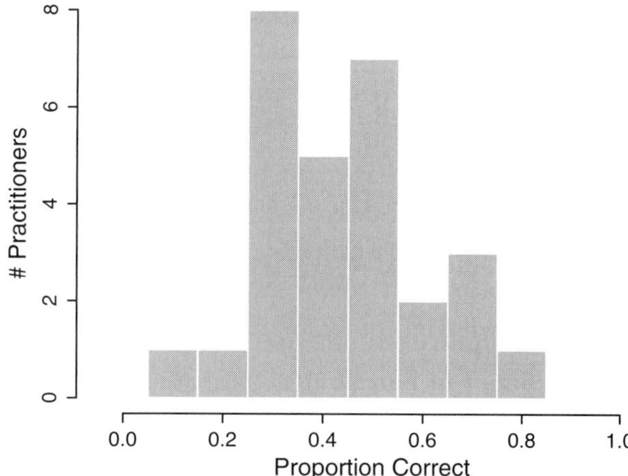

Figure 9.9 Data from the therapeutic touch experiment of Rosa et al. (1998). Histogram of proportion correct for the 28 practitioners.

the consistency of the group. Moreover, the distribution of proportions correct across subjects (Figure 9.9) is essentially unimodal and can be meaningfully described as a beta distribution. With 28 subjects, there are a total of 30 parameters being estimated.

Below is a script for running the analysis on the therapeutic touch data. The script has a structure directly analogous to the scripts used previously, such as the one described in Section 8.3. The complete script is the file named `Jags-Ydich-XnomSsubj-MbernBetaOmegaKappa-Example.R`. Please read through the excerpt below, paying attention to the comments:

```
# Read the data file:
myData = read.csv("TherapeuticTouchData.csv")
# Load the relevant model functions into R's working memory:
source("Jags-Ydich-XnomSsubj-MbernBetaOmegaKappa.R")
# Generate the MCMC chain:
mcmcCoda = genMCMC( data=myData , sName="s" , yName="y" ,
                    numSavedSteps=20000 , thinSteps=10 )
# Display diagnostics of chain, for specified parameters:
diagMCMC( codaObject=mcmcCoda , parName="omega" )
diagMCMC( codaObject=mcmcCoda , parName="kappa" )
diagMCMC( codaObject=mcmcCoda , parName="theta[1]" )
# Get summary statistics of chain:
smryMCMC( mcmcCoda ,
          compVal=0.5 , diffIdVec=c(1,14,28), compValDiff=0.0 )
# Display posterior information:
plotMCMC( mcmcCoda , data=myData , sName="s" , yName="y" ,
          compVal=0.5 , diffIdVec=c(1,14,28), compValDiff=0.0 )
```

The data file must be arranged with one column that contains the outcome of the trials (i.e., 0 or 1 in each row), and another column that contains the identifier of the subject who generated each outcome. The subject identifier could be a numerical integer or a character string. The first row of the data file must contain the names of the columns, and those column names are included as the arguments `sName` and `yName` in the functions `genMCMC` and `plotMCMC`. If you look at the file TherapeuticTouchData.csv, you will see that it has two columns, one column named "y" consisting of 0's and 1's, and a second column named "s" consisting of subject identifiers such as S01, S02, and so on. When the file is read by the `read.csv` function, the result is a data frame that has a vector named "y" and a factor named "s."

In the call to the `genMCMC` function, you can see that there are 20,000 saved steps and thinning of 10 steps. These values were chosen after a short initial run with no thinning (i.e., `thinSteps=1`) showed strong autocorrelation that would have required a very long chain to achieve an effective sample size of 10,000 for the ω parameter. To keep the saved MCMC sample down to a modest file size (under 5 MB), I chose to set `thinSteps=10` and to save 20,000 of the thinned steps. This still required waiting through 200,000 steps, however, and throwing away information. If computer memory is of no concern, then thinning is neither needed nor recommended. When you run the script, you will notice that it takes a minute for JAGS to generate the long chain. Section 9.4 discusses ways to speed up the processing.

The diagnostic plots produced by calls to the `diagMCMC` function all showed adequate convergence. They are not shown here, but you can see them by running the script yourself. Only the `kappa` parameter showed high autocorrelation, which is typical of higher-level parameters that control the variance of lower-level parameters. Nevertheless, the chains for κ show good overlap and we are not concerned with a very accurate estimate of its value. Moreover, we will see that the marginal posterior on κ spans a wide range of values, and the estimates of the other parameters are not hugely affected by small changes in κ.

A numerical summary of the posterior distribution is produced by the `smryMCMC` function. An argument of the function that is unique to this model is `diffIdVec`. It specifies a vector of subject indices that should have their posterior differences summarized. For example, `diffIdVec=c(1,14,28)` produces summaries of the posterior distributions of $\theta_1 - \theta_{14}$, $\theta_1 - \theta_{28}$, and $\theta_{14} - \theta_{28}$. The argument defaults to showing no differences instead of all differences (which, in this application, would yield $28(28-1)/2 = 378$ differences).

Finally, a graphical summary of the posterior distribution is produced by the `plotMCMC` function. It also has the `diffIdVec` argument to specify which differences should be displayed. The result is shown in Figure 9.10. The posterior distribution reveals several interesting implications, described next.

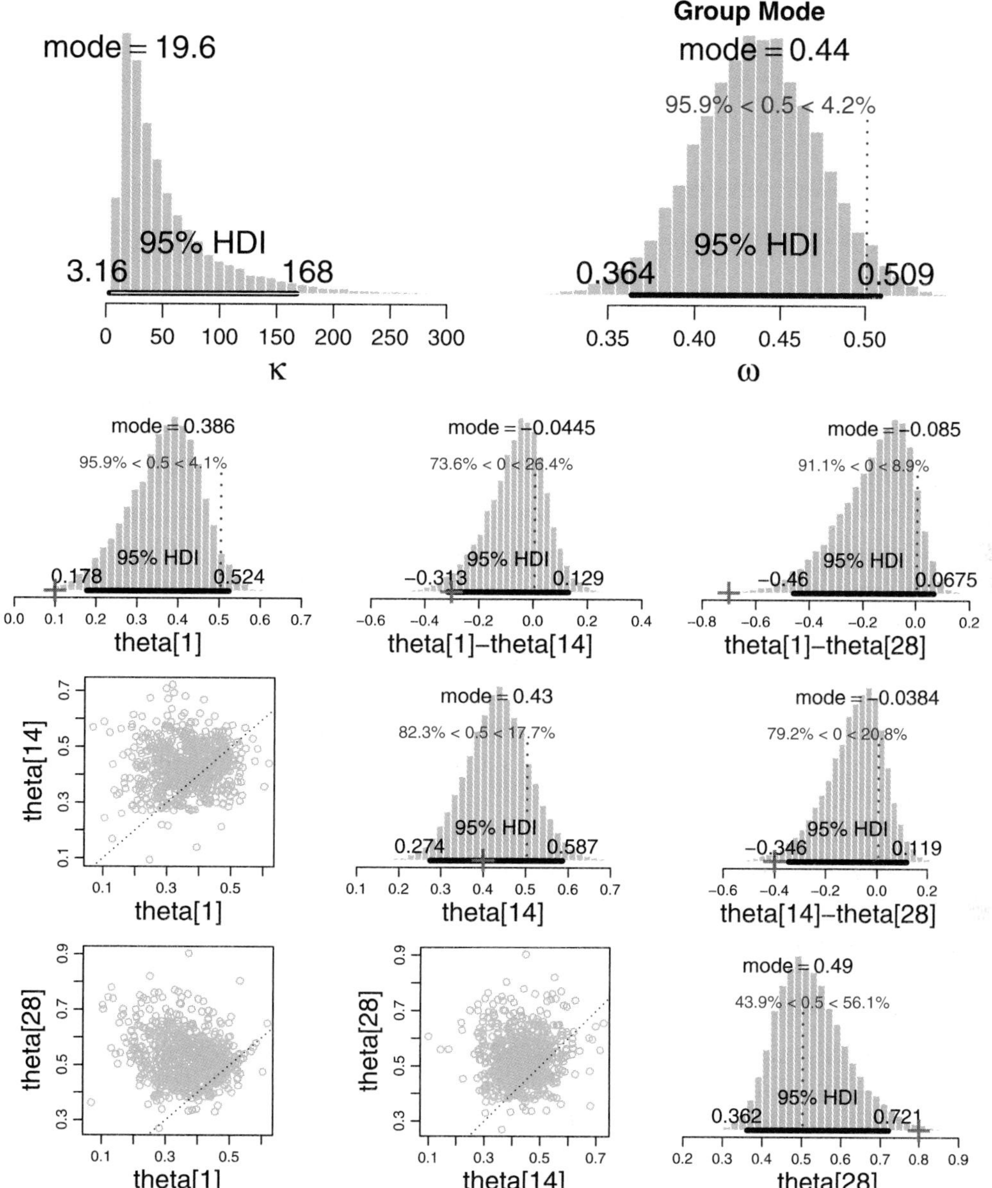

Figure 9.10 Marginal posterior distributions for the therapeutic touch data.

The upper-right panel of Figure 9.10 shows the marginal posterior distribution on the group-level mode, ω. It's most credible value is less than 0.5, and its 95% HDI includes the chance value of 0.5, so we would certainly not want to conclude that the group of practitioners, as a class, detects the energy field of a nearby hand better than chance. (If the group mode had been credibly *less* than 0.5, we might have inferred that the practitioners as a group could detect the experimenter's hand but then systematically misinterpreted the energy field to give a contrary response.)

The lower three rows of Figure 9.10 show marginal posteriors for θ_s of selected individuals. The data conveniently ordered the subject identifiers so that subject 1 had the worst performance and subject 28 had the best performance, with the others between, such that subject 14 was near the 50th percentile of the group. Down the diagonal panels are the estimates of the individual parameters. You can see that the estimate for theta[1] has its peak around 0.4, even though this subject got only 1 out of 10 trials correct, as indicated by the "+" marked on the horizontal axis at 0.10. The estimate is pulled above the subject's actual performance because there are 27 other subjects all providing data that indicate typical performance is higher. In other words, the 27 other subjects inform the estimate of group-level parameters, and those group-level parameters pull the estimate of the individual toward what is typical for the group. The same phenomenon can be seen in the lower-right panel, where it can be seen the estimate for the best subject peaks around 0.5, even though this subject got 8 out of 10 trials correct, as shown by the "+" marked at 0.8 on the horizontal axis. The 95% HDI's of even the worst and best subjects include 0.5, so we cannot conclude that any of the subjects were performing credibly differently than chance.

The lower three rows of Figure 9.10 also show pairs of individual estimates and their differences. We might be interested, for example, in whether the worst and best subjects performed very differently. The answer is provided by the marginal posterior on the difference $\theta_1 - \theta_{28}$, plotted in the right column. The most credible difference is near -0.1 (which is nowhere near the difference of proportions exhibited by the subjects, indicated by the "+" on the horizontal axis), and the 95% HDI includes a difference of zero.

In conclusion, the posterior distribution indicates that the most credible values for the group as a whole and for all the individuals include chance performance. Either the experiment procedure was not appropriate for detecting the abilities of therapeutic touch practitioners, or much more data would be needed to detect a very subtle effect detectable by the procedure, or there is no actual sensing ability in the practitioners. Notice that these conclusions are tied to the particular descriptive model we used to analyze the data. We used a hierarchical model because it is a reasonable way to capture individual differences and group-level tendencies. The model assumed that all individuals were representative of the same overarching group, and therefore all individuals mutually informed each other's estimates.

It can be very useful to view the prior distribution in hierarchical models, as sometimes the prior on mid-level parameters, implied by the top-level prior, is not actually what was intuited or intended. It is easy to make JAGS produce an MCMC sample from the prior by running the analysis without the data, as was explained in Section 8.5 (p. 211).

The prior for the therapeutic touch data is shown in Figure 9.11. It can be seen that the prior is uniform on the group-level mode, ω, and uniform on each of the individual level biases, θ_s. This is good because it was what was intended, namely, to be noncommittal and to give equal prior credibility to any possible bias.[4]

The priors on the differences of individual biases are also shown in Figure 9.11. They are triangular distributions, peaked at a difference of zero. You can understand why this shape arises by examining the scatter plots of the joint distributions in the lower left. Those scatter plots are uniform across the two dimensions, so when they are collapsed along the diagonal that marks a difference of zero, you can see that there are many more points along the diagonal, and the number of points drops off linearly toward the corners. This triangular prior on the differences might not have been what you imagined it would be, but it is the natural consequence of independent uniform priors on the individual parameters. Nevertheless, the prior does show a modest preference for differences of zero between individuals.

9.3. SHRINKAGE IN HIERARCHICAL MODELS

In typical hierarchical models, the estimates of low-level parameters are pulled closer together than they would be if there were not a higher-level distribution. This pulling together is called *shrinkage* of the estimates. We have seen two cases of shrinkage, one in Figure 9.10 and another previously in Figure 9.6 (p. 234). In Figure 9.10, the most credible values of individual-level biases, θ_s, were closer to the group-level mode, ω, than the individual proportions correct, z_s/N_s. Thus, the variance between the estimated values θ_s is less than the variance between the data values z_s/N_s.

This reduction of variance in the estimators, relative to the data, is the general property referred to by the term "shrinkage." But the term can be misleading because shrinkage leads to reduced variance only for *uni*modal distributions of parameters. In general, shrinkage in hierarchical models causes low-level parameters to shift toward the modes of the higher-level distribution. If the higher-level distribution has multiple modes, then the low-level parameter values cluster more tightly around those multiple modes, which might actually pull some low-level parameter estimates apart instead of together. Examples are presented at the book's blog (http://doingbayesiandataanalysis.blogspot.com/2013/03/shrinkage-in-bimodal-hierarchical.html) and in Kruschke and Vanpaemel (in press).

[4] The first edition of this book presented this model parameterized by the group-level mean instead of the group-level mode as is done here. There is nothing logically or mathematically wrong with parameterizing by the group-level mean, but it produces an unintended prior distribution on the individual θ_s values. The parameterization by the group-level mode, used in this edition, supersedes the previous parameterization.

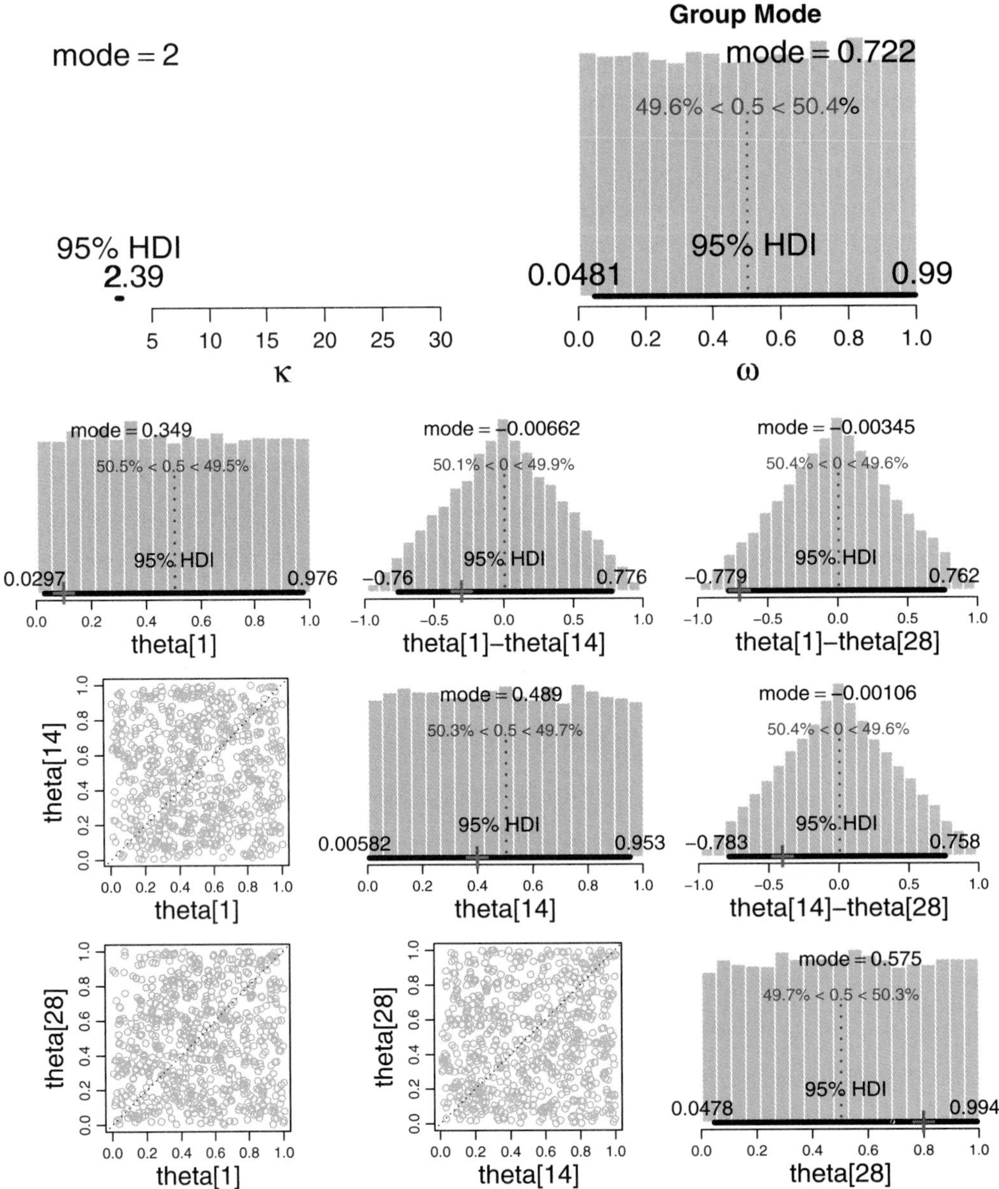

Figure 9.11 Marginal prior distributions for the therapeutic touch data. The upper-left panel, showing κ, does not plot well because it is a tall narrow peak near 2, with a long short tail extending far right. The estimated modal values of uniform distributions should be disregarded, as they are merely marking whatever random ripple happens to be a little higher than the other random ripples.

Shrinkage is a rational implication of hierarchical model structure, and is (usually) desired by the analyst because the shrunken parameter estimates are less affected by random sampling noise than estimates derived without hierarchical structure. Intuitively, shrinkage occurs because the estimate of each low-level parameter is influenced from two sources: (1) the subset of data that are directly dependent on the low-level parameter, and (2) the higher-level parameters on which the low-level parameter depends. The higher-level parameters are affected by all the data, and therefore the estimate of a low-level parameter is affected indirectly by all the data, via their influence on the higher-level parameters.

It is important to understand that shrinkage is a consequence of hierarchical model structure, not Bayesian estimation. To better understand shrinkage in hierarchical models, we will consider the maximum likelihood estimate (MLE) for the hierarchical model of Figure 9.7. The MLE is the set of parameter values that maximizes the probability of the data. To find those values, we need the formula for the probability of the data, given the parameter values. For a single datum, $y_{i|s}$, that formula is

$$\begin{aligned} & p\left(y_{i|s} \mid \theta_s, \omega, \kappa \right) \\ & = \text{bern}(y_{i|s}|\theta_s) \cdot \text{beta}(\theta_s \mid \omega(\kappa-2)+1, (1-\omega)(\kappa-2)+1) \end{aligned} \tag{9.9}$$

For the whole set of data, $\{y_{i|s}\}$, because we assume independence across data values, we take the product of that probability across all the data:

$$\begin{aligned} & p\left(\{y_{i|s}\} \mid \{\theta_s\}, \omega, \kappa \right) \\ & = \prod_s \prod_{i|s} p\left(y_{i|s} \mid \theta_s, \omega, \kappa \right) \\ & = \prod_s \prod_{i|s} \text{bern}(y_{i|s}|\theta_s) \cdot \text{beta}(\theta_s \mid \omega(\kappa-2)+1, (1-\omega)(\kappa-2)+1) \end{aligned} \tag{9.10}$$

In that formula, remember that the $y_{i|s}$ values are constants, namely the 0's and 1's in the data. Our goal here is to find the values of the parameters in the formula that will maximize the probability of the data. You can see that θ_s participates in both the low-level data probability, $\prod_{i|s} \text{bern}(y_{i|s}|\theta_s)$, and in the high-level group probability, $\text{beta}(\theta_s \mid \omega(\kappa-2)$ $+1, (1-\omega)(\kappa-2)+1)$. The value of θ_s that maximizes the low-level data probability is the data proportion, z_s/N_s. The value of θ_s that maximizes the high-level group probability is the higher-level mode, ω. The value of θ_s that maximizes the product of low-level and high-level falls between the data proportion, z_s/N_s, and the higher-level mode, ω. In other words, θ_s is pulled toward the higher-level mode.

As a concrete numerical example, suppose we have five individuals, each of whom went through 100 trials, and had success counts of 30, 40, 50, 60, and 70. It might seem reasonable that a good description of these data would be $\theta_1 = z_1/N_1 = 30/100 = 0.30$, $\theta_2 = z_2/N_2 = 40/100 = 0.40$, and so on, with a higher-level beta distribution that is nearly flat. This description is shown in the left panel of Figure 9.12. The probability of the data, given this choice of parameter values, is the likelihood computed from Equation 9.10 and is displayed at the top of the panel. Although this choice of parameter values is reasonable, we ask: are there other parameter values that increase the likelihood? The answer is yes, and the parameter values that maximize the likelihood are displayed in the right panel of Figure 9.12. Notice that the likelihood (shown at the top of the panel) has increased by a factor of more than 20. The θ_s values in the MLE are shrunken closer to the mode of the higher-level distribution, as emphasized by the arrows in the plot.

Intuitively, shrinkage occurs because the data from all individuals influence the higher-level distribution, which in turn influences the estimates of each individual. The estimate of θ_s is a compromise between the data z_s/N_s of that individual and the group-level distribution informed by all the individuals. Mathematically, the compromise is expressed by having to jointly maximize $\prod_{i|s} \text{bern}(y_{i|s}|\theta_s)$ and $\text{beta}(\theta_s \mid \omega(\kappa-2)+1, (1-\omega)(\kappa-2)+1)$. The first term is maximized by $\theta_s = z_s/N_s$ and the second term is maximized by $\theta_s = \omega$. As θ_s is pulled away from z_s/N_s toward ω, the first term gets smaller but the second term gets larger. The MLE finds the best compromise between the two factors.

In summary, shrinkage is caused by hierarchical structure. We have seen how shrinkage occurs for the MLE. The MLE involves no prior on the top-level parameters,

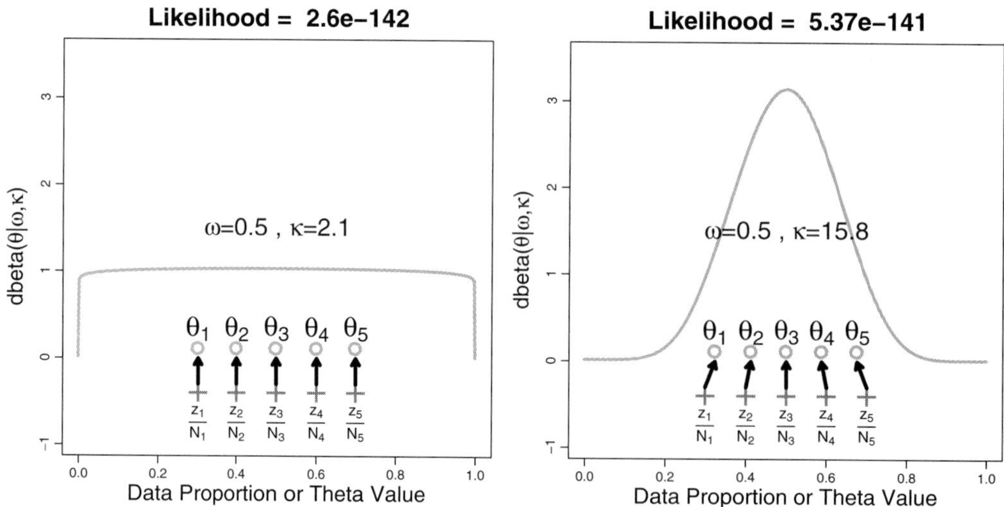

Figure 9.12 Shrinkage of individual parameter values in maximum likelihood estimation (MLE). Within each panel, the data proportion (z_s/N_s) from each individual is plotted as a "+" symbol, and the candidate value of θ_s is plotted as a circle, and the overarching beta distribution is also plotted with its candidate values of mode (ω) and concentration (κ) annotated. The left panel shows the choice of $\theta_s = z_s/N_s$ with a nearly flat beta distribution. The right panel shows the MLE, which exhibits shrinkage. Arrows highlight the shrinkage. The likelihood has improved by a factor greater than 20.

of course. When we use Bayesian estimation, we supply a prior distribution on the top-level parameters, and infer an entire posterior distribution across the joint parameter space. The posterior distribution explicitly reveals the parameter values that are credible and their uncertainty, given the data. Exercise 9.3 explores the Bayesian analysis of the data in Figure 9.12.

9.4. SPEEDING UP JAGS

In this section, we consider two ways to speed up processing of JAGS. One method changes the likelihood function in the JAGS model specification. A second method uses the runjags package to run chains in parallel on multicore computers. The two methods are discussed in turn.

When there are many trials per subject, generating the chains in JAGS can become time consuming. One reason for slowness is that the model specification (recall Section 9.2.3) tells JAGS to compute the Bernoulli likelihood for every single trial:

```
for ( i in 1:Ntotal ) {
  y[i] ~ dbern( theta[s[i]] )
}
```

Although that statement (above) looks like a `for`-loop in R that would be run in a sequential procedure, it is really just telling to JAGS to set up many copies of the Bernoulli relation. Even though JAGS is not cycling sequentially through the copies, there is extra computational overhead involved. It would be nice if, instead of evaluating a Bernoulli distribution N_s times, we could tell JAGS to compute $\theta_s^{z_s}(1-\theta_s)^{(N_s-z_s)}$ just once.

We can accomplish this by using the binomial likelihood function instead of the Bernoulli likelihood. The binomial likelihood is $p(z_s|\theta_s, N_s) = \binom{N_s}{z_s}\theta_s^{z_s}(1-\theta_s)^{(N_s-z_s)}$, where $\binom{N_s}{z_s}$ is a constant called the binomial coefficient. It will be explained later, accompanying Equation 11.5 on p. 303, but for now all you need to know is that it is a constant. Therefore, when the binomial likelihood is put into Bayes' rule, the constant appears in the numerator and in the denominator and consequently cancels out, having no influence. Thus, we "trick" JAGS into computing $\theta_s^{z_s}(1-\theta_s)^{(N_s-z_s)}$ by using its built-in binomial likelihood.[5]

[5] Notice that we are using the binomial likelihood only as a computational short cut, not as a claim about sampling intentions in data collection. The formula for the binomial distribution is derived by assuming that the sample size, N_s, is fixed in advance and defines the scope of a single data-collection event. Then the outcome for a single trial can take on values of $z_s \in \{0, 1, \dots, N_s\}$ and the binomial likelihood function gives a probability for each of the possible outcome values. In our applications, we do not necessarily assume that N_s is fixed, and instead we treat a single "flip" as the elementary data-collection event. See also footnote 2 on p. 126.

The necessary changes in the program are simple and few. First, we compute the z_s and N_s values from the data. The code below assumes that y is a vector of 0's and 1's, and that s is vector of integer subject identifiers. The code uses the aggregate function, which was explained in Section 3.6.

```
z = aggregate( y , by=list(s) , FUN=sum )$x
N = aggregate( rep(1,length(y)) , by=list(s) , FUN=sum )$x
dataList = list(
  z = z ,
  N = N ,
  Nsubj = length(unique(s))
)
```

Notice that the dataList (above) no longer includes any mention of y or Ntotal. The only information sent to JAGS is z_s, N_s, and the overall number of subjects, Nsubj.

Then we modify the model specification so it uses the binomial distribution, which is called dbin in JAGS:

```
model {
  for ( s in 1:Nsubj ) {
    z[s] ~ dbin( theta[s] , N[s] )
    theta[s] ~ dbeta( omega*(kappa-2)+1 , (1-omega)*(kappa-2)+1 )
  }
  omega ~ dbeta( 1 , 1 )
  kappa <- kappaMinusTwo + 2
  kappaMinusTwo ~ dgamma( 0.01 , 0.01 )  # mean=1 , sd=10 (generic vague)
}
```

The only change in the model specification (above) was removing the loop for y[i]~dbern(theta[s[i]]) and putting the new binomial likelihood inside the subject loop.

The changes described above have been implemented in programs with file names beginning with Jags-Ydich-XnomSsubj-MbinomBetaOmegaKappa. Notice in the file name that the part starting with "-M" says binom instead of bern. Notice also that the part starting with "-Y" still says dich because the data are still dichotomous 0's and 1's, only being converted internally to z's and N's for computational purposes. To run the binomial version, all you have to do is tell R to load it instead of the Bernoulli version:

```
source("Jags-Ydich-XnomSsubj-MbinomBetaOmegaKappa.R")
```

For the modest amount of data in the therapeutic-touch experiment, the acceleration achieved by using the binomial instead of the Bernoulli is slight. But for larger data sets, the reduction in duration can be noticeable. The duration of the MCMC sampling can be measured in R using the `proc.time` function that was explained in Section 3.7.5, p. 66. Exercise 9.4 shows an example.

An important way to speed the processing of the MCMC is to run them in parallel with the runjags package, as was explained in Section 8.7, p. 215. The program `Jags-Ydich-XnomSsubj-MbinomBetaOmegaKappa.R` has been set up with runjags with parallel chains as its default. You can alter the setting inside the program, then save the program and re-`source` it, and then see how much slower it runs in nonparallel mode. I have found that it takes about half the time to run with three parallel chains as for one long chain, achieving about the same ESS either way. Exercise 9.4 provides more prompts for you to try this yourself.

9.5. EXTENDING THE HIERARCHY: SUBJECTS WITHIN CATEGORIES

Many data structures invite hierarchical descriptions that may have multiple levels. Software such as JAGS makes it easy to implement hierarchical models, and Bayesian inference makes it easy to interpret the parameter estimates, even for complex nonlinear hierarchical models. Here, we take a look at one type of extended hierarchical model.

Suppose our data consist of many dichotomous values for individual subjects from different categories, with all the categories under an overarching common distribution. For example, consider professional baseball players who, over the course of a year of games, have many opportunities at bat, and who sometimes get a hit. Players have different fielding positions (e.g., pitcher, catcher, and first base), with different specialized skills, so it is meaningful to categorize players by their primary position. Thus, we estimate batting abilities for individual players, and for positions, and for the overarching group of professional players. You may be able to think of a number of analogous structures from other domains.

This sort of hierarchical model structure is depicted in Figure 9.13. It is the same as the diagram in Figure 9.7 but with an extra layer added for the category level, along with the subscripts needed to indicate the categories. At the bottom of Figure 9.13, the individual trials are denoted as $y_{i|s,c}$ to indicate instances within subjects s and categories c. The underlying bias of subject s within category c is denoted $\theta_{s|c}$. The biases of subjects within category c are assumed to be distributed as a beta density with mode ω_c and concentration κ_c. Thus, each category has its own modal bias ω_c, from which all subject biases in the category are assumed to be drawn. Moving up the diagram, the model assumes that all the category modes come from a higher-level beta distribution that describes the variation across categories. The modal bias across categories is denoted ω (without a subscript), and the concentration of the category biases is denoted κ (without a subscript). When κ is large, the category biases ω_c are tightly concentrated. Because we are estimating ω and κ, we must specify prior distributions for them, indicated at the top of the diagram.

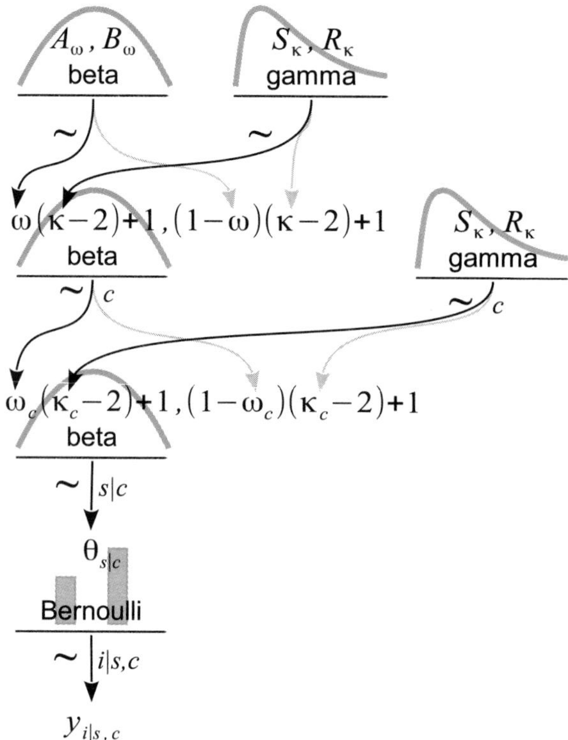

Figure 9.13 A model of hierarchical dependencies for data from several coins (indexed by subscript *s*) created by more than one category of mint (indexed by subscript *c*), with an overarching distribution across categories.

This model has an over-arching distribution over the category modes, ω_c, that has its central tendency ω and scale κ estimated. But, purely for simplicity, the model does not have an over-arching distribution on the category concentrations κ_c that has its central tendency and scale estimated. In other words, the prior on κ_c applies independently to each κ_c in a manner fixed by the prior constants S_κ and R_κ, and the κ_c's do not mutually inform each other via that part of the hierarchy. The model could be extended to include such an overarching distribution, as was done in a version on the book's blog (http://doingbayesiandataanalysis.blogspot.com/2012/11/shrinkage-in-multi-level-hierarchical.html).

Whereas the data are really 0's and 1's from individual trials, we will assume the data file has total results for each subject, and is arranged with one row per subject, with each row indicating the number of successes (or heads), $z_{s|c}$, the number of attempts (or flips), $N_{s|c}$, and the category of the subject, c. Each row could also contain a unique subject identifier for ease of reference. The categories could be indicated by meaningful textual labels, but we will assume that they are converted by the program into consecutive integers for indexing purposes. Therefore, in the JAGS code below, we will use the notation c[s] for the integer category label of subject s.

Expressing the hierarchical model of Figure 9.13 in JAGS is straight forward. For every arrow in the diagram, there is a corresponding line of code in the JAGS model specification:

```
model {
  for ( s in 1:Nsubj ) {
    z[s] ~ dbin( theta[s] , N[s] )
    theta[s] ~ dbeta( omega[c[s]]*(kappa[c[s]]-2)+1 ,
                      (1-omega[c[s]])*(kappa[c[s]]-2)+1 )
  }
  for ( c in 1:Ncat ) {
    omega[c] ~ dbeta( omega0*(kappa0-2)+1 ,
                      (1-omega0)*(kappa0-2)+1 )
    kappa[c] <- kappaMinusTwo[c] + 2
    kappaMinusTwo[c] ~ dgamma( 0.01 , 0.01 ) # mean=1 , sd=10 (generic vague)
  }
  omega0 ~ dbeta( 1.0 , 1.0 )
  kappa0 <- kappaMinusTwo0 + 2
  kappaMinusTwo0 ~ dgamma( 0.01 , 0.01 )  # mean=1 , sd=10 (generic vague)
}
```

Because the data have been totaled, the expression of the likelihood uses a binomial (dbin) instead of Bernoulli distribution, without looping over trials i, as was explained previously in Section 9.4. Notice the nested indexing for categories in omega[c[s]], which uses the mode of the appropriate category for describing the individual theta[s]. The overarching ω and κ (with no subscript) in Figure 9.13 are coded as omega0 and kappa0.

This example is intended to demonstrate how easy it is to specify a complex hierarchical model in JAGS. If you can make a coherent model diagram for whatever application you are considering, then you can probably implement that model in JAGS. In fact, that is the method I use when I am building a new JAGS program: I start by making a diagram to be sure I fully understand the descriptive model, then I create the JAGS code arrow by arrow, starting at the bottom of the diagram. In the next section, we consider a specific application and see the richness of the information produced by Bayesian inference.

9.5.1. Example: Baseball batting abilities by position

Consider the sport of baseball. During a year of games, different players have different numbers of opportunities at bat, and on some of these opportunities a player might actually hit the ball. In American Major League Baseball, the ball is pitched very fast, sometimes at speeds exceeding 90 miles (145 km) per hour, and batters typically hit the ball on about only 23% of their opportunities at bat. That ratio, of hits divided by opportunities at bat, is called the batting average of each player. We can think of it as an indicator of the underlying probability that the player will hit the ball for any opportunity at bat. We would like to estimate that underlying probability, as one indicator of a player's ability. Players also must play a position in the field when the other team is at bat. Different fielding positions have different specialized skills, and those players are expected to focus on those skills, not necessarily on hitting. In particular, pitchers typically are not expected to be strong hitters, and catchers might also not be expected to focus so much on hitting. Most players have a primary fielding position, although many players perform different fielding positions at different times. For purposes of simplifying the present example, we will categorize each player into a single primary fielding position.

If you, like me, are not a big sports fan and do not really care about the ability of some guy to hit a ball with a stick, then translate the situation into something you do care about. Instead of opportunities at bat, think of opportunities to cure a disease, for treatments categorized into types. We are interested in estimating the probability of cure for each treatment and each type of treatment and overall across treatments. Or think of opportunities for students to graduate from high school, with schools categorized by districts. We are interested in estimating the probability of graduation for each school and each district and overall. Or, if that's boring for you, keep thinking about guys swinging sticks.

The data consist of records from 948 players in the 2012 regular season of Major League Baseball.[6] For player s, we have his (it was an all-male league) number of opportunities at bat, $N_{s|c}$, his number of hits, $z_{s|c}$, and his primary position when in the field, c_s, which was one of nine possibilities (e.g., pitcher, catcher, and first base). All players for whom there were zero at-bats were excluded from the data set. To give some sense of the data, there were 324 pitchers with a median of 4.0 at-bats, 103 catchers with a median of 170.0 at-bats, and 60 right fielders with a median of 340.5 at-bats, along with 461 players in six other positions. As you can guess from these data, pitchers are not often at bat because they are prized for their pitching abilities not necessarily for their batting abilities.

The CSV data file is named BattingAverage.csv and looks like this:

```
Player,PriPos,Hits,AtBats,PlayerNumber,PriPosNumber
Fernando Abad,Pitcher,1,7,1,1
Bobby Abreu,Left Field,53,219,2,7
Tony Abreu,2nd Base,18,70,3,4
Dustin Ackley,2nd Base,137,607,4,4
Matt Adams,1st Base,21,86,5,3
...[943 more rows]...
```

[6] Data retrieved 22 December 2012 from http://www.baseball-reference.com/leagues/MLB/2012-standard-batting.shtml.

The first line (above) is the column names. Notice that there are six columns. The first four columns are the player's name (Player), primary position (PriPos), hits (Hits), and at-bats (AtBats). The last two columns are redundant numerical recodings of the player name and primary position; these columns will not be used in our analysis and could be completely omitted from the data file.

A script for analyzing the data follows the same sequence of steps as previous scripts, and looks like this:

```
# Read the data
myData = read.csv("BattingAverage.csv")
# Load the relevant model into R's working memory:
source("Jags-Ybinom-XnomSsubjCcat-MbinomBetaOmegaKappa.R")
# Generate the MCMC chain:
mcmcCoda = genMCMC( data=myData ,
                    zName="Hits", NName="AtBats", sName="Player",
                    cName="PriPos", numSavedSteps=11000 , thinSteps=20 )
# Display diagnostics of chain, for specified parameters:
for ( parName in c("omega[1]","omega0","kappa[1]","kappa0", "theta[1]") ) {
  diagMCMC( codaObject=mcmcCoda , parName=parName ,
            saveName=fileNameRoot , saveType=graphFileType )
}
# Get summary statistics of chain:
summaryInfo = smryMCMC( mcmcCoda , compVal=NULL )
# Display posterior information:
plotMCMC( mcmcCoda , data=myData ,
          zName="Hits", NName="AtBats", sName="Player", cName="PriPos",
          compVal=NULL ,
          diffCList=list( c("Pitcher","Catcher") ,
                          c("Catcher","1st Base") ) ,
          diffSList=list( c("Kyle Blanks","Bruce Chen") ,
                          c("Mike Leake","Wandy Rodriguez") ,
                          c("Andrew McCutchen","Brett Jackson") ,
                          c("ShinSoo Choo","Ichiro Suzuki") ) ,
          compValDiff=0.0 )
```

Notice that the name of the function-definition file includes "`Ybinom`" because the data file codes the data as z and N, and the file name includes "`XnomSsubjCcat`" because both subject and category identifiers are used as predictors. There are a few arguments in the function calls that are specific to this form of data and model. In particular, we have to tell the functions which columns of the data frame correspond to the variables $z_{s|c}$, $N_{s|c}$, s, and c. For example, the argument `zName="Hits"` indicates that the column name of the $z_{s|c}$ data is `Hits`, and the argument `cName="PriPos"` indicates that the column name of the category labels is `PriPos`.

The `genMCMC` command indicates that JAGS should save 11,000 steps thinned by 20. These values were chosen after an initial short run showed fairly high autocorrelation.

The goal was to have an effective sample size (ESS) of at least 10,000 for the key parameters (and differences of parameters) such as θ_s and ω_c, while also keeping the saved file size as small as possible. With 968 parameters, 11,000 steps takes more than 77 MB of computer memory. The `genMCMC` function uses three parallel chains in runjags by default, but you can change that by altering the innards of the `genMCMC` function. Even when using three parallel chains, it takes about 11 min to run on my modest desktop computer (plus additional time for making diagnostic graphs, etc.).

The last command in the script, `plotMCMC`, uses new arguments `diffCList` and `diffSList`. These arguments take lists of vectors that indicate which specific categories or subjects you want to compare. The example above produces plots of the marginal posterior of the difference between the modes for category Pitcher and category Catcher, and for subjects Kyle Blanks versus Bruce Chen, etc. The plots are shown in figures described next.

JAGS produces combinations of parameter values in the 968-dimensional parameter space that are jointly credible, given the data. We could look at the results many different ways, based on whatever question we might ponder. For every pair of players, we could ask how much their estimated batting abilities differ. For every pair of positions, we can ask how much their batting abilities differ. (We could also ask questions about combinations of positions, such as, Do outfielders have different batting averages than basemen? But we will not pursue that sort of question here.) Selected illustrative results are shown in Figures 9.14–9.16. (We cannot get a sense of the difference of two parameters merely by looking at the marginal distributions of those parameters, as will be explained in Section 12.1.2.1, p. 340.)

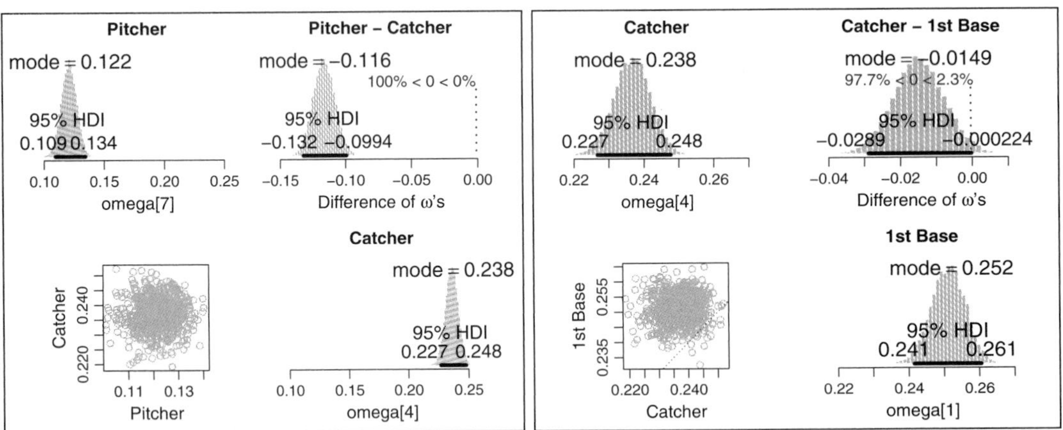

Figure 9.14 Marginal posterior distributions for baseball batting data. Left quartet shows that the pitchers have far lower batter abilities than the catchers. Right quartet shows that the catchers have slightly lower batting abilities than 1st-base men.

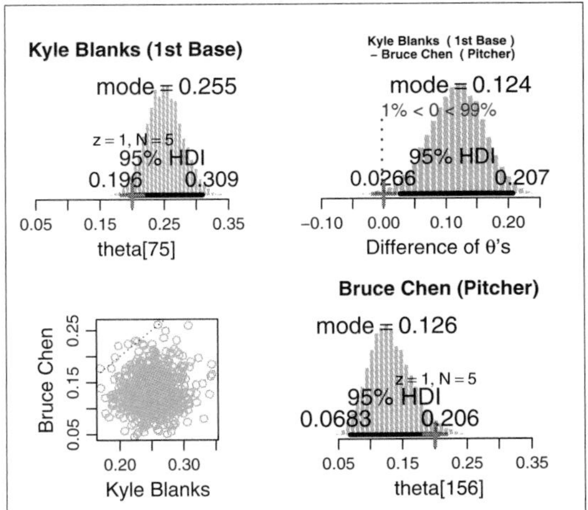

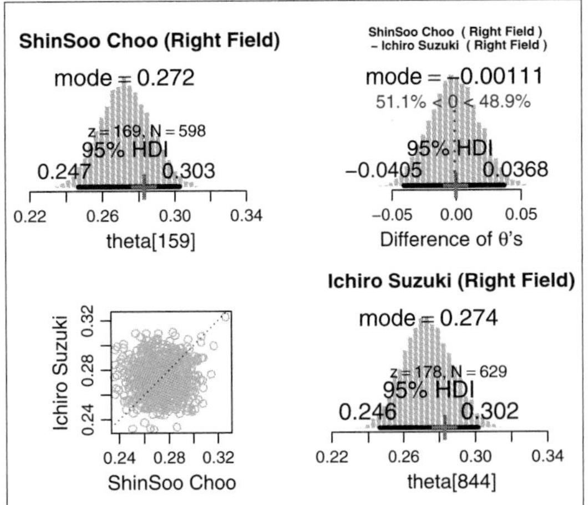

Figure 9.15 Marginal posterior distributions for baseball batting data. Left quartet: individual estimates for two players with identical records of 1 hit in only 5 at-bats, but from two different positions. Although the batting records are identical, the estimated batting abilities are very different. Right quartet: individual estimates for two right fielders with large numbers of at-bats. The posterior distributions of their individual performances have narrow HDIs compared with the left quartet, and are shrunken slightly toward the position-specific mode (which is about 0.247). The posterior distribution of their difference is essentially zero and the 95% HDI of the difference is very nearly contained within a ROPE from −0.04 to +0.04 (except for MCMC instability).

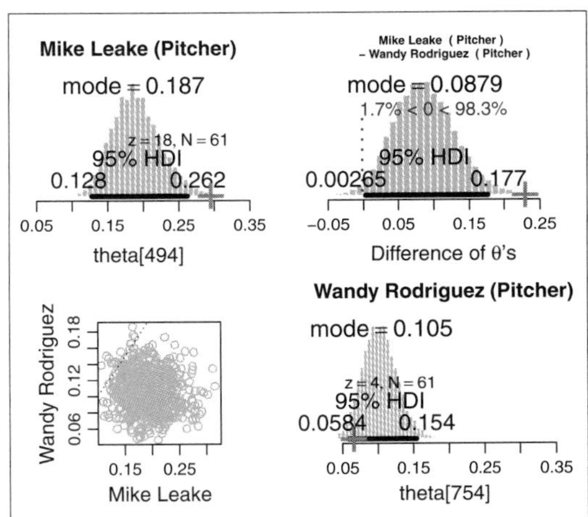

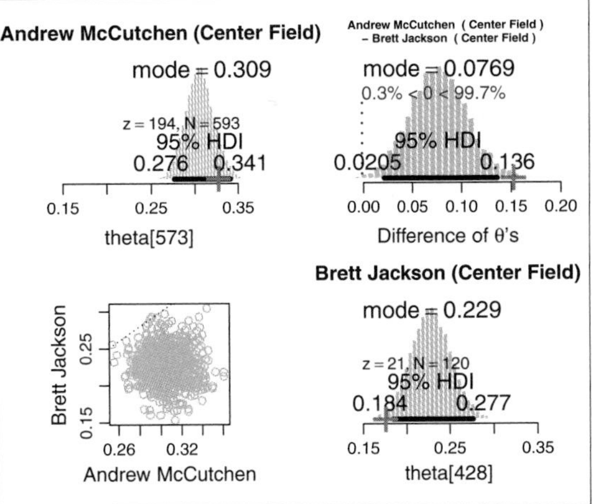

Figure 9.16 Marginal posterior distributions for baseball batting data. Left quartet: two pitchers each with 61 at-bats but very different numbers of hits. Despite the difference in performance, shrinkage toward the position-specific mode leaves the posterior distribution of their difference nearly spanning zero. Right quartet: two center fielders with very different batting averages, and moderately large at-bats. Despite some shrinkage toward the position-specific mode, the larger set of data makes the posterior distribution of their difference notably exclude zero.

Figure 9.14 shows estimates of position-level batting abilities (the ω_c parameters in the model). Clearly the modal batting ability of pitchers is lower than that of catchers. The modal ability of catchers is slightly lower than that of first-base men. Because the individual-level parameters ($\theta_{s|c}$) are modeled as coming from position-specific modes, the individual estimates are shrunken toward the position-specific modes.

Figure 9.15 shows estimated abilities of selected individuals. The left panel shows two players with identical batting records but with very different estimated batting abilities because they play different positions. These two players happened to have very few opportunities at bat, and therefore their estimated abilities are dominated by the position information. This makes sense because if all we knew about the players was their positions, then our best guess would come from what we knew about other players from those positions. Notice that if we did not incorporate position information into our hierarchical model, and instead put all 948 players under a single over-arching distribution, then the estimates for two players with identical batting records would be identical regardless of the position they play.

The right panel of Figure 9.15 shows estimated abilities of two right fielders with many at-bats (and very strong batting records). Notice that the widths of their 95% HDIs are smaller than those in the left panel because these players have so much more data. Despite the large amount of data contributing to the individual estimates, there is still noticeable shrinkage of the estimates away from the proportion of hits (marked by the "+" signs on the horizontal axis) toward the modal value of all right fielders, which is about 0.247. The difference between the estimated abilities of these two players is centered almost exactly on zero, with the 95% HDI of the difference essentially falling within a ROPE of -0.04 to $+0.04$.

Figure 9.16 shows two comparisons of players from the same positions but with very different batting records. The left panel shows two pitchers, each with a modest number of at-bats. Despite the very different proportions of hits attained by these players, the posterior estimate of the difference of their abilities only marginally excludes zero because shrinkage pulls their individual estimates toward the mode of pitchers. The right panel shows two center fielders, one of whom has a large number of at-bats and an exceptional batting record. Because of the large amount of data from these individuals, the posterior distribution of the difference excludes zero despite within-position shrinkage.

This application was initially presented at the book's blog (http://doingbayesiandataanalysis.blogspot.com/2012/11/shrinkage-in-multi-level-hierarchical.html) and subsequently described by Kruschke and Vanpaemel (in press). In those reports, the models parameterized the beta distributions by means instead of by modes, and the models also put an estimated higher-level distribution on the κ_c parameters, instead of using a fixed prior. Despite those differences, they came to similar conclusions as those shown here.

Summary of example. This example has illustrated many important concepts in hierarchical modeling that are recapitulated and amplified in the following paragraphs.

The example has dramatically illustrated the phenomenon of shrinkage in hierarchical models. In particular, we saw shrinkage of individual-ability estimates based on category (fielding position). Because there were so many players contributing to each position, the position information had strong influence on the individual estimates. Players with many at-bats (large N_s) had somewhat less shrinkage of their individual estimates than players with few at-bats (small N_s), who had estimates dominated by the position information.

Why should we model the data with hierarchical categories at all? Why not simply put all 948 players under one group, namely, major-leaguers, instead of in nine subcategories within major leaguers? Answer: the hierarchical structure is an expression of how you think the data should be meaningfully modeled, and the model description captures aspects of the data that you care about. If we did not categorize by position, then the estimated abilities of the pitchers would be much more noticeably pulled up toward the modal abilities of all the nonpitchers. And, if we did not categorize by position, then the estimated abilities of any two players with identical batting records would also be identical regardless of their positions; for example, the first-base man and pitcher in Figure 9.15 would have identical estimates. But if you think that position information is relevant for capturing meaningful aspects of the data, then you should categorize by position. Neither model is a uniquely "correct" description of the data. Instead, like all models, the parameter estimates are meaningful descriptions of the data only in the context of the model structure.

An important characteristic of hierarchical estimation that has not yet been discussed is illustrated in Figure 9.17. The certainty of the estimate at a level in the hierarchy depends (in part) on how many parameter values are contributing to that level. In the present application, there are nine position parameters (ω_c) contributing to the overall mode (ω), but dozens or hundreds of players (θ_s) contributing to each position (ω_c). Therefore, the certainty of estimate at the overall level is less than the certainty of estimate within each position. You can see this in Figure 9.17 by the widths of the 95% HDIs. At the overall level, the 95% HDI on ω has a width of about 0.09. But for the specific positions, the 95% HDI on ω_c has a width of about only 0.02. Other aspects of the data contribute to the widths and shapes of the marginal posterior distributions at the different levels, but one implication is this: if there are only a few categories, the overall level typically is not estimated very precisely. Hierarchical structure across categories works best when there are many categories to inform the higher-level parameters.

Finally, we have only looked at a tiny fraction of the relations among the 968 parameters. We could investigate many more comparisons among parameters if we were specifically interested. In traditional statistical testing based on p-values (which will be discussed in Chapter 11), we would pay a penalty for even intending to make

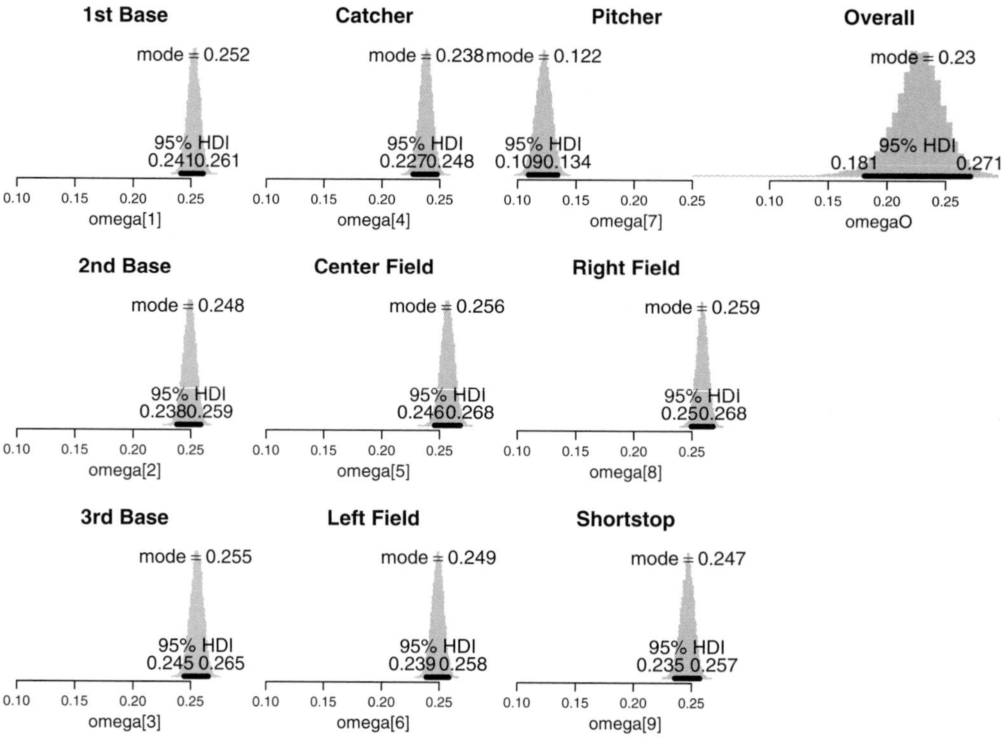

Figure 9.17 Marginal posterior distributions for baseball batting data. Notice that the estimate of the overall mode `omegaO` is less certain (wider HDI) than the estimate of position modes `omega[c]`. One reason for the different certainties is that there are dozens or hundreds of individuals contributing to each position, but only nine positions contributing to the overall level.

more comparisons. This is because a p value depends on the space of counter-factual possibilities created from the testing intentions. In a Bayesian analysis, however, decisions are based on the posterior distribution, which is based only on the data (and the prior), not on the testing intention. More discussion of multiple comparisons can be found in Section 11.4.

9.6. EXERCISES

Look for more exercises at https://sites.google.com/site/doingbayesiandataanalysis/

Exercise 9.1. [Purpose: Try different priors on κ to explore the role of κ in shrinkage.] Consider the analysis of the therapeutic touch data in Figure 9.10, p. 243. The analysis used a generic gamma distributed prior on κ that had a *mean* of 1.0 and a standard deviation of 10.0. We assumed that the prior had minimal influence on the results; here, we examine the robustness of the posterior when we change the prior to other reasonably vague and noncommittal distributions. In particular, we will examine a gamma distributed prior on κ that had a *mode* of 1.0 and a standard deviation of 10.0.

(A) What are the shape and rate parameters for a gamma distribution that has mean of 1.0 and standard deviation of 10.0? What are the shape and rate parameters for a gamma distribution that has mode of 1.0 and standard deviation of 10.0? *Hint:* use the utility functions `gammaShRaFromMeanSD` and `gammaShRaFromModeSD`.

(B) Plot the two gamma distributions, superimposed, to see which values of κ they emphasize. If you like, make the graphs with this R code:

```
openGraph(height=7,width=7)
layout(matrix(1:3,ncol=1))
k=seq(0,200,length=10001)
plot( k , dgamma(k,1.105125,0.105125) , ylab="dgamma(k)" ,
      type="l" , main="Gamma Distrib's (SD=10)" )
lines( k , dgamma(k,0.01,0.01) , col="skyblue" )
legend( "topright" , c("Mode 1","Mean 1") ,
        lty=c(1,1) , col=c("black","skyblue") , text.col=c("black", "skyblue") )
plot( k , dgamma(k,1.105125,0.105125) , ylab="dgamma(k)" ,
      type="l" , ylim=c(.07,.08) , main="Gamma Distrib's (SD=10), zoomed in" )
lines( k , dgamma(k,0.01,0.01) , col="skyblue" )
legend( "topright" , c("Mode 1","Mean 1") ,
        lty=c(1,1) , col=c("black","skyblue") , text.col=c("black", "skyblue") )
plot( k , dgamma(k,1.105125,0.105125) , ylab="dgamma(k)" ,
      type="l" , ylim=c(0,8.0e-5) , main="Gamma Distrib's (SD=10), zoomed in" )
lines( k , dgamma(k,0.01,0.01) , col="skyblue" )
legend( "topright" , c("Mode 1","Mean 1") ,
        lty=c(1,1) , col=c("black","skyblue") , text.col=c("black", "skyblue") )
```

The result is shown in Figure 9.18. Relative to each other, which gamma distribution favors values of κ between about 0.1 and 75? Which gamma distribution favors values of κ that are tiny or greater than 75?

(C) In the program Jags-Ydich-XnomSsubj-MbinomBetaOmegaKappa.R, find the line in the model specification for the prior on `kappaMinusTwo`. Run the program once using a gamma with mean of 1.0, and run the program a second time using a gamma with a mode of 1.0. Show the graphs of the posterior distribution. *Hints:* in the model specification, just comment out one or the other of the lines:

```
#kappaMinusTwo ~ dgamma( 0.01 , 0.01 )  # mean=1 , sd=10 (generic vague)
kappaMinusTwo ~ dgamma( 1.105125 , 0.1051249 )  # mode=1 , sd=10
```

Be sure to save the program before calling it from the script! In the script, you might want to change the file name root that is used for saved graph files.

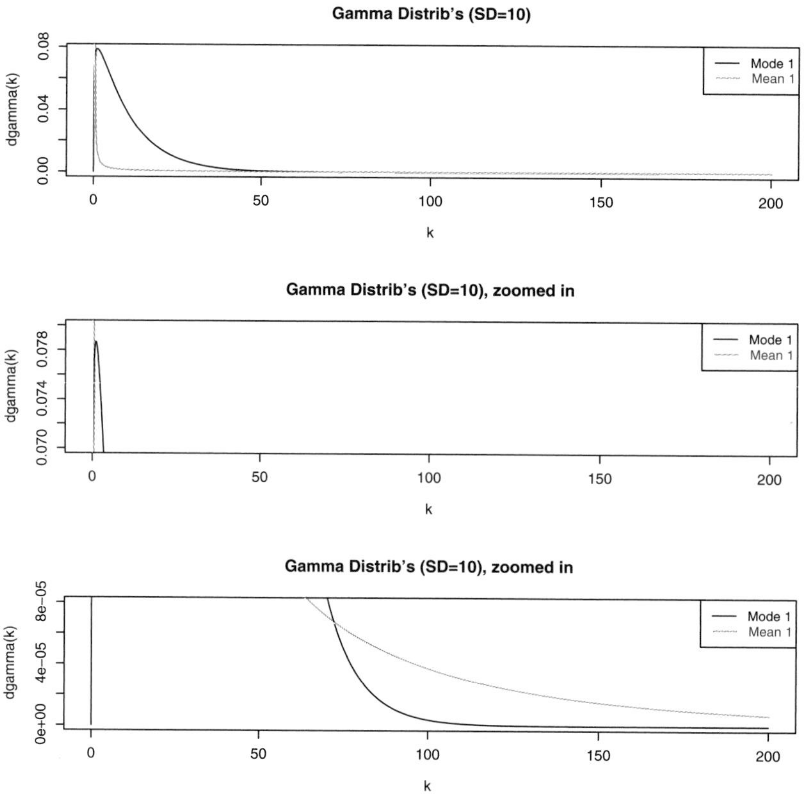

Figure 9.18 Two gamma distributions superimposed, for use with Exercise 9.1.

(D) Does the posterior distribution change much when the prior is changed? In particular, for which prior does the marginal posterior distribution on κ have a bigger large-value tail? When κ is larger, what effect does that have on shrinkage of the θ_s values?

(E) Which prior do you think is more appropriate? To properly answer this question, you should do the next exercise!

Exercise 9.2. [Purpose: Examine the prior on θ_s implied by the prior constants at higher levels.] To sample from the prior in JAGS, we just comment out the data, as was explained in Section 8.5. In the program Jags-Ydich-XnomSsubj-MbinomBetaOmegaKappa.R, just comment out the line that specifies z, like this:

```
  dataList = list(
#    z = z ,
     N = N ,
     Nsubj = Nsubj
  )
```

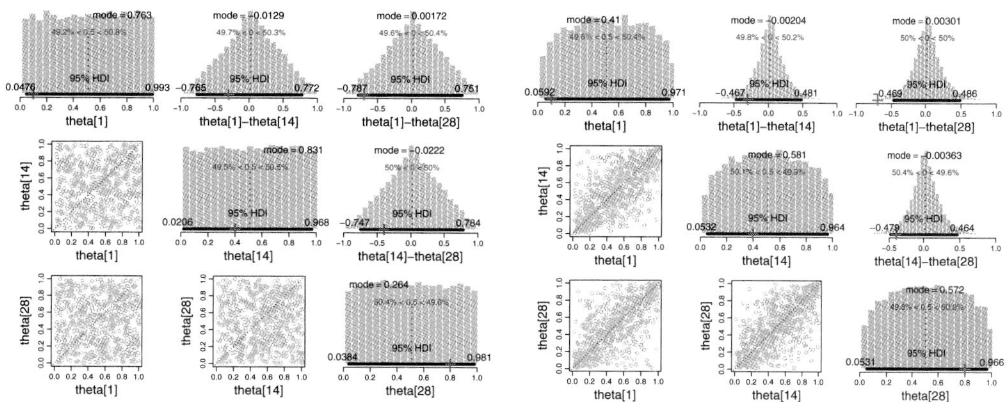

Figure 9.19 Priors on θ_s implied by different gamma distributions on κ. For use with Exercise 9.2.

Save the program, and run it with the two priors on κ discussed in the previous exercise. You may want to change the file name root for the saved graphics files. For both priors, include the graphs of the prior distributions on θ_s and the differences of θ_s's such as `theta[1]-theta[28]`. See Figure 9.19.

(A) Explain why the implied prior distribution on individual θ_s has rounded shoulders (instead of being essentially uniform) when using a prior on κ that has a mode of 1 (instead of a mean of 1).

(B) Which prior do you think is more appropriate?

Exercise 9.3. [Purpose: Compare Bayesian shrinkage with MLE shrinkage] Construct a data set like the data in Figure 9.12 and do a Bayesian analysis like that done for the therapeutic touch data. Compare the Bayesian parameter estimates with the MLE estimates (gleaned from Figure 9.12). What does the Bayesian analysis provide that is not provided by the MLE?

Exercise 9.4. [Purpose: Explore duration of processing by JAGS] Consider the therapeutic touch data of Figure 9.9, as analyzed by the program `Jags-Ydich-XnomSsubj-MbinomBetaOmegaKappa.R`. Open that program in RStudio and find the section that calls runjags or rjags.

(A) Set the program (if it is not already) to use three parallel chains with runjags. Be sure to save the program after any changes. Then run the high-level script, `Jags-Ydich-XnomSsubj-MbinomBetaOmegaKappa-Example.R`. Be sure it has `proc.time()` before and after `genMCMC` so that you can monitor how long it takes to generate the MCMC chain. Report the elapsed time of the chain generation, and also include the chain-diagnostic graph of omega, which includes its ESS.

(B) Set the program `Jags-Ydich-XnomSsubj-MbinomBetaOmegaKappa.R` to use rjags (with a single long chain). Be sure to save the program after any changes. Then run the high-level script, `Jags-Ydich-XnomSsubj-Mbinom BetaOmega Kappa-Example.R`. Be sure it has `proc.time()` before and after `genMCMC` so that you can monitor how long it takes to generate the MCMC chain. Report the elapsed time of the chain generation, and also include the chain-diagnostic graph of omega, which includes its ESS.

(C) Compare the two runs. What is the difference in run durations? Are the ESSs about the same?

CHAPTER 10

Model Comparison and Hierarchical Modeling

Contents

The magazine model comparison game
Leaves all of us wishing that we looked like them.
But they have mere fantasy's bogus appeal,
'Cause none obeys fact or respects what is real.[1]

There are situations in which different models compete to describe the same set of data. In one classic example, the data consist of the apparent positions of the planets and sun against the background stars, during the course of many months. Are these data best described by an earth-centric model or by a solar-centric model? In other words, how should we allocate credibility across the models?

As was emphasized in Chapter 2 (specifically Section 2.1 and following), Bayesian inference is reallocation of credibility over possibilities. In model comparison, the focal

[1] This chapter is about Bayesian model comparison, which is about relative credibility of models. The poem refers to comparing ourselves to models of the human type used in advertising, and their relative credibility. Word play: Listen carefully and you will hear "Bayes factor" sounded out in the verse. Making that happen took effort!

Doing Bayesian Data Analysis, 2nd Edition
http://dx.doi.org/10.1016/B978-0-12-405888-0.00010-6

possibilities are the models, and Bayesian model comparison reallocates credibility across the models, given the data. In this chapter, we explore examples and methods of Bayesian inference about the relative credibilities of models. We will see that Bayesian model comparison is really just a case of Bayesian parameter estimation applied to a hierarchical model in which the top-level parameter is an index for the models. A central technical point is that a "model" consists of both its likelihood function and prior distribution, and model comparison can be extremely sensitive to the choice of prior, even if the prior is vague, unlike continuous parameter estimation within models.

10.1. GENERAL FORMULA AND THE BAYES FACTOR

In this section, we will consider model comparison in a general, abstract way. This will establish a framework and terminology for the examples that follow. If this abstract treatment does not make complete sense to you on a first reading, do not despair, because simple examples quickly follow that illustrate the ideas.

Recall the generic notation we have been using, in which data are denoted as D or y, and the parameters of a model are denoted θ. The likelihood function is denoted $p(y|\theta)$, and the prior distribution is denoted $p(\theta)$. We are now going to expand that notation so we can make reference to different models. We will include a new indexical parameter m that has value $m=1$ for model 1, $m=2$ for model 2, and so on. Then the likelihood function for model m is denoted $p_m(y|\theta_m, m)$ and the prior is denoted $p_m(\theta_m|m)$. Notice that the parameter is given a subscript because each model might involve different parameters. Notice also that the probability density is given a subscript, because different models might involve different distributional forms.

Each model can be given a prior probability, $p(m)$. Then we can think of the whole system of models as a large joint parameter space spanning $\theta_1, \theta_2, \ldots$, and m. Notice the joint space includes the indexical parameter m. Then, on this joint space, Bayes' rule becomes

$$p(\theta_1, \theta_2, \ldots, m|D) = \frac{p(D|\theta_1, \theta_2, \ldots, m)\, p(\theta_1, \theta_2, \ldots, m)}{\sum_m \int \mathrm{d}\theta_m p(D|\theta_1, \theta_2, \ldots, m)\, p(\theta_1, \theta_2, \ldots, m)} \tag{10.1}$$

$$= \frac{\prod_m p_m(D|\theta_m, m)\, p_m(\theta_m|m)\, p(m)}{\sum_m \int \mathrm{d}\theta_m \prod_m p_m(D|\theta_m, m)\, p_m(\theta_m|m)\, p(m)}. \tag{10.2}$$

(As usual in Bayes' rule, the variables in the numerator refer to specific values, and the variables in the denominator take on all possible values for the integral or summation.) Notice the key change in going from Equation 10.1 to Equation 10.2 was converting the numerator from an expression about the joint parameter space to an expression of dependencies among parameters. Thus, $p(D|\theta_1, \theta_2, \ldots, m)\, p(\theta_1, \theta_2, \ldots, m)$ became $\prod_m p_m(D|\theta_m, m)\, p_m(\theta_m|m)\, p(m)$. The factoring of the likelihood-times-prior into a chain of dependencies is the hallmark of a hierarchical model.

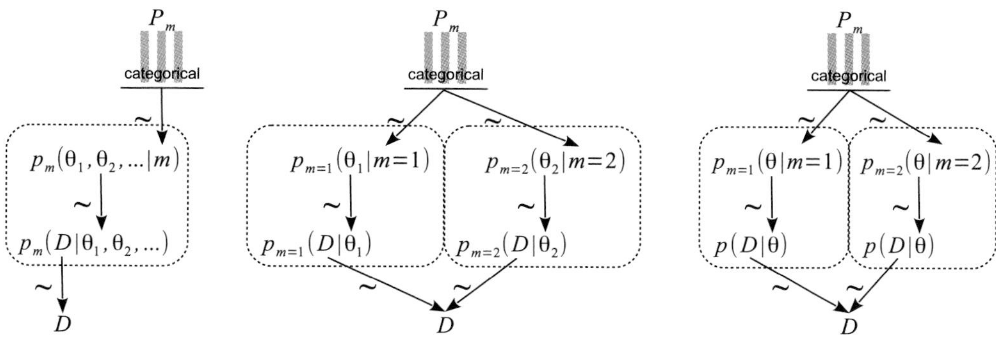

Figure 10.1 Model comparison as a single hierarchical model. Dashed boxes enclose the sub-models being compared. Left panel shows general conception, with parameters θ_m for all submodels in a joint space. Middle panel shows the usual case in which the likelihood and prior reduce to functions of only θ_m for each m. Right panel shows the special case in which the likelihood function is the same for all m, and only the form of the prior is different for different m. (Middle and right panels depict only two sub-models, but there can be many.)

A hierarchical diagram illustrating these relations appears in Figure 10.1. The left panel shows the general, joint parameter space across all the models, as expressed in Equation 10.1. The middle panel shows the factorization into model subspaces, as expressed in Equation 10.2. The middle panel is meant to emphasize that each submodel has its own parameters and distributions, suggested by the dashed boxes, but the submodels are under a higher-level indexical parameter. The prior on the model index, $p(m)$, is depicted as a bar graph over candidate indexical values $m = 1, 2, \ldots$ (just as the graphical icon for a Bernoulli distribution includes bars over candidate values of $y = 0, 1$).

The aim of this discussion has been to show that this situation is just like any other application of Bayes' rule, but with specific dependencies among parameters, and with the top-level parameter having a discrete, indexical value. Bayesian inference reallocates credibility across the values of the parameters, simultaneously across values of the indexical parameter and across values of the parameters within the component models.

We can also consider Bayes' rule applied to the model index alone, that is, marginalized across the parameters within the component models. This framing is useful when we want to know the relative credibilities of models overall. Bayes' rule says that the posterior probability of model m is[2]

$$p(m|D) = \frac{p(D|m)\,p(m)}{\sum_m p(D|m)\,p(m)} \tag{10.3}$$

[2] Equation 10.3 stands on its own directly from Bayes' rule, without reference to Equation 10.2. But if you want to, you can start with Equation 10.2 and take its integral, $\int d\theta_m$, to arrive at Equation 10.3.

where the probability of the data, given the model m, is, by definition, the probability of the data within that model marginalized across all possible parameter values:

$$p(D|m) = \int d\theta_m \, p_m(D|\theta_m, m) \, p_m(\theta_m|m) \tag{10.4}$$

Thus, to get from the prior probability of a model, $p(m)$, to its posterior probability, $p(m|D)$, we must multiply by the probability of the data given that model, $p(D|m)$. *Notice that the probability of the data for a model, $p(D|m)$, is computed by marginalizing across the prior distribution of the parameters in the model, $p_m(\theta_m|m)$. This role of the prior distribution in $p(D|m)$ is a key point for understanding Bayesian model comparison.* It emphasizes that a model is not only the likelihood function, but also includes the prior on its parameters. This fact, that a model refers to both the likelihood and the prior, is explicit in the notation that includes the model index in the likelihood and prior, and is also marked in Figure 10.1 as the dashed lines that enclose the likelihood and prior for each model. As we will see, one consequence is that Bayesian model comparison can be very sensitive to the choice of priors within models, even if those priors are thought to be vague.

Suppose we want to know the relative posterior probabilities of two models. We can write Equation 10.3 once with $m=1$ and again with $m=2$ and take their ratio to get:

$$\frac{p(m=1|D)}{p(m=2|D)} = \underbrace{\frac{p(D|m=1)}{p(D|m=2)}}_{\text{BF}} \, \frac{p(m=1)}{p(m=2)} \, \underbrace{\frac{/\sum_m p(D|m)\,p(m)}{/\sum_m p(D|m)\,p(m)}}_{=1} \tag{10.5}$$

The ratio at the end, underbraced with "$=1$," equals 1 and can be removed from the equation. The ratio underbraced and marked with "BF" is the *Bayes factor* for models 1 and 2. The Bayes factor (BF) is the ratio of the probabilities of the data in models 1 and 2. Those probabilities are marginalized across the parameters within each model, as expressed by Equation 10.4. Thus, Equation 10.5 says that the so-called "posterior odds" of two models are the "prior odds" times the Bayes factor. The Bayes factor indicates how much the prior odds change, given the data. One convention for converting the magnitude of the BF to a discrete decision about the models is that there is "substantial" evidence for model $m=1$ when the BF exceeds 3.0 and, equivalently, "substantial" evidence for model $m=2$ when the BF is less than 1/3 (Jeffreys, 1961; Kass & Raftery, 1995; Wetzels et al., 2011).

10.2. EXAMPLE: TWO FACTORIES OF COINS

We will consider a simple, contrived example in order to illustrate the concepts and mechanisms. Suppose we have a coin, embossed with the phrase, "Acme Novelty and Magic Company, Patent Pending." We know that Acme has two coin factories. One factory generates tail-biased coins, and the other factory generates head-biased coins. The tail-biased factory generates coins with biases distributed around a mode of $\omega_1 = 0.25$, with a consistency (or concentration) of $\kappa = 12$, so that the biases of coins from the factory are distributed as $\theta \sim \text{beta}(\theta | \omega_1(\kappa{-}2){+}1\,,\ (1{-}\omega_1)(\kappa{-}2){+}1\,) = \text{beta}(\theta|3.5, 8.5)$. The head-biased factory generates coins with biases distributed around a mode of $\omega_2 = 0.75$, again with a consistency of $\kappa = 12$, such that the biases of coins from the factory are distributed as $\theta \sim \text{beta}(\theta | \omega_2(\kappa - 2) + 1\,,\ (1 - \omega_2)(\kappa - 2) + 1\,) = \text{beta}(\theta|8.5, 3.5)$.

A diagram representing the two factories is shown in Figure 10.2. The two models are shown within the two dashed boxes. The left dashed box shows the tail-biased model, as suggested by the beta-distributed prior that has its mode at 0.25. The right dashed box shows the head-biased model, as suggested by the beta-distributed prior that has its mode at 0.75. At the top of the diagram is the indexical parameter m, and we want to compute the posterior probabilities of $m = 1$ and $m = 2$, given the observed flips of the coin, y_i. The diagram in Figure 10.2 is a specific case of the right panel of Figure 10.1, because the likelihood function is the same for both models and only the priors are different.

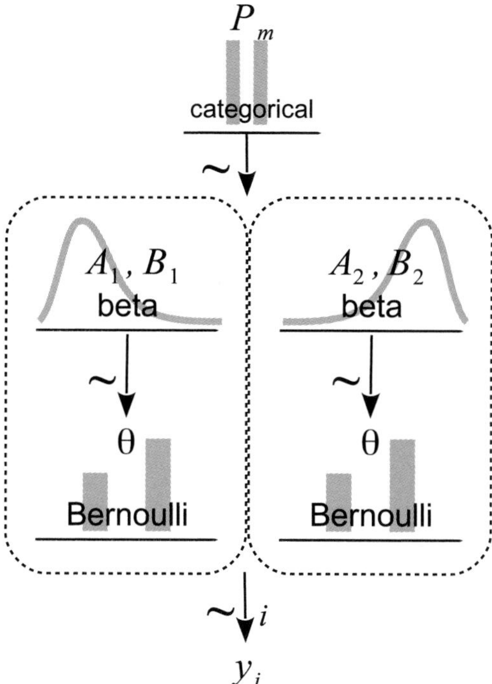

Figure 10.2 Hierarchical diagram for two models of a coin. Model 1 is a tail-biased mint; model 2 is a head-biased mint. This diagram is a specific case of the right panel of Figure 10.1, because the likelihood function is the same for both models and only the priors are different.

Suppose we flip the coin nine times and get six heads. Given those data, what are the posterior probabilities of the coin coming from the head-biased or tail-biased factories? We will pursue the answer three ways: via formal analysis, grid approximation, and MCMC.

10.2.1. Solution by formal analysis

Both the tail-biased and head-biased models have prior distributions that are beta density functions, and both models use the Bernoulli likelihood function. This is exactly the form of model for which we derived Bayes' rule analytically in Equation 6.8 (p. 132). In that equation, we denoted the data as z, N to indicate z heads in N flips. The derivation showed that if we start with a beta$(\theta|a, b)$ prior, we end up with a beta$(\theta|z+a, N-z+b)$ posterior. In particular, it was pointed out in Footnote 5 on p. 132 that the denominator of Bayes' rule, that is, the marginal likelihood, is

$$p(z, N) = \frac{B(z + a, N - z + b)}{B(a, b)} \tag{10.6}$$

Equation 10.6 provides an exact formula for the value of $p(D|m)$. Here is a function in *R* that implements Equation 10.6:

```
pD = function(z,N,a,b) { beta(z+a,N-z+b) / beta(a,b) }
```

That function works fine for modest values of its arguments, but unfortunately it suffers underflow errors when its arguments get too big. A more robust version converts the formula into logarithms, using the identity that

$$\begin{aligned} \mathrm{B}(z+a, N-z+b)/\mathrm{B}(a, b) &= \exp\left(\log\left(\mathrm{B}(z+a, N-z+b)/\mathrm{B}(a, b)\right)\right) \\ &= \exp\left(\log\left(\mathrm{B}(z+a, N-z+b)\right) - \log\left(\mathrm{B}(a, b)\right)\right) \end{aligned}$$

The logarithmic form of the beta function is built into *R* as the function `lbeta`. Therefore, we instead use the following form:

```
pD = function(z,N,a,b) { exp( lbeta(z+a,N-z+b) - lbeta(a,b) ) }
```

Let's plug in the specific priors and data to compute exactly the posterior probabilities of the two factories. First, consider the tail-biased factory (model index $m=1$). For its prior distribution, we know $\omega_1 = 0.25$ and $\kappa = 12$, hence $a_1 = 0.25 \cdot 10 + 1 = 3.5$ and $b_1 = (1 - 0.25) \cdot 10 + 1 = 8.5$. Then, from Equation 10.6, $p(D|m=1) = p(z, N|m=1) = B(z + a_1, N - z + b_1)/B(a_1, b_1) \approx 0.000499$, which you can verify in R using `pD(z=6,N=9,a=3.5,b=8.5)`. Next, consider the head-biased factory (model index $m=2$). For its prior distribution, we know $\omega_2 = 0.75$ and $\kappa = 12$, hence $a_2 = 0.75 \cdot 10 + 1 = 8.5$ and $b_2 = (1 - 0.75) \cdot 10 + 1 = 3.5$. Then, from Equation 10.6, $p(D|m=2) = p(z, N|m=2) = B(z + a_2, N - z + b_2)/B(a_2, b_2) \approx 0.002339$. The Bayes factor is $p(D|m=1)/p(D|m=2)=0.000499/0.002339=0.213$. In other words, if the prior odds for the two factories were 50/50, then the posterior odds are 0.213 against the tail-biased factory, which is to say 4.68 (i.e., 1/0.213) in favor of the head-biased factory.

We can convert the Bayes factor to posterior probabilities, assuming specific prior probabilities, as follows: we start with Equation 10.5 and plug in the values of the Bayes factor and the prior probabilities. Here, we assume that the prior probabilities of the models are $p(m=1)=p(m=2)=0.5$.

$$\frac{p(m=1|D)}{p(m=2|D)}=\frac{p(D|m=1)}{p(D|m=2)}\frac{p(m=1)}{p(m=2)}=\frac{0.000499}{0.002339}\frac{0.5}{0.5}=0.213$$

$$\frac{p(m=1|D)}{1-p(m=1|D)}=0.213 \quad \text{because } p(m=1|D)+p(m=2|D) \text{ must be } 1$$

$$p(m=1|D)=\frac{0.213}{1+0.213}=17.6\% \quad \text{by algebraic rearrangement}$$

And therefore $p(m=2|D)=82.4\%$. (These percentages were computed by retaining more significant digits than were presented in the rounded values above.) In other words, given the data, which showed six heads in nine flips, we reallocated beliefs from a prior head-biased credibility of $p(m=2)=50\%$ to a posterior head-biased credibility of $p(m=2|D)=82.4\%$.

Exercise 10.1 has you explore variations of this scheme. In particular, when the models make very different predictions, which in this case means the factories produce very different biases, then only a small amount of data is needed to strongly prefer one model over the other.

Notice that the model comparison only indicates the relative probabilities of the models. It does not, by itself, provide a posterior distribution on the coin bias. In the above calculations, there was nothing derived for $p(\theta|D,m)$. Thus, while we have estimated which factory might have produced the coin, we have not estimated what the bias of the coin might be!

10.2.2. Solution by grid approximation

The analytical solution of the previous section provides an exact answer, but reproducing it with a grid approximation provides additional insights when we visualize the parameter space.

In our current scenario, the model index, m, determines the value of the factory mode, ω_m. Therefore, instead of thinking of a discrete indexical parameter m, we can think of the continuous mode parameter ω being allowed only two discrete values by the prior. The overall model has only one other parameter, namely the coin bias θ. We can graph the prior distribution on the ω, θ parameter space as shown in the upper panels of Figure 10.3. Notice that the prior is like two dorsal fins, each having the profile of a

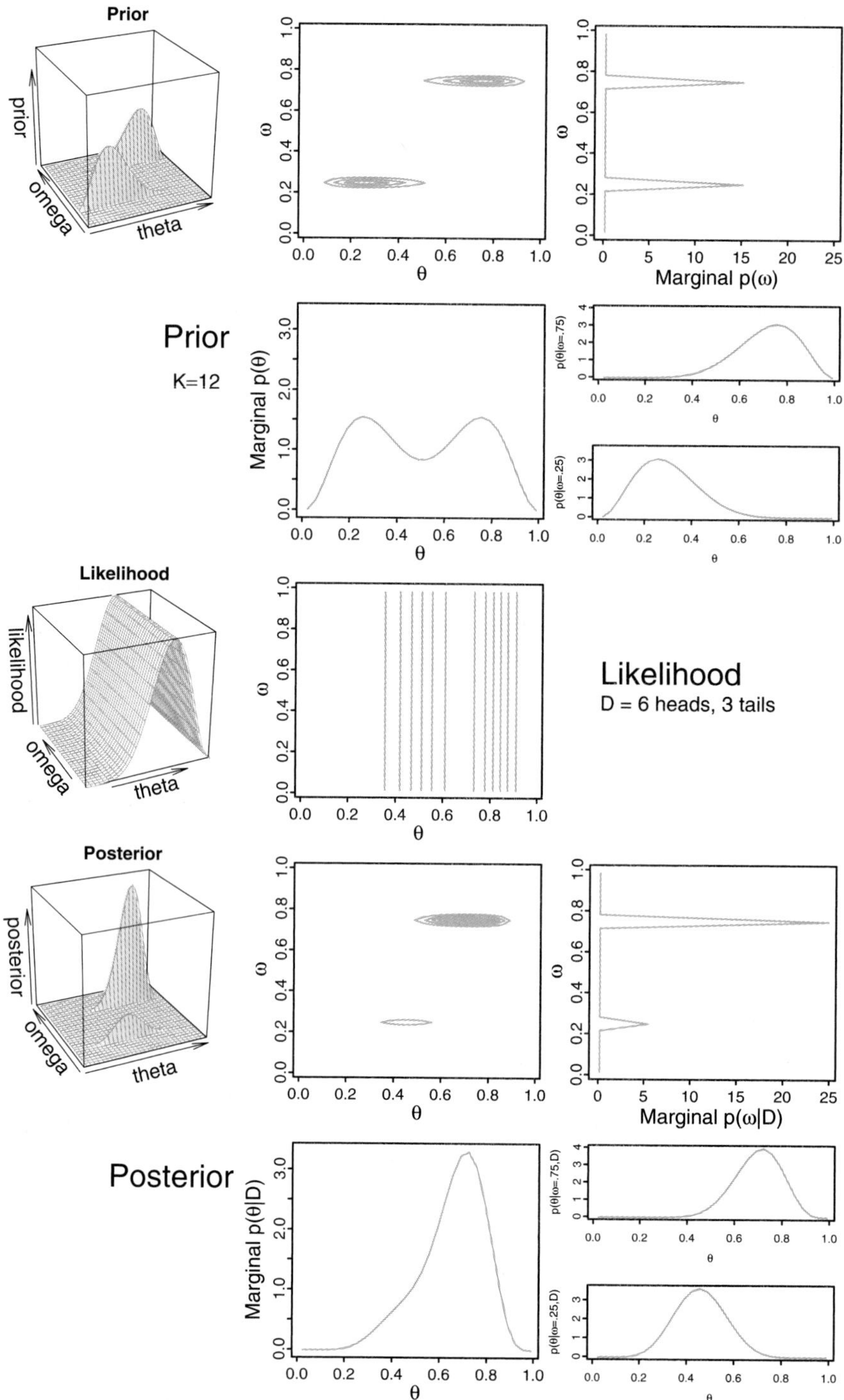

Figure 10.3 A representation of the joint ω, θ parameter space when the mode parameter, ω, is allowed only two discrete values. (For an example with a continuous distribution on ω, compare with Figure 9.2, p. 227.)

beta distribution, with one fin at $\omega = 0.25$ and the other fin at $\omega = 0.75$. These two fins correspond to the two beta distributions shown in Figure 10.2.

Importantly, notice the top-right panel of Figure 10.3, which shows the marginal prior distribution on ω. You can see that it is two spikes of equal heights, which means that the prior puts equal probability on the two candidate values of ω for the two models. (The absolute scale on $p(\omega)$ is irrelevant because it is the probability density for an arbitrary choice of grid approximation.) The two-spike prior on ω represents the discrete prior on the model index, m.

One advantage of the visualization in Figure 10.3 is that we can see both models simultaneously and see the marginal prior distribution on θ when averaging across models. Specifically, the middle panel of the second row shows $p(\theta)$, where you can see it is a bimodal distribution. This illustrates that the overall model structure, as a whole, asserts that biases are probability near 0.25 or 0.75.

The likelihood is shown in the middle row of Figure 10.3, again for data consisting of $z = 6$ heads in $N = 9$ flips. Notice that the level contours of the likelihood depend only on the value of θ, as they should.

The posterior distribution is shown in the lower two rows of Figure 10.3. The posterior distribution was computed at each point of the $\langle\theta, \omega\rangle$ grid by multiplying prior times likelihood, and then normalizing (exactly as done for previous grid approximations such as in Figure 9.2, p. 227).

Notice the marginal posterior distribution on ω, shown in the right panel of the fourth row. You can see that the spike over $\omega = 0.75$ is much taller than the spike over $\omega = 0.25$. In fact, *visual inspection suggests that the ratio of the heights is about 5 to 1, which matches the Bayes factor of 4.68 that we computed exactly in the previous section.* (As mentioned above, the absolute scale on $p(\omega)$ is irrelevant because it is the probability density for an arbitrary choice of grid approximation.) Thus, at the level of model comparison, the grid approximation has duplicated the analytical solution of the previous section. The graphs of the spikes on discrete levels of ω provide a visual representation of model credibilities.

The visualization of the grid approximation provides additional insights, however. In particular, the bottom row of Figure 10.3 shows the posterior distribution of θ within each discrete value of ω, and the marginal distribution of θ across values of ω. If we restrict our attention to the model index alone, we do not see the implications for θ. One way of verbally summarizing the posterior distribution is like this: Given the data, the head-biased factory (with $\omega = 0.75$) is about five times more credible than the tail-biased factory (with $\omega = 0.25$), and the bias of the coin is near $\theta = 0.7$ with uncertainty shown by the oddly-skewed distribution[3] in the bottom-middle panel of Figure 10.3.

[3] The marginal distribution on θ is just a weighted sum of conditional posterior distributions for each discrete value of ω. Those conditional posterior distributions are simply beta distributions computed with the beta updating rule of Equation 6.8. The weights are simply the posterior probabilities $p(\omega|D)$.

10.3. SOLUTION BY MCMC

For large, complex models, we cannot derive $p(D|m)$ analytically or with grid approximation, and therefore we will approximate the posterior probabilities using MCMC methods. We will consider two approaches. The first approach computes $p(D|m)$ using MCMC for individual models. The second approach puts the models together into a hierarchy as diagrammed in Figure 10.1, and the MCMC process visits different values of the model index proportionally to their posterior probabilities.

10.3.1. Nonhierarchical MCMC computation of each model's marginal likelihood

In this section, we consider how to compute $p(D|m)$ for a single model m. This section is a little heavy on the math, and the method has limited usefulness for complex models, and may be of less immediate value to readers who wish to focus on applications in the latter part of the book. Therefore this section may be skipped on a first reading, without loss of dignity. But you will have less to talk about at parties. The method presented in this section is just one of many; Friel and Wyse (2012) offer a recent review.

To explain how the method works, we first need to establish a basic principle of how to approximate functions on probability distributions: For any function $f(\theta)$, the integral of that function, weighted by the probability distribution $p(\theta)$, is approximated by the average of the function values at points sampled from the probability distribution. In math:

$$\int \mathrm{d}\theta\, f(\theta)\, p(\theta) \approx \frac{1}{N} \sum_{\theta_i \sim p(\theta)}^{N} f(\theta_i) \tag{10.7}$$

To understand why that is true, consider discretizing the integral in the left side of Equation 10.7, so that it is approximated as a sum over many small intervals: $\int \mathrm{d}\theta\, f(\theta)\, p(\theta) \approx \sum_j [\Delta\theta\, p(\theta_j)] f(\theta_j)$, where θ_j is a representative value of θ in the jth interval. The term in brackets, $\Delta\theta\, p(\theta_j)$, is the probability mass of the small interval around θ_j. That probability mass is approximated by the relative number of times we happen to get a θ value from that interval when sampling from $p(\theta)$. Denote the number of times we get a θ value from the jth interval as n_j, and the total number of sampled values as N. With a large sample, notice that $n_j/N \approx \Delta\theta\, p(\theta_j)$. Then $\int \mathrm{d}\theta\, f(\theta)\, p(\theta) \approx \sum_j [\Delta\theta\, p(\theta_j)] f(\theta_j) \approx \sum_j [n_j/N] f(\theta_j) = \frac{1}{N} \sum_j n_j f(\theta_j)$. In other words, every time we sample a θ value from the jth interval, we add into the summation another iteration of the interval's representative value, $f(\theta_j)$. But there is no need to use the interval's representative value; just use the value of $f(\theta)$ at the sampled value of θ, because the sampled θ already is in the jth interval. So the approximation becomes $\int \mathrm{d}\theta\, f(\theta)\, p(\theta) \approx \frac{1}{N} \sum_j n_j f(\theta_j) \approx \frac{1}{N} \sum_{\theta_i \sim p(\theta)}^{N} f(\theta_i)$, which is Equation 10.7.

For the goal of model comparison, we want to compute the marginal likelihood for each model, which is $p(D) = \int d\theta\, p(D|\theta)\, p(\theta)$, where $p(\theta)$ is the prior on the parameters in the model. In principle, we could just apply Equation 10.7 directly:

$$\begin{aligned} p(D) &= \int d\theta\, p(D|\theta)\, p(\theta) \\ &\approx \frac{1}{N} \sum_{\theta_i \sim p(\theta)}^{N} p(D|\theta_i) \end{aligned}$$

This means that we are getting random values from the prior. But in practice, the prior is very diffuse, and for nearly all of its sampled values, $p(D|\theta)$ is nearly zero. Therefore, we might need a ginormous number of samples for the approximation to converge to a stable value.

Instead of sampling from the prior, we will use our MCMC sample from the posterior distribution, in a clever way. First, consider Bayes' rule:

$$p(\theta|D) = \frac{p(D|\theta)p(\theta)}{p(D)}$$

We can rearrange it to get

$$\frac{1}{p(D)} = \frac{p(\theta|D)}{p(D|\theta)p(\theta)}$$

Now a trick (due to Gelfand & Dey, 1994; summarized by Carlin & Louis, 2009): for any probability density function $h(\theta)$, it is the case that it integrates to 1, which expressed mathematically is $\int d\theta\, h(\theta) = 1$. We will multiply the rearranged Bayes' rule by 1:

$$\begin{aligned} \frac{1}{p(D)} &= \frac{p(\theta|D)}{p(D|\theta)p(\theta)} \\ &= \frac{p(\theta|D)}{p(D|\theta)p(\theta)} \int d\theta\, h(\theta) && \text{by multiplying by 1} \\ &= \int d\theta\, \frac{h(\theta)}{p(D|\theta)p(\theta)}\, p(\theta|D) && \text{by magic} \\ &\approx \frac{1}{N} \sum_{\theta_i \sim p(\theta|D)}^{N} \frac{h(\theta_i)}{p(D|\theta_i)p(\theta_i)} && (10.8) \end{aligned}$$

where the last line is simply applying Equation 10.7 to the penultimate line. Now I reveal the magic in the transition from second to third lines of Equation 10.8. The θ in $h(\theta)$ varies over the range of the integral, but the θ in $p(\theta|D)/p(D|\theta)p(\theta)$ is a specific value. Therefore, we cannot treat the two θ's as interchangeable without further justification. However, the value of the ratio $p(\theta|D)/p(D|\theta)p(\theta)$ is the same value for *any* value of θ, because the ratio always equals the constant $1/p(D)$, and therefore we can let the value of θ in $p(\theta|D)/p(D|\theta)p(\theta)$ equal the value of θ in $h(\theta)$ as the latter varies. In other words, although the θ in $p(\theta|D)/p(D|\theta)p(\theta)$ began as a single value, it transitioned into being a value that varies along with the θ in $h(\theta)$. This magic trick is great entertainment at parties.

All there is yet to do is specify our choice for the function $h(\theta)$. It would be good for $h(\theta)$ to be similar to $p(D|\theta)p(\theta)$, so that their ratio, which appears in Equation 10.8, will not get too extremely large or extremely small for different values of θ. If their ratio did get too big or small, that would upset the convergence of the sum as N grows.

When the likelihood function is the Bernoulli distribution, it is reasonable that $h(\theta)$ should be a beta distribution with mean and standard deviation corresponding to the mean and standard deviation of the samples from the posterior. The idea is that the posterior will tend to be beta-ish, especially for large-ish data sets, regardless of the shape of the prior, because the Bernoulli likelihood will overwhelm the prior as the amount of data gets large. Because I want $h(\theta)$ to be a beta distribution that mimics the posterior distribution, I will set the beta distribution's mean and standard deviation to the mean and standard deviation of the θ values sampled from the posterior. Equation 6.7, p. 131, provides the corresponding shape parameters for the beta distribution. To summarize: We approximate $p(D)$ by using Equation 10.8 with $h(\theta)$ being a beta distribution with its a and b values set to imitate the posterior distribution. Note that Equation 10.8 yields the reciprocal of $p(D)$, so we have to invert the result to get $p(D)$ itself.

In general, there might not be strong theoretical motivations to select a particular $h(\theta)$ density. No matter. All that's needed is any density that reasonably mimics the posterior. In many cases, this can be achieved by first generating a representative sample of the posterior, and then finding an "off-the-shelf" density that describes it reasonably well. For example, if the parameter is limited to the range $[0, 1]$, we might be able to mimic its posterior with a beta density that has the same mean and standard deviation as the sampled posterior, even if we have no reason to believe that the posterior really is exactly a beta distribution. If the parameter is limited to the range $[0, +\infty)$, then we might be able to mimic its posterior with a gamma density (see Figure 9.8 , p. 237) that has the same mean and standard deviation as the sampled posterior, even if we have no reason to believe that the posterior really is exactly a gamma distribution.

For complex models with many parameters, it may be difficult to create a suitable $h(\theta)$, where θ here represents a vector of many parameters. One approach is to mimic the marginal posterior distribution of each individual parameter, and to let $h(\theta)$ be the product of the mimicked marginals. This approach assumes that there are no strong correlations of parameters in the posterior distribution, so that the product of marginals is a good representation of the joint distribution. If there are strong correlations of parameters in the posterior distribution, then this approach may produce summands in Equation 10.8 for which $h(\theta_i)$ is large but $p(D|\theta_i)p(\theta_i)$ is close to zero, and their ratio "explodes." Even if the mimicry by $h(\theta)$ is good, the probability densities of the various terms might get so small that they exceed the precision of the computer. Thus, for complex models, this method might not be tractable. For the simple application here, however, the method works well, as demonstrated in the next section.

10.3.1.1 Implementation with JAGS

We have previously implemented the Bernoulli-beta model in JAGS, as the introductory example in Section 8.2, in the program named `Jags-Ydich-Xnom1subj-MbernBeta.R`. We will adapt the program so it uses the priors from the current model comparison, and then use its MCMC output to compute $p(D)$ in R from Equation 10.8.

In the model section of `Jags-Ydich-Xnom1subj-MbernBeta.R`, change the prior to the appropriate model. For example, for $m = 2$, which has $\omega = 0.75$ and $\kappa = 12$, we simply change the constants in the `dbeta` prior as follows:

```
model {
  for ( i in 1:Ntotal ) {
    y[i] ~ dbern( theta )
  }
  theta ~ dbeta( 0.75*(12-2)+1 , (1-0.75)*(12-2)+1 )
}
```

Be sure to save the altered file, perhaps with a different file name than the original. Then generate the MCMC posterior sample with these commands:

```
source("Jags-Ydich-Xnom1subj-MbernBeta.R") # or whatever file name you saved as
myData = c(rep(0,9-6),rep(1,6))            # 9 flips with 6 heads
mcmcCoda = genMCMC( data=myData , numSavedSteps=10000 )
theta = as.matrix(mcmcCoda)[,"theta"]      # convert from coda object to vector
```

Having thereby generated the MCMC chain and put it in a vector named `theta`, we then compute $p(D)$ by implementing Equation 10.8 in R:

```
# Compute mean and standard deviation of MCMC values:
meanTheta = mean(theta)
sdTheta = sd(theta)
# Convert to a,b shape parameters for use in h(theta) function:
aPost = meanTheta   * ( meanTheta*(1-meanTheta)/sdTheta^2 - 1 )
bPost = (1-meanTheta) * ( meanTheta*(1-meanTheta)/sdTheta^2 - 1 )
# Compute 1/p(D):
oneOverPD = mean( dbeta( theta , aPost , bPost ) /
                  ( theta^6*(1-theta)^(9-6)
                    * dbeta( theta , 0.75*(12-2)+1 , (1-0.75)*(12-2)+1 ) ) )
PD = 1/oneOverPD
show(PD)
```

The result is PD= 0.002338 for one JAGS run, which is remarkably close to the value computed from formal analysis. Repeating the above, but with the model statement and formula for oneOverPD using $\omega = 0.25$, yields PD= 0.000499 for one JAGS run, which is also remarkably close to the value computed from formal analysis.

10.3.2. Hierarchical MCMC computation of relative model probability

In this section, we implement the full hierarchical structure depicted by Figure 10.2 (p. 269), in which the top-level parameter is the index across models. The specific example is implemented in the R script Jags-Ydich-Xnom1subj-MbernBetaModelComp.R. (The program is a script run on its own, not packaged as a function and called from another program, because this example is a specialized demonstration that is unlikely to be used widely with different data sets.)

The data are coded in the usual way, with y being a vector of 0's and 1's. The variable N is the total number of flips. The model of Figure 10.2 (p. 269) can then be expressed in JAGS as follows:

```
model {
  for ( i in 1:N ) {
    y[i] ~ dbern( theta )
  }
  theta ~ dbeta( omega[m]*(kappa-2)+1 , (1-omega[m])*(kappa-2)+1 )
  omega[1] <- .25
  omega[2] <- .75
  kappa <- 12
  m ~ dcat( mPriorProb[] )
  mPriorProb[1] <- .5
  mPriorProb[2] <- .5
}
```

As you read through the model specification (above), note that the Bernoulli likelihood function and beta prior distribution are only specified once, even though they participate in both models. This is merely a coding convenience, not a necessity. In fact, a later section shows a version in which distinct beta distributions are used. Importantly, notice that the value of omega[m] in the beta distribution depends on the model index, m. The next lines in the model specification assign the particular values for omega[1] and omega[2]. At the end of the model specification, the top-level prior on the model index is a categorical distribution, which is denoted dcat in JAGS. The argument of the dcat distribution is a vector of probabilities for each category. JAGS does not allow the vector constants to be defined inside the argument, like this: m ~ dcat(c(.5,.5)). With the model specified as shown above, the rest of the calls to JAGS are routine.

When evaluating the output, it is important to keep in mind that the sample of `theta` values is a mixture of values combining cases of `m=1` and `m=2`. At every step in the MCMC chain, JAGS provides representative values of θ and m that are jointly credible, given the data. At any step in the chain, JAGS lands on $m = 1$ or $m = 2$ proportionally to each model's posterior probability, and, for whichever value of m is sampled at that step, JAGS produces a credible value of θ for the corresponding ω_m. If we plot a histogram of all the `theta` values, collapsed across model indices, the result looks like the bottom-middle panel of Figure 10.3. But if we want to know the posterior distribution of θ values within each model, we have to separate the steps in the chain for which $m = 1$ or $m = 2$. This is trivial to do in R; see the script.

Figure 10.4 shows the prior and posterior distributions of m, θ_1, and θ_2. The prior was generated from JAGS in the usual way; see Section 8.5 (p. 211). Notice that the prior is as intended, with a 50-50 categorical distribution on the model index, and the appropriate tail-biased and head-biased priors on θ_1 and θ_2. The lower frame of Figure 10.4 shows the posterior distribution, where it can be seen that the results match those of the analytical solution, the grid approximation, and the one-model-at-a-time MCMC implementation of previous sections.

In particular, the posterior probability of $m=1$ is about 18%. This means that during the MCMC random walk, the value $m=1$ was visited on only about 18% of the steps. Therefore the histogram of θ_1 is based on only 18% of the chain, while the histogram of θ_2 is based on the complementary 82% of the chain. If the data were to favor one model over the other very strongly, then the losing model would have very few representative values in the MCMC sample.

10.3.2.1 Using pseudo-priors to reduce autocorrelation

Here we consider an alternative way to specify the model of the previous section, using two distinct beta distributions that generate distinct θ_m values. This method is more faithful to the diagram in Figure 10.2, and it is also more general because it allows different functional forms for the priors, if wanted. For realistic, complex model comparison, the implementation techniques described in this section are needed (or some other technique altogether, as opposed to the simplification of the previous section).

In this implementation, there are three parameters in the model, namely m, θ_1, and θ_2. At every step in the MCMC random walk, JAGS generates values of all three parameters that are jointly credible, given the data. However, at any step in the chain, only the θ_m of the sampled model index m is actually used to describe the data. If, at some step in the chain, the sampled value of m is 1, then θ_1 is used to describe the data, while θ_2 floats free of the data, constrained only by its prior. If, at another step in the chain, the sampled value of m is 2, then θ_2 is used to describe the data, and θ_1 is constrained only by its prior.

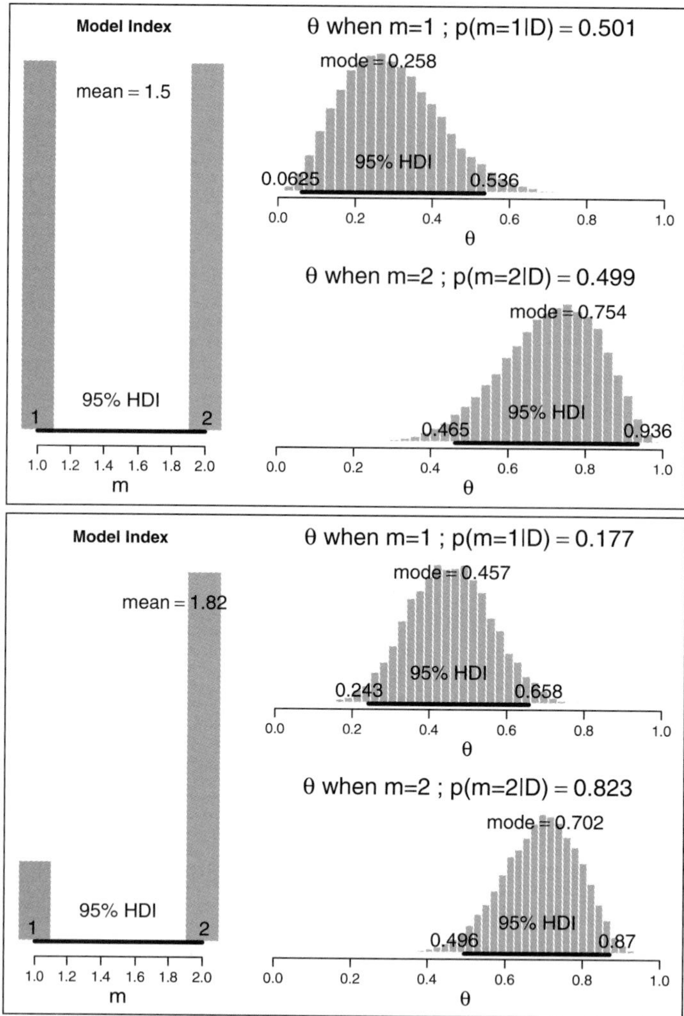

Figure 10.4 The prior and posterior distributions for script `Jags-Ydich-Xnom1subj-MbernBetaModelComp.R`. The upper frame, which shows the prior distribution, has labels that indicate $p(\theta|D)$ but the data set, D, is empty. The lower frame shows the posterior distribution.

Let's take a look at this model structure in JAGS, to see how it works. One novel requirement is that we must somehow tells JAGS to use θ_1 when $m=1$ but to use θ_2 when $m = 2$. We cannot do this with an `if` statement, as we might use in R, because JAGS is not a procedural language. The JAGS model specification is a "static" declaration of structure, not a sequential script of commands. Therefore, we use the JAGS `equals(,)` function, which returns `TRUE` (or 1) if its arguments are equal, and returns `FALSE` (or 0) if its arguments are not equal. Thus, the JAGS statement

```
theta <- equals(m,1)*theta1 + equals(m,2)*theta2
```

sets `theta` to the value of `theta1` when `m` equals 1 and sets `theta` to the value of `theta2` when `m` equals 2.

Below is the complete model specification for JAGS. Notice that `theta1` and `theta2` have different `dbeta` priors. The priors are given ω and κ values that match the examples of previous sections.

```
model {
  for ( i in 1:N ) {
    y[i] ~ dbern( theta )
  }
  theta <- equals(m,1)*theta1 + equals(m,2)*theta2
  theta1 ~ dbeta( omega1*(kappa1-2)+1 , (1-omega1)*(kappa1-2)+1 )
  omega1 <- .25
  kappa1 <- 12
  theta2 ~ dbeta( omega2*(kappa2-2)+1 , (1-omega2)*(kappa2-2)+1 )
  omega2 <- .75
  kappa2 <- 12
  m ~ dcat( mPriorProb[] )
  mPriorProb[1] <- .5
  mPriorProb[2] <- .5
}
```

The model specified above will run in JAGS without any problems in principle. In practice, however, for model structures of this type, the chain for the model index can be highly autocorrelated: The chain will linger in one model for a long time before jumping to the other model. In the very long run, the chain will visit each model proportionally to its posterior probability, but it might take a long time for the visiting proportion to represent the probability accurately and stably. Therefore, we will implement a trick in JAGS to help the chain jump between models more efficiently. The trick uses so-called "pseudopriors" (Carlin & Chib, 1995).

To motivate the use of pseudopriors, we must understand why the chain may have difficulty jumping between models. As was mentioned above, at each step in the chain, JAGS generates values for all three parameters. At a step for which $m=1$, then θ_1 is used to describe the data, while θ_2 floats unconstrained by the data and is sampled randomly from its prior. At a step for which $m=2$, then θ_2 is used to describe the data, while θ_1 floats unconstrained by the data and is sampled randomly from its prior. The problem is that on the next step, when JAGS considers a new random value for m while leaving the other parameter values unchanged, JAGS is choosing between the current value of m, which has a value for $\theta_{m=\text{current}}$ that is credible for the data, and the other value m, which has a value for $\theta_{m=\text{other}}$ from its prior that might be far away from posterior credible values. Because $\theta_{m=\text{other}}$ might be a poor description of the data, the chain rarely jumps to it, instead lingering on the current model.

A solution to this problem is to make the unused, free-floating parameter values remain in their zone of posterior credibility. Recall that parameter values are generated randomly from their prior when not being used to describe data. If we make the prior mimic the posterior when the parameter is not being used to describe data, then the randomly generated value will remain in the credible zone. Of course, when the parameter *is* being used to describe the data, then we must use its true prior. The prior that is used when the parameter is not describing the data is called the pseudoprior. Thus, when the model index is $m = 1$, θ_1 uses its true prior and θ_2 uses its pseudoprior. When the model index is $m = 2$, then θ_2 uses its true prior and θ_1 uses its pseudoprior.

To implement pseudopriors in JAGS, we can establish indexed values for the prior constants. Instead of using a single value `omega1` for the mode of the prior beta on `theta1`, we will instead use a mode indexed by the model, `omega1[m]`, which has different values depending on `m`. Thus, instead of specifying a single prior for `theta1` like this,

```
theta1 ~ dbeta( omega1*(kappa1-2)+1 , (1-omega1)*(kappa1-2)+1 )
omega1 <- .25 # true prior value
kappa1 <- 12 # true prior value
```

we will specify a prior that depends on the model index, like this:

```
theta1 ~ dbeta( omega1[m]*(kappa1[m]-2)+1 , (1-omega1[m])* (kappa1[m]-2)+1 )
omega1[1] <- .25 # true prior value
omega1[2] <- .45 # pseudo prior value
kappa1[1] <- 12 # true prior value
kappa1[2] <- 21 # pseudo prior value
```

In particular, notice that `omega1[1]` refers to the true prior value because it is the mode for `theta1` when `m=1`, but `omega1[2]` refers to the pseudo-prior value because it is the mode for `theta1` when `m=2`. The point of the above example is for you to understand the indexing. How to set the constants for the pseudo-priors is explained in the example below.

To show clearly the phenomenon of autocorrelation in the model index, the example will use data and prior constants that are a little different than the previous example. The data consist of $z = 17$ heads in $N = 30$ flips. Model 1 has $\omega_1 = 0.10$ with $\kappa_1 = 20$, and model 2 has $\omega_2 = 0.90$ with $\kappa_2 = 20$. The JAGS model specification therefore has true priors specified as follows:

```
model {
  for ( i in 1:N ) {
    y[i] ~ dbern( theta )
  }
  theta <- equals(m,1)*theta1 + equals(m,2)*theta2
  theta1 ~ dbeta( omega1[m]*(kappa1[m]-2)+1 , (1-omega1[m])*(kappa1[m]-2)+1 )
  omega1[1] <- .10 # true prior value
```

```
  omega1[2] <- .40 # pseudo prior value
  kappa1[1] <- 20 # true prior value
  kappa1[2] <- 50 # pseudo prior value
  theta2 ~ dbeta( omega2[m]*(kappa2[m]-2)+1 , (1-omega2[m])*(kappa2[m]-2)+1 )
  omega2[1] <- .70 # pseudo prior value
  omega2[2] <- .90 # true prior value
  kappa2[1] <- 50 # pseudo prior value
  kappa2[2] <- 20 # true prior value
  m ~ dcat( mPriorProb[] )
  mPriorProb[1] <- .5
  mPriorProb[2] <- .5
}
```

The complete script for this example has file name `Jags-Ydich-Xnom1subj-MbernBetaModelCompPseudoPrior.R`.

The values for the pseudo-priors are determined as follows:

1. Do an initial run of the analysis *with the pseudo-prior set to the true prior.* Note the characteristics of the marginal posterior distributions on the parameters.
2. Set the pseudo-prior constants to values that mimic the currently estimated posterior. Run the analysis. Note the characteristics of the marginal posterior distributions on the parameters. Repeat this step if the pseudo-prior distributions are very different from the posterior distributions.

As an example, Figure 10.5 shows an initial run of the analysis, using the true prior for the pseudoprior values. The MCMC chain had 10,000 steps altogether. The top part of the figure shows the convergence diagnostics on the model index. In particular, you can see from the top-right panel that the effective sample size (ESS) of the chain is very small, in this case less than 500. The top-left panel shows that that within each chain the model index lingered at one value for many steps before eventually jumping to the other value.

The lower part of Figure 10.5 shows the posterior distribution on the parameters. Consider the panel in the left column of the penultimate row, which shows θ_1 when $m = 1$. This represents the actual posterior distribution on θ_1. Below it, in the bottom-left panel, is θ_1 when $m = 2$. These values of θ_1 are not describing the data, because they were generated randomly from the prior when JAGS was instead using θ_2 to describe the data. Thus, the lower-left panel shows the prior on θ_1, which has its mode very close to the model-specified value of $\omega_1 = 0.10$. Notice that the (prior) values of θ_1 when $m = 2$ are not very close to the (posterior) values of θ_1 when $m = 1$. Analogously, the bottom-right panel shows the posterior distribution of θ_2 when $m = 2$, and the panel immediately above it shows the distribution of θ_2 when $m = 1$, which is the prior distribution of θ_2 (which has a mode of $\omega_2 = 0.90$). Again, notice that the (prior) values of θ_2 when $m = 1$ are not very close to the (posterior) values of θ_2 when $m = 2$.

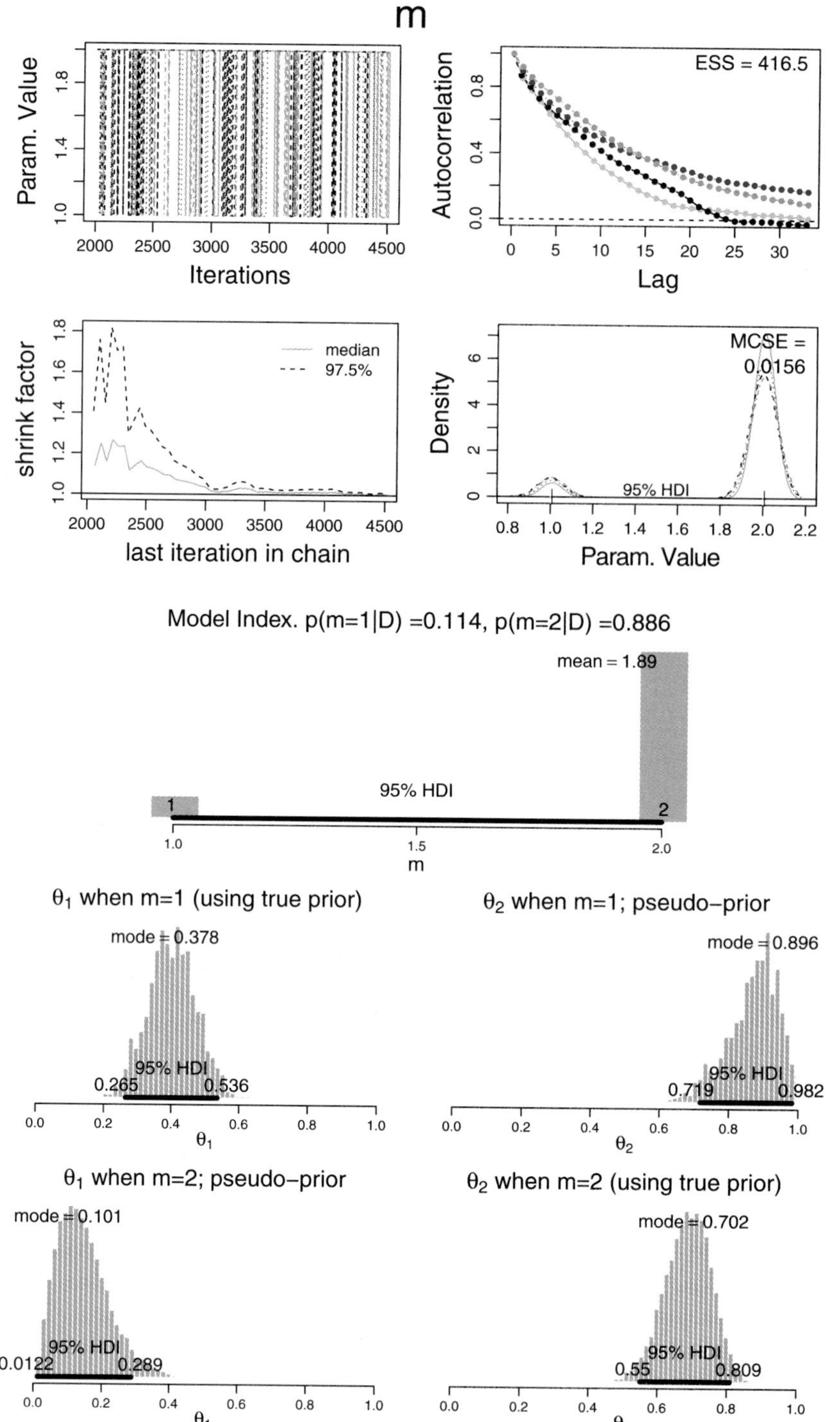

Figure 10.5 Poor jumping between models when not using pseudopriors; that is, when the implemented pseudoprior is the true prior. Compare with Figure 10.6.

With an eye on the lower two rows of Figure 10.5, consider the MCMC steps in JAGS. Suppose that at some step it is the case that $m = 2$, which means we are in the bottom row of Figure 10.5, so θ_1 has a value from the bottom-left distribution and θ_2 has a value from the bottom-right distribution. Now JAGS considers jumping from $m = 2$ to $m = 1$, using its current values of θ_1 and θ_2. Unfortunately, θ_1 is (probably) set to a value that is not among the credible values, and therefore JAGS has little incentive to jump to $m = 1$. Analogous reasoning applies when $m = 1$: The present value of θ_2 is (probably) not among the credible values, so JAGS is unlikely to jump to $m = 2$.

Despite the severe autocorrelation in the model index, the initial run has provided an indication of the posterior distributions of θ_1 and θ_2. We use their characteristics to set their pseudopriors. In particular, the posterior mode of θ_1 is near 0.40, so we set `omega1[2] <- .40`, and the posterior mode of θ_2 is near 0.70, so we set `omega2[1] <- .70`. The κ values were determined by remembering that the concentration parameter increases by N (recall the updating rule for Bernoulli-beta model in Section 6.3). Because the prior κ is 20, and the data have $N = 30$, the pseudoprior κ is 50, which is specified in the model statement as `kappa1[2] <- 50` and `kappa2[1] <- 50`.

Figure 10.6 shows the results when running the model with the new specification of the pseudopriors. The top-right panel shows that the chain for the model index has virtually no autocorrelation, and the ESS is essentially equal to the total length of the chain. The bottom two rows show the distributions of θ_1 and θ_2. Notice that the pseudo-priors nicely mimic the actual posteriors.

Because the ESS of the model index is far better than in the initial run, we trust its estimate in this run more than in the initial run. In this second run, with the pseudopriors appropriately set, the posterior probability of $m = 2$ is very nearly 92%. In the initial run, the estimate was about 89%. While these estimates are not radically different in this case, the example does illustrate what is at stake.

When examining the output of JAGS, it is important to realize that the chain for θ_m is a mixture of values from the true posterior and the pseudoprior. The values for the true posterior are from those steps in the chain when the model index corresponds to the parameter. Below is an example of R code, from the script `Jags-Ydich-Xnom1subj-MbernBetaModelCompPseudoPrior.R`, that extracts the relevant steps in the chain. The MCMC output of JAGS is the coda object named `codaSamples`.

```
# Convert coda-object codaSamples to matrix object for easier handling:
mcmcMat = as.matrix( codaSamples )
# Pull out the model index for easier handling:
m = mcmcMat[,"m"]
# Extract theta values for each model index:
theta1M1 = mcmcMat[,"theta1"][ m == 1 ] # true theta1
```

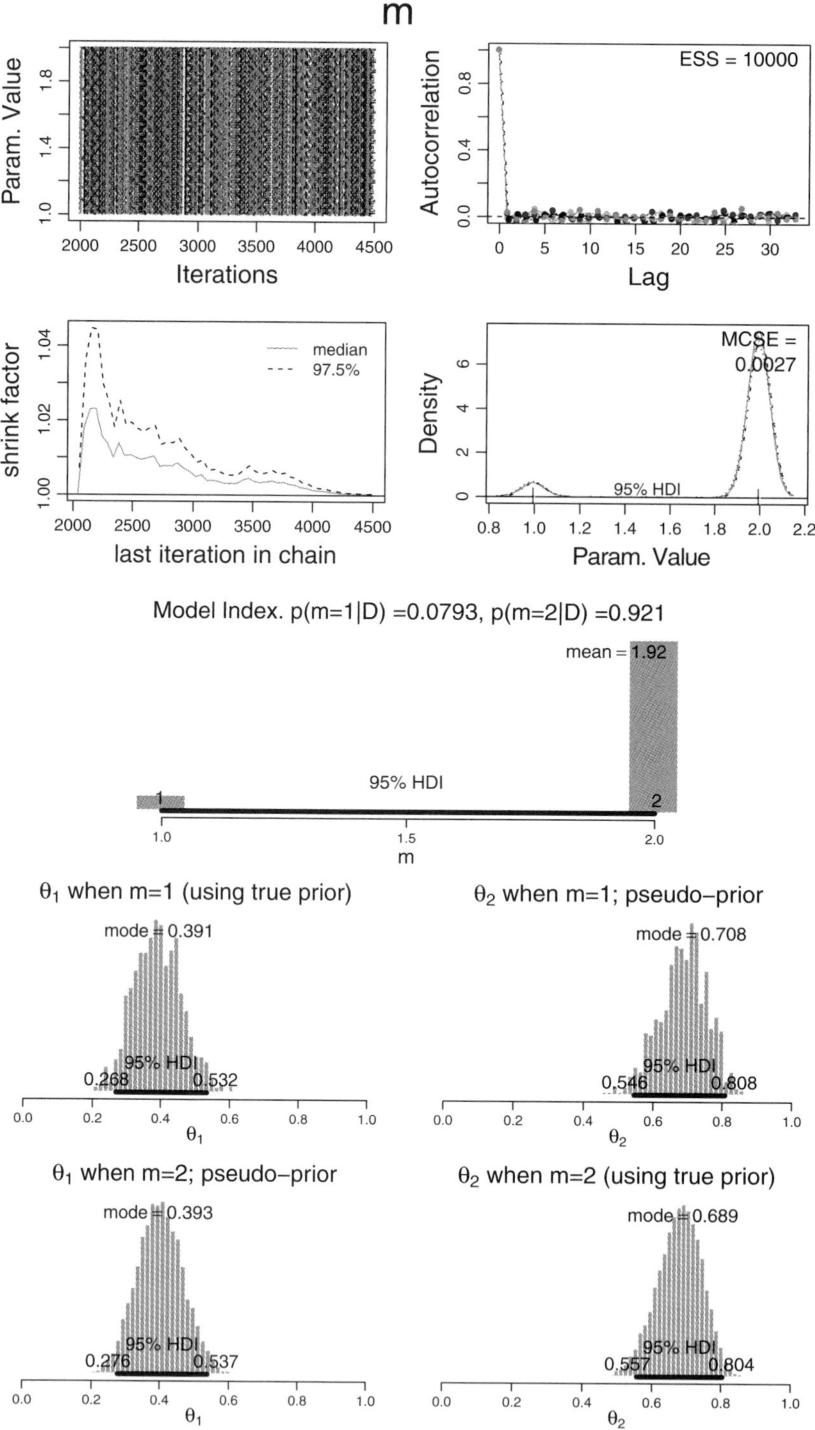

Figure 10.6 Better jumping between models when using pseudopriors; that is, when the implemented pseudoprior mimics the posterior. Compare with Figure 10.5.

```
theta1M2 = mcmcMat[,"theta1"][ m == 2 ] # pseudo theta1
theta2M1 = mcmcMat[,"theta2"][ m == 1 ] # pseudo theta2
theta2M2 = mcmcMat[,"theta2"][ m == 2 ] # true theta2
```

In the above R code, `mcmcMat` is a matrix with columns named `m`, `theta1`, and `theta2`, because those were the monitored variables in the call to JAGS that created the MCMC chain. The vector `m` contains a sequence of 1's and 2's indicating which model is used at each step. Then `mcmcMat[,"theta1"][m==1]` consists of the rows of `theta1` for which the model index is 1. These are the values of `theta1` for the true posterior. Analogously, `mcmcMat[,"theta2"][m==2]` contains the true posterior values for `theta2`.

The focus of model comparison is the posterior probability of the model index, m, not the estimates of the parameters in each model. If, however, you are interested in the parameters within each model, then you will want a sufficiently large MCMC sample within each model. In the present example, the chain visited $m = 1$ in only about 8% of its steps, yielding only about 800 representative values instead of the typically desired 10,000. (The small MCMC sample explains the choppy appearance of the histograms for $m = 1$.) To increase the number of values, we could simply increase the length of the chain, and wait. In large models with lots of data, however, increasing the chain length might exceed our patience. Instead, we can make JAGS sample more equally between the models by changing the prior on the models to compensate for their relative credibilities. From the initial run using 50-50 priors, we note that $p(m = 1|D) \approx 0.11$ and $p(m = 2|D) \approx 0.89$. Therefore, in the next run, we set `mPriorProb[1] <- .89` and `mPriorProb[2] <- .11`. The result is that the chain visits the two models about equally often, so we get a better representation of the parameters within each model. This may also provide a better estimate of the relative probabilities of the models, especially if one model is much better than the other.

But, you might ask, if we don't *use* 50-50 priors, how do we know the posterior probabilities of the models for 50-50 priors? The answer is revealed through the miracle of algebra. Let BF denote the value of the Bayes factor for the models (it is a constant). We know from Equation 10.5, p. 268, that $p(m = 1|D)/p(m = 2|D) = BF \cdot p(m = 1)/p(m = 2)$. In particular, for 50-50 priors, that equation reduces to $p(m = 1|D)/p(m = 2|D) = BF$. Therefore, if we want the posterior odds when the prior odds are 50/50, all we need is BF, which is given by rearranging the general equation: $BF = [p(m = 1|D)/p(m = 2|D)] \cdot [p(m = 2)/p(m = 1)]$.

Another example of model comparison using pseudopriors is presented in Section 12.2.2.1, p. 351. For more information about pseudopriors in transdimensional MCMC, see the tutorial article by Lodewyckx et al. (2011), and the original article by Carlin and Chib (1995), among other useful examples presented by Dellaportas, Forster, and Ntzoufras (2002), Han and Carlin (2001), and Ntzoufras (2002).

10.3.3. Models with different "noise" distributions in JAGS

All of the examples to this point have been cases of the structure in the far right panel of Figure 10.1. In that structure, the probability density function, $p(D|\theta)$, is the same for all models. This probability distribution is sometimes called the "noise" distribution because it describes the random variability of the data values around the underlying trend. In more general applications, different models can have different noise distributions. For example, one model might describe the data as log-normal distributed, while another model might describe the data as gamma distributed. The distinct noise distributions are suggested by the middle panel of Figure 10.1, where the likelihood functions have their probability functions subscripted with different indices: $p_1(D|\theta_1, m_1)$ vs $p_2(D|\theta_2, m_2)$.

To implement the general structure in JAGS, we would like to tell JAGS to use different probability density functions depending on the value of an indexical parameter. Unfortunately, we cannot use syntax such as `y[i] ~ equals(m,1) *pdf1(param1) + equals(m,2)*pdf2(param2)`. Instead, we can use the ones trick from Section 8.6.1, p. 214. The general form is shown here:

```
data {
  C <- 10000 # JAGS does not warn if too small!
  for (i in 1:N) {
    ones[i] <- 1
  }
}
model {
  for (i in 1:N) {
    spy1[i] <-  pdf1( y[i] , parameters1 )/C # where pdf1 is a formula
    spy2[i] <-  pdf2( y[i] , parameters2 )/C # where pdf2 is a formula
    spy[i] <- equals(m,1)*spy1[i] + equals(m,2)*spy2[i]
    ones[i] ~ dbern( spy[i] )
  }
  parameters1 ~ dprior1...
  parameters2 ~ dprior2...
  m ~ dcat( mPriorProb[] )
  mPriorProb[1] <- .5
  mPriorProb[2] <- .5
}
```

Notice above that each likelihood function is defined as an explicit formula for its probability density function, abbreviated above as `pdf1` or `pdf2`. Then one of the probability density values, `pdf1` or `pdf2`, is selected for the current MCMC step by using the `equals` construction. Each likelihood function should be specified in its entirety and both likelihood functions must be scaled by the same constant, `C`. The prior distributions on the parameters, including the model index, are then set up in JAGS as usual. Of course, the prior distributions on the parameters should use pseudopriors to facilitate efficient sampling of the model index.

10.4. PREDICTION: MODEL AVERAGING

In many applications of model comparison, the analyst wants to identify the best model and then base predictions of future data on that single best model, denoted with index b. In this case, predictions of future $\hat{y}$ are based exclusively on the likelihood function $p_b(\hat{y}|\theta_b, m=b)$ and the posterior distribution $p_b(\theta_b|D, m=b)$ of the winning model:

$$p(\hat{y}|D, m=b) = \int d\theta_b \, p_b(\hat{y}|\theta_b, m=b) \, p_b(\theta_b|D, m=b)$$

But the full model of the data is actually the complete hierarchical structure that spans all the models being compared, as indicated in Figure 10.1 (p. 267). Therefore, if the hierarchical structure really expresses our prior beliefs, then the most complete prediction of future data takes into account all the models, weighted by their posterior credibilities. In other words, we take a weighted average across the models, with the weights being the posterior probabilities of the models. Instead of conditionalizing on the winning model, we have

$$\begin{aligned} p(\hat{y}|D) &= \sum_m p(\hat{y}|D, m)p(m|D) \\ &= \sum_m \int d\theta_m \, p_m(\hat{y}|\theta_m, m) \, p_m(\theta_m|D, m)p(m|D) \end{aligned}$$

This is called model averaging.

The difference between $p_b(\theta_b|D, m=b)$ and $\sum_m p_m(\theta_m|D, m)p(m|D)$ is illustrated by the bottom row of Figure 10.3, p. 272. Recall that there were two models of mints that created the coin, with one mint being tail-biased with mode $\omega = 0.25$ and one mint being head-biased with mode $\omega = 0.75$. The two subpanels in the lower-right illustrate the posterior distributions on θ within each model, $p(\theta|D, \omega=0.25)$ and $p(\theta|D, \omega=0.75)$. The winning model was $\omega = 0.75$, and therefore the predicted value of future data, based on the winning model alone, would use $p(\theta|D, \omega=0.75)$. But the overall model included $\omega = 0.25$, and if we use the overall model, then the predicted value of future data should be based on the complete posterior summed across values of ω. The complete posterior distribution, $p(\theta|D)$, is shown in the lowest-middle panel. You can see the contribution of $p(\theta|D, \omega=0.25)$ as the extended leftward tail.

10.5. MODEL COMPLEXITY NATURALLY ACCOUNTED FOR

One of the nice qualities of Bayesian model comparison is that it naturally compensates for model complexity. A complex model (usually) has an inherent advantage over a simpler model because the complex model can find some combination of its parameter values that match the data better than the simpler model. There are so many more parameter options in the complex model that one of those options is likely to fit the data better than any of the fewer options in the simpler model. The problem is that data are contaminated by random noise, and we do not want to always choose the more complex model merely because it can better fit noise. Without some way of accounting for model complexity, the presence of noise in data will tend to favor the complex model.

Bayesian model comparison compensates for model complexity by the fact that each model must have a prior distribution over its parameters, and more complex models must dilute their prior distributions over larger parameter spaces than simpler models. Thus, even if a complex model has some particular combination of parameter values that fit the data well, the prior probability of that particular combination must be small because the prior is spread thinly over the broad parameter space.

An example should clarify. Consider again the case of two factories that mint coins. One model for a factory is simple: it consistently creates coins that are fair. Its modal coin bias is $\omega_s = 0.5$ and its concentration is $\kappa_s = 1,000$. This model is simple insofar as it has limited options for describing data: all the θ values for individual coins must be very nearly 0.5. I will call this the "must-be-fair" model. The other model for a factory is relatively complex: it creates coins of any possible bias, all with equal probability. While its central coin bias is also $\omega_c = 0.5$, its consistency is slight, with $\kappa_c = 2$. This model is complex because it has many options for describing data: the θ values for individual coins can be anything between 0 and 1. I will call this the "anything's-possible" model.

We saw how to solve the Bayes factor for this situation back in Equation 10.6, p. 270. An implementation of the formula in R was also provided, as the "`pD(z,N,a,b)`" function. For the present example, the logarithmic implementation is required.

Suppose that we flip a coin $N = 20$ times and observe $z = 15$ heads. These data seem to indicate the coin is biased, and indeed the complex, anything's-possible model is favored by the Bayes factor:

```
> z=15 ; N=20 ; pD(z,N,a=500,b=500)/pD(z,N,a=1,b=1)
[1] 0.3229023
```

In other words, the probability of the simple, must-be-fair model is only about a third of the probability of the complex, anything's-possible model. The must-be-fair model loses because it has no θ value sufficiently close to the data proportion. The anything's-possible model has available θ values that exactly match the data proportion.

Suppose instead that we flip the coin $N = 20$ times and observe $z = 11$ heads. These data seem pretty consistent with the simpler, must-be-fair model, and indeed it is favored by the Bayes factor:

```
> z=11 ; N=20 ; pD(z,N,a=500,b=500)/pD(z,N,a=1,b=1)
[1] 3.337148
```

In other words, the probability of the simple, must-be-fair model is more than three times the probability of the complex, anything's-possible model. How could this favoring of the simple model happen? After all, the anything's-possible model has available to it the value of θ that exactly matches the data proportion. Should not the anything's-possible model win? The anything's-possible model loses because it pays the price of having a small prior probability on the values of θ near the data proportion, while the must-be-fair model has large prior probability on θ values sufficiently near the data proportion to be credible. Thus, in Bayesian model comparison, a simpler model can win if the data are consistent with it, even if the complex model fits just as well. The complex model pays the price of having small prior probability on parameter values that describe simple data. For additional reading on this topic, see, for example, the general-audience article by Jefferys and Berger (1992) and the application to cognitive modeling by Myung and Pitt (1997).

10.5.1. Caveats regarding nested model comparison

A frequently encountered special case of comparing models of different complexity occurs when one model is "nested" within the other. Consider a model that implements all the meaningful parameters we can contemplate for the particular application. We call that the full model. We might consider various restrictions of those parameters, such as setting some of them to zero, or forcing some to be equal to each other. A model with such a restriction is said to be nested within the full model. Notice that the full model is always able to fit data at least as well as any of its restricted versions, because the restricted model is merely a particular setting of the parameter values in the full model. But Bayesian model comparison will be able to prefer the restricted model, if the data are well described by the restricted model, because the full model pays the price of diluting its prior over a larger parameter space.

As an example, recall the hierarchical model of baseball batting abilities in Figure 9.13, p. 252. The full model has a distinct modal batting ability, ω_c, for each of the nine fielding positions. The full model also has distinct concentration parameters for each of the nine positions. We could consider restricted versions of the model, such as a version in which all infielders (first base, second base, etc.) are grouped together versus all outfielders (right field, center field, and left field). In this restricted model, we are forcing the modal batting abilities of all the outfielders to be the same, that is, $\omega_{\text{left field}} = \omega_{\text{center field}} = \omega_{\text{right field}}$ and analogously for the other parameters. Effectively, we are using a single parameter where there were three parameters (and so on for the other restrictions). The full model will always be able to fit the data at least as well as the restricted model, because the parameter values in the restricted model are also available in the full model. But the full model must also spread its prior distribution thinly over a larger parameter space. The restricted model piles up its prior distribution on points where $\omega_{\text{left field}}$ equals $\omega_{\text{center field}}$, with zero allocation to the vast space of unequal combinations, whereas the full model must dilute its prior over all those unequal combinations. If the data happen to be well described by the parameter values of the restricted model, then the restricted model will be favored by its higher prior probability on those parameter values.

There is a tendency in statistical modeling to systematically test all possible restrictions of a full model. For the baseball batting example, we could test all possible subsets of equalities among the nine positions, which results in 21,147 restricted models.[4] This is far too many to meaningfully test, of course. But there are other, more important reasons to avoid testing restricted models merely because you could. The main reason is that many restricted models have essentially zero prior probability, and you would not want to accept them even if they "won" a model comparison.

This caveat follows essentially from Bayes' rule for model comparison as formulated in Equation 10.5, p. 268. If the prior probability of a model is zero, then its posterior probability is zero, even if the Bayes factor favors the model. In other words, do not confuse the Bayes factor with the posterior odds.

As an example, suppose that a restricted model, that groups together the seven positions other than pitchers and catchers, wins a model comparison against the full model with nine separate positions, in the sense that the Bayes factor favors the restricted model. Does that mean we should believe that the seven positions (such as first base and right field) have literally identical batting abilities? Probably not, because the prior probability of such a restriction is virtually nil. Instead, it is more meaningful to use the full model for parameter estimation, as was done in Section 9.5. The parameter estimation will reveal that the differences between the seven positions may be small relative to the uncertainty of the estimates, but will retain distinct estimates. We will see an example in Section 12.2.2 where a model comparison favors a restricted model that collapses some groups, but explicit parameter estimation shows differences between those groups.

10.6. EXTREME SENSITIVITY TO PRIOR DISTRIBUTION

In many realistic applications of Bayesian model comparison, the theoretical emphasis is on the difference between the models' likelihood functions. For example, one theory predicts planetary motions based on elliptical orbits around the sun, and another theory predicts planetary motions based on circular cycles and epicycles around the earth. The two models involve very different parameters. In these sorts of models, the form of the prior distribution on the parameters is not a focus, and is often an afterthought. But, when doing Bayesian model comparison, the form of the prior is crucial because the Bayes factor integrates the likelihood function weighted by the prior distribution.

[4] The number of restricted models is, in this example, the number of possible partitions of the nine positions. The number of partitions of a set is called its Bell number, which has been extensively studied in combinatorics (e.g., Rota, 1964).

As we have seen repeatedly, Bayesian model comparison involves marginalizing across the prior distribution in each model. Therefore, the posterior probabilities of the models, and the Bayes factors, can be extremely sensitive to the choice of prior distribution. If the prior distribution happens to place a lot of probability mass where the likelihood distribution peaks, then the marginal likelihood (i.e., $p(D|m)$) will be large. But if the prior distribution happens to place little probability mass where the likelihood distribution is, then the marginal likelihood will be small. The sensitivity of Bayes factors to prior distributions is well known in the literature (e.g., Kass & Raftery, 1995; Liu & Aitkin, 2008; Vanpaemel, 2010).

When doing Bayesian model comparison, different forms of vague priors can yield very different Bayes factors. As an example, consider again the must-be-fair versus anything's-possible models of the previous section. The must-be-fair model was characterized as a beta prior with shape parameters of $a = 500$ and $b = 500$ (i.e., mode $\omega = 0.5$ and concentration $\kappa = 1000$). The anything's-possible model was defined as a beta prior with shape parameters of $a = 1$ and $b = 1$. Suppose we have data with $z = 65$ and $N = 100$. Then the Bayes factor is

```
> z=65 ; N=100 ; pD(z,N,a=500,b=500)/pD(z,N,a=1,b=1)
[1] 0.125287
```

This means that the anything's-possible model is favored. But why did we choose those particular shape parameter values for the anything's-possible model? It was merely that intuition suggested a uniform distribution. On the contrary, many mathematical statisticians recommend a different form of prior to make it uninformative according to a particular mathematical criterion (Lee & Webb, 2005; Zhu & Lu, 2004). The recommended prior is the so-called Haldane prior, which uses shape constants that are very close to zero, such as $a = b = 0.01$. (See Figure 6.1, p. 128, for an example of a beta distribution with shape parameters less than 1.) Using a Haldane prior to express the anything's-possible model, the Bayes factor is

```
> z=65 ; N=100 ; pD(z,N,a=500,b=500)/pD(z,N,a=0.01,b=0.01)
[1] 5.728066
```

This means that the must-be-fair model is favored. Notice that we reversed the Bayes factor merely by changing from a "vague" $\text{beta}(\theta|1, 1)$ prior to a "vague" $\text{beta}(\theta|.01, .01)$ prior.

Unlike Bayesian model comparison, when doing Bayesian estimation of continuous parameters within models and using realistically large amounts of data, the posterior distribution on the continuous parameters is typically robust against changes in vague priors. It does not matter if the prior is extremely vague or only a little vague (and yes, what I meant by "extremely vague" and "only a little vague" is vague, but the point is that it doesn't matter).

As an example, consider the two versions of the anything's-possible model, using either a "vague" beta(θ|1, 1) prior or a "vague" beta(θ|.01, .01) prior. Using the data $z = 65$ and $N = 100$, we can compute the posterior distribution on θ. Starting with the beta(θ|1, 1) yields a beta(θ|66, 36) posterior, which has a 95% HDI from 0.554 to 0.738. (The HDI was computed by using the `HDIofICDF` function that comes with the utilities program accompanying this book.) Starting with the beta(θ|.01, .01) yields a beta(θ|65.01, 35.01) posterior, which has a 95% HDI from 0.556 to 0.742. The HDIs are virtually identical. In particular, for either prior, the posterior distribution rules out $\theta = 0.5$, which is to say that the must-be-fair hypothesis is not among the credible values. For additional discussion and related examples, see Kruschke (2011a) and Section 12.2 of this book.

10.6.1. Priors of different models should be equally informed

We have established that seemingly innocuous changes in the vagueness of a vague prior can dramatically change a model's marginal likelihood, and hence its Bayes factor in comparison with other models. What can be done to ameliorate the problem? One useful approach is to inform the priors of all models with a small set of representative data (the same for all models). The idea is that even a small set of data overwhelms any vague prior, resulting in a new distribution of parameters that is at least "in the ballpark" of reasonable parameter values for that model. This puts the models on an equal playing field going into the model comparison.

Where do the data come from, that will act as the small representative set for informing the priors of the models? They could come from previous research. They could be fictional but representative of previous research, as long as the audience of the analysis agrees that the fictional data are valid. Or, the data could be a small percentage of the data from the research at hand. For example, a random 10% of the data could inform the priors of the models, and the remaining 90% used for computing the Bayes factor in the model comparison. In any case, the data used for informing the priors should be representative of real data and large enough in quantity to usefully overwhelm any reasonable vague prior. Exactly what that means will depend on the details of the model, but the following simple example illustrates the idea.

Recall, from the previous section, the comparison of the must-be-fair model and the anything's-possible model. When $z = 65$ with $N = 100$, the Bayes factor changed dramatically depending on whether the "vague" anything's-possible model used a beta(θ|1, 1) prior or a beta(θ|.01, .01) prior. Now let's compute the Bayes factors after informing both models with just 10% of the data. Suppose that the 10% subset has 6 heads in 10 flips, so the remaining 90% of the data has $z = 65 - 6$ and $N = 100 - 10$.

Suppose we start with beta(θ|1, 1) for the prior of the anything's-possible model. We inform it, and the must-be-fair model, with the 10% subset. Therefore, the anything's-possible model becomes a beta($\theta|1+6, 1+10-6$) prior, and the must-be-fair model becomes a beta($\theta|500+6, 500+10-6$) prior. The Bayes factor is

```
> z=65-6 ; N=100-10 ; pD(z,N,a=500+6,b=500+10-6)/pD(z,N,a=1+6,b=1+10-6)
[1] 0.05570509
```

Now let's instead start with beta($\theta|.01, .01$) for the anything's-possible model. The Bayes factor using the weakly informed priors is

```
> z=65-6 ; N=100-10 ; pD(z,N,a=500+6,b=500+10-6)/pD(z,N,a=0.01+6,b=0.01+10-6)
[1] 0.05748123
```

Thus, the Bayes factor has barely changed at all. With the two models equally informed by a small amount of representative data, the Bayes factor is stable.

The idea of using a small amount of training data to inform the priors for model comparison has been discussed at length in the literature and is an active research topic. A selective overview was provided by J. O. Berger and Pericchi (2001), who discussed conventional default priors (e.g., Jeffreys, 1961), "intrinsic" Bayes factors (e.g., J. O. Berger & Pericchi, 1996), and "fractional" Bayes factors (e.g., O'Hagan, 1995, 1997), among others.

10.7. EXERCISES

Look for more exercises at https://sites.google.com/site/doingbayesiandataanalysis/

Exercise 10.1. [Purpose: To illustrate the fact that models with more distinctive predictions can be more easily discriminated.] Consider the scenario of Section 10.2.1, in which there were two coin factories, one of which was tail-biased and the other head-biased. Suppose we flip a coin that we know is from one of the two factories but we do not know which factory, and the prior probabilities of the factories are 50/50. The results show $z = 7$ heads in $N = 10$ flips.

(A) If $\omega_1 = 0.25$, $\omega_2 = 0.75$, and $\kappa = 6$, what are the posterior probabilities of the factories?

(B) If $\omega_1 = 0.25$, $\omega_2 = 0.75$, and $\kappa = 202$, what are the posterior probabilities of the factories?

(C) Why are the posterior probabilities so different in parts A and B, even though the modes of the factories are the same?

Exercise 10.2. [Purpose: To be sure you really understand the JAGS program for Figure 10.4.]

(A) Use the script Jags-Ydich-Xnom1subj-MbernBetaModelComp.R to reproduce Figure 10.4, including both the prior and the posterior. Explain how you generated the MCMC sample from the prior. Include the graphical output in your answer, which will be slightly different than Figure 10.4 because of randomness in the MCMC chain.

(B) Make a histogram of the θ values collapsed across both models. It should look like the bottom-middle panel of Figure 10.3. Explain why.

(C) Use the script to reproduce the previous exercise. That is, change the data to $z = 7$ heads in $N = 10$, and run the script once with $\kappa = 6$ and once with $\kappa = 202$. Do the MCMC results match the analytical results?

Exercise 10.3. [Purpose: To get some hands-on experience with pseudo-priors.]

(A) Use the script Jags-Ydich-Xnom1subj-MbernBetaModelCompPseudoPrior.R to reproduce Figures 10.5 and 10.6. That is, run the script once with the pseudo-prior set to the true prior, and then again with the pseudo-prior set to the values shown in the text. Include the graphical output of the chain diagnostics on the model index. Your results will differ slightly from Figures 10.5 and 10.6 because of randomness in the MCMC chain.

(B) Change the pseudo-prior to broad distributions, with `omega1[2]=omega2[1]=0.5` and `kappa1[2]=kappa2[1]=2.1`. Run the script and report the results, including the chain diagnostic on the model index. Discuss.

CHAPTER 11

Null Hypothesis Significance Testing

Contents

My baby don't value what I really do.
She only imagines who else might come through.
She'll only consider my worth to be high
If she can't conceive of some much bigger guy.[1]

In the previous chapters, we have seen a thorough introduction to Bayesian inference. It is appropriate now to compare Bayesian inference with NHST. In NHST, the goal of inference is to decide whether a particular value of a parameter can be rejected. For

[1] This chapter is about p values in null hypothesis significance testing (NHST). p values are computed by considering imaginary, counterfactual possibilities—things that could have happened but didn't. The poem brings this idea to an everyday domain, pregnant with possibilities.

Doing Bayesian Data Analysis, 2nd Edition
http://dx.doi.org/10.1016/B978-0-12-405888-0.00011-8

example, we might want to know whether a coin is fair, which in NHST becomes the question of whether we can reject the "null" hypothesis that the bias of the coin has the specific value $\theta = 0.50$.

The logic of conventional NHST goes like this. Suppose the coin is fair (i.e., $\theta = 0.50$). Then, when we flip the coin, we expect that about half the flips should come up heads. If the actual number of heads is far greater or fewer than half the flips, then we should reject the hypothesis that the coin is fair. To make this reasoning precise, we need to figure out the exact probabilities of all possible outcomes, which in turn can be used to figure out the probability of getting an outcome as extreme as (or more extreme than) the actually observed outcome. This probability, of getting an outcome from the null hypothesis that is as extreme as (or more extreme than) the actual outcome, is called a "p value." If the p value is very small, say less than 5%, then we decide to reject the null hypothesis.

Notice that this reasoning depends on defining a space of all possible outcomes from the null hypothesis, because we have to compute the probabilities of each outcome relative to the space of all possible outcomes. The space of all possible outcomes is based on how we intend to collect data. For example, was the intention to flip the coin exactly N times? In that case, the space of possible outcomes contains all sequences of exactly N flips. Was the intention to flip until the zth head appeared? In that case, the space of possible outcomes contains all sequences for which the zth head appears on the last flip. Was the intention to flip for a fixed duration? In that case, the space of possible outcomes contains all combinations of N and z that could be obtained in that fixed duration. Thus, a more explicit definition of a p value is the probability of getting a sample outcome from the hypothesized population that is as extreme as or more extreme than the actual outcome *when using the intended sampling and testing procedures*.

Figure 11.1 illustrates how a p value is defined. The actual outcome is an observed constant, so it is represented by a solid block. The space of all possible outcomes is represented by a cloud generated by the null hypothesis with a particular sampling intention. For example, from the hypothesis of a fair coin, we would expect to get 50% heads, but sometimes we would get a greater or lesser percentage of heads in the sample of random flips. The center of the cloud in Figure 11.1 is most dense, with typical outcomes we would expect from the null hypothesis. The fringe of the cloud is less dense, with unusual outcomes obtained by chance. The dashed line indicates how far the actual outcome is from the expected outcome. The proportion of the cloud beyond the dashed line is the p value: The probability that possible outcomes would meet or exceed the actual outcome. The left panel of Figure 11.1 shows the cloud of possibilities with sampling intention A, which might be to stop collecting data when N reaches a particular value. Notice that the p value is relatively large. The right panel of Figure 11.1 shows the cloud of possibilities with sampling intention B, which might

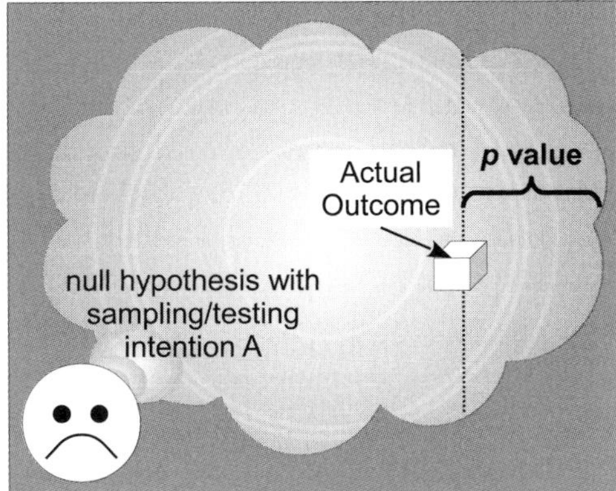

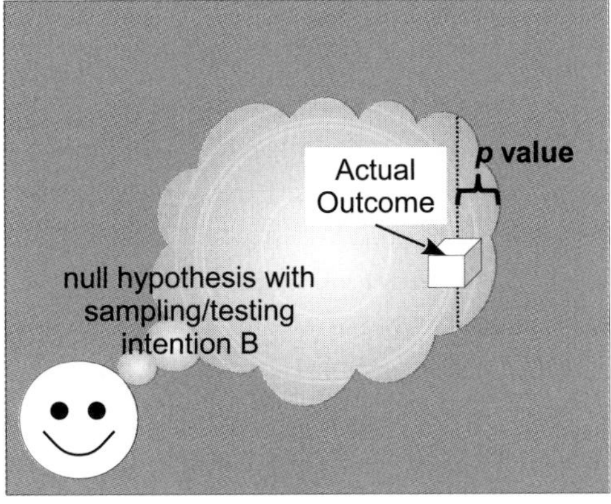

Figure 11.1 The null hypothesis generates a cloud of imaginary outcomes, most of which fall in the center of the cloud, but some of which fall beyond the actual outcome marked by block. The *p* value is the proportion of the cloud as or more extreme than the actual outcome. Left panel: With sampling intention A, the cloud of imaginary possibilities has a large proportion that exceeds the actual outcome, hence the *p* value is large. Right panel: With sampling intention B, the cloud of imaginary possibilities has a small proportion that exceeds the actual outcome, hence the *p* value is small.

be to stop collecting data when the duration reaches a particular value. Notice that the cloud of possible outcomes is different, and the *p* value is relatively small.

Compare the two panels and notice that the single actual outcome has different *p* values depending on the sampling intention. This belies the common parlance of talking about "*the*" *p* value for a set of data, as if there were only one unique *p* value for the data. There are, in fact, many possible *p* values for any set of data, depending on how the cloud of imaginary outcomes is generated. The cloud depends not only on the intended stopping criterion, but also on the intended tests that the analyst wants to make, because intending to make additional tests expands the cloud with additional possibilities from the additional tests.

Do the actually observed data depend on the intended stopping rule or intended tests? No, not in appropriately conducted research. A good experiment (or observational survey) is founded on the principle that the data are insulated from the experimenter's intentions. The coin "knows" only that it was flipped so many times, regardless of what the experimenter had in mind while doing the flipping. Therefore our conclusion about the coin should not depend on what the experimenter had in mind while flipping it, nor what tests the experimenter had in mind before or after flipping it.

The essential constraint on the stopping rule is that it should not bias the data that are obtained. Stopping at a fixed number of flips does not bias the data (assuming that the procedure for a *single* flip does not bias the outcome). Stopping at a fixed duration does not bias the data. Stopping at a fixed number of heads (or tails) *does* bias the data, because a random sequence of unrepresentative flips can cause the data collection to stop prematurely, preventing the subsequent collection of compensatory representative data. In general, peeking at the data as it accumulates, and continuing to collect additional data only if there is not yet an extreme outcome, does bias the data, because a random extreme outcome can stop data collection with no subsequent opportunity for compensatory data. A focus of this chapter is the effect of the stopping intention on the p value; the effect of the stopping intention on the content of the data is discussed en route.

This chapter explains some of the gory details of NHST, to bring mathematical rigor to the above comments, and to bring rigor mortis to NHST. You'll see how NHST is committed to the notion that the covert intentions of the experimenter are crucial to making decisions about the data, even though the data are not supposed to be influenced by the covert intentions of the experimenter. We will also discuss what the cloud of possibilities *is* good for, namely prospective planning of research and predicting future data.

11.1. PAVED WITH GOOD INTENTIONS

In this section, we will derive exact p values for observed coin flips under different sampling intentions. To make the calculations concrete, suppose we have a coin that we want to test for fairness. We have an assistant to flip the coin. The assistant is not told the hypothesis we are testing. We ask the assistant to flip the coin a few times as we watch. Here is the sequence of results:

$$TTHHTTHTTTTTTTTTHTTHHTTH \tag{11.1}$$

We observe that of $N = 24$ flips, there were $z = 7$ heads, that is, a proportion of $7/24$. It seems that there were fewer heads than what we would expect from the hypothesis of fairness. We would like to derive the probability of getting a proportion of heads that is $7/24$ or smaller if the null hypothesis is true.

11.1.1. Definition of *p* value

To derive that probability, it can help to be clear about the general form of what we are trying to derive. Here's the idea. The null hypothesis in NHST starts with a likelihood function and specific parameter value that describes the probabilities of outcomes for single observations. This is the same likelihood function as in Bayesian analysis. In the case of a coin, the likelihood function is the Bernoulli distribution, with its parameter θ that describes the probability of getting the outcome "head" on a single flip. Typically the null value of θ is 0.5, as when testing whether a coin is fair, but the hypothesized value of θ could be different.

To derive a p value from the null hypothesis, we must also specify how to generate full samples of data. The sample generation process should reflect the way that the real data were actually collected. Perhaps the data collection stopped when the sample size N reached a predetermined limit. Or perhaps the data were collected for a fixed duration, like a pollster standing on a street corner for 1 hour, asking random passers-by. In this case, the sample size is random, but there is a typical average sample size based on the rate of passers-by per unit time. And the sample-generation process must also reflect other sources of hypothetical samples, such as other tests that might be run. We need to include those hypothetical samples in the cloud of possibilities that surrounds our actually observed outcome.

In summary, the likelihood function defines the probability for a single measurement, and the intended sampling process defines the cloud of possible sample outcomes. The null hypothesis is the likelihood function with its specific value for parameter θ, and the cloud of possible samples is defined by the stopping and testing intentions, denoted I. Each imaginary sample generated from the null hypothesis is summarized by a descriptive statistic, denoted $D_{\theta,I}$. In the case of a sample of coin flips, the descriptive summary statistic is z/N, the proportion of heads in the sample. Now, imagine generating infinitely many samples from the null hypothesis using stopping and testing intention I; this creates a cloud of possible summary values $D_{\theta,I}$, each of which has a particular probability. The probability distribution over the cloud of possibilities is the *sampling distribution*: $p(D_{\theta,I}|\theta, I)$.

To compute the p value, we want to know how much of that cloud is as extreme as, or more extreme than, the actually observed outcome. To define "extremeness" we must determine the typical value of $D_{\theta,I}$, which is usually defined as the expected value, $E[D_{\theta,I}]$ (recall Equations 4.5 and 4.6). This typical value is the center of the cloud of possibilities. An outcome is more "extreme" when it is farther away from the central tendency. The p value of the actual outcome is the probability of getting a hypothetical outcome that is as or more extreme. Formally, we can express this as

$$p \text{ value} = p(D_{\theta,I} \succcurlyeq D_{\text{actual}}|\theta, I) \tag{11.2}$$

where "$\succcurlyeq$" in this context means "as extreme as or more extreme than, relative to the expected value from the hypothesis." Most introductory applied statistics textbooks suppress the sampling intention I from the definition, but precedents for making the sampling intention explicit can be found in Wagenmakers (2007, Online Supplement A) and additional references cited therein. For the case of coin flips, in which the sample summary statistic is z/N, the p value becomes

$$p \text{ (right tail)} = p\left((z/N)_{\theta,I} \geq (z/N)_{\text{actual}} \,|\, \theta, I\right) \tag{11.3}$$

$$p \text{ (left tail)} = p\left((z/N)_{\theta,I} \leq (z/N)_{\text{actual}} \,|\, \theta, I\right) \tag{11.4}$$

Those p values are called "one tailed" because they indicate the probability of hypothetical outcomes more extreme than the actual outcome in only one direction. Typically we care about the right tail when $(z/N)_{\text{actual}}$ is greater than $E[(z/N)_{\theta,I}]$, and we care

z \ N	1	2	3	4	5	6	7	8	...
0									...
1									...
2	-								...
3	-	-							...
4	-	-	-						...
5	-	-	-	-					...
6	-	-	-	-	-				...
7	-	-	-	-	-	-			...
8	-	-	-	-	-	-	-		...
...	-	-	-	-	-	-	-	-	...

z \ N	1	2	3	4	5	6	7	8	...
0									...
1									...
2	-								...
3	-	-							...
4	-	-	-						...
5	-	-	-	-					...
6	-	-	-	-	-				...
7	-	-	-	-	-	-			...
8	-	-	-	-	-	-	-		...
...	-	-	-	-	-	-	-	-	...

z \ N	1	2	3	4	5	6	7	8	...
0									...
1									...
2	-								...
3	-	-							...
4	-	-	-						...
5	-	-	-	-					...
6	-	-	-	-	-				...
7	-	-	-	-	-	-			...
8	-	-	-	-	-	-	-		...
...	-	-	-	-	-	-	-	-	...

Figure 11.2 Sample space for flips of a coin, in which columns show candidate values for N and rows show candidate values for z. Left table: Space of possibilities when N is considered fixed, highlighted by shaded column (at $N = 5$). Middle table: Space of possibilities when z is considered fixed, highlighted by shaded row (at $z = 4$). Right table: Space of possibilities when duration is considered fixed, with probabilities of sample sizes suggested by differential shading of columns.

about the left tail when $(z/N)_{\text{actual}}$ is less than $E[(z/N)_{\theta,I}]$. The "two-tailed" p value can be defined in various ways, but for our purposes we will define the two-tailed p value as simply two times the one-tailed p value.

To compute the p value for any specific situation, we need to define the space of possible outcomes. Figure 11.2 shows the full space of possible outcomes when flipping a coin, showing all possible combinations of z and N. The figure shows only small values of z and N, but both values extend to infinity. Each cell represents the proportion z/N. For compactness, each cell of Figure 11.2 collapses across different specific sequences of heads and tails that have the same z and N. (For example, the cell for $z = 1$ and $N = 2$ refers to the sequences H, T and T, H.) Each combination of z and N has a particular probability of occurring from a particular null hypothesis and sampling intention. In our example, the null hypothesis of a fair coin means that $\theta = 0.5$ in the Bernoulli likelihood function, and therefore the expected outcome should be about $0.5 \cdot N$. With an actual outcome of $z = 7$ and $N = 24$ (recall the sequence of flips in Equation 11.1), that means we want to compute the probability of landing in a cell of the table for which $(z/N)_{\theta,I}$ is *less* than $7/24$. And to do that, we must specify the stopping and testing intention, I.

11.1.2. With intention to fix *N*

Suppose we ask the assistant why she stopped flipping the coin. She says that her lucky number is 24, so she decided to stop when she completed 24 flips of the coin. This means the space of possible outcomes is restricted to combinations of z and N for which N is fixed at $N = 24$. This corresponds to a single column of the z, N space as shown in the left panel of Figure 11.2 (which shows $N = 5$ highlighted instead of $N = 24$ because of lack of space). The computational question then becomes, what is the probability of the actual proportion, or a proportion more extreme than expected, *within that column of the outcome space*?

What is the probability of getting a particular number of heads when N is fixed? The answer is provided by the *binomial probability distribution*, which states that the probability of getting z heads out of N flips is

$$p(z|N,\theta) = \binom{N}{z}\theta^z(1-\theta)^{N-z} \tag{11.5}$$

where the notation $\binom{N}{z}$ will be defined below. The binomial distribution is derived by the following logic. Consider any specific sequence of N flips with z heads. The probability of that specific sequence is simply the product of the individual flips, which is the product of Bernoulli probabilities $\prod_i \theta^{y_i}(1-\theta)^{1-y_i} = \theta^z(1-\theta)^{N-z}$, which we first saw in Section 6.1, p. 124. But there are many different specific sequences with z heads. Let's count how many there are. Consider allocating z heads to N flips in the sequence. The first head could go in any one of the N slots. The second head could go in any one of the remaining $N-1$ slots. The third head could go in any one of the remaining $N-2$ slots. And so on, until the zth head could go in any one of the remaining $N-(z-1)$ slots. Multiplying those possibilities together means that there are $N \cdot (N-1) \cdot \ldots \cdot (N-(z-1))$ ways of allocating z heads to N flips. As an algebraic convenience, notice that $N \cdot (N-1) \cdot \ldots \cdot (N-(z-1)) = N!/(N-z)!$, where "!" denotes factorial. In this counting of the allocations, we've counted different orderings of the same allocation separately. For example, putting the first head in the first slot and the second head in the second slot was counted as a different allocation than putting the first head in the second slot and the second head in the first slot. There is no meaningful difference in these allocations, because they both have a head in the first and second slots. Therefore, we remove this duplicate counting by dividing out by the number of ways of permuting the z heads among their z slots. The number of permutations of z items is $z!$. Putting this all together, the number of ways of allocating z heads among N flips, without duplicate counting of equivalent allocations, is $N!/[(N-z)!z!]$. This factor is also called the number of ways of choosing z items from N possibilities, or "N choose z" for short, and is denoted $\binom{N}{z}$. Thus, the overall probability of getting z heads in N flips is the probability of any particular sequence of z heads in N flips times the number of ways of choosing z slots from among the N possible flips. The product appears in Equation 11.5.

A graph of a binomial probability distribution is provided in the right panel of Figure 11.3, for $N = 24$ and $\theta = 0.5$. Notice that the graph contains 25 spikes, because there are 25 possible proportions, from 0/24, 1/24, 2/24, through 24/24. The binomial probability distribution in Figure 11.3 is also called a *sampling distribution*. This terminology stems from the idea that any set of N flips is a representative sample of the behavior of the coin. If we were to repeatedly run experiments with a fair coin, such that in every experiment we flip the coin exactly N times, then, in the long run, the probability of getting each possible z would be the distribution shown in Figure 11.3. To

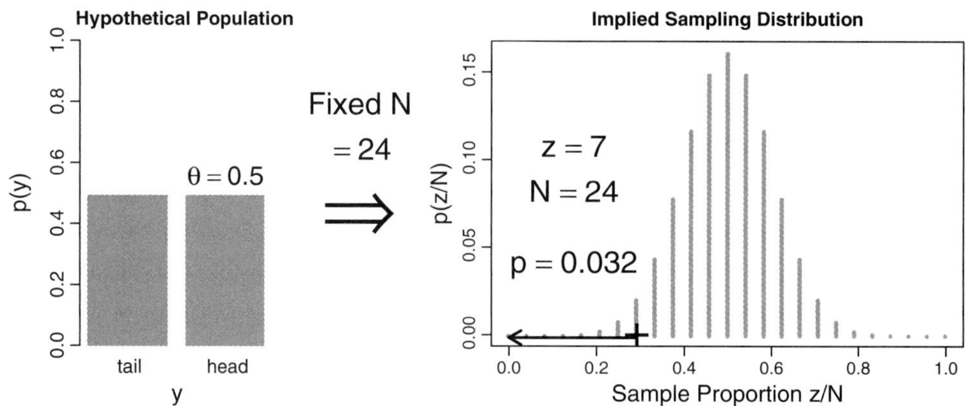

Figure 11.3 The imaginary cloud of possible outcomes when N is fixed. The null hypothesis likelihood distribution and parameter are shown on the left. The stopping intention is shown in the middle. The sampling distribution and p value are shown on the right. Compare with Figures 11.4 and 11.5.

describe it carefully, we would call it "the probability distribution of the possible sample outcomes," but that's usually just abbreviated as "the sampling distribution."

Terminological aside: Statistical methods that rely on sampling distributions are sometimes called *frequentist* methods. A particular application of frequentist methods is NHST.

Figure 11.3 is a specific case of the general structure shown in Figure 11.1. The left side of Figure 11.3 shows the null hypothesis as the probability distribution for the two states of the coin, with $\theta = 0.5$. This corresponds to the face in the lower-left corner of Figure 11.1, who is thinking of a particular hypothesis. The middle of Figure 11.3 shows an arrow marked with the sampling intention. This arrow indicates the intended manner by which random samples will be generated from the null hypothesis. This sampling intention also corresponds to the face in the lower-left corner of Figure 11.1, who is thinking of the sampling intention. The right side of Figure 11.3 shows the resulting probability distribution of possible outcomes. This sampling distribution corresponds to the cloud of imaginary possibilities in Figure 11.1.

It is important to understand that the sampling distribution is a probability distribution over samples of data, and is *not* a probability distribution over parameter values. The right side of Figure 11.3 has the sample proportion, z/N, on its abscissa, and does *not* have the parameter value, θ, on its abscissa. Notice that the parameter value, θ, is fixed at a specific value and appears in the left panel of the figure.

Our goal, as you might recall, is to determine whether the probability of getting the observed result, $z/N = 7/24$, is tiny enough that we can reject the null hypothesis. By using the binomial probability formula in Equation 11.5, we determine that the probability of getting *exactly* $z = 7$ heads in $N = 24$ flips is 2.063%. Figure 11.3 shows this probability as the height of the bar at $z/N = 7/24$ (where the "+" is plotted). However, we do not want to determine the probability of only the actually observed result. After all, for large N, *any* specific result z can be very improbable. For example, if we flip a fair coin $N = 1000$ times, the probability of getting exactly $z = 500$ heads is only 2.5%, even though $z = 500$ is precisely what we would expect if the coin were fair.

Therefore, instead of determining the probability of getting exactly the result z/N from the null hypothesis, we determine the probability of getting z/N *or a result even more extreme than expected from the null hypothesis.* The reason for considering more extreme outcomes is this: If we would reject the null hypothesis because the result z/N is too far from what we would expect, then any other result that has an even more extreme value would also cause us to reject the null hypothesis. Therefore we want to know the probability of getting the actual outcome *or an outcome more extreme* relative to what we expect. This total probability is referred to as "the p value." The p value defined at this point is the "one-tailed" p value, because it sums the extreme probabilities in only one tail of the sampling distribution. (The term "tail" here refers to the end of a sampling distribution, not to the side of a coin.) In practice, the one-tailed p value is multiplied by 2, to get the two-tailed p value. We consider both tails of the sampling distribution because the null hypothesis could be rejected if the outcome were too extreme in either direction. If this p value is less than a critical amount, then we reject the null hypothesis.

The critical two-tailed probability is conventionally set to 5%. In other words, we will reject the null hypothesis whenever the total probability of the observed z/N or an outcome more extreme is less than 5%. Notice that this decision rule will cause us to reject the null hypothesis 5% of the time *when the null hypothesis is true*, because the null hypothesis itself generates those extreme values 5% of the time, just by chance. The critical probability, 5%, is the proportion of false alarms that we are willing to tolerate in our decision process. When considering a single tail of the distribution, the critical probability is half of 5%, that is, 2.5%.

Here's the conclusion for our particular case. The actual observation was $z/N = 7/24$. The one-tailed probability is $p = 0.032$, which was computed from Equation 11.4, and is shown in Figure 11.3. Because the p value is not less than 2.5%, we do *not* reject the null hypothesis that $\theta = 0.5$. In NHST parlance, we would say that the result "has failed to reach significance." This does not mean we *accept* the null hypothesis; we merely suspend judgment regarding rejection of this particular hypothesis. Notice that we have not determined any degree of belief in the hypothesis that $\theta = 0.5$. The hypothesis might be true or might be false; we suspend judgment.

11.1.3. With intention to fix *z*

You'll recall from the previous section that when we asked the assistant why she stopped flipping the coin, she said it was because N reached her lucky number. Suppose instead that when we ask her why she stopped she says that her favorite number is 7, and she stopped when she got $z = 7$. Recall the sequence of flips in Equation 11.1 that the 7th head occurred on the final flip. In this situation, z is fixed in advance and N is the random variable. We don't talk about the probability of getting z heads out of N flips, we instead talk about the probability of taking N flips to get z heads. If the coin is head biased, it will tend to take relatively few flips to get z heads, but if the coin is tail biased, it will tend to take relatively many flips to get z heads. This means the space of possible outcomes is restricted to combinations of z and N for which z is fixed at $z = 7$ (and the 7th head occurs on the final flip). This corresponds to a single row of the z, N space as shown in the middle panel of Figure 11.2 (which shows $z = 4$ highlighted merely for ease of visualization). The computational question then becomes, what is the probability of the actual proportion, or a proportion more extreme than expected, *within that row of the outcome space*? (Actually, it is not entirely accurate to say that the sample space corresponds to a row of Figure 11.2. The sample space must have the zth head occur on the Nth flip, but the cells in Figure 11.2 do not have this requirement. A more accurate depiction would refer to the $z-1$th row, suffixed with one more flip that must be a head.) Notice that the set of possible outcomes in the fixed z space are quite different than in the fixed N space, as is easy to see by comparing the left and middle panels of Figure 11.2.

What is the probability of taking N flips to get z heads? To answer this question, consider this: We know that the Nth flip is the zth head, because that is what caused flipping to stop. Therefore the previous $N - 1$ flips had $z - 1$ heads in some random sequence. The probability of getting $z-1$ heads in $N-1$ flips is $\binom{N-1}{z-1}\theta^{z-1}(1-\theta)^{N-z}$. The probability that the last flip comes up heads is θ. Therefore, the probability that it takes N flips to get z heads is

$$\begin{aligned} p(N|z,\theta) &= \binom{N-1}{z-1}\theta^{z-1}(1-\theta)^{N-z} \cdot \theta \\ &= \binom{N-1}{z-1}\theta^{z}(1-\theta)^{N-z} \\ &= \frac{z}{N}\binom{N}{z}\theta^{z}(1-\theta)^{N-z} \end{aligned} \tag{11.6}$$

(This distribution is sometimes called the "negative binomial" but that term sometimes refers to other formulations and can be confusing, so I will not use it here.) This is a sampling distribution, like the binomial distribution, because it specifies the relative probabilities of all the possible data outcomes for the hypothesized fixed value of θ and the intended stopping rule.

An example of the sampling distribution is shown in the right panel of Figure 11.4. The distribution consists of vertical spikes at all the possible values of z/N. The spike at 1.0 indicates the probability of $N = 7$ (for which $z/N = 7/7 = 1.0$). The spike at 0.875

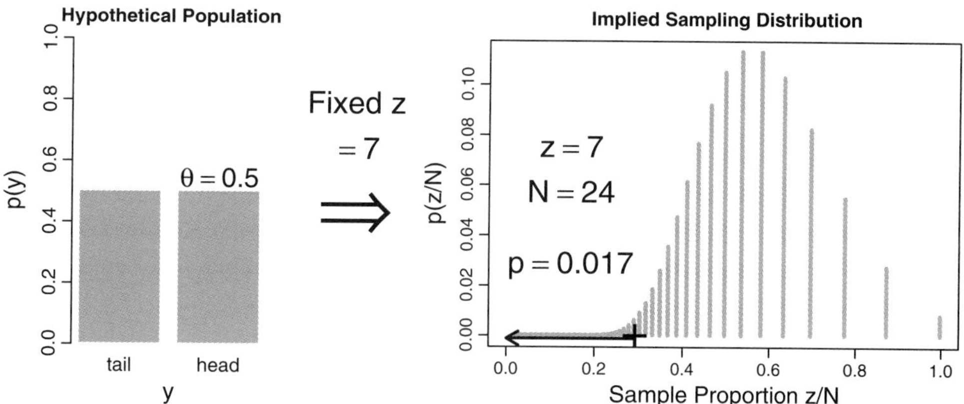

Figure 11.4 The imaginary cloud of possible outcomes when z is fixed. The null hypothesis likelihood distribution and parameter are shown on the left. The stopping intention is shown in the middle. The sampling distribution and p value are shown on the right. Compare with Figures 11.3 and 11.5.

indicates the probability of $N = 8$ (for which $z/N = 7/8 = 0.875$). The spike near 0.78 indicates the probability of $N = 9$ (for which $z/N = 7/9 \approx 0.78$). The spikes in the left tail become infinitely dense and infinitesimally short as N approaches infinity.

Now for the dramatic conclusion: Figure 11.4 shows that the p value is 0.017, which is less than the decision threshold of 2.5%, and therefore we reject the null hypothesis.[2] Compare this with Figure 11.3, for which the p value was greater than 2.5% and we did not reject the null hypothesis. The data are the same in the two analyses. All that has changed is the cloud of imaginary possibilities in the sample space. If the coin flipper intended to stop because N reached 24, then we do *not* reject the null hypothesis. If the coin flipper intended to stop because z reached 7, then we *do* reject the null hypothesis. The point here is not about where to set the limit for declaring significance (e.g., 5%, 2.5%, 1%, or whatever). The point is that the p values are different even though the data are the same.

The focus of this section has been that the p value is different when the stopping intention is different. It should also be mentioned, however, that if data collection stops when the number of heads (or tails) reaches a threshold, then the data are biased. Any stopping rule that is triggered by extremity of data can produce a biased sample, because an accidental sequence of randomly extreme data will cause data collection to stop and thereby prevent subsequent collection of compensatory data that are more representative. Thus, a person could argue that stopping at threshold z is not good practice because it biases the data; but, that does not change the fact that stopping at threshold z implies a

[2] The total probability in the left tail of Figure 11.4 is an infinite sum. It is easily computed by considering the finite complement to its right. In particular, $\sum_{n=24}^{\infty} p(n|z,\theta) = 1 - \sum_{n=z}^{n=24-1} p(n|z,\theta)$.

different p value than stopping at threshold N. Many practitioners do stop collecting data when an extreme is exceeded; this issue is discussed at greater length in Section 13.3.

11.1.4. With intention to fix duration

The previous two sections explained the imaginary sampling distributions for stopping at threshold N or stopping at threshold z. These cases have been discussed many times in the literature, including the well-known and accessible articles by Lindley and Phillips (1976) and J. O. Berger and Berry (1988). Derivation of the binomial coefficients, as in the fixed N scenario, is attributed to Jacob Bernoulli (1655-1705). The threshold-z scenario is attributed to the geneticist J. B. S. Haldane (1892-1964) by Lindley and Phillips (1976, p. 114), as was also alluded by Anscombe (1954, p. 89).

In this section we consider another variation. Suppose, when we ask the assistant why she stopped flipping the coin, she replies that she stopped because 2 min had elapsed. In this case, data collection stopped not because of reaching threshold N, nor because of reaching threshold z, but because of reaching threshold duration. Neither N nor z is fixed. Lindley and Phillips (1976, p. 114) recognized that this stopping rule would produce yet a different sampling distribution, but said, "In fact, in the little experiment with the [coin] I continued tossing until my wife said "Coffee's ready." Exactly how a significance test is to be performed in these circumstances is unclear to me."[3] In this section I fill in the details and perform the test. This scenario was discussed in the first edition of this book in its Exercise 11.3 (Kruschke, 2011b, p. 289). For fixed-duration tests of metric data (not dichotomous data) see Kruschke (2010, p. 659) and Kruschke (2013a, p. 588).

The key to analyzing this scenario is specifying how various combinations of z and N can arise when sampling for a fixed duration. There is no single, uniquely "correct" specification, because there are many different real-world constraints on sampling through time. But one approach is to think of the sample size N as a random value: If the 2 min experiment is repeated, sometimes N will be larger, sometimes smaller. What is the distribution of N? A convenient formulation is the Poisson distribution. The Poisson distribution is a frequently used model of the number of occurrences of an event in a fixed duration (e.g., Sadiku & Tofighi, 1999). It is a probability distribution over integer values of N from 0 to $+\infty$. The Poisson distribution has a single parameter, λ, that controls its mean (and also happens to be its variance). The parameter λ can have any non-negative real value; it is not restricted to integers. Examples of the Poisson distribution are provided in Chapter 24. According to the Poisson distribution, if $\lambda = 24$, the value of N will typically be near 24, but sometimes larger and sometimes smaller.

[3] Sorry for the old-fashioned gender roles in that quote; it was the 1970s. Remember the proposed Equal Rights Amendment in the USA? It passed both houses of Congress in 1972, but failed to be ratified by a sufficient number of states.

A schematic of this distribution on the sample space appears in the right panel of Figure 11.2 (p. 302). Because the table does not have room for $N = 24$, the figure highlights sample sizes near 5 instead of near 24. The shading of columns suggests that $N = 5$ occurs often for the fixed duration, but other columns can also occur. *For whichever column of N happens to occur, the distribution of z is the binomial distribution for that N. Thus, the overall distribution of $\langle z, N\rangle$ combinations is a weighted mixture of binomial distributions.*

An example of the sampling distribution appears in right side of Figure 11.5. In principle, the distribution has spikes at every possible sample proportion, including $0/N, 1/N, 2/N, \ldots, N/N$ for all $N \geq 0$. But only values of N near λ have a visible appearance in the plot because the Poisson distribution makes values far from λ very improbable. In other words, the sampling distribution shown in the right side of Figure 11.5 is a weighted mixture of binomial distributions for Ns near λ. (The parameter λ was chosen to be 24 in this example merely because it matches N of the observed data and makes the plots most comparable to Figures 11.3 and 11.4.)

The p value for this fixed-duration stopping rule is shown in Figure 11.5 to be $p = 0.024$. This p value barely squeaks beneath the rejection limit of 2.5%, and might be reported as "marginally significant." Compare this conclusion with the conclusion from assuming fixed N, which was "not significant," and the conclusion from assuming fixed z, which was "significant." The key point, however, is not the decision criterion. The point is that the p value changes when the imaginary sample space changes, while the data are unchanged.

In the example of Figure 11.5, λ was set to 24 merely to match N and make the example most comparable to the preceding examples. But the value of λ should

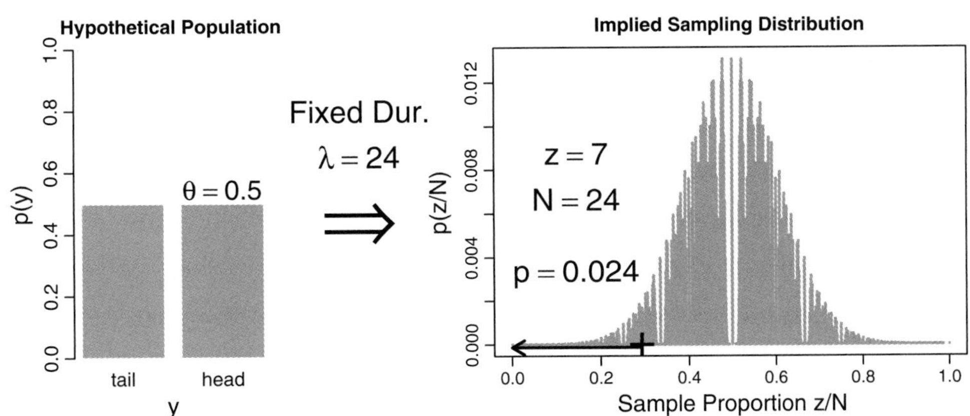

Figure 11.5 The imaginary cloud of possible outcomes when duration is fixed. The null hypothesis likelihood distribution and parameter are shown on the left. The stopping intention is shown is the middle. The sampling distribution and p value are shown on the right. Sample sizes are drawn randomly from a Poisson distribution with mean λ. Compare with Figures 11.3 and 11.4.

really be chosen on the basis of real-world sampling constraints. Thus, if it takes about 6 s to manually toss a coin and record its outcome, the typical number of flips in 2 min is $\lambda = 20$, not 24. The sampling distribution for $\lambda = 20$ is different than the sampling distribution for $\lambda = 24$, resulting in a p value of 0.035 (not 0.024 as in Figure 11.5). Real-world sampling constraints are often complex. For example, some labs have limited resources (lab benches, computers, assistants, survey administrators) for collecting data, and can run only some maximum number of observation events at any one time. Therefore the distribution of N is truncated at that maximum for any one session of observations. If there are multiple sessions, then the distribution of N is a mixture of truncated distributions. Each of these constraints produces a different cloud of imaginary sample outcomes from the null hypothesis, and hence the possibility of a different p value. As another example, in mail surveys, the number of respondents is a random value that depends on how many surveys actually get to the intended recipients (because of erroneous or obsolete addresses) and how many respondents take the effort to complete the survey and return it. As another example, in observational field research such as wildlife ecology, the number of observations of a species during any session is a random value.

11.1.5. With intention to make multiple tests

In the preceding sections we have seen that when a coin is flipped $N = 24$ times and comes up $z = 7$ heads, the p value can be 0.032 or 0.017 or 0.024 or other values. The change in p is caused by the dependence of the imaginary cloud of possibilities on the stopping intention. Typical NHST textbooks never mention the dependency of p values on the stopping intention, but they often do discuss the dependency of p values on the testing intention. In this section we will see how testing intentions affect the imaginary cloud of possibilities that determines the p value.

Suppose that we want to test the hypothesis of fairness for a coin, and we have a second coin in the same experiment that we are also testing for fairness. In real biological research, for example, this might correspond to testing whether the male/female ratio of babies differs from 50/50 in each of two species of animal. We want to monitor the false alarm rate overall, so we have to consider the probability of a false alarm from *either* coin. Thus, the p value for the first coin is the probability that a proportion, equal to or more extreme than its actual proportion, could arise by chance from *either* coin. Thus, the left-tail p value (cf. Equation 11.4) is

$$p\Big((z_1/N_1)_{\theta_1,I_1} \leq (z_1/N_1)_{\text{actual}} \text{ OR } (z_2/N_2)_{\theta_2,I_2} \leq (z_1/N_1)_{\text{actual}} \Big| \theta_1, \theta_2, I_1, I_2\Big)$$

and the right-tail p value (cf. Equation 11.3) is

$$p\Big((z_1/N_1)_{\theta_1,I_1} \geq (z_1/N_1)_{\text{actual}} \text{ OR } (z_2/N_2)_{\theta_2,I_2} \geq (z_1/N_1)_{\text{actual}} \Big| \theta_1, \theta_2, I_1, I_2\Big)$$

Because that notation is a little unwieldy, I introduce an abbreviated notation that I hope simplifies more than it obscures. I will use the expression $\text{Extrem}\{z_1/N_1, z_2/N_2\}$ to denote the lesser of the proportions when computing the low tail, but the higher of proportions when computing the high tail. Then the left-tail p value is

$$p\left(\text{Extrem}\{(z_1/N_1)_{\theta_1,I_1}, (z_2/N_2)_{\theta_2,I_2}\} \leq (z_1/N_1)_{\text{actual}} \middle| \theta_1, \theta_2, I_1, I_2\right)$$

and the right-tail p value is

$$p\left(\text{Extrem}\{(z_1/N_1)_{\theta_1,I_1}, (z_2/N_2)_{\theta_2,I_2}\} \geq (z_1/N_1)_{\text{actual}} \middle| \theta_1, \theta_2, I_1, I_2\right)$$

For concreteness, suppose we flip both coins $N_1 = N_2 = 24$ times, and the first coin comes up heads $z_1 = 7$ times. This is the same result as in the previous examples. Suppose that we intended to stop when the number of flips reached those limits. The p value for the outcome of the first coin is the probability that either z_1/N_1 or z_2/N_2 would be as extreme, or more extreme, than 7/24 when the null hypothesis is true. This probability will be larger than when considering a single coin, because even if the imaginary flips of the first coin do not exceed 7/24, there is still a chance that the imaginary flips of the second coin will. For every imaginary sample of flips from the two coins, we need to consider the sample proportion, either z_1/N_1 or z_2/N_2, that is most extreme relative to θ. The p value is the probability that the extreme proportion meets or exceeds the actual proportion.

Figure 11.6 shows the numerical details for this situation. The upper-right panel shows the sampling distribution of the extreme of z_1/N_1 or z_2/N_2, when the null hypothesis is true and the stopping intention is fixed N for both coins. You can see that the p value for $z_1/N_1 = 7/24$ is $p = 0.063$. This p value is almost twice as big as when considering the coin by itself, as was shown in Figure 11.3 (p. 304). The actual outcome of the coin is the same in the two figures; what differs is the cloud of imaginary possibilities relative to which the outcome is judged.[4]

Notice that we do not need to know the result of the second coin (i.e., z_2) to compute the p value of the first coin. In fact, we do not need to flip the second coin at all. All we need for computing the p value for the first coin is the *intention* to flip the second coin N_2 times. *The cloud of imaginary possibilities is determined by sampling intentions, not by observed data.*

[4] There is a direct mathematical relation between the p values of Figures 11.6 and 11.3. For the single-coin test in Figure 11.3, we had $p = 0.03195733$ for one tail (displayed rounded to 0.032). We use that p value to compute the probability that two independent coins, each tossed 24 times, would have at least one of them meet or exceed 7/24. This probability is the same as 1.0 minus the probability that both coins do *not* meet or exceed 7/24, which is $1.0 - (1 - 0.03195733)^2 = 0.06289339$. This should match the p value for the two-coin scenario in Figure 11.6. Its one-tailed p is, in fact, 0.06289339 (displayed rounded as 0.063).

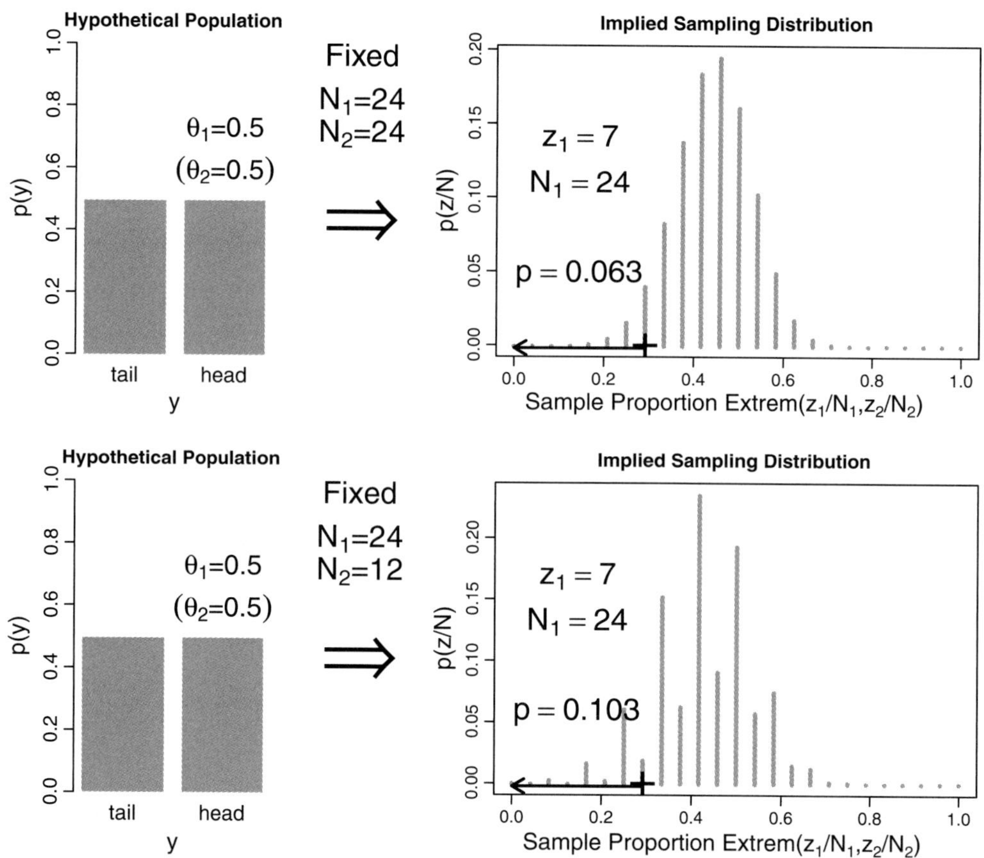

Figure 11.6 The imaginary cloud of possible outcomes when N is fixed and there are two independent tests. Upper is for $N_2 = N_1$. Lower is for $N_2 < N_1$. The null hypothesis likelihood distribution and parameter are shown on the left. The stopping and testing intentions are shown in the middle. The sampling distribution and p value are shown on the right. Compare with Figure 11.3. Extrem$\{z_1/N_1, z_2/N_2\}$ is the smaller (for left-tailed) of the hypothetically sampled proportions.

Figure 11.6 shows another scenario in its lower half, wherein the second coin is flipped only $N_2 = 12$ times instead of 24 times. With so few flips, it is relatively easy for the second coin to exhibit an extreme outcome by chance alone, even if the null hypothesis is true. Therefore the p value for the first coin gets even larger than before, here up to $p = 10.3\%$, because the space of possible outcomes from both coins now includes many more extreme outcomes.

It might seem artificial to compute a p value for one outcome based on possible results from other outcomes, but this is indeed a crucial concern in NHST. There is a vast literature on "correcting" the p value of one outcome when making multiple tests of other outcomes. As mentioned earlier, typical textbooks in NHST describe at least some of these corrections, usually in the context of analysis of variance (which involves multiple groups of metric data, as we will see in Chapter 19) and rarely in the context of proportions for dichotomous data.[5]

11.1.6. Soul searching

A defender of NHST might argue that I'm quibbling over trivial differences in the p values (although going from $p = 1.7\%$ in Figure 11.4 to $p = 10.3\%$ in Figure 11.6 is hard to ignore), and, for the examples I've given above, the differences get smaller when N gets larger. There are two flaws in this argument. First, it does not deny the problem, and gives no solution when N is small. Second, it understates the problem because there are many other examples in which the p values differ dramatically across sampling intentions, even for large N. In particular, as we will see in more detail later, when a researcher merely intends to make multiple comparisons across different conditions or parameters, the p value for any single comparison increases greatly.

Defenders of NHST might argue that the examples regarding the stopping intention are irrelevant, because it is okay to compute a p value for every set of data *as if N were fixed in advance*. The reason it is okay to conditionalize on N (that is, to consider only the observed N column in the z, N outcome space) is that doing so will result in exactly 5% false alarms in the long run when the null hypothesis is true. One expression of this idea comes from Anscombe (1954, p. 91), who stated, "In any experiment or sampling inquiry where the number of observations is an uncertain quantity but does not depend on the observations themselves, it is always legitimate to treat the observations in the statistical analysis as if their number had been fixed in advance. We are then in fact using perfectly correct conditional probability distributions." The problem with this argument is that it applies equally well to the other stopping rules for computing p values. It is important to notice that Anscombe did *not* say it is *uniquely* or *only* legitimate to treat N as fixed in advance. Indeed, if we treat z as fixed in advance, or if we treat duration as fixed in advance, then we still have perfectly correct conditional probability distributions that result in 5% false alarms in the long run when the null hypothesis is true. The p values for individual data sets may differ, but across many data sets the long-run false alarm rate is 5%. Note also that Anscombe concluded that this whole issue would be avoided if we did Bayesian analysis instead. Anscombe (1954, p. 100) said, "All risk of error is avoided if the method of analysis uses the observations only in the form of their likelihood function, since the likelihood function (given the observations) is independent of the sampling rule. One such method of analysis is provided by the classical theory of rational belief, in which a distribution of posterior probability is deduced, by Bayes' theorem, from the likelihood function of the observations and a distribution of prior probability."

[5] In particular, I do not recall having previously seen in the literature graphs of sampling distributions such as those shown in Figures 11.6 and 11.11.

Within the context of NHST, the solution is to establish the true intention of the researcher. This is the approach taken explicitly when applying corrections for multiple tests. The analyst determines what the truly intended tests are, and determines whether those testing intentions were honestly conceived *a priori* or *post hoc* (i.e., motivated only after seeing the data), and then computes the appropriate p value. The same approach should be taken for stopping rules: The data analyst should determine what the truly intended stopping rule was, and then compute the appropriate p value. Unfortunately, determining the true intentions can be difficult. Therefore, perhaps researchers who use p values to make decisions should be required to publicly pre-register their intended stopping rule and tests, before collecting the data. (There are other motivations for pre-registering research, such as preventing selective inclusion of data or selective reportage of results. In the present context, I am focusing on pre-registration as a way to establish p values.) But what if an unforeseen event interrupts the data collection, or produces a windfall of extra data? What if, after the data have been collected, it becomes clear that there should have been other tests? In these situations, the p values must be adjusted despite the pre-registration. Fundamentally, the intentions should not matter to the interpretation of the data because the propensity of the coin to come up heads does not depend on the intentions of the flipper (except when the stopping rule biases the data collection). Indeed, we carefully design experiments to insulate the coins from the intentions of the experimenter.[6]

11.1.7. Bayesian analysis

The Bayesian interpretation of data does not depend on the covert sampling and testing intentions of the data collector. In general, for data that are independent across trials (and uninfluenced by the sampling intention), the probability of the set of data is simply the product of the probabilities of the individual outcomes. Thus, for $z = \sum_{i=1}^{N} y_i$ heads in N flips, the likelihood is $\prod_{i=1}^{N} \theta^{y_i}(1-\theta)^{1-y_i} = \theta^z(1-\theta)^{N-z}$. The likelihood function captures everything we assume to influence the data. In the case of the coin, we assume that the bias (θ) of the coin is the only influence on its outcome, and that the flips are independent. The Bernoulli likelihood function completely captures those assumptions.

In summary, the NHST analysis and conclusion depend on the covert intentions of the experimenter, because those intentions define the probabilities over the space of all possible (unobserved) data, as depicted by the imaginary clouds of possibilities in Figure 11.1, p. 299. This dependence of the analysis on the experimenter's intentions

[6] A more exuberant statement comes from Howson and Urbach (2006, pp. 158-159): "*We suggest that such information about experimenters' subjective intentions ...has no inductive relevance whatever in this context, and that in practice it is never sought or even contemplated. The fact that significance tests and, indeed, all classical inference models require it is a decisive objection to the whole approach.*" (italics in original)

conflicts with the opposite assumption that the experimenter's intentions have no effect on the observed data. The Bayesian analysis, on the other hand, does not depend on the imaginary cloud of possibilities. The Bayesian analysis operates only with the actual data obtained.

11.2. PRIOR KNOWLEDGE

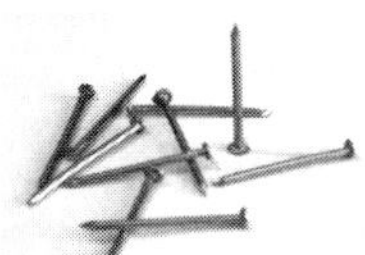

Suppose that we are not flipping a coin, but we are flipping a flat-headed nail. In a social science setting, this is like asking a survey question about left or right handedness of the respondent, which we know is far from 50/50, as opposed to asking a survey question about male or female sex of the respondent, which we know is close to 50/50. When we flip the nail, it can land with its pointy tail touching the ground, an outcome I'll call tails, or the nail can land balanced on its head with its pointy tail sticking up, an outcome I'll call heads. We believe, just by looking at the nail and thinking of our previous experience with nails, that it will *not* come up heads and tails equally often. Indeed, the nail will very probably come to rest with its point touching the ground. In other words, we have a strong prior belief that the nail is tail-biased. Suppose we flip the nail 24 times and it comes up heads on only 7 flips. Is the nail "fair"? Would we use it to determine which team gets to kick off at the Superbowl?

11.2.1. NHST analysis

The NHST analysis does not care if we are flipping coins or nails. The analysis proceeds the same way as before. As we saw in the previous sections, if we declare that the intention was to flip the nail 24 times, then an outcome of 7 heads means we do *not* reject the hypothesis that the nail is fair (recall Figure 11.3, where $p > 2.5\%$). Let me say that again: We have a nail for which we have a strong prior belief that it is tail biased. We flip the nail 24 times, and find it comes up heads 7 times. We conclude, therefore, that we cannot reject the null hypothesis that the nail can come up heads or tails 50/50. Huh? This is a *nail* we're talking about. How can you not reject the null hypothesis?

11.2.2. Bayesian analysis

The Bayesian statistician starts the analysis with an expression of the prior knowledge. We know from prior experience that the narrow-headed nail is biased to show tails, so we express that knowledge in a prior distribution. In a scientific setting, the prior is established by appealing to publicly accessible and reputable previous research. In our present fictional example involving a nail, suppose that we represent our prior beliefs by a fictitious previous sample that had 95% tails in a sample size of 20. That translates into a beta$(\theta|2, 20)$ prior distribution if the "proto-prior," before the fictional data, was beta(θ|1, 1). If we wanted to go through the trouble, we could instead derive a prior from established theories regarding the mechanics of such objects, after making physical measurements of the nail such as its length, diameter, mass, rigidity, etc. In any case, to make the analysis convincing to a scientific audience, the prior must be cogent to that audience. Suppose that the agreed prior for the nail is beta(θ|2, 20), then the posterior distribution is beta(θ|2 + 7, 20 + 17), as shown in the left side of Figure 11.7. The posterior 95% highest density interval (HDI) clearly does not include the nail being fair.

On the other hand, if we have prior knowledge that the object is fair, such as a coin, then the posterior distribution is different. For example, suppose that we represent our prior beliefs by a fictitious previous sample that had 50% tails in a sample size of 20. That translates into a beta(θ|11, 11) prior distribution if the "proto-prior," before the fictional data, was beta(θ|1, 1). Then the posterior distribution is beta(θ|11+7, 11+17), as shown in the right side of Figure 11.7. The posterior 95% HDI includes the nail being fair.

The differing inferences for a coin and a nail make good intuitive sense. Our posterior beliefs about the bias of the object *should* depend on our prior knowledge of the object:

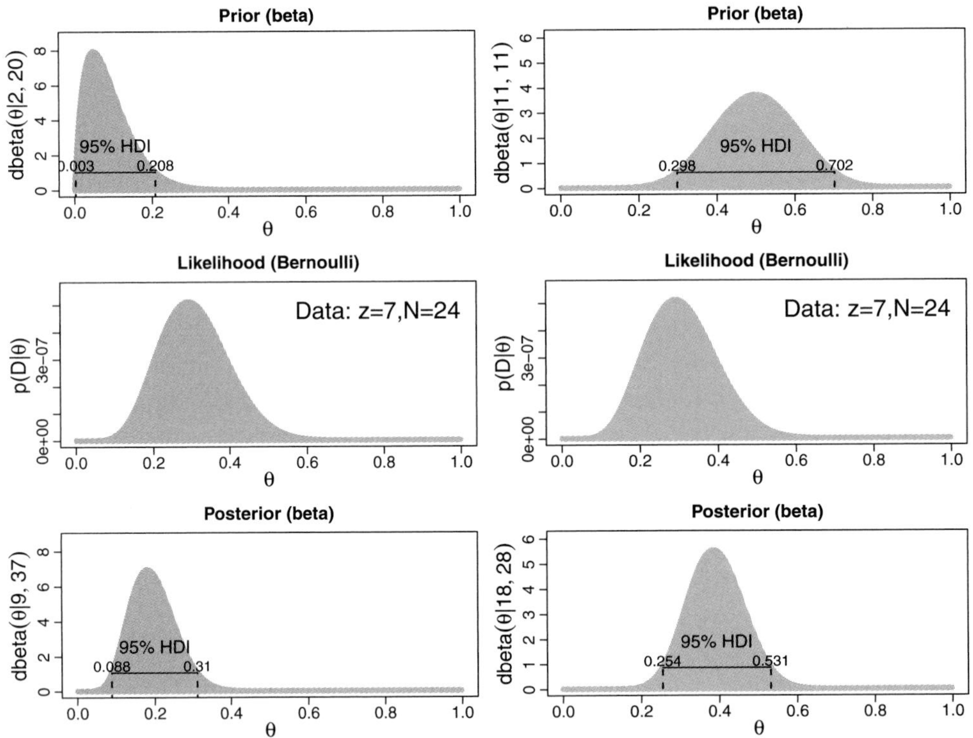

Figure 11.7 Posterior HDI for the bias of a Bernoulli process, when the prior assumes a tail-strong nail (left column) or a fair coin (right column).

7 heads in 24 flips of narrow-headed nail *should* leave us with a different opinion than 7 heads in 24 flips of a coin. For additional details and a practical example, see Lindley and Phillips (1976). Despite the emphasis here on the important and appropriate role of prior knowledge, please remember that the prior distributions are overwhelmed with sufficiently large data sets, and the posterior distributions converge to the same result.

11.2.2.1 Priors are overt and relevant

Some people might have the feeling that prior beliefs are no less mysterious than the experimenter's stopping and testing intentions. But prior beliefs are not capricious and idiosyncratic. Prior beliefs are overt, explicitly debated, and founded on publicly accessible previous research. A Bayesian analyst might have personal priors that differ from what most people think, but if the analysis is supposed to convince an audience, then the analysis must use priors that the audience finds palatable. It is the job of the Bayesian analyst to make cogent arguments for the particular prior that is used. The research will not get published if the reviewers and editors think that that prior is untenable. Perhaps the researcher and the reviewers will have to agree to disagree about the prior, but even in that case the prior is an explicit part of the argument, and the analysis should be conducted with both priors to assess the robustness of the posterior. Science is a cumulative process, and new research is presented always in the context of previous research. A Bayesian analysis can incorporate this obvious fact.

Some people might wonder, if informed priors are allowed for Bayesian analyses, then why not allow subjective intentions for NHST? Because the subjective stopping and testing intentions in the data collector's mind only influence the cloud of possible data that were not actually observed. Informed prior beliefs, on the other hand, are not about what didn't happen, but about how the data influence subsequent beliefs: Prior beliefs are the starting point from which we move in the light of new data. Indeed, it can be a blunder *not* to use prior knowledge, as was discussed in Section 5.3.2 (p. 113) with regard to random disease or drug testing. Bayesian analysis tells us how much we should re-allocate credibility from our prior allocation. Bayesian analysis does not tell us what our prior allocation should be. Nevertheless, the priors are overt, public, cumulative, and overwhelmed as the amount of data increases. Bayesian analysis provides an intellectually coherent method for determining the degree to which beliefs should change.

11.3. CONFIDENCE INTERVAL AND HIGHEST DENSITY INTERVAL

Many people have acknowledged perils of p values, and have suggested that data analysis would be better if practitioners used *confidence intervals* (CIs). For example, in a well-known article in a respected medical journal, Gardner and Altman (1986, p. 746) stated, "Overemphasis on hypothesis testing and the use of p values to dichotomize significant or non-significant results has detracted from more useful approaches to interpreting study results, such as estimation and confidence intervals." As another example, from the perspective a professor in a department of management in a college of business, Schmidt (1996, p. 116) said in another well-known article, "My conclusion is that we must abandon the statistical significance test. In our graduate programs we must teach that ... the appropriate statistics are point estimates of effect sizes and confidence intervals around these point estimates. ... I am not the first to reach the conclusion that significance testing should be replaced by point estimates and confidence intervals." Schmidt then listed several predecessors, going back to 1955. Numerous recent articles and books have been published that recommend use of CIs (e.g., Cumming, 2012).

Those recommendations have important and admirable goals, a primary one being to get people to understand the uncertainty of estimation instead of only a yes/no decision about a null hypothesis. Within the context of frequentist analysis, the CI is a device for addressing the goal. Unfortunately, as will be explained in this section, the goals are not well accomplished by frequentist CIs. Instead, the goals are well achieved by Bayesian analysis. This section defines CIs and provides examples. It shows that, while CIs ameliorate some of the problems of p values, ultimately CIs suffer the same problems as p values because CIs are defined in terms of p values. Bayesian posterior distributions, on the other hand, provide the needed information.

11.3.1. CI depends on intention

The primary goal of NHST is determining whether a particular "null" value of a parameter can be rejected. One can also ask what *range* of parameter values would not be rejected. This range of nonrejectable parameter values is called the CI. (There are different ways of defining an NHST CI; this one is conceptually the most general and coherent with NHST precepts.) The 95% CI consists of all values of θ that would not be rejected by a (two-tailed) significance test that allows 5% false alarms.

For example, in a previous section we found that $\theta = 0.5$ would not be rejected when $z = 7$ and $N = 24$, for a data collector who intended to stop when $N = 24$. The question is, which other values of θ would we not reject? Figure 11.8 shows the sampling distribution for different values of θ. The upper row shows the case of $\theta = 0.126$, for which the sampling distribution has a p value of almost exactly 2.5%. In fact, if θ is nudged any smaller, p becomes smaller than 2.5%, which means that smaller values of θ can be rejected. The lower row of Figure 11.8 shows the case of $\theta = 0.511$, for which the sampling distribution shows p is almost exactly 2.5%. If θ is nudged any larger, then p falls below 2.5%, which means that larger values of θ can be rejected. In summary, the range of θ values we would not reject is $\theta \in [0.126, 0.511]$. This is the 95% confidence interval when $z = 7$ and $N = 24$, for a data collector who intended to stop when $N = 24$. Exercise 11.2 has you examine this "hands-on."

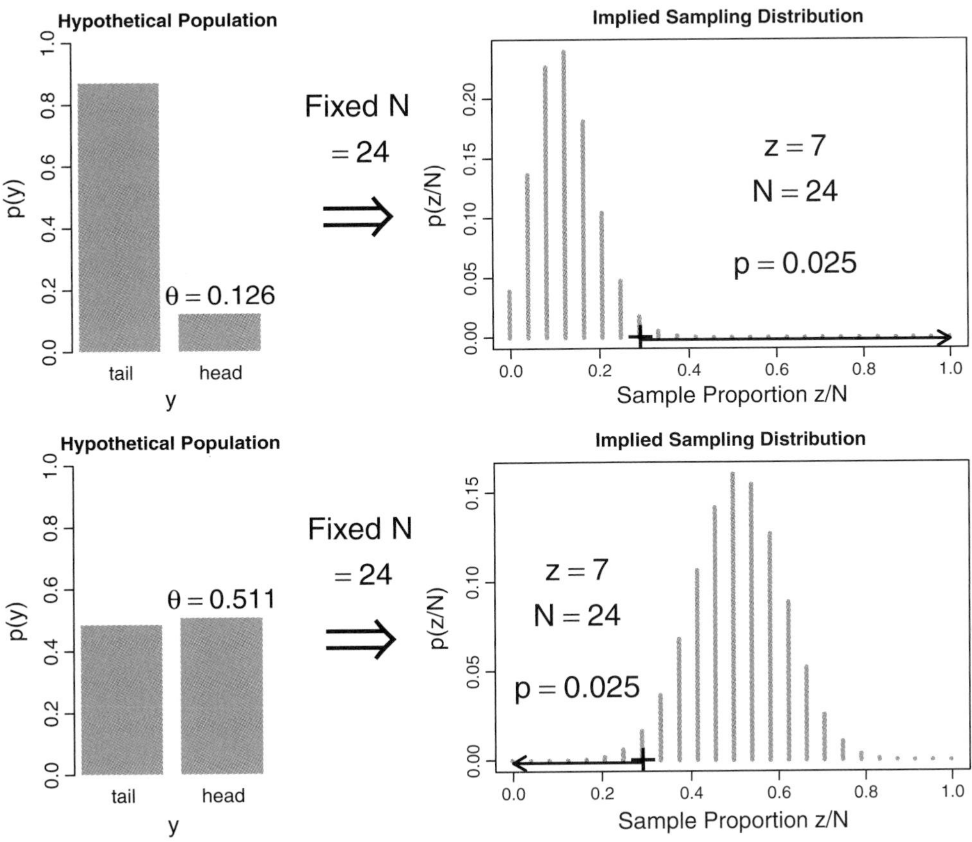

Figure 11.8 95% confidence interval when N is fixed by the experimenter's intention extends from $\theta = 0.126$ (top row) to $\theta = 0.511$ (bottom row). Compare with Figures 11.9 and 11.10.

Notice that the CI describes the limits of the θ value, which appear in the *left* side of Figure 11.8. The parameter θ describes the hypothetical population. Although θ exists on a range from 0 to 1, it is quite different than the sample proportion, z/N, on the right of the figure. Notice also that the sampling distribution is a distribution over the sample proportion; the sampling distribution is not a distribution over the parameter θ. The CI is simply the smallest and largest values of θ that yield $p \geq 2.5\%$.

We can also determine the CI for the experimenter who intended to stop when $z = 7$. Figure 11.9 shows the sampling distribution for different values of θ. The upper row shows the case of $\theta = 0.126$, for which the sampling distribution has $p = 2.5\%$. In fact, if θ is nudged any smaller, p is less than 2.5%, which means that smaller values of θ can be rejected. The lower row of Figure 11.9 shows the case of $\theta = 0.484$, for which the sampling distribution has $p = 2.5\%$. If θ is nudged any larger, p falls below 2.5%,

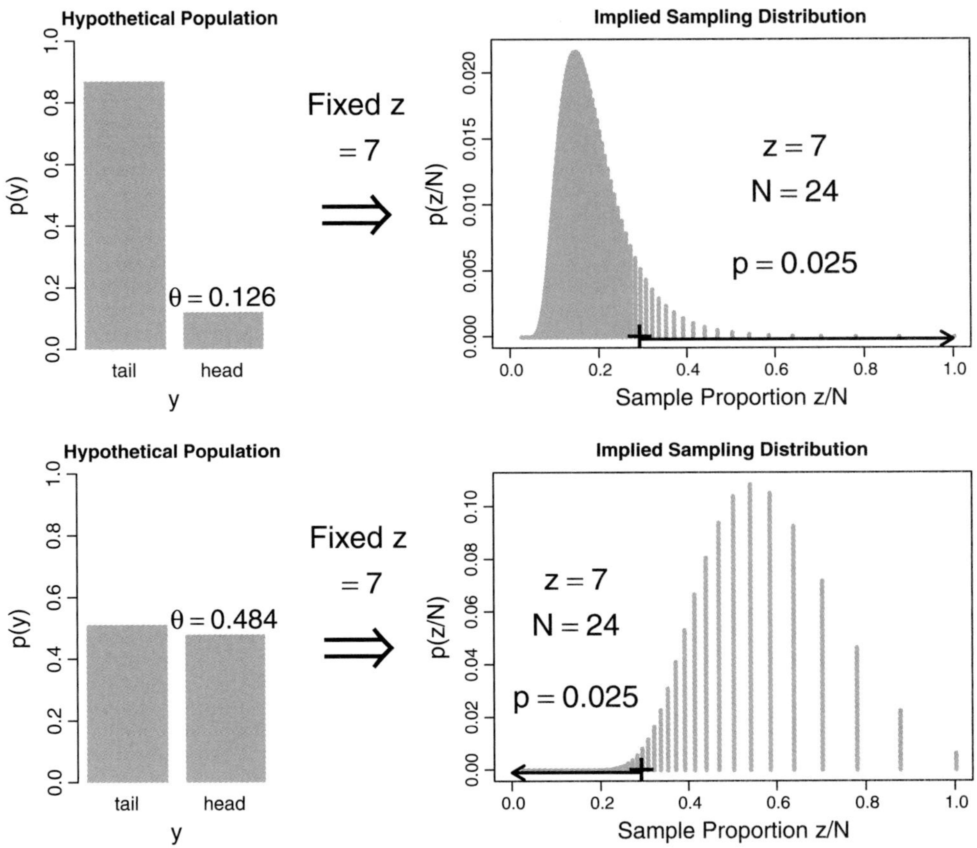

Figure 11.9 95% confidence interval when z is fixed by the experimenter's intention extends from $\theta = 0.126$ (top row) to $\theta = 0.484$ (bottom row). Compare with Figures 11.8 and 11.10.

which means that larger values of θ can be rejected. In summary, the range of θ values we would not reject is $\theta \in [0.126, 0.484]$. This is the 95% CI when $z = 7$ and $N = 24$, for a data collector who intended to stop when $z = 7$.

Furthermore, we can determine the CI for the experimenter who intended to stop when a fixed duration expired. Figure 11.10 shows the sampling distribution for different values of θ. The upper row shows the case of $\theta = 0.135$, for which the sampling distribution has $p = 2.5\%$. If θ is nudged any smaller, p is less than 2.5%, which means that smaller values of θ can be rejected. The lower row of Figure 11.9 shows the case of $\theta = 0.497$, for which the sampling distribution has $p = 2.5\%$. If θ is nudged any larger, p falls below 2.5%, which means that larger values of θ can be rejected. In summary, the range of θ values we would not reject is $\theta \in [0.135, 0.497]$. This is the 95% CI when $z = 7$ and $N = 24$, for a data collector who intended to stop when time expired.

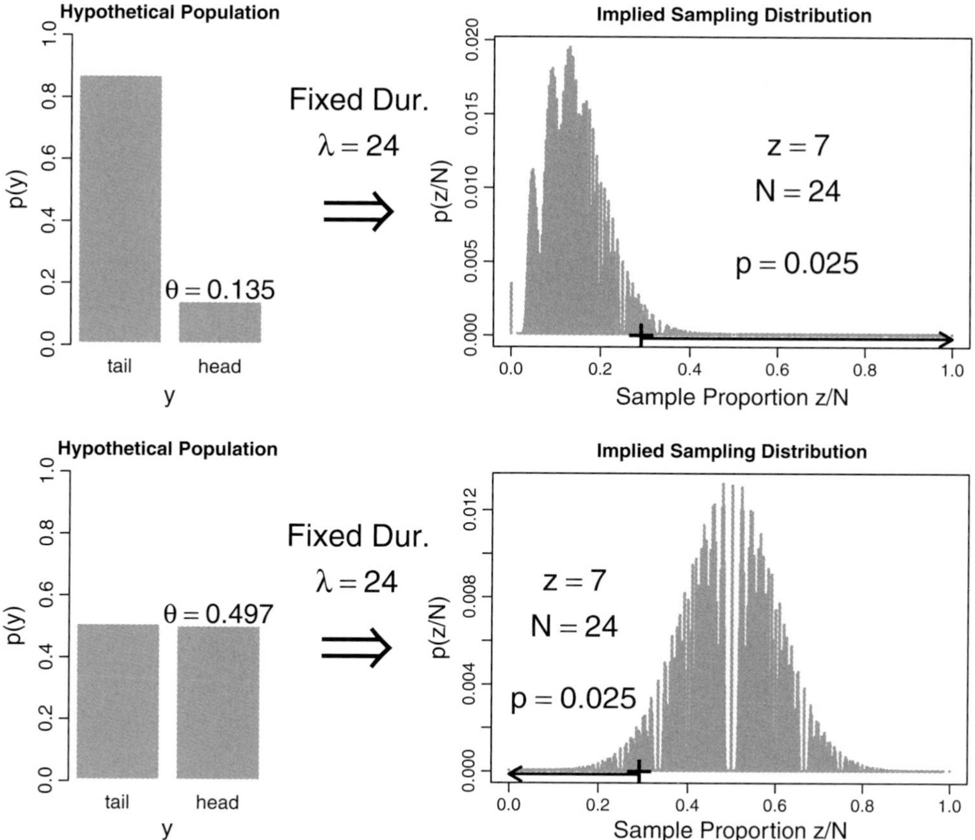

Figure 11.10 95% confidence interval when duration is fixed by the experimenter's intention extends from $\theta = 0.135$ (top row) to $\theta = 0.497$ (bottom row). Compare with Figures 11.8 and 11.9.

Finally, we can determine the CI for the experimenter who intended to test two coins, with a fixed N for both coins. Figure 11.11 shows the sampling distribution for different values of θ.[7] The upper row shows the case of $\theta = 0.110$, for which the sampling distribution has $p = 2.5\%$. If θ is nudged any smaller, p is less than 2.5%, which means that smaller values of θ can be rejected. The lower row of Figure 11.11 shows the case of $\theta = 0.539$, for which the sampling distribution has $p = 2.5\%$. If θ

[7] The multiple-test confidence interval for Figure 11.11 is conventionally computed as follows. First, the overall false alarm rate for two independent tests is $\alpha_{EW} = 1 - (1 - \alpha_{PT})^2$, where α_{PT} is the per-test false alarm rate. Therefore, when $\alpha_{EW} = 0.05$, $\alpha_{PT} = 1 - (1 - \alpha_{EW})^{1/2} = 0.0253$. For a two-tailed test, that means $\alpha_{PT} = 0.0253/2 = 0.0127$ in each tail. We then use this tail probability to find the confidence interval limits. Assuming fixed N as in Figure 11.8, the resulting limits are 0.110 to 0.539, just as shown in Figure 11.11.

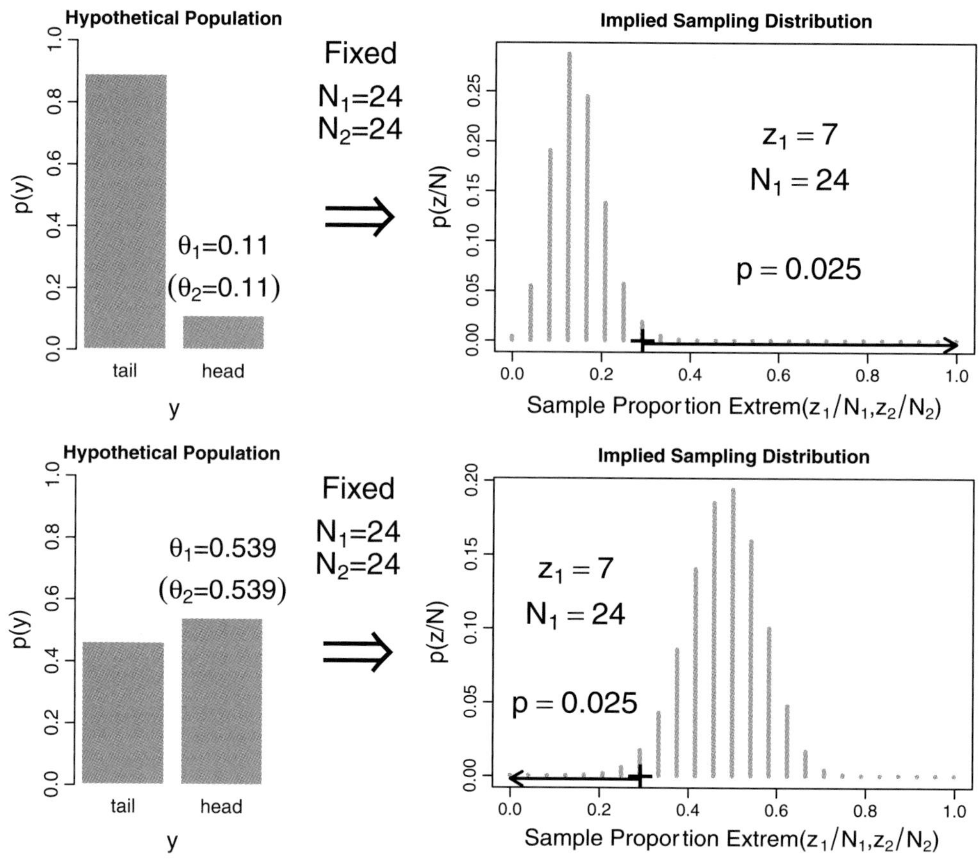

Figure 11.11 95% confidence interval, when N is fixed by the experimenter's intention and there are two tests, extends from $\theta = 0.110$ (top row) to $\theta = 0.539$ (bottom row). Compare with Figure 11.8. Extrem$\{z_1/N_1, z_2/N_2\}$ is the hypothetical sampled proportion that is more extreme relative to θ.

is nudged any larger, p falls below 2.5%, which means that larger values of θ can be rejected. In summary, the range of θ values we would not reject is $\theta \in [0.110, 0.539]$. This is the 95% CI when $z = 7$ and $N = 24$, for a data collector who intended to test two coins and stop when N reached a fixed value.

We have just seen that the NHST CI depends on the stopping and testing intentions of the experimenter. When the intention was to stop when $N = 24$, then the range of biases that would not be rejected is $\theta \in [0.126, 0.511]$. But when the intention was to stop when $z = 7$, then the range of biases that would not be rejected is $\theta \in [0.126, 0.484]$. And when the intention was to stop when time was up, then the range of biases that would not be rejected is $\theta \in [0.135, 0.497]$. And when the intention was to test two coins, stopping each at fixed N, then the CI was $\theta \in [0.110, 0.539]$. The CI depends on the experimenter's intentions because the intentions determine the cloud of imaginary possibilities relative to which the actually observed data are judged. Thus, the interpretation of the NHST CI is as cloudy as the interpretation of NHST itself, because the CI is merely the significance test conducted at every candidate value of θ. Because CIs are defined by p values, for every misconception of p values, there can be a corresponding misconception of CIs. Although motivated somewhat differently, a similar conclusion was colorfully stated by Abelson (1997, p. 13): "under the Law of Diffusion of Idiocy, every foolish application of significance testing will beget a corresponding foolish practice for confidence limits."

11.3.1.1 CI is not a distribution

A CI is merely two end points. A common misconception of a confidence interval is that it indicates some sort of probability distribution over values of θ. It is very tempting to think that values of θ in the middle of a CI should be more believable than values of θ at or beyond the limits of the CI.

There are various ways of imposing some form of distribution over a CI. One way is to superimpose the sampling distribution onto the CI. For example, in the context of estimating the mean of normally distributed data, Cumming and Fidler (2009, p. 18) stated, "In fact, the [sampling distribution of M_{diff} from the null hypothesis], if centered on our [actual] M_{diff}, rather than on μ as in [the sampling distribution], gives the relative likelihood of the various values in and beyond our CI being the true value of μ. Values close to the point estimate M_{diff} are the most plausible for μ. Values inside our [confidence] interval but out toward either limit are progressively less plausible for μ. Values just outside the [confidence] interval are relatively implausible ..." Cumming and Fidler (2009) were careful to say that "plausibility" means relative likelihood, and that μ is not a variable but has an unknown fixed value. Nevertheless, it is all too easy for readers to interpret "plausibility" of a parameter value as the posterior probability of the parameter value. The distinction becomes especially evident when the method of Cumming and Fidler (2009) is applied to estimating the bias of a coin. The sampling distribution is a set of spikes over discrete values of z/N. When that sampling distribution is transferred to the CI for θ, we get a *discrete* set of "plausible" values for θ, which, of course, is very misleading because the (unknown but fixed) value of θ could be anywhere on the *continuum* from 0 to 1.

There have been other proposals to display distributional information on CIs. For example, one could plot the p value as a function of the parameter value θ (e.g., Poole, 1987; Sullivan & Foster, 1990). The θ values at which the curve hits 2.5% are the limits of the 95% CI. But notice these curves are not probability distributions: They do not integrate to 1.0. The curves are plots of $p(D_{\theta,I} \geq D_{\text{actual}}|\theta, I)$ where I is the stopping and testing intentions, and where D refers to a summary description of the data such as z/N. Different stopping or testing intentions produce different p curves and different CIs. Contrast the p curve with a Bayesian posterior distribution, which is $p(\theta|D_{\text{actual}})$. Some theorists have explored normalized p curves, which do integrate to 1, and are called *confidence distributions* (e.g., Schweder & Hjort, 2002; Singh, Xie, & Strawderman, 2007). But these confidence distributions are still sensitive to sampling and testing intentions. Under some specific assumptions, special cases are equivalent to a Bayesian posterior distribution (Schweder & Hjort, 2002). See further discussion in Kruschke (2013a).

All these methods for imposing a distribution upon a CI seem to be motivated by a natural Bayesian intuition: Parameter values that are consistent with the data should be more credible than parameter values that are not consistent with the data (subject to prior credibility). If we were confined to frequentist methods, then the various proposals outlined above would be expressions of that intuition. But we are not confined to frequentist methods. Instead, we can express our natural Bayesian intuitions in fully Bayesian formalisms.

11.3.2. Bayesian HDI

A concept in Bayesian inference, that is somewhat analogous to the NHST CI, is the HDI, which was introduced in Section 4.3.4, p. 87. The 95% HDI consists of those values of θ that have at least some minimal level of posterior credibility, such that the total probability of all such θ values is 95%.

Let's consider the HDI when we flip a coin and observe $z = 7$ and $N = 24$. Suppose we have a prior informed by the fact that the coin appears to be authentic, which we express here, for illustrative purposes, as a beta$(\theta|11, 11)$ distribution. The right side of Figure 11.7 shows that the 95% HDI goes from $\theta = 0.254$ to $\theta = 0.531$. These limits span the 95% most credible values of the bias. Moreover, the posterior density shows exactly how credible each bias is. In particular, we can see that $\theta = 0.5$ is within the 95% HDI. Rules for making discrete decisions are discussed in Chapter 12.

There are at least three advantages of the HDI over an NHST CI. First, the HDI has a direct interpretation in terms of the credibilities of values of θ. The HDI is explicitly about $p(\theta|D)$, which is exactly what we want to know. The NHST CI, on the other hand, has no direct relationship with what we want to know; there's no clear relationship between the probability of rejecting the value θ and the credibility of θ. Second, the HDI has no dependence on the sampling and testing intentions of the experimenter, because the likelihood function has no dependence on the sampling and testing intentions of the experimenter.[8] The NHST confidence interval, in contrast, tells us about probabilities of data relative to imaginary possibilities generated from the experimenter's intentions.

[8] The likelihood function is, actually, defined in terms of a sampling intention, namely, the intention to stop when $N = 1$. This means merely that we must agree on the operationalization for making a single observation, and the possible outcomes of a single observation. Both Bayesian and NHST approaches

Third, the HDI is responsive to the analyst's prior beliefs, as it should be. The Bayesian analysis indicates how much the new data should alter our beliefs. The prior beliefs are overt and publicly decided. The NHST analysis, on the contrary, does not incorporate prior knowledge.

11.4. MULTIPLE COMPARISONS

In many research situations, there are multiple conditions or treatments being compared. Recall, for example, the estimation of baseball batting abilities for players from different fielding positions, in Section 9.5.1 (p. 253). We examined several comparisons of different positions (e.g., Figure 9.14) and of different individual players (e.g., Figure 9.15). With 9 positions and 948 players, there are hundreds if not thousands of meaningful comparisons we might want to make. In experimental research with several conditions, researchers can make many different comparisons across conditions and combinations of conditions.

When comparing multiple conditions, a key goal in NHST is to keep the overall false alarm rate down to a desired maximum such as 5%. Abiding by this constraint depends on the number of comparisons that are to be made, which in turn depends on the intentions of the experimenter. In a Bayesian analysis, however, there is just one posterior distribution over the parameters that describe the conditions. That posterior distribution is unaffected by the intentions of the experimenter, and the posterior distribution can be examined from multiple perspectives however is suggested by insight and curiosity. The next two sections expand on frequentist and Bayesian approaches to multiple comparisons.

11.4.1. NHST correction for experimentwise error

When there are multiple groups, it often makes sense to compare each group to every other group. With nine fielding positions, for example, there are 36 different pairwise comparisons we can make. The problem is that each comparison involves a decision with the potential for false alarm, and the p values for all comparisons increase. We already saw, in Section 11.1.5 (p. 310), that the intention to make multiple tests increases the p value because the imaginary cloud of possible outcomes expands. In NHST, we have to take into account all comparisons we intend for the whole experiment. Suppose we set a criterion for rejecting the null such that each decision has a "per-comparison" (PC) false alarm rate of α_{PC}, e.g., 5%. Our goal is to determine the overall false alarm

---

rely on this foundation. It is explicit in the likelihood function of Bayes' rule. It is also explicit as the population hypothesis in NHST, as shown, for example, as the Bernoulli distribution in the left sides of Figures 11.3–11.5. In this sense, even a Bayesian analysis is based on the stopping intention that operationalizes a single measurement.

rate when we conduct several comparisons. To get there, we do a little algebra. First, suppose the null hypothesis is true, which means that the groups are identical, and we get apparent differences in the samples by chance alone. This means that we get a false alarm on a proportion α_{PC} of replications of a comparison test. Therefore, we do *not* get a false alarm on the complementary proportion $1 - \alpha_{PC}$ of replications. If we run c independent comparison tests, then the probability of not getting a false alarm on *any* of the tests is $(1 - \alpha_{PC})^{c}$. Consequently, the probability of getting at least one false alarm is $1 - (1 - \alpha_{PC})^{c}$. We call that probability of getting at least one false alarm, across all the comparisons in the experiment, the "experimentwise" false alarm rate, denoted α_{EW}. Here's the rub: α_{EW} is greater than α_{PC}. For example, if $\alpha_{PC} = .05$ and $c = 36$, then $\alpha_{EW} = 1 - (1 - \alpha_{PC})^{c} = 0.84$. Thus, even when the null hypothesis is true, and there are really no differences between groups, if we conduct 36 independent comparisons, we have an 84% chance of falsely rejecting the null hypothesis for at least one of the comparisons. Usually not all comparisons are structurally independent of each other, so the false alarm rate does not increase so rapidly, but it does increase whenever additional comparison tests are conducted.

One way to keep the experimentwise false alarm rate down to 5% is by reducing the permitted false alarm rate for the individual comparisons, i.e., setting a more stringent criterion for rejecting the null hypothesis in individual comparisons. One often-used re-setting is the *Bonferonni correction*, which sets $\alpha_{PC} = \alpha_{EW}^{desired}/c$. For example, if the desired experimentwise false alarm rate is 0.05, and there are 36 comparisons planned, then we set each individual comparison's false alarm rate to 0.05/36. This is a conservative correction, because the actual experimentwise false alarm rate will usually be much less than $\alpha_{EW}^{desired}$.

There are many different corrections available to the discerning NHST aficionado (e.g., Maxwell & Delaney, 2004, chap. 5). Not only do the correction factors depend on the structural relationships of the comparisons, but the correction factors also depend on whether the analyst intended to conduct the comparison before seeing the data, or was provoked into conducting the comparison only after seeing the data. If the comparison was intended in advance, it is called a *planned* comparison. If the comparison was thought of only after seeing a trend in the data, it is called a *post hoc* comparison. Why should it matter whether a comparison is planned or *post hoc*? Because even when the null hypothesis is true, and there are no real differences between groups, there will always be a highest and lowest random sample among the groups. If we don't plan in advance which groups to compare, but do compare whichever two groups happen to be farthest apart, we have an inflated chance of declaring groups to be different that aren't truly different.

The point, for our purposes, is not which correction to use. The point is that the NHST analyst must make some correction, and the correction depends on the number and type of comparisons that the analyst *intends* to make. This creates a problem because two analysts can come to the same data but leave with different conclusions because of the variety of comparisons that they find interesting enough to conduct, and what provoked their interest. The creative and inquisitive analyst, who wants to conduct many comparisons either because of deep thinking about implications of theory, or because of provocative unexpected trends in the data, is penalized for being thoughtful. A large set of comparisons can be conducted only at the cost of using a more stringent threshold for each comparison. The uninquisitive analyst is rewarded with an easier criterion for achieving significance. This seems to be a counterproductive incentive structure: You have a higher chance of getting a "significant" result, and getting your work published, if you feign narrow mindedness under the pretense of protecting the world from false alarms.

To make this concrete, consider again the estimation of baseball batting abilities for players from different fielding positions, in Section 9.5.1 (p. 253). A basic question might be whether the batting ability of infielders differs from the batting ability of outfielders. Therefore an uninquisitive analyst might plan to make the single comparison of the average of the six non-outfield positions against the average of the three outfield positions. A more inquisitive or knowledgeable analyst might plan additional comparisons, suspecting that pitchers and catchers might be different from basemen because of the different skills demanded for the pitcher-catcher duo. This analyst might plan four comparisons: outfielders versus non-outfielders, outfielders versus basemen, outfielders versus average of catchers and pitchers, and basemen versus average of catchers and pitchers. The more inquisitive and knowledgeable analyst is punished with a more stringent criterion for declaring significance, even on the comparison of outfielders versus non-outfielders that the uninquisitive analyst also made.

Suppose that, upon seeing the data, the detail-oriented analyst discovers that catchers actually have about the same batting average as basemen, and therefore comparison should be made between catchers and basemen and between catchers and pitchers. The analyst should treat this as a *post hoc*, not planned, comparison. But wait—upon reflection, it is clear merely from knowledge of the game that catchers have much different demands than pitchers, and therefore these comparisons should have been considered from the start. So, perhaps these comparisons should be considered *planned*, not *post hoc*, after all. Suppose also that, upon seeing the data, the analyst notices that the basemen don't differ much from the outfielders. Therefore it seems superfluous to compare them. This lack of comparison is *post hoc*, because a comparison had been planned. But, in retrospect, it's clear from background knowledge about the game that basemen and outfielders actually have similar demands: they all must catch fly balls and throw to basemen. Therefore the lack of comparison should have been planned after all.

All this leaves the NHST analyst walking on the quicksand of soul searching. Was the comparison truly planned or *post hoc*? Did the analyst commit premeditated exclusion of comparisons that should have been planned, or was the analyst merely superficial, or was the exclusion *post hoc*? This problem is not solved by picking a story and sticking to it, because any story still presumes that the analyst's testing intentions should influence the data interpretation.

11.4.2. Just one Bayesian posterior no matter how you look at it

The data from an experiment, or from an observational study, are carefully collected so to be totally insulated from the experimenter's intentions regarding subsequent tests. In experiments, each datum should be uninfluenced by the presence or absence of any other condition or subject in the experiment. For properly conducted experiments, in which subjects are kept uninformed about the goals or structure of the experiment (until after the experiment is done), there is no way for an individual in one experimental group to be influenced by the presence or absence of any groups or subjects, before or after. Moreover, the data are uninfluenced by the experimenter's intentions regarding the other groups and sample size.

In a Bayesian analysis, the interpretation of the data is not influenced by the experimenter's stopping and testing intentions (assuming that those intentions do not affect the data). A Bayesian analysis yields a posterior distribution over the parameters of the model. The posterior distribution is the complete implication of the data. The posterior distribution can be examined in as many different ways as the analyst deems interesting; various comparisons of groups are merely different perspectives on the posterior distribution.

For example, in the baseball data we examined several comparisons of different positions (e.g., Figure 9.14) and of different individual players (e.g., Figure 9.15). Those marginal distributions merely summarize the posterior distribution from various perspectives. The posterior distribution itself is unchanged by how we look at it. We can examine any other comparison of parameters without worrying about what motivated us to consider it, because the posterior distribution is unchanged by those motivations, unlike the cloud of imaginary possibilities from the null hypothesis.

In summary, the Bayesian posterior distribution is appropriately *in*sensitive to the experimenter's stopping and sampling intentions to compare or not compare various groups. The Bayesian posterior also directly tells us the credibilities of the magnitudes of differences, unlike NHST which tells us only about whether a difference is extreme in a cloud of possibilities determined by the experimenter's intentions.

11.4.3. How Bayesian analysis mitigates false alarms

No analysis is immune to false alarms, because randomly sampled data will occasionally contain accidental coincidences of outlying values. Bayesian analysis eschews the use of p values as a criterion for decision making, however, because p values control false alarms on the basis of the analyst's intentions, not on the basis of the data. Bayesian analysis instead accepts the fact that the posterior distribution is the best inference we can make, given the observed data and the prior knowledge.

How, then, does a Bayesian analysis address the problem of false alarms? By incorporating prior knowledge into the structure of the model. Specifically, if we know that different groups have some overarching commonality, even if their specific treatments are different, we can nevertheless describe the different group parameters as having been drawn from an overarching distribution that expresses the commonality. Examples of hierarchical models were given in Figures 9.7 (p. 236, re therapeutic touch) and 9.13 (p. 252, re baseball batting ability). If several of the groups yield similar data, this similarity informs the overarching distribution, which in turn implies that any outlying groups should be estimated to be a little more similar than they would be otherwise. In other words, just as there can be shrinkage of individual estimates toward the group central tendency, there can be shrinkage of group estimates toward the overall central tendency. Shrinkage was described extensively in Section 9.3 (p. 245). The shrinkage pulls in the estimates of accidental outliers and reduces false alarms (e.g., D. A. Berry & Hochberg, 1999; Gelman, 2005; Gelman, Hill, & Yajima, 2009; Lindquist & Gelman, 2009; Meng & Dempster, 1987). This shrinkage is not an arbitrary "correction" like those applied in NHST. The shrinkage is a rational consequence of the prior knowledge expressed in the model structure. Hierarchical structure can be put into models that are assessed using NHST, but Bayesian estimation is especially seamless and straightforward for implementing and evaluating hierarchical models.

11.5. WHAT A SAMPLING DISTRIBUTION *IS* GOOD FOR

I hope to have made it clear that sampling distributions (the cloud of imaginary possibilities) aren't as useful as posterior distributions for making inferences from a set of observed data. The reason is that sampling distributions tell us the probabilities of possible data if we run an intended experiment given a particular hypothesis, rather than the credibilities of possible hypotheses given that we have a particular set of data. That is, sampling distributions tell us the probability of imaginary outcomes given a parameter value and an intention, $p(D_{\theta,I}|\theta, I)$, instead of the probability of parameter values given the actual data, $p(\theta|D_{\text{actual}})$. Nevertheless, sampling distributions are appropriate and useful for other applications. Two of those applications are described in the following sections.

11.5.1. Planning an experiment

So far in this book, we have only considered the analysis of data that have already been obtained. But a crucial part of conducting research is planning the study before actually obtaining the data. When planning research, we have some hypothesis about how the world might be, and we want to gather data that will inform us about the viability of that hypothesis. Typically we have some notion already about the experimental treatments or observational settings, and we want to plan how many observations we'll probably need to make, or how long we'll need to run the study, in order to have reasonably reliable evidence one way or the other.

For example, suppose that my theory suggests a coin should be biased with θ around 0.60, perhaps a little higher and perhaps a little lower. The coin might represent a population of voters, hence flipping the coin means polling a person in the population, and the outcome heads means preference for candidate A. The theory regarding the bias may have come from previous polls regarding political attitudes. We would like to plan a survey of the population that will give us precise posterior beliefs about the true preference for candidate A. Suppose we contemplate a survey that polls 500 people. By simulating the experiment over and over, using the hypothesized random $\theta \approx 0.60$ and $N = 500$, we can generate simulated data, and then derive a Bayesian posterior distribution for every set of simulated data. For every posterior distribution, we determine some measure of accuracy, such as the width of the 95% HDI. From many simulated experiments, we get a sampling distribution of HDI widths. From the sampling distribution of HDI widths, we can decide whether $N = 500$ typically yields high enough accuracy for our purposes. If not, we repeat the simulation with a larger N. Once we know how big N needs to be to get the accuracy we seek, we can decide whether or not it is feasible to conduct such a study.

Notice that we used the intended experiment to generate a space of possible data in order to anticipate what is likely to happen *when the data are analyzed with Bayesian methods*. For any single set of data (simulated or actual), we recognize that the individual data points in the set are insulated from the intentions of the design, and we conduct a Bayesian analysis of the data set. The use of a distribution of possible sample data, from an intended experiment, is perfectly appropriate here because it is exactly the implications of this hypothetical data distribution that we want to discover.

The issues of research design will be explored in depth in Chapter 13. You might want to glance at Figure 13.1, p. 363, to see how a cloud of possible data is used as a "dress rehearsal" for planning actual data collection. In particular, notice that a Bayesian analysis of actual data, illustrated in the top panel of Figure 13.1, does *not* use a cloud of possibilities from a hypothesis, and is quite different from the construction of p values illustrated in Figure 11.1, p. 299.

11.5.2. Exploring model predictions (posterior predictive check)

A Bayesian analysis only indicates the *relative* credibilities of the various parameter values or models under consideration. The posterior distribution only tells us which parameter values are relatively less bad than the others. The posterior does not tell us whether the least bad parameter values are actually any good.

For example, suppose we believe that a coin is a heavily biased trick coin, and either comes up heads 99% of the time, or else comes up tails 99% of the time; we just don't know which direction of bias it has. Now we flip the coin 40 times and it comes up heads 30 of those flips. It turns out that the 99%-head model has a far bigger posterior probability than the 99%-tail model. But it is also the case that the 99%-head model is a terrible model of a coin that comes up heads 30 out of 40 flips!

One way to evaluate whether the least unbelievable parameter values are any good is via a posterior predictive check. A posterior predictive check is an inspection of patterns in simulated data that are generated by typical posterior parameters values. The idea of a posterior predictive check is as follows: If the posterior parameter values really are good descriptions of the data, then the predicted data from the model should actually "look like" real data. If the patterns in the predicted data do not mirror the patterns in the actual data, then we are motivated to invent models that can produce the patterns of interest.

This use of the posterior predictive check is suspiciously like NHST: We start with a hypothesis (i.e., the least unbelievable parameter values), and we generate simulated data as if we were repeating our intended experiment over and over. Then we see if the actual data are typical or atypical in the space of simulated data. If we were to go further, and determine critical values for false alarm rates and then reject the model if the actual data fall in its extreme tails, then we would indeed be doing something tantamount to NHST. Some authors do promote this sort of "Bayesian p value." But I prefer to keep posterior predictive checks fully Bayesian. The goal of the posterior predictive check is to drive intuitions about the qualitative manner in which the model succeeds or fails, and about what sort of novel model formulation might better capture the trends in the data. Once we invent another model, then we can use Bayesian methods to quantitatively compare it with the other models. For further discussion, see Section 17.5.1 and Kruschke (2013b).

11.6. EXERCISES

Look for more exercises at https://sites.google.com/site/doingbayesiandataanalysis/

Exercise 11.1. [Purpose: To compute p values for stopping at fixed N and fixed z.] We have a six-sided die, and we want to know whether the probability that the six-dotted face comes up is fair. Thus, we are considering two possible outcomes: six-dots or not six-dots. If the die is fair, the probability of the six-dotted face is 1/6.

(A) Suppose we roll the die $N = 45$ times, intending to stop at that number of rolls. Suppose we get 3 six-dot rolls. What is the two-tailed p value?

Hints: Use Equation 11.5 (p. 303) to compute the tail probability of the binomial sampling distribution in R. R has various relevant functions built in, such as `factorial`, choose, and even dbinom.[9] To maintain correspondence with Equation 11.5, I will not use dbinom. Try this script:

```
N = 45 ; z = 3 ; theta = 1/6
lowTailZ = 0:z
sum( choose(N,lowTailZ) * theta^lowTailZ * (1-theta)^(N-lowTailZ) )
```

Explain carefully what each line of the script does. Why does it consider the low tail and not the high tail? Explain the meaning of the final result.

(B) Suppose that instead of stopping at fixed N, we stop when we get 3 six-dot outcomes. It takes 45 rolls. (Notice this is the same result as the previous part.) What is the two-tailed p value?

Hint: Use Equation 11.6 (p. 306). Try this:

```
sum( (lowTailZ/N) * choose(N,lowTailZ) * theta^lowTailZ * (1-theta) ^(N-lowTailZ) )
```

Explain carefully what that code does and what its result means.

Exercise 11.2. [Purpose: To determine NHST CIs, and notice that they depend on the experimenter's intention.] We continue with the scenario of the previous exercise: A dichotomous outcome, with $N = 45$ and $z = 3$.

(A) If the intention is to stop when $N = 45$, what is the 95% CI?

Hints: Try this continuation of the R script from the previous exercise:

```
for ( theta in seq( 0.170 , 0.190 , 0.001) ) {
  show( c(
    theta ,
    2*sum( choose(N,lowTailZ) * theta^lowTailZ * (1-theta)^(N-lowTailZ) )
  ))
}

highTailZ = z:N
for ( theta in seq( 0.005 , 0.020 , 0.001) ) {
  show( c(
    theta ,
    2*sum( choose(N,highTailZ) * theta^highTailZ * (1-theta)^(N-highTailZ) )
  ))
}
```

Explain carefully what the code does and what it means!

[9] You might find it helpful to use R's dbinom(x,size,prob) function, where x corresponds to z (a vector from 0 to N) and size corresponds to N (a constant) in Equation 11.6. R also has a function for the negative binomial, dnbinom(x,size,prob). Be very careful if you use the negative binomial density, because the argument x corresponds to $N-z$ (a vector starting with 0) and the argument size corresponds to z (a constant) in Equation 11.6.

(B) If the intention is to stop when $z = 3$, what is the 95% CI? Is the CI the same as for stopping when $N = 45$?

Hint: Modify the R script of the previous part for use with stopping at z, like the second part of the previous exercise.

Exercise 11.3. [Purpose: To determine the p value when data collection stops at a fixed duration.] (For another example of NHST for fixed-duration samples, see Kruschke, 2010.) We continue with the scenario of the previous exercises: A dichotomous outcome, with $N = 45$ and $z = 3$. Suppose that the die-roller of the previous exercises stopped rolling because time expired at 6 min. For simplicity, suppose that during a 6-min interval, the roller could have rolled $N = 40$, or $N = 41$, or $N = 42$, through $N = 50$, with equal probability. What is the p value for the observed outcome? Is it the same p value as when assuming fixed N or fixed z?

Hints: We need to compute the p value for each possible N, and then average them according to the probability they would happen. For each N, the low tail consists of outcomes that are a proportion less than or equal to the observed $z/N = 3/45$. Examine the follow R script. Explain exactly what it does and interpret its output.

```
N = 45 ; z = 3 ; theta = 1/6
# Specify possible N values:
Nposs = 40:50
# Specify probability of each N (here all equal):
Nprob = rep(1,length(Nposs)) ; Nprob = Nprob/sum(Nprob)
# For each possible N, compute p value, and compute the weighted total p:
totalP = 0
for ( i in 1:length(Nposs) ) {
  thisN = Nposs[i]
  # For this N, determine the max z that is in the low tail:
  thisZ =  max( (0:thisN)[ (0:thisN)/thisN <= z/N ] )
  lowTailZ = 0:thisZ
  thisP = 2*sum( choose(thisN,lowTailZ) * theta^lowTailZ * (1-theta) ^(thisN-lowTailZ) )
  totalP = totalP + Nprob[i] * thisP
  show( c( thisN , thisP ) )
}
show( totalP )
```

CHAPTER 12

Bayesian Approaches to Testing a Point ("Null") Hypothesis

Contents

Is he a loser or is he real great?
Words without actions make bad estimates.
I need big outcomes when going gets rough, 'cause
Better than nothing just ain't good enough.[1]

Suppose that you have collected some data, and now you want to answer the question, Is there a non-zero effect or not? Is the coin fair or not? Is there better-than-chance accuracy or not? Is there a difference between groups or not? In the previous chapter, I argued that answering this type of question via null hypothesis significance testing (NHST) has deep problems. This chapter describes Bayesian approaches to the question.

In the context of coin flipping, the question we are asking is whether the probability of heads has some particular value. For example, if we are asking whether the coin is fair, we are asking whether a head probability of 0.5 is credible. There are two different ways of formalizing this question in a Bayesian framework. One way to pose the question is to ask whether the value of interest ($\theta = 0.5$) falls among the most credible values

[1] This chapter explains two Bayesian approaches to evaluating null values, one approach based on parameter estimation and the other approach based on comparing nothing with something. The poem suggests the need for estimating magnitudes in real life, as opposed to merely saying something is better than nothing.

Doing Bayesian Data Analysis, 2nd Edition
http://dx.doi.org/10.1016/B978-0-12-405888-0.00012-X

in the posterior. A different way to pose the question sets up a dichotomy between, on the one hand, a prior distribution that allows *only* the value of interest, and, on the other hand, a prior distribution that allows a broad range of all possible values. The posterior believability of the two priors is assessed via Bayesian model comparison. This chapter will explore the two approaches in some detail. The conclusions drawn by the two approaches do not yield the same information, so it is important to choose the approach that is most appropriate to the situation and the question you want to ask of your data.

12.1. THE ESTIMATION APPROACH

Throughout this book, we have used Bayesian inference to derive a posterior distribution over a parameter of interest, such as the bias θ of a coin. We can then use the posterior distribution to discern the credible values of the parameter. If the null value is far from the credible values, then we reject the null value as not credible. But if all the credible values are virtually equivalent to the null value, then we can accept the null value. This intuitive decision procedure is given formal expression in this section.

12.1.1. Region of practical equivalence

A *region of practical equivalence* (ROPE) indicates a small range of parameter values that are considered to be practically equivalent to the null value for purposes of the particular application. For example, if we wonder whether a coin is fair, for purposes of determining which team will kick off at a football game, then we want to know if the underlying bias in the coin is reasonably close to 0.50, and we don't really care if the true bias is 0.473 or 0.528, because those values are practically equivalent to 0.50 for our application. Thus, the ROPE on the bias might go from 0.45 to 0.55. As another example, if we are assessing the efficacy of a drug versus a placebo, we might only consider using the drug if it improves the probability of cure by at least 5 percentage points. Thus, the ROPE on the difference of cure probabilities could have limits of ± 0.05. There will be more discussion later of how to set the ROPE limits.

Once a ROPE is set, we make a decision to reject a null value according the following rule:

> *A parameter value is declared to be* not *credible, or rejected, if its entire ROPE lies outside the 95% highest density interval (HDI) of the posterior distribution of that parameter.*

For example, suppose that we want to know whether a coin is fair, and we establish a ROPE that goes from 0.45 to 0.55. We flip the coin 500 times and observe 325 heads. If the prior is uniform, the posterior has a 95% HDI from 0.608 to 0.691, which falls completely outside the ROPE. Therefore we declare that the null value of 0.5 is rejected for practical purposes. *Notice that when the HDI excludes the ROPE, we are not rejecting all values within the ROPE; we are rejecting only the null value.*

Because the ROPE and HDI can overlap in different ways, there are different decisions that can be made. In particular, we can decide to "accept" a null value:

A parameter value is declared to be accepted for practical purposes if that value's ROPE completely contains the 95% HDI of the posterior of that parameter.

With this decision rule, a null value of a parameter can be accepted only when there is sufficient precision in the estimate of the parameter. For example, suppose that we want to know whether a coin is fair, and we establish a ROPE that goes from 0.45 to 0.55. We flip the coin 1,000 times and observe 490 heads. If the prior is uniform, the posterior has a 95% HDI from 0.459 to 0.521, which falls completely within the ROPE. Therefore we declare that the null value of 0.5 is confirmed for practical purposes, because all of the most credible values are practically equivalent to the null value.

In principle, though rarely in practice, a situation could arise in which a highly precise HDI does not include the null value but still falls entirely within a wide ROPE. According to the decision rule, we would "accept" the null value despite it having low credibility according to the posterior distribution. This strange situation highlights the difference between the posterior distribution and the decision rule. *The decision rule for accepting the null value says merely that the most credible values are practically equivalent to the null value according to the chosen ROPE, not necessarily that the null value has high credibility.* If this situation actually arises, it could be a sign that the ROPE is too large and has been ill-conceived or is outdated because the available data are so much more precise than the ROPE.

When the HDI and ROPE overlap, with the ROPE not completely containing the HDI, then neither of the above decision rules is satisfied, and we withhold a decision. This means merely that the current data are insufficient to yield a clear decision one way or the other, according to the stated decision criteria. The posterior distribution provides complete information about the credible parameter values, regardless of the subsequent decision procedure. There are other types of decisions that could be declared, if the HDI and ROPE overlap in different ways. We will not be pursuing those other types of decisions here, but they can be useful in some applied situations. For further discussion of the ROPE, under somewhat different appellations of "range of equivalence" and "indifference zone," see, for example, Carlin and Louis (2009), Freedman, Lowe, and Macaskill (1984), Hobbs and Carlin (2008), and Spiegelhalter, Freedman, and Parmar (1994).

Aside from the intuitive appeal of using a ROPE to declare practical equivalence, there are sound logical reasons from the broader perspective of scientific method. Serlin and Lapsley (1985, 1993) pointed out that using a ROPE to *affirm* a predicted value is essential for scientific progress, and is a solution to Meehl's paradox (e.g., Meehl, 1967, 1978, 1997). Meehl started with the premise that all theories must be wrong, in the sense that they must oversimplify some aspect of any realistic scenario. The magnitude of discrepancy between theory and reality might be small, but there must be some discrepancy. Therefore, as measurement precision improves (e.g., with collection of more data), the probability of detecting the discrepancy and disproving the theory must increase. This is how it should be: More precise data should be a more challenging test of the theory. But the logic of null hypothesis testing paradoxically yields the opposite result. In null hypothesis testing, all that it takes to "confirm" a (non-null) theory is to reject the null value. Because the null hypothesis is certainly wrong, at least slightly, increased precision implies increased probability of rejecting the null, which means increased probability of "confirming" the theory. What is needed instead is a way to affirm substantive theories, not a way to disconfirm straw-man null hypotheses. Serlin and Lapsley (1985, 1993) showed that by using a ROPE around the predicted value of a theory, the theory can be affirmed. Crucially, as the precision of data increases, and the width of the ROPE decreases, then the theory is tested more stringently.[2]

How is the size of the ROPE determined? In some domains such as medicine, expert clinicians can be interviewed, and their opinions can be translated into a reasonable consensus regarding how big of an effect is useful or important for the application. Serlin and Lapsley (1993, p. 211) said, "Admittedly, specifying [ROPE limits] is difficult. ... The width of the [ROPE] depends on the state of the art of the theory and of the best measuring device available. It depends on the state of the art of the theory ... [because] a historical look at one's research program or an examination of a competing research program will help determine how accurately one's theory should predict in order that it be competitive with other theories." In other words, the limits of the ROPE depend on the practical purpose of the ROPE. If the purpose is to assess the equivalence of drug-treatment outcomes, then the ROPE limits depend on the real-world costs and benefits of the treatment and the ability to measure the outcome. If the purpose is to affirm a scientific theory, then the ROPE limits depend on what other theories need to be distinguished.

The ROPE limits, by definition, cannot be uniquely "correct," but instead are established by practical aims, bearing in mind that wider ROPEs yield more decisions to accept the ROPEd value and fewer decision to reject the ROPEd value. In many situations, the exact limit of the ROPE can be left indeterminate or tacit, so that the audience of the analysis can use whatever ROPE is appropriate at the time, as competing theories and measuring devices evolve. When the HDI is far from the ROPEd value, the exact ROPE is inconsequential because the ROPEd value would be rejected for any reasonable ROPE. When the HDI is very narrow and overlaps the target value, the HDI might again fall within any reasonable ROPE, again rendering the exact ROPE inconsequential. When, however, the HDI is only moderately narrow and near the target value, the analysis can report how much of the posterior falls

[2] Serlin and Lapsley (1985, 1993) called a ROPE a "good-enough belt" and used frequentist methods, but the logic of their argument still applies.

within a ROPE as a function of different ROPE widths. An example is shown at this book's blog at http://doingbayesiandataanalysis.blogspot.com/2013/08/how-much-of-bayesian-posterior.html.

It is important to be clear that any discrete decision about rejecting or accepting a null value does *not* exhaustively capture our knowledge about the parameter value. Our knowledge about the parameter value is described by the full posterior distribution. When making a binary decision, we have merely compressed all that rich detail into a single bit of information. The broader goal of Bayesian analysis is conveying an informative summary of the posterior, and where the value of interest falls within that posterior. Reporting the limits of an HDI region is more informative than reporting the declaration of a reject/accept decision. By reporting the HDI and other summary information about the posterior, different readers can apply different ROPEs to decide for themselves whether a parameter is practically equivalent to a null value. The decision procedure is separate from the Bayesian inference. The Bayesian part of the analysis is deriving the posterior distribution. The decision procedure uses the posterior distribution, but does not itself use Bayes' rule.

In applications when the Bayesian posterior is approximated with an MCMC sample, it is important to remember the instability of the HDI limits. Recall the discussion accompanying Figure 7.13, p. 185, which indicated that the standard deviation of a 95% HDI limit for a normal distribution, across repeated runs with an MCMC sample that has an effective sample size (ESS) of 10,000, is roughly 5% of the standard deviation of parameter posterior. Thus, if the MCMC HDI limit is very near the ROPE limit, be cautious in your interpretation because the HDI limit has instability due to MCMC randomness. Analytically derived HDI limits do not suffer this problem, of course.

It may be tempting to try to adopt the use of the ROPE in NHST because an NHST confidence interval (CI) has some properties analogous to the Bayesian posterior HDI. The analogous decision rule in NHST would be to accept a null hypothesis if a 95% CI falls completely inside the ROPE. This approach goes by the name of *equivalence testing* in NHST (e.g., Rogers, Howard, & Vessey, 1993; Westlake, 1976, 1981). While the spirit of the approach is laudable, it has two main problems. One problem is technical: CIs can be difficult to determine. For example, with the goal of equivalence testing for two proportions, Dunnett and Gent (1977) devoted an entire article to various methods that approximate the CI for that particular case. With modern Bayesian methods, on the other hand, HDIs can be computed seamlessly for arbitrary complex models; indeed the case of two proportions was addressed in Section 7.4.5, p. 176, and the more complex case of hundreds of grouped proportions was addressed in Section 9.5.1, p. 253. The second problem with frequentist equivalence testing is foundational: A CI does not actually indicate the most credible parameter values. In a Bayesian approach, the 95% HDI actually includes the 95% of parameter values that are most credible. Therefore, when the 95% HDI falls within the ROPE, we can conclude that 95% of the credible parameter values are practically equivalent to the null value. But a 95% CI from NHST says nothing directly about the credibility of parameter values. Crucially, even if a 95% CI falls within the ROPE, a change of stopping or testing intention will change the CI and the CI may no longer fall within the ROPE. For example, if the two groups being compared are intended to be compared to other groups, then the 95% CI is much wider and may no longer fall inside the ROPE. This dependency of equivalence testing on the set of tests to be conducted is pointed out by Rogers et al. (1993, p. 562). The Bayesian HDI, on the other hand, is not affected by the intended tests.

12.1.2. Some examples

The left side of Figure 9.14 (p. 256) shows the difference of batting abilities between pitchers and catchers. The 95% HDI goes from −0.132 to −0.0994. If we use a ROPE from −0.05 to +0.05 (somewhat arbitrarily), the HDI falls far outside the ROPE, and we reject the null, even taking into account the MCMC instability of the HDI limits.

The right side of Figure 9.14 (p. 256) shows the difference of batting abilities between catchers and first basemen. The 95% HDI goes from −0.0289 to 0.0 (essentially). With any non-zero ROPE, and taking into account the MCMC instability of the HDI limits, we would not want to declare that a difference of zero is rejected, instead saying that the difference is only marginally non-zero. The posterior gives the full information, indicating that there is a suggestion of difference, but the difference is small relative to the uncertainty of its estimate.

The right side of Figure 9.15 (p. 257) shows the difference of batting abilities between two individual players who provided lots of data. The 95% HDI goes from −0.0405 to +0.0368. This falls completely inside a ROPE of −0.05 to +0.05 (even taking MCMC instability into account). Thus, we could declare that a difference of zero is accepted for practical purposes. This example also illustrates that it can take a lot of data to achieve a narrow HDI; in this case both batters had approximately 600 opportunities at bat.

The `plotPost` function, in the utilities that accompany this book, has options for displaying a null value and ROPE. Details are provided in Section 8.2.5.1 (p. 205), and an example graph is shown in Figure 8.4 (p. 205). In particular, the null value is specified with the `compVal` argument (which stands for comparison value), and the ROPE is specified as a two-element vector with the `ROPE` argument. The resulting graph displays the percentage of the posterior on either side of the comparison value, and the percentage of the posterior within the ROPE and on either side of the ROPE limits.

12.1.2.1 Differences of correlated parameters

It is important to understand that the marginal distributions of two parameters do not reveal whether or not the two parameter values are different. Figure 12.1, in its left quartet, shows a case in which the posterior distribution for two parameter values has a strong positive correlation. Two of the panels show the marginal distributions of the single parameters. Those two marginal distributions suggest that there is a lot of overlap between the two parameters values. Does this overlap imply that we should not believe that they are very different? No! The histogram of the differences shows that the true difference between parameters is credibly greater than zero, with a difference of zero falling well outside the 95% HDI, even taking into account MCMC sampling instability and a small ROPE. The upper left panel shows why: The credible values of the two parameters are highly correlated, such that when one parameter value is large, the other parameter value is also large. Because of this high correlation, the points in the joint distribution fall almost all on one side of the line of equality.

Figure 12.1 shows, in its right quartet, a complementary case. Here, the marginal distributions of the single parameters are exactly the same as before: Compare the histograms of the marginal distributions in the two quartets. Despite the fact that the marginal distributions are the same as before, the bottom right panel reveals that the difference of parameter values now straddles zero, with a difference of zero firmly in the midst of the HDI. The plot of the joint distribution shows why: Credible values of the two parameter are negatively correlated, such that when one parameter value is large, the other parameter value is small. The negative correlation causes the joint distribution to straddle the line of equality.

In summary, the marginal distributions of two parameters do not indicate the relationship between the parameter values. The joint distribution of the two parameters

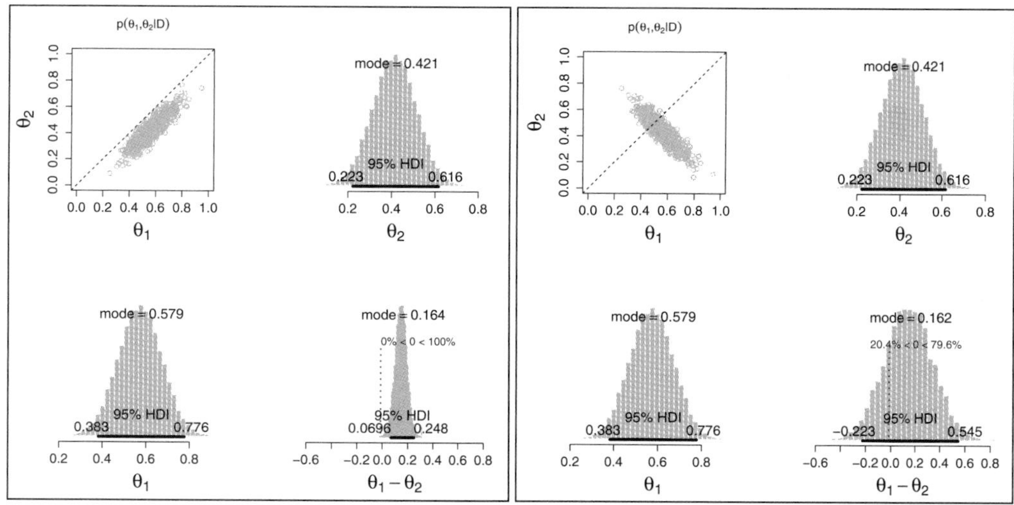

Figure 12.1 When there is a positive correlation between parameters, as shown in the left quartet, the distribution of differences is narrower than when there is a negative correlation, as shown in the right quartet.

might have positive or negative correlation (or even a non-linear dependency), and therefore the difference of the parameter values should be explicitly examined.

12.1.2.2 Why HDI and not equal-tailed interval?

I have advocated using the HDI as the summary credible interval for the posterior distribution, also used in the decision rule along with a ROPE. The reason for using the HDI is that it is very intuitively meaningful: All the values inside the HDI have higher probability density (i.e., credibility) than any value outside the HDI. The HDI therefore includes the most credible values.

Some other authors and software use an *equal-tailed interval* (ETI) instead of an HDI. A 95% ETI has 2.5% of the distribution on either side of its limits. It indicates the 2.5th percentile and the 97.5th percentile. One reason for using an ETI is that it is easy to compute.

In symmetric distributions, the ETI and HDI are the same, but not in skewed distributions. Figure 12.2 shows an example of a skewed distribution with its 95% HDI and 95% ETI marked. (It is a gamma distribution, so its HDI and ETI are easily computed to high accuracy.) Notice on the right there is a region, marked by an arrow, that is outside the HDI but inside the ETI. On the left there is another region marked by an arrow, that is inside the HDI but outside the ETI. The ETI has the strange property that parameter values in the region marked by the right arrow are *included* in the ETI, even though they have *lower credibility* than parameter values in the region marked by the left arrow that are *excluded* from the ETI. This property seems undesirable as a summary of the credible values in a distribution.

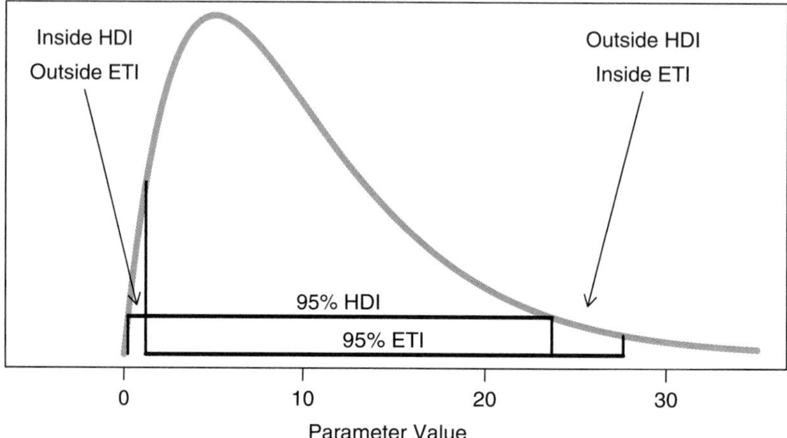

Figure 12.2 A skewed distribution has different 95% highest density interval (HDI) than 95% equal-tailed interval (ETI).

The strange property of the ETI also leads to weirdness when using it as a decision tool. If a null value and ROPE were in the region marked by the right arrow, it would be rejected by the HDI, but not by the ETI. Which decision makes more sense? I think the decision by HDI makes more sense, because it is saying that the values outside its limits have low credibility. But the decision by ETI says that values in this region are *not* rejected, even though they have low credibility. The complementary conflict happens in the region marked by the left arrow. If a null value and ROPE overlap that region, the decision by HDI would be *not* to reject, but the decision by ETI *would* be to reject. Again, I think the decision by HDI makes more sense, because these values have high credibility, even though they are in the extreme tail of the distribution.

Proponents of using the ETI point out that the ETI limits are invariant under nonlinear transformations of the parameter. The ETI limits of the transformed parameter are just the transformed limits of the original scale. This is not the case for HDI limits (in general). This property is handy when the parameters are arbitrarily scaled in abstract model derivations, or in some applied models for which parameters might be nonlinearly transformed for different purposes. But in most applications, the parameters are meaningfully defined on the canonical scale of the data, and the HDI has meaning relative to that scale. Nevertheless, it is important to recognize that if the scale of the parameter is nonlinearly transformed, the HDI limits will change relative to the percentiles of the distribution.

12.2. THE MODEL-COMPARISON APPROACH

Recall that the motivating issue for this chapter is the question, Is the null value of a parameter credible? The previous section answered the question in terms of parameter estimation. In that approach, we started with a possibly informed prior distribution and examined the posterior distribution.

In this section we take a different approach. Some researchers prefer instead to pose the question in terms of model comparison. In this framing of the question, the focus is not on estimating the magnitude of the parameter. Instead, the focus is on deciding which of two hypothetical prior distributions is least incredible. One prior expresses the hypothesis that the parameter value is exactly the null value. The alternative prior expresses the hypothesis that the parameter could be any value, according to some form of broad distribution. In some formalizations, the alternative prior is a default uninformed distribution, chosen according to mathematical desiderata. This lack of being informed is often taken as a desirable aspect of the approach, not a defect, because the method might obviate disputes about prior knowledge. We will see, however, that the model-comparison method can be extremely sensitive to the choice of "uninformed" prior for the alternative hypothesis. The model comparison is not necessarily meaningful unless both hypotheses are viable in the first place.

Recall the model-comparison framework of Figure 10.1, p. 267, where the right panel shows a special case of model comparison in which the only difference between models is their prior distribution. It is this special case that is used to express the comparison of null and alternative hypotheses. The null hypothesis is expressed as a "spike" prior on the parameter of interest, such that only the null value has non-zero prior credibility. The alternative hypothesis is expressed as a broad prior, allowing a wide range of non-null values of the parameter. We previously saw an example similar to this in Section 10.5, p. 289, when we compared the must-be-fair model with the anything's-possible model. In that case, the must-be-fair model was almost a spike-shaped prior for the null hypothesis, and the anything's-possible model was a form of alternative hypothesis.

12.2.1. Is a coin fair or not?

For the null hypothesis, the prior distribution is a "spike" at the null value. The prior probability is zero for all values of θ other than the null value. The probability of the data for the null hypothesis is

$$p(z, N|M_{\text{null}}) = \theta_{\text{null}}^{z}(1-\theta_{\text{null}})^{(N-z)} \tag{12.1}$$

where M_{null} denotes the null model, that is, the null hypothesis. For the alternative hypothesis, we assume a broad beta distribution. Recall from Footnote 5 on p. 132 that for a single coin with a beta prior, the marginal likelihood is

$$p(z, N|M_{\text{alt}}) = B(z + a_{\text{alt}}, N - z + b_{\text{alt}})/B(a_{\text{alt}}, b_{\text{alt}}) \tag{12.2}$$

This equation was expressed as functions in R immediately after Equation 10.6 on p. 270. Combining Equations 12.2 and 12.1 we get the Bayes' factor for the alternative hypothesis relative to the null hypothesis:

$$\frac{p(z, N|M_{\text{alt}})}{p(z, N|M_{\text{null}})} = \frac{B(z + a_{\text{alt}}, N - z + b_{\text{alt}})/B(a_{\text{alt}}, b_{\text{alt}})}{\theta_{\text{null}}^{z}(1-\theta_{\text{null}})^{(N-z)}} \tag{12.3}$$

For a default alternative prior, the beta distribution is supposed to be uninformed, according to particular mathematical criteria. Intuition might suggest that a uniform distribution suits this requirement, that is, beta$(\theta|1, 1)$. Instead, some argue that the most appropriate uninformed beta distribution is beta$(\theta|\epsilon, \epsilon)$, where ϵ is a small number approaching zero (e.g., Lee & Webb, 2005; Zhu & Lu, 2004). This is called the *Haldane prior* (as was mentioned in Section 10.6). A Haldane prior is illustrated in the top-left plot of Figure 12.3, using $\epsilon = 0.01$.

Let's compute the value of the Bayes' factor in Equation 12.3, for the data used repeatedly in the previous chapter on NHST, namely $z = 7$ and $N = 24$. Here are the results for various values of a_{alt} and b_{alt}:

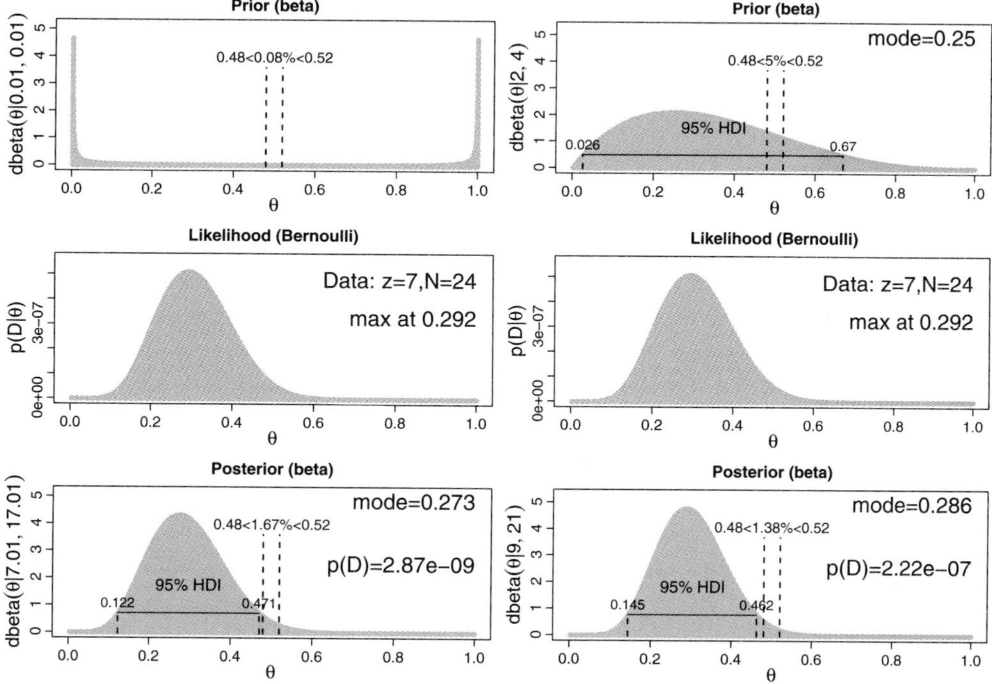

Figure 12.3 Left column: Haldane prior. Right column: Mildly informed prior. Vertical dashed lines mark a ROPE from 0.48 to 0.52. Annotation above the dashed lines indicates the percentage of the distribution within the ROPE.

$$\frac{p(z,N|M_{\text{alt}})}{p(z,N|M_{\text{null}})} = \begin{cases} 3.7227 & \text{for } a_{\text{alt}} = 2, b_{\text{alt}} = 4 \\ 1.9390 & \text{for } a_{\text{alt}} = b_{\text{alt}} = 1.000 \\ 0.4211 & \text{for } a_{\text{alt}} = b_{\text{alt}} = 0.100 \\ 0.0481 & \text{for } a_{\text{alt}} = b_{\text{alt}} = 0.010 \\ 0.0049 & \text{for } a_{\text{alt}} = b_{\text{alt}} = 0.001 \end{cases} \tag{12.4}$$

The first case, $a_{\text{alt}} = 2, b_{\text{alt}} = 4$, will be discussed later. For now, notice that when the alternative prior is uniform, with $a_{\text{alt}} = b_{\text{alt}} = 1.000$, the Bayes' factor shows a (small) preference for the alternative hypothesis, but when the alternative prior approximates the Haldane, the Bayes' factor shows a strong preference for the null hypothesis. As the alternative prior gets closer to the Haldane limit, the Bayes' factor changes by orders of magnitude. Thus, as we have seen before (e.g. Section 10.6, p. 292), the Bayes' factor is very sensitive to the choice of prior distribution.

You can see why this happens by considering Figure 12.3, which shows two priors in its two columns. The Haldane prior, in the left column, puts extremely small prior credibility on the parameter values that are most consistent with the data. Because the marginal likelihood used in the Bayes' factor is the product of the prior and the likelihood, the marginal likelihood from the Haldane prior will be relatively small. The lower left plot of Figure 12.3 indicates that the marginal likelihood is $p(D) = 2.87 \times 10^{-9}$, which is small compared to the probability of the data from the null hypothesis, $p(D) = \theta^{z}_{\text{null}}(1 - \theta_{\text{null}})^{(N-z)} = 5.96 \times 10^{-8}$. The right column of Figure 12.3 uses a mildly informed prior (discussed below) that puts modestly high probability on the parameter values that are most consistent with the data. Because of this, the marginal likelihood is relatively high, at $p(D) = 2.22 \times 10^{-7}$.

If we consider the posterior distribution instead of the Bayes' factor, we see that the posterior distribution on θ within the alternative model is only slightly affected by the prior. With $z = 7$ and $N = 24$, for the uniform $a_{\text{alt}} = b_{\text{alt}} = 1.00$, the 95% HDI is $[0.1407, 0.4828]$. For the approximate Haldane $a_{\text{alt}} = b_{\text{alt}} = 0.01$, the 95% HDI is $[0.1222, 0.4710]$, as shown in the lower-left plot of Figure 12.3. And for the mildly informed prior $a_{\text{alt}} = 2, b_{\text{alt}} = 4$, the 95% HDI is $[0.1449, 0.4624]$, as shown in the lower-right plot of Figure 12.3. (These HDI limits were accurately determined by function optimization using `HDIofICDF` in `DBDA2E-utilities.R`, not by MCMC.) The lower and upper limits vary by only about 2 percentage points. In all cases, the 95% HDI excludes the null value, although a wide ROPE might overlap the HDI. Thus, the explicit estimation of the bias parameter robustly indicates that the null value should be rejected, but perhaps only marginally. This contrasts with the Bayes' factor, model-comparison approach, which rejected the null or accepted the null depending on the alternative prior.

Of the Bayes' factors in Equation 12.4, which is most appropriate? If your analysis is driven by the urge for a default, uninformed alternative prior, then the prior that best approximates the Haldane is most appropriate. Following from that, we should strongly prefer the null hypothesis to the Haldane alternative. While this is mathematically correct, it is meaningless for an applied setting because the Haldane alternative represents nothing remotely resembling a credible alternative hypothesis. The Haldane prior sets prior probabilities of virtually zero at all values of θ except $\theta = 0$ and $\theta = 1$. There are very few applied settings where such a U-shaped prior represents a genuinely meaningful theory.

I recommended back in Section 10.6.1, p. 294, that the priors of the models should be equivalently informed. In the present application, the null-hypothesis prior is, by definition, fixed. But the alternative prior should be whatever best represents a meaningful and credible hypothesis, not a meaningless default. Suppose, for example, that we have some mild prior information that the coin is tail-biased. We express this as fictional prior data containing 1 head in 4 flips. With a uniform "proto-prior," that implies the alternative prior should be $\text{beta}(\theta|1{+}1, 3{+}1) = \text{beta}(\theta|2, 4)$, as shown in the upper-right plot of Figure 12.3. The Bayes' factor for this meaningful alternative prior is given as the first case in Equation 12.4, where is can be seen that the null is rejected. This agrees with the conclusion from explicit estimation of the parameter value in the posterior distribution. Exercise 12.1 has you generate these examples for yourself. More extensive discussion, and an example from extrasensory perception, can be found in Kruschke (2011a).

12.2.1.1 Bayes' factor can accept null with poor precision

Figure 12.4 shows two examples in which the Bayes' factor favors the null hypothesis. In these cases, the data have a proportion of 50% heads, which is exactly consistent with the null value $\theta = 0.5$. The left column of Figure 12.4 uses a Haldane prior (with $\epsilon = 0.01$), and the data comprise only a single head in two flips. The Bayes' factor is 51.0 in favor of the null hypothesis! But should we really believe therefore that $\theta = 0.5$? No, I don't think so, because the posterior distribution on θ has a 95% HDI from 0.026 to 0.974.

The right column of Figure 12.4 uses a uniform prior. The data show 7 heads in 14 flips. The resulting Bayes' factor is 3.14 in favor of the null. But should we really believe therefore that $\theta = 0.5$? No, I don't think so, because the posterior distribution on θ has a 95% HDI from 0.266 to 0.734.

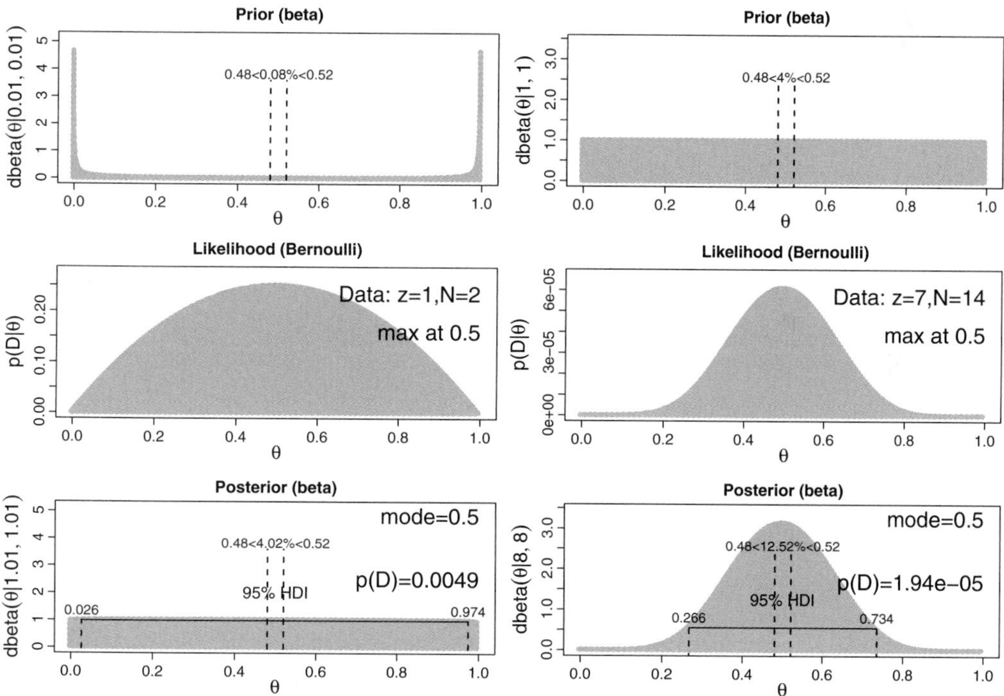

Figure 12.4 Bayes' factor (model comparison) approach can accept the null even with low precision on estimate. Left column: Haldane prior. Bayes' factor is 51.0 in favor of null, but 95% HDI extends from 0.026 to 0.974 (!). Right column: Uniform prior. Bayes' factor is 3.14 in favor of null, but the 95% HDI extends from 0.266 to 0.734 (!).

Thus, the Bayes' factor can favor the null hypothesis when the data are consistent with it, even for small quantities of data. For another example, involving continuous instead of dichotomous data, see Appendix D of Kruschke (2013a). The problem is that there is very little precision in the estimate of θ for small quantities of data. It seems inappropriate, even contradictory, to accept a particular value of θ when there is so much uncertainty about it.

When using the estimation approach instead of the model-comparison approach, accepting the null value demands that the HDI falls entirely inside a ROPE, which typically demands high precision. Narrower ROPEs require higher precision to accept the null value. For example, if we use the narrow ROPE from $\theta = 0.48$ to $\theta = 0.52$ as in Figure 12.4, we would need $N = 2400$ and $z = 1200$ for the 95% HDI to fall inside the ROPE. There is no way around this inconvenient statistical reality: high precision demands a large sample size (and a measurement device with minimal possible noise). But when we are trying to accept a specific value of θ, is seems logically appropriate that we should have a reasonably precise estimate indicating that specific value.

12.2.2. Are different groups equal or not?

In many research applications, data are collected from subjects in different conditions, groups, or categories. We saw an example with baseball batting abilities in which players were grouped by fielding position (e.g., Figure 9.14, p. 256), but there are many other situations. Experiments often have different subjects in different treatment conditions. Observational studies often measure subjects from different classifications, such as gender, location, etc. Researchers often want to ask the question, Are the groups different or not?

As a concrete example, suppose we conduct an experiment about the effect of background music on the ability to remember. As a simple test of memory, each person tries to memorize the same list of 20 words (such as "chair," "shark," "radio," etc.). They see each word for a specific time, and then, after a brief retention interval, recall as many words as they can. For simplicity, we assume that all words are equally memorable, and a person's ability to recall words is modeled as a Bernoulli distribution, with probability θ_{ij} for the ith person in the jth condition. The individual recall propensities θ_{ij} depends on condition-level parameters, ω_j and κ_j, that describe the overarching recall propensity in each condition, because $\theta_{ij} \sim \text{dbeta}(\omega_j(\kappa_j-2)+1\,,\ (1-\omega_j)(\kappa_j-2)+1\,)$.

The only difference between the conditions is the type of music being played during learning and recall. For the four groups, the music comes from, respectively, the death-metal band "Das Kruschke"[3], Mozart, Bach, and Beethoven. For the four conditions, the mean number of words recalled is 8.0, 10.0, 10.2, and 10.4.

[3] To find information regarding the death metal band Das Kruschke, search www.metal-archives.com. Appropriate to its genre, the band was short-lived. The author has no relation to the band, other than, presumably, some unknown common ancestor many generations in the past. The author was, however,

The most straight-forward way to find out whether the different types of music produce different memory abilities is to estimate the condition-level parameters and then examine the posterior differences of the parameter estimates. The histograms in the upper part of Figure 12.5 show the distributions of differences between the ω_j parameters. It can be seen that ω_1 is quite different than ω_3 and ω_4, and possibly ω_2. A difference of zero falls well outside the 95% HDI intervals, even taking into account MCMC instability and a small ROPE. From this we would conclude that Das Kruschke produces poorer memory than the classical composers.

A model-comparison approach addresses the issue a different way. It compares the full model, which has distinct ω_j parameters for the four conditions, against a restricted model, which has a shared ω_0 parameter to describe all the conditions simultaneously. The two models have equal (50/50) prior probabilities. The bottom panel of Figure 12.5 shows the results of a model comparison. The single ω_0 model is preferred over the distinct ω_j model, by about 85% to 15%. In other words, from the model comparison we might conclude that there is *no* difference in memory between the groups.

Which analysis should we believe? Is condition 1 different from some other conditions, as the parameter estimate implies, or are all the conditions the same, as the model comparison seems to imply? Consider carefully what the model comparison actually says: Given the choice between one shared mode and four different group modes, the one-mode model is less improbable. But that does not mean that the one-mode model is the best possible model. In fact, if a different model comparison is conducted, that compares the one-mode model against a different model that has one mode for group 1 and a second mode that is shared for groups 2 through 4, then the comparison favors the two-mode model. Exercise 12.2 has you carry out this alternative comparison.

In principle, we could consider all possible models formed by partitioning the four groups. For four groups, there are 15 distinct partitions. We could, in principle, put a prior probability on each of the 15 models, and then do a comparison of the 15 models (Gopalan & Berry, 1998). From the posterior probabilities of the models, we could ascertain which partition was most credible, and decide whether it is more credible than other nearly-as-credible partitions. (Other approaches have been described by D. A. Berry & Hochberg, 1999; Mueller, Parmigiani, & Rice, 2007; Scott & Berger, 2006). Suppose that we conducted such a large-scale model comparison, and found that the most credible model partitioned groups 2-4 together, separate from group 1. Does this mean that we should truly believe that there is zero difference between groups 2, 3, and 4? Not necessarily. If the group treatments are different, such as the four types of music in the present scenario, then there is almost certainly at least some small difference between their outcomes. (In fact, the simulated data do come from groups

---

in a garage band as a teenager. That band did not think it was playing death metal, although the music may have sounded that way to the critters fleeing the area.

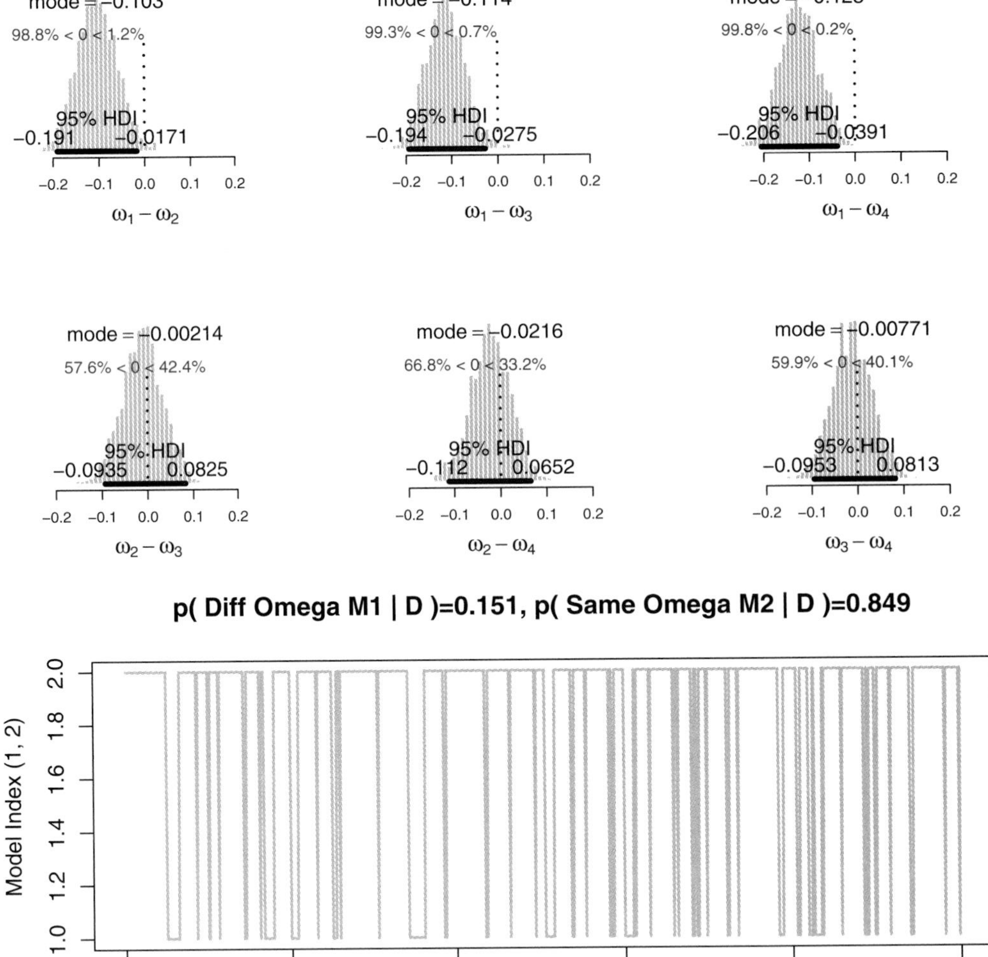

Figure 12.5 Top: Differences of posterior ω_j values for the four groups in the different-omega model. Notice that ω_1 *is credibly different from* ω_3 *and* ω_4, and possibly different from ω_2. The histograms are a bit choppy because the MCMC chain visits the different-omega model relatively rarely. Bottom: Trace plot of the model index shows that the model with a *single* omega parameter ("Same Omega M2") is *preferred* to a model with a separate omega parameter for each group ("Diff Omega M1").

with all different means.) We may still want to estimate the magnitude of those small differences, even if they are small. An explicit posterior estimate will reveal the magnitude and uncertainty of those estimates. Thus, unless we have a viable reason to believe that different group parameters may be literally identical, an estimation of distinct group parameters will tell us what we want to know, without model comparison.

12.2.2.1 Model specification in JAGS

Although it is somewhat tangential to the conceptual points of this section, here is the complete model specification for the results presented in Figure 12.5. These implementation details are relevant for Exercise 12.2 and they provide a review of pseudopriors that were introduced in Section 10.3.2.1.

The data structure has one row per subject, with the number of trials (words) for subject `s` denoted `nTrlOfSubj[s]`, the number correctly recalled for subject `s` denoted `nCorrOfSubj[s]`, and the condition of subject `s` denoted `CondOfSubj[s]`. The model specification begins with saying that each subject has an individual ability `theta[s]` from a condition-specific beta distribution:

```
model {
  for ( s in 1:nSubj ) {
    nCorrOfSubj[s] ~ dbin( theta[s] , nTrlOfSubj[s] )
    theta[s] ~ dbeta( aBeta[CondOfSubj[s]] , bBeta[CondOfSubj[s]] )
  }
```

The shape parameters of the beta distribution are then re-written in terms of the mode and concentration. Model 1 uses condition-specific `omega[j]`, while Model 2 uses the same `omega0` for all conditions. The JAGS function `equals(mdlIdx,...)` is used to select the appropriate model for index `mdlIdx`:

```
for ( j in 1:nCond ) {
  # Use omega[j] for model index 1, omega0 for model index 2:
  aBeta[j] <-       ( equals(mdlIdx,1)*omega[j]
                    + equals(mdlIdx,2)*omega0   )    * (kappa[j]-2)+1
  bBeta[j] <- ( 1 - ( equals(mdlIdx,1)*omega[j]
                    + equals(mdlIdx,2)*omega0   ) ) * (kappa[j]-2)+1
  omega[j] ~ dbeta( a[j,mdlIdx] , b[j,mdlIdx] )
}
omega0 ~ dbeta( a0[mdlIdx] , b0[mdlIdx] )
```

The priors on the concentration parameters are then specified:

```
for ( j in 1:nCond ) {
  kappa[j] <- kappaMinusTwo[j] + 2
  kappaMinusTwo[j] ~ dgamma( 2.618 , 0.0809 ) # mode 20 , sd 20
}
```

Notice that the groups have distinct concentration parameters under either model, even the single-mode model. This is merely a simplification for purposes of presentation. The concentration parameters could be structured differently across groups, analogous to the mode parameters.

Finally, the true and pseudoprior constants are set for the condition-level mode and concentration priors. (Pseudopriors were discussed in Section 10.3.2.1, p. 279.) The pseudoprior constants were selected to mimic the posterior reasonably well:

```
  # Constants for prior and pseudoprior:
  aP <- 1
  bP <- 1
  # a0[model] and b0[model]
  a0[1] <- 0.48*500        # pseudo
  b0[1] <- (1-0.48)*500    # pseudo
  a0[2] <- aP              # true
  b0[2] <- bP              # true
  # a[condition,model] and b[condition,model]
  a[1,1] <- aP             # true
  a[2,1] <- aP             # true
  a[3,1] <- aP             # true
  a[4,1] <- aP             # true
  b[1,1] <- bP             # true
  b[2,1] <- bP             # true
  b[3,1] <- bP             # true
  b[4,1] <- bP             # true
  a[1,2] <- 0.40*125        # pseudo
  a[2,2] <- 0.50*125        # pseudo
  a[3,2] <- 0.51*125        # pseudo
  a[4,2] <- 0.52*125        # pseudo
  b[1,2] <- (1-0.40)*125    # pseudo
  b[2,2] <- (1-0.50)*125    # pseudo
  b[3,2] <- (1-0.51)*125    # pseudo
  b[4,2] <- (1-0.52)*125    # pseudo
  # Prior on model index:
  mdlIdx ~ dcat( modelProb[] )
  modelProb[1] <- 0.5
  modelProb[2] <- 0.5
}
```

The pseudoprior constants used above are not uniquely correct. Other values might reduce autocorrelation even better.

12.3. RELATIONS OF PARAMETER ESTIMATION AND MODEL COMPARISON

We have now seen several examples of Bayesian approaches to assessing a null value, using a model-comparison approach or a parameter-estimation approach. In the model-comparison approach, a decision is made by putting a threshold on the Bayes' factor. In the parameter-estimation approach, a decision is made by putting a threshold on the parameter (involving the HDI and ROPE). In other words, both approaches involve decision rules applied to some aspect of a Bayesian posterior distribution.

One key relationship between the two approaches was emphasized in the introduction to model comparison, back in the hierarchical diagram of Figure 10.1, p. 267. Recall that model comparison is really a single hierarchical model in which the submodels fall under a top-level indexical parameter. The model-comparison approach to null assessment focuses on the top-level model-index parameter. The parameter-estimation approach to null assessment focuses on the parameter distribution within the alternative model that has a meaningful prior. Thus, both approaches are logically coherent and can be applied simultaneously. Their conclusions about the null value do not have to agree, however, because they are assessing the null value in different ways, posing the question at different levels of the model. Neither level is the uniquely "correct" level, but one or the other level can be more or less meaningful in different applications.

A second relationship between the two approaches is that the Bayes' factor in the model comparison approach can be discerned in the estimation approach by noting how much the credibility of the null value increases or decreases in the parameter estimation. Consider, for example, the left side of Figure 12.3, p. 345, which shows parameter estimation using a Haldane prior. The top-left plot shows a fairly small ROPE around the null value, indicating that 0.08% of the prior distribution falls within the ROPE. The bottom-left plot shows that the posterior distribution has 1.67% of the distribution within the ROPE. *The ratio of proportions is (approximately) the Bayes' factor in favor of the null value.* The intuition for this fact is straightforward: The prior puts a certain probability on the null value, and Bayes' rule re-allocates probability, producing greater or lesser probability on the null value. If the null value gets more probability than it had in the prior, then the null hypothesis is favored, but if the null value gets less than it had in the prior, then the alternative is favored. In the present example case, the ratio of posterior probability in the ROPE to prior probability in the ROPE is $1.67/0.08 = 20.9$. Compare that with the analytically computed Bayes' factor from Equation 12.4: $1/0.0481 = 20.8$.

The right side of Figure 12.3 shows another example. The ratio of posterior probability in the ROPE to prior probability in the ROPE is $1.38/5.00 = 0.28$. Compare that with the analytically computed Bayes' factor from Equation 12.4: $1/3.7227 = 0.27$. Figure 12.4, p. 347, provides more examples. In its left side, the ratio of posterior to prior probabilities in the ROPE is $4.02/0.08 \approx 50$, which is nearly the same as the analytically computed Bayes' factor of 51. In the right side of Figure 12.4, the ratio of posterior to prior probabilities in the ROPE is $12.52/4.00 = 3.13$, which is nearly the same as the analytically computed Bayes' factor of 3.14.

Visualizing the Bayes' factor as the ratio of posterior to prior probabilities within a narrow ROPE helps us intuit the apparent contradictions between the conclusions of the model comparison and the parameter estimation. In the left side of Figure 12.3, p. 345, the proportion of the distribution inside the ROPE increases by a large ratio, even though most of the posterior distribution falls outside the ROPE. Similarly in the right side of Figure 12.4, p. 347, which again shows a large increase in the proportion of the distribution inside the ROPE, but with most of the distribution outside the ROPE. Thus, the model comparison focuses on the null value and whether its local probability increases from prior to posterior. The parameter estimation considers the entire posterior distribution, including the uncertainty (i.e., HDI) of the parameter estimate relative to the ROPE.

The derivation of the Bayes' factor by considering the null value in parameter estimation is known as the Savage-Dickey method. A lucid explanation is provided by Wagenmakers, Lodewyckx, Kuriyal, and Grasman (2010), who also provide some historical references and applications to MCMC analysis of hierarchical models.

12.4. ESTIMATION OR MODEL COMPARISON?

As mentioned above, neither method for null value assessment (parameter estimation or model comparison) is uniquely "correct." The two approaches merely pose the question of the null value in different ways. In typical situations, I find the estimation approach to be more transparently meaningful and informative because parameter estimation provides an explicit posterior distribution on the parameter, while the Bayes' factor by itself does not provide that information. As I have emphasized, the two methods can be applied together in an integrated hierarchical model, but when their conclusions agree, the parameter estimation provides the information I typically want, and when their conclusions disagree, the parameter estimation still usually provides the more meaningful information.

The model-comparison approach to null-value assessment has two requirements for it to be meaningful. First, it must be theoretically meaningful for the parameter value to be exactly the null value. Second, the alternative-hypothesis prior must be meaningfully informed. These two requirements simply demand that both priors should be genuinely meaningful and viable, and they are merely an instance of the general requirement made back in Section 10.6.1 (p. 294) that the priors of both models should be meaningful and equally informed. Regarding the null-hypothesis prior, if it is not really plausible for two different treatment groups to be exactly the same, is it meaningful to give credibility to a null model that describes them as exactly the same? Maybe only as an approximate description, but perhaps not literally. Regarding the alternative-hypothesis prior, if it does not represent a plausible theory, is it meaningful to say it is more or less credible than any other prior? Much of the effort in pursuing the model-comparison approach to null-hypothesis testing goes into justifying an "automatic" prior for the alternative model that has desirable mathematical properties (e.g., in the psychological methods literature, Dienes, 2008, 2011; Edwards, Lindman, & Savage, 1963; Gallistel, 2009; Rouder, Speckman, Sun, Morey, & Iverson, 2009; Wagenmakers, 2007). But, in my opinion, a default prior is only useful to the extent that it happens to express a meaningful informed theory.

The estimation approach to null-value assessment uses a decision rule that involves the HDI and ROPE. Unfortunately, the ROPE has no automatic default limits, and the decision maker must make a case for a reasonable ROPE. In some applications, the ROPE can be specified with respect to the magnitude of conventionally "small" effect sizes for the domain of research (Cohen, 1988). We have not yet had the need to discuss a technical definition of effect size, but the topic will arise in Chapter 16. But just as the labeling of effect sizes as "small" depends on conventional practice, the setting of a ROPE must be made in the context of current theory, measurement abilities, and the practical purpose of the decision. Recall that Serlin and Lapsley (1985, 1993) argued that research should affirm quantitative predictions, not merely reject null values, and affirmation of a theory is always relative to currently competing theories and the current state of measurement noise. Thus, the ROPE should be argued reasonably with the understanding that it may subsequently change.

12.5. EXERCISES

Look for more exercises at https://sites.google.com/site/doingbayesiandataanalysis/

Exercise 12.1. [Purpose: To make sure you understand the Bayes' factors regarding a single coin in Figure 12.3 and Equation 12.4, including the Savage-Dickey method.] Find the file BernBeta.R in the programs that accompany this book. Open RStudio with the folder of that file as R's working directory. Source the file so that R knows about the function `BernBeta`:

```
source("BernBeta.R")
```

Now, suppose we have a coin that is flipped 24 times and shows 7 heads. Enter these data into R, like this:

```
z=7 ; N=24
```

(A) According to the spike null hypothesis, for which the only credible value of θ is 0.5, what is the probability of the data? *Hint:* It is $\theta^z(1-\theta)^{(N-z)}$. Compute the value.

(B) Verify the result of the previous part by approximating a spike prior with a narrow beta distribution. Use the `BernBeta` function with a beta$(\theta|2000, 2000)$ prior, like this:

```
a=2000 ; b=2000
openGraph(width=5,height=7)
BernBeta( c(a,b) , c(rep(0,N-z),rep(1,z)) ,  ROPE=c(0.48,0.52) ,
          plotType="Bars" , showCentTend="Mode" , showHDI=TRUE , showpD=TRUE )
```

Include the resulting graph in your report. What is the value of $p(D)$ for this prior? Is it very close to the value computed for the exact spike prior in the previous part of this exercise? (It should be.) Explain why they are not exactly equal.

(C) Show the result when using a nearly Haldane prior, like this:

```
a=0.01 ; b=0.01
openGraph(width=5,height=7)
BernBeta( c(a,b) , c(rep(0,N-z),rep(1,z)) ,  ROPE=c(0.48,0.52) ,
          plotType="Bars" , showCentTend="Mode" , showHDI=TRUE ,showpD=TRUE )
```

Include the resulting graph in your report. What is the value of $p(D)$ for this prior? Compute and report the Bayes' factor of this prior relative to the spike (null) prior, using the formula $p(D|\text{Haldane})/p(D|\text{null})$.

(D) Continuing with the Haldane prior from the previous part, compute the approximate Bayes' factor using the Savage-Dickey method. That is, compute and report the ratio of percentage of prior within the ROPE over percentage of posterior with the ROPE.

(E) Suppose we have previous knowledge that in this application there tend to be more tails than heads. Show the result when using a mildly informed prior, like this:

```
a=2 ; b=4
openGraph(width=5,height=7)
BernBeta( c(a,b) , c(rep(0,N-z),rep(1,z)) ,  ROPE=c(0.48,0.52) ,
          plotType="Bars" , showCentTend="Mode" , showHDI=TRUE , showpD=TRUE )
```

Include the resulting graph in your report. What is the value of $p(D)$ for this prior? Compute and report the Bayes' factor of this prior relative to the spike (null) prior, using the formula $p(D|\text{informed})/p(D|\text{null})$.

(F) Continuing with the mildly informed prior from the previous part, compute the approximate Bayes' factor using the Savage-Dickey method. That is, compute and report the ratio of percentage of prior within the ROPE over percentage of posterior with the ROPE.

(G) Report the 95% HDIs when starting with the Haldane prior and the mildly informed prior. Are the HDIs very different? Were the Bayes' factors very different?

(H) Which approach, model comparison or estimation, seems most informative? Why? Within the model-comparison approach, which prior, uninformed Haldane or mildly informed, seems most meaningful? Why?

Exercise 12.2. [Purpose: Model comparison for different partitions of group modes, using the script of Section 12.2.2.1.] Open the script `OneOddGroupModelComp2E.R`, making sure that R's working directory includes the various utility programs used with this book.

(A) For this part of the exercise, the goal is to reproduce the findings presented in Section 12.2.2.1. First, be sure that the prior probabilities on the models are set to 50/50:

```
modelProb[1] <- 0.5
modelProb[2] <- 0.5
```

Run the script and report the results, including the graphs for the model index and the modes and differences of modes. State what the two models are, and state which model is preferred and by how much. (*Hint:* Model 2 is the single-mode model, and it is preferred.)

(B) Continuing with the previous part, consider the graphs of differences of modes. What do they imply about differences between groups? Does this conclusion agree or disagree with the conclusion from the model comparison? How do you reconcile the conclusions? (*Hint:* The model index and the groups modes are all parameters being simultaneously estimated, so there is no contradiction. The parameters answer different questions; which questions?)

(C) For this part of the exercise, the goal is to compare the single-mode model against a different partitioning of the group modes. Instead of letting each group have its own distinct mode, we will allow a distinct mode for the first group, but restrict groups 2 through 4 to use a single mode. One way to accomplish this is to change this part of the model specification:

```
for ( j in 1:nCond ) {
  # Use omega[j] for model index 1, omega0 for model index 2:
  aBeta[j] <-       ( equals(mdlIdx,1)*omega[j]
                    + equals(mdlIdx,2)*omega0   )    * (kappa[j]-2)+1
  bBeta[j] <- ( 1 - ( equals(mdlIdx,1)*omega[j]
                    + equals(mdlIdx,2)*omega0   ) ) * (kappa[j]-2)+1
  omega[j] ~ dbeta( a[j,mdlIdx] , b[j,mdlIdx] )
}
```

to this:

```
for ( j in 1:nCond ) {
  # Use omega[j] for model index 1, omega0 for model index 2:
  aBeta[j] <-       ( equals(mdlIdx,1)*omega[j]
                    + equals(mdlIdx,2)*omega0   )    * (kappa[j]-2)+1
  bBeta[j] <- ( 1 - ( equals(mdlIdx,1)*omega[j]
                    + equals(mdlIdx,2)*omega0   ) ) * (kappa[j]-2)+1
}
for ( j in 1:2 ) {
  omega[j] ~ dbeta( a[j,mdlIdx] , b[j,mdlIdx] )
}
omega[3] <- omega[2]
omega[4] <- omega[2]
```

In your report, *carefully explain what the change does*. Make the change, and run the script (with the prior probabilities on the models set to 50/50). Report the results, including the graphs for the model index and the modes and differences of modes. State what the two models are, and state which model is preferred and by how much. (*Hint:* Model 2 is the single-mode model, and it is *not* preferred.)

(D) Continuing with the previous part, consider the graphs of differences of modes. What do they imply about differences between groups? Does this conclusion agree or disagree with the conclusion from the model comparison? Even though Model 1 is preferred, is it really meaningful?

(E) Considering the results of the previous parts, what seems to be the most meaningful approach to analyzing differences of groups? (I'm hoping you'll say parameter estimation, not model comparison. But you might give arguments to the contrary.) Can you think of other applications in which model comparison might be more useful?

CHAPTER 13

Goals, Power, and Sample Size

Contents

Just how many times must I show her I care,
Until she believes that I'll always be there?
Well, while she denies that my value's enough,
I'll have to rely on the power of love.[1]

Researchers collect data in order to achieve a goal. Sometimes the goal is to show that a suspected underlying state of the world is credible; other times the goal is to achieve a minimal degree of precision on whatever trends are observed. Whatever the goal, it can only be probabilistically achieved, as opposed to definitely achieved, because data are replete with random noise that can obscure the underlying state of the world. Statistical *power* is the probability of achieving the goal of a planned empirical study, if a suspected underlying state of the world is true. Scientists don't want to waste time and resources pursuing goals that have a small probability of being achieved. In other words, researchers desire power.

[1] The power of "luff"? Sailors know that there's not much power in luffing.

Doing Bayesian Data Analysis, 2nd Edition
http://dx.doi.org/10.1016/B978-0-12-405888-0.00013-1

13.1. THE WILL TO POWER[2]

In this section, a framework for research and data analysis will be described which leads to a more precise definition of power and how to compute it.

13.1.1. Goals and obstacles

There are many possible goals for an experimental or observational study. For example, we might want to show that the rate of recovery for patients who take a drug is higher than the rate of recovery for patients who take a placebo. This goal involves showing that a null value (zero difference) is not tenable. We might want to confirm a specific effect size predicted by a quantitative theory, such as the curvature of light around a massive object predicted by general relativity. This goal involves showing that a specific value *is* tenable. We might want merely to measure accurately whatever effect is present, for example when measuring voter preferences in a political poll. This goal involves establishing a minimal degree of precision.

Any goal of research can be formally expressed in various ways. In this chapter I will focus on the following goals formalized in terms of the highest density interval (HDI):

- *Goal:* Reject a null value of a parameter.
  - *Formal expression:* Show that a region of practical equivalence (ROPE) around the null value excludes the posterior 95% HDI.
- *Goal:* Affirm a predicted value of a parameter.
  - *Formal expression:* Show that a ROPE around the predicted value includes the posterior 95% HDI.
- *Goal:* Achieve precision in the estimate of a parameter.
  - *Formal expression:* Show that the posterior 95% HDI has width less than a specified maximum.

There are other mathematical formalizations of the various goals, and they will be mentioned later. This chapter focuses on the HDI because of its natural interpretation for purposes of parameter estimation and measurement of precision.

If we knew the benefits of achieving our goal, and the costs of pursuing it, and if we knew the penalties for making a mistake while interpreting the data, then we could express the results of the research in terms of the long-run expected payoff. When we know the costs and benefits, we can conduct a full decision-theoretic treatment of the situation, and plan the research and data interpretation accordingly (e.g., Chaloner & Verdinelli, 1995; Lindley, 1997). In our applications we do not have access to those costs and benefits, unfortunately. Therefore we rely on goals such as those outlined above.

[2] Regarding the title of this section: Other than the fact that researchers desire statistical power, the notion of statistical power might have profound connections with concepts from Friedrich Nietzsche's work, *The Will to Power*. See Exercise 13.1.

The crucial obstacle to the goals of research is that a random sample is only a probabilistic representation of the population from which it came. Even if a coin is actually fair, a random sample of flips will rarely show exactly 50% heads. And even if a coin is not fair, it might come up heads 5 times in 10 flips. Drugs that actually work no better than a placebo might happen to cure more patients in a particular random sample. And drugs that truly are effective might happen to show little difference from a placebo in another particular random sample of patients. Thus, a random sample is a fickle indicator of the true state of the underlying world. Whether the goal is showing that a suspected value is or isn't credible, or achieving a desired degree of precision, random variation is the researcher's bane. Noise is the nemesis.

13.1.2. Power

Because of random noise, the goal of a study can be achieved only probabilistically. The probability of achieving the goal, given the hypothetical state of the world and the sampling plan, is called the *power* of the planned research. In traditional null hypothesis significance testing (NHST), power has only one goal (rejecting the null hypothesis), and there is one conventional sampling plan (stop at predetermined sample size) and the hypothesis is only a single specific value of the parameter. In traditional statistics, that is *the* definition of power. That definition is generalized in this book to include other goals, other sampling plans, and hypotheses that involve an entire distribution on parameters.

Scientists go to great lengths to try to increase the power of their experiments or observational studies. There are three primary methods by which researchers can increase the chances of detecting an effect. First, we reduce measurement noise as much as possible. For example, if we are trying to determine the cure rate of a drug, we try to reduce other random influences on the patients, such as other drugs they might be stopping or starting, changes in diet or rest, etc. Reduction of noise and control of other influences is the primary reason for conducting experiments in the lab instead of in the maelstrom of the real world. The second method, by which we can increase the chance of detecting an effect, is to amplify the underlying magnitude of the effect if we possibly can. For example, if we are trying to show that a drug helps cure a disease, we will want to administer as large a dose as possible (assuming there are no negative side effects). In non-experimental research, in which the researcher does not have the luxury of manipulating the objects being studied, this second method is unfortunately unavailable. Sociologists, economists, and astronomers, for example, are often restricted to observing events that the researchers cannot control or manipulate.

Once we have done everything we can to reduce noise in our measurements, and to amplify the effect we are trying to measure, the third way to increase power is to increase the sample size. The intuition behind this method is simple: With more and more measurements, random noises will tend to cancel themselves out, leaving on average a clear signature of the underlying effect. In general, as sample size increases, power increases. Increasing the sample size is an option in most experimental research, and in a lot of observational research (e.g., more survey respondents can be polled), but not in some domains where the population is finite, such as comparative studies of the states or provinces of a nation. In this latter situation, we cannot create a larger sample size, but Bayesian inference is still valid, and perhaps uniquely so (Western & Jackman, 1994).

In this chapter we precisely calculate power. We compute the probability of achieving a specific goal, given (i) a hypothetical distribution of effects in the population being measured, and (ii) a specified data-sampling plan, such as collecting a fixed number of observations. Power calculations are very useful for planning an experiment. To anticipate likely results, we conduct "dress rehearsals" before the actual performance. We repeatedly simulate data that we suspect we might get, and conduct Bayesian analyses on the simulated data sets. If the goal is achieved for most of the simulated data sets, then the planned experiment has high power. If the goal is rarely achieved in the analyses of simulated data, then the planned experiment is likely to fail, and we must do something to increase its power.

The upper part of Figure 13.1 illustrates the flow of information in an actual Bayesian analysis. The real world provides a single sample of actually observed data (illustrated as concrete blocks). We use Bayes' rule, starting with a prior suitable for a skeptical audience, to derive an actual posterior distribution. The illustration serves as a reference for the lower part of Figure 13.1, which shows the flow of information in a power analysis. Starting at the leftmost cloud:

1. *From the hypothetical distribution of parameter values, randomly generate representative values.* In many cases, the hypothesis is a posterior distribution derived from previous research or idealized data. The specific parameter values serve as one representative description.
2. *From the representative parameter values, generate a random sample of data, using the planned sampling method.* The sample should be generated according to how actual data will be gathered in the eventual real experiment. For example, typically it is assumed that the number of data points is fixed at N. It might instead be assumed that data will be collected for a fixed interval of time T, during which data points appear randomly at a known mean rate n/T. Or there might be some other sampling scheme.
3. *From the simulated sample of data, compute the posterior estimate, using Bayesian analysis with audience-appropriate priors.* This analysis should be the same as used for the actual data. The analysis must be convincing to the anticipated audience of the research, which presumably includes skeptical scientists.

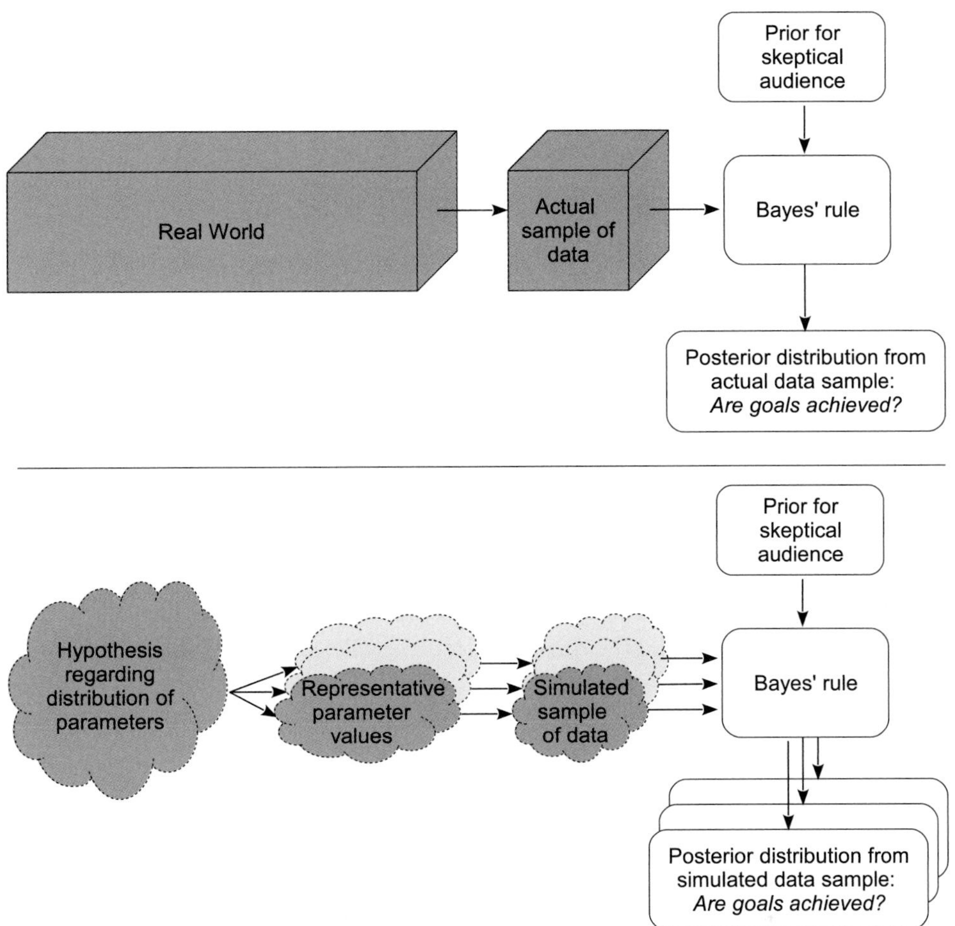

Figure 13.1 Upper diagram illustrates the flow of information in an actual Bayesian analysis, in which the data come from the real world. Lower diagram illustrates flow of information in a power analysis, in which simulated data come from random hypothetical parameters.

4. *From the posterior estimate, tally whether or not the goals were attained.* The goals could be any of those outlined previously, such as having a ROPE exclude or include the 95% HDI, or having the 95% HDI be narrower than a desired width, applied to a variety of parameters.
5. *Repeat the above steps many times, to approximate the power.* The repetition is indicated in Figure 13.1 by the layers of anticipated data samples and anticipated posteriors. Power is, by definition, the long-run proportion of times that the goal is attained. As we have only a finite number of simulations, we use Bayesian inference to establish a posterior distribution on the power.

Details of these steps will be provided in examples. Notice that if the data-sampling procedure uses a fixed sample of size N, then the process determines power as a function of N. If the data-sampling procedure uses a fixed sampling duration T, then the process determines power as a function of T.

13.1.3. Sample size

Power increases as sample size increases (usually). Because gathering data is costly, we would like to know the minimal sample size, or minimal sampling duration, that is required to achieve a desired power.

The goal of precision in estimation can always be attained with a large enough sample size. This is because the likelihood of the data, thought of graphically as a function of the parameter, tends to get narrower and narrower as the sample size increases. This narrowing of the likelihood is also what causes the data eventually to overwhelm the prior distribution. As we collect more and more data, the likelihood function gets narrower and narrower on average, and therefore the posterior gets narrower and narrower. Thus, with a large enough sample, we can make the posterior distribution as precise as we like.

The goal of showing that a parameter value is different from a null value might not be attainable with a high enough probability, however, no matter how big the sample size. Whether or not this goal is attainable with high probability depends on the hypothetical data-generating distribution. *The best that a large sample can do is exactly reflect the data-generating distribution.* If the data-generating distribution has considerable mass straddling the null value, then the best we can do is get estimates that include and straddle the null value. As a simple example, suppose that we think that a coin may be biased, and the data-generating hypothesis entertains four possible values of θ, with $p(\theta = 0.5) = 25\%$, $p(\theta = 0.6) = 25\%$, $p(\theta = 0.7) = 25\%$, and $p(\theta = 0.8) = 25\%$. Because 25% of the simulated data come from a fair coin, the maximum probability of excluding $\theta = 0.5$, even with a huge sample, is 75%.

Therefore, when planning the sample size for an experiment, it is crucial to decide what a realistic goal is. If there are good reasons to posit a highly certain data-generating hypothesis, perhaps because of extensive previous results, then a viable goal may be to exclude a null value. On the other hand, if the data-generating hypothesis is somewhat vague, then a more reasonable goal is to attain a desired degree of precision in the posterior. As will be shown in Section 13.3, the goal of precision is also less biased in sequential testing. In a frequentist setting, the goal of precision has been called *accuracy in parameter estimation* (AIPE; e.g., Maxwell, Kelley, & Rausch, 2008). Kelley (2013, p. 214) states:

> *The goal of the AIPE approach to sample size planning is [that] the confidence interval for the parameter of interest will be sufficiently narrow, where "sufficiently narrow" is necessarily context-specific. Sample size planning with the goal of obtaining a narrow confidence interval dates back to at least Guenther (1965) and Mace (1964), yet the AIPE approach to sample size planning has taken on a more important role in the research design literature recently. This is the case due to the increased emphasis on effect sizes, their confidence intervals, and the undesirable situation of 'embarrassingly wide' [(Cohen, 1994)] confidence intervals.*

The Bayesian approach to the goal of precision, as described in this chapter, is analogous to frequentist AIPE, but Bayesian HDIs do not suffer from the instability of frequentist confidence intervals when testing or sampling intentions change (as was discussed in Section 11.3.1, p. 318).

13.1.4. Other expressions of goals

There are other ways to express mathematically the goal of precision in estimation. For example, another way of using HDIs was described by Joseph, Wolfson, and du Berger (1995a, 1995b). They considered an "average length criterion," which requires that the *average* HDI width, across repeated simulated data, does not exceed some maximal value L. There is no explicit mention of power, i.e., the probability of achieving the goal, because the sample size is chosen so that the goal is definitely achieved. The goal itself is probabilistic, however, because it regards an average: While some data sets will have HDI width less than L, many other data sets will not have an HDI width greater than L. Another goal considered by Joseph et al. (1995a) was the "average coverage criterion." This goal starts with a specified width for the HDI, and requires its mass to exceed 95% (say) on average across simulated data. The sample size is chosen to be large enough to achieve that goal. Again, power is not explicitly mentioned, but the goal is probabilistic: Some data sets will have an L-width HDI mass greater than 95%, and other data sets will not have an L-width HDI mass less than 95%. Other goals regarding precision are reviewed by Adcock (1997) and by De Santis (2004, 2007). The methods emphasized in this chapter focus on limiting the worst precision, instead of the average precision.

A rather different mathematical expression of precision is the *entropy* of a distribution. Entropy describes how spread out a distribution is, such that smaller entropy connotes a narrower distribution. A distribution, that consists of an infinitely dense spike with infinitesimally narrow width, has zero entropy. At the opposite extreme, a uniform distribution has maximal entropy. The goal of high precision in the posterior distribution might be re-expressed as a goal of small entropy in the posterior distribution. For an overview of this approach, see Chaloner and Verdinelli (1995). For an introduction to how minimization of expected entropy might be used spontaneously by people as they experiment with the world, see Kruschke (2008). Entropy may be a better measure of posterior precision than HDI width especially in cases of multimodal distributions, for which HDI width is more challenging to determine. I will not further explicate the use of entropy because I think that HDI width is a more intuitive quantity than entropy, at least for most researchers in most contexts.

There are also other ways to express mathematically the goal of excluding a null value. In particular, the goal could be expressed as wanting a sufficiently large Bayes' factor (BF) in a model comparison between the spike-null prior and the automatic alternative prior (e.g., Wang & Gelfand, 2002; Weiss, 1997). I will not further address this approach, however, because the goal of a criterial BF for untenable caricatured priors has problems as discussed in the previous chapter. However, the procedure diagrammed in Figure 13.1 is directly applicable to BFs for those who wish to pursue the idea. In the remainder of this chapter, it will be assumed that the goal of the research is estimation of the parameter values, starting with a viable prior. The resulting posterior distribution is then used to assess whether the goal was achieved.

13.2. COMPUTING POWER AND SAMPLE SIZE

As our first worked-out example, consider the simplest case: Data from a single coin. Perhaps we are polling a population and we want to precisely estimate the preferences for candidates A or B. Perhaps we want to know if a drug has more than a 50% cure rate. We will go through the steps listed in Section 13.1.2 to compute the exact sample size needed to achieve various degrees of power for different data-generating hypotheses.

13.2.1. When the goal is to exclude a null value

Suppose that our goal is to show that the coin is unfair. In other words, we want to show that the 95% HDI excludes a ROPE around $\theta = 0.50$.

We must establish the hypothetical distribution of parameter values from which we will generate simulated data. Often, the most intuitive way to create a parameter distribution is as the posterior distribution inferred from prior data, which could be actual or idealized. Usually it is more intuitively accessible to get prior data, or to think of idealized prior data, than to directly specify a distribution over parameter values. For example, based on knowledge about the application domain, we might have 2000 actual or idealized flips of the coin for which the result showed 65% heads. Therefore we'll describe the data-generating hypothesis as a beta distribution with a mode of 0.65 and concentration based on 2000 flips after a uniform "proto-prior": $\text{beta}(\theta \mid 0.65 \cdot (2000-2)+1,\ (1-0.65) \cdot (2000-2)+1)$. This approach to creating a data-generating parameter distribution by considering previous data is sometimes called the *equivalent prior sample* method (Winkler, 1967).

Next, we randomly draw a representative parameter value from the hypothetical distribution, and with that parameter value generate simulated data. We do that repeatedly, and tally how often the HDI excludes a ROPE around $\theta = 0.50$. One iteration of the process goes like this: First, select a value for the "true" bias in the coin, from the hypothetical distribution that is centered on $\theta = 0.65$. Suppose that the selected value is 0.638. Second, simulate flipping a coin with that bias N times. The simulated data have z heads and $N - z$ tails. The proportion of heads, z/N, will tend to be around 0.638, but will be higher or lower because of randomness in the flips.

Third, using the audience-appropriate prior for purposes of data analysis, determine the posterior beliefs regarding θ if z heads in N flips were observed. Tally whether or not the 95% HDI excludes a ROPE around the null value of $\theta = 0.50$. Notice that even though the data were generated by a coin with bias of 0.638, the data might, by chance, show a proportion of heads near 0.5, and therefore the 95% HDI might not exclude a ROPE around $\theta = 0.50$. This process is repeated many times to estimate the power of the experiment.

Table 13.1 shows the minimal sample size needed for the 95% HDI to exclude $\theta = 0.5$ when flipping a single coin. As an example of how to read the table, suppose you have a data-generating hypothesis that the coin has a bias very near $\theta = 0.65$. This hypothesis is implemented, for purposes of Table 13.1, as a beta distribution with shape parameters of $0.65 \cdot (2000 - 2) + 1$ and $(1 - 0.65) \cdot (2000 - 2) + 1$. The value of 2000 is arbitrary; as described in the previous paragraph, it's as if the generating mean of 0.65 was based on fictitious previous data containing 2000 flips. The table indicates that if we desire a 90% probability of obtaining a 95% HDI that excludes a ROPE from $\theta = 0.48$ to $\theta = 0.52$, we need a sample size of $N = 150$, i.e., we need to flip the coin at least 150 times in order to have a 90% chance that the 95% HDI falls outside the ROPE.

Notice, in Table 13.1, that as the generating mode increases, the required sample size decreases. This makes sense intuitively: When the generating mode is large, the sample proportion of heads will tend to be large, and so the HDI will tend to fall toward the high end of the parameter domain. In other words, when the generating mode is large, it doesn't take a lot of data for the HDI to fall consistently above $\theta = 0.5$. On the other hand, when the generating mode is only slightly above $\theta = 0.5$, then it takes a large sample for the sample proportion of heads to be consistently above 0.5, and for the HDI to be consistently entirely above 0.5 (and the ROPE).

Notice also, in Table 13.1, that as the desired power increases, the required sample size increases quite dramatically. For example, if the data-generating mode is 0.6, then as the desired power rises from 0.7 to 0.9, the minimal sample size rises from 238 to 430.

Table 13.1 Minimal sample size required for 95% HDI to exclude a ROPE from 0.48 to 0.52, when flipping a single coin.

	Generating Mode ω					
Power	**0.60**	**0.65**	**0.70**	**0.75**	**0.80**	**0.85**
0.7	238	83	40	25	16	7
0.8	309	109	52	30	19	14
0.9	430	150	74	43	27	16

Note. The data-generating distribution is a beta density with mode ω, as indicated by the column header, and concentration $\kappa = 2000$. The audience prior is a uniform distribution.

13.2.2. Formal solution and implementation in R

For this simple situation, the exact power can be computed analytically, without need for Monte Carlo simulation. In this section we derive the relevant formulas and compute the power using a program in R (without using MCMC). The program was used to generate Tables 13.1 and 13.2.

The key idea for the analytical derivation is that, for this application, there are only a finite number of possible data sets, namely $z \in \{0, \ldots, N\}$, and each data set completely determines the posterior distribution (because the audience prior is fixed). Therefore all we have to do is figure out the probability of each possible outcome, and sum up the probabilities of the outcomes that achieve the desired goal.

The hypothetical data are generated by first sampling a θ value according to the data-generating prior, which is a beta distribution that we will denote as beta$(\theta|a, b)$, where the shape constants could be determined by converting from a specified mode and concentration. This sampling from the distribution on θ is illustrated in Figure 13.1 by the arrows from the hypothesis to the representative parameter value. Next, we generate N flips of the coin according to the binomial distribution. This is illustrated in Figure 13.1 by the arrows from representative parameter value to simulated sample of data. We need to integrate this procedure across the entire hypothetical distribution to determine the probability of getting z heads. Thus, the probability of getting z heads in the simulated sample of N flips is

$$
\begin{aligned}
p(z|N) &= \int_0^1 \mathrm{d}\theta \; p(z|N,\theta)\, p(\theta) \\
&= \int_0^1 \mathrm{d}\theta \; \text{binomial}(z|N,\theta)\;\, \text{beta}(\theta|a,b) \\
&= \int_0^1 \mathrm{d}\theta \binom{N}{z} \theta^{z}(1-\theta)^{(N-z)}\, \theta^{(a-1)}(1-\theta)^{(b-1)}/B(a,b) \\
&= \binom{N}{z} \int_0^1 \mathrm{d}\theta \; \theta^{(z+a-1)}(1-\theta)^{(N-z+b-1)}\;/B(a,b) \\
&= \binom{N}{z} B(z+a, N-z+b)/B(a,b)
\end{aligned}
\tag{13.1}
$$

The transition to the final line, above, was made by the definition of the beta function, explained back in Equation 6.4, p. 127. The probability of possible data is sometimes called the "preposterior marginal distribution of z" (cf. Equation 5 of Pham-Gia & Turkkan, 1992). For each possible outcome, z, we update the audience-agreeable prior to render a posterior distribution, and then we assess whether the goal has been achieved for that outcome. Because the decision is determined by the outcome z, the probability of the decisions is determined by the probability of the outcomes.

Equation 13.1 is implemented in the R function `minNforHDIpower`, in the set of programs accompanying this book, but in logarithmic form to prevent underflow errors. The function has several arguments, including the generating distribution mode and concentration, called `genPriorMode` and `genPriorN`. The audience prior has mode and concentration specified by `audPriorMode` and `audPriorN`. The function allows specification of a maximum HDI width, `HDImaxwid`, or a null value and ROPE, `nullVal` and `ROPE`, but not both. The function does not check whether the ROPE fully *contains* the HDI, but could be expanded to do so. The function finds the required sample size by trying a small sample size, checking the power, and incrementing the sample size repeatedly until a sufficient size is found. The initial sample size is specified by the argument `initSampSize`. Take a look at the function definition now, attending to the embedded comments:

```
minNforHDIpower = function( genPriorMode , genPriorN ,
                            HDImaxwid=NULL , nullVal=NULL ,
                            ROPE=c(max(0,nullVal-0.02),min(1,nullVal+0.02)) ,
                            desiredPower=0.8 , audPriorMode=0.5 ,  audPriorN=2 ,
                            HDImass=0.95 , initSampSize=20 , verbose=TRUE ) {
  # Check for argument consistency:
  if ( !xor( is.null(HDImaxwid) , is.null(nullVal) ) ) {
    stop("One and only one of HDImaxwid and nullVal must be specified.")
  }
  # Load HDIofICDF function if not already present:
  if ( !exists("HDIofICDF") ) source("DBDA2E-utilities.R")
  # Convert prior mode and N to a,b parameters of beta distribution:
  genPriorA = genPriorMode * (genPriorN-2) + 1
  genPriorB = ( 1.0 - genPriorMode ) * (genPriorN-2) + 1
  audPriorA = audPriorMode * (audPriorN-2) + 1
  audPriorB = ( 1.0 - audPriorMode ) * (audPriorN-2) + 1
  # Initialize loop for incrementing sampleSize:
  sampleSize = initSampSize
  notPowerfulEnough = TRUE
  # Increment sampleSize until desired power is achieved:
  while( notPowerfulEnough ) {
    zvec = 0:sampleSize # vector of all possible z values for N flips.
    # Compute probability of each z value for data-generating prior:
    pzvec = exp( lchoose( sampleSize , zvec )
                 + lbeta( zvec + genPriorA , sampleSize-zvec + genPriorB )
                 - lbeta( genPriorA , genPriorB ) )
    # For each z value, compute posterior HDI:
    # hdiMat will hold HDI limits for each z:
    hdiMat = matrix( 0 , nrow=length(zvec) , ncol=2 )
    for ( zIdx in 1:length(zvec) ) {
      z = zvec[zIdx]
      hdiMat[zIdx,] = HDIofICDF( qbeta ,
                                 shape1 = z + audPriorA ,
```

```
                                    shape2 = sampleSize - z + audPriorB ,
                                    credMass = HDImass )
    }
    # Compute HDI widths:
    hdiWid = hdiMat[,2] - hdiMat[,1]
    # Sum the probabilities of outcomes with satisfactory HDI widths:
    if ( !is.null( HDImaxwid ) ) {
      powerHDI = sum( pzvec[ hdiWid < HDImaxwid ] )
    }
    # Sum the probabilities of outcomes with HDI excluding ROPE:
    if ( !is.null( nullVal ) ) {
      powerHDI = sum( pzvec[ hdiMat[,1] > ROPE[2] | hdiMat[,2] < ROPE[1] ] )
    }
    if ( verbose ) {
      cat( " For sample size = ", sampleSize , ", power = " , powerHDI ,
           "\n" , sep="" ) ; flush.console()
    }
    if ( powerHDI > desiredPower ) {  # If desired power is attained,
      notPowerfulEnough = FALSE       # set flag to stop,
    } else {                          # otherwise
      sampleSize = sampleSize + 1     # increment the sample size.
    }
  } # End while( notPowerfulEnough ).
  # Return the sample size that achieved the desired power:
  return( sampleSize )
} # End of function.
```

An example of calling the function looks like this:

```
source("minNforHDIpower.R") # only needed once per R session
sampSize = minNforHDIpower( genPriorMode=0.75, genPriorN=2000,
                            HDImaxwid=NULL, nullVal=0.5, ROPE=c (0.48,0.52),
                            desiredPower=0.8,
                            audPriorMode=0.5, audPriorN=2,
                            HDImass=0.95, initSampSize=5, verbose=TRUE )
```

In that function call, the data-generating distribution has a mode of 0.75 and concentration of 2000, which means that the hypothesized world is pretty certain that coins have a bias of 0.75. The goal is to exclude a null value of 0.5 with a ROPE from 0.48 to 0.52. The desired power if 80%. The audience prior is uniform. When the function is executed, it displays the power for increasing values of sample size, until stopping at $N = 30$ (as shown in Table 13.1).

13.2.3. When the goal is precision

Suppose you are interested in assessing the preferences of the general population regarding political candidates A and B. In particular, you would like to have high confidence in estimating whether the preference for candidate A exceeds $\theta = 0.5$. A recently conducted poll by a reputable organization found that of 10 randomly selected voters, 6 preferred candidate A, and 4 preferred candidate B. If we use a uniform pre-poll prior, our post-poll estimate of the population bias is a beta$(\theta|7, 5)$ distribution. As this is our best information about the population so far, we can use the beta$(\theta|7, 5)$ distribution as a data-generating distribution for planning the follow-up poll. Unfortunately, a beta$(\theta|7, 5)$ distribution has a 95% HDI from $\theta = 0.318$ to $\theta = 0.841$, which means that $\theta = 0.5$ is well within the data-generating distribution. How many more people do we need to poll so that 80% of the time we would get a 95% HDI that falls *above* $\theta = 0.5$?

It turns out, in this case, that we can never have a sample size large enough to achieve the goal of 80% of the HDIs falling above $\theta = 0.5$. To see why, consider what happens when we sample a particular value θ from the data-generating distribution, such as $\theta = 0.4$. We use that θ value to simulate a random sample of votes. Suppose N for the sample is huge, which implies that the HDI will be very narrow. What value of θ will the HDI focus on? Almost certainly it will focus on the value $\theta = 0.4$ that was used to generate the data. To reiterate, when N is very large, the HDI essentially just reproduces the θ value that generated it. Now recall the data-generating hypothesis of our example: The beta$(\theta|7, 5)$ distribution has only about 72% of the θ values above 0.5. Therefore, even with an extremely large sample size, we can get at most 72% of the HDIs to fall above 0.5.

There is a more useful goal, however. Instead of trying to reject a particular value of θ, we set as our goal a desired degree of precision in the posterior estimate. For example, our goal might be that the 95% HDI has width less than 0.2, at least 80% of the time. This goal implies that regardless of what values of θ happen to be emphasized by the posterior distribution, the width of the posterior is usually narrow, so that we have attained a suitably high precision in the estimate.

Table 13.2 shows the minimal sample size needed for the 95% HDI to have maximal width of 0.2. As an example of how to read the table, suppose you have a data-generating hypothesis that the coin has a bias roughly around $\theta = 0.6$. This hypothesis

Table 13.2 Minimal sample size required for 95% HDI to have maximal width of 0.2, when flipping a single coin.

	Generating Mode ω					
Power	**0.60**	**0.65**	**0.70**	**0.75**	**0.80**	**0.85**
0.7	91	90	88	86	81	75
0.8	92	92	91	90	87	82
0.9	93	93	93	92	91	89

Note. The data-generating distribution is a beta density with mode ω, as indicated by the column header, and with concentration $\kappa = 10$. The audience-agreeable prior is uniform.

is implemented, for purposes of Table 13.2, as a beta distribution with mode of 0.6 and concentration of 10. The value of 10 is arbitrary; it's as if the generating distribution were based on fictitious previous data containing only 10 flips. The table indicates that if we desire a 90% probability of obtaining an HDI with maximal width of 0.2, we need a sample size of 93.

Notice in Table 13.2 that as the desired power increases, the required sample size increases only slightly. For example, if the data-generating mean is 0.6, then as the desired power rises from 0.7 to 0.9, the minimal sample size rises from 91 to 93. This is because the distribution of HDI widths, for a given sample size, has a very shunted high tail, and therefore small changes in N can quickly pull the high tail across a threshold such as 0.2. On the other hand, as the desired HDI width decreases (not shown in the table), the required sample size increases rapidly. For example, if the desired HDI width is 0.1 instead of 0.2, then the sample size needed for 80% power is 378 instead of 92.

An example of using the R function, defined in the previous section, for computing minimum sample sizes for achieving desired precision, looks like this:

```
source("minNforHDIpower.R") # only needed once per R session
sampSize = minNforHDIpower( genPriorMode=0.75, genPriorN=10,
                            HDImaxwid=0.20, nullVal=NULL, ROPE=NULL,
                            desiredPower=0.8,
                            audPriorMode=0.5, audPriorN=2,
                            HDImass=0.95, initSampSize=50, verbose=TRUE )
```

In that function call, the data-generating distribution has a mode of 0.75 and concentration of 10, which means that the hypothesized world is *un*certain that coins have a bias of 0.75. The goal is to have a 95% HDI with width less than 0.20. The desired power is 80%. The audience prior is uniform. When the function is executed, it displays the power for increasing values of sample size, until stopping at $N = 90$ (as shown in Table 13.2).

13.2.4. Monte Carlo approximation of power

The previous sections illustrated the ideas of power and sample size for a simple case in which the power could be computed by mathematical derivation. In this section, we approximate the power by Monte Carlo simulation. The R script for this simple case serves as a template for more realistic applications. The R script is named `Jags-Ydich-Xnom1subj-MbernBeta-Power.R`, which is the name for the JAGS program for dichotomous data from a single "subject" suffixed with the word "Power." As you read through the script, presented below, remember that you can find information about any general R command by using the help function in R, as explained in Section 3.3.1 (p. 39).

The script has three main parts. The first part defines a function that does a JAGS analysis of a set of data and checks the MCMC chain for whether the desired goals have been achieved. The function takes a data vector as input, named `data`, and returns a list, named goalAchieved, of TRUE or FALSE values for each goal. Notice the comments that precede each command in the function definition:

```
# Load the functions genMCMC, smryMCMC, and plotMCMC:
# (This also sources DBDA2E-utilities.R)
source("Jags-Ydich-Xnom1subj-MbernBeta.R")

# Define function that assesses goal achievement for a single set of data:
goalAchievedForSample = function( data ) {
  # Generate the MCMC chain:
  mcmcCoda = genMCMC( data=data , numSavedSteps=10000 , saveName=NULL )
  # Check goal achievement. First, compute the HDI:
  thetaHDI = HDIofMCMC( as.matrix(mcmcCoda[,"theta"]) )
  # Define list for recording results:
  goalAchieved = list()
  # Goal: Exclude ROPE around null value:
  thetaROPE = c(0.48,0.52)
  goalAchieved = c( goalAchieved ,
                    "ExcludeROPE"=( thetaHDI[1] > thetaROPE[2]
                                    | thetaHDI[2] < thetaROPE[1] ) )
  # Goal: HDI less than max width:
  thetaHDImaxWid = 0.2
  goalAchieved = c( goalAchieved ,
                    "NarrowHDI"=( thetaHDI[2]-thetaHDI[1] < thetaHDImaxWid ) )
  # More goals can be inserted here if wanted...
  # Return list of goal results:
  return(goalAchieved)
}
```

The function above accomplishes the lower-right side of Figure 13.1, involving the white rectangles (not the shaded clouds). The audience prior is specified inside the genMCMC function, which is defined in the file Jags-Ydich-Xnom1subj- MbernBeta.R. The function above (goalAchievedForSample) simply runs the JAGS analysis of the data and then checks whether various goals were achieved. Inside the function, the object goalAchieved is initially declared as an empty list so that you can append as many different goals as you want. Notice that each goal is named (e.g., "ExcludeROPE" and "NarrowHDI") when it is appended to the list. Be sure that you use distinct names for the goals.

The next part of the script accomplishes the lower-left side of Figure 13.1, involving the shaded clouds. It loops through many simulated data sets, generated from a hypothetical distribution of parameter values. The value of genTheta is the randomly generated representative value of the parameter drawn from the hypothetical beta distribution. The value of sampleZ is the number of heads in the randomly generated data based on genTheta. Take a look at the comments before each command in the simulation loop:

```
# Specify mode and concentration of hypothetical parameter distribution:
omega = 0.70
kappa = 2000
# Specify sample size for each simulated data set:
sampleN = 74
# Run a bunch of simulated experiments:
nSimulatedDataSets = 1000 # An arbitrary large number.
for ( simIdx in 1:nSimulatedDataSets ) {
  # Generate random value from hypothesized parameter distribution:
  genTheta = rbeta( 1 , omega*(kappa-2)+1 , (1-omega)*(kappa-2)+1 )
  # Generate random data based on parameter value:
  sampleZ = rbinom( 1 , size=sampleN , prob=genTheta )
  # Convert to vector of 0's and 1's for delivery to JAGS function:
  simulatedData = c(rep(1,sampleZ),rep(0,sampleN-sampleZ))
  # Do Bayesian analysis on simulated data:
  goalAchieved = goalAchievedForSample( simulatedData )
  # Tally the results:
  if (!exists("goalTally")) { # if goalTally does not exist, create it
    goalTally=matrix( nrow=0 , ncol=length(goalAchieved) )
  }
  goalTally = rbind( goalTally , goalAchieved )
}
```

The object goalTally is a matrix that stores the results of each simulation in successive rows. The matrix is created after the first analysis inside the loop, instead of before the loop, so that it can decide how many columns to put in the matrix based on how many goals are returned by the analysis. The simulation loop could also contain two additional but optional lines at the end. These lines save the ongoing goalTally matrix at each iteration of the simulated data sets. This saving is done in case each iteration takes a long time and there is the possibility that the run would be interrupted before reaching nSimulatedDataSets. In the present application, each data set is analyzed very quickly in real time and therefore there is little need to save interim results. For elaborate models of large data sets, however, each simulated data set might take minutes.

After all the simulated data sets have been analyzed, the final section of the script computes the proportion of successes for each goal, and the Bayesian HDI around each proportion. The function HDIofICDF, used below, is defined in DBDA2E-utilities.R. The comments before each line below explain the script:

```
# For each goal...
for ( goalIdx in 1:NCOL(goalTally) ) {
  # Extract the goal name for subsequent display:
  goalName = colnames(goalTally)[goalIdx]
  # Compute number of successes:
  goalHits = sum(unlist(goalTally[,goalIdx]))
  # Compute number of attempts:
  goalAttempts = NROW(goalTally)
```

```
    # Compute proportion of successes:
    goalEst = goalHits/goalAttempts
    # Compute HDI around proportion:
    goalEstHDI = HDIofICDF( qbeta ,
                            shape1=1+goalHits ,
                            shape2=1+goalAttempts-goalHits )
    # Display the result:
    show( paste0( goalName,
                  ": Est.Power=" , round(goalEst,3) ,
                  "; Low Bound=" , round(goalEstHDI[1],3) ,
                  "; High Bound=" , round(goalEstHDI[2],3) ) )
}
```

On a particular run of the full script above, the results were as follows:

```
[1] "ExcludeROPE: Est.Power=0.896; Low Bound=0.876; High Bound=0.914"
[1] "NarrowHDI: Est.Power=0.38; Low Bound=0.35; High Bound=0.41"
```

When you run the script, the results will be different because you will create a different set of random data sets. Compare the first line above, which shows that the power for excluding the ROPE is 0.896, with Table 13.1, p. 367, in which the minimum N needed to achieve a power of 0.9 (when $\omega = 0.7$ and $\kappa = 2000$) is shown as 74. Thus, the analytical and Monte Carlo results match. The output of the exact result from `minNforHDIpower.R` shows that the power for N=74 is 0.904.

If the script is run again, but with `kappa=10` and `sampleN=91`, we get

```
[1] "ExcludeROPE: Est.Power=0.651; Low Bound=0.621; High Bound=0.68"
[1] "NarrowHDI: Est.Power=0.863; Low Bound=0.841; High Bound=0.883"
```

The second line above indicates that the power for having an HDI width less than 0.2 is 0.863. This matches Table 13.2, which shows that a sample size of 91 is needed to achieve a power of at least 0.8 for the HDI to have width less than 0.2. In fact, the output of the program `minNforHDIpower.R` says that the power is 0.818, which implies that the Monte Carlo program has somewhat overestimated the power. This might be because Monte Carlo HDI widths tend to be slightly underestimated, as was described with Figure 7.13, p. 185.

In general, the script presented here can be used as a template for power calculations of complex models. Much of the script remains the same. The most challenging part for complex models is generating the simulated data, in the second part of the script. Generating simulated data is challenging from a programming perspective merely to get all the details right; patience and perseverance will pay off. But it is also conceptually challenging in complex models because it is not always clear how to express hypothetical parameter distributions. The next section provides an example and a general framework.

13.2.5. Power from idealized or actual data

Recall the example of therapeutic touch from Section 9.2.4. Practitioners of therapeutic touch were tested for ability to sense the presence of the experimenter's hand near one of their own. The data were illustrated in Figure 9.9 (p. 241). The hierarchical model for the data was shown in Figure 9.7 (p. 236). The model had a parameter ω for the modal ability of the group, along with parameters θ_s for the individual abilities of each subject, and the parameter κ for the concentration of the individual abilities across the group.

To generate simulated data for power computation, we need to generate random values of ω and κ (and subsequently θ_s) that are representative of the hypothesis. One way to do that is to explicitly hypothesize the top-level constants that we think capture our knowledge about the true state of the world. In terms of Figure 9.7, that means we specify values of A_ω, B_ω, S_κ, and R_κ that directly represent our uncertain hypothesis about the world. This can be done in principle, although it is not immediately intuitive in practice because it can be difficult to specify the uncertainty (width) of the distribution on the group mode and the uncertainty (width) on the group concentration.

In practice, it is often more intuitive to specify actual or idealized *data* that express the hypothesis, than it is to specify top-level parameter properties. The idea is that we start with the actual or idealized data and then use Bayes' rule to generate the corresponding distribution on parameter values. Figure 13.2 illustrates the process. In the top-left of Figure 13.2, the real or hypothetical world creates an actual or an

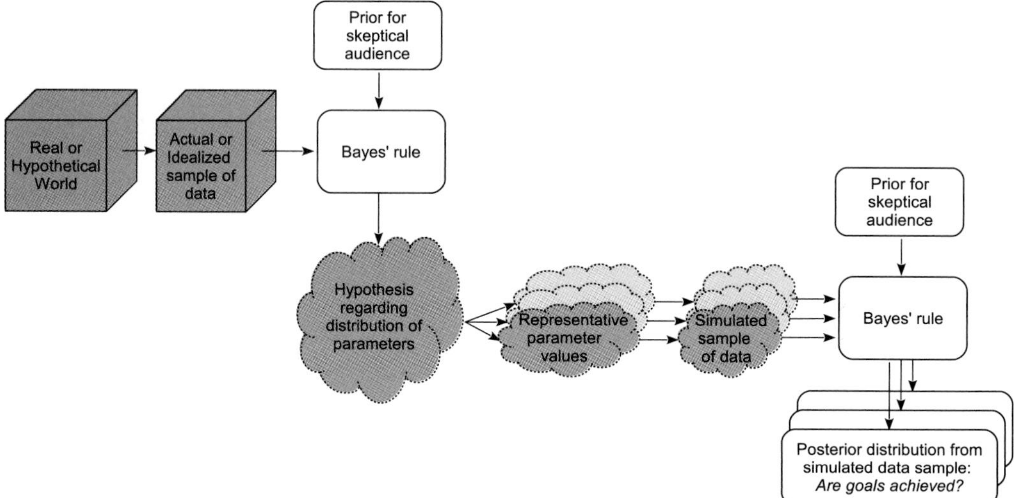

Figure 13.2 Flow of information in a power analysis when the hypothesis regarding the distribution of parameters is a posterior distribution from a Bayesian analysis on real or idealized previous data. Compare with Figure 13.1, p. 363.

idealized data set. Bayes' rule is the applied, which creates a posterior distribution. The posterior distribution is used as the hypothetical distribution of parameter values for power analysis. Specifying actual or idealized data that represent the hypothesis is typically very intuitive because it's concrete. The beauty of the this approach is that the hypothesis is expressed as concrete, easily intuited data, and the Bayesian analysis converts it into the corresponding parameter distribution. Importantly, our confidence in the hypothesis is expressed concretely by the amount of data in the actual or idealized sample. The bigger the actual or idealized sample, the tighter will be the posterior distribution. Thus, instead of having to explicitly specify the tightness of the parameter distribution, we let Bayes' rule do it for us.

Another benefit of this approach is that appropriate correlations of parameters are automatically created in the posterior distribution of Bayes' rule, rather than having to be intuited and explicitly specified (or inappropriately ignored). The present application does not involve parameters with strong correlations, but there are many applications in which correlations do occur. For example, when estimating the standard deviation (scale) and normality (kurtosis) of set of metric data, those two parameters are correlated (see σ and ν in Figure 16.8, p. 378, and Kruschke (2013a)). As another example, when estimating the slope and intercept in linear regression, those two parameters are usually correlated (see β_1 and β_0 in Figure 17.3, p. 393, and Kruschke, Aguinis, and Joo (2012)).

The script `Jags-Ydich-XnomSsubj-MbinomBetaOmegaKappa-Power.R` provides an example of carrying out this process in R. The first step is merely loading the model and utility functions for subsequent use:

```
# Load the functions genMCMC, smryMCMC, and plotMCMC:
# (This also sources DBDA2E-utilities.R)
source("Jags-Ydich-XnomSsubj-MbinomBetaOmegaKappa.R")
```

Next, we generate some idealized data. Suppose we believe, from anecdotal experiences, that therapeutic-touch practitioners as a group have a 65% probability of correctly detecting the experimenter's hand. Suppose also we believe that different practitioners will have accuracies above or below that mean, with a standard deviation of 7% points. This implies that the worst practitioner will be at about chance, and the best practitioner will be at about 80% correct. This hypothesis is expressed in the next two lines:

```
# Specify idealized hypothesis:
idealGroupMean = 0.65
idealGroupSD = 0.07
```

We then specify how much (idealized) data we have to support that hypothesis. The more data we have, the more confident we are in the hypothesis. Suppose we are fairly confident in our idealized hypothesis, such that we imagine we have data from 100 practitioners, each of whom contributed 100 trials. This is expressed in the next two lines:

```
idealNsubj = 100        # more subjects => higher confidence in hypothesis
idealNtrlPerSubj = 100 # more trials => higher confidence in hypothesis
```

Next we generate data consistent with the values above.

```
# Generate random theta values for idealized subjects:
betaAB = betaABfromMeanSD( idealGroupMean , idealGroupSD )
theta = rbeta( idealNsubj , betaAB$a , betaAB$b )
# Transform the theta values to exactly match idealized mean, SD:
theta = ((theta-mean(theta))/sd(theta))*idealGroupSD + idealGroupMean
theta[ theta >= 0.999 ] = 0.999 # must be between 0 and 1
theta[ theta <= 0.001 ] = 0.001 # must be between 0 and 1
# Generate idealized data very close to theta's:
z = round( theta*idealNtrlPerSubj )
# Convert to data format needed by JAGS function:
# Set up an empty matrix for holding the data:
dataMat=matrix(0,ncol=2,nrow=0,dimnames=list(NULL,c("y","s")))
# For each simulated subject,
for ( sIdx in 1:idealNsubj ) {
  # Create vector of 0's and 1's matching the z values generated above:
  yVec = c(rep(1,z[sIdx]),rep(0,idealNtrlPerSubj-z[sIdx]))
  # Bind the subject data to the bottom of the matrix:
  dataMat = rbind( dataMat , cbind( yVec , rep(sIdx, idealNtrlPerSubj) ) )
}
# Make it a data frame:
idealDatFrm = data.frame(dataMat)
```

Then we run the Bayesian analysis on the idealized data. We are trying to create a set of representative parameter values that can be used for subsequent power analysis. Therefore we want each successive step of joint parameter values to be clearly distinct, which is to say that we want chains with very small autocorrelation. To achieve this in the present model, we must thin the chains. On the other hand, we are not trying to create a high-resolution impression of the posterior distribution because we are not using the posterior to estimate the parameters. Therefore we only generate as many steps in the chain as we may want for subsequent power analysis.

```
# Run Bayesian analysis on idealized data:
mcmcCoda = genMCMC( data=idealDatFrm , saveName=NULL ,
                    numSavedSteps=2000 , thinSteps=20 )
# Convert coda object to matrix for convenience:
mcmcMat = as.matrix(mcmcCoda)
```

The above code creates a posterior distribution on parameters ω and κ (as well as all the individual θ_s). Now we have a distribution of parameter values consistent with our idealized hypothesis, but we did not have to figure out the top-level constants in the model. We merely specified the idealized tendencies in the data and expressed our confidence by its amount. The above part of the script accomplished the left side of Figure 13.2, so we now have a large set of representative parameter values for conducting a power analysis. The representative values are shown in the upper part of Figure 13.3. Notice that the hypothesized values of ω are centered around the idealized data mean. What was not obvious from the idealized data alone is how much uncertainty there should be on ω; the posterior distribution in Figure 13.3 reveals the answer. Analogous remarks apply to the concentration parameter, κ.

The function that assesses goal achievement for a single set of data is structured the same as before, with only the specific goals changed. In this example, I considered goals for achieving precision and exceeding a ROPE around the null value, at both the group level and individual level. For the group level, the goals are for the 95% HDI on the group mode, ω, to fall above the ROPE around the null value, and for the width of the HDI to be less than 0.2. For the individual level, the goals are for at least one of the θ_ss

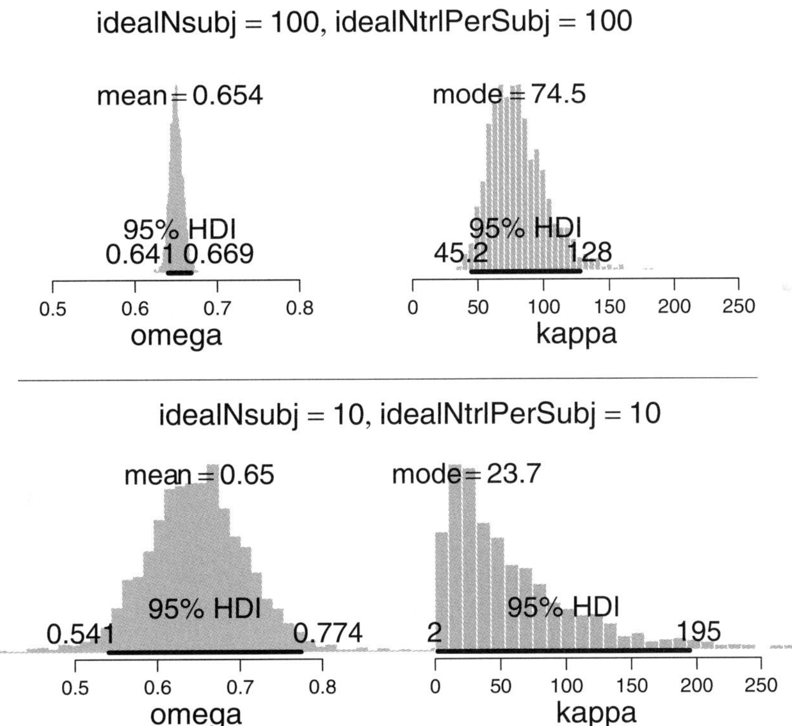

Figure 13.3 Distributions of parameters consistent with idealized data. Upper panel is for large amounts of idealized data, lower panel is for small amounts.

95% HDIs to exceed the ROPE with none that fall below the ROPE, and for all the θ_ss 95% HDIs to have widths less than 0.2. Here is the code that specifies these goals:

```
# Define function that assays goal achievement for a single set of data:
goalAchievedForSample = function( data ) {
  # Generate the MCMC chain:
  mcmcCoda = genMCMC( data=data , saveName=NULL ,
                      numSavedSteps=5000 , thinSteps=2 )
  # Convert coda object to matrix for convenience:
  mcmcMat = as.matrix(mcmcCoda)
  # Specify criteria for goals:
  nullROPE = c(0.48,0.52)
  HDImaxWid = 0.2
  # Compute HDIs:
  HDImat = apply( mcmcMat , 2 , "HDIofMCMC" )
  show( HDImat[,1:5] )
  # Define list for recording results:
  goalAchieved = list()
  # Goal: omega greater than ROPE:
  goalAchieved = c( goalAchieved ,
                    "omegaAboveROPE"=unname( HDImat[1,"omega"] > nullROPE[2] ) )
  # Goal: omega HDI width less than max width:
  goalAchieved = c( goalAchieved ,
                    "omegaNarrowHDI"=unname( HDImat[2,"omega"]-HDImat[1,"omega"]
                                             < HDImaxWid ) )
  # Goal: at least one theta greater than ROPE with none below:
  thetaCols = grep("theta",colnames(HDImat)) # column indices of thetas
  goalAchieved = c( goalAchieved ,
                    "thetasAboveROPE"= (any(HDImat[1,thetaCols] > nullROPE[2])
                                        & !any(HDImat[2,thetaCols] < nullROPE[1])))
  # Goal: all theta's HDI width less than max width:
  goalAchieved = c( goalAchieved ,
                    "thetasNarrowHDI"= all( HDImat[2,thetaCols]
                                            - HDImat[1,thetaCols]
                                            < HDImaxWid ) )
  # More goals can be inserted here if wanted...
  # Return list of goal results:
  return(goalAchieved)
}
```

The above function is then called repeatedly for simulated data sets, created from the hypothetical distribution of parameter values. Importantly, notice that we use only the ω and κ values from the hypothesized distribution of parameters, *not* the θ_s values. The reason is that the θ_s values are "glued" to the data of the idealized subjects. In particular, the number of θ_s's is `idealNsubj`, but our simulated data could use more or fewer simulated subjects. To generate simulated data, we must specify the number of simulated subjects and the number of trials per subject. I will run the power analysis twice, using different selections of subjects and trials. In both cases there is a total of 658 trials, but in the first case there are 14 subjects with 47 trials per subject, and in the second case there are seven subjects with 94 trials per subject. Please consider the R code, attending to the comments:

```
# Specify sample size for each simulated data set:
Nsubj = 2*7 ; NtrlPerSubj = 47  # 658 flips total
#Nsubj = 7 ; NtrlPerSubj = 2*47  # 658 flips total
# Specify the number of simulated experiments:
nSimulatedDataSets = min(500,NROW(mcmcMat)) # An arbitrary large number.
# Run the simulated experiments:
simCount=0
for ( simIdx in ceiling(seq(1,NROW(mcmcMat), length=nSimulatedDataSets)) ) {
  simCount=simCount+1
  cat( "\n\n==================== Simulation",simCount,"of", nSimulatedDataSets,
       "====================\n\n" )
  # Generate random omega and kappa for group distribution:
  genOmega = mcmcMat[simIdx,"omega"]
  genKappa = mcmcMat[simIdx,"kappa"]
  # Generate random theta's for individuals:
  genTheta = rbeta( Nsubj , genOmega*(genKappa-2)+1 , (1-genOmega)*(genKappa-2)+1 )
  # Generate random data based on parameter value:
  dataMat=matrix(0,ncol=2,nrow=0,dimnames=list(NULL,c("y","s")))
  for ( sIdx in 1:Nsubj ) {
    z = rbinom( 1 , size=NtrlPerSubj , prob=genTheta[sIdx] )
    yVec = c(rep(1,z),rep(0,NtrlPerSubj-z))
    dataMat = rbind( dataMat , cbind( yVec , rep(sIdx,NtrlPerSubj) ) )
  }
  # Do Bayesian analysis on simulated data:
  goalAchieved = goalAchievedForSample( data.frame(dataMat) )
  # Tally the results:
  if (!exists("goalTally")) { # if goalTally does not exist, create it
    goalTally=matrix( nrow=0 , ncol=length(goalAchieved) )
  }
  goalTally = rbind( goalTally , goalAchieved )
}
```

The next and final section of the script, that tallies the goal achievement across repeated simulations, is unchanged from the previous example, and therefore will not be shown again here.

We now consider the results of running the script, using 500 simulated experiments. When

```
Nsubj = 2*7 ; NtrlPerSubj = 47  # 658 flips total
```

then

```
[1] "omegaAboveROPE: Est.Power=0.996; Low Bound=0.987; High Bound=0.999"
[1] "omegaNarrowHDI: Est.Power=0.99; Low Bound=0.978; High Bound=0.996"
[1] "thetasAboveROPE: Est.Power=1; Low Bound=0.994; High Bound=1"
[1] "thetasNarrowHDI: Est.Power=0.266; Low Bound=0.229; High Bound=0.306"
```

Notice that there is extremely high power for achieving the group-level goals, but there is low probability that the subject-level estimates will be precise. If we use fewer subjects with more trials per subject, with

```
Nsubj = 7 ; NtrlPerSubj = 2*47  # 658 flips total
```

then

```
[1] "omegaAboveROPE: Est.Power=0.642; Low Bound=0.599; High Bound=0.683"
[1] "omegaNarrowHDI: Est.Power=0.524; Low Bound=0.48; High Bound=0.568"
[1] "thetasAboveROPE: Est.Power=0.996; Low Bound=0.987; High Bound=0.999"
[1] "thetasNarrowHDI: Est.Power=0.906; Low Bound=0.878; High Bound=0.929"
```

Notice that now there is lower probability of achieving the group-level goals, but a much higher probability that the subject-level estimates will achieve the desired precision. This example illustrates a general trend in hierarchical estimates. If you want high precision at the individual level, you need lots of data within individuals. If you want high precision at the group level, you need lots of individuals (without necessarily lots of data per individual, but more is better).

As another important illustration, suppose that our idealized hypothesis was less certain, which we express by having idealized data with fewer subjects and fewer trials per subject, while leaving the group mean and standard deviation unchanged. Thus,

```
# Specify idealized hypothesis:
idealGroupMean = 0.65
idealGroupSD = 0.07
idealNsubj = 10        # instead of 100
idealNtrlPerSubj = 10 # instead of 100
```

Notice that the idealized group mean and group standard deviation are the same as before. Only the idealized amount of data contributing to the ideal is reduced. The resulting hypothetical parameter distribution is shown in the lower part of Figure 13.3. Notice that the parameter distribution is more spread out, reflecting the uncertainty inherent in the small amount of idealized data. For the power analysis, we use the simulated same sample size as the first case above:

```
Nsubj = 2*7 ; NtrlPerSubj = 47  # 658 flips total
```

The resulting power for each goal is

```
[1] "omegaAboveROPE: Est.Power=0.788; Low Bound=0.751; High Bound=0.822"
[1] "omegaNarrowHDI: Est.Power=0.816; Low Bound=0.781; High Bound=0.848"
[1] "thetasAboveROPE: Est.Power=0.904; Low Bound=0.876; High Bound=0.928"
[1] "thetasNarrowHDI: Est.Power=0.176; Low Bound=0.144; High Bound=0.211"
```

Notice that *the less certain hypothesis has reduced the power* for all goals, even though the group-level mean and standard deviation are unchanged. This influence of the uncertainty of the hypothesis is a central feature of the Bayesian approach to power analysis.

The classical definition of power in NHST assumes a specific value for the parameters without any uncertainty. The classical approach can compute power for different specific parameter values, but the approach does not weigh the different values by their credibility. One consequence is that for the classical approach, retrospective power is extremely uncertain, rendering it virtually useless, because the estimated powers at the two ends of the confidence interval are close to the baseline false alarm rate and 100% (Gerard, Smith, & Weerakkody, 1998; Miller, 2009; Nakagawa & Foster, 2004; O'Keefe, 2007; Steidl, Hayes, & Schauber, 1997; Sun, Pan, & Wang, 2011; L. Thomas, 1997).

You can find another complete example of using idealized data for power analysis, applied to comparing two groups of metric data, in Kruschke (2013a). The software accompanying that article, also included in this book's programs, is called "BEST" for Bayesian estimation. See the Web site http://www.indiana.edu/~kruschke/BEST/ for links to videos, a web app, an implementation in Python, and an enhanced implementation in R.

13.3. SEQUENTIAL TESTING AND THE GOAL OF PRECISION

In classical power analysis, it is assumed that the goal is to reject the null hypothesis. For many researchers, the *sine qua non* of research is to reject the null hypothesis. The practice of NHST is so deeply institutionalized in scientific journals that it is difficult to get research findings published without showing "significant" results, in the sense of $p < 0.05$. As a consequence, many researchers will monitor data as they are being collected and stop collecting data only when $p < 0.05$ (conditionalizing on the current sample size) or when their patience runs out. This practice seems intuitively not to be problematic because the data collected after testing previous data are not affected by the previously collected data. For example, if I flip a coin repeatedly, the probability of heads on the next flip is not affected by whether or not I happened to check whether $p < 0.05$ on the previous flip.

Unfortunately, that intuition about independence across flips only tells part of story. What's missing is the realization that the stopping procedure biases which data are sampled, because the procedure stops only when extreme values happen to be randomly sampled. After stopping, there is no opportunity to sample compensatory values from the opposite extreme. In fact, as will be explained in more detail, in NHST the null hypothesis will always be rejected even if it is true, when doing sequential testing with infinite patience. In other words, under sequential testing in NHST, the true probability of false alarm is 100%, not 5%.

Moreover, any stopping rule based on getting extreme outcomes will provide estimates that are too extreme. Regardless of whether we use NHST or Bayesian decision criteria, if we stop collecting data only when a null value has been rejected, then the sample will tend to be biased too far away from the null value. The reason is that the stopping rule caused data collection to stop as soon as there were enough accidentally extreme values, cutting off the opportunity to collect compensating representative values. We saw an example of stopping at extreme outcomes back in Section 11.1.3, p. 305. The section discussed flipping a coin until reaching a certain number of heads (as opposed to stopping at a certain number of flips). This procedure will tend to overestimate the probability of getting a head, because if a random subsequence of several heads happens to reach the stopping criterion, there will be no opportunity for subsequent flips with tails to compensate. Of course, if the coin is flipped until reaching a certain number of tails (instead of heads), then the procedure will tend to overestimate the probability of getting a tail.

One solution to these problems is not to make rejecting the null value be the goal. Instead, we make precision the goal. For many parameters, precision is unaffected by the true underlying value of the parameter, and therefore stopping when a criterial precision is achieved does not bias the estimate. The goal of achieving precision thereby seems to be motivated by a desire to learn the true value, or, more poetically, by love of the truth, regardless of what it says about the null value. The goal of rejecting a null value, on the other hand, seems too often to be motivated by fear: fear of not being published or not being approved if the null fails to be rejected. The two goals for statistical power might be aligned with different core motivations, love or fear. The Mahatma Gandhi noted that "Power is of two kinds. One is obtained by the fear of punishment and the other by acts of love. Power based on love is a thousand times more effective and permanent than the one derived from fear of punishment."[3]

The remainder of this section shows examples of sequential testing with different decision criteria. We consider decisions by p values, BFs, HDIs with ROPEs, and precision. We will see that decisions by p values not only lead to 100% false alarms (with infinite patience), but also lead to biased estimates that are more extreme than the true value. The two Bayesian methods both can decide to accept the null hypothesis, and therefore do not lead to 100% false alarms, but both do produce biased estimates

[3] I have seen this quote attributed to Gandhi on many web pages, but I have been unable to find an original source.

because they stop when extreme values are sampled. Stopping when precision is achieved produces accurate estimates.

13.3.1. Examples of sequential tests

Figures 13.4 and 13.5 show two sequences of coins flips. In Figure 13.4, the true bias of the coin is $\theta = 0.50$, and therefore a correct decision would be to accept the null hypothesis. In Figure 13.5, the true bias of the coin is $\theta = 0.65$, and therefore a correct decision would be to reject the null hypothesis. The top panel in each figure shows the proportion of heads (z/N) in the sequence plotted against the flip number, N. You can see that the proportion of heads eventually converges to the underlying bias of the coin, which is indicated by a horizontal dashed line.

For both sequences, at every step we compute the (two-tailed) p value, conditionalized on the N. The second panel in each figure plots the p values. The plots also show a dashed line at $p = 0.05$. If $p < 0.05$, the null hypothesis is rejected. You can see in Figure 13.4 that the early trials of this particular sequence happen to have a preponderance of tails, and p is less than 0.05 for many of the early trials. Thus, the null value is falsely rejected. Subsequently the p value rises above 0.05, but eventually it would cross below 0.05 again, by chance, even though the null value is true in this case. In Figure 13.5, you can see that the early trials of the particular sequence happen to hover around $z/N \approx 0.5$. Eventually the value of z/N converges to the true generating value of $\theta = 0.65$, and the p value drops below 0.05 and stays there, correctly rejecting the null.

The third panels of Figures 13.4 and 13.5 show the BF for each flip of the coin. The BF is computed as in Equation 12.3, p. 344. One conventional decision threshold states that BF > 3 or BF $< 1/3$ constitutes "substantial" evidence in favor or one hypothesis or the other (Jeffreys, 1961; Kass & Raftery, 1995; Wetzels et al., 2011). For visual and numerical symmetry, we take the logarithm and the decision rule is $\log(\text{BF}) > \log(3) \approx 1.1$ or $\log(\text{BF}) < \log(1/3) \approx -1.1$. These decision thresholds are plotted as dashed lines. You can see in Figure 13.4 that the BF falsely rejects the null in the early trials (similar to the p value). If the sequence had not been stopped at that point, the BF would have changed eventually to the *accept null* region and stayed there. In Figure 13.5, the BF falsely accepts the null in the early trials of this particular sequence. If the sequence had not been stopped, the BF would have changed to the *reject null* zone and stayed there.

The fourth panels of Figures 13.4 and 13.5 show the 95% HDIs at every flip, assuming a uniform prior. The y-axis is θ, and at each N the 95% HDI of the posterior, $p(\theta|z, N)$, is plotted as a vertical line segment. The dashed lines indicate the limits of a ROPE from 0.45 to 0.55, which is an arbitrary but reasonable choice for illustrating the behavior of the decision rule. You can see in Figure 13.4 that the HDI eventually falls within the ROPE, thereby correctly accepting the null value for practical purposes. Unlike

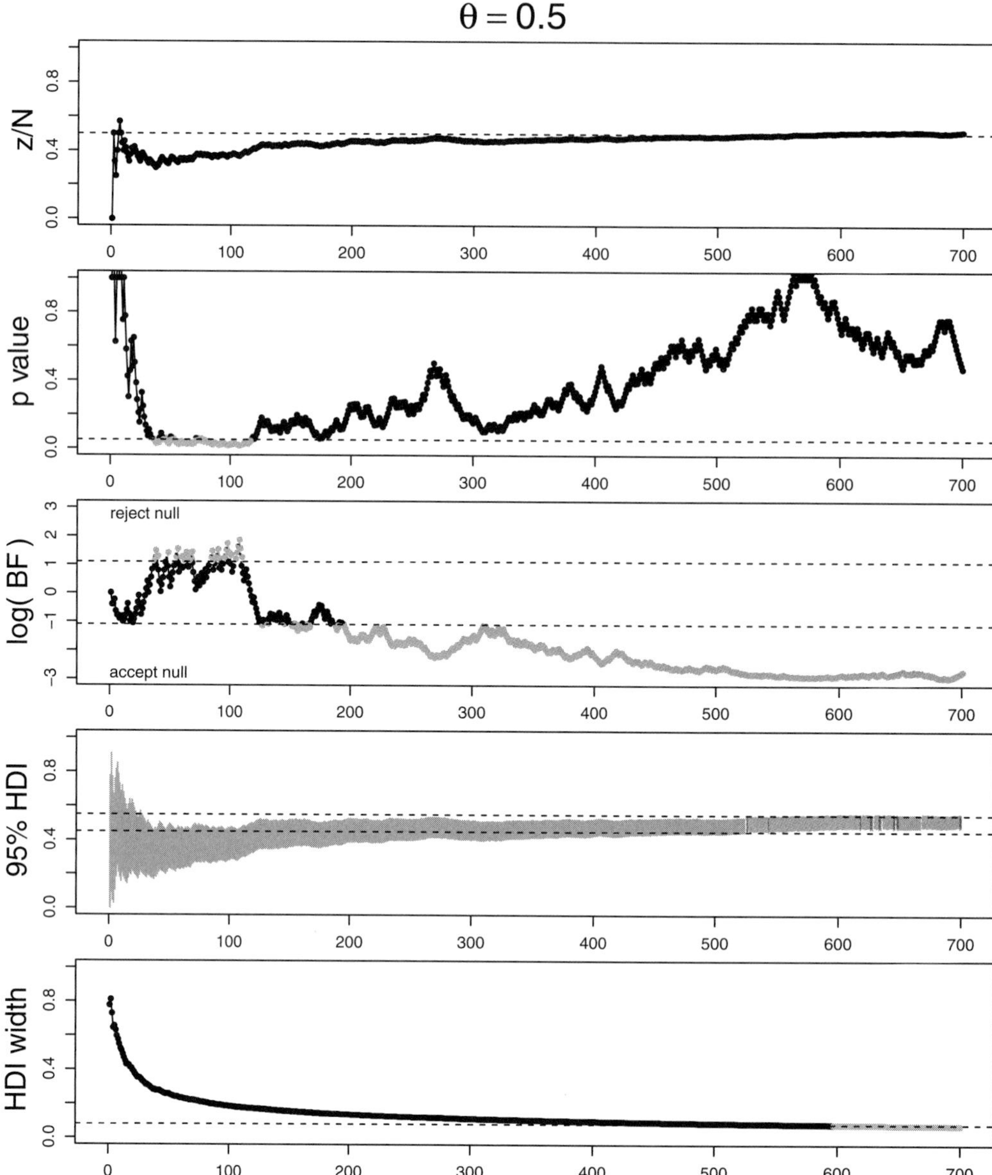

Figure 13.4 An example of a sequence of flips with testing of cumulative data at every flip. The abscissa is *N*. In this case the null hypothesis is true (i.e., $\theta = 0.50$). This sequence of flips (top panel) happens to show a preponderance of tails early in the sequence, hence both the *p* value and Bayes' factor (BF) reject the null early on.

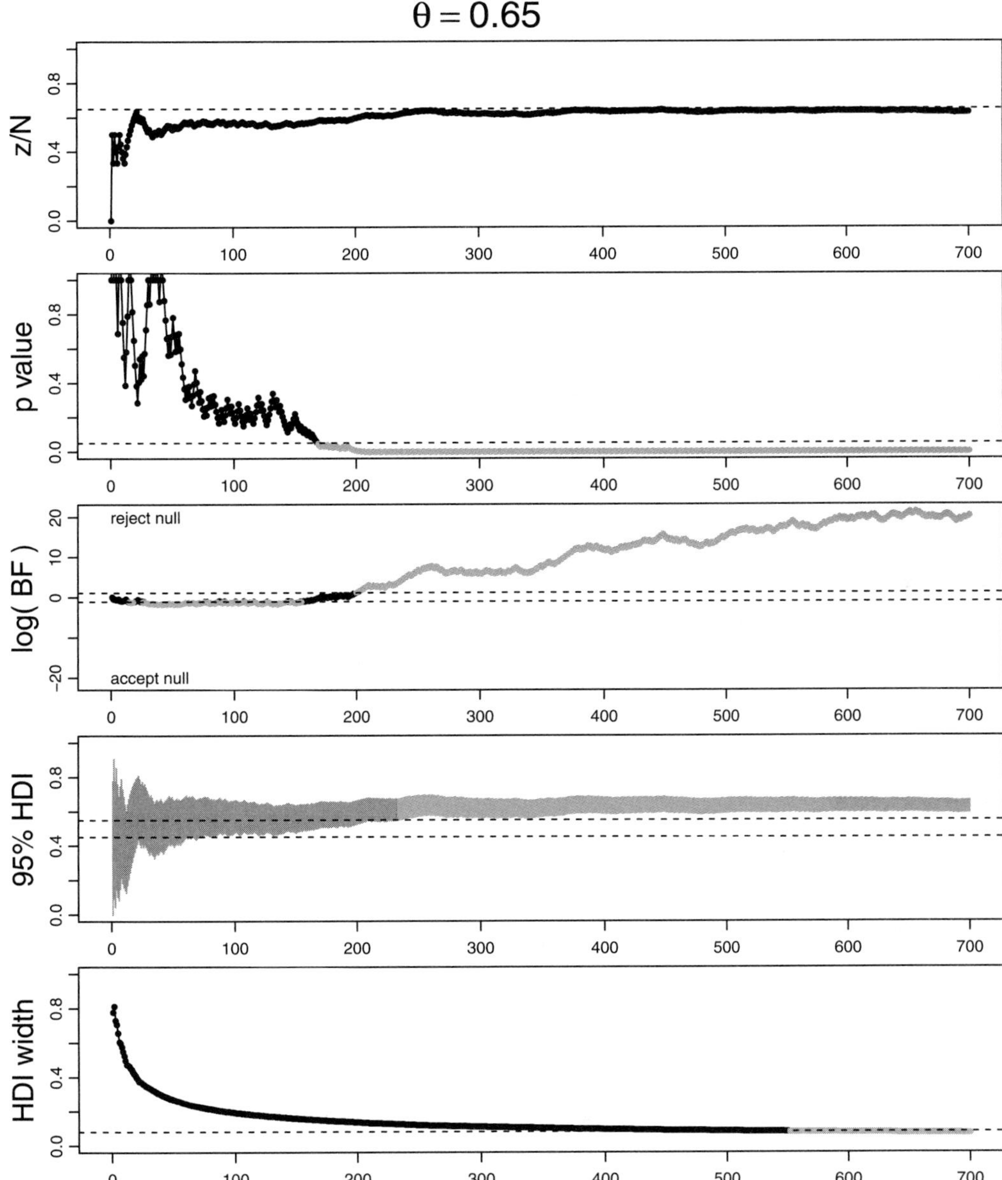

Figure 13.5 An example of a sequence of flips with testing of cumulative data at every flip. The abscissa is *N*. In this case the null hypothesis is *not* true (i.e., $\theta = 0.65$). This sequence of flips (top panel) happens to show proportions near 0.5 early in the sequence, hence the Bayes' factor (BF) accepts the null early on.

the p value and BF, the HDI does not falsely reject the null in the early trials, for this particular sequence. In Figure 13.5, the HDI eventually falls completely outside the ROPE, thereby correctly rejecting the null value. Unlike the BF, the HDI cannot accept the null early in the sequence because there is not sufficient precision with so few flips.

The fifth (lowermost) panels of Figures 13.4 and 13.5 show the width of the 95% HDI at every flip. There is no decision rule to accept or reject the null hypothesis on the basis of the width of the HDI. But there is a decision to stop when the width falls to 80% of the width of the ROPE. The 80% level is arbitrary and chosen merely so that there is some opportunity for the HDI to fall entirely within the ROPE when data collection stops. As the ROPE in these examples extends from 0.45 to 0.55, the critical HDI width is 0.08, and a dashed line marks this height in the plots. Under this stopping rule, data collection continues until the HDI width falls below 0.08. At that point, if desired, the HDI can be compared to the ROPE and a decision to reject or accept the null can also be made. In these examples, when the HDI reaches critical precision, it also happens to fall entirely within or outside the ROPE and yields correct decisions.

13.3.2. Average behavior of sequential tests

The previous examples (Figures 13.4 and 13.5) were designed to illustrate the behavior of various stopping rules in sequential testing. But those two examples were merely for specific sequences that happened to show correct decisions for the HDI-with-ROPE method and wrong decisions by the BF method. There are other sequences in which the opposite happens. The question then is, what is the average behavior of these methods?

The plots in Figures 13.6 and 13.7 were produced by running 1000 random sequences like those shown in Figures 13.4 and 13.5. For each of the 1000 sequences, the simulation kept track of where in the sequence each stopping rule would stop, what decision it would make at that point, and the value of z/N at that point. Each sequence was allowed to continue up to 1500 flips. Figure 13.6 is for when the null hypothesis is true, with $\theta = 0.50$. Figure 13.7 is for when the null hypothesis is false with $\theta = 0.65$. Within each figure, the upper row plots the proportion of the 1000 sequences that have come to each decision by the Nth flip. One curve plots the proportion of sequences that have stopped and decided to accept the null, another curve plots the proportion of sequences that have stopped and decided to reject the null, and a third curve plots the remaining proportion of undecided sequences. The lower rows plot histograms of the 1000 values of z/N at stopping. The true value of θ is plotted as a black triangle, and the mean of the z/N values is plotted as an outline triangle.

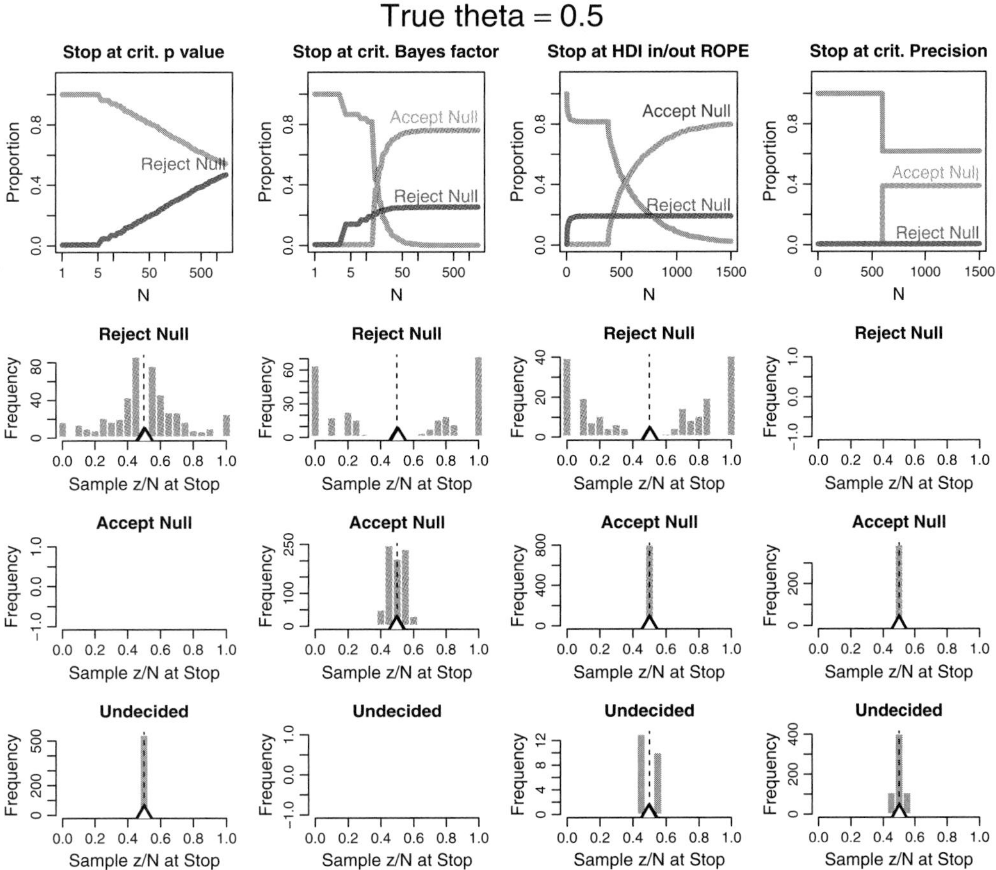

Figure 13.6 Behaviors of the four stopping rules are in the four columns, for $\theta = 0.50$. The top row shows the proportion of 1000 sequences that make each decision (accept, reject, undecided) at each flip. The lower rows show the value of z/N when data collection stops. Black triangle marks true θ and outline triangle marks mean z/N at stopping.

Consider Figure 13.6 for which the null hypothesis is true, with $\theta = 0.50$. The top left plot shows decisions by the p value. You can see that as N increases, more and more of the sequences have falsely rejected the null. The abscissa shows N on a logarithmic scale, so you see that the proportion of sequences that falsely rejects the null rises linearly on $\log(N)$. If the sequences had been allowed to extend beyond 1500 flips, the proportion of false rejections would continue to rise. This phenomenon has been called "sampling to reach a foregone conclusion" (Anscombe, 1954).

The second panel of the top row (Figure 13.6) shows the decisions reached by the Bayes' factor (BF). Unlike the p value, the BF reaches an asymptotic false alarm rate far less than 100%; in this case the asymptote is just over 20%. The BF correctly accepts the

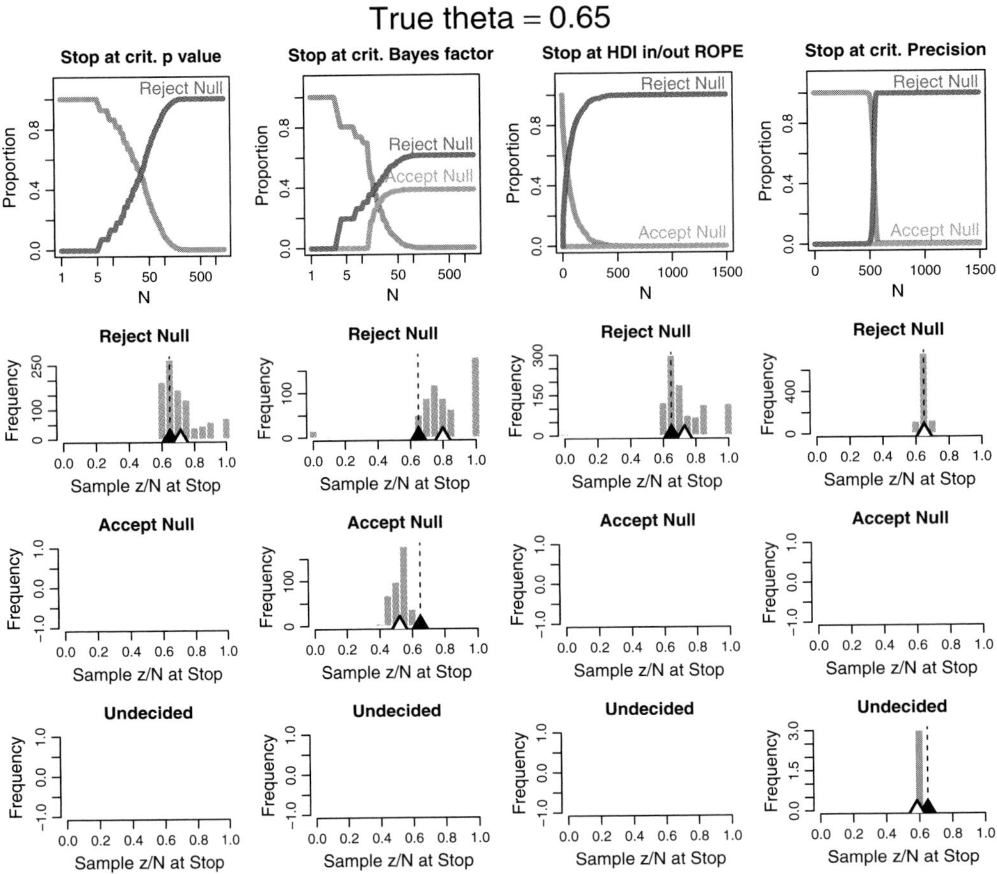

Figure 13.7 Behaviors of the four stopping rules are in the four columns, for $\theta = 0.65$. The top row shows the proportion of 1000 sequences that make each decision (accept, reject, undecided) at each flip. The lower rows show the value of z/N when data collection stops. Black triangle marks true θ and outline triangle marks mean z/N at stopping.

null, eventually, for the remaining sequences. The abscissa is displayed on a logarithmic scale because most of the decisions are made fairly early in the sequence.

The third panel of the top row (Figure 13.6) shows the decisions reached by the HDI-with-ROPE criterion. Like the BF, the HDI-with-ROPE rule reaches an asymptotic false alarm rate far below 100%, in this case just under 20%. The HDI-with-ROPE rule eventually accepts the null in all the remaining sequences, although it can take a large N to reach the required precision. As has been emphasized in Figure 12.4, p. 347, the HDI-with-ROPE criterion only accepts the null value when there is high precision in the estimate, whereas the BF can accept the null hypothesis even when there is little precision in the parameter estimate. (And, of course, the BF by itself does not provide an estimate of the parameter.)

The fourth panel of the top row (Figure 13.6) shows the decisions reached by stopping at a criterial precision. Nearly all sequences reach the criterial decision at about the same N. At that point, about 40% of the sequences have an HDI that falls within the ROPE, whence the null value is accepted. None of the HDIs falls outside the ROPE because the estimate has almost certainly converged to a value near the correct null value when the precision is high. In other words, there is a 0% false alarm rate.

The lower rows (Figure 13.6) show the value of z/N when the sequence is stopped. In the left column, you can see that for stopping at $p < 0.05$ when the null is rejected, the sample z/N can only be significantly above or below the true value of $\theta = 0.5$. For stopping at the limiting N of 1500, before encountering $p < 0.05$ and remaining undecided, the sample z/N tends to be very close to the true value of $\theta = 0.5$.

The second column (Figure 13.6), for the BF, shows that the sample z/N is quite far from $\theta = 0.5$ when the null hypothesis is rejected. Importantly, the sample z/N can also be noticeably off of $\theta = 0.5$ when the null hypothesis is *accepted*. The third column, for the HDI-with-ROPE, shows similar outcomes when rejecting the null value, but gives very accurate estimates when accepting the null value. This makes sense, of course, because the HDI-with-ROPE rule only accepts the null value when it is precisely estimated within the ROPE. The fourth column, for stopping at criterial precision, of course shows accurate estimates.

Now consider Figure 13.7 for which the null hypothesis is false, with $\theta = 0.65$. The top row shows that the null is eventually rejected for all stopping rules except the BF. In this case, the BF falsely accepts the null hypothesis almost 40% of the time. The second row shows that when the decision is to reject the null, only the criterial-precision rule does not noticeably overestimate θ. The BF overestimates θ the most. The third row shows that when the BF accepts the null, θ is underestimated. The bottom row shows those rare cases, only 3 in 1000 sequences, that the criterial-precision stopping rule remains undecided, in which case θ is slightly underestimated.

In summary, Figures 13.6 and 13.7 have shown that when testing sequentially, p values (conditionalizing on N) will always eventually reject the null, even when it is true, and the resulting estimates of θ tend to be too extreme. Stopping at criterial BF prevents 100% false alarms when the null is true, but also often falsely accepts the null when it is not true. The BF also results in estimates of θ that are even more extreme. The HDI-with-ROPE rule prevents 100% false alarms when the null is true, and does not falsely accept the null when it is not true (in this case). The HDI-with-ROPE rule results in estimates of θ that are extreme, but not as badly as the BF stopping rule. The price paid by the HDI-with-ROPE rule is that it tends to require larger sample sizes. Stopping at criterial precision never falsely rejected the null when it was true, and never falsely accepted the null when it was not true, but did remain undecided in some cases. It gave virtually unbiased estimates of θ. To achieve the criterial precision, a large sample was required.

The key point is this: If the sampling procedure, such as the stopping rule, biases the data in the sample then the estimation can be biased whether it's Bayesian estimation or not. A stopping rule based on getting extreme values will automatically bias the sample toward extreme estimates, because once some extreme data appear by chance, sampling stops. A stopping rule based on precision will not bias the sample unless the measure of precision depends on the value of the parameter (which actually is the case here, just not very noticeably for parameter values that aren't very extreme).

Thus, if one wants unbiased estimates and low error rates, then sampling to achieve a criterial precision is a method to consider. The down side of stopping at criterial precision is that it can require a large sample. There may be some situations in which the costs of a large sample are too high, such as medical trials in which lives are at stake. A full treatment of these trade-offs falls under the rubrics of Bayesian *adaptive design* and Bayesian *decision theory*. Book-length discussions of these topics are available, and there is not space here to delve deeply into them. A fundamental idea of decision theory is that each decision or action is assigned a *utility*, which is a measure of its cost or benefit. For example, what is the cost of falsely rejecting the null hypothesis that the new drug is the same as the old drug? What is the cost of falsely deciding that the new drug is better than the old drug? What is the cost of overestimating the effectiveness of the drugs? What is the cost of sampling more data, especially when those data come from sick patients waiting for effective treatment? Conversely, what are the benefits of correct decisions, accurate estimations, and less data collection?

In closing this brief section on sequential testing, it is appropriate to recapitulate its relation to power, which is the main topic of this chapter. Power (defined generally, not traditionally) is the probability of achieving a specified goal, when the world is described by a hypothesized distribution over parameter values, and when the data are sampled by a specified stopping rule. For power analysis, the typically assumed stopping rule is fixed sample size. What has changed in this section is the stopping rule: We do not stop when the sample size N reaches a threshold, but instead when a summary statistic of the sampled data (such as p, BF, HDI, or HDI width) reaches a threshold (respectively 0.05, 3, ROPE limits, and 0.8 ROPE width). Under these stopping rules, the probabilities of achieving the goals of rejecting or accepting the null are the asymptotic decision proportions in Figures 13.6 and 13.7. In other words, those figures illustrate the power of these various stopping rules in sequential testing. The figures also emphasized that the resulting parameter estimates are biased when stopping is based on extremeness in the data rather than precision in the data. The sequential-testing examples also differed from the previous power examples in the shape of the hypothetical distribution over parameter values. The examples using sequential testing posited "spike" shaped distributions, such as $\theta = 0.50$ with no uncertainty, so that the issue of false-alarm rates could be addressed.

13.4. DISCUSSION

13.4.1. Power and multiple comparisons

In NHST, the overall p value for any particular test is increased when the test is considered in the space of all other intended tests. An example was discussed in Section 11.1.5, p. 310. This must be taken into account in frequentist power analysis. When there are multiple tests, the power of any one test is reduced.

In a frequentist approach, multiple tests must also be taken into account when the goal is precision instead of rejecting a null hypothesis. In the frequentist approach, precision can be measured by a confidence interval (CI). We saw in Section 11.3.1, specifically Figure 11.11, p. 322, that CIs depend on the intended tests because p values depend on the intended tests. Pan and Kupper (1999) discuss frequentist power when the goal is achieving precision (as opposed to rejecting the null) and when multiple comparisons are intended.

Bayesian power analysis is not affected by intending multiple tests. In Bayesian analysis, the decision is based on the posterior distribution, which is determined by the data in hand, whether actual or simulated, and not by what other tests are intended. In Bayesian analysis, the probability of achieving a goal, that is the power, is determined only by the data-generating process (which includes the stopping rule) and not by the cloud of counterfactual samples (which includes other tests).

13.4.2. Power: prospective, retrospective, and replication

There are different types of power analysis, depending on the source of the hypothetical distribution over parameter values and the prior used for analyzing the simulated data. The most typical and useful type is *prospective* power analysis. In prospective power analysis, research is being planned for which there has not yet been any data collected. The hypothetical distribution over parameter values comes from either theory or idealized data or actual data from related research. De Santis (2007) describes how to combine results of previous experiments to construct a data-generating distribution. An example of prospective power analysis was presented in Section 13.2.5, which showed how to use idealized data to create representative hypothetical parameter values.

On the other hand, *retrospective* power analysis refers to a situation in which we have already collected data from a research project, and we want to determine the power of the research we conducted. In this case, we can use the posterior distribution, derived from the actual data, as the representative parameter values for generating new simulated data. (This is tantamount to a posterior predictive check.) In other words, at a step in the posterior MCMC chain, the parameter values are used to generate simulated data. The simulated data are then analyzed with the same Bayesian model as the actual data, and the posterior from the simulated data is examined for whether or not the goals are achieved.

In traditional power analysis, for which the hypothesis is a spike and the only goal is to reject the null based on a p value, it is well known that the estimate of retrospective power has a direct correspondence with the p value, and therefore retrospective power is not useful for additional inference beyond the p value (Gerard et al., 1998; Hoenig & Heisey, 2001; Nakagawa & Foster, 2004; O'Keefe, 2007; Steidl et al., 1997; Sun et al., 2011; L. Thomas, 1997). Retrospective power analysis does make the power explicit however. And, in the generalized Bayesian setting described in this book, retrospective power analysis can reveal the probabilities of achieving other goals.

Finally, suppose that we have already collected some data, and we want to know the probability that we would achieve our goal if we exactly replicated the experiment. In other words, if we were simply to collect a new batch of data, what is the probability that we would achieve our goal in the replicated study, also taking into account the results of the first set of data? This is the *replication* power. As with retrospective power analysis, we use the actual posterior derived from the first sample of data as the data generator. But for analysis of the simulated data, we again use the actual posterior from first sample of data, because that is the best-informed prior for the follow-up experiment. An easy way to execute this analysis by MCMC is as follows: Use the actual set of data with a skeptical-audience prior to generate representative parameter values and representative simulated data. Then, *concatenate the original data with the novel simulated data* and update the original skeptical-audience prior with the enlarged data set. This technique is tantamount to using the posterior of the original data set as the prior for the novel simulated data. Computation of replication power is natural in a Bayesian setting, but is difficult or impossible for traditional NHST (Miller, 2009). NHST has trouble when addressing replication probability because it has no good way to model a data generator: It has no access to the posterior distribution from the initial analysis.

13.4.3. Power analysis requires verisimilitude of simulated data

Power analysis is only useful when the simulated data imitate actual data. We generate simulated data from a descriptive model that has uncertainty in its parameter values, but we assume that the model is a reasonably good description of the actual data. If the model is instead a poor description of the actual data, then the simulated data do not imitate actual data, and inferences from the simulated data are not very meaningful. It is advisable, therefore, to check that the simulated data accurately reflect the actual data.

When simulated data differ from actual data, strange results can arise in power analysis. Consider an analysis of replication probability in which the simulated data are quite different to the actual data. The novel simulated data are combined with the original data to conduct the replication analysis. The combined data are a mixture of two different trends (i.e., the actual trend and the different simulated trend), and therefore the estimates of the parameters become more *un*certain than for the original data alone. It is only when the simulated sample size becomes large, relative to the original sample size, that the simulated trend overwhelms the actual trend, and the replication uncertainty becomes smaller again. If you find in your analyses of replication power that parameter uncertainty initially gets larger as the simulated sample size increases, then you may have a situation in which the model does not faithfully mimic the actual data.

13.4.4. The importance of planning

Conducting a power analysis in advance of collecting data is very important and valuable. Often in real research, a fascinating theory and clever experimental manipulation imply a subtle effect. It can come as a shock to the researcher when power analysis reveals that detecting the subtle effect would take many hundreds of subjects! But the shock of power analysis is far less than the pain of actually running dozens of subjects and finding highly uncertain estimates of the sought-after effect.

Power analysis can reduce research pain in other ways. Sometimes in real research, an experiment or observational study is conducted merely to objectively confirm what is anecdotally known to be a strong effect. A researcher may be tempted to conduct a study using the usual large sample size that is typical of related research. But a power analysis may reveal that the strong effect can be easily detected with a much smaller sample size.

Power analysis is also important when proposing research to funding agencies. Proposals in basic research might have fascinating theories and clever research designs, but if the predicted effects are subtle, then reviewers of the proposal may be justifiably dubious, and want to be reassured by a power analysis. Proposals in applied research are even more reliant on power analysis, because the costs and benefits are more immediate and tangible. For example, in clinical research (e.g., medicine, pharmacology, psychiatry, counseling), it can be very costly to test patients, and therefore it is important to anticipate the probable sample size or sampling duration.

While it is important to plan sample size in advance, it can also be important, especially in clinical applications, to monitor data as they are collected and to stop the research as soon as possible. It behooves the researcher to discontinue an experiment as soon as the data clearly indicate a positive or negative outcome: It would be unethical to slavishly continue treating patients with an experimental treatment that is clearly detrimental, and it would be unethical to slavishly continue running patients in a placebo condition when the experimental treatment is clearly having positive effects. The decision regarding when to stop collecting data is a topic of much investigation, and goes under the name of Bayesian *optimal* or *adaptive design*. It will not be discussed further here, but the interested reader is referred to books such as the one by S. M. Berry, Carlin, Lee, and Müller (2011), and books in decision theory such as the ones by J. O. Berger

(1985) and DeGroot (2004), and various articles, for example D. A. Berry (2006, 2011), Cavagnaro, Myung, Pitt, and Kujala (2010), and those cited by Roy, Ghosal, and Rosenberger (2009, p. 427).

13.5. EXERCISES

Look for more exercises at https://sites.google.com/site/doingbayesiandataanalysis/

Exercise 13.1. [Purpose: Comic relief.] Read the complete oeuvre of Friedrich Nietzsche, with special attention to his posthumous work, *The Will to Power* (Nietzsche, 1967). Provide a mathematical formalization of the Nietzschian concepts of will and power, using Bayesian probability theory. Show that the notion of statistical power is a special case of formalized Nietzschian power, *and vice versa.* Post your answer on your personal blog. If this exercise does not destroy you, it will make you stronger.

Exercise 13.2. [Purpose: Understanding power for flipping a single coin, in Tables 13.1 and 13.2.] For this exercise, consider flipping a single coin and inferring its bias.

(A) Table 13.2 indicates that when the data-generating distribution is vague, with $\kappa = 10$ and $\omega = 0.80$, then 87 flips are needed for an 80% chance of getting the 95% HDI width to be less than 0.2. What is the minimal N needed if the data-generating distribution is very certain, with $\kappa = 2000$? Show the command you used, and report the exact power for the smallest N that has power greater than 0.8. *Hint:* Change the appropriate argument(s) in `minNforHDIpower (genPriorMode=0.80, genPriorN=10, HDI-maxwid=0.2, nullVal=NULL, ROPE=c(0.48,0.5 2), desiredPower=0.8, aud PriorMode=0.5, audPriorN=2, HDImass=0.95, initSampSize=5, verbose=TRUE )`. Don't forget to `source` the function first.

(B) Regarding the previous part, why might a researcher pursue a goal of precision if the data-generating hypothesis is already very precise? *Hint:* The audience prior may be different than the data-generating hypothesis. Discuss briefly, perhaps with an example.

(C) Table 13.1 indicates that when the data-generating distribution is highly certain, with $\kappa = 2000$ and $\omega = 0.80$, then 19 flips are needed for an 80% chance of getting the 95% HDI to exclude a small ROPE around $\theta = 0.5$. What is the minimal N needed if the data-generating distribution is vague, with $\kappa = 2$? Show the command you used, and report the exact power for the smallest N that has power greater than 0.8.

(D) For the previous part, the goal was for the HDI to exclude the null value (i.e., 0.5). Notice that the goal can be satisfied if the HDI is above the null value *or* if the HDI is below the null value. (i) When the data-generating prior is a beta distribution with $\mu = 0.8$ and $\kappa = 2$, as in the previous part, what proportion of the data-generating biases are greater than the null value? (ii) If the goal is for the HDI to fall entirely *above* the null value, what sample size is needed to achieve a power of 0.8? *Hint:* Use `minNforHDIpower.R` with the argument `ROPE=c(0,0.5)`. Watch the sample size increase indefinitely, with the power creeping toward an asymptote. Why does the power never exceed the proportion you computed for (i)?

Exercise 13.3. [Purpose: Hands on experience with Monte Carlo power simulation in Section 13.2.5.] The script `Jags-Ydich-XnomSsubj-MbinomBetaOmegaKappa-Power.R` in Section 13.2.5 was used to estimate power for the therapeutic-touch experiment of Figure 9.9.

(A) Run the script using `nSimulatedDataSets` of 50. Show which line of code you changed to accomplish this. Report the final power estimates. How do your results compare with the results shown in Section 13.2.5? *Hint:* The power estimates should be about the same, but because you used a smaller number of simulated data sets, the bounds on your power estimate should be wider (less certain).

(B) Now you will run the power simulation starting with idealized data that mimic the actual data. Refer to the posterior distribution from the analysis of the actual data in Figure 9.10, p. 243. Notice the central tendency and HDI on the group-level mode. We will use those characteristics for the idealized data generating hypothesis. Specifically, near the beginning of the script, set `idealGroupMean = 0.44`, `idealGroupSD = 0.04`, `idealNsubj = 28`, and `idealNtrlPerSubj = 10`. Explain what each of those settings does and explain why those values were chosen.

(C) Because the idealized data have central tendency near chance performance, we cannot have high hopes for rejecting the null, and therefore our goal might be high precision. In the function `goalAchievedForSample`, set the `HDImaxwid` to 0.1. Also, for high precision, we will need more data than was obtained in the original experiment, so try setting `NSubj` to 40 and `NtrlPerSubj` to 100. Because this is an exercise, not real research, change the number of simulated data sets to only 20. Report the lines of code you changed (and any you deleted or commented out). Now run the simulation and report the final estimated power for each of the goals. Why does the goal `omegaNarrowHDI` have high power but the goal `thetaNarrowHDI` have low power?

(D) For those who want a simple programming exercise in R, try this: Instead of using idealized data to create hypothetical data-generating parameter values, use the actual data from the original experiment. In the first part of the script, just comment out or delete the lines that create idealized data. Instead, use the actual data in the `genMCMC` function. Then repeat the previous part. Are the power estimates about the same?

Exercise 13.4. [Purpose: To explore sequential testing of coin flips.] For this exercise, your job is to create your own versions of the graphs in Figure 13.4. Ready? Go! *Hints:* (continued on next page)

- For a reminder of how to create and plot z/N, see Figure 4.1, p. 75, and its accompanying description in the text.
- To compute a p value for a proportion, assuming the intention was to stop at N, you could use R's function `binom.test`. For example: `z=9 ; N=10 ; theta=0.5 ; binom.test( x=z , n=N , p=theta , alternative="two.sided" )$p.value` returns the p value. Or you could compute it "from scratch" by using the definitions of the binomial distribution.
- The BF can be computed from Equation 12.3, p. 344. Be careful to use the beta *function* and not the beta *distribution*.
- The HDI can be computed with the function `HDIofICDF` that has been used in some of the power scripts and functions. It is defined in the utilities functions that come with this book, and therefore that file must be sourced before the function can be used. For this application, its basic format is `HDIofICDF`(qbeta , shape1=1+z , shape2=1+N-z).

CHAPTER 14

Stan

Contents

Fools lob proposals on random trajectories,
Finding requital just once in a century.
True love homes in on a heart that is radiant,
Guided by stars from Sir Hamilton's gradient.[1]

Stan is the name of a software package that creates representative samples of parameter values from a posterior distribution for complex hierarchical models, analogous to JAGS. Take a look at Figure 8.1, p. 194, which shows the relation of R to JAGS and Stan. Just as we can specify models for JAGS and communicate with JAGS from R via rjags, we can specify models for Stan and communicate with Stan from R via RStan.

According to the Stan reference manual, Stan is named after Stanislaw Ulam (1909-1984), who was a pioneer of Monte Carlo methods. (Stan is not named after the slang term referring to an overenthusiastic or psychotic fanatic, formed by a combination of the words "stalker" and "fan.") The name of the software package has also been unpacked as the acronym, Sampling Through Adaptive Neighborhoods (Gelman et al., 2013, p. 307), but it is usually written as Stan not STAN.

Stan uses a different method than JAGS for generating Monte Carlo steps. The method is called *Hamiltonian Monte Carlo* (HMC). HMC can be more effective than the

[1] This chapter is about an MCMC sampling scheme that creates proposal distributions that are pulled toward the mode(s) of the posterior distribution instead of being symmetrical around the current position. The proposals use trajectories based on the gradient of the posterior, using a mathematical scheme named after the physicist Sir William Hamilton.

Doing Bayesian Data Analysis, 2nd Edition
http://dx.doi.org/10.1016/B978-0-12-405888-0.00014-3

various samplers in JAGS and BUGS, especially for large complex models. Moreover, Stan operates with compiled C++ and allows greater programming flexibility, which again is especially useful for unusual or complex models. For large data sets or complex models, Stan can provide solutions when JAGS (or BUGS) takes too long or fails. However, Stan is not universally faster or better (at this stage in its development). For some of the applications in this book, JAGS works as fast or faster, and there are some models that cannot (yet) be directly expressed in Stan.

Stan involves a little extra programming overhead than JAGS, so Stan is a little harder to learn from scratch than JAGS. But once you know JAGS, it is fairly easy to learn the additional details of Stan. Stan also has extensive documentation and a number of programming abilities not available in JAGS. Because Stan is undergoing rapid development at the time of this writing, the goal of this book is not to present a complete library of programs for Stan. Instead, this chapter presents the ideas, a complete example, and guidelines for composing programs in Stan. Later chapters include a variety of additional Stan programs. By the time you learn Stan, some of its details may have changed.

14.1. HMC SAMPLING

Stan generates random representative samples from a posterior distribution by using a variation of the Metropolis algorithm called HMC. To understand it, we briefly review the Metropolis algorithm, which was explained in Section 7.3, p. 156. In the Metropolis algorithm, we take a random walk through parameter space, favoring parameter values that have relatively high posterior probability. To take the next step in the walk, there is a proposed jump from the current position, with the jump sampled randomly from a proposal distribution. The proposed jump is accepted or rejected probabilistically, according to the relative densities of the posterior at the proposed position and the current position. If the posterior density is higher at the proposed position than at the current position, the jump is definitely accepted. If the posterior density is lower at the proposed position than the current position, the jump is accepted only with probability equal to the ratio of the posterior densities.

The key feature I want to emphasize here is the shape of the proposal distribution: In the vanilla Metropolis algorithm, the proposal distribution is symmetrically centered on the current position. In multidimensional parameter spaces, the proposal distribution could be a multivariate Gaussian distribution, with its variances and covariances tuned for the particular application. But the multivariate Gaussian is always centered on the current position and is always the same shape regardless of where the walk roams in parameter space. This fixedness can lead to inefficiencies. For example, in the tails of the posterior distribution, the proposals will just as often go away from a posterior mode as go toward it, and therefore proposals will often be rejected. As another example, if the posterior distribution curves through parameter space, a fixed-shape proposal distribution that is well tuned for one part of the posterior may be poorly tuned for another part of the posterior.

HMC instead uses a proposal distribution that changes depending on the current position. HMC figures out the direction in which the posterior distribution increases, called its gradient, and warps the proposal distribution toward the gradient. Consider Figure 14.1. The top panel shows a simple posterior distribution on a single parameter called theta. The current position in the Markov chain is indicated by a large dot on the abscissa. The two columns of Figure 14.1 show two different current positions. From either current position, a jump is proposed. In vanilla Metropolis, the proposal distribution would be a Gaussian centered on the current position, such that jumps above or below the current position would be equally likely to be proposed. But HMC generates proposals quite differently.

HMC generates a proposal by analogy to rolling a marble on the posterior distribution turned upside down. The second row of Figure 14.1 illustrates the upside-down posterior distribution. Mathematically, the upside-down posterior is the negative logarithm of the posterior density, and it is called the "potential" function for reasons to be revealed shortly. Wherever the posterior is tall, the potential is low, and wherever the posterior is short, the potential is high. The large dot, that represents the current position, is resting on the potential function like a marble waiting to roll downhill. The proposed next position is generated by flicking the marble in a random direction and letting it roll around for a certain duration. In this simple one-parameter example, the direction of the initial flick is randomly to the right or to the left, and the magnitude of the flick is randomly sampled from a zero-mean Gaussian. The flick imparts a random initial *momentum* to the ball, as suggested by the annotation in the second row of Figure 14.1. When time is up, the ball's new position is the proposed position for the Metropolis jump. You can imagine that the marble will tend to be caught at positions that are downhill on the potential function relative to the starting position. In other words, the proposed position will tend to be in regions of higher posterior probability.

The terminology comes from analogy to the physics of objects in gravity. In theoretical physics, a moving object has *kinetic* energy that trades off with *potential* energy: A stationary marble at the top of a hill has lots of potential energy but no kinetic energy, but when it is rolling to the bottom of the hill the marble has exchanged some of its potential energy for kinetic energy. In idealized frictionless systems, the sum of potential and kinetic energy is constant, and the dynamics of the system conserve the total energy. Real rolling balls violate the ideal because they incur friction (and change their angular momentum as they roll, which complicates the dynamics). A better analogy to the ideal is a puck sliding (not rolling) on a virtually frictionless surface such as smooth ice or a cushion of pressurized air. But ice feels cold and cushions of air involve noisy machines, so let's just imagine a pleasantly perfect marble on a smooth hill that has no friction and no spin.

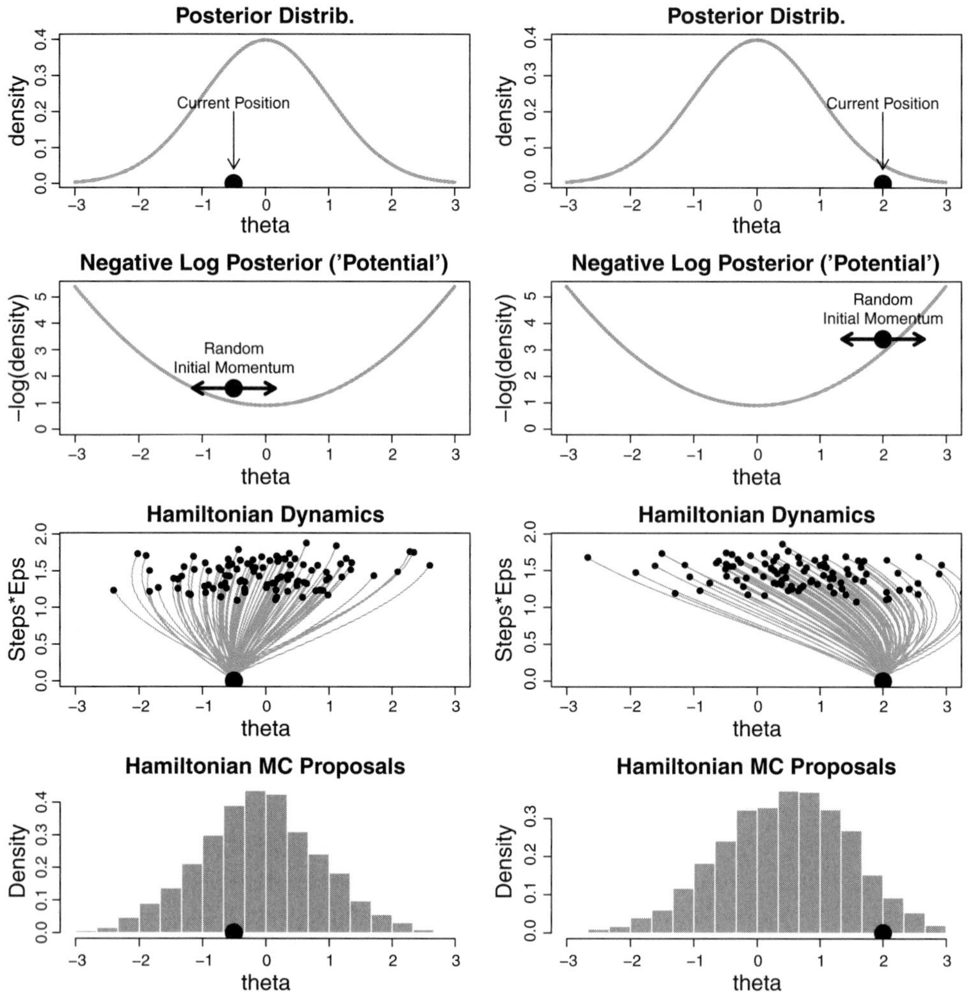

Figure 14.1 Examples of a Hamiltonian Monte Carlo proposal distributions. Two columns show two different current parameter values, marked by the large dots. First row shows posterior distribution. Second row shows the potential energy, with a random impulse given to the dot. Third row shows trajectories, which are the theta value (*x*-axis) as a function of time (*y*-axis marked Steps*Eps). Fourth row shows histograms of the proposals.

The third row of Figure 14.1 shows examples of many trajectories taken by balls that start on the current position and are given random impulses. The ordinate (*y*-axis) is mysteriously labeled as "Steps*Eps." Its exact meaning will be revealed below, but you can think of it simply as duration. Time increases as Steps*Eps increases. The trajectories in Figure 14.1 have random durations constrained within a fairly narrow range of possibilities. The trajectories show the theta value as a function of time as the ball rolls after it receives the random initial impulse. The end point of each trajectory is marked with a small dot; this is the proposed position.

The bottom row of Figure 14.1 shows histograms of all the proposed positions. In particular, you can see that the proposal distribution is *not* centered on the current position. Instead, the proposal distribution is shifted toward the mode of the posterior distribution. Notice that the proposal distributions are quite different for the different current positions. In both cases, however, the proposals are shifted toward the mode of the posterior distribution. You can imagine that *for high-dimensional posterior distributions that have narrow diagonal valleys and even curved valleys, the dynamics of HMC will find proposed positions that are much more promising than a vanilla symmetric proposal distribution, and more promising than Gibbs sampling which can get stuck at diagonal walls (as was described at the very end of Section 7.4.4)*.

Once the proposed jump is established, then the proposal is accepted or rejected according to the Metropolis decision rule as in Equation 7.1, p. 151, *except* that the terms involve not only the relative posterior density, but also the momentum at the current and proposed positions. The initial momentum applied at the current position is drawn randomly from a simple probability distribution such as a normal (Gaussian). Denote the momentum as ϕ. Then the Metropolis acceptance probability for HMC becomes

$$p_{\text{accept}} = \min\left(\frac{p(\theta_{\text{proposed}}|D)\, p(\phi_{\text{proposed}})}{p(\theta_{\text{current}}|D)\, p(\phi_{\text{current}})}, 1\right) \tag{14.1}$$

In an idealized continuous system, the sum of potential and kinetic energy [corresponding to $-\log(p(\theta|D))$ and $-\log(p(\phi))$] is constant, and therefore the ratio in Equation 14.1 would be 1, and the proposal would never be rejected. But in practical simulations, the continuous dynamics are discretized into small intervals of time, and the calculations are only approximate. Because of the discretization noise, the proposals will not always be accepted.

If the discrete steps of the trajectory are very small, then the approximation to the true continuous trajectory will be relatively good. But it will take many steps to go very far from the original position. Conversely, larger individual steps will make a poorer approximation to the continuous mathematics, but will take fewer steps to move far from the original position. Therefore, the proposal distribution can be "tuned" by adjusting the step size, called epsilon or "eps" for short, and by adjusting the number of steps. We think of the step size as the time it takes to make the step, therefore the total duration of the trajectory is the number of steps multiplied by the step size, or "Steps*Eps" as displayed on the y-axis of the trajectories in Figure 14.1. Practitioners of HMC typically strive for an acceptance rate of approximately 65% (Neal, 2011, p. 142). If the acceptance rate of a simulation is too low, then epsilon is reduced, and if the acceptance rate of a simulation is too high, then epsilon is increased, with compensating changes in the number of steps to maintain the trajectory duration.

The step size controls the smoothness or jaggedness of the trajectory. The overall duration, Steps*Eps, controls how far the proposal ventures from the current position. This duration is important to tune, because we want the proposal to be closer to a mode, without overshooting, and without rolling all the way back to the starting point. Figure 14.2 shows a wide range of trajectories for the same starting positions as Figure 14.1. Notice that the long trajectories overshoot the mode and return to the current position. To prevent inefficiencies that would arise from letting the trajectories make a U-turn, Stan incorporates an algorithm that generalizes the notion of U-turn to high-dimensional parameter spaces and estimates when to stop the trajectories before they make a U-turn back toward the starting position. The algorithm is called the "no U-turn sampler" (NUTS; M. Hoffman & Gelman, 2014). By comparing the proposal distributions in Figures 14.1 and 14.2, you can intuit that the ones in Figure 14.1 will explore the posterior distribution more efficiently.

Other than step size and number of steps, there is a third tuning knob on the proposal distribution, namely, the standard deviation of the distribution from which the initial momentum is selected. Figure 14.3 shows examples of proposal distributions starting at the same current position, with the same trajectory duration, but with different standard deviations for the random initial momentum. You can see that when the standard deviation of the momentum distribution is wider, the proposal distribution is wider. As you can see by comparing Figures 14.1 and 14.3, the most efficient standard deviation is not too wide and not too narrow. The standard deviation of the momentum distribution is typically set by adaptive algorithms in Stan to match the standard deviation of the posterior.

Computing a proposal trajectory involves simulating the rolling of the marble down the potential hillside. In other words, we must be able to compute the gradient (i.e., derivative) of the posterior density, at any value of the parameter. This is done efficiently on high-dimensional parameter spaces only with explicit formulas for the gradient (as opposed to using numerical approximation by finite differentials). The formula could be derived by a human being, but for complex models with hundreds of parameters the formulas are derived algorithmically by symbolic-math syntactical engines. In simulating a proposal trajectory with discrete steps, the usual method is to first take a half step along the gradient to update the momentum, before alternating full steps along the gradients of potential and momentum, and finishing with another half step of momentum. This is called a "leapfrog" procedure because of the half steps.

The algorithms for computing gradients and for tuning proposal trajectories are sophisticated and complex. There are many technical hurdles, including a general-purpose system for analytically finding derivatives, dealing with limited-range parameters

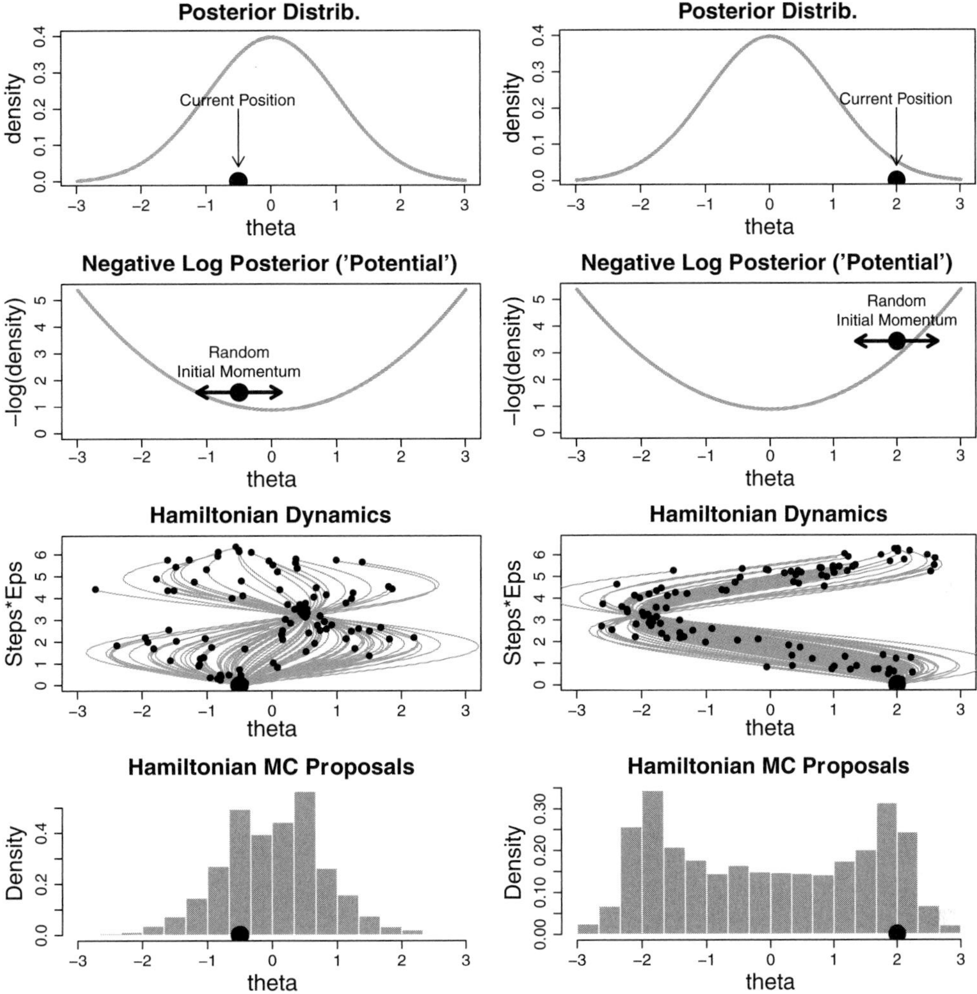

Figure 14.2 Examples of a Hamiltonian Monte Carlo proposal distributions for two different current parameter values, marked by the large dots, in the two columns. For this figure, a large range of random trajectory lengths (Steps*Eps) is sampled. Compare with Figure 14.1.

of different types, and various ways to tune the discretization of the Hamiltonian dynamics in high-dimensional parameter spaces. As Gelman et al. (2013, p. 307) suggest in this understatement, "Hamiltonian Monte Carlo takes a bit of effort to program and tune." Fortunately, the Stan system makes it relatively easy for the user.

Mathematical theories that accurately describe the dynamics of mechanical systems have been worked out by physicists. The formulation here, in terms of kinetic and

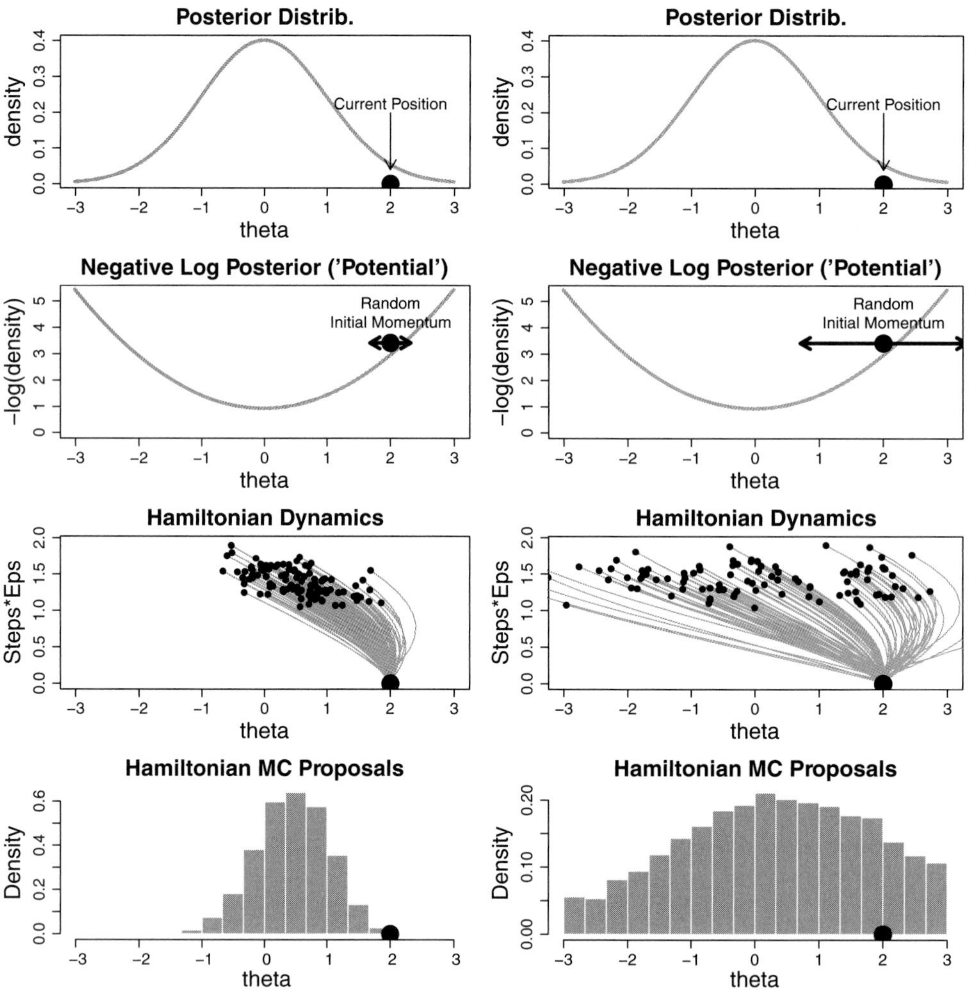

Figure 14.3 Examples of a Hamiltonian Monte Carlo proposal distributions for two different variances of the initial random momentum, indicated in the second row. Compare with Figure 14.1, which shows an intermediate variance of the initial random momentum.

potential energy, is named after William Rowan Hamilton (1805-1865). HMC was described in the physics literature by Duane, Kennedy, Pendleton, and Roweth (1987) (who called it "hybrid" Monte Carlo), and HMC was applied to statistical problems by Neal (1994). A brief mathematical overview of HMC is presented by MacKay (2003, chap. 30). A more thorough mathematical review of HMC is provided by Neal (2011). Details of how HMC is implemented in Stan can be found in the Stan reference manual and in the book by Gelman et al. (2013).

14.2. INSTALLING STAN

Go to the Stan home page at http://mc-stan.org/. (Web site addresses occasionally change. If the address stated here does not work, please search the web for "Stan language." Be sure the site is legitimate before downloading anything to your computer.) On that page there is a link to RStan. Click that link, and you will see extensive installation instructions. Be careful to follow every step in detail. Be sure that your versions of R and RStudio are up to date. I will not include specific details here because those details will change rapidly as Stan continues to be developed. Be sure to download the Stan modeling language user's guide and reference manual. When the RStan library is loaded in R, most of its functions have help pages. For example, if you want to learn more about the RStan `sampling` command, you can type `?sampling` at R's command line.

14.3. A COMPLETE EXAMPLE

The best way to explain programming in Stan with RStan is to show an example. We begin with the simple script named `Stan-BernBeta-Script.R`. The script implements the model of Figure 8.2, p. 196. The model estimates the bias of a single coin. The data are flips of a coin, $y_i \in \{0, 1\}$, described by a Bernoulli distribution, $y_i \sim \text{dbern}(\theta)$, with a beta prior, $\theta \sim \text{beta}(A, B)$. Like JAGS with rjags, the model for Stan with RStan is specified as a string in R. Unlike JAGS, however, the model specification begins with explicit declarations of which variables are data and which variables are parameters, in separately marked blocks before the model statement. Each variable declaration also states what numeric type the variable is, and any restrictions on its domain. For example, the declaration

```
int<lower=0> N ;
```

means that N is an integer value that has a lower bound of zero. Here is the complete model specification, enclosed as a string in R:

```
modelString = "
  data {
   int<lower=0> N ;
   int y[N] ; // y is a length-N vector of integers
  }
  parameters {
    real<lower=0,upper=1> theta ;
  }
  model {
    theta ~ beta(1,1) ;
    y ~ bernoulli(theta) ;
  }
" # close quote for modelString
```

You will have noticed that every line within a block ends with an explicit end-of-command marker, which is a semi-colon. This is the syntax used in C++, which is the underlying language used by Stan. R also interprets a semi-colon as an end-of-command marker, which can be useful for typing several short commands on a single line. But whereas R recognizes either a carriage return or a semi-colon as the end of a command, Stan and C++ recognize only a semi-colon as the end of a command.

Another difference from JAGS is that comments in Stan are indicated by a double slash "//" instead of by a number sign "#." This difference again stems from the fact that Stan specifications are compiled into C++ code, and comments in C++ are indicated by a double slash.

In the model specification, above, the statements in the model block look analogous to the form used by JAGS and BUGS because that was a conscious design principle for Stan. But the Stan model specification is not identical to JAGS. For example, the probability densities are denoted as `beta` instead of `dbeta` and as `bernoulli` instead of `dbern`.

Another important difference is that Stan allows, in fact encourages, vectorization of operations. Thus, in Stan we can write a single line to indicate that every y_i value comes from the Bernoulli distribution:

```
y ~ bernoulli(theta) ;
```

But in JAGS we would have to write an explicit `for` loop:

```
for ( i in 1:N ) {
  y[i] ~ dbern(theta)
}
```

Stan does have `for` loops, but processing is faster when operations can be vectorized.

With the model specified as discussed above, the next step is to translate the model into C++ code and compile the C++ code into an executable *dynamic shared object* (DSO). The command in RStan for doing this is `stan_model`. Before it is called, however, the RStan library must be loaded into R:

```
library(rstan)
stanDso = stan_model( model_code=modelString )
```

If there were errors in the model, Stan would tell you about them here. This translation and compilation step can take a while, depending on how complex the model is. One of the key things that Stan is doing at this point is figuring out the gradient functions for the Hamiltonian dynamics. The resulting DSO is assigned, above, to the variable named `stanDso`.

Once the DSO is created, it can be used for generating a Monte Carlo sample from the posterior distribution. First we specify the data exactly as we did for JAGS, and then we generate the MC sample with the `sampling` command:

```
# Create some fictitious data:
N = 50 ; z = 10 ; y = c(rep(1,z),rep(0,N-z))
dataList = list( y = y , N = N )

stanFit = sampling( object=stanDso , data=dataList ,
                    chains=3 , iter=1000 , warmup=200 , thin=1 )
```

The arguments of the `sampling` command start with telling Stan what DSO to use. The next arguments should look familiar from rjags or runjags, except that Stan uses "`warmup`" instead of "`burnin`." In Stan, `iter` is the total number of steps per chain, including `warmup` steps in each chain. Thinning merely marks some steps as not to be used; thinning does not increase the number of steps taken. Thus, the total number of steps that Stan takes is `chains·iter`. Of those steps, the ones actually used as representative have a total count of `chains·(iter−warmup)/thin`. Therefore, if you know the desired total steps you want to keep, and you know the warm-up, chains, and thinning, then you can compute that the necessary `iter` equals the desired total multiplied by `thin/chains+warmup`.

We did not specify the initial values of the chains in the example above, instead letting Stan randomly initialize the chains by default. The chains can be initialized by the user with the argument `init`, analogous to JAGS. For more information, type "`?sampling`" at R's command line (after previously loading RStan with `library(rstan)`).

The `sampling` command returns more information than only the MC sample of representative parameter values. Also included is the DSO (again), along with information about the run details. There are various ways to examine the MC sample itself. RStan has methods for the standard R `plot` and `summary` commands, and RStan also has its own version of the `traceplot` command (for which there is a different version in the `coda` package used by JAGS). You can experiment with them easily from the command line in R. The RStan versions of `traceplot` and `plot` have an argument, `pars`, which takes a vector of strings that specify which parameters you want to plot. Here, we will convert that output of Stan into a `coda` object so that we can view the results using the same graphical format as we have been using for JAGS:

```
# Load rjags, coda, and DBDA2E functions:
source("DBDA2E-utilities.R")
# Convert stan format to coda format:
mcmcCoda = mcmc.list( lapply( 1:ncol(stanFit) ,
                              function(x) { mcmc(as.array(stanFit)[,x,]) } ) )
# Graph chain diagnostics using DBDA2E function:
diagMCMC( mcmcCoda , parName=c("theta") )
```

The resulting graph, not shown here, has the same format as Figure 8.3, p. 204.

14.3.1. Reusing the compiled model

Because model compilation can take a while in Stan, it is convenient to store the DSO of a successfully compiled model and use it repeatedly for different data sets. It is trivial to do this; just use the `sampling` command again with whatever data set is appropriate. This ability is especially useful for power analysis, which runs the analysis on many simulated data sets. Exercise 14.2 suggests how.

14.3.2. General structure of Stan model specification

The example presented above is very simple for the purpose of a first introduction to Stan. For more complex models, additional components of Stan model specification will be needed. The general structure of model specifications in Stan consist of six blocks, as suggested in the following outline:

```
data {
... declarations ...
}
transformed data {
... declarations ... statements ...
}
parameters {
... declarations ...
}
transformed parameters {
... declarations ... statements ...
}
model {
... declarations ... statements ...
}
generated quantities {
... declarations ... statements ...
}
```

Notice, for example, that after the `data` block there can be a `transformed data` block, in which statements can be placed that transform the data from the `data` statement into new values that can be used in the subsequently specified model. Analogously, parameters declared in the `parameters` block can be transformed to other parameters via statements in the `transformed parameters` block. At the end, if you want to monitor quantities generated from the model at each step, such as predictive values, you can create these in the `generated quantities` block.

The various blocks are optional (except the model block itself), but must be in the order shown above. As explained in detail in Section 14.4, the lines in a Stan model specification are processed in order. Therefore a variable must be declared before it is used in a statement. In particular, that explains why the `transformed parameters` block must be placed after the `parameters` block. For an example, see Exercise 14.1.

14.3.3. Think log probability to think like Stan

If you think a bit about the algorithm used by Stan, you realize that most of Stan's effort is spent on computing the trajectory for Hamiltonian dynamics. From a current parameter position, the algorithm randomly generates an initial momentum and jittered step size and number of steps. Then the Hamiltonian dynamics are deterministically computed from the gradient of the potential function, that is, from the gradient of the (negative) logarithm of the posterior density. This repeats for many steps. At the end of the trajectory, the ratio of the posterior density at proposed and current positions (along with the momentum) is used to deterministically compute an acceptance probability. A random number from a uniform distribution is sampled if the acceptance probability is less than 1. Notice that the probability of acceptance in Equation 14.1 (p. 403) could be re-written by taking the logarithm of both sides, so that the acceptance probability also involves computing the logarithm of the posterior density. Thus, the essence of computation in Stan is dealing with the logarithm of the posterior probability density and its gradient; there is no direct random sampling of parameters from distributions.

If there is no random sampling of parameters from distributions, then what could a model specification like `y ~ normal(mu,sigma)` actually mean to Stan? It means to multiply the current posterior probability by the density of the normal distribution at the datum value `y`. Equivalently, it means to increment the current log-probability by the log-density of the normal at the datum. In fact, in Stan, you could replace the "sampling statement"

```
y ~ normal(mu,sigma)
```

with a corresponding command to increment the log-probability:

```
increment_log_prob( normal_log( y,mu,sigma ))
```

and get the same result. In the command above, `normal_log` is the logarithm of the normal density, and `increment_log_prob` is the function that adds the result to the running total of the log-probability.

The two Stan statements above, while yielding the same posterior MCMC sample, are not completely equivalent, however. The sampling statement assumes that all you want is a representative sample from the posterior distribution, for which all that is needed is the relative not absolute posterior density, so Stan cleverly removes all the constants from the formula for the normal density to improve efficiency in computation. The explicit log-probability form retains the exact posterior density.

One benefit of this computational method for Stan is that you have great flexibility in specifying whatever distribution you want for your model. All you need to do is express the log-probability inside the `increment_log_prob` function. At least, in principle. Stan must also be able to figure out a gradient for the Hamiltonian dynamics. My point of mentioning this here is for you to understand a bit better why log-probability and gradients are so central to the architecture of Stan. For advanced programming details, please see the Stan reference manual.

14.3.4. Sampling the prior in Stan

As was discussed in Section 8.5, p. 211, there are several reasons why we might want to examine a sample from the prior distribution of a model. This can be especially useful for viewing the implied prior on mid-level parameters in a hierarchical model, or for viewing the implied prior on derived parameters such as differences of means.

In the current version of Stan, data cannot have missing values. (Unlike JAGS, which imputes missing values as if they were parameters to be estimated.) Therefore the prior distribution cannot be sampled in Stan merely by commenting out the data as we did with JAGS in Section 8.5. Instead, we comment out the likelihood from the model specification. Here is an example:

```
modelString = "
  data {
    int<lower=0> N ;
    int y[N] ;
  }
  parameters {
    real<lower=0,upper=1> theta ;
  }
  model {
    theta ~ beta(1,1) ;
//     y ~ bernoulli(theta) ;  // likelihood commented out
  }
" # close quote for modelString
```

You must recompile the model with the `stan_model` command. Then sample from the model exactly as before, with the same data as before.

You can understand why this works by thinking about what the sampling statement (now commented out) really means. As mentioned in the previous section, the sampling statement is just incrementing the log-probability according to the data. If we leave that line out, then the log-probability is influenced only by the other statements in the model, namely, the statements that specify the prior.

Stan can have convergence problems when sampling from very diffuse, "flat" distributions because there is such a small gradient. If Stan has trouble sampling from the prior of your model, you can experiment with the priors to see if less diffuse priors solve the convergence problem, and still get a sense of the qualitative nature of the implied priors on derived parameters. JAGS does not have this problem.

14.3.5. Simplified scripts for frequently used analyses

As you will fondly recall from Section 8.3, p. 206, I have wrapped the JAGS scripts for frequently used analyses in functions that can be called using a consistent sequence of commands across different analyses. That way, for example, the single command `plotMCMC` will display different sets of graphs for different applications. I have done the same wrapping for analogous Stan scripts. The file names start with "Stan-" instead of with "Jags-."

For example, the simple Bernoulli-beta model described above is called with the script named `Stan-Ydich-Xnom1subj-MbernBeta-Example.R`. The file name convention was explained in Section 8.3. The script is virtually identical to the JAGS version, except that "Jags" is replaced with "Stan" and there are a few extra lines at the end to illustrate the use of RStan plotting functions:

```
# Load The data
myData = read.csv("z15N50.csv")
# Load the functions genMCMC, smryMCMC, and plotMCMC:
source("Stan-Ydich-Xnom1subj-MbernBeta.R")
# Specify filename root and graphical format for saving output.
fileNameRoot = "Stan-Ydich-Xnom1subj-MbernBeta-"
graphFileType = "eps" # or "png" or "pdf" etc.
# Generate the MCMC chain:
mcmcCoda = genMCMC( data=myData , numSavedSteps=10000 , saveName=fileNameRoot )
# Display diagnostics of chain, for specified parameters:
parameterNames = varnames(mcmcCoda) # get all parameter names
for ( parName in parameterNames ) {
  diagMCMC( mcmcCoda , parName=parName ,
              saveName=fileNameRoot , saveType=graphFileType )
}
# Get summary statistics of chain:
summaryInfo = smryMCMC( mcmcCoda , compVal=0.5 , rope=c(0.45,0.55) ,
                        saveName=fileNameRoot )
# Display posterior information:
plotMCMC( mcmcCoda , data=myData , # compVal=0.5 , rope=c(0.45,0.55) ,
          saveName=fileNameRoot , saveType=graphFileType )
# Use Stan display functions instead of DBDA2E functions:
# Load the stanFit object that was saved by genMCMC:
load("Stan-Ydich-Xnom1subj-MbernBeta-StanFit.Rdata")
# Display information:
show(stanFit)
openGraph()
```

```
traceplot(stanFit,pars=c("theta"))
openGraph()
plot(stanFit,pars=c("theta"))
```

14.4. SPECIFY MODELS TOP-DOWN IN STAN

For humans, descriptive models begin, conceptually, with the data that are to be described. We first know the measurement scale of the data and their structure. Then we conceive of a likelihood function for the data. The likelihood function has meaningful parameters, which we might want to re-express in terms of other data (called covariates, predictors, or regressors). Then we build a meaningful hierarchical prior on the parameters. Finally, at the top level, we specify constants that express our prior knowledge, which might be vague or noncommittal.

A nice feature of JAGS is that models can be specified in that conceptual order: from data, to likelihood, to successively higher levels of the prior. This format was discussed in Section 8.2.2, p. 198, where it was also explained that JAGS does not care about the order in which the dependencies are specified in its model statement. JAGS does not execute the lines in order as a procedure; instead JAGS examines the specifications of the dependencies, checks them for consistency, and assembles MCMC samplers for the corresponding model structure.

This bottom-up ordering is *not* appropriate for model specifications in Stan, however. Stan translates the model specification directly into corresponding C++ commands, which are processed in order. For example, we cannot start a model specification with `y ~ bernoulli(theta)` because we have not yet told Stan the value of `theta`. Stan would try to fill in the value of `theta` with a default value from the variable's declaration or with the value from the previous MCMC step, neither of which is want we intend. Therefore, *in Stan, model specifications usually begin with the top level of the prior, and then lower-level dependencies are filled in, finishing with the likelihood of the data.*

The fact that Stan executes the model specification in order can be very useful for flow control in complex models. For example, unlike JAGS, Stan has `while` loops, which keep cycling until a condition has been met. Stan also has `if else` structures, not available in JAGS.

I find that when I am initially typing a new model in Stan ("typing" in the sense of manually poking fingers on a computer keyboard, not "typing" in the sense of declaring variable types), I type the model in bottom-up conceptual order, moving the cursor around to different blocks of the specification as I go. Before any typing, I make a (hand sketched) diagram of the model as in Figure 8.2, p. 196. Then I start by typing the likelihood function, such as `y ~ bernoulli(theta)` in the model block. Then I move the cursor to the data block and declare the `y` and `N` variables, then I move the cursor to the parameter block and declare the `theta` variable. Then I move, conceptually, to the next level up the model structure, visually moving up the hierarchical diagram (as in Figure 8.2, p. 196). I place the cursor at the start of the model block, *above* the previously specified likelihood function, and type the corresponding specification, `theta ~ beta(1,1)`. If this statement had introduced new higher-level parameters or data, I would then declare them in the appropriate blocks. This process repeats until reaching the top of the hierarchical structure. After the model has been typed-in using bottom-up conceptual order, I then check the model specification for accuracy and efficiency in procedural order, reading the text from top to bottom. I ask myself, Are variables processed in correct logical order? Are loops and if-else statements really in the correct order? Could the processing efficiency be improved by using different flow structure or vectorization?

14.5. LIMITATIONS AND EXTRAS

At the time of this writing, one of the main limitations of Stan is that it does not allow discrete (i.e., categorical) parameters. The reason for this limitation is that Stan has HMC as its foundational sampling method, and HMC requires computing the gradient (i.e., derivative) of the posterior distribution with respect to the parameters. Of course, gradients are undefined for discrete parameters. The Stan 2.1.0 reference manual states, "Plans are in place to add full discrete sampling in Stan 2.0" (p. 4, Footnote 2). Perhaps by the time you read this book, Stan will allow discrete parameters.

In particular, the lack of discrete parameters in Stan means that we cannot do model comparison as a hierarchical model with an indexical parameter at the top level, as was done in Section 10.3.2. There might be ways to work around this restriction by using clever programming contrivances, but presently there is nothing as straight forward as the model specification in JAGS.

A facility of Stan that is not present in JAGS is the ability to find the mode of the posterior distribution in the joint parameter space, directly from the formulation of the model and not from an MCMC sample. If the posterior distribution is curved, the multidimensional mode is not necessarily the same as the marginal modes. We will not use this facility for the applications in this book, however.

14.6. EXERCISES

Look for more exercises at https://sites.google.com/site/doingbayesiandataanalysis/

Exercise 14.1. [Purpose: Transformed parameters in Stan, and comparison with JAGS.] For this exercise, we analyze in Stan the therapeutic-touch data from Section 9.2.4, p. 240. The model structure is depicted in Figure 9.7, p. 236. The relevant Stan scripts are included in the online suite of programs that accompany the book; see

`Stan-Ydich-XnomSsubj-MbernBetaOmegaKappa-Example.R`. Compare it with the JAGS version, which is the same file name but starting with Jags-.

(A) Consider the model block of the Stan code called by the script. What does the `transformed parameters` block do?

(B) Consider the model block of the Stan code called by the script. Why is `theta` vectorized but not `y`?

(C) Run the Stan program, and note how long it takes. Run the Jags version and note how long it takes. Do they produce the same posterior distribution with the same effective sample size (ESS)? Which is faster? You might find that Stan is no faster, or even much slower. Thus, there is no guarantee that Stan will work better. Please note that to make comparisons between Stan and Jags, the ESS of the resulting chains must be matched, not the number of steps. The summaries and graphs created by the `coda`-based functions are convenient for displaying the ESS computed with the same ESS algorithm for the outputs of both Stan and JAGS. For a fair comparison of Stan and JAGS, they also must be matched for their required burn-in phases, as they might need different amounts of burn-in to reach converged chains.

Exercise 14.2. [Purpose: Power analysis in Stan.] It was briefly mentioned in Section 14.3.1 that the DSO created by the Stan command `stan_model` could be reused in multiple calls of the Stan `sampling` command. For this exercise, your goal is to implement in Stan the Monte Carlo approximation of power described in Section 13.2.4, p. 372. It is straight forward to implement it in Stan using the exactly analogous function definitions, but this approach, while logically correct, is inefficient because it will recompile the Stan model every time a new simulated data set is analyzed. The real exercise of this exercise is to modify the function that does the Stan analysis so that it can either compile the Stan model afresh or reuse an existing DSO. In your modified version, *compile the Stan model outside the loop that runs the analyses on simulated data, and pass the compiled DSO as an argument into the function that runs an analysis on a simulated data set.* Show your code, with explanatory comments.

PART III

The Generalized Linear Model

In this part of the book, the methods and concepts explained in the previous parts will be applied to cases of the generalized linear model (GLM). The GLM encompasses multiple linear regression, logistic regression, and Bayesian analogues to classical analysis of variance (ANOVA) and frequency-table analysis, among other cases. Many realistic examples will be discussed in detail. The accompanying computer programs are high-level scripts that can be readily adapted to your own data files, because the scripts use only the few commands needed for loading data and specifying summaries of results.

CHAPTER 15

Overview of the Generalized Linear Model

Contents

Straight and proportionate, deep in your core
All is orthogonal, ceiling to floor.
But on the outside the vines creep and twist
'round all the parapets shrouded in mist.[1]

The previous part of the book explored all the basic concepts of Bayesian analysis applied to a simple likelihood function, namely the Bernoulli distribution. The focus on a simple likelihood function allowed the complex concepts of Bayesian analysis, such as MCMC

[1] The poem is a metaphorical description of the generalized linear model (GLM). The core of the GLM is a linear combination of predictors; the resulting value is proportional to the magnitudes of the predictors, as described in the poem. The GLM can have a nonlinear inverse link function; this is the twisting vine in the poem. The GLM has a random noise distribution that obscures the underlying trend; this is the shrouding mist of the poem.

Doing Bayesian Data Analysis, 2nd Edition
http://dx.doi.org/10.1016/B978-0-12-405888-0.00015-5

methods and hierarchical priors, to be developed without interference from additional complications of elaborate likelihood functions with multiple parameters.

In this new part of the book, we apply all the concepts to a more complex but versatile family of models known as the generalized linear model (GLM; McCullagh & Nelder, 1989; Nelder & Wedderburn, 1972). This family of models comprises the traditional "off the shelf" analyses such as *t* tests, analysis of variance (ANOVA), multiple linear regression, logistic regression, log-linear models, etc. Because we now know from previous chapters the concepts and mechanisms of Bayesian analysis, we can focus on applications of this versatile model. The present chapter is important for understanding subsequent chapters because it lays out the framework for all the models in the remainder of the book.

15.1. TYPES OF VARIABLES

To understand the GLM and its many specific cases, we must build up a variety of component concepts regarding relationships between variables and how variables are measured in the first place.

15.1.1. Predictor and predicted variables

Suppose we want to predict someone's weight from their height. In this case, weight is the predicted variable and height is the predictor. Or, suppose we want to predict high school grade point average (GPA) from Scholastic Aptitude Test (SAT) score and family income. In this case, GPA is the predicted variable, while SAT score and income are predictor variables. Or, suppose we want to predict the blood pressure of patients who either take drug A, or take drug B, or take a placebo, or merely wait. In this case, the predicted variable is blood pressure, and treatment category is the predictor.

The key mathematical difference between predictor and predicted variables is that the likelihood function expresses the probability of values of the predicted variable as a function of values of the predictor variable. The likelihood function does not describe the probabilities of values of the predictor variable. The value of the predictor variable comes from outside the system being modeled, whereas the value of the predicted variable depends on the value of the predictor variable.

Because the predicted variable depends on the predictor variable, at least mathematically in the likelihood function if not causally in the real world, the predicted variable can also be called the "dependent" variable. The predictor variables are sometimes called "independent" variables. The key conceptual difference between independent and dependent variables is that the value of the dependent variable depends on the value of the independent variable. The term "independent" can be confusing because it can be used strictly or loosely. In experimental settings, the variables that are actually manipulated and set by the experimenter are the independent variables. In this context of experimental manipulation, the values of the independent variables truly are (in principle, at least) independent of the values of other variables, because the experimenter has intervened to arbitrarily set the values of the independent variables. But sometimes a non-manipulated variable is also referred to as "independent," merely as a way to indicate that it is being used as a predictor variable.

Among non-manipulated variables, the roles of predicted and predictor are arbitrary, determined only by the interpretation of the analysis. Consider, for example, people's weights and heights. We could be interested in predicting a person's weight from his or her height, or we could be interested in predicting a person's height from his or her weight. Prediction is merely a mathematical dependency, not necessarily a description of underlying causal relationship. Although height and weight tend to co-vary across people, the two variables are not directly causally related. When a person slouches, thereby getting shorter, she does not lose weight. And when a person drinks a glass of water, thereby weighing more, she does not get taller.

Just as "prediction" does not imply causation, "prediction" also does not imply any temporal relation between the variables. For example, we may want to predict a person's sex, male or female, from his or her height. Because males tend to be taller than females, this prediction can be made with better-than-chance accuracy. But a person's sex is not caused by his or her height, nor does a person's sex occur only after their height is measured. Thus, we can "predict" a person's sex from his or her height, but this does not mean that the person's sex occurred later in time than his or her height.

In summary, all manipulated independent variables are predictor variables, not predicted. Some dependent variables can take on the role of predictor variables, if desired. All predicted variables are dependent variables. The likelihood function specifies the probability of values of the predicted variables as a function of the values of the predictor variables.

Why we care: We care about these distinctions between predicted and predictor variables because the likelihood function is a mathematical description of the dependency of the predicted variable on the predictor variable. The first thing we have to do in statistical inference is identify what variables we are interested in predicting, on the basis of what predictors. As you should recall from Section 2.3, p. 25, the first step of Bayesian data analysis is to identify the data relevant to the analysis, and which variables are predictors and which variable is predicted.

15.1.2. Scale types: metric, ordinal, nominal, and count

Items can be measured on different scales. For example, the participants in a foot race can be measured either by the time they took to run the race, or by their placing in the race (first, second, third, etc.), or by the name of the team they represent. These three measurements are examples of metric, ordinal, and nominal scales, respectively (Stevens, 1946).

Examples of *metric* scales include response time (i.e., latency or duration), temperature, height, and weight. Those are actually cases of a specific type of metric scale called a *ratio* scale, because they have a natural zero point on the scale. The zero point on the scale corresponds to there being a complete absence of the stuff being measured. For example, when the duration is zero, there has been no time elapsed, and when the weight is zero, there is no downward force. Because these scales have a natural zero point, it is meaningful to talk about ratios of amounts being measured, and that is why they are called ratio scales. For example, it is meaningful to say that taking 2 min to solve a problem is twice as long as taking 1 min to solve the problem. On the other hand, the scale of historical time has no known absolute zero. We cannot say, for example, that there is twice as much time in January 2nd as there is time in January 1st. We can refer to the duration since some arbitrary reference point, but we cannot talk about the absolute amount of time in any given moment. Scales that have no natural zero are called *interval* scales because all we know about them is the amount of stuff in an interval on the scale, not the amount of stuff at a point on the scale. Despite the conceptual difference between ratio and interval scales, I will lump them together into the category of metric scales.

A special case of metric-scaled data is *count* data, also called *frequency* data. For example, the number of cars that pass through an intersection during an hour is a count. The number of poll respondents who say they belong to a particular political party is a count. Count data can only have values that are nonnegative integers. Distances between counts have meaning, and therefore the data are metric, but because the data cannot be negative and are not continuous, they are treated with different mathematical forms than continuous, real-valued metric data.

Examples of *ordinal* scales include placing in a race, or rating of degree of agreement. When we are told that, in a race, Jane came in first, Jill came in second, and Jasmine came in third, we only know the order. We do not know whether Jane beat Jill by a nose or by a mile. There is no distance or metric information in an ordinal scale. As another example, many polls have ordinal response scales: Indicate how much you agree with this statement: "Bayesian statistical inference is better than null hypothesis significance testing," with 5 = strongly agree, 4 = mildly agree, 3 = neither agree nor disagree, 2 = mildly disagree, and 1 = strongly disagree. Notice that there is no metric information in the response scale, because we cannot say the difference between ratings of 5 and 4 is the same amount of difference as between ratings of 4 and 3.

Examples of *nominal*, a.k.a. categorical, scales include political party affiliation, the face of a rolled die, and the result of a flipped coin. For nominal scales, there is neither distance between categories nor order between categories. For example, suppose we measure the political party affiliation of a person. The categories of the scale might be Green, Democrat, Republican, Libertarian, and Other (in the United States, with different parties in different countries). While some political theories might suggest that the parties fall on an underlying liberal-conservative scale, there is no such scale in the actual categorical values themselves. In the actual categorical labels there is neither distance nor ordering. As another example of values on a nominal scale, consider eye color and hair color, as in Table 4.1, p. 90. Both eye color and hair color are nominal variables because they have neither distance between levels nor ordering of levels. The cells of Table 4.1 show count data that may be predicted from the nominal color predictors, as will be explored in Chapter 24.

In summary, if two items have different nominal values, all we know is that the two items are different (and what categories they are in). On the other hand, if two items have different ordinal values, we know that the two items are different and we know which one is "larger" than the other, but not how much larger. If two items have different metric values, then we know that they are different, which one is larger, and how much larger.

Why we care: We care about the scale type because the likelihood function must specify a probability distribution on the appropriate scale. If the scale has two nominal values, then a Bernoulli likelihood function may be appropriate. If the scale is metric, then a normal distribution may be appropriate as a probability distribution to describe the data. Whenever we are choosing a model for data, we must answer the question, What kind of scale are we dealing with? As you should recall from Section 2.3, p. 25, the first step of Bayesian data analysis includes identifying the measurement scales of the predictor and predicted variables.

In the following sections, we will first consider the case of a metric-predicted variable with metric predictors. In that context of all metric variables, we will develop the concepts of linear functions and interactions. Once those concepts are established for metric predictors, the notions will be extended to nominal predictors.

15.2. LINEAR COMBINATION OF PREDICTORS

The core of the GLM is expressing the combined influence of predictors as their weighted sum. The following sections build this idea by scaffolding from the simplest intuitive cases.

15.2.1. Linear function of a single metric predictor

Suppose we have identified one variable to be predicted, which we'll call y, and one variable to be the predictor, which we'll call x. Suppose we have determined that both variables are metric. The next issue we need to address is how to model a relationship between x and y. This is now Step 2 of Bayesian data analysis from Section 2.3, p. 25. There are many possible dependencies of y on x, and the particular form of the dependency is determined by the specific meanings and nature of the variables. But in general, across all possible domains, what is the most basic or simplistic dependency of y on x that we might consider? The usual answer to this question is: a linear relationship. A linear function is the generic, "vanilla," off-the-shelf dependency that is used in statistical models.

Linear functions preserve proportionality (relative to an appropriate baseline). If you double the input, then you double the output. If the cost of a candy bar is a linear function of its weight, then when the weight is reduced 10%, the cost should be reduced 10%. If automobile speed is a linear function of fuel injection to the engine, then when you press the gas pedal 20% further, the car should go 20% faster. Nonlinear functions do not preserve proportionality. For example, in actuality, car speed is not a linear function of the amount of fuel consumed. At higher and higher speeds, it takes proportionally more and more gas to make the car go faster. Despite the fact that many real-world dependencies are nonlinear, most are at least approximately linear over moderate ranges of the variables. For example, if you have twice the wall area, it takes approximately twice the amount of paint. It is also the case that linear relationships are intuitively prominent (Brehmer, 1974; P. J. Hoffman, Earle, & Slovic, 1981; Kalish, Griffiths, & Lewandowsky, 2007). Linear relationships are the easiest to think about: Turn the steering wheel twice as far, and we believe that the car should turn twice as sharp. Turn the volume knob 50% higher, and we believe that the loudness should increase 50%.

The general mathematical form for a linear function of a single variable is

$$y = \beta_0 + \beta_1 x \tag{15.1}$$

When values of x and y that satisfy Equation 15.1 are plotted, they form a line. Examples are shown in Figure 15.1. The value of parameter β_0 is called the y-intercept because it is where the line intersects the y-axis when $x = 0$. The left panel of Figure 15.1 shows two lines with different y-intercepts. The value of parameter β_1 is called the slope because it indicates how much y increases when x increase by 1. The right panel of Figure 15.1 shows two lines with the same intercept but different slopes.

In strict mathematical terminology, the type of transformation in Equation 15.1 is called *affine*. When $\beta_0 \neq 0$, the transformation does not preserve proportionality. For example, consider $y = 10 + 2x$. When x is doubled from $x = 1$ to $x = 2$, y increases from $y = 12$ to $y = 14$, which is not doubling y. Nevertheless, *the rate of increase in y is the same for all values of x: Whenever x increases by 1, y increases by β_1*. Moreover, Equation 15.1 can have either y or x shifted so that proportionality is achieved. If y is simply shifted by β_0 and the shifted value is called y^*, then proportionality is achieved: $y^* = y - \beta_0 = \beta_1 x$. Or, if x is shifted by $-\beta_0/\beta_1$ and the shifted value is called x^*, then proportionality is achieved: $y = \beta_1(x - \beta_0/\beta_1) = \beta_1 x^*$. This form of the equation is called x-intercept form or x-threshold form, because $-\beta_0/\beta_1$ is where the line crosses the x-axis, and is the threshold between negative and positive values of y.

Summary of why we care. The likelihood function includes the form of the dependency of y on x. When y and x are metric variables, the simplest form of dependency, both mathematically and intuitively, is one that preserves proportionality.

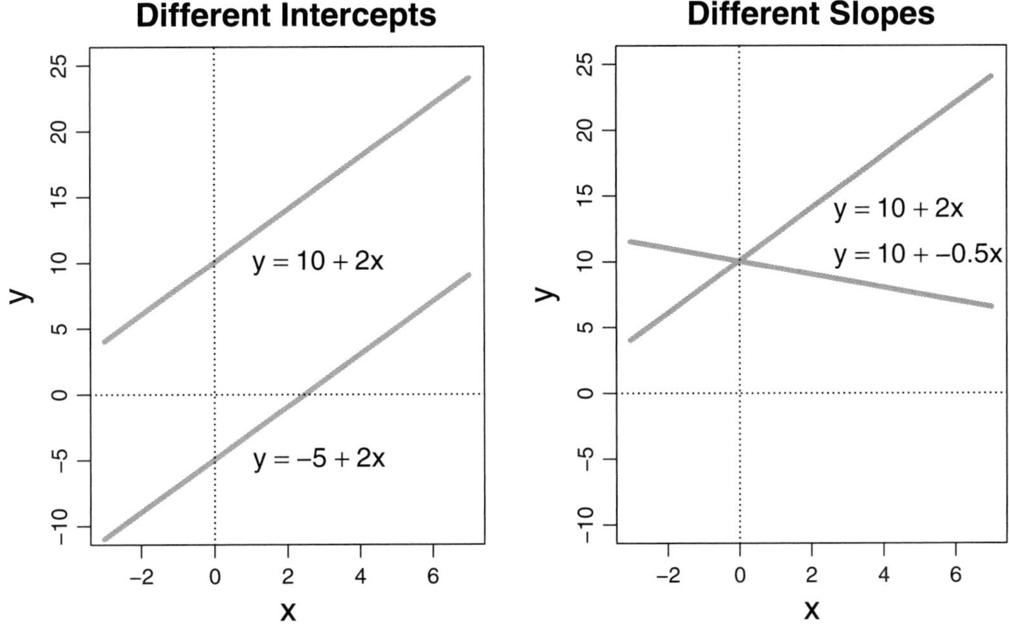

Figure 15.1 Examples of linear functions of a single x variable. The left panel shows examples of two lines with the same slope but different intercepts. The right panel shows examples of two lines with the same intercept but different slopes.

The mathematical expression of this relation is a so-called linear function. The usual mathematical expression of a line is the y-intercept form, but sometimes a more intuitive expression is the x threshold form. Linear functions form the core of the GLM.

15.2.2. Additive combination of metric predictors

If we have more than one predictor variable, what function should we use to combine the influences of all the predictor variables? If we want the combination to be linear in each of the predictor variables, then there is just one answer: Addition. In other words, if we want an increase in one predictor variable to predict the *same* increase in the predicted variable *for any value of the other predictor variables*, then the predictions of the individual predictor variables must be added.

In general, a linear combination of K predictor variables has the form

$$\begin{aligned} y &= \beta_0 + \beta_1 x_1 + \cdots + \beta_K x_K \\ &= \beta_0 + \sum_{k=1}^{K} \beta_k x_k \end{aligned} \tag{15.2}$$

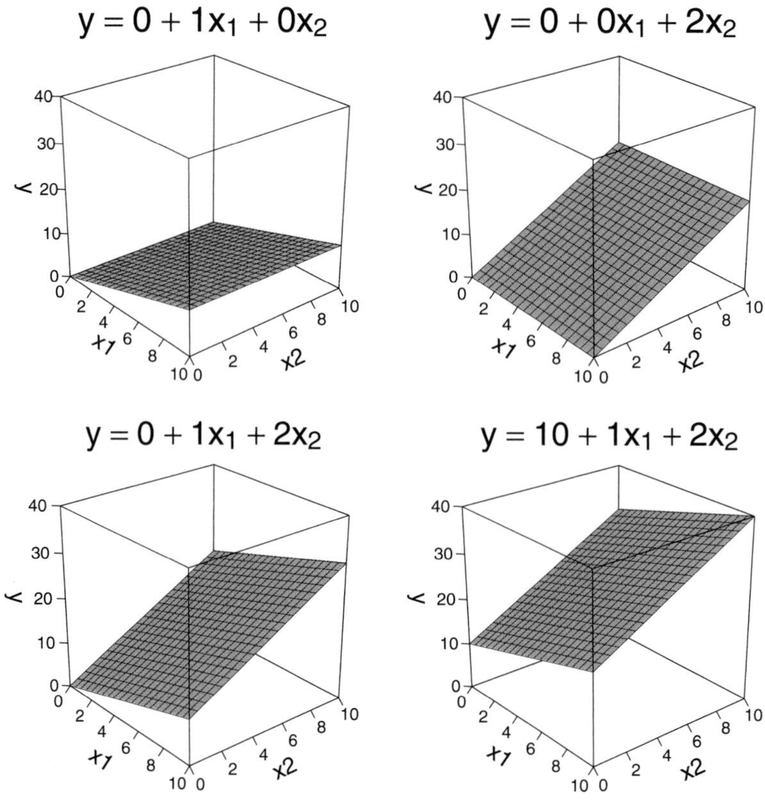

Figure 15.2 Examples of linear functions of two variables, x_1 and x_2. Upper left: Only x_1 has an influence on y. Upper right: Only x_2 has an influence on y. Lower left: x_1 and x_2 have an additive influence on y. Lower right: Nonzero intercept is added.

Figure 15.2 shows examples of linear functions of *two* variables, x_1 and x_2. The graphs show y plotted only over a the domain with $0 \leq x_1 \leq 10$ and $0 \leq x_2 \leq 10$. It is important to realize that the plane extends from minus to plus infinity, and the graphs only show a small region. Notice in the upper left panel, where $y = 0 + 1x_1 + 0x_2$, the plane tilts upward in the x_1 direction, but the plane is horizontal in the x_2 direction. For example, when $x_1 = 10$ then $y = 10$ regardless of the value of x_2. The opposite is true in the upper right panel, where $y = 0 + 0x_1 + 2x_2$. In this case, the plane tilts upward in the x_2 direction, but the plane is horizontal in the x_1 direction. The lower left panel shows the two influences added: $y = 0 + 1x_1 + 2x_2$. Notice that the slope in the x_2 direction is steeper than in the x_1 direction. Most importantly, notice that the slope in the x_2 direction is the same at any specific value of x_1. For example, when x_1 is fixed at 0, then y rises from $y = 0$ to $y = 20$ when x_2 goes from $x_2 = 0$ to $x_2 = 10$. When x_1 is fixed at 10, then y again rises 20 units, from $y = 10$ to $y = 30$.

Summary of section: When the influence of every individual predictor is unchanged by changing the values of other predictors, then the influences are additive. The combined influence of two or more predictors can be additive even if the individual influences are nonlinear. But if the individual influences are linear, and the combined influence is additive, then the overall combined influence is also linear. The formula of Equation 15.2 is one expression of a linear model, which forms the core of the GLM.

15.2.3. Nonadditive interaction of metric predictors

The combined influence of two predictors does not have to be additive. Consider, for example, a person's self-rating of happiness, predicted from his or her overall health and annual income. It's likely that if a person's health is very poor, then the person is not happy, regardless of his or her income. And if the person has zero income, then the person is probably not happy, regardless of his or her health. But if the person is both healthy and rich, then the person has a higher probability of being happy (despite celebrated counter-examples in the popular media).

A graph of this sort of nonadditive interaction between predictors appears in the upper left panel of Figure 15.3. The vertical axis, labeled y, is happiness. The horizontal axes, x_1 and x_2, are health and income. Notice that if either $x_1 = 0$ or $x_2 = 0$, then $y = 0$. But if both $x_1 > 0$ and $x_2 > 0$, then $y > 0$. The specific form of interaction plotted here is *multiplicative*: $y = 0 + 0x_1 + 0x_2 + 0.2x_1x_2$. For comparison, the upper right panel of Figure 15.3 shows a non-interactive (i.e., additive) combination of x_1 and x_2. Notice that the graph of the interaction has a twist or curvature in it, but the graph of the additive combination is flat.

The lower left panel of Figure 15.3 shows a multiplicative interaction in which the individual predictors increase the outcome, but the combined variables decrease the outcome. A real-world example of this occurs with some drugs: Individually, each of two drugs might improve symptoms, but when taken together, the two drugs might interact and cause a decline in health. As another example, consider lighter-than-air travel in the form of ballooning. The levity of a balloon can be increased by fire, as in hot air balloons. And the levity of a balloon can be increased by hydrogen, as in many early twentieth century blimps and dirigibles. But the levity of a balloon is dramatically decreased by the combination of fire and hydrogen.

The lower right panel of Figure 15.3 shows a multiplicative interaction in which the direction of influence of one variable depends on the magnitude of the other variable. Notice that when $x_2 = 0$, then an increase in the x_1 variable leads to a decline in y. But when $x_2 = 10$, then an increase in the x_1 variable leads to an increase in y. Again, the graph of the interaction shows a twist or curvature; the surface is not flat.

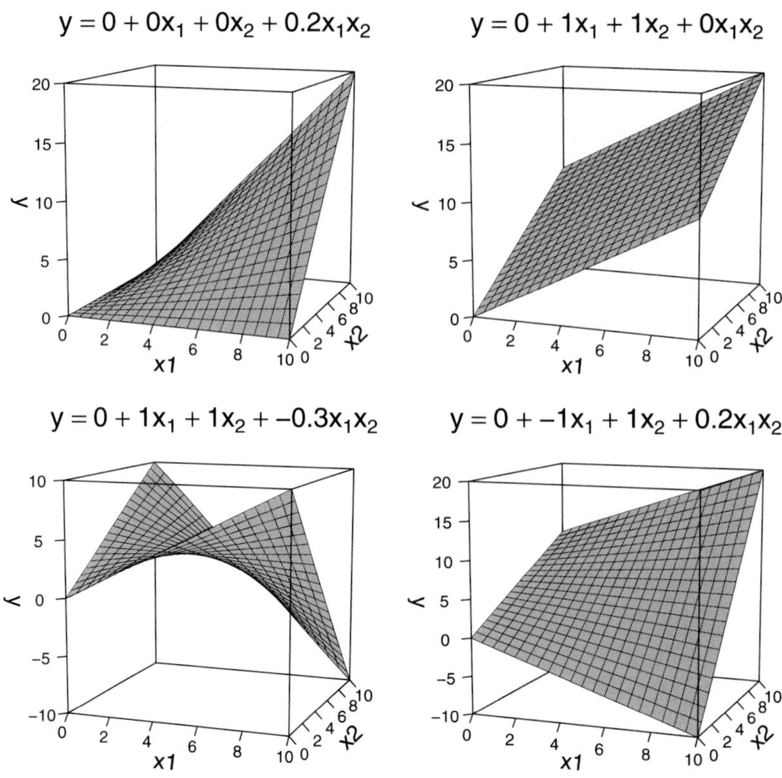

Figure 15.3 Multiplicative interaction of two variables, x_1 and x_2. Upper right panel shows *zero* interaction, for comparison. Figure 18.8, p. 526, provides additional perspective and insight.

There is a subtlety in the use of the term "linear" that can sometimes cause confusion in this context. The interactions shown in Figure 15.3 are *not* linear on the *two* predictors x_1 and x_2. But if the product of the two predictors, x_1x_2, is thought of as a third predictor, then the model *is* linear on the *three* predictors, because the predicted value of y is a weighted additive combination of the three predictors. This reconceptualization can be useful for implementing nonlinear interactions in software for linear models, but we will not be making that semantic leap to a third predictor, and instead we will think of a nonadditive combination of two predictors.

A nonadditive interaction of predictors does not have to be multiplicative. Other types of interaction are possible. The type of interaction is motivated by idiosyncratic theories for different variables in different application domains. Consider, for example, predicting the magnitude of gravitational force between two objects from three predictor variables: the mass of object one, the mass of object two, and the distance between the objects. The force is proportional to the multiplicative product of the two masses *divided by* the *squared* distance between them.

15.2.4. Nominal predictors

15.2.4.1 Linear model for a single nominal predictor

The previous sections assumed that the predictor was metric. But what if the predictor is nominal, such as political party affiliation or sex? A convenient formulation has each value of the nominal predictor generate a particular deflection of y away from its baseline level. For example, consider predicting height from sex (male or female[2]). We can consider the overall average height across both sexes as the baseline height. When an individual has the value "male," that adds an upward deflection to the predicted height. When an individual has the value "female," that adds a downward deflection to the predicted height.

Expressing that idea in mathematical notation can get a little awkward. First consider the nominal predictor. We can't represent it appropriately as a single scalar value, such as 1 through 5, because that would suggest that level 1 is closer to level 2 than it is to level 5, which is not true of nominal values. Therefore, instead of representing the value of the nominal predictor by a single scalar value x, we will represent the nominal predictor by a vector $\vec{x} = \langle x_{[1]}, \ldots, x_{[J]} \rangle$, where J is the number of categories that the predictor has. As you may have just noticed, I will use a subscript in square brackets to indicate a particular element of the vector, by analogy to indices in R. Thus, the first component of $\vec{x}$ is denoted $x_{[1]}$ and the jth component is denoted $x_{[j]}$. When an individual has level j of the nominal predictor, this is represented by setting $x_{[j]} = 1$ and $x_{[i \neq j]} = 0$. For example, suppose x is sex, with level 1 being male and level 2 being female. Then male is represented as $\vec{x} = \langle 1, 0 \rangle$ and female is represented as $\vec{x} = \langle 0, 1 \rangle$. As another example, suppose that the predictor is political party affiliation, with Green as level 1, Democrat as level 2, Republican as level 3, Libertarian as level 4, and Other as level 5. Then Democrat is represented as $\vec{x} = \langle 0, 1, 0, 0, 0 \rangle$, and Libertarian is represented as $\vec{x} = \langle 0, 0, 0, 1, 0 \rangle$.

Now that we have a formal representation for the nominal predictor variable, we can create a formal representation for the generic model of how the predictor influences the predicted variable. As mentioned above, the idea is that there is a baseline level of the predicted variable, and each category of the predictor indicates a deflection above or below that baseline level. We will denote the baseline value of the prediction as β_0. The deflection for the jth level of the predictor is denoted $\beta_{[j]}$. Then the predicted value is

$$\begin{aligned} y &= \beta_0 + \beta_{[1]} x_{[1]} + \cdots + \beta_{[J]} x_{[J]} \\ &= \beta_0 + \vec{\beta} \cdot \vec{x} \end{aligned} \tag{15.3}$$

where the notation $\vec{\beta} \cdot \vec{x}$ is sometimes called the "dot product" of the vectors.

[2] For simplicity of discussion, I will consider the two conventional sex categories of male and female, acknowledging here that there are other biological and gender-identity categories.

Notice that Equation 15.3 has a form similar to the basic linear form of Equation 15.1. The conceptual analogy is this: In Equation 15.1 for a metric predictor, the coefficient β_1 indicates how much y changes when x changes from 0 to 1. In Equation 15.3 for a nominal predictor, the coefficient $\beta_{[j]}$ indicates how much y changes when x changes from neutral to category j.

There is one more consideration when expressing the influence of a nominal predictor as in Equation 15.3: How should the baseline value be set? Consider, for example, predicting height from sex. We could set the baseline height to be zero. Then the deflection from baseline for male might be 1.76 m, and the deflection from baseline for female might be 1.62 m. On the other hand, we could set the baseline height to be 1.69 m. Then the deflection from baseline for male would be +0.07 m, and the deflection from baseline for female would be −0.07 m. The second way of setting the baseline is a typical form in generic statistical modeling. In other words, the baseline is constrained so that the deflections sum to zero across the categories:

$$\sum_{j=1}^{J} \beta_{[j]} = 0 \tag{15.4}$$

The expression of the model in Equation 15.3 is not complete without the constraint in 15.4. Figure 15.4 shows examples of a nominal predictor expressed in terms of Equations 15.3 and 15.4. Notice that the deflections from baseline sum to zero, as demanded by the constraint in Equation 15.4.

15.2.4.2 Additive combination of nominal predictors

Suppose we have two (or more) nominal predictors of a metric value. For example, we might be interested in predicting income as a function of political party affiliation and sex. Figure 15.4 showed examples of each of those predictors individually (for different predicted variables). Now we consider the joint influence of those predictors. If the two influences are additive, then the model from Equation 15.3 becomes

$$\begin{aligned} y &= \beta_0 + \vec{\beta}_1 \cdot \vec{x}_1 + \vec{\beta}_2 \cdot \vec{x}_2 \\ &= \beta_0 + \sum_j \beta_{1[j]} x_{1[j]} + \sum_k \beta_{2[k]} x_{2[k]} \end{aligned} \tag{15.5}$$

with the constraints

$$\sum_j \beta_{1[j]} = 0 \quad \text{and} \quad \sum_k \beta_{2[k]} = 0 \tag{15.6}$$

The left panel of Figure 15.5 shows an example of two nominal predictors that have additive effects on the predicted variable. In this case, the overall baseline is $y = 5$. When

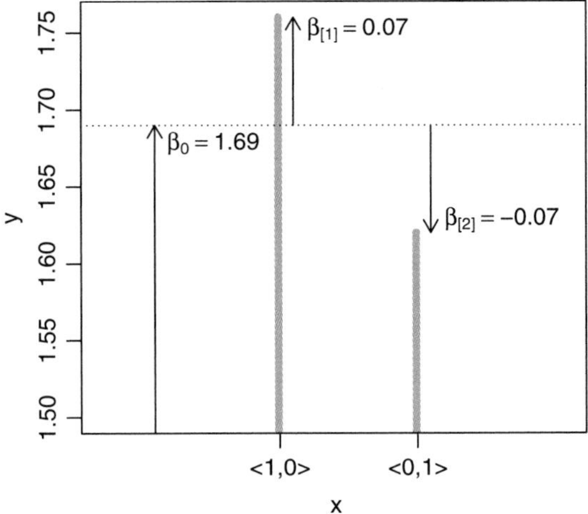

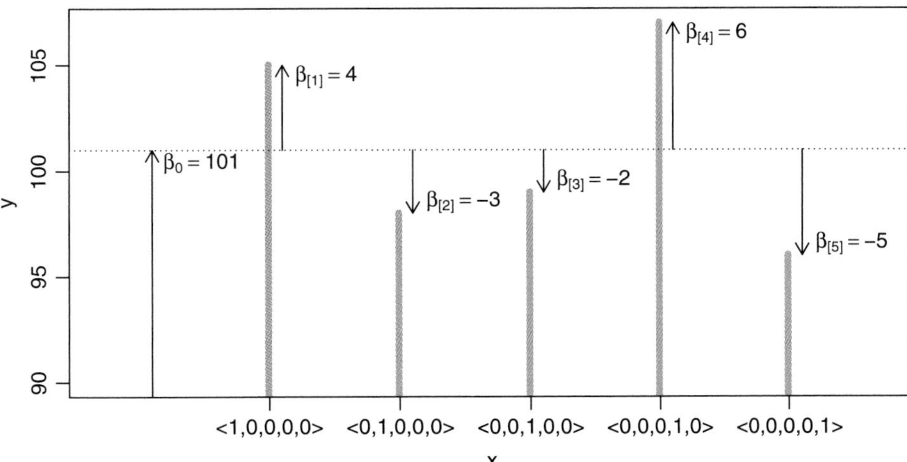

Figure 15.4 Examples of a nominal predictor (Equations 15.3 and 15.4). Upper panel shows a case with $J = 2$, lower panel shows a case with $J = 5$. In each panel, the baseline value of y is on the far left. Notice that the deflections from baseline sum to zero.

$x_1 = \langle 1, 0 \rangle$, there is a deflection in y of -1, and when $x_1 = \langle 0, 1 \rangle$, there is a deflection in y of $+1$. These deflections by x_1 are the same at every level of x_2. The deflections for the three levels of x_2 are $+3$, -2, and -1. These deflections by x_2 are the same at all levels of x_1. Formally, the left panel of Figure 15.5 is expressed mathematically by the additive combination:

$$y = 5 + \langle -1, 1 \rangle \cdot \vec{x}_1 + \langle 3, -2, -1 \rangle \cdot \vec{x}_2 \tag{15.7}$$

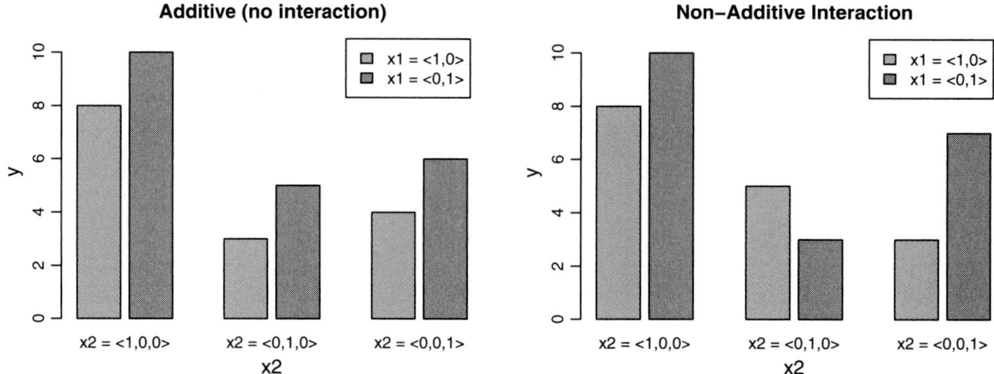

Figure 15.5 Combinations of two nominal variables. *Left:* Additive combination. Notice that the difference between levels of x_1 is the same for every level of x_2. *Right:* Nonadditive interaction. Notice that the difference between levels of x_1 is *not* the same for every level of x_2. The labels elevate subscripts for readability; thus x_1 is displayed as x1 and x_2 is displayed as x2. Figure 20.1, p. 585, provides additional perspective and insight.

For example, consider $\overrightarrow{x}_1 = \langle 0, 1\rangle$ and $\overrightarrow{x}_2 = \langle 0, 0, 1\rangle$. According to Equation 15.7, the value of the predicted variable is $y = 5 + \langle -1, 1\rangle \cdot \langle 0, 1\rangle + \langle 3, -2, -1\rangle \cdot \langle 0, 0, 1\rangle = 5 + 1 - 1 = 5$, which does indeed match the corresponding bar in the left panel of Figure 15.5.

15.2.4.3 Nonadditive interaction of nominal predictors

In many applications, an additive model is not adequate for describing the combined influence of two predictors. For example, consider predicting annual income from political party affiliation and sex (in the contemporary United States). Men, on average, have a higher income than women. Republicans, on average, have a higher income than Democrats. But it may be that the influences of sex and political party combine nonadditively: Perhaps people who are both Republican and male have a higher average income than would be predicted by merely adding the average income boosts for being Republican and for being male. (This nonadditive interaction is not claimed to be true; it is being used only as a hypothetical example.)

We need new notation to formalize the nonadditive influence of a combination of nominal values. Just as $\overrightarrow{x}_1$ refers to the value of predictor 1, and $\overrightarrow{x}_2$ refers to the value of predictor 2, the notation $\overrightarrow{x}_{1\times 2}$ will refer to a particular *combination* of values of predictors 1 and 2. If there are J levels of predictor 1 and K levels of predictor 2, then there are $J \times K$ combinations of the two predictors. To indicate a particular combination of levels from predictors 1 and 2, the corresponding component of $\overrightarrow{x}_{1\times 2}$ is set to 1 while all other components are set to 0. A nonadditive interaction of predictors is formally represented by including a term for the influence of combinations of predictors, beyond the additive influences, as follows: $y = \beta_0 + \overrightarrow{\beta}_1 \cdot \overrightarrow{x}_1 + \overrightarrow{\beta}_2 \cdot \overrightarrow{x}_2 + \overrightarrow{\beta}_{1\times2} \cdot \overrightarrow{x}_{1\times2}$.

The right panel of Figure 15.5 shows a graphical example of two nominal predictors that have interactive (i.e., nonadditive) effects on the predicted variable. Notice, in the left pair of bars ($x_2 = \langle 1, 0, 0 \rangle$), that a change from $x_1 = \langle 1, 0 \rangle$ to $x_1 = \langle 0, 1 \rangle$ produces an increase of $+4$ in y, from $y = 6$ to $y = 10$. But for the middle pair of bars ($x_2 = \langle 0, 1, 0 \rangle$), a change from $x_1 = \langle 1, 0 \rangle$ to $x_1 = \langle 0, 1 \rangle$ produces a change of -2 in y, from $y = 4$ down to $y = 2$. Thus, the influence of x_1 is *not* the same at all levels of x_2. Formally, the right panel of Figure 15.5 is expressed mathematically by the combination:

$$\begin{aligned} y &= 5 + \langle -1, 1 \rangle \cdot \overrightarrow{x}_1 + \langle 3, -2, -1 \rangle \cdot \overrightarrow{x}_2 \\ &\quad + \left\langle \begin{matrix} -1, & +2, & -1, \\ +1, & -2, & +1 \end{matrix} \right\rangle \cdot \overrightarrow{x}_{1\times2} \end{aligned} \tag{15.8}$$

The interaction coefficients, $\overrightarrow{\beta}_{1\times2}$, have been displayed in two rows like a matrix to make it easier to understand which component corresponds to which combination of levels. The two rows correspond to the two levels of $\overrightarrow{x}_1$ and the three columns correspond to the three levels of $\overrightarrow{x}_2$. For example, consider $\overrightarrow{x}_1 = \langle 0, 1 \rangle$ and $\overrightarrow{x}_2 = \langle 0, 0, 1 \rangle$. According to Equation 15.8, the value of the predicted variable is

$$\begin{aligned} y &= 5 + \langle -1, 1 \rangle \cdot \langle 0, 1 \rangle + \langle 3, -2, -1 \rangle \cdot \langle 0, 0, 1 \rangle \\ &\quad + \left\langle \begin{matrix} -1, & +2, & -1, \\ +1, & -2, & +1 \end{matrix} \right\rangle \cdot \left\langle \begin{matrix} 0, & 0, & 0, \\ 0, & 0, & 1 \end{matrix} \right\rangle \\ &= 5 + 1 - 1 + 1 \\ &= 6 \end{aligned}$$

which does indeed match the corresponding bar in the right panel of Figure 15.5.

An interesting aspect of the pattern in the right panel of Figure 15.5 is that the *average* influences of x_1 and x_2 are the same as in the left panel. Overall, on average, going from $x_1 = \langle 1, 0 \rangle$ to $x_1 = \langle 0, 1 \rangle$ produces a change of $+2$ in y, in both the left and right panels. And overall, on average, for both panels it is the case that $x_2 = \langle 1, 0, 0 \rangle$ is $+3$ above baseline, $x_2 = \langle 0, 1, 0 \rangle$ is -2 below baseline, and $x_2 = \langle 0, 0, 1 \rangle$ is -1 below baseline. The only difference between the two panels is that the combined influence of the two predictors equals the sum of the individual influences in the left panel, but the combined influence of the two predictors does not equal the sum of the individual influences in the right panel. This appears explicitly in Equations 15.7 and 15.8: The only difference between them is the interaction coefficients, which are (tacitly) zero in Equation 15.7.

In general, the expression that includes an interaction term can be written as

$$y = \beta_0 + \vec{\beta}_1 \cdot \vec{x}_1 + \vec{\beta}_2 \cdot \vec{x}_2 + \vec{\beta}_{1\times2} \cdot \vec{x}_{1\times2} \tag{15.9}$$
$$= \beta_0 + \sum_j \beta_{1[j]} x_{1[j]} + \sum_k \beta_{2[k]} x_{2[k]} + \sum_{j,k} \beta_{1\times2[j,k]} x_{1\times2[j,k]}$$

with the constraints

$$\sum_j \beta_{1[j]} = 0 \quad \text{and} \quad \sum_k \beta_{2[k]} = 0 \quad \text{and}$$
$$\sum_j \beta_{1\times2[j,k]} = 0 \ \text{ for all } k \quad \text{and} \quad \sum_k \beta_{1\times2[j,k]} = 0 \ \text{ for all } j \tag{15.10}$$

Notice that these constraints were satisfied in the example of Equation 15.8. In particular, within every row and every column of the matrix representation of $\vec{\beta}_{1\times2}$, the coefficients summed to zero.

The notation used here is a bit unwieldy, so do not fret if is not clear to you yet. That's my fault, not yours, because I'm only presenting an overview at this point. When we implement these ideas in Chapter 20 there will be more examples and different notation for computer programs. The main point to understand now is that *the term "interaction" refers to a nonadditive influence of the predictors on the predicted, regardless of whether the predictors are measured on a nominal scale or a metric scale.*

Summary: We have now seen how predictors of different scale types are weighted and added together to form an underlying trend for the predicted variable. Table 15.1 shows a summary of the cases we have covered. Each column of Table 15.1 corresponds to a type of predictor, and the cells of the table show the mathematical form of the linear combination. As a general notation to refer to these linear functions of predictors, I will use the expression lin(x). For example, when there is a single metric predictor,

Table 15.1 For the generalized linear model: typical linear functions lin(x) of the predictor variables x, for various scale types of x

Scale type of predictor x					
Metric				Nominal	
Single group	Two groups	Single predictor	Multiple predictors	Single factor	Multiple factors
β_0	$\beta_{x=1}$ $\beta_{x=2}$	β_0 $+\beta_1 x$	β_0 $+\sum_k \beta_k x_k$ $+\sum_{j,k} \beta_{j\times k} x_j x_k$ $+\left[\text{higher order interactions}\right]$	β_0 $+\vec{\beta} \cdot \vec{x}$	β_0 $+\sum_k \vec{\beta}_k \cdot \vec{x}_k$ $+\sum_{j,k} \vec{\beta}_{j\times k} \cdot \vec{x}_{j\times k}$ $+\left[\text{higher order interactions}\right]$

The value lin(x) is mapped to the predicted data by functions shown in Table 15.2.

$\text{lin}(x) = \beta_0 + \beta_1 x$. All of the cells in Table 15.1 show forms of $\text{lin}(x)$. Even if there are multiple predictors, the notation $\text{lin}(x)$ refers to a linear combination of all of them.

Table 15.1 also indicates a few special cases and generalizations not previously mentioned. In the left columns of the table, the special cases of one or two groups are shown. When there is a single group, then the predictor x is merely a single value, namely, an indicator of being in the group. It does not matter if we think of this single value as nominal or ordinal or metric because all we are describing with the linear core is the central tendency of the group, denoted β_0. When there are two groups, the predictor x is a nominal indicator of group membership. We could subsume the two-group case under the column for a nominal, single-factor predictor (as was shown, for example, in the top panel of Figure 15.4), but the two-group case is encountered so often that we give it a column of its own. Finally, in the columns of Table 15.1 for multiple predictors, the formulas for $\text{lin}(x)$ include optional terms for higher order interactions. Just as any two predictors may have combined effects that are not captured by merely summing their separate influences, there may be effects of multiple predictors that are not captured by two-way combinations. In other words, the magnitudes of the two-way interactions might depend on the levels of other predictors. Higher order interactions are routinely incorporated into models of nominal predictors, but relatively rarely used in models of metric predictors.

15.3. LINKING FROM COMBINED PREDICTORS TO NOISY PREDICTED DATA

15.3.1. From predictors to predicted central tendency

After the predictor variables are combined, they need to be mapped to the predicted variable. This mathematical mapping is called the *(inverse) link* function, and is denoted by $f()$ in the following equation:

$$y = f\,(\text{lin}(x)) \tag{15.11}$$

Until now, we have been assuming that the link function is merely the identity function, $f(\text{lin}(x)) = \text{lin}(x)$. For example, in Equation 15.9, y *equals* the linear combination of the predictors; there is no transformation of the linear combination before mapping the result to y.

Before describing different link functions, it is important to make some clarifications of terminology and corresponding concepts. First, the function $f()$ in Equation 15.11 is called the *inverse* link function, because the link function is traditionally thought of as transforming the value y into a form that can be linked to the linear model. That is, the link function goes from y to the predictors, not from the predictors to y. I may occasionally abuse convention and refer to either $f()$ or $f^{-1}()$ as "the" link function, and rely on context to disambiguate which direction of linkage is intended. The reason for this terminological sadism is that the arrows in hierarchical diagrams of Bayesian models will flow from the linear combination toward the data, and therefore it is natural for the functions to map toward the predicted values, as in Equation 15.11. But repeatedly referring to this function as the "inverse" link would strain my patience and violate my aesthetic sensibilities. Second, the value y that results from the link function $f(\text{lin}(x))$ is not a data value *per se*. Instead, $f(\text{lin}(x))$ is the value of a parameter that expresses a central tendency of the data, usually their mean. Therefore the function $f()$ in Equation 15.11 is sometimes called the *mean* function (instead of the inverse link function), and is written $\mu = f()$ instead of $y = f()$. I will use this notation temporarily for some summary tables below, but then abandon it because $f()$ does not always refer to the mean of the predicted data.

There are situations in which a non-identity link function is appropriate. Consider, for example, predicting response time as a function of amount of caffeine consumed. Response time decreases as caffeine dosage increases (e.g., Smit & Rogers, 2000, Fig. 1; although most of the decrease is produced by even a small dose). Therefore a linear prediction of RT (y) from dosage (x) would have a negative slope. This negative slope on a linear function implies that for a very large dosage of caffeine, response time would become negative, which is impossible unless caffeine causes precognition (i.e., foreseeing events before they occur). Therefore a linear function cannot be used for extrapolation to large doses, and we might instead want to use a link function that asymptotes above zero, such as an exponential function with $y = \exp\left(\beta_0 + \beta_1 x\right)$. In the next sections we will consider some frequently used link functions.

15.3.1.1 The logistic function

A frequently used link function is the *logistic*:

$$y = \text{logistic}(x) = 1 \big/ \big(1 + \exp(-x)\big) \tag{15.12}$$

Notice the negative sign in front of the x. The value y of the logistic function ranges between 0 and 1. The logistic is nearly 0 when x is large negative, and is nearly 1 when x is large positive. In our applications, x is a linear combinations of predictors. For a single metric predictor, the logistic can be written:

$$y = \text{logistic}(x; \beta_0, \beta_1) = 1 \big/ \big(1 + \exp\left(-(\beta_0 + \beta_1 x)\right)\big) \tag{15.13}$$

The logistic function is often most conveniently parameterized in a form that shows the gain γ (Greek letter gamma) and threshold θ:

$$y = \text{logistic}\,(x; \gamma, \theta) = 1 \big/ \big(1 + \exp\left(-\gamma\,(x - \theta)\right)\big) \tag{15.14}$$

Examples of Equation 15.14 are shown in Figure 15.6. *Notice that the threshold is the point on the x-axis for which $y = 0.5$. The gain indicates how steeply the logistic rises through that point.*

Figure 15.7 shows examples of a logistic of two predictor variables. Above each panel is the equation for the corresponding graph. The equations are parameterized

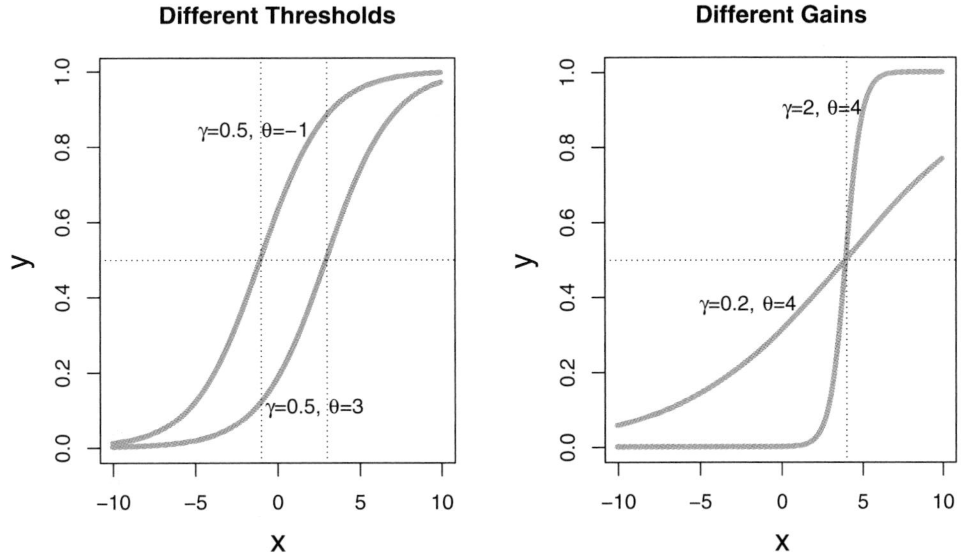

Figure 15.6 Examples of logistic functions of a single variable. The left panel shows logistics with the same gain but different thresholds. The right panel shows logistics with the same threshold but different gains.

in normalized-threshold form: $y = \text{logistic}\left(\gamma\left(\sum_k w_k x_k - \theta\right)\right)$, with $\left(\sum_k w_k^2\right)^{1/2} = 1$. Notice, in particular, that the coefficients of x_1 and x_2 in the plotted equations do indeed have Euclidean length of 1.0. For example, in the upper right panel, $\left(0.71^2 + 0.71^2\right)^{1/2} = 1.0$, except for rounding error. If you start with the linear part parameterized as $\beta_0 + \sum_{k=1}^{K} \beta_k x_k$, you can convert to the equivalent normalized-threshold form as follows: Let $\gamma = \left(\sum_{k=1}^{K} \beta_k^2\right)^{1/2}$, $\theta = -\beta_0/\gamma$, and $w_k = \beta_k/\gamma$ for $k \neq 0$. Then $\beta_0 + \sum_{k=1}^{K} \beta_k x_k = \gamma(w_k x_k - \theta)$ and it is true that $\left(\sum_k w_k^2\right)^{1/2} = 1$.

The coefficients of the x variables determine the *orientation* of the logistical "cliff." For example, compare the two top panels in Figure 15.7, which differ only in the coefficients, not in gain or threshold. In the top left panel, the coefficients are $w_1 = 0$ and $w_2 = 1$, and the cliff rises in the x_2 direction. In the top right panel, the coefficients are $w_1 = 0.71$ and $w_2 = 0.71$, and the cliff rises in the positive diagonal direction.

The threshold determines the *position* of the logistical cliff. In other words, the threshold determines the x values for which $y = 0.5$. For example, compare the two left panels of Figure 15.7. The coefficients are the same, but the thresholds (and gains) are different. In the upper left panel, the threshold is zero, and therefore the mid-level of the cliff is over $x_2 = 0$. In the lower left panel, the threshold is -3, and therefore the mid-level of the cliff is over $x_2 = -3$.

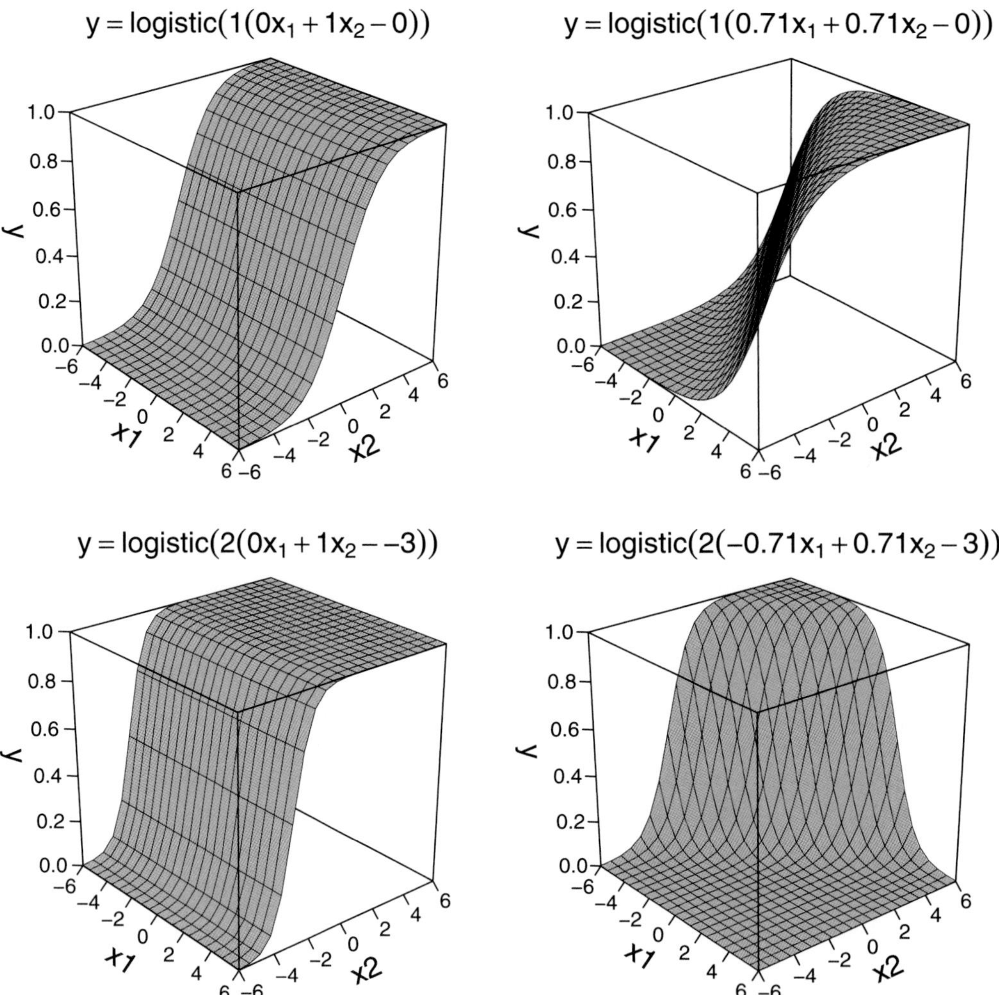

Figure 15.7 Examples of logistic functions of two variables. Top two panels show logistics with the same gain and threshold, but different coefficients on the predictors. The left two panels show logistics with the same coefficients on the predictors, but different gains and thresholds. The lower right panel shows a case with a negative coefficient on the first predictor.

The gain determines the *steepness* of the logistical cliff. Again compare the two left panels of Figure 15.7. The gain of the upper left is 1, whereas the gain of the lower left is 2.

Terminology: The *logit* function. The inverse of the logistic function is called the logit function. For $0 < p < 1$, $\text{logit}(p) = \log\big(p/(1-p)\big)$. It is easy to show (try it!) that $\text{logit}(\text{logistic}(x)) = x$, which is to say that the logit is indeed the inverse of the logistic.

Some authors, and programmers, prefer to express the connection between predictors and predicted in the opposite direction, by first transforming the predicted variable to match the linear model. In other words, you may see the link expressed either of these ways:

$$\begin{aligned} y &= \text{logistic}\,(\text{lin}(x)) \\ \text{logit}(y) &= \text{lin}(x) \end{aligned} \tag{15.15}$$

The two expressions achieve the same result, mathematically. The difference between them is merely a matter of emphasis. In the first expression, the combination of predictors is transformed so it maps onto y. In the second expression, y is transformed onto a new scale, and that transformed value is modeled as a linear combination of predictors. I find that the logistic formulation is usually more intuitive than the logit formulation, but we will see in Chapter 21 that the logit formulation will be useful for interpreting the β coefficients.

15.3.1.2 The cumulative normal function

Another frequently used link function is the cumulative normal distribution. It is qualitatively very similar to the logistic function. Modelers will use the logistic or the cumulative normal depending on mathematical convenience or ease of interpretation. For example, when we consider ordinal predicted variables (in Chapter 23), it will be natural to model the responses in terms of a continuous underlying variable that has normally distributed variability, which leads to using the cumulative normal as a model of response probabilities.

The cumulative normal is denoted $\Phi(x;\mu,\sigma)$, where x is a real number and where μ and σ are parameter values, called the mean and standard deviation of the normal distribution. The parameter μ governs the point at which the cumulative normal, $\Phi(x)$, equals 0.5. In other words, μ plays the same role as the threshold in the logistic. The parameter σ governs the steepness of the cumulative normal function at $x=\mu$, but inversely, such that a *smaller* value of σ corresponds to a steeper cumulative normal. A graph of a cumulative standardized normal appears in Figure 15.8.

Terminology: The inverse of the cumulative normal is called the *probit* function. ("Probit" stands for "probability unit"; Bliss, 1934). The probit function maps a value p, for $0.0 \le p \le 1.0$, onto the infinite real line, and a graph of the probit function looks very much like the logit function. You may see the link expressed either of these ways:

$$\begin{aligned} y &= \Phi\,(\text{lin}(x)) \\ \text{probit}(y) &= \text{lin}(x) \end{aligned}$$

Traditionally, the transformation of y (in this case, the probit function) is called the link function, and the transformation of the linear combination of x (in this case, the Φ

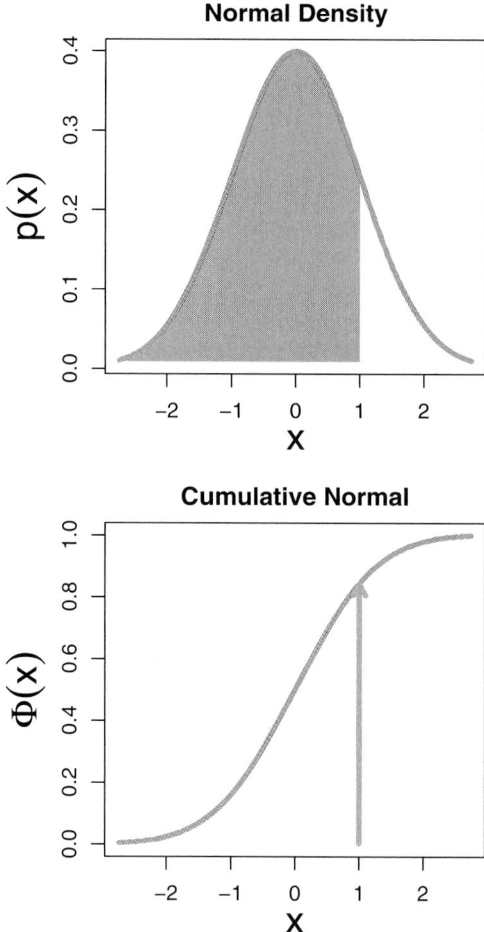

Figure 15.8 Top: A standardized normal density (i.e., with mean zero and standard deviation one). Bottom: The corresponding standardized cumulative normal function. The area under the normal distribution to the left of x (top panel) is the height of the cumulative normal at x (lower panel).

function) is called the inverse link function. As mentioned before, I abuse the traditional terminology and call either one a link function, relying on context to disambiguate. In later applications, we will be using the Φ function, not the probit.

15.3.2. From predicted central tendency to noisy data

In the real world, there is always variation in y that we cannot predict from x. This unpredictable "noise" in y might be deterministically caused by sundry factors we have neither measured nor controlled, or the noise might be caused by inherent non-determinism in y. It does not matter either way because in practice the best we can do is predict the *probability* that y will have any particular value, dependent upon x.

Therefore we use the deterministic value predicted by Equation 15.11 as the predicted *tendency* of y as a function of the predictors. We do not predict that y is exactly $f(\text{lin}(x))$ because we would surely be wrong. Instead, we predict that y tends to be *near* $f(\text{lin}(x))$.

To make this notion of probabilistic tendency precise, we need to specify a probability distribution for y that depends on $f(\text{lin}(x))$. To keep the notation tractable, first define $\mu = f(\text{lin}(x))$. The value μ represents the central tendency of the predicted y values, which might or might not be the mean. With this notation, we then denote the probability distribution of y as some to-be-specified probability density function, abbreviated as "pdf":

$$y \sim \text{pdf}\big(\mu, [\text{scale,shape,etc.}]\big)$$

As indicated by the bracketed terms after μ, the pdf might have various additional parameters that control the distribution's scale (i.e., standard deviation), shape, etc.

The form of the pdf depends on the measurement scale of the predicted variable. If the predicted variable is metric and can extend infinitely in both positive and negative directions, then a typical pdf for describing the noise in the data is a normal distribution. A normal distribution has a mean parameter μ and a standard deviation parameter σ, so we would write $y \sim \text{normal}(\mu, \sigma)$ with $\mu = f(\text{lin}(x))$. In particular, if the link function is the identity, then we have a case of conventional *linear regression*. Examples of linear regression are shown in Figure 15.9. The upper panel of Figure 15.9 shows a case in which there is a single metric predictor, and the lower panel shows a case with two metric predictors. For both cases, the black dots indicate data that are normally distributed around a linear function of the predictors.

If the predicted variable is dichotomous, with $y \in \{0, 1\}$, then a typical pdf is the Bernoulli distribution, $y \sim \text{Bernoulli}(\mu)$, where $\mu = f(\text{lin}(x))$ is the probability that $y = 1$. In other words, μ is playing the role of the parameter that was called θ in Equation 5.10, p. 109. Because μ must be between 0 and 1, the link function must convert $\text{lin}(x)$ to a value between 0 and 1. A typical link function for this purpose is the logistic function, and when the predictor variables are metric, we have a case of conventional *logistic regression*. An example of logistic regression is shown in Figure 15.10. The black dots indicate data that are Bernoulli distributed, at 0 or 1, with probability indicated by the logistic surface.

Table 15.2 shows a summary of typical link functions and pdfs for various types of predicted variables. The previous paragraphs provided examples for the first two rows of Table 15.2, which correspond to predicted variables that are metric or dichotomous. The remaining rows of Table 15.2 are for predicted variables that are nominal, ordinal, or count. Each of these cases will be explained at length in later chapters. The point now is for you to notice that in each case, there is a linear combination of the predictors converted by an inverse link function in the right column to a predicted central tendency μ, which is then used in a pdf in the middle column.

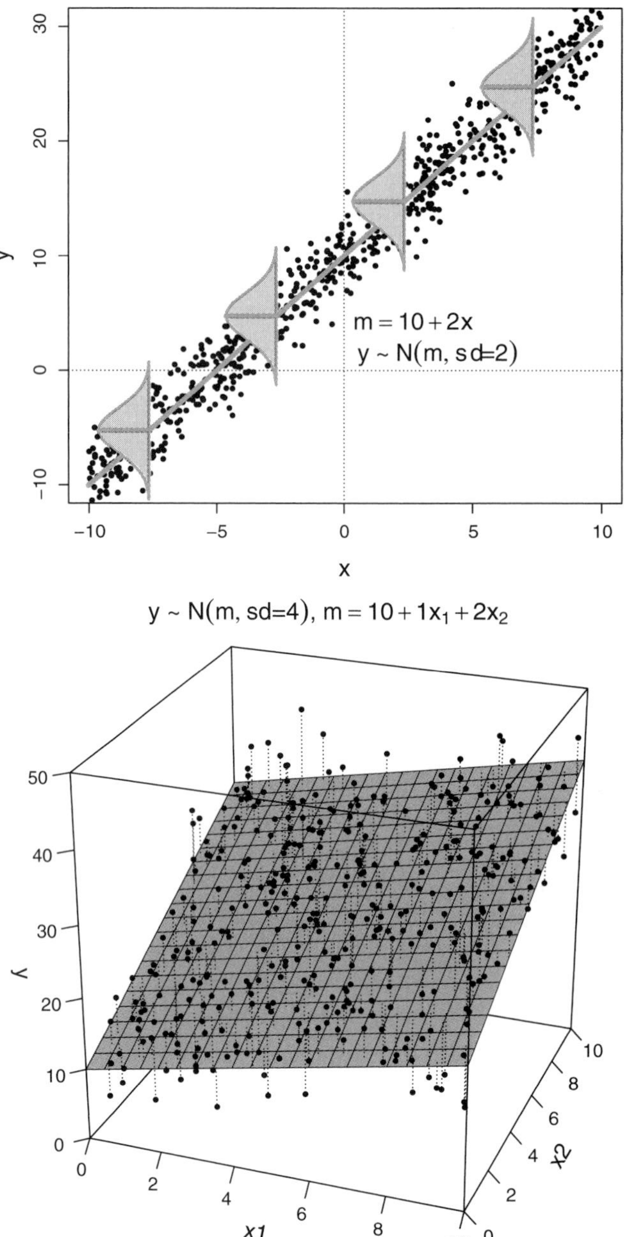

Figure 15.9 Examples of points normally distributed around a linear function. The upper panel shows normal distributions superimposed on the linear function to emphasize that the random variability is vertical, along the *y*-axis, and centered on the line. The lower panel shows each datum connected to the plane by a dotted line, to again emphasize the vertical displacement from the plane.

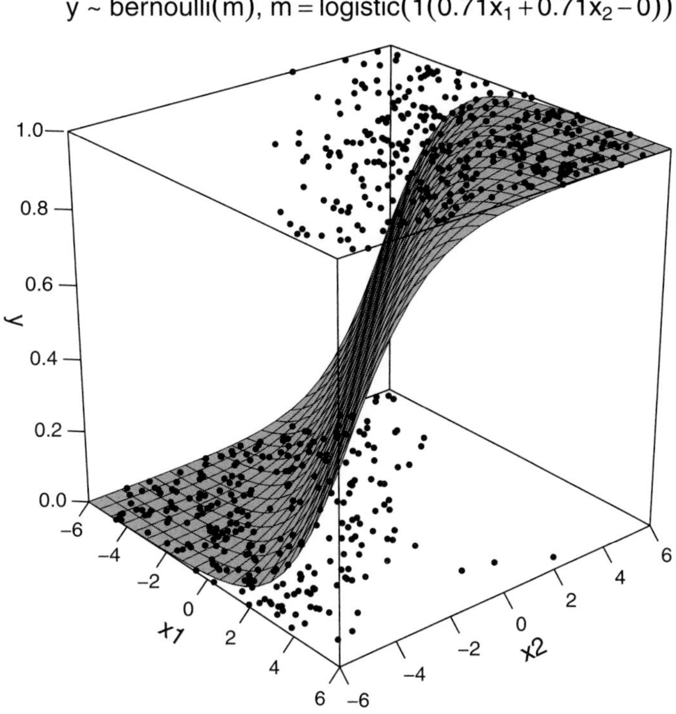

Figure 15.10 Examples of points that are Bernoulli distributed around a logistic function of two predictors. All the points are either at $y = 1$ or $y = 0$; intermediate values such as $y = 0.6$ cannot occur.

Table 15.2 For the generalized linear model: typical noise distributions and inverse link functions for describing various scale types of the predicted variable y

Scale type of predicted y	Typical noise distribution $y \sim \text{pdf}(\mu, [\text{parameters}])$	Typical inverse link function $\mu = f(\text{lin}(x), [\text{parameters}])$
Metric	$y \sim \text{normal}(\mu, \sigma)$	$\mu = \text{lin}(x)$
Dichotomous	$y \sim \text{bernoulli}(\mu)$	$\mu = \text{logistic}\,(\text{lin}(x))$
Nominal	$y \sim \text{categorical}(\ldots, \mu_k, \ldots)$	$\mu_k = \dfrac{\exp\,(\text{lin}_k(x))}{\sum_c \exp\,(\text{lin}_c(x))}$
Ordinal	$y \sim \text{categorical}(\ldots, \mu_k, \ldots)$	$\mu_k = \Phi\left(\left(\theta_k - \text{lin}(x)\right)/\sigma\right) - \Phi\left(\left(\theta_{k-1} - \text{lin}(x)\right)/\sigma\right)$
Count	$y \sim \text{poisson}(\mu)$	$\mu = \exp\,(\text{lin}(x))$

The value μ is a central tendency of the predicted data (not necessarily the mean). The predictor variable is x, and $\text{lin}(x)$ is a linear function of x, such as those shown in Table 15.1.

The forms shown in Table 15.2 are merely typical and not necessary. For example, if metric data are skewed or kurtotic (i.e., have heavy tails) then a non-normal pdf could be used that better describes the noise in the data. Furthermore, you might have realized that the case of a dichotomous scale is subsumed by the case of a nominal scale. But the dichotomous case is encountered often, and its model is generalized to the nominal scale, so the dichotomous case is treated separately.

15.4. FORMAL EXPRESSION OF THE GLM

The GLM can be written as follows:

$$\mu = f\left(\text{lin}(x), [\text{parameters}]\right) \tag{15.16}$$

$$y \sim \text{pdf}\left(\mu, [\text{parameters}]\right) \tag{15.17}$$

As has been previously explained, the predictors x are combined in the linear function lin(x), and the function f in Equation 15.16 is called the inverse link function. The data, y, are distributed around the central tendency μ according to the probability density function labeled "pdf."

The GLM covers a large range of useful applications. In fact, the GLM is even more general than indicated in Table 15.2 because there can be multiple predicted variables. These cases are also straight forward to address with Bayesian methods, but they will not be covered in this book.

15.4.1. Cases of the GLM

Table 15.3 displays the various cases of the GLM that are considered in this book, along with the chapters that address them. Each cell of Table 15.3 corresponds to a combination of lin(x) from Table 15.1 and link with pdf from Table 15.2. The chapter for each combination provides conceptual explanation of the descriptive mathematical model and of how to do Bayesian estimation of the parameters in the model.

Table 15.3 Book chapters that discuss combinations of scale types for predicted and predictor variables of Tables 15.1 and 15.2

<table>
<tr><th></th><th colspan="6">Scale type of predictor x</th></tr>
<tr><th></th><th></th><th></th><th colspan="2">Metric</th><th colspan="2">Nominal</th></tr>
<tr><th>Scale type of predicted y</th><th>Single group</th><th>Two groups</th><th>Single predictor</th><th>Multiple predictors</th><th>Single factor</th><th>Multiple factors</th></tr>
<tr><td>Metric</td><td colspan="2">Chapter 16</td><td>Chapter 17</td><td>Chapter 18</td><td>Chapter 19</td><td>Chapter 20</td></tr>
<tr><td>Dichotomous</td><td colspan="2">Chapters 6–9</td><td colspan="4">Chapter 21</td></tr>
<tr><td>Nominal</td><td colspan="6">Chapter 22</td></tr>
<tr><td>Ordinal</td><td colspan="6">Chapter 23</td></tr>
<tr><td>Count</td><td colspan="6">Chapter 24</td></tr>
</table>

The first row of Table 15.3 lists cases for which the predicted variable is metric. Moving from left to right within this row, the first two columns are for when there are one or two groups of data. Classical NHST would apply a single-group or two-group t test to these cases. This case is described in its Bayesian setting in Chapter 16. Moving to the next column, there is a single metric predictor. This corresponds to so-called "simple linear regression," and is explored in Chapter 17. Moving rightward to the next column, we come to the scenario involving two or more metric predictors, which corresponds to "multiple linear regression" and is explored in Chapter 18.

The next two columns involve nominal predictors, instead of metric predictors, with the penultimate column devoted to a single predictor and the final column devoted to two or more predictors. The last two columns correspond to what NHST calls "one-way ANOVA" and "multifactor ANOVA." Bayesian approaches are explained in Chapters 19 and 20. For both the Bayesian approaches to linear regression and the Bayesian approaches to ANOVA, we will use hierarchical models that are not used in classical approaches.

In the second row of Table 15.3, the predicted variable is dichotomous. This simplest of data scales was used to develop all the foundational concepts of Bayesian data analysis in Chapters 6–9. When the predictors are more elaborate, and especially when the predictors are metric, this situation is referred to as "logistic regression" because of the logistic (inverse) link function. It is discussed in Chapter 21.

The next two rows of Table 15.3 are for nominal and ordinal scales on the predicted variable. Both of these will use a categorical noise distribution, but with probabilities computed through different link functions. For nominal predicted values, Chapter 22 will explain how logistic regression is generalized to address multiple categories. For ordinal predicted values, Chapter 23 will explain how an underlying metric scale is mapped to an ordinal scale using thresholds on a cumulative normal distribution.

Finally, the bottom row of Table 15.3 refers to a predicted variable that measures counts. We have previously seen this sort of data in the counts of eye color and hair color in Table 4.1, p. 90. For this situation, we will consider a new pdf called the Poisson distribution, which requires a positive-valued mean parameter. After all, counts are nonnegative. Because the mean must be positive, the inverse link function must provide a positive value. A natural function that satisfies that requirement is the exponential, as shown in Table 15.2. Because the inverse link function is the exponential, the link function is the logarithm, and therefore these are called "log-linear models." Chapter 24 provides an introduction.

How to be your own statistical consultant: The key point to understand is that each cell of Table 15.3 corresponds to a combination of lin(x) from Table 15.1 and link with pdf from Table 15.2. This organization is well known to every statistical consultant. When a client brings an application to a consultant, one of the first things the consultant does is find out from the client which data are supposed to be predictors and which data are supposed to be predicted, and the measurement scales of the data. Very often the situation falls into one of the cells of Table 15.3, and then the standard models can be applied. A benefit of Bayesian analysis is that it is easy to create nonstandard variations of the models, as appropriate for specific situations. For example, if the data have many outliers, it is trivial to use a heavy-tailed pdf for the noise distribution. If there are lots of data for each individual, and there are many individuals, perhaps in groups, it is usually straight forward to create hierarchical structure on the parameters of the GLM. When you are considering how to analyze data, your first task is to be your own consultant and find out which data are predictors, which are predicted, and what measurement scales they are. Then you can determine whether one of the cases of the GLM applies to your situation. This also constitutes the first steps of Bayesian data analysis, described back in Section 2.3, p. 25.

15.5. EXERCISES

Look for more exercises at https://sites.google.com/site/doingbayesiandataanalysis/

Exercise 15.1. [Purpose: For real-world examples of research, identify which statistical model is relevant.] For each of the examples below, identify the predicted variable and its scale type, identify the predictor variable(s) and its scale type, and identify the cell of Table 15.3 to which the example belongs.

(A) Guber (1999) examined average performance by public high school students on the SAT as a function of how much money was spent per pupil by the state, and what percentage of eligible students actually took the exam.

(B) Hahn, Chater, and Richardson (2003) were interested in perceived similarity of simple geometric patterns. Human observers rated pairs of patterns for how similar the patterns appeared, by circling one of the digits 1–7 printed on the page, where 1 meant "very dissimilar" and 7 meant "very similar." The authors presented a theory of perceived similarity, in which patterns are perceived to be dissimilar to the extent that it takes more geometric transformations to produce one pattern from the other. The theory specified the exact number of transformations needed to get from one pattern to the other.

(C) R. L. Berger, Boos, and Guess (1988) were interested in the longevity of rats, measured in days, as a function of the rat's diet. One group of rats fed freely, another group of rats had a very low calorie diet.

(D) McIntyre (1994) was interested in predicting the tar content of a cigarette (measured in milligrams) from the weight of the cigarette.

(E) You are interested in predicting the gender of a person, based on the person's height and weight.

(F) You are interested in predicting whether a respondent will agree or disagree with the statement, "The United States needs a federal health care plan with a public option," on the basis of the respondent's political party affiliation.

Exercise 15.2. [Purpose: Find *student-relevant* real-world examples of each type of situation in Table 15.3.] For each cell of Table 15.3, provide an example of research involving that cell's model structure. Do this by finding published articles that describe research with the corresponding structure. The articles do *not* need to have Bayesian data analysis; the articles *do* need to report research that involves the corresponding types of predictor and predicted variables. Because it might be overly time consuming to find published examples of all the cells, please find published articles of at least six cells spanning at least three different rows. For each example, specify the following:

- The full citation to the published article (see the references of this book for examples of how to cite articles),
- The predictor and predicted variables. Describe their meaning and their type of scale. Briefly describe the meaningful context for the variables, that is, the goal of the research.

CHAPTER 16

Metric-Predicted Variable on One or Two Groups

Contents

It's normal to want to fit in with your friends,
Behave by their means and believe all their ends.
But I'll be high tailing it, fast and askew,
Precisely 'cause I can't abide what you do.[1]

In this chapter, we consider a situation in which we have a metric-predicted variable that is observed for items from one or two groups. For example, we could measure the blood pressure (i.e., a metric variable) for people randomly sampled from first-year university students (i.e., a single group). In this case, we might be interested in how much the group's typical blood pressure differs from the recommended value for people of that age as published by a federal agency. As another example, we could measure the IQ (i.e., a metric variable) of people randomly sampled from everyone self-described as vegetarian (i.e., a single group). In this case, we could be interested in how much this group's IQ differs from the general population's average IQ of 100.

In the context of the generalized linear model (GLM) introduced in the previous chapter, this chapter's situation involves the most trivial cases of the linear core of the GLM, as indicated in the left cells of Table 15.1 (p. 434), with a link function that is the

[1] This chapter describes data with a normal distribution, which is parameterized by its mean and precision. But data can have outliers, which demand descriptive distributions with high tails or skewed ends. The poem plays with alternative meanings of the words normal, fit, mean, end, high tail, skew, precise, and believe.

http://dx.doi.org/10.1016/B978-0-12-405888-0.00016-7

identity along with a normal distribution for describing noise in the data, as indicated in the first row of Table 15.2 (p. 443). We will explore options for the prior distribution on parameters of the normal distribution, and methods for Bayesian estimation of the parameters. We will also consider alternative noise distributions for describing data that have outliers.

16.1. ESTIMATING THE MEAN AND STANDARD DEVIATION OF A NORMAL DISTRIBUTION

The normal probability density function was introduced in Section 4.3.2.2, p. 83. The normal distribution specifies the probability density of a value y, given the values of two parameters, the mean μ and standard deviation σ:

$$p(y|\mu,\sigma) = \frac{1}{Z}\exp\left(-\frac{1}{2}\frac{(y-\mu)^2}{\sigma^2}\right) \tag{16.1}$$

where Z is the normalizer, which is a constant that makes the probability density integrate to 1. It turns out that $Z = \sigma\sqrt{2\pi}$, but we won't need to use this fact in the derivations below.

To get an intuition for the normal distribution as a likelihood function, consider three data values $y_1 = 85$, $y_2 = 100$, and $y_3 = 115$, which are plotted as large dots in Figure 16.1. The probability density of any single datum, given particular parameter values, is $p(y|\mu,\sigma)$ as specified in Equation 16.1. The probability of the entire set of independent data values is the multiplicative product, $\prod_i p(y_i|\mu,\sigma) = p(D|\mu,\sigma)$, where $D = \{y_1, y_2, y_3\}$. Figure 16.1 shows $p(D|\mu,\sigma)$ for different values of μ and σ. As you can see, there are values of μ and σ that make the data most probable, but other nearby values also accommodate the data reasonably well. (For another example, see Figure 2.4, p. 23.) The question is, given the data, how should we allocate credibility to combinations of μ and σ?

The answer is provided by Bayes' rule. Given a set of data, D, we estimate the parameters with Bayes' rule:

$$p(\mu,\sigma|D) = \frac{p(D|\mu,\sigma)\,p(\mu,\sigma)}{\iint d\mu\, d\sigma\, p(D|\mu,\sigma)\,p(\mu,\sigma)} \tag{16.2}$$

Figure 16.1 shows examples of $p(D|\mu,\sigma)$ for a particular data set at different values of μ and σ. The prior, $p(\mu,\sigma)$, specifies the credibility of each combination of μ,σ values in the two-dimensional joint parameter space, without the data. Bayes' rule says that the posterior credibility of each combination of μ,σ values is the prior credibility times the likelihood, normalized by the marginal likelihood. Our goal now is to evaluate Equation 16.2 for reasonable choices of the prior distribution, $p(\mu,\sigma)$.

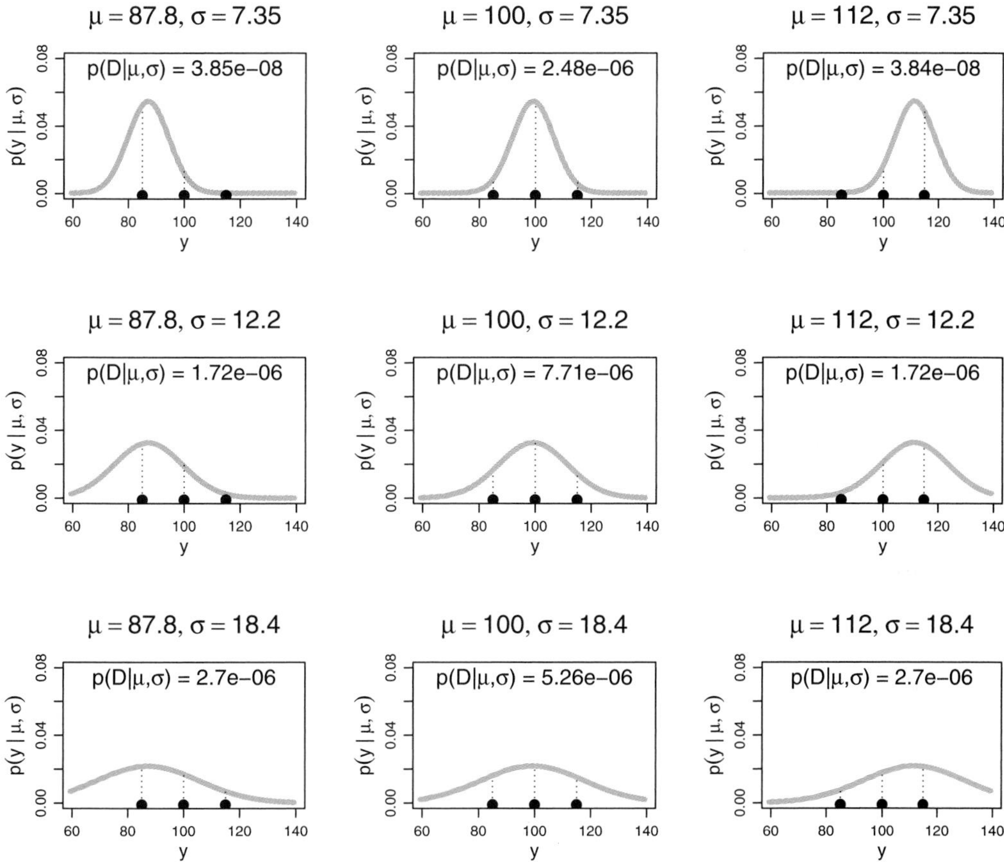

Figure 16.1 The likelihood $p(D|\mu, \sigma)$ for three data points, $D = \{85, 100, 115\}$, according to a normal likelihood function with different values of μ and σ. Columns show different values of μ, and rows show different values of σ. The probability density of an individual datum is the height of the dotted line over the point. The probability of the set of data is the product of the individual probabilities. The middle panel shows the μ and σ that maximize the probability of the data. (For another example, see Figure 2.4, p. 23.)

16.1.1. Solution by mathematical analysis

Because we are already familiar with JAGS and Stan, we could go directly to an MCMC solution. But this case is simple enough that it can be solved analytically, and the resulting formulas motivate some of the traditional parameterizations and priors for normal distributions. Therefore, we take a short algebraic tour before moving on to MCMC implementations.

It is convenient first to consider the case in which the standard deviation of the likelihood function is fixed at a specific value. In other words, the prior distribution on σ is a spike over that specific value. We'll denote that fixed value as $\sigma = S_y$. With this simplifying assumption, we are only estimating μ because we are assuming perfectly certain prior knowledge about σ.

When σ is fixed, then the prior distribution on μ in Equation 16.2 can be easily chosen to be conjugate to the normal likelihood. (The term "conjugate prior" was defined in Section 6.2, p. 126.) It turns out that the product of normal distributions is again a normal distribution; in other words, if the prior on μ is normal, then the posterior on μ is normal. It is easy to derive this fact, as we do next.

Let the prior distribution on μ be normal with mean M_μ and standard deviation S_μ. Then the likelihood times the prior (i.e., the numerator of Bayes' rule) is

$$
\begin{aligned}
p(y|\mu,\sigma)\,p(\mu,\sigma) &= p(y|\mu,S_y)\,p(\mu) \\
&\propto \exp\left(-\frac{1}{2}\frac{(y-\mu)^2}{S_y^2}\right)\exp\left(-\frac{1}{2}\frac{\left(\mu-M_\mu\right)^2}{S_\mu^2}\right) \qquad \text{notice } \propto \text{ not } = \\
&= \exp\left(-\frac{1}{2}\left[\frac{(y-\mu)^2}{S_y^2}+\frac{\left(\mu-M_\mu\right)^2}{S_\mu^2}\right]\right) \\
&= \exp\left(-\frac{1}{2}\left[\frac{S_\mu^2\,(y-\mu)^2+S_y^2\left(\mu-M_\mu\right)^2}{S_y^2S_\mu^2}\right]\right) \\
&= \exp\left(-\frac{1}{2}\left[\frac{S_y^2+S_\mu^2}{S_y^2S_\mu^2}\left(\mu^2-2\frac{S_y^2M_u+S_\mu^2y}{S_y^2+S_\mu^2}\mu+\frac{S_y^2M_u^2+S_\mu^2y^2}{S_y^2+S_\mu^2}\right)\right]\right) \\
&= \exp\left(-\frac{1}{2}\left[\frac{S_y^2+S_\mu^2}{S_y^2S_\mu^2}\left(\mu^2-2\frac{S_y^2M_u+S_\mu^2y}{S_y^2+S_\mu^2}\mu\right)\right]\right) \\
&\quad\times\exp\left(-\frac{1}{2}\left[\frac{S_y^2+S_\mu^2}{S_y^2S_\mu^2}\left(+\frac{S_y^2M_u^2+S_\mu^2y^2}{S_y^2+S_\mu^2}\right)\right]\right) \\
&\propto \exp\left(-\frac{1}{2}\left[\frac{S_y^2+S_\mu^2}{S_y^2S_\mu^2}\left(\mu^2-2\frac{S_y^2M_u+S_\mu^2y}{S_y^2+S_\mu^2}\mu\right)\right]\right)
\end{aligned}
\tag{16.3}
$$

where the transition to the last line is valid because the term that was dropped was merely a constant. This result, believe it or not, is progress. Why? Because we ended up, in the innermost parentheses, with a quadratic expression in μ. Notice that the normal prior is also a quadratic expression in μ. All we have to do is "complete the square" inside the parentheses, and do the same trick that got us to the last line of Equation 16.3:

$$
\begin{aligned}
p(y|\mu,S_y)p(\mu) &\propto \exp\left(-\frac{1}{2}\left[\frac{S_y^2+S_\mu^2}{S_y^2S_\mu^2}\left(\mu^2-2\frac{S_y^2M_u+S_\mu^2y}{S_y^2+S_\mu^2}\mu+\left(\frac{S_y^2M_\mu+S_\mu^2y}{S_y^2+S_\mu^2}\right)^2\right)\right]\right) \\
&= \exp\left(-\frac{1}{2}\left[\frac{S_y^2+S_\mu^2}{S_y^2S_\mu^2}\left(\mu-\frac{S_y^2M_\mu+S_\mu^2y}{S_y^2+S_\mu^2}\right)^2\right]\right)
\end{aligned}
\tag{16.4}
$$

Equation 16.4 is the numerator of Bayes' rule. When it is normalized, it becomes a probability density function. What is the shape of the function? You can see that Equation 16.4 has the same form as a normal distribution on μ, such that the mean is $(S_y^2 M_\mu + S_\mu^2 y)/(S_y^2 + S_\mu^2)$ and the standard deviation is $\sqrt{S_y^2 S_\mu^2/(S_y^2 + S_\mu^2)}$.

That formula is rather unwieldy! It becomes more compact if the normal density is re-expressed in terms of $1/\sigma^2$ instead of σ. The reciprocal of the squared standard deviation is called the *precision*. To get an intuition for precision, notice that a very narrow distribution is highly precise. When the standard deviation gets smaller, the precision gets bigger. Now, because the posterior standard deviation is $\sqrt{S_y^2 S_\mu^2/(S_y^2 + S_\mu^2)}$, the posterior precision is

$$\frac{S_y^2 + S_\mu^2}{S_y^2 S_\mu^2} = \frac{1}{S_\mu^2} + \frac{1}{S_y^2} \tag{16.5}$$

Thus, the posterior precision is the sum of the prior precision and the likelihood precision.

The posterior mean can also be compactly re-expressed in terms of precisions. The posterior mean is $(S_y^2 M_\mu + S_\mu^2 y)/(S_y^2 + S_\mu^2)$, which can be re-arranged as

$$\frac{1/S_\mu^2}{1/S_y^2 + 1/S_\mu^2} M_\mu + \frac{1/S_y^2}{1/S_y^2 + 1/S_\mu^2} y \tag{16.6}$$

In other words, the posterior mean is a weighted average of the prior mean and the datum, with the weighting corresponding to the relative precisions of the prior and the likelihood. When the prior is highly precise compared to the likelihood, that is when $1/S_\mu^2$ is large compared to $1/S_y^2$, then the prior is weighed heavily and the posterior mean is near the prior mean. But when the prior is imprecise and very uncertain, then the prior does not get much weight and the posterior mean is close to the datum. We have previously seen this sort of relative weighting of prior and data in the posterior. It showed up in the case of updating a beta prior, back in Equation 6.9, p. 133.

The formulas for the mean and precision of the posterior normal can be naturally extended when there are N values of y in a sample, instead of only a single value of y. The formulas can be derived from the defining formulas, as was done above, but a short cut can be taken. It is known from mathematical statistics that when a set of values y_i are generated from a normal likelihood function, the mean of those values, $\overline{y}$, is also distributed normally, with the same mean as the generating mean, and with a standard deviation of $\sigma/\sqrt{N}$. Thus, instead of conceiving of this situation as N scores y_i sampled from the likelihood, normal$(y_i|\mu, \sigma)$, we conceive of this as a single score, $\overline{y}$, sampled from the likelihood, normal$(\overline{y}|\mu, \sigma/\sqrt{N})$. Then we just apply the updating formulas we previously derived. Thus, for N scores y_i generated from a normal$(y_i|\mu, S_y)$ likelihood and a normal$(\mu|M_\mu, S_\mu)$ prior, the posterior distribution on μ is also normal with mean

$$\frac{1/S_\mu^2}{N/S_y^2 + 1/S_\mu^2} M_\mu + \frac{N/S_y^2}{N/S_y^2 + 1/S_\mu^2} \overline{y}$$

and precision

$$\frac{1}{S_\mu^2} + \frac{N}{S_y^2}.$$

Notice that as the sample size N increases, the posterior mean is dominated by the data mean.

In the derivations above, we estimated the μ parameter when σ is fixed. We can instead estimate the σ parameter when μ is fixed. Again the formulas are more conveniently expressed in terms of precision. It turns out that when μ is fixed, a conjugate prior for the precision is the gamma distribution (e.g., Gelman et al., 2013, p. 43). The gamma distribution was described in Figure 9.8, p. 237. For our purposes, it is not important to state the updating formulas for the gamma distribution in this situation. But it is important to understand the meaning of a gamma prior on precision. Consider a gamma distribution that is loaded heavily over very small values, but has a long shallow tail extending over large values. This sort of gamma distribution on precision indicates that we believe most strongly in small precisions, but we admit that large precisions are possible. If this is a belief about the precision of a normal likelihood function, then this sort of gamma distribution expresses a belief that the data will be more spread out, because small precisions imply large standard deviations. If the gamma distribution is instead loaded over large values of precision, it expresses a belief that the data will be tightly clustered.

Because of its role in conjugate priors for the normal likelihood function, the gamma distribution is routinely used as a prior on precision. But there is no logical necessity to do so, and modern MCMC methods permit more flexible specification of priors. Indeed, because precision is less intuitive than standard deviation, it can be more useful instead to give the standard deviation a uniform prior that spans a wide range.

Summary. We have assumed that the data are generated by a normal likelihood function, parameterized by a mean μ and standard deviation σ, and denoted $y \sim \text{normal}(y|\mu, \sigma)$. For purposes of mathematical derivation, we made unrealistic assumptions that the prior distribution is either a spike on σ or a spike on μ, in order to make three main points:

1. A natural way to express a prior on μ is with a normal distribution, because this is conjugate with the normal likelihood when its standard deviation is fixed.
2. A way to express a prior on the precision $1/\sigma^2$ is with a gamma distribution, because this is conjugate with the normal likelihood when its mean is fixed. However in practice the standard deviation can instead be given a uniform prior (or anything else that reflects prior beliefs, of course).
3. The formulas for Bayesian updating of the parameter distribution are more conveniently expressed in terms of precision than standard deviation. Normal distributions are described sometimes in terms of standard deviation and sometimes in terms of precision, so it is important to glean from context which is being referred to. *In R and Stan, the normal distribution is parameterized by mean and standard deviation. In JAGS and BUGS, the normal distribution is parameterized by mean and precision.*

A joint prior, on the combination of μ and σ parameter values, can also be specified, in such a way that the posterior has the same form as the prior. We will not pursue these mathematical analyses here, because our purpose is merely to justify and motivate typical expressions for the prior distributions on the parameters, so that they can then be used in MCMC sampling.

Various other sources describe conjugate priors for the joint parameter space (e.g., Gelman et al., 2013).

16.1.2. Approximation by MCMC in JAGS

It is easy to estimate the mean and standard deviation in JAGS. Figure 16.2 illustrates two options for the model. The data, y_i, are assumed to be generated by a normal likelihood function with mean μ and standard deviation σ. In both models, the prior on μ is a normal distribution with mean M and standard deviation S. For our applications, we will assume a noncommittal prior that places M in the midst of typical data and sets S to an extremely large value so that the prior has minimal influence on the posterior. For the prior on the standard deviation, the left panel of Figure 16.2 re-expresses σ as precision: $\tau = 1/\sigma^2$ hence $\sigma = 1/\sqrt{\tau}$. Then a noncommittal gamma prior is placed on the precision. The conventional noncommittal gamma prior has shape and rate constants that are close to zero, such as $Sh = 0.01$ and $R = 0.01$. The right panel of Figure 16.2 instead puts a broad uniform distribution directly on σ. The low and high values of the uniform distribution are set to be far outside any realistic value for the data, so that the prior has minimal influence on the posterior. The uniform prior on σ is easier to intuit than a gamma prior on precision, but the priors are not equivalent.

How are the constants in the prior actually determined? In this application we seek broad priors relative to typical data, so that the priors have minimal influence on the posterior. One way to discover the constants is by asking an expert in the domain being studied. But in lieu of that, we will use the data themselves to tell us what the typical scale of the data is. We will set M to the mean of the data, and set S to a huge multiple (e.g., 100) of the standard deviation of the data. This way, no matter what the scale of the data is, the prior will be vague. Similarly, we will set the high value H of the uniform prior on σ to a huge multiple of the standard deviation in the data, and set the low value L to a tiny fraction of the standard deviation in the data. Again, this means that the prior is vague no matter what the scale of the data happens to be.

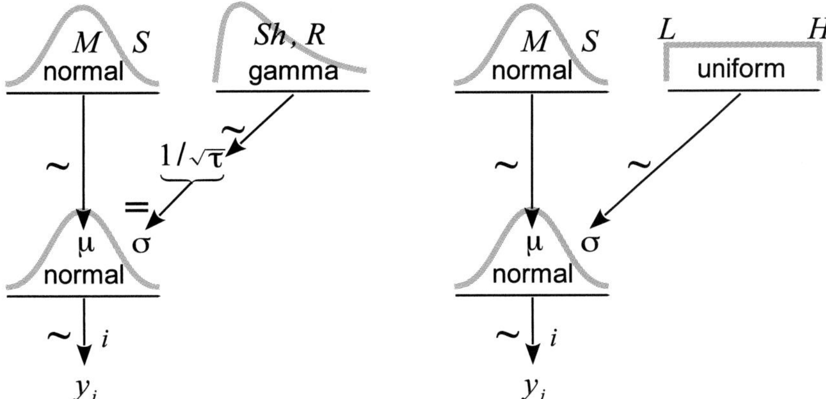

Figure 16.2 Dependencies of variables for metric data described by a normal distribution. The left panel puts a gamma prior on the precision $\tau = 1/\sigma^2$. The right panel puts a uniform prior on σ.

Before we examine the model specification for JAGS, consider the following specification of the data. The data themselves are in the vector named `y`. Then we define:

```
dataList = list(
  y = y ,
  Ntotal = length(y) ,
  meanY = mean(y) ,
  sdY = sd(y)
)
```

Notice above that the mean and standard deviation of the data are packaged into the list that gets shipped to JAGS, so JAGS can use those constants in the model specification. The model in the right side of Figure 16.2 is expressed in JAGS as follows:

```
model {
  for ( i in 1:Ntotal ) {
    y[i] ~ dnorm( mu , 1/sigma^2 )      # JAGS uses precision
  }
  mu ~ dnorm( meanY , 1/(100*sdY)^2 )  # JAGS uses precision
  sigma ~ dunif( sdY/1000 , sdY*1000 )
}
```

Notice that each arrow in Figure 16.2 has a corresponding line in the JAGS code. Notice that JAGS parameterizes `dnorm` by mean and *precision*, not by mean and standard deviation. Notice also that we can put the expression `1/sigma^2` in the argument of `dnorm`, instead of having to define precision as a separate explicit variable.

I have packaged the model and supporting functions in a program that is called from the high-level script, `Jags-Ymet-Xnom1grp-Mnormal-Example.R`. If the file name seems mysterious, please review the file name system explained back in Section 8.3, p. 206. The file name used here indicates that the predicted variable is metric (`Ymet`) and the predictor is nominal with a single group (`Xnom1grp`) and the model is a normal likelihood (`Mnormal`). The program produces graphical output that is specialized for this model.

For purposes of illustration, we use fictitious data. The data are IQ (intelligence quotient) scores from a group of people who have consumed a "smart drug." We know that IQ tests have been normed to the general population so that they have an average score of 100 and a standard deviation of 15. Therefore, we would like to know how differently the smart-drug group has performed relative to the general population average.

The aforementioned script loads the data file, runs JAGS, produces MCMC diagnostics, and graphs the posterior. Before considering the posterior, it is important to check that the chains are well behaved. The diagnostics (not shown here, but you can run the script and see for yourself) indicate converged chains with an ESS on both parameters of at least 10,000. Figure 16.3 shows aspects of the posterior distribution, along with a histogram of the data in the upper right panel.

The upper left panel of Figure 16.3 shows the posterior on the mean parameter. The estimated 95% HDI extends from 101.35 to 114.21. This HDI barely excludes a somewhat arbitrary ROPE from 99 to 101, especially if we also consider MCMC variability in the

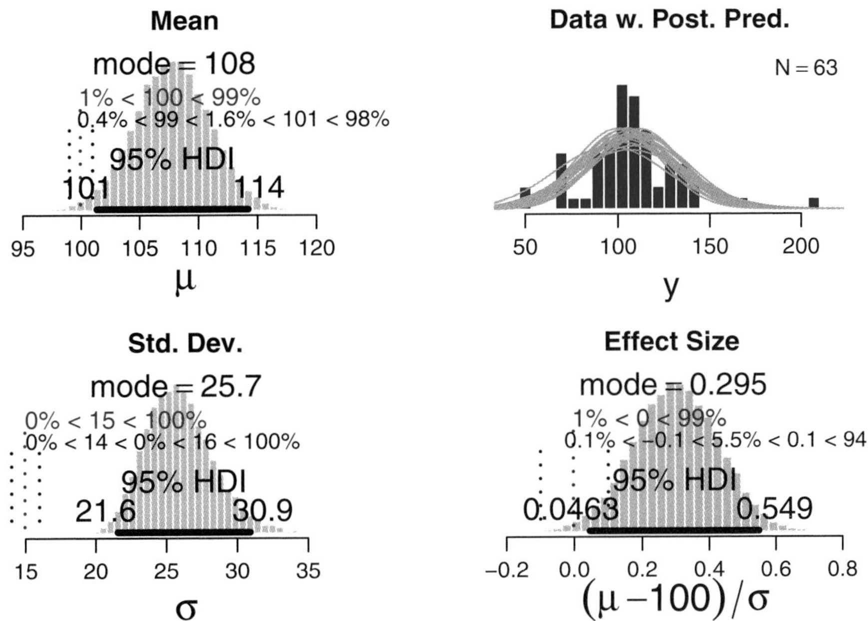

Figure 16.3 Posterior distribution of `Jags-Ymet-Xnom1grp-Mnormal-Example.R` applied to fictitious IQ data from a "smart drug" group.

HDI limit (recall Figure 7.13, p. 185). Thus it appears that the smart drug increases IQ performance somewhat, but the improvement is not large relative to the uncertainty in the estimate. Therefore, we might not want to make any strong decision, from these data, that IQ is credibly increased by the smart drug. This conservative conclusion is reinforced by considering the effect size in the lower right panel of Figure 16.3. *Effect size* is simply the amount of change induced by the treatment relative to the standard deviation: $(\mu - 100)/\sigma$. In other words, effect size is the "standardized" change. The posterior on the effect size has a 95% HDI barely excluding zero, and clearly not excluding a ROPE from −0.1 to 0.1. A conventionally "small" effect size in psychological research is 0.2 (Cohen, 1988), and the ROPE limits are set at half that size for purposes of illustration.

However, the lower left panel of Figure 16.3 suggests that the smart drug did have an effect on the standard deviation of the group. The posterior of the standard deviation is far outside any reasonable ROPE around the general-population value of 15. One interpretation of this result is that the smart drug caused some people to increase their IQ test performance but caused other people to decrease their IQ test performance. Such an effect on variance has real-world precedents; for example, stress can increase variance across people (Lazarus & Eriksen, 1952). In the next section we will model the data with a distribution that accommodates outliers, and we will have to modify this interpretation of an increased standard deviation.

Finally, the upper right panel of Figure 16.3 shows a histogram of the data superimposed with a smattering of normal curves that have credible μ and σ values from the MCMC sample. This constitutes a form of posterior-predictive check, by which we check whether the model appears to be a reasonable description of the data. With such a small amount of data, it is difficult to visually assess whether normality is badly violated, but there appears to be a hint that the normal model is straining to accommodate some outliers: The peak of the data protrudes prominently above the normal curves, and there are gaps under the shoulders of the normal curves.

16.2. OUTLIERS AND ROBUST ESTIMATION: THE *t* DISTRIBUTION

When data appear to have outliers beyond what would be accommodated by a normal distribution, it would be useful to be able to describe the data with a more appropriate distribution that has taller or heavier tails than the normal. A well-known distribution with heavy tails is the *t* distribution. The *t* distribution was originally invented by Gosset (1908), who used the pseudonym "Student" because his employer (Guinness Brewery) prohibited publication of any research that might be proprietary or imply problems with their product (such as variability in quality). Therefore the distribution is often referred to as the *Student t* distribution.

Figure 16.4 shows examples of the *t* distribution. Like the normal distribution, it has two parameters that control its mean and its width. The standard deviation is controlled indirectly via the *t* distribution's "scale" parameter. In Figure 16.4, the mean is set to zero and the scale is set to one. The *t* distribution has a third parameter that controls the heaviness of its tails,

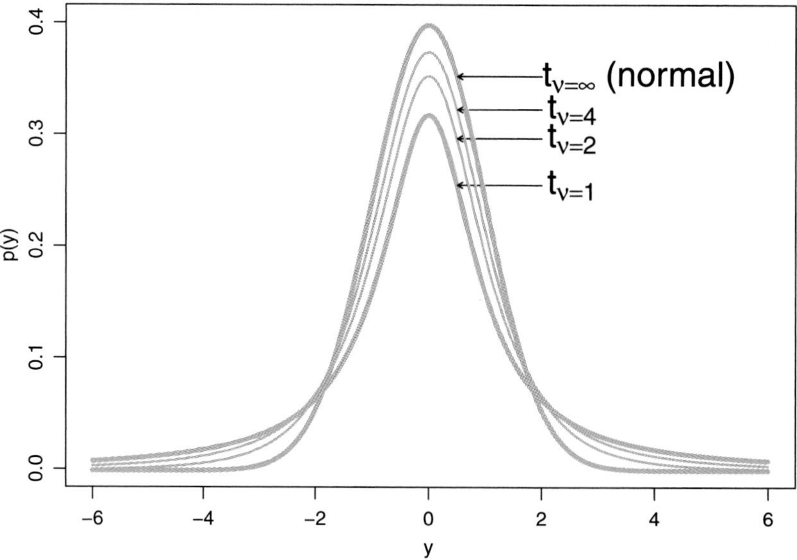

Figure 16.4 Examples of *t* distributions. In all cases, $\mu = 0$ and $\sigma = 1$. The normality parameter, ν, controls the heaviness of the tails. Curves for different values of ν are superimposed for easy comparison. The abscissa is labeled as *y* (not *x*) because the distribution is intended to describe predicted data.

which I will refer to as the "normality" parameter, ν (Greek letter nu). Many people might be familiar with this parameter as the "degrees of freedom" from its use in NHST. But because we will not be using the t distribution as a sampling distribution, and instead we will be using it only as a descriptive shape, I prefer to name the parameter by its effect on the distribution's shape. The normality parameter can range continuously from 1 to ∞. As can be seen in Figure 16.4, when $\nu = 1$ the t distribution has very heavy tails, and when ν approaches ∞ the t distribution becomes normal.

Although the t distribution is usually conceived as a sampling distribution for the NHST t test, we will use it instead as a convenient descriptive model of data with outliers (as is often done; e.g., Damgaard, 2007; M. C. Jones & Faddy, 2003; Lange, Little, & Taylor, 1989; Meyer & Yu, 2000; Tsionas, 2002; Zhang, Lai, Lu, & Tong, 2013). Outliers are simply data values that fall unusually far from a model's expected value. Real data often contain outliers relative to a normal distribution. Sometimes the anomalous values can be attributed to extraneous influences that can be explicitly identified, in which case the affected data values can be corrected or removed. But usually we have no way of knowing whether a suspected outlying value was caused by an extraneous influence, or is a genuine representation of the target being measured. Instead of deleting suspected outliers from the data according to some arbitrary criterion, we retain all the data but use a noise distribution that is less affected by outliers than is the normal distribution.

Figure 16.5 shows examples of how the t distribution is robust against outliers. The curves show the maximum likelihood estimates (MLEs) of the parameters for the t and normal distributions. More formally, for the given data $D = \{y_i\}$, parameter values were found for the normal that maximized $p(D|\mu, \sigma)$, and parameter values were found for the t distribution that maximized $p(D|\mu, \sigma, \nu)$. The curves for those MLEs are plotted with the data. The upper panel of Figure 16.5 shows "toy" data to illustrate that the normal is strongly influenced by an outlier, while the t distribution remains centered over the bulk of the data. For the t distribution, the mean μ is 0.12, which is very close to the center of the cluster of five data points. The outlying datum is accommodated by setting the normality to a very small value. The normal distribution, on the other hand, can accommodate the outlier only by using a mean that is shifted toward the outlier and a standard deviation that is bloated to span the outlier. The t distribution appears to be a better description of the data.

It is important to understand that the scale parameter σ in the t distribution is not the standard deviation of the distribution. (Recall that the standard deviation is the square root of the variance, which is the expected value of the squared deviation from the mean, as defined back in Equation 4.8, p. 86.) The standard deviation is actually larger than σ because of the heavy tails. In fact, when ν drops below 2 (but is still ≥ 1), the standard deviation of the mathematical t distribution goes to infinity. For example, in the upper panel of Figure 16.5, ν is only 1.14, which means that the standard deviation of the mathematical t distribution is infinity, even though the sample standard deviation of the data is finite. At the same time, the scale parameter of the t distribution has value $\sigma = 1.47$. While this value of the scale parameter is not the standard deviation of the distribution, it does have an intuitive relation to the spread of the data. Just as the range $\pm\sigma$ covers the middle 68% of a *normal* distribution, the range $\pm\sigma$ covers the middle 58% of a t distribution when $\nu = 2$, and the middle 50%

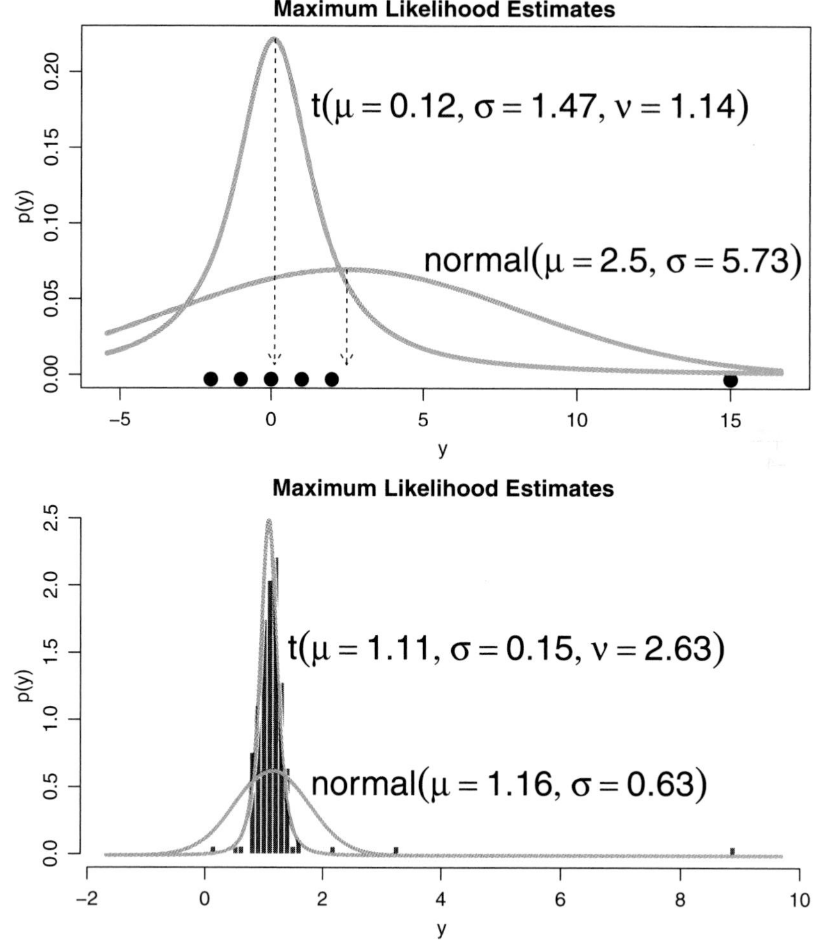

Figure 16.5 The maximum likelihood estimates of normal and t distributions fit to the data shown. Upper panel shows "toy" data to illustrate that the normal accommodates an outlier only by enlarging its standard deviation and, in this case, by shifting its mean. Lower panel shows actual data (Holcomb & Spalsbury, 2005) to illustrate realistic effect of outliers on estimates of the normal.

when $\nu = 1$. These areas are illustrated in the left column of Figure 16.6. The right column of Figure 16.6 shows the width under the middle of a t distribution that is needed to span 68.27% of the distribution, which is the area under a normal distribution for $\sigma = \pm 1$.

The lower panel of Figure 16.5 uses realistic data that indicates levels of inorganic phosphorous, measured in milligrams per deciliter, in 177 human subjects aged 65 or older. The authors of the data (Holcomb & Spalsbury, 2005) intentionally altered a few data points to reflect typical transcription errors and to illustrate methods for detecting and correcting such errors. We instead assume that we no longer have access to records of the original individual measurements, and must model the uncorrected data set. The t distribution accommodates the outliers and fits the distribution of data much better than the normal.

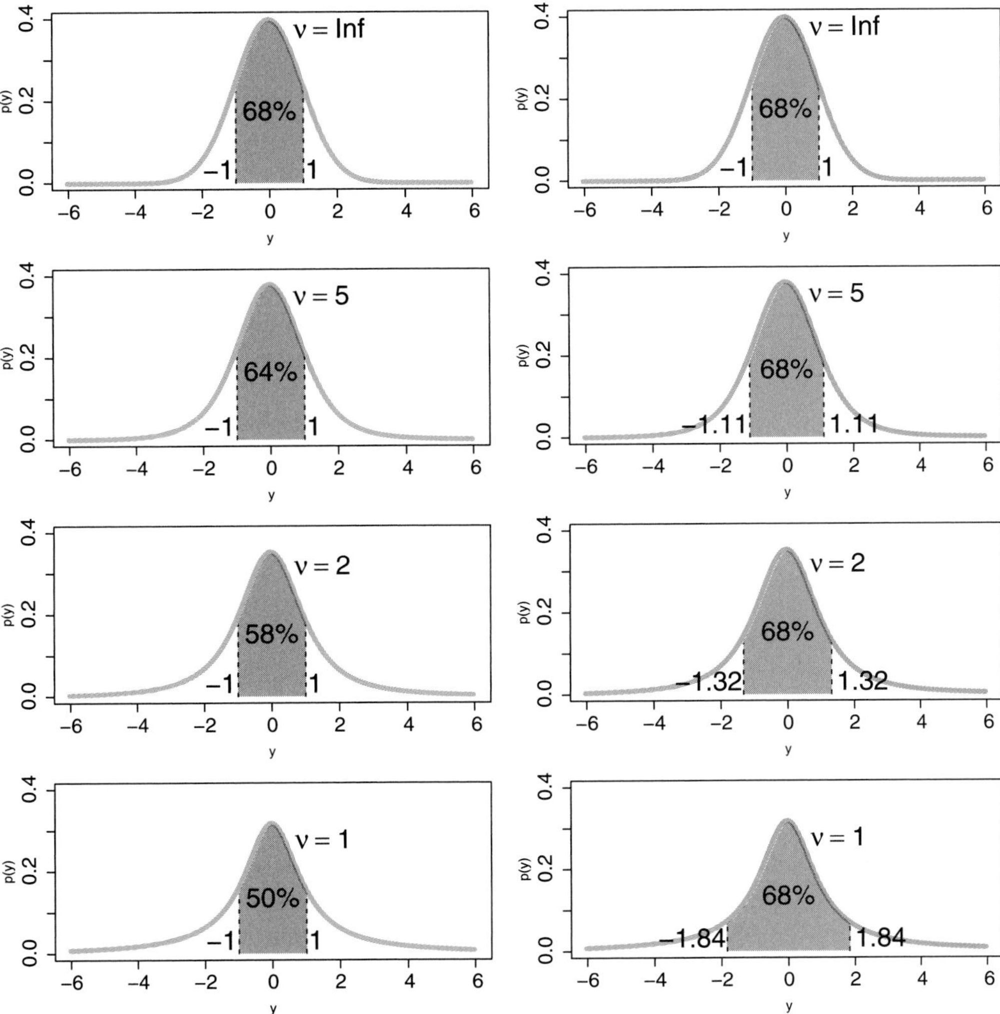

Figure 16.6 Examples of *t* distributions with areas under the curve. In all cases, $\mu = 0$ and $\sigma = 1$. Rows show different values of the normality parameter, ν. Left column shows area under the *t* distribution from $y = -1$ to $y = +1$. Right column shows values of $\pm y$ needed for an area of 68.27%, which is the area under a standardized normal curve from $y = -1$ to $y = +1$. The abscissa is labeled as *y* (not *x*) because the distribution is intended to describe predicted data.

The use of a heavy-tailed distribution is often called *robust estimation* because the estimated value of the central tendency is stable, that is, "robust," against outliers. The *t* distribution is useful as a likelihood function for modeling outliers at the level of observed data. But the *t* distribution is also useful for modeling outliers at higher levels in a hierarchical prior. We will encounter several applications.

16.2.1. Using the *t* distribution in JAGS

It is easy to incorporate the *t* distribution into the model specification for JAGS. The likelihood changes from `dnorm` to `dt` with the inclusion of its normality parameter, as follows:

```
model {
  for ( i in 1:Ntotal ) {
    y[i] ~ dt( mu , 1/sigma^2 , nu )      # JAGS: dt uses precision
  }
  mu ~ dnorm( meanY , 1/(100*sdY)^2 )
  sigma ~ dunif( sdY/1000 , sdY*1000 )
  nu <- nuMinusOne+1                      # nu must be >= 1
  nuMinusOne ~ dexp(1/29)                 # prior on nu-1
}
```

Notice above that `dt` has three parameters, the first two of which are just like the parameters in `dnorm`. The third parameter is ν. The final lines of the model specification implement the prior on ν. The motivation for the prior is as follows. Look at the family of *t* distributions in Figure 16.4, and notice that nearly all the variation happens when ν is fairly small. In fact, once ν gets up to about 30, the *t* distribution is essentially normal. Thus, we would like a prior that gives equal opportunity to small values of ν (less than 30) and larger values of ν (greater than 30). A convenient distribution that captures this allocation is the exponential distribution. It is defined over positive values. It has a single parameter that specifies the reciprocal of its mean. Thus, the JAGS expression `nuMinusOne` $\sim$ `dexp(1/29)` says that the variable named `nuMinusOne` is exponentially distributed with a mean of 29. The value of `nuMinusOne` ranges from zero to infinity, so we must add 1 to make it range from 1 to infinity, with a mean of 30. This is accomplished by the penultimate line in the model specification. Figure 16.7, upper panel, shows the prior distribution on ν. Notice that it does indeed put considerable probability mass on small values of ν, but the mean of the distribution is 30 and large values of ν are also entertained. The lower panel of Figure 16.7 shows the same prior distribution on a logarithmic scale. The logarithmic scale is useful for displaying the distribution because it is extremely skewed on the original scale.

It should be emphasized that this choice of prior for ν is not uniquely "correct." While it is well motivated and has reasonable operational characteristics in many applications, there may be situations in which you would want to put more or less prior mass on small values of ν. The prior on ν can have lingering influence on the posterior even for fairly large data sets because the estimate of ν is influenced strongly by data in the tails of the distribution which, by definition, are relatively rare. To make small values of ν credible in the posterior, the data must contain some extreme outliers or many moderate outliers. Because outliers are rare, even in samples from truly heavy-tailed distributions, the prior on ν must have a fair amount of credibility on small values.

The script for running this model is called `Jags-Ymet-Xnom1grp-Mrobust-Example.R`. Figure 16.8 shows pair-wise plots of the three parameters. Notice that the credible values of σ are positively correlated with credible values of $\log10(\nu)$. This means that if the distribution is more normal with larger ν, then the distribution must also be wider with larger σ. This correlation is a signature of the data having outliers. To accommodate

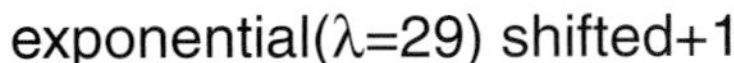

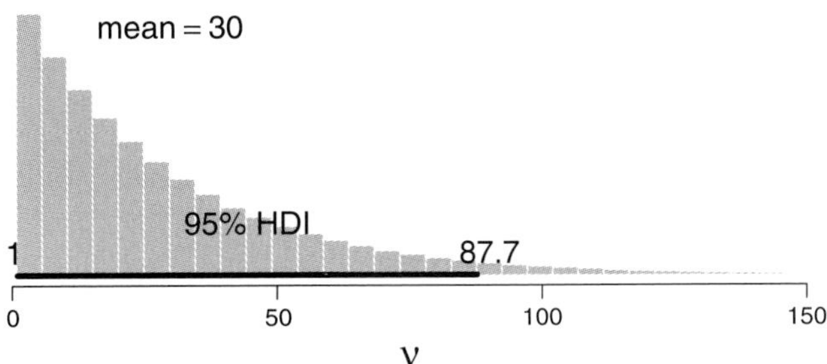

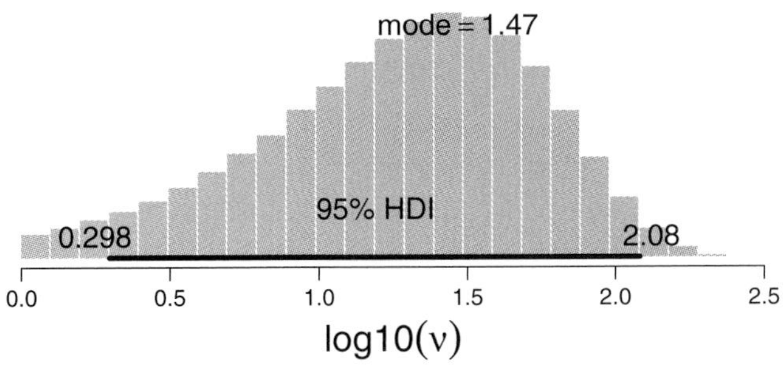

Figure 16.7 The prior on the normality parameter. Upper panel shows the shifted exponential distribution on the original scale of ν. Lower panel shows the same distribution on a logarithmic scale.

the outliers, either ν must be small to provide heavy tails or σ must be large to provide a wide distribution. We saw this same trade-off in the MLE examples of Figure 16.5.

Figure 16.9 shows other aspects of the posterior distribution. Notice the marginal posterior on the normality parameter in the lower left panel. Its mode (on the log10 scale) is only 0.68, which is noticeably reduced from the prior mode of 1.47 in the lower panel of Figure 16.7. This indicates that the data are better fit by small-ish values of ν because there are outliers.

Typically we are not interested in the exact estimated value of ν. We merely want to give the model flexibility to use heavy tails if the data demand them. If the data are normally distributed, then the posterior of ν will emphasize large values. Any value of ν greater than about 30 ($\approx$1.47 on the log10 scale) represents a nearly normal distribution, so its exact value does not have strong consequences for interpretation. On the other hand, small values of ν can be estimated accurately only by data in the extreme tails of the distribution, which, by their very nature, are rare. Therefore, we cannot anticipate much certainty in the estimate of ν, and we settle for broad statements about whether the posterior emphasizes small-ish values of log10(ν) (e.g., under 1.47) or large-ish values of log10(ν) (e.g., over 1.47).

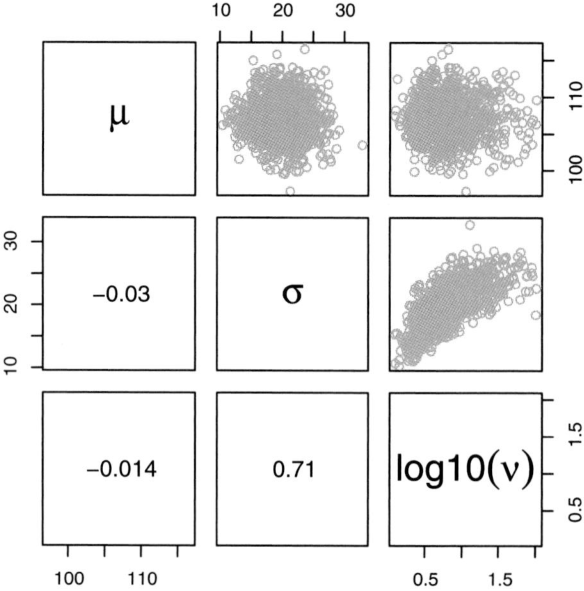

Figure 16.8 Posterior distribution of `Jags-Ymet-Xnom1grp-Mrobust-Example.R` applied to fictitious IQ data from a "smart drug" group. Off-diagonal cells show scatter plots and correlations of parameters indicated in the corresponding diagonal cells. Notice the strong positive correlation of σ and log10(ν).

The upper right panel of Figure 16.9 shows that the posterior predictive t distributions appear to describe the data better than the normal distribution in Figure 16.3, insofar as the data histogram does not poke out at the mode and the gaps under the shoulders are smaller.

Detailed comparison of Figure 16.9 with Figure 16.3 also reveals that the marginal posteriors on μ and effect size are a little bit tighter, with a little more of the distributions falling above the ROPEs. More prominently, σ in the robust estimate is much smaller than in the normal estimate. What we had interpreted as increased standard deviation induced by the smart drug might be better described as increased outliers. Both of these differences, that is, μ more tightly estimated and σ smaller in magnitude, are a result of there being outliers in the data. The only way a normal distribution can accommodate the outliers is to use a large value for σ. In turn, that leads to "slop" in the estimate of μ because there is a wider range of μ values that reasonably fit the data when the standard deviation is large, as we can see by comparing the upper and lower rows of Figure 16.1.

16.2.2. Using the *t* distribution in Stan

When you run the JAGS program yourself, you will see that it uses many steps to produce a posterior sample for σ that has an ESS exceeding 10,000. You will also see that the ESS for

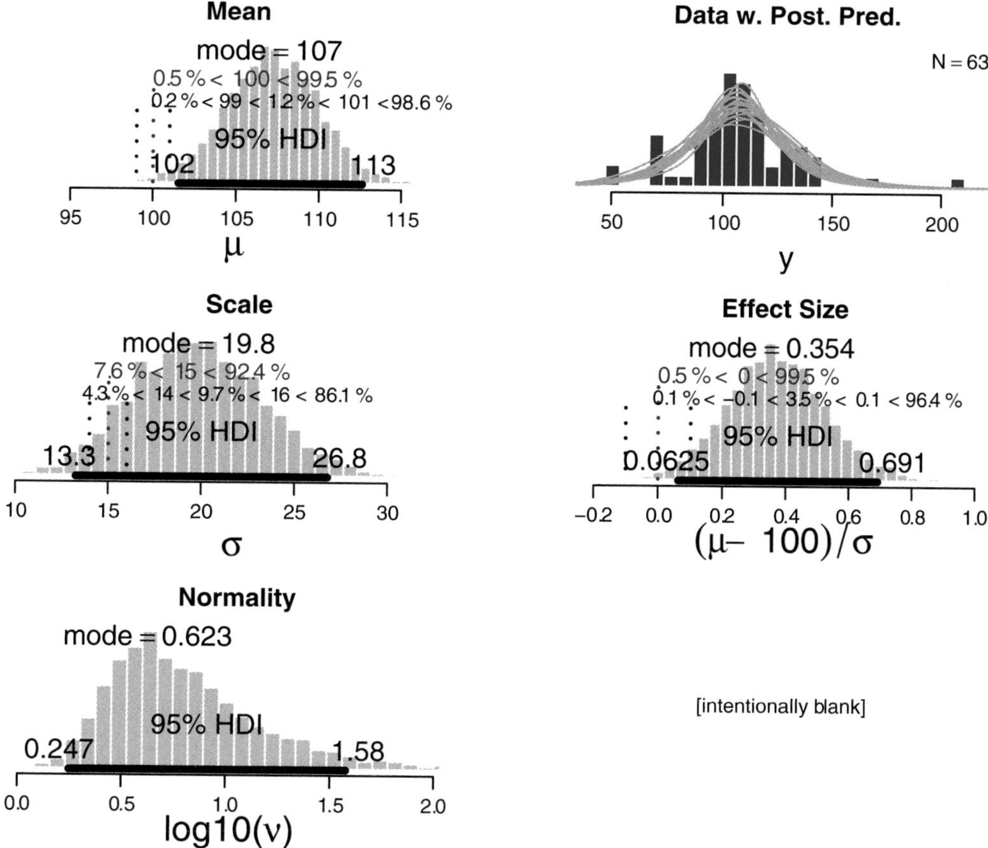

Figure 16.9 Posterior distribution of `Jags-Ymet-Xnom1grp-Mrobust` applied to fictitious IQ data from a "smart drug" group. Compare with Figure 16.3.

ν is less than 10,000 despite the long chain. In other words, there is high autocorrelation in the chains in JAGS.

We do not care that the chain for ν has a relatively small ESS because (a) we do not care about the exact value of ν when interpreting the posterior, as explained above, and (b) the exact value of ν has relatively little effect on the estimates of the other parameters. To be sure, the posterior sample of ν must be converged and truly representative of the posterior, but it does not need to be as finely detailed as the other parameters. Nevertheless, it would be less worrying if ν had a larger ESS.

The autocorrelation of the MCMC sampling in JAGS requires a long chain, which requires us to have patience while the computer chugs along. We have discussed two options for improving the efficiency of the sampling. One option is to run parallel chains on multiple cores using runjags (Section 8.7, p. 215). Another option is to implement the model in Stan, which may explore the parameter space more efficiently with its HMC sampling

(Chapter 14). This section shows how to run the model in Stan. Its results do indeed show a posterior sample of ν with higher ESS than JAGS.

In Stan, the sampling statement for a t distribution has the form `y ~ student_t (nu,mu,sigma)`. Notice that the normality parameter is the *first* argument, not the last as in JAGS (and BUGS). Notice also that the scale parameter is entered directly, not indirectly as the precision $1/\sigma^2$ as in JAGS (and BUGS). Here is the complete model specification in Stan, which begins with declaring all the variables for the data and parameters before getting to the model at the end:

```
data {
  int<lower=1> Ntotal ;
  real y[Ntotal] ;
  real meanY ;
  real sdY ;
}
transformed data { // compute the constants for the priors
  real unifLo ;
  real unifHi ;
  real normalSigma ;
  real expLambda ;
  unifLo <- sdY/1000 ;
  unifHi <- sdY*1000 ;
  normalSigma <- sdY*100 ;
  expLambda <- 1/29.0 ;
}
parameters {
  real<lower=0> nuMinusOne ;
  real mu ;
  real<lower=0> sigma ;
}
transformed parameters {
  real<lower=0> nu ;
  nu <- nuMinusOne + 1 ;
}
model {
  sigma ~ uniform( unifLo , unifHi ) ;
  mu ~ normal( meanY , normalSigma ) ;
  nuMinusOne ~ exponential( expLambda ) ;
  y ~ student_t( nu , mu , sigma ) ;  // vectorized
}
```

The script for running this model is called `Stan-Ymet-Xnom1grp-Mrobust-Example.R`. It uses a chain length and thinning that match the specifications for the corresponding JAGS script merely for purposes of comparison, but it turns out that Stan does not need such long chains as JAGS to produce the same ESS. Figure 16.10 shows the chain diagnostics for the ν parameter in JAGS and Stan. For both runs, there were 20,000 steps with a thinning of 5. The thinning was done merely to keep the saved file size down to a modest size for JAGS; thinning is not recommended if computer memory is not an

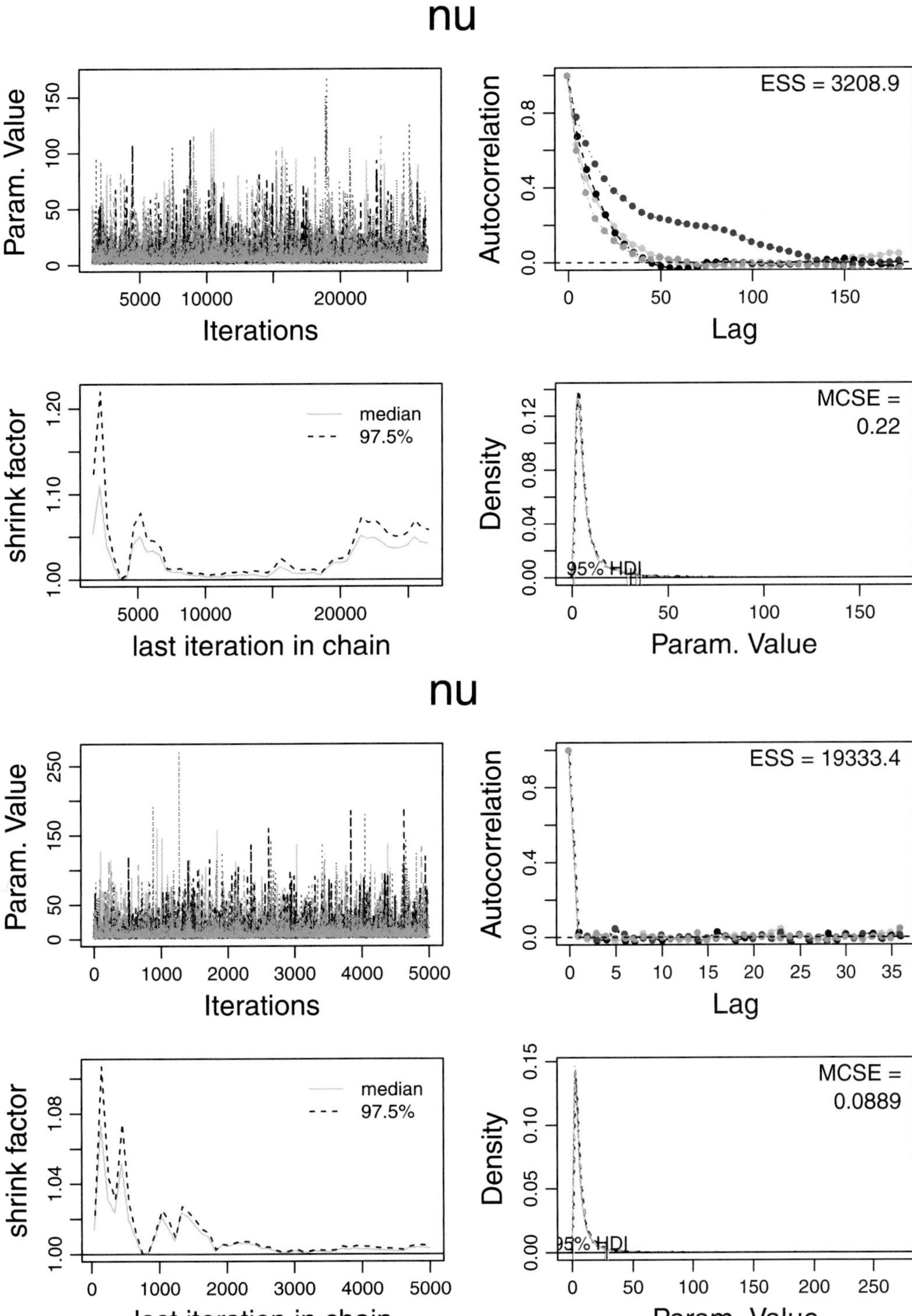

Figure 16.10 Chain diagnostics for JAGS (above) and Stan (below). Notice difference in autocorrelation in the upper right panels, and the resulting ESS.

issue. As you can see, Stan does an excellent job of sampling the normality parameter ν. Presumably this is because the Hamiltonian dynamics create proposed moves that efficiently jump around the parameter space. Stan also does a better job than JAGS in sampling σ and μ.

16.3. TWO GROUPS

An often-used research design is a comparison of two groups. For example, in the context of assaying the effect of a "smart drug" on IQ, instead of comparing the mean of the treatment group against an assumed landmark such as 100 (see Figure 16.3), it would make more sense to compare the treatment group against an identically handled placebo group. When there are two groups, we estimate the mean and scale for each group. When using t distributions for robust estimation, we could also estimate the normality of each group separately. But because there usually are relatively few outliers, we will use a single normality parameter to describe both groups, so that the estimate of the normality is more stably estimated.

Figure 16.11 illustrates the model structure. At the bottom of the diagram, the ith datum within group j is denoted $y_{i|j}$. The data within group j come from a t distribution with mean μ_j and scale σ_j. The normality parameter ν has no subscript because it is used for both groups. The prior for the parameters is exactly what was used for robust estimation of a single group in the previous section.

For two groups, there are only a few changes to the Stan model specification for a single group. The data now include a group identifier. The ith row of the data file specifies the IQ score as `y[i]` and the group identity as `x[i]`. In the data block of the model specification, the group membership variable `x` must be declared. The transformed data block is unchanged. The parameter block merely makes the mu and sigma variables vectors of 2 elements (for the

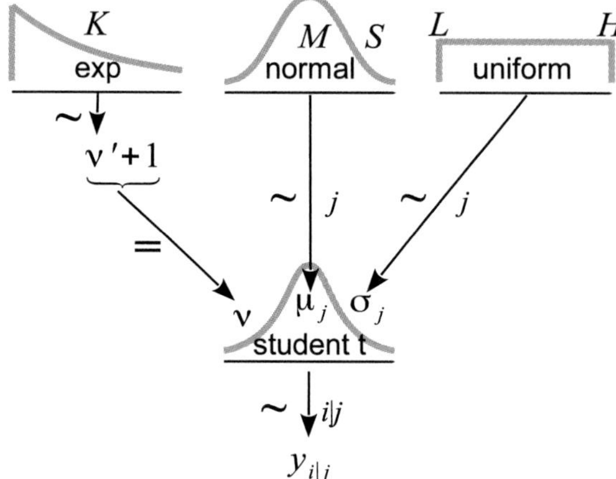

Figure 16.11 Dependency diagram for robust estimation of two groups. At the bottom of the diagram, $y_{i|j}$ is the ith datum within the jth group.

2 groups). Finally, in the model block, the likelihood function is put in a loop so that nested indexing of the group identifier can be used. Here is the complete model specification, with the changed lines marked by comments:

```
data {
  int<lower=1> Ntotal ;
  int x[Ntotal] ;                                          // group identifier
  real y[Ntotal] ;
  real meanY ;
  real sdY ;
}
transformed data {
  real unifLo ;
  real unifHi ;
  real normalSigma ;
  real expLambda ;
  unifLo <- sdY/1000 ;
  unifHi <- sdY*1000 ;
  normalSigma <- sdY*100 ;
  expLambda <- 1/29.0 ;
}
parameters {
  real<lower=0> nuMinusOne ;
  real mu[2] ;                                             // 2 groups
  real<lower=0> sigma[2] ;                                 // 2 groups
}
transformed parameters {
  real<lower=0> nu ;
  nu <- nuMinusOne + 1 ;
}
model {
  sigma ~ uniform( unifLo , unifHi ) ;                     // vectorized 2 groups
  mu ~ normal( meanY , normalSigma ) ;                     // vectorized 2 groups
  nuMinusOne ~ exponential( expLambda ) ;
  for ( i in 1:Ntotal ) {
    y[i] ~ student_t( nu , mu[x[i]] , sigma[x[i]] ) ; // nested index of group
  }
}
```

Notice in the model block that there is essentially one line of code for each arrow in Figure 16.11. The only arrow that is not in the model block is the additive shift of ν by $+1$, which appears in Stan's transformed parameters block. When the diagram is implemented in JAGS, all the arrows appear in the model block.

A script for running the program is `Stan-Ymet-Xnom2grp-MrobustHet-Example.R`, and its format is minimally different from the other high-level scripts included with this book. The substantive changes are in the function definitions in `Stan-Ymet-Xnom2grp-MrobustHet.R`, which includes the Stan model specification (shown in the previous paragraph) and the specialized output graphs for this application.

The posterior distribution is shown in Figure 16.12. You can see the marginal posterior distributions on the five parameters (μ_1, μ_2, σ_1, σ_2, and ν) and the posterior distributions of the difference of means $\mu_2 - \mu_1$, the difference of scales $\sigma_2 - \sigma_1$, and the effect size, which is defined as the difference of means relative to the average scale: $(\mu_2 - \mu_1)/\sqrt{(\sigma_1^2 + \sigma_2^2)/2}$. The posterior distribution indicates that the difference of means is about 7.7 IQ points, but the 95% HDI of the difference barely excludes a small ROPE around zero. The posterior distribution on the effect size also suggests a nonzero difference, but the 95% HDI slightly overlaps the ROPE spanning ±0.1. The difference of scales (i.e., $\sigma_2 - \sigma_1$) shows a credible nonzero difference, suggesting that the smart drug causes greater variance than the placebo.

Recall from Section 12.1.2.1, p. 340, that the differences of parameter values are computed from their jointly credible values at every step in the MCMC chain. This is worth remembering because the credible values of σ_2 and σ_1 are positively correlated in the posterior distribution, and therefore their difference cannot be accurately gleaned by considering their separate marginal distributions. The two scale parameters are positively correlated because they both trade off with the normality parameter. When the normality parameter is large, then both scale parameters must be large to accommodate the outliers in the two groups. When the normality parameter is small, then both scale parameters are better off being small to give higher probability to the centrally located data.

The posterior predictive check in the upper right panels of Figure 16.12 suggests that the t distribution is a reasonably good description of both groups. Neither group's data show clear departures from the smattering of credible t distributions.

16.3.1. Analysis by NHST

In traditional NHST, metric data from two groups would be submitted to a t-test. The t test is part of the standard R facilities; learn about it by typing `?t.test` at R's command line. When applied to the IQ data, here are the results:

```
> myDataFrame = read.csv( file="TwoGroupIQ.csv" )
> t.test( Score ~ Group , data=myDataFrame )

     Welch Two Sample t test
data:  Score by Group
t = -1.958, df = 111.441, p-value = 0.05273
alternative hypothesis: true difference in means is not equal to 0
95 percent confidence interval:
 -15.70602585   0.09366161
sample estimates:
   mean in group Placebo mean in group Smart Drug
              100.0351                 107.8413
```

Notice that the p value is greater than 0.05, which means that the conventional decision would be *not* to reject the null hypothesis. This conclusion conflicts with the Bayesian analysis in Figure 16.12, unless we use a conservatively wide ROPE. The reason that the t test is less sensitive than the Bayesian estimation in this example is that the t test assumes normality and therefore its estimate of the within-group variances is too large when there are outliers.

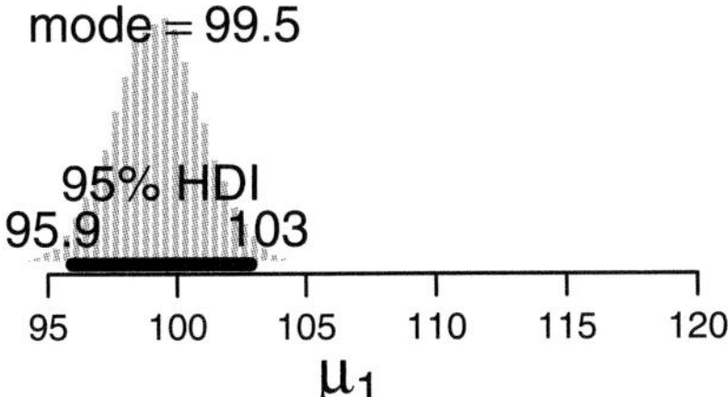

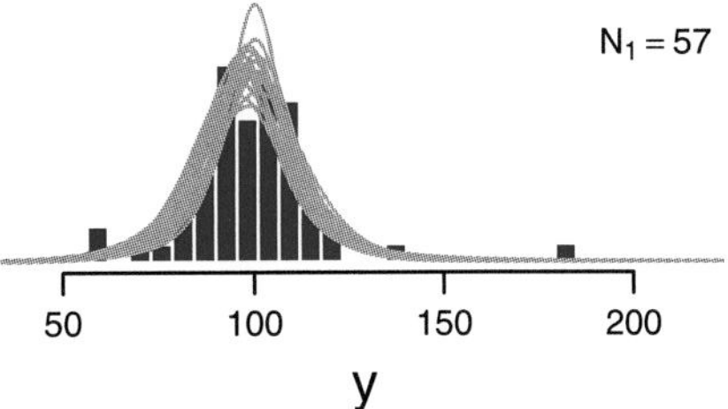

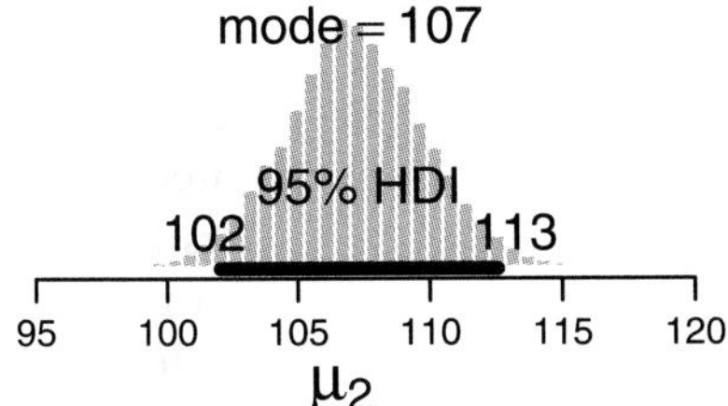

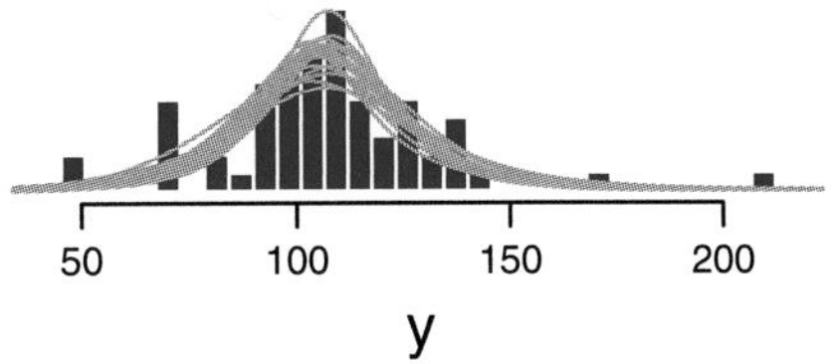

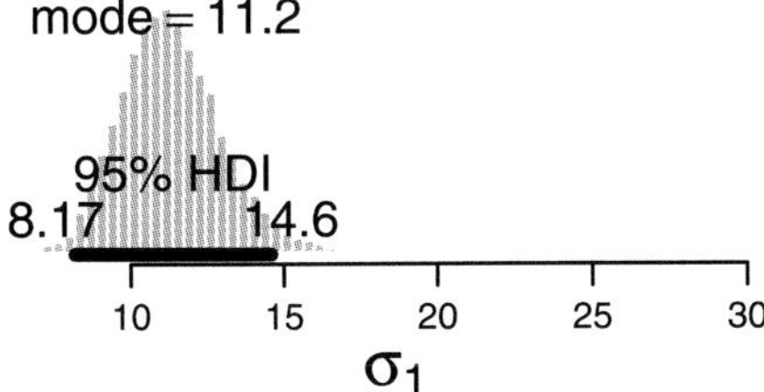

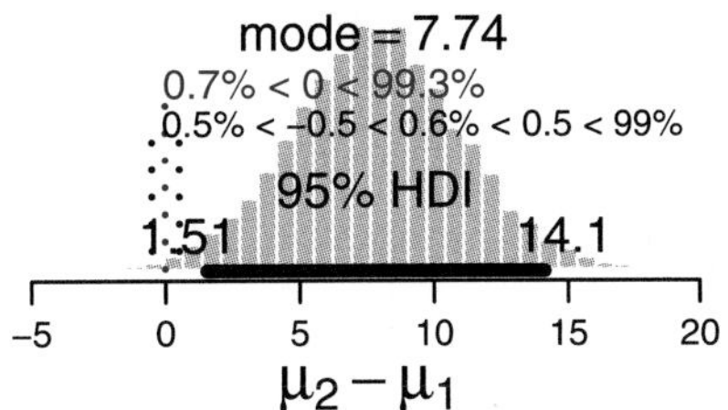

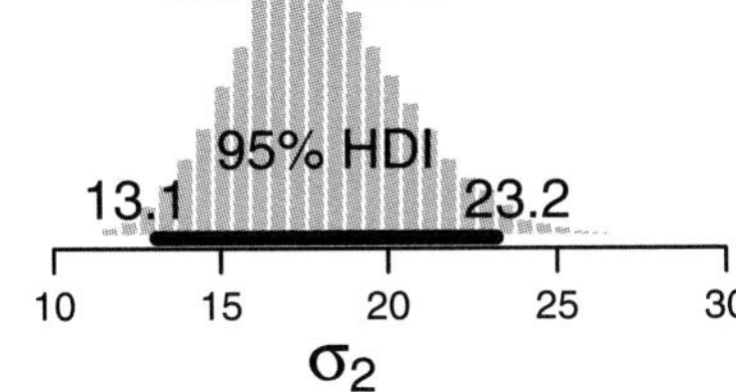

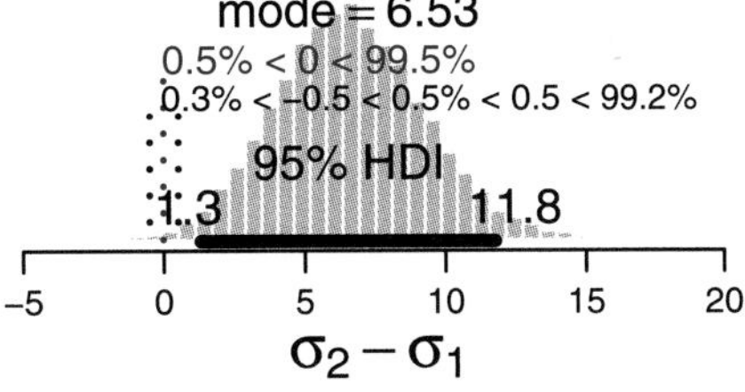

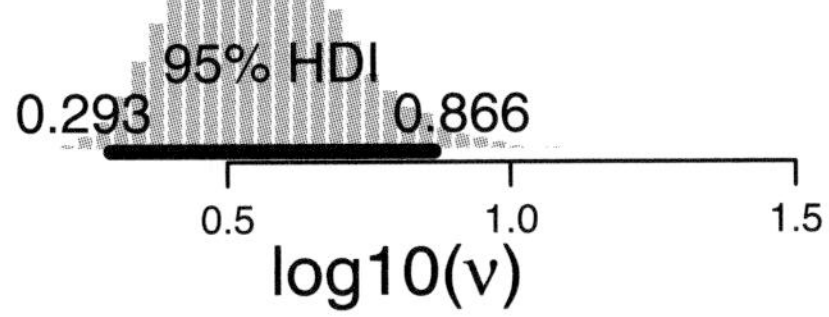

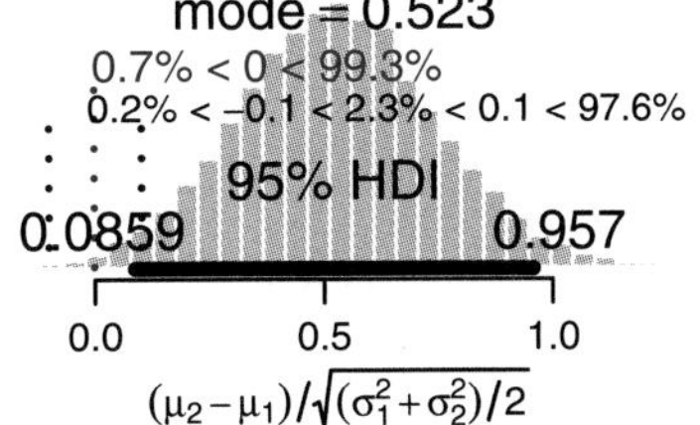

Figure 16.12 Posterior distribution for two groups.

The t test has other problems. Unlike the Bayesian analysis, the t test provides only a test of the equality of means, without a test of the equality of variances. To test equality of variances, we need to run an additional test, namely an F test of the ratio of variances, which would inflate the p values of both tests. Moreover, both tests compute p values based on hypothetical normally distributed data, and the F test is particularly sensitive to violations of this assumption. Therefore it would be better to use resampling methods to compute the p values (and correcting them for multiple tests).

I have previously written an extensive but self-contained article about this case of estimating two groups, with other examples comparing conclusions from NHST and Bayesian analysis (Kruschke, 2013a). The article includes programs for JAGS that have been translated into other formats by friendly enthusiasts; you can find links to their work at the Web site http://www.indiana.edu/kruschke/BEST/ (where BEST stands for Bayesian estimation). The article also explains Bayesian power analysis and some of the perils of p values, and, in an appendix, Bayes' factor approaches to assessing null values. If you are looking for a compact introduction to Bayesian data analysis, perhaps as a gift for a loved one, or as something to bring to the host of a party you are attending, that article might be just what you need.[2]

16.4. OTHER NOISE DISTRIBUTIONS AND TRANSFORMING DATA

When the data are not distributed like the assumed noise distribution, then the interpretation of the parameters can be problematic. For example, if the data have many outliers, and the assumed noise distribution is normal, then the estimate of the standard deviation parameter is artificially inflated by the outliers. If the data are skewed but the assumed distribution is symmetric, then the estimate of the mean parameter is artificially pulled by the skewed data values. In general, we want the noise distribution to accurately mimic the data, so that the parameters are meaningful.

If the initially assumed noise distribution does not match the data distribution, there are two ways to pursue a better description. The preferred way is to use a better noise distribution. The other way is to transform the data to a new scale so that they tolerably match the shape of the assumed noise distribution. In other words, we can either change the shoe to fit the foot, or we can squeeze the foot to fit in the shoe. Changing the shoe is preferable to squeezing the foot. In traditional statistical software, users were stuck with the pre-packaged noise distribution, and had no way to change it, so they transformed their data and squeezed them into the software. This practice can lead to confusion in interpreting the parameters because they are describing the transformed data, not the data on the original scale. In software such as JAGS and Stan, however, there is great flexibility in specifying various noise distributions (and higher level structure). We have seen one example in this chapter, in which an initially assumed normal distribution was changed to a t distribution. Many other distributions are available in JAGS and Stan, and we can also specify noise distributions by using the Bernoulli ones trick that was explained in Section 8.6.1, p. 214.

[2] Of course, the article can also be used for lining the bottom of bird cages, or for wrapping fish at the market.

As another example of non-normal noise distributions, consider models of response time. Response time data are typically positively skewed, because response times can only be so fast, but can often be very slow. There is a debate in the scientific literature regarding what sort of distribution best describes response times and why. As one recent example, Rouder, Lu, Speckman, Sun, and Jiang (2005) used the Weibull distribution in a hierarchical Bayesian model to describe response times. Whatever the preferred descriptive distribution is, probably it can be implemented in JAGS and Stan. For example, both JAGS and Stan have the Weibull distribution built in.

16.5. EXERCISES

Look for more exercises at https://sites.google.com/site/doingbayesiandataanalysis/

Exercise 16.1. [Purpose: Practice using different data files in the high-level script, with an interesting real example about alcohol preference of sexually frustrated males.] Shohat-Ophir et al. (2012) were interested in alcohol preferences of sexually deprived males. The procedure is illustrated in Figure 16.13, and was described as follows: "One cohort, rejected-isolated, was subjected to courtship conditioning; they experienced 1-h sessions of sexual rejection by mated females, three times a day, for 4 days. …Flies in the mated-grouped cohort experienced 6-h sessions of mating with multiple receptive virgin females (ratio 1:5) for 4 days. Flies from each cohort were then tested in a two-choice preference assay, in which they voluntarily choose to consume food with or without 15% ethanol supplementation. (Shohat-Ophir et al., 2012, p. 1351, citations and figure reference removed)" For each fly, the amount of each type of food consumed was

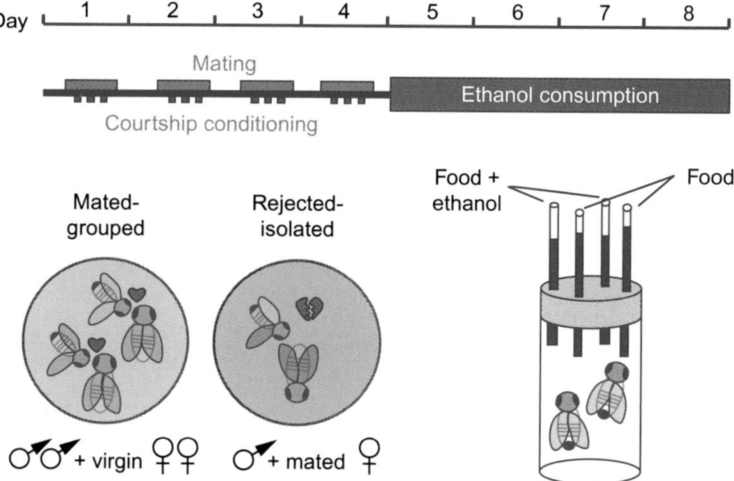

Figure 16.13 Procedure used for investigation of alcohol preference of sexually frustrated males (*Drosophila melanogaster*). From Figure 1A of Shohat-Ophir, Kaun, Azanchi, Mohammed, and Heberlein (2012). Reprinted with permission from AAAS.

converted to a *preference ratio*: the amount of ethanol-supplemented food minus the amount of regular food divided by the total of both. I constructed 3-day summary preference scores for each individual fruit fly by summing the consumption of ethanol and non-ethanol across days 6–8. The amounts of food consumed and the preference ratios are in the data file named ShohatOphirKAMH2012dataReduced.csv. My thanks to Dr. Galit Shohat-Ophir for providing the data.

(A) Run `Jags-Ymet-Xnom2grp-MrobustHet-Example.R` on the preference scores. Make sure that the ROPE on the means and standard deviation is scaled appropriately to the data. How big are differences between groups relative to the uncertainty of the estimate? What do you conclude? (If this result interests you, then you will also be intrigued by the results in Section 19.3.2, p. 561.)

(B) Instead of focusing on the *relative* amounts of ethanol and regular food consumed, we might also be interested in the absolute total amount of food consumed. Run the analysis on the total consumption data, which has column name GrandTotal in the data file. What do you conclude? In particular, would you want to make an argument to accept the null hypothesis of no difference? (Review Section 12.1.1, beginning on p. 336.)

Exercise 16.2. [Purpose: More practice using different data files in the high-level script, using a real example, with skewed data.] The typical lifespan of a laboratory rat that eats *ad lib* is approximately 700 days. When rats are placed on a restricted diet, their longevity can increase, but there is a lot of variability in lifespans across different individual rats. Restricting the diet might not only affect the typical lifespan, but restricting the diet might also affect the variance of the lifespan across rats. We consider data from R. L. Berger, Boos, and Guess (1988), as reported in Hand, Daly, Lunn, McConway, and Ostrowski (1994, data set #242), and which are available in the file named `RatLives.csv`.

(A) Run the two-group analysis on the rat longevity data. Use JAGS or Stan as you prefer (report which one you used). Report the code you used to read in the data file, specify the column names for the data, and the ROPEs appropriate to the scale of the data. Do the groups appear to differ in their central tendencies and variances? Does the value of the normality parameter suggest that the data have outliers relative to a normal distribution?

(B) The data within each group appear to be skewed to the left. That is, within each group, there are many rats that died relatively young, but there are fewer outliers on the high end. We could try to implement a skewed noise distribution, or we could try to transform the data so they are approximately symmetric within each group. We will try the latter approach here. To get rid of leftward skew, we need a transformation that expands the rightward values. We will try squaring the data. Read in the data and append a transformed data column like this:

```
myDataFrame = read.csv( file="RatLives.csv" )
myDataFrame = cbind( myDataFrame , DaysLiveSq = myDataFrame$ DaysLive^2 )
yName="DaysLiveSq"
```

Change the specification of the ROPEs to be appropriate to the transformed data. Do the groups appear to differ in their central tendencies and variances on the days-squared scale?

Does the value of the normality parameter suggest that the data have outliers relative to a normal distribution on the days-squared scale? Is the posterior effect size on the days-squared scale much different than the posterior effect size on the days scale from the previous part?

Exercise 16.3. [Purpose: For two groups, examine the implied prior on the effect size and on the differences of means and scales.]

(A) Modify the script `Stan-Ymet-Xnom2grp-MrobustHet-Example.R` and functions in `Stan-Ymet-Xnom2grp-MrobustHet.R` as needed to display the prior distribution. See Section 14.3.4, p. 412, for details. Explain what changes you made, and include the graphical output. Does Stan have convergence problems?

(B) Modify the JAGS version of the program to sample from the prior. Refer to Section 8.5, p. 211, for a reminder of how to sample from the prior in JAGS.

(C) From the previous two parts, is the prior on the effect size and differences suitably "noncommittal"? Briefly discuss.

CHAPTER 17

Metric Predicted Variable with One Metric Predictor

Contents

The agri-bank's threatnin' to revoke my lease
If my field's production don't rapid increase.
Oh Lord how I wish I could divine the trend,
Will my furrows deepen? and Will my line end?[1]

In this chapter, we consider situations such as predicting a person's weight from their height, or predicting their blood pressure from their weight, or predicting their income from years of education. In these situations, the predicted variable is metric, and the single predictor is also metric.

[1] This chapter discusses linear regression (among other topics). Analysts who use linear regression are sometimes interested in extrapolating a trend line into the future. In the poem, deeper furrows in the field might mean fresh tilling, while deeper furrows on the forehead would mean worsening worries. The "end of the line" means one thing for simple linear regression, but it can also refer to family lineage.

Doing Bayesian Data Analysis, 2nd Edition
http://dx.doi.org/10.1016/B978-0-12-405888-0.00017-9

We will initially describe the relationship between the predicted variable, y, and predictor, x, with a simple linear model and normally distributed residual randomness in y. This model is often referred to as "simple linear regression." We will generalize the model in three ways. First, we will give it a noise distribution that accommodates outliers, which is to say that we will replace the normal distribution with a t distribution as we did in the previous chapter. The model will be implemented in both JAGS and Stan. Next, we will consider differently shaped relations between the predictor and the predicted, such as quadratic trend. Finally, we will consider hierarchical models of situations in which every individual has data that can be described by an individual trend, and we also want to estimate group-level typical trends across individuals.

In the context of the generalized linear model (GLM) introduced in Chapter 15, this chapter's situation involves a linear function of a single metric predictor, as indicated in the third column of Table 15.1 (p. 434), with a link function that is the identity along with a normal distribution for describing noise in the data, as indicated in the first row of Table 15.2 (p. 443). For a reminder of how this chapter's combination of predicted and predictor variables relates to other combinations, see Table 15.3 (p. 444).

17.1. SIMPLE LINEAR REGRESSION

Figure 17.1 shows examples of simulated data generated from the model for simple linear regression. First a value of x is arbitrarily generated. At that value of x, the mean

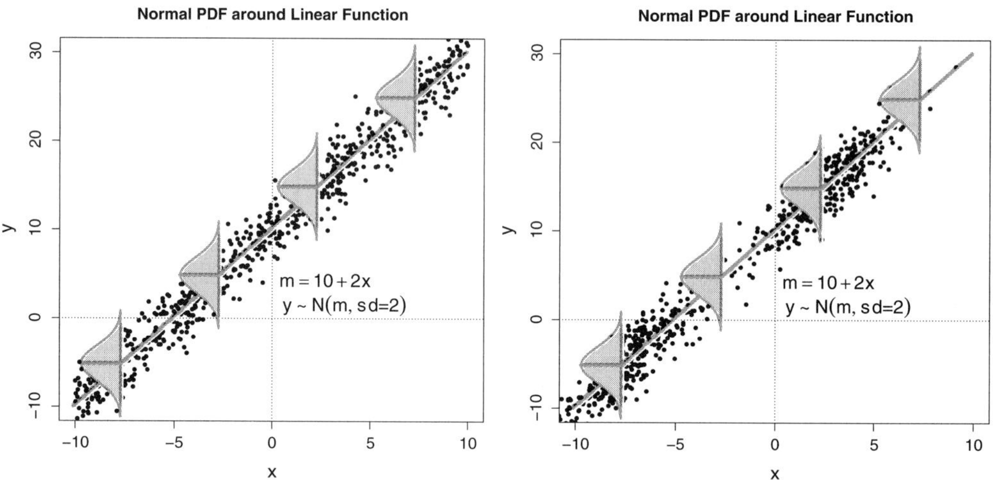

Figure 17.1 Examples of points normally distributed around a linear function. (The left panel repeats Figure 15.9, p. 442.) The model assumes that the data y are normally distributed vertically around the line, as shown. Moreover, the variance of y is the same at all values of x. The model puts no constraints on the distribution of x. The right panel shows a case in which x are distributed bimodally, whereas in the left panel the x are distributed uniformly. In both panels, there is homogeneity of variance.

predicted value of y is computed as $\mu = \beta_0 + \beta_1 x$. Then, a random value for the datum y is generated from a normal distribution centered at μ with standard deviation σ. For a review of how to interpret the slope, β_1, and intercept, β_0, see Figure 15.1 (p. 425).

Note that the model only specifies the dependency of y on x. The model does not say anything about what generates x, and there is no probability distribution assumed for describing x. The x values in the left panel of Figure 17.1 were sampled randomly from a uniform distribution, merely for purposes of illustration, whereas the x values in the right panel of Figure 17.1 were sampled randomly from a bimodal distribution. Both panels show data from the same model of the dependency of y on x.

It is important to emphasize that the model assumes *homogeneity of variance*: At every value of x, the variance of y is the same. This homogeneity of variance is easy to see in the left panel of Figure 17.1: The smattering of data points in the vertical, y, direction appears visually to be the same at all values of x. But that is only because the x values are uniformly distributed. Homogeneity of variance is less easy to identify visually when the x values are not uniformly distributed. For example, the right panel of Figure 17.1 displays data that may appear to violate homogeneity of variance, because the apparent vertical spread of the data seems to be larger at $x = 2.5$ than at $x = 7.5$ (for example). Despite this deceiving appearance, the data do respect homogeneity of variance. The reason for the apparent violation is that for regions in which x is sparse, there is less opportunity for the sampled y values to come from the tails of the noise distribution. In regions where x is dense, there are many opportunities for y to come from the tails.

In applications, the x and y values are provided by some real-world process. In the real-world process, there might or might not be any direct causal connection between x and y. It might be that x causes y, or y causes x, or some third factor causes both x and y, or x and y have no causal connection, or some combination of those from multiple types of causes. The simple linear model makes no claims about causal connections between x and y. The simple linear model merely describes a tendency for y values to be linearly related to x values, hence "predictable" from the x values. When describing data with this model, we are starting with a scatter plot of points generated by an unknown process in the real world, and estimating parameter values that would produce a smattering of points that might mimic the real data. Even if the descriptive model mimics the data well (and it might not), the mathematical "process" in the model may have little if anything to do with the real-world process that created the data. Nevertheless, the parameters in the descriptive model are meaningful because they describe tendencies in the data.

17.2. ROBUST LINEAR REGRESSION

There is no requirement to use a normal distribution for the noise distribution. The normal distribution is traditional because of its relative simplicity in mathematical derivations. But real data may have outliers, and the use of (optionally) heavy-tailed noise

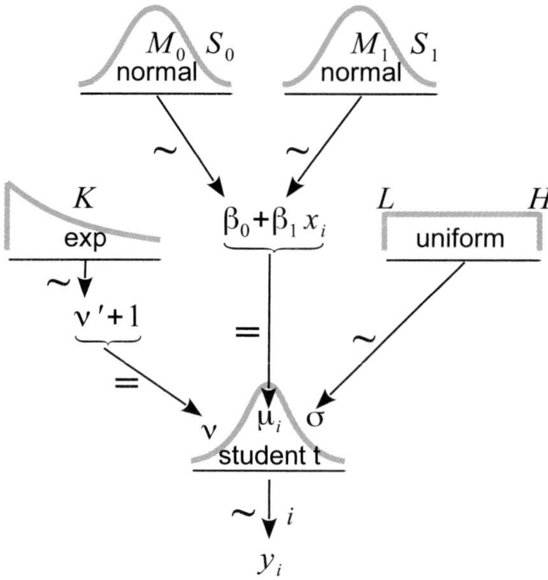

Figure 17.2 A model of dependencies for robust linear regression. The datum, y_i at the bottom of the diagram, is distributed around the central tendency μ_i, which is a linear function of x_i. Compare with Figure 16.11 on p. 468.

distributions is straight forward in contemporary Bayesian software. Section 16.2 (p. 458) explained the t distribution and its usefulness for describing data that might have outliers.

Figure 17.2 illustrates the hierarchical dependencies in the model. At the bottom of the diagram, the datum y_i is a t-distributed random value around the central tendency $\mu_i = \beta_0 + \beta_1 x_i$. The rest of the diagram illustrates the prior distribution on the four parameters. The scale parameter σ is given a noncommittal uniform prior, and the normality parameter ν is given a broad exponential prior, as they were in the previous chapter. The intercept and slope are given broad normal priors that are vague on the scale of the data. Figure 17.2 is just like the diagram for robust estimation of the central tendency of groups, shown in Figure 16.11 on p. 468, except here the two parameters β_0 and β_1 are shown with separate normal priors, whereas in Figure 16.11, the two parameters μ_1 and μ_2 had their normal priors superimposed.

As an example, suppose we have measurements of height, in inches, and weight, in pounds, for some randomly selected people. Figures 17.3 and 17.4 show two examples, for data sets of $N = 30$ and $N = 300$. At this time, focus your attention only on the scatter of data points in the top panels of each figure. The superimposed lines and curves will be explained later. You can see that the axes of the top panels are labeled height and weight, so each point represents a person with a particular combination of height and weight. The data were generated from a program I created (HtWtDataGenerator.R) that uses realistic population parameters (Brainard & Burmaster, 1992). The simulated data

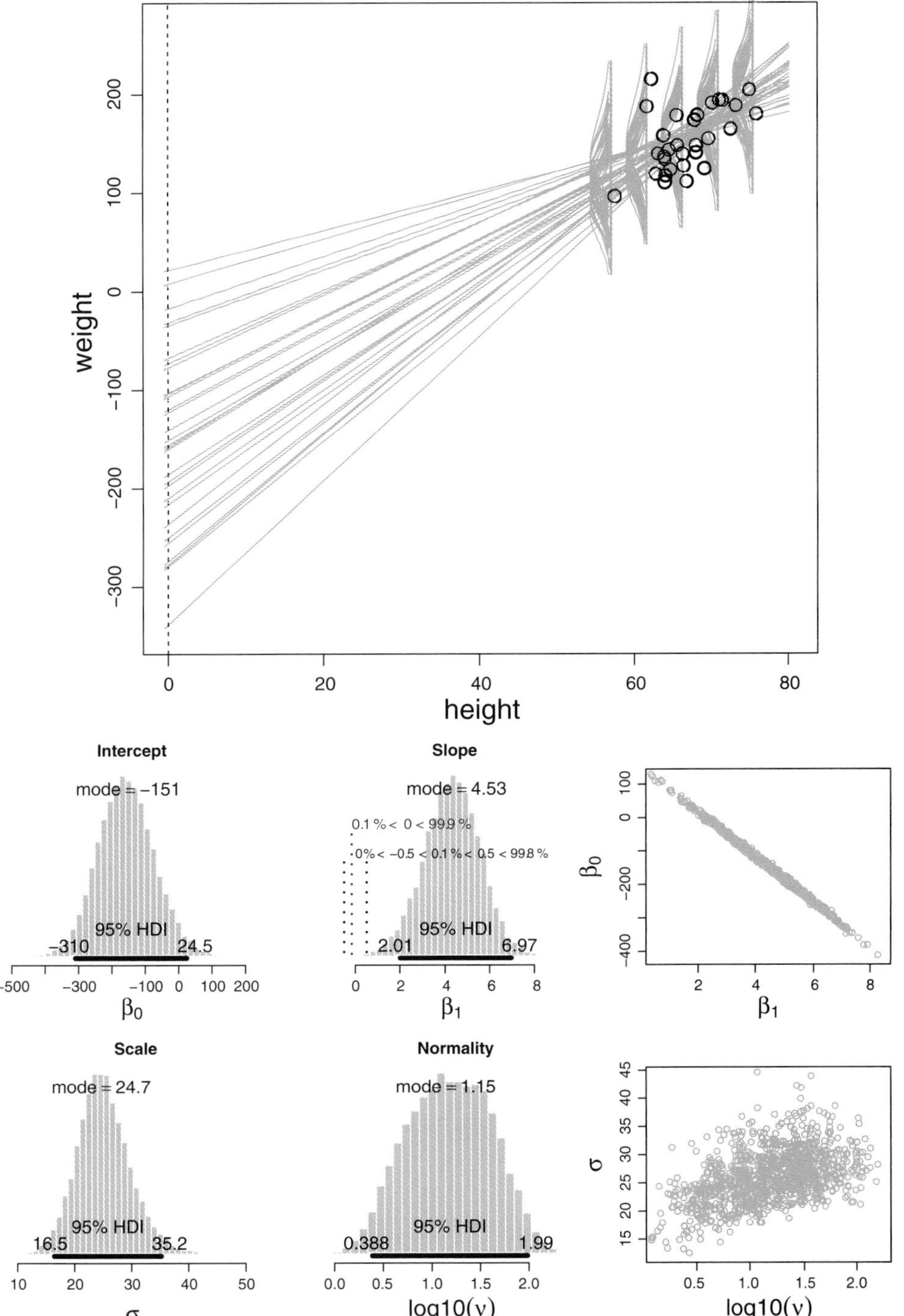

Figure 17.3 Upper panel: Data ($N = 30$) with a smattering of credible regression lines and t-noise distributions superimposed. Lower panels: Marginal posterior distribution on parameters. Compare with Figure 17.4.

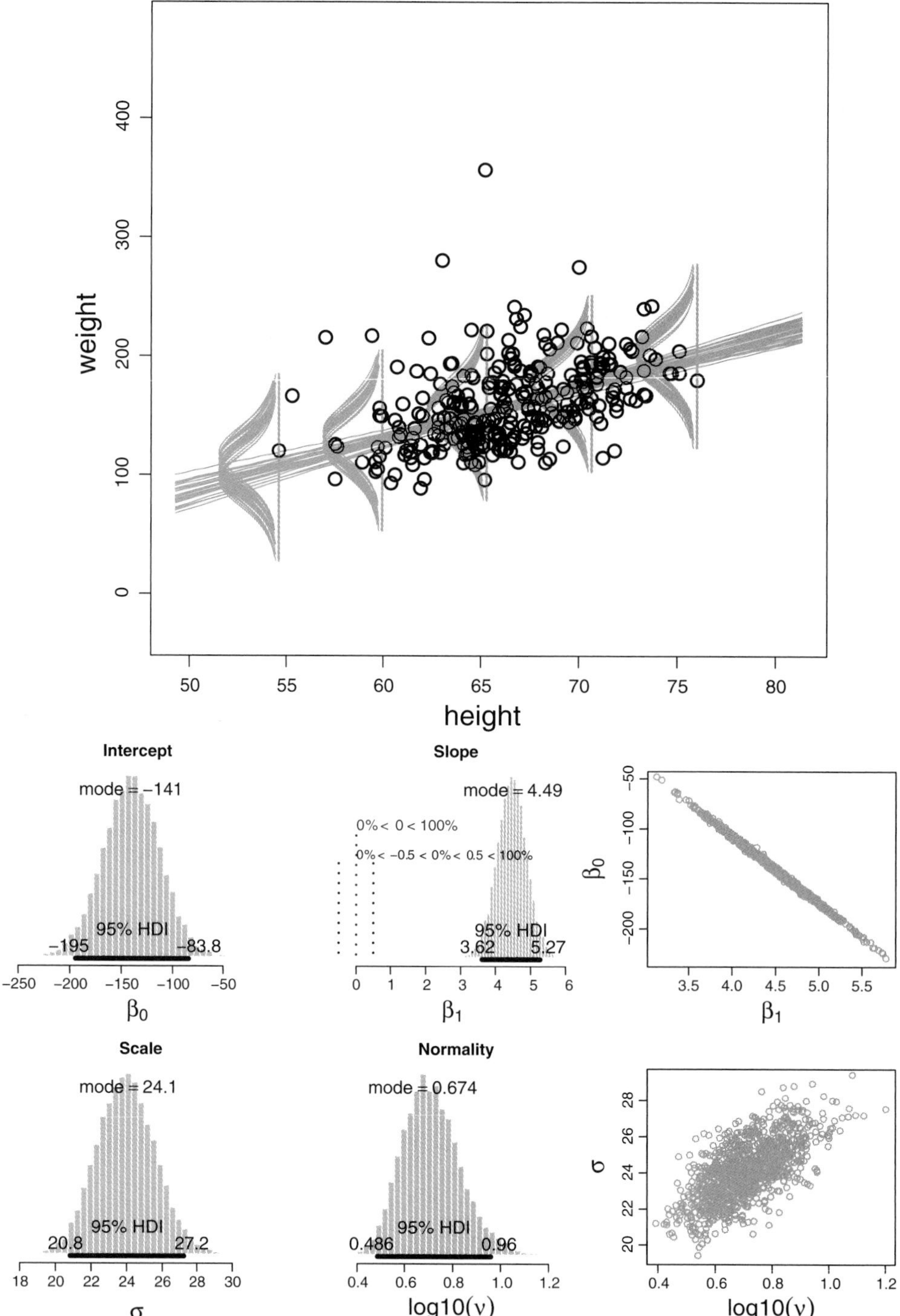

Figure 17.4 Upper panel: Data ($N = 300$) with a smattering of credible regression lines and t-noise distributions superimposed. Lower panels: Marginal posterior distribution on parameters. Compare with Figure 17.3.

are actually based on three bivariate Gaussian clusters, not a single linear dependency. But we can still attempt to describe the data with linear regression. The scatter plots in Figures 17.3 and 17.4 do suggest that as height increases, weight also tends to increase.

Our goal is to determine what combinations of β_0, β_1, σ, and ν are credible, given the data. The answer comes from Bayes' rule:

$$p(\beta_0, \beta_1, \sigma, \nu|D) = \frac{p(D|\beta_0, \beta_1, \sigma, \nu)\, p(\beta_0, \beta_1, \sigma, \nu)}{\iiiint d\beta_0\, d\beta_1\, d\sigma\, d\nu\, p(D|\beta_0, \beta_1, \sigma, \nu)\, p(\beta_0, \beta_1, \sigma, \nu)}$$

Fortunately, we do not have to worry much about analytical derivations because we can let JAGS or Stan generate a high resolution picture of the posterior distribution. Our job, therefore, is to specify sensible priors and to make sure that the MCMC process generates a trustworthy posterior sample that is converged and well mixed.

17.2.1. Robust linear regression in JAGS

As has been emphasized (e.g., at the end of Section 8.2), hierarchical dependency diagrams like Figure 17.2 are useful not only for conceptualizing the structure of a model but also for programming it. Every arrow in Figure 17.2 has a corresponding line of code in JAGS or Stan. The following sections describe implementations of the model. The core of the model specification for JAGS is virtually an arrow-to-line transliteration of the diagram. The seven arrows in Figure 17.2 have seven corresponding lines of code:

```
model {
  for ( i in 1:Ntotal ) {
    zy[i] ~ dt( mu[i] , 1/zsigma^2 , nu )
    mu[i] <- zbeta0 + zbeta1 * zx[i]
  }
  zbeta0 ~ dnorm( 0 , 1/(10)^2 )
  zbeta1 ~ dnorm( 0 , 1/(10)^2 )
  zsigma ~ dunif( 1.0E-3 , 1.0E+3 )
  nu <- nuMinusOne+1
  nuMinusOne ~ dexp(1/29.0)
}
```

Some of the parameter names in the code above are preceded by the letter "z" because the data will be standardized before being sent to the model, as will be described later, and standardized data often are referred to by the letter "z."

One of the nice features of JAGS (and Stan) is that arguments in distributions are functionally evaluated, so we can put a statement with operations in the place of an argument instead of only a variable name. For example, in the `dt` distribution earlier, the precision argument has the expression `1/zsigma^2` instead of just a variable name. We could do the same thing for `dt`'s other arguments. In particular, instead of assigning a value to `mu[i]` on a separate line, we could simply insert `zbeta0 + zbeta1 * zx[i]` in place of `mu[i]` in the mean argument of `dt`. Thus, assignment arrows in diagrams like

Figure 17.2 do not necessarily need a separate line in the JAGS code. However, when the expression is put directly into an argument, JAGS cannot record the value of the expression throughout the chain. Thus, in the form shown earlier, `mu[i]` is available for recording along with the parameter values at every step. We are not interested in `mu[i]` at every step; therefore, the revised code (exhibited later) will put its expression directly into the mean argument of `dt`. On the other hand, the expression for `nu` will remain on a separate line (as shown above) because we do want an explicit record of its value.

17.2.1.1 Standardizing the data for MCMC sampling

In principle, we could run the JAGS code, shown earlier, on the "raw" data. In practice, however, the attempt often fails. There's nothing wrong with the mathematics or logic; the problem is that credible values of the slope and intercept parameters tend to be strongly correlated, and this narrow diagonal posterior distribution is difficult for some sampling algorithms to explore, resulting in extreme inefficiency in the chains.

You can get an intuition for why this happens by considering the data in the upper panel of Figure 17.3. The superimposed lines are a smattering of credible regression lines from the posterior MCMC chain. All the regression lines must go through the midst of the data, but there is uncertainty in the slope and intercept. As you can see, if the slope is steep, then the y intercept (i.e., the y value of the regression line when $x = 0$) is low, at a large negative value. But if the slope is not so steep, then the y intercept is not so low. The credible slopes and intercepts trade off. The middle-right panel of Figure 17.3 shows the posterior distribution of the jointly credible values of the slope β_1 and the intercept β_0. You can see that the correlation of the credible values of the slope and intercept is extremely strong.

MCMC sampling from such a tightly correlated distribution can be difficult. Gibbs sampling gets stuck because it keeps "bumping into the walls" of the distribution. Recall how Gibbs sampling works: One parameter is changed at a time. Suppose we want to change β_0 while β_1 is constant. There is only a very small range of credible values for β_0 when β_1 is constant, so β_0 does not change much. Then we try changing β_1 while β_0 is constant. We run into the same problem. Thus, the two parameter values change only very slowly, and the MCMC chain is highly autocorrelated. Other sampling algorithms can suffer the same problem if they are not able to move quickly along the long axis of the diagonal distribution. There is nothing wrong with this behavior in principle, but in practice, it requires us to wait too long before getting a suitably representative sample from the posterior distribution.

There are (at least) two ways to make the sampling more efficient. One way is to change the sampling algorithm itself. For example, instead of using Gibbs sampling (which is one of the main methods in JAGS), we could try Hamiltonian Monte Carlo (HMC) in Stan. We will explore this option in a later section. A second way is to transform the data so that credible regression lines do not suffer such strong correlation between their slopes and intercepts. It is this option we explore presently.

Look again at the data with the superimposed regression lines in Figure 17.3. The problem arises because the arbitrary position of zero on the x axis is far away from the data. If we simply slid the axis so that zero fell under the mean of the data, then the regression lines could tilt up or down without any big changes on the y intercept. Sliding the axis so that zero falls under the mean is equivalent to subtracting the mean of the x values from each x value. Thus, we can transform the x values by "*mean centering*" them. This alone would solve the parameter-correlation problem.

But if we're going to bother to mean center the x values, we might as well completely standardize the data. This will allow us to set vague priors for the standardized data that will be valid no matter the scale of the raw data. Standardizing simply means re-scaling the data relative to their mean and standard deviation:

$$z_x = \frac{(x - M_x)}{\text{SD}_x} \quad \text{and} \quad z_y = \frac{(y - M_y)}{\text{SD}_y} \tag{17.1}$$

where M_x is the mean of the x values and SD_x is the standard deviation of the x values. It is easy to prove, using simple algebra, that the mean of the resulting z_x values is zero, and the standard deviation of the resulting z_x values is one, for any initial data set. Thus, standardized values are not only mean centered, but are also stretched or shrunken so that their standard deviation is one.

We will send the standardized data to the JAGS model, and JAGS will return credible parameter values for the standardized data. We then need to convert the parameter values back to the original raw scales. Denote the intercept and slope for standardized data as ζ_0 and ζ_1 (Greek letter "zeta"), and denote the predicted value of y as $\hat{y}$. Then:

$$\begin{aligned} z_{\hat{y}} &= \zeta_0 + \zeta_1 z_x && \text{by definition of the model} \\ \frac{(\hat{y} - M_y)}{\text{SD}_y} &= \zeta_0 + \zeta_1 \frac{(x - M_x)}{\text{SD}_x} && \text{from Equation 17.1} \\ \hat{y} &= \underbrace{\zeta_0 \text{SD}_y + M_y - \zeta_1 M_x \text{SD}_y/\text{SD}_x}_{\beta_0} + \underbrace{\zeta_1 \text{SD}_y/\text{SD}_x}_{\beta_1} x && (17.2) \end{aligned}$$

Thus, for every combination of ζ_0, ζ_1 values, there is a corresponding combination of β_0, β_1 values specified by Equation 17.2. To get from the σ value of the standardized scale to the σ value on the original scale, we simply multiply by SD_y. The normality parameter remains unchanged because it refers to the relative shape of the distribution and is not affected by the scale parameter.[2]

[2] The fact that ν is unchanged by standardization comes from the formula for the t distribution, which shows that the standardized score of x is isolated in its core, with no other role: $p(x|\nu, \mu, \sigma) = \frac{\Gamma\left(\frac{\nu+1}{2}\right)}{\Gamma\left(\frac{\nu}{2}\right)\sqrt{\pi \nu}\,\sigma} \left(1 + \frac{1}{\nu}\left(\frac{x-\mu}{\sigma}\right)^2\right)^{-\frac{\nu+1}{2}}$.

There is one more choice to make before implementing this method: Should we use R to standardize the data, then run JAGS on the standardized data, and then in R transform the parameters back to the original scale? Or, should we send the original data to JAGS, and do the standardization and transformation back to original scale, all in JAGS? Either method can work. We will do the transformation in JAGS. The reason is that this method keeps an integrated record of the standardized and original-scale parameters in JAGS' coda format, so convergence diagnostics can be easily run. This method also gives me the opportunity to show another aspect of JAGS programming that we have not previously encountered, namely, the `data` block.

The data are denoted as vectors `x` and `y` on their original scale. No other information is sent to JAGS. In the JAGS model specification, we first standardize the data by using a `data` block, like this:

```
data {
  Ntotal <- length(y)
  xm <- mean(x)
  ym <- mean(y)
  xsd <- sd(x)
  ysd <- sd(y)
  for ( i in 1:length(y) ) {
    zx[i] <- ( x[i] - xm ) / xsd
    zy[i] <- ( y[i] - ym ) / ysd
  }
}
```

All the values computed in the `data` block can be used in the `model` block. In particular, the model is specified in terms of the standardized data, and the parameters are preceded by a letter `z` to indicate that they are for the standardized data. Because the data are standardized, we know that the credible slope cannot greatly exceed the range ± 1, so a standard deviation of 10 on the slope makes it extremely flat relative to its possible credible values. Similar considerations apply to the intercept and noise, such that the following model block has extremely vague priors no matter what the original scale of the data:

```
model {
  for ( i in 1:Ntotal ) {
    zy[i] ~ dt( zbeta0 + zbeta1 * zx[i] , 1/zsigma^2 , nu )
  }
  # Priors vague on standardized scale:
  zbeta0 ~ dnorm( 0 , 1/10^2 )
  zbeta1 ~ dnorm( 0 , 1/10^2 )
  zsigma ~ dunif( 1.0E-3 , 1.0E+3 )
  nu <- nuMinusOne+1
  nuMinusOne ~ dexp(1/29.0)
  # Transform to original scale:
  beta1 <- zbeta1 * ysd / xsd
```

```
  beta0 <- zbeta0 * ysd  + ym -  zbeta1 * xm * ysd / xsd
  sigma <- zsigma * ysd
}
```

Notice at the end of the model block, above, the standardized parameters are transformed to the original scale, using a direct implementation of the correspondences in Equation 17.2.

A high-level script for using the JAGS model above is named `Jags-Ymet-Xmet-Mrobust-Example.R`, with functions themselves defined in `Jags-Ymet-Xmet-Mrobust.R`. Running the script produces the graphs shown in Figures 17.3 and 17.4. Before discussing the results, let's explore how to implement the model in Stan.

17.2.2. Robust linear regression in Stan

Recall from Section 14.1 (p. 400) that Stan uses Hamiltonian dynamics to find proposed positions in parameter space. The trajectories use the gradient of the posterior distribution to move large distances even in narrow distributions. Thus, HMC by itself, without data standardization, should be able to efficiently generate a representative sample from the posterior distribution.

17.2.2.1 Constants for vague priors

The only new aspect of the Stan implementation is the setting of the priors. Because the data will not be standardized, we cannot use the constants from the priors in the JAGS model. Instead, we need the constants in the priors to be vague on whatever scale the data happen to fall on. We could query the user to supply a prior, but instead we will let the data themselves suggest what is typical, and we will set the priors to be extremely broad relative to the data (just as we did in the previous chapter).

A regression slope can take on a maximum value of SD_y/SD_x for data that are perfectly correlated. Therefore, the prior on the slope will be given a standard deviation that is large compared to that maximum. The biggest that an intercept could be, for data that are perfectly correlated, is $M_x\text{SD}_y/\text{SD}_x$. Therefore, the prior on the intercept will have a standard deviation that is large compared to that maximum. The following model specification for Stan implements these ideas:

```
data {
  int<lower=1> Ntotal ;
  real x[Ntotal] ;
  real y[Ntotal] ;
  real meanY ;
  real sdY ;
  real meanX ;
  real sdX ;
}
transformed data { // for constants in priors
  real unifLo ;
```

```
  real unifHi ;
  real expLambda ;
  real beta0sigma ;
  real beta1sigma ;
  unifLo <- sdY/1000 ;
  unifHi <- sdY*1000 ;
  expLambda <- 1/29.0 ;
  beta1sigma <- 10*fabs(sdY/sdX) ; // fabs is absolute value
  beta0sigma <- 10*fabs(meanX*sdY/sdX) ;
}
parameters {
  real beta0 ;
  real beta1 ;
  real<lower=0> nuMinusOne ;
  real<lower=0> sigma ;
}
transformed parameters {
  real<lower=0> nu ; // actually lower=1
  nu <- nuMinusOne + 1 ;
}
model {
  sigma ~ uniform( unifLo , unifHi ) ;
  nuMinusOne ~ exponential( expLambda ) ;
  beta0 ~ normal( 0 , beta0sigma ) ;
  beta1 ~ normal( 0 , beta1sigma ) ;
  for ( i in 1:Ntotal ) {
    y[i] ~ student_t( nu , beta0 + beta1 * x[i] , sigma ) ;
  }
}
```

A script for the Stan model above is named `Stan-Ymet-Xmet-Mrobust-Example.R`, with functions defined in `Stan-Ymet-Xmet-Mrobust.R`. Running it produces graphs just like those shown in Figures 17.3 and 17.4. Before discussing the results, let's briefly compare the performance of Stan and JAGS.

17.2.3. Stan or JAGS?

The Stan and JAGS implementations were both run on the data in Figure 17.4, and they produced very similar results in terms of the details of the posterior distribution. For a comparison, both implementations were run with 500 adaptation steps, 1000 burn-in steps, and 20,000 saved steps with no thinning, using four chains. On my modest desktop computer, a run in JAGS took about 180 seconds, while a run in Stan (including compilation) took about 485 seconds (2.7 times as long). The ESS in JAGS was quite different for different parameters, being about 5000 for ν and σ but about 16,000 for β_0 and β_1. The effective sample size (ESS) in Stan was more consistent across parameters but not as high as JAGS for some, being about 8000 for ν and σ but only about 7000 for β_0 and β_1.

Thus, Stan does indeed go a long way toward solving the correlated-parameter problem for which JAGS needed the intervention of standardizing the data. Stan also samples the normality and scale parameters with less autocorrelation than JAGS, as we saw previously in Figure 16.10 (p. 467). HMC deftly navigates through posterior distributions where other samplers may have troubles. But Stan took longer in real time and yielded a smaller ESS for the main parameters of interest, namely the slope and intercept. Stan's efficiency with this model might be improved if we used standardized data as we did for JAGS, but a main point of this demonstration was to show that HMC is able to navigate difficult posterior distributions. Stan's efficiency (and JAGS') might also be improved if the t distribution were reparameterized, as explained in the Stan Reference Manual. Although subsequent programs in this section focus on JAGS, keep in mind that it is straight forward to translate the programs into Stan.

17.2.4. Interpreting the posterior distribution

Now that we are through all the implementation details, we will discuss the results of the analysis shown in Figures 17.3 and 17.4. Recall that both figures show results from the same data generator, but Figure 17.3 has only 30 data points while Figure 17.4 has 300. By comparing the marginal distributions of the posterior across the two figures, you can see that the modal estimates of the slope, intercept, and scale are about the same, but the certainty of the estimate is much tighter for 300 data points than for 30 data points. For example, with $N = 30$, the slope has modal estimate of about 4.5 (pounds per inch), with a 95% HDI that extends from about 2.0 to 7.0. When $N = 300$, again the slope has modal estimate of about 4.5, but the 95% HDI is narrower, extending from about 3.6 to 5.2.

It is interesting to note that the normality parameter does not have the same estimate for the two data sets. As the data set gets bigger, the estimate of the normality parameter gradually gets smaller, overcoming the prior which remains fairly dominant with $N = 30$. (Recall the discussion of the exponential prior on ν in Section 16.2.1, p. 462.) With larger data sets, there is more opportunity for outliers to appear, which are inconsistent with large values of ν. (Of course, if the data are genuinely normally distributed, then larger data sets just reinforce large values of the normality parameter.)

In some applications, there is interest in extrapolating or interpolating trends at x values sparsely represented in the current data. For instance, we might want to predict the weight of a person who is 50 inches tall. A feature of Bayesian analysis is that we get an entire distribution of credible predicted values, not only a point estimate. To get the distribution of predicted values, we step through the MCMC chain, and at each step, we randomly generate simulated data from the model using the parameter values at that step. You can think of this by looking at the vertically plotted noise distributions in Figure 17.4. For any given value of height, the noise distributions show the credible predicted values of weight. By simulating data from all the steps in the chain, we integrate over all those credible noise distributions. An interesting aspect of the predictive distributions is that they are naturally wider, that is, less certain, the farther away we probe from the actual data. This can be seen most easily in Figure 17.3. As we probe farther from the data, the spread between credible regression lines gets larger, and thus the predicted value gets less certain.

The upper panels of Figures 17.3 and 17.4 show the data with a smattering of credible regression lines and t distributions superimposed. This provides a visual posterior predictive check, so we can qualitatively sense whether there are systematic deviations of the data from the form implied by the model. For example, we might look for nonlinear trends in the data or asymmetry in the distribution of weights. The data suggest that there might be some positive skew in the distribution of weights relative to the regression lines. If this possibility had great theoretical importance or were strong enough to cast doubt on the interpretation of the parameters, then we would want to formulate a model that incorporated skew in the noise distribution. For more information about posterior predictive checks, see Kruschke (2013b). Curiously, in the present application, the robust linear-regression model does a pretty good job of mimicking the data, even though the data were actually generated by three bivariate normal distributions. Remember, though, that the model only mimics the dependency of y on x, and the model does not describe the distribution of x.

17.3. HIERARCHICAL REGRESSION ON INDIVIDUALS WITHIN GROUPS

In the previous applications, the jth individual contributed a single x_j, y_j pair. But suppose instead that every individual, j, contributes multiple observations of $x_{i|j}$, $y_{i|j}$ pairs. (The subscript notation $i|j$ means the ith observation within the jth individual.) With these data, we can estimate a regression curve for every individual. If we also assume that the individuals are mutually representative of a common group, then we can estimate group-level parameters too.

One example of this scenario comes from Walker, Gustafson, and Frimer (2007), who measured reading-ability scores of children across several years. Thus, each child contributed several age and reading-score pairs. A regression line can describe each child's reading ability through time, and higher-level distributions can describe the credible intercepts and slopes for the group as a whole. Estimates of each individual are mutually informed by data from all other individuals, by virtue of being linked indirectly through the higher-level distribution. In this case, we might be interested in assessing individual reading ability and rate of improvement, but we might also or instead be interested in estimating the group-level ability and rate of improvement.

As another example, suppose we are interested in how family income varies with the size of the family, for different geographical regions. The U.S. Census Bureau has published data that show the median family income as a function of the number of persons in a family, for all 50 states, the District of Columbia, and Puerto Rico. In this case, each territory is a "subject" and contributes several pairs of data: Median income for family size 2, median income for family size 3, and so on, up to family size 6 or 7. If we had the actual income for each family, we could use those data in the analysis instead of the median. In this application, we might be interested in estimates for each individual region, or in the overall estimate. The estimated trend in each region is informed by the data from all the other regions indirectly through the higher-level distribution that is informed by all the regions.

Figure 17.5 shows some fictitious data for a worked example. The scales for x and y are roughly comparable to the height and weight example from the previous section, but the data are not intended to be realistic in this case. In these data, each subject contributes several x, y pairs. You might think of these as growth curves for individual people, but with the caveat that they are not realistic. The top panel of Figure 17.5 shows the data from all the individuals superimposed, with subsets of data from single individuals connected by line segments. The lower panels of Figure 17.5 show the data separated into individuals. Notice that different individuals contribute different numbers of data points, and the x values for different individuals can be completely different.

Our goal is to describe each individual with a linear regression, and simultaneously to estimate the typical slope and intercept of the group overall. A key assumption for our analysis is that each individual is representative of the group. Therefore, every individual informs the estimate of the group slope and intercept, which in turn inform the estimates of all the individual slopes and intercepts. Thereby we get sharing of information across individuals, and shrinkage of individual estimates toward the overarching mode. We will expand our discussion of shrinkage after twisting our minds around the model specification.

17.3.1. The model and implementation in JAGS

Figure 17.6 shows a diagram of the model. It might seem a bit daunting, but if you compare it with Figure 17.2 on p. 480 you will see that it merely adds a level to the basic model we are already very familiar with. The diagram is useful for understanding all the dependencies in the model, and for programming in JAGS (and Stan) because every arrow in the diagram has a corresponding line in the JAGS model specification.

Starting at the bottom of Figure 17.6, we see that the ith datum within the jth subject comes from a t distribution that has a mean of $\mu_{i|j} = \beta_{0,j} + \beta_{1,j} x_{i|j}$. Notice that the intercept and slope are subscripted with j, meaning that every subject has its own slope and intercept. Moving up the hierarchical diagram, we see that the slopes across the individuals all come from a higher-level normal distribution that has mean denoted μ_1 and standard deviation of σ_1, both of which are estimated. In other words, the model assumes that $\beta_{1,j} \sim \text{normal}(\mu_1, \sigma_1)$, where μ_1 describes the typical slope of the individuals and σ_1 describes the variability of those individual slopes. If every

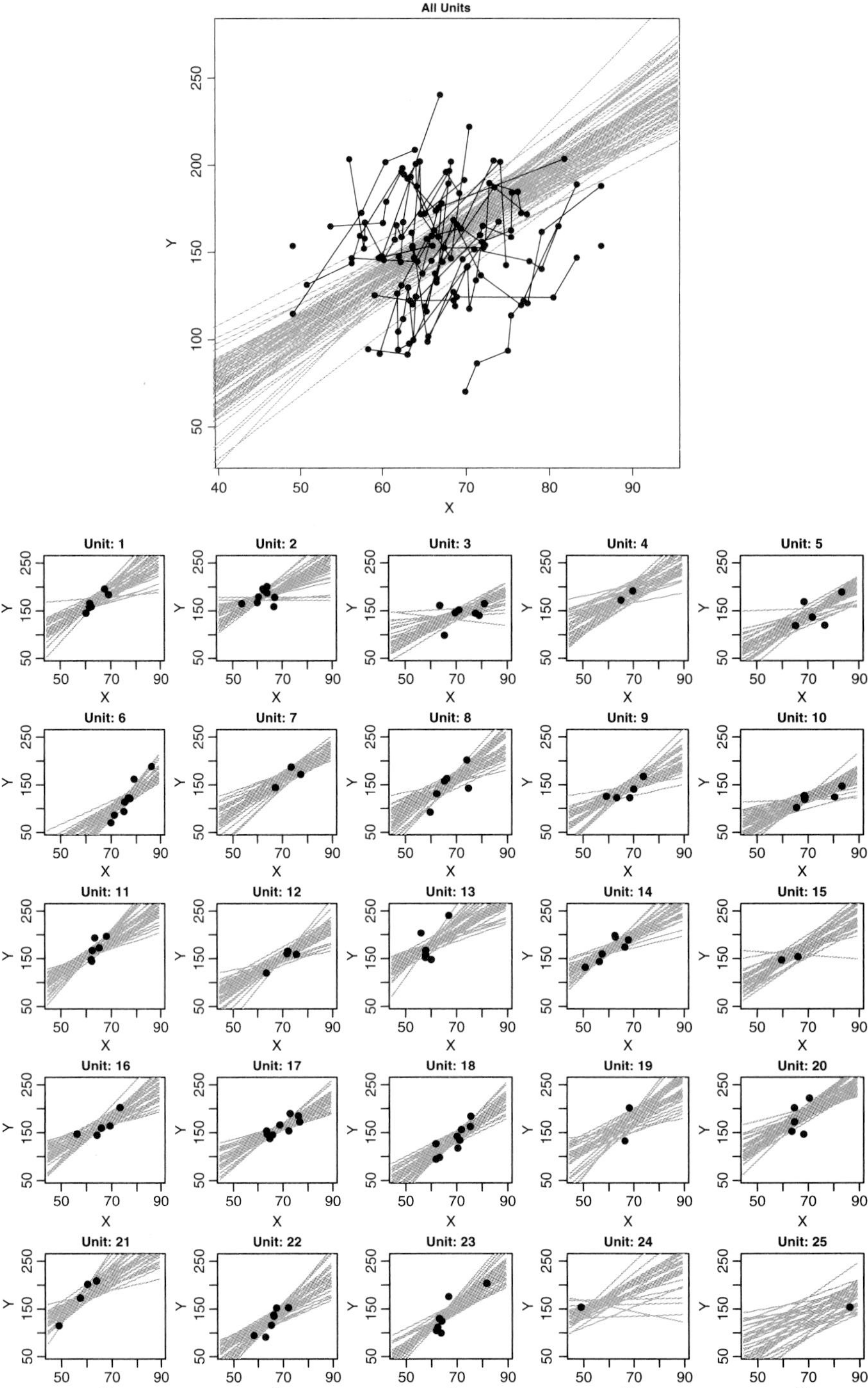

Figure 17.5 Fictitious data for demonstrating hierarchical linear regression, with posterior predicted lines superimposed. Upper panel: All data together, with individuals represented by connected segments. Lower panels: Plots of individual data. *Notice that the final two subjects have only single data points, yet the hierarchical model has fairly tight estimates of the individual slopes and intercepts.*

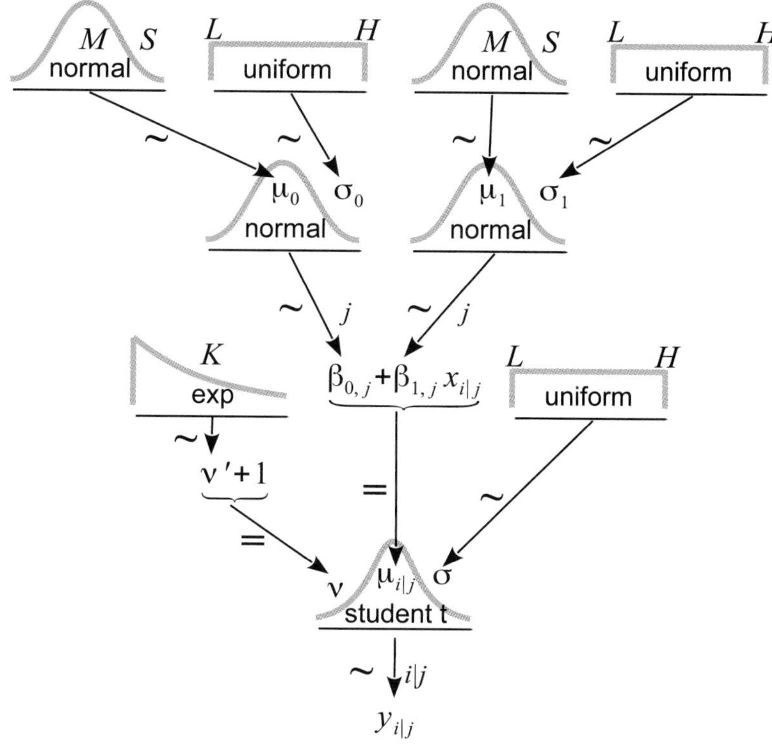

Figure 17.6 A model of dependencies for robust hierarchical linear regression. Compare with Figure 17.2 on p. 480.

individual seems to have nearly the same slope, then σ_1 is estimated to be small, but if different individuals have very different slopes, then σ_1 is estimated to be large. Analogous considerations apply to the intercepts. At the top level of the hierarchy, the group-level parameters are given generic vague priors.

This model assumes that the standard deviation of the noise within each subject is the same for all subjects. In other words, there is a single σ parameter, as shown at the bottom of Figure 17.6. This assumption could be relaxed by providing subjects with distinct noise parameters, all under a higher level distribution. We will see our first example of this approach later, in Figure 19.6 (p. 574).

To understand the JAGS implementation of the model, we must understand the format of the data file. The data format consists of three columns. Each row corresponds to one point of data and specifies the x, y values of the point and the subject who contributed that point. For the ith row of the data file, the x value is `x[i]`, the y value is `y[i]`, and the subject index j is `s[i]`. Notice that the index `i` in the JAGS model specification is the row of overall data file, not the ith value within subject j. The model specification assumes that the subject indices are consecutive integers, but the data file can use any sort of unique subject identifiers because the identifiers get converted to consecutive integers inside the program. The subject indices will be used to keep track of individual slopes and intercepts.

The JAGS implementation of this model begins by standardizing the data, exactly like the nonhierarchical model of the previous section:

```
data {
  Ntotal <- length(y)
  xm <- mean(x)
  ym <- mean(y)
  xsd <- sd(x)
  ysd <- sd(y)
  for ( i in 1:length(y) ) {
    zx[i] <- ( x[i] - xm ) / xsd
    zy[i] <- ( y[i] - ym ) / ysd
  }
}
```

Then we get to the novel part, where the hierarchical model itself is specified. Notice below that the intercept and slope use nested indexing. Thus, to describe the standardized y value `zy[i]`, the model uses the slope, `zbeta1[s[i]]`, of the subject who contributed that y value. As you read the model block, below, compare each line with the arrows in Figure 17.6:

```
model {
  for ( i in 1:Ntotal ) {
    zy[i] ~ dt( zbeta0[s[i]] + zbeta1[s[i]] * zx[i], 1/zsigma^2, nu)
  }
  for ( j in 1:Nsubj ) {
    zbeta0[j] ~ dnorm( zbeta0mu , 1/(zbeta0sigma)^2 )
    zbeta1[j] ~ dnorm( zbeta1mu , 1/(zbeta1sigma)^2 )
  }
  # Priors vague on standardized scale:
  zbeta0mu ~ dnorm( 0 , 1/(10)^2 )
  zbeta1mu ~ dnorm( 0 , 1/(10)^2 )
  zsigma ~ dunif( 1.0E-3 , 1.0E+3 )
  zbeta0sigma ~ dunif( 1.0E-3 , 1.0E+3 )
  zbeta1sigma ~ dunif( 1.0E-3 , 1.0E+3 )
  nu <- nuMinusOne+1
  nuMinusOne ~ dexp(1/29.0)
  # Transform to original scale:
  for ( j in 1:Nsubj ) {
    beta1[j] <- zbeta1[j] * ysd / xsd
    beta0[j] <- zbeta0[j] * ysd  + ym - zbeta1[j] * xm * ysd / xsd
  }
  beta1mu <- zbeta1mu * ysd / xsd
```

```
  beta0mu <- zbeta0mu * ysd  + ym - zbeta1mu * xm * ysd / xsd
  sigma <- zsigma * ysd
}
```

17.3.2. The posterior distribution: Shrinkage and prediction

A complete high-level script for running the analysis is in `Jags-Ymet-XmetSsubj-MrobustHier-Example.R`, with the corresponding functions defined in `Jags-Ymet-XmetSsubj-MrobustHier.R` You will notice that there is nonzero thinning used; this is merely to keep the saved file size relatively small while attaining an ESS of at least 10,000 for relevant parameters. The MCMC file can get large because there are a lot of parameters to store. For example, with 25 subjects, and tracking standardized and original-scale parameters, there are 107 variables recorded from JAGS. The model uses three parallel chains in runjags to save time, but it can still take a few minutes. Occasionally, a run will produce a chain that is well behaved for all parameters except the normality parameter ν. This chain nevertheless explores reasonable values of ν and is well behaved on other parameters. If this is worrisome, simply run JAGS again until all chains are well behaved (or translate to Stan).

Figure 17.5 shows the data with a smattering of credible regression lines superimposed. The lines on the overall data, in the top panel of Figure 17.5, plot the group-level slope and intercept. Notice that the slopes are clearly positive, even though the collective of data points, disconnected from their individual sources, show no obvious upward trend.

The lower panels of Figure 17.5 show individual data with a smattering of credible regression lines superimposed, using individual-level slopes and intercepts. Because of information sharing across individuals, via the higher-level distribution, there is notable shrinkage of the estimates of the individuals. That is to say, the estimates of the individual slopes are pulled together. This shrinkage is especially clear for the final two subjects, who each have only a single data point. Despite this dearth of data, the estimates of the slope and intercept for these subjects are surprisingly tightly constrained, as shown by the fact that the smattering of credible lines is a fairly tight bundle that looks like a blurry version of the posterior predictive lines in the other subjects. Notice that the estimated intercepts of the two final subjects are pulled toward the data values, so that predictions are different for the different individuals. This is another illustration of shrinkage of estimation in hierarchical models, which was introduced in Section 9.3 (p. 245).

17.4. QUADRATIC TREND AND WEIGHTED DATA

The U.S. Census Bureau publishes information from the American Community Survey and Puerto Rico Community Survey, including data about family income and family size.[3] Suppose we are interested in the median family income for different family sizes, in each of the 50 states, Puerto Rico, and the District of Columbia. The data are displayed in Figure 17.7. You can see that the incomes appear to have a nonlinear, inverted-U trend as family size increases. Indeed, if you run the data through the linear-regression model of the previous section, you will notice obvious systematic deviations away from the posterior predicted lines.

Because a linear trend appears to be an inadequate description of the data, we will extend the model to include a quadratic trend. This is not the only way to express a curve mathematically, but it is easy and conventional. A quadratic has the form $y = b_0 + b_1 x + b_2 x^2$. When b_2 is zero, the form reduces to a line. Therefore, this extended model can produce any fit that the linear model can. When b_2 is positive, a plot of the curve is a parabola that opens upward. When b_2 is negative, the curve is a parabola that opens downward. We have no reason to think that the curvature in the family-income data is exactly a parabola, but the quadratic trend might describe the data much better than a line alone. If the posterior distribution on the parameters indicates that credible values of b_2 are far less than zero, we have evidence that the linear model is not adequate (because the linear model would have $b_2 = 0$).

We will expand the model to include a quadratic component for every individual (i.e., state) and for the group level. To make the MCMC more efficient, we will standardize the data and then transform the parameters back to the original scale, just as we did for the linear model (see Equation 17.2, p. 485). Because of the quadratic component, the transformation involves a little more algebra:

$$
\begin{aligned}
z_{\hat{y}} &= \zeta_0 + \zeta_1 z_x + \zeta_2 z_x^2 && \text{by definition of the model} \\
\frac{(\hat{y} - M_y)}{\text{SD}_y} &= \zeta_0 + \zeta_1 \frac{(x - M_x)}{\text{SD}_x} + \zeta_2 \frac{(x - M_x)^2}{\text{SD}_x^2} && \text{from Equation 17.1} \\
\hat{y} &= \underbrace{\zeta_0 \text{SD}_y + M_y - \zeta_1 M_x \text{SD}_y/\text{SD}_x + \zeta_2 M_x^2 \text{SD}_y/\text{SD}_x^2}_{\beta_0} \\
&\quad + \underbrace{\left(\zeta_1 \text{SD}_y/\text{SD}_x - 2\zeta_2 M_x \text{SD}_y/\text{SD}_x^2\right)}_{\beta_1} x + \underbrace{\zeta_2 \text{SD}_y/\text{SD}_x^2}_{\beta_2} x^2 && (17.3)
\end{aligned}
$$

The correspondences in Equation 17.3 are implemented in JAGS, so JAGS keeps track of the standardized coefficients and the original-scale coefficients.

There is one other modification we will make for the family-income data. The data report the median income at each family size, but the median is based on different numbers of families at each size. Therefore, every median has a different amount of

[3] Data are from http://www.census.gov/hhes/www/income/data/Fam_Inc_SizeofFam1.xls, retrieved December 11, 2013. Median family income for years 2009-2011.

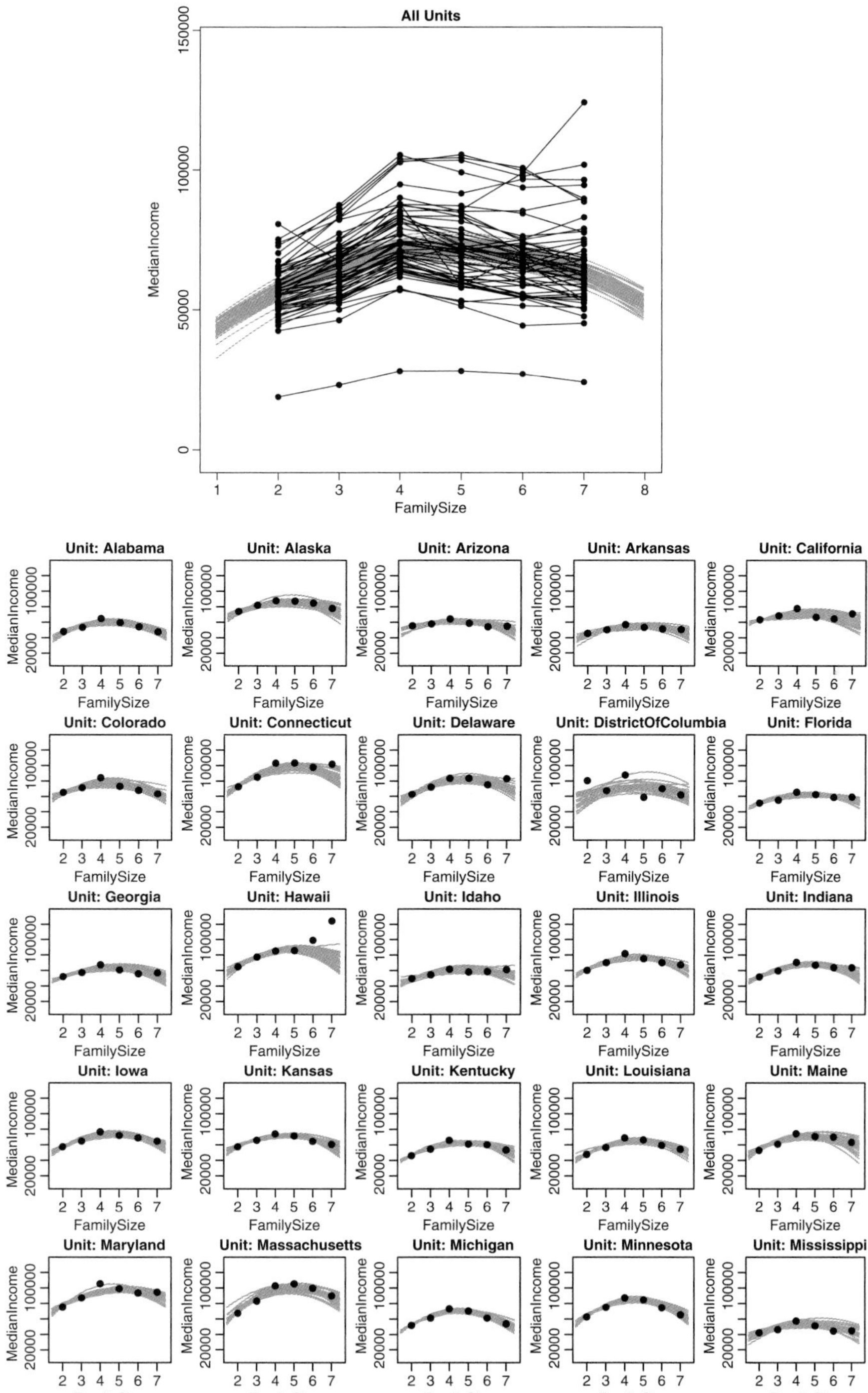

Figure 17.7 Upper panel: Median family income for various family sizes, with separate line for each of 50 states and the District of Columbia and Puerto Rico. Lower panels show data of a subset of individual states. Credible quadratic trends are superimposed.

sampling noise. Typically, there are fewer large-sized families than small-sized families, and therefore, the medians for the large-size families are noisier than the medians for the small-size families. The model should take this into account, with parameter estimates being more tightly constrained by the less noisy data points.

Fortunately, the U.S. Census Bureau reported each median with a "margin of error" that reflects the standard error of the data collected for that point. We will use the margin of error to modulate the noise parameter in the model. If the margin of error is high, then the noise parameter should be increased proportionally. If the margin of error is small, then the noise parameter should be decreased proportionally. More formally, instead of using $y_i \sim \text{normal}(\mu_i, \sigma)$ with the same σ for every datum, we use $y_i \sim \text{normal}(\mu_i, w_i\sigma)$, where w_i reflects the relative standard error for the ith datum.[4] When w_i is small, then μ_i has relatively little wiggle room around its best fitting value. When w_i is large, then μ_i has can deviate farther from its best fitting value without greatly changing the likelihood of the datum. Overall, then, this formulation finds parameter values that better match the data points with small margin of error than the data points with large margin of error. In the JAGS specification below, the original-scale standard errors are divided by their mean, so that a weight of 1.0 implies the mean standard error.

The quadratic component and weighted noise are implemented in the JAGS model as follows:

```
# Standardize the data:
data {
  Ntotal <- length(y)
  xm <- mean(x)
  ym <- mean(y)
  wm <- mean(w)    # standard error of y
  xsd <- sd(x)
  ysd <- sd(y)
  for ( i in 1:length(y) ) {
    zx[i] <- ( x[i] - xm ) / xsd
    zy[i] <- ( y[i] - ym ) / ysd
    zw[i] <- w[i] / wm    # set noise weights relative to mean noise
  }
}
# Specify the model for standardized data:
model {
  for ( i in 1:Ntotal ) {
    zy[i]~dt( zbeta0[s[i]]+zbeta1[s[i]]*zx[i]+zbeta2[s[i]]*zx[i]^2 ,
              1/(zw[i]*zsigma)^2 , nu )
```

[4] In this formulation, the weights w_i represent noise or error. In other formulations of weighted regression, the weights represent precision or inverse error variance, and therefore, σ is divided by the weight instead of multiplied.

```
  }
  for ( j in 1:Nsubj ) {
    zbeta0[j] ~ dnorm( zbeta0mu , 1/(zbeta0sigma)^2 )
    zbeta1[j] ~ dnorm( zbeta1mu , 1/(zbeta1sigma)^2 )
    zbeta2[j] ~ dnorm( zbeta2mu , 1/(zbeta2sigma)^2 )
  }
  # Priors vague on standardized scale:
  zbeta0mu ~ dnorm( 0 , 1/(10)^2 )
  zbeta1mu ~ dnorm( 0 , 1/(10)^2 )
  zbeta2mu ~ dnorm( 0 , 1/(10)^2 )
  zsigma ~ dunif( 1.0E-3 , 1.0E+3 )
  zbeta0sigma ~ dunif( 1.0E-3 , 1.0E+3 )
  zbeta1sigma ~ dunif( 1.0E-3 , 1.0E+3 )
  zbeta2sigma ~ dunif( 1.0E-3 , 1.0E+3 )
  nu <- nuMinusOne+1
  nuMinusOne ~ dexp(1/29.0)
  # Transform to original scale:
  for ( j in 1:Nsubj ) {
    beta2[j] <- zbeta2[j]*ysd/xsd^2
    beta1[j] <- zbeta1[j]*ysd/xsd - 2*zbeta2[j]*xm*ysd/xsd^2
    beta0[j] <- zbeta0[j]*ysd  + ym - zbeta1[j]*xm*ysd/xsd + zbeta2[j]*xm^2*ysd/xsd^2
  }
  beta2mu <- zbeta2mu*ysd/xsd^2
  beta1mu <- zbeta1mu*ysd/xsd - 2*zbeta2mu*xm*ysd/xsd^2
  beta0mu <- zbeta0mu*ysd  + ym - zbeta1mu*xm*ysd/xsd + zbeta2mu*xm^2*ysd/xsd^2
  sigma <- zsigma * ysd
}
```

In the JAGS specification above, notice that the mean argument of the student t distribution includes the new quadratic component. And notice that the precision argument of the student t distribution multiplies `zsigma` by the datum-specific noise weight, `zw[i]`. After the likelihood function, the remaining additions to the model specification merely specify the prior on the quadratic terms and implement the new transformation to the original scale (Equation 17.3). The complete script and functions for the model are in `Jags-Ymet-XmetSsubj-MrobustHierQuadWt-Example.R` and `Jags-Ymet-XmetSsubj-MrobustHierQuadWt.R`. The script is set up such that if you have a data file without standard errors for each data point, the weight name is simply omitted (or explicitly set as `wName=NULL`).

The Stan version of the specification above is provided in Exercise 17.3.

17.4.1. Results and interpretation

As you can see from the superimposed credible regression curves in Figure 17.7, the curvature is quite prominent. In fact, the 95% HDI on the quadratic coefficient (not shown here, but displayed by the script) goes from −2200 to −1700, which is far from zero. Thus, there is no doubt that there is a nonlinear trend in these data.

Interestingly, shrinkage from the hierarchical structure informs the estimates for each state. For example, inspect the subpanel for Hawaii in Figure 17.7. Its credible trend lines are curved downward even though its data in isolation curve upward. Because the vast majority of states show downward curvature in their data, and because Hawaii is assumed to be like the other states, the most credible estimates for Hawaii have a downward curvature. If Hawaii happened to have had a ginormous amount of data, so that its noise weights were extremely small, then its individual credible curves would more closely match its individual trend. In other words, shrinkage from the group level is a compromise with data from the individual.

Care must be taken when interpreting the linear component, that is, the slope. The slope is only meaningful in the context of adding the quadratic component. In the present application, the 95% HDI on the slope goes from 16,600 to 21,800. But this slope, by itself, greatly overshoots the data because the quadratic component *subtracts* a large value.

The weighting of data by their standard error is revealed by greater uncertainty at large family sizes than small. For example, consider the subpanel for California (or almost every other state) in Figure 17.7. The credible curves have a narrow spread at family size 2 but large spread at family size 7. This is caused in part by the fact that most of the data for large family sizes have large standard errors, and therefore, the parameters have more "slop" or flexibility at the large family sizes. (Another cause of spread in credible curves is variability within and between units.)

Finally, the posterior distribution reveals that these data have outliers within individual states because the normality parameter is estimated to be very small. Almost all of the posterior distribution is below $\nu = 4$. This suggests that most of the data points within each individual state fall fairly close to a quadratic-trend line, and the remaining points are accommodated by large tails of the t distribution.

17.4.2. Further extensions

Even this model has many simplifications for the purpose of clarity. One simplification was the use of only linear and quadratic trends. It is easy in Bayesian software to include any sort of non-linear trend, such as higher-order polynomial terms, or sinusoidal trends, or exponential trends or any other mathematically defined trend.

This model assumed a single underlying noise for all individuals. The noise could be modulated by the relative standard error of each datum, but the reference quantity, σ, was the same for all individuals. We do not need to make this assumption. Different individuals might have different inherent amounts of noise in their data. For example, suppose we were measuring a person's blood pressure at various times of day. Some people might have relatively consistent blood pressure while others might have greatly varying blood pressure. When there is sufficient data for each individual, it could be useful to expand the model to include subject-specific standard deviations or "noise" parameters. This lets the less noisy individuals have a stronger influence on the group-level estimate than the noisier individuals. The individual noise parameters could be modeled as coming from a group-level gamma distribution: $\sigma_i \sim \text{gamma}(r, s)$, where r and s are estimated, so that the gamma distribution describes the variation of the noise parameter across individuals. We will see our first example of this approach later in Figure 19.6 (p. 574).

Another simplification was assuming that intercepts, slopes, and curvatures are normally distributed across individuals. But it could be that there are outliers: Some individuals might have very unusual intercepts, slopes, or curvatures. Therefore, we could use t distributions at the group level. It is straight forward in Bayesian software to change the distributions from normal to student t. The model thereby becomes robust to outliers at multiple levels. But remember that estimating the normality parameter relies on data in the tails, so if the data set has few individuals and no extreme individuals, there will be nothing gained by using t distributions at the group level.

Another simplification in the model was that it had no explicit parameters to describe covariation of the intercepts, slopes, and curvatures across individuals. For example, rat pups that are born relatively small tend to grow at a smaller rate than rat pups that are born relatively larger: The intercept (weight at birth) and slope (growth rate) naturally covary. In other applications, the correlation could be the opposite, of course. The point is that the present model has no explicit parameter to capture such a correlation. It is straight forward in Bayesian software to use a multivariate normal prior on the intercept, slope, and curvature parameters. The multivariate normal has explicit covariance parameters, which are estimated along with the other parameters. (See, e.g., the "birats" example from WinBUGS Examples Volume 2 at http://www.mrc-bsu.cam.ac.uk/wp-content/uploads/WinBUGS_Vol2.pdf.)

17.5. PROCEDURE AND PERILS FOR EXPANDING A MODEL

17.5.1. Posterior predictive check

Comparing the data to the posterior predictions of the model is called a *posterior predictive check*. When there appear to be systematic discrepancies that would be meaningful to address, you should consider expanding or changing the model so it may be a better description of the data.

But what extension or alternative model should you try? There is no uniquely correct answer. The model should be both meaningful and computationally tractable. Meaning can come from mere familiarity in the catalog of functions learned in traditional math courses. This familiarity and simplicity was what motivated the use of a quadratic trend for the family-income data. Alternatively, meaning can come from theories of the application domain. For example, for family income, we might imagine that total income rises as the square root of the number of adults in the family (because each adult has the potential to bring in money to the family) and declines exponentially with the number of children in the family (because each child requires more time away from income generation by the adults). That model is completely fictitious and is mentioned merely for illustration.

The extended model is intended to better describe the data. One way to extend a model is to leave the original model "nested" inside the extended model, such that all the original parameters and mathematical form can be recovered from the extended model by setting the extended parameters to specific constants, such as zero. An example is extending the linear model to a quadratic-trend model; when the curvature is set to zero, the nested linear model is recovered. In the case of nested models, it is easy to check whether the extension fits the data better merely by inspecting the posterior distribution on the new parameters. If the new parameter values are far from the setting that produces the original model, then we know that the new parameters are useful for describing the data.

Instead of adding new parameters to a nested existing model, we might be inspired to try a completely different model form. To assess whether the new model describes the data better than the original model, Bayesian model comparison could be used, in principle. The priors on the parameters within the two models would need to be equivalently informed, as was discussed in Section 10.6.1 (p. 294).

Some people caution that looking at the data to inspire a model form is "double dipping" into the data, in the sense that the data are being used to change the prior distribution on the model space. Of course, we can consider any model space we care to, and the motivation for that model space can evolve through time and come from many different sources, such as different theories and different background literature. If a set of data suggests a novel trend or functional form, presumably the prior probability on that novel trend was small unless the analyst realizes the same trend had occurred unheralded in previous results. The analyst should keep the low prior probability in mind, either formally or informally. Moreover, novel trends are retrospective descriptions of particular data sets, and the trends need to be confirmed by subsequent, independently collected data.

Another approach to a posterior predictive check is to create a posterior predictive sampling distribution of a measure of discrepancy between the predictions and the data. In this approach, a "Bayesian p value" indicates the probability of getting the data's discrepancy, or something more extreme, from the model. If the Bayesian p value is too small, then the model is rejected. From my discussion of p values in Chapter 11, you might imagine that I find this approach to be problematic. And you would be right. I discuss this issue in more detail in an article (Kruschke, 2013b).

17.5.2. Steps to extend a JAGS or Stan model

One of the great benefits of Bayesian software is its tremendous flexibility for specifying models that are theoretically meaningful and appropriate to the data. To take advantage of this flexibility, you will want to be able to modify existing programs. An example of such a modification was presented in this chapter in the extension from the linear-trend model to the quadratic-trend model. The steps in such a modification are routine, as outlined here:

- *Carefully specify the model with its new parameters.* It can help to sketch a diagram like Figure 17.6 to make sure that you really understand all the parameters and their priors. The diagram can also help coding the model, because there is usually a correspondence of arrows to lines of code. For example when extending a linear model to a quadratic, the diagram gets just one more branch, for the quadratic coefficient, that is structurally identical to the branch for the linear coefficient. The expression for the mean must be extended with the quadratic term: $\mu_{i|j} = \cdots + \beta_{2,j}\, x^2_{i|j}$, using notation in the code that is consistent with the already-established code. The group-level distribution over the unit-level quadratic coefficients must also be included. Again, this is easy because the model structure of the quadratic coefficient is exactly analogous to the model structure of the linear coefficient.
- *Be sure all the new parameters have sensible priors.* Be sure the priors for the old parameters still make sense in the new model.
- *If you are defining your own initial values for the chains, define initial values for all the new parameters*, and make sure that the initial values of the old parameters still make sense for the extended model. JAGS will initialize any stochastic variables that are not given explicit initial values. It is often easiest to let JAGS initialize parameters automatically.
- *Tell JAGS to track the new parameters.* JAGS records only those parameters that you explicitly tell it to track. Stan tracks all nonlocal variables (such as parameters and transformed parameters) by default.
- *Modify the summary and graphics output to properly display the extended model.* I have found that this step is the most time-consuming and often the most error-prone. Because the graphics are displayed by R, not by JAGS, you must modify all of your R graphics code to be consistent with the modified model in JAGS. Depending on the role of the parameter in the model and its meaning for interpreting the data, you might want a plot of the parameter's marginal posterior distribution, and perhaps `pairs` plots of the parameter crossed with other parameters to see if the parameters are correlated in the posterior, and probably specialized displays of the posterior predictive that show trends and predicted spread of data superimposed on the data, and possibly differences between parameters or other comparisons and contrasts, etc.

17.5.3. Perils of adding parameters

When including new parameters, the extended model has new flexibility in fitting the data. There may be a far wider range of credible values for the previous parameters. For example, consider again the data that were introduced for the hierarchical linear model, back in Figure 17.5 (p. 492). We can easily use the hierarchical *quadratic*-trend model on these data, which yields the posterior credible trends shown in Figure 17.8.[5] Notice that there are many positive or negative curvatures that are consistent with the data. The curvature and slope trade-off strongly in their fit to the original-scale data, such that a positive curvature with small slope fits the data as well as a negative curvature with large slope. Thus, the introduction of the curvature parameter makes the slope parameter less certain, even though the posterior distribution of the curvature parameter is centered nearly at zero. Specifically, in the linear-trend fit, wherein the curvature is implicitly forced to be exactly zero, the 95% HDI on the slope goes from +2.2 to +4.1 (not shown here, but produced in the output of the script). In the quadratic-trend fit, the curvature has a mode very near zero with a 95% HDI from about −0.10 to +0.07, but the 95% HDI of the slope goes from about −7.0 to +16.5. In other words, even though the quadratic-trend model indicates that there is little if any quadratic trend in the data, the estimate of the slope is blurred. The magnitude of this apparent blurring of the slope depends strongly on the data scale, however. Using standardized data, the 95% HDI on the slope parameter changes only slightly.

This increase in uncertainty of a parameter estimate is a form of penalizing complexity: For nested models, in which the simpler model is the more complex model with one of the complex model's parameters fixed (e.g., to zero), then the estimates of the nested parameters will tend to be less precise in the more complex model. The degree to which a parameter estimate broadens, when expanding the model, depends on the model structure, the data, and the parameterization. The "blurring" can be especially pronounced if the parameter values can trade-off and fit the data equally well.

A different way of penalizing complexity comes from Bayesian model comparison. The prior on the larger parameter space dilutes the prior probability of any particular combination of parameter values, thereby favoring the simpler model unless the data are much better fit by a combination of parameter values that is not available to the simpler model. It is worth recapitulating that in Bayesian model comparison, the priors for the two models would need to be equivalently informed, as was discussed in Section 10.6.1 (p. 294).

17.6. EXERCISES

Look for more exercises at https://sites.google.com/site/doingbayesiandataanalysis/

[5] In JAGS, many runs take a long burn-in to converge, and even after that, most runs yield highly autocorrelated chains for the normality parameter. But all chains show similar smooth marginal distributions on the normality parameter, and all chains are well behaved on the other parameters. In Stan, burn-in is rapid, and the normality parameter is smoothly sampled, but most runs yield chains that get temporarily stuck and show unrepresentative bumps in the group-level means. Exercise 17.3 has you try this yourself.

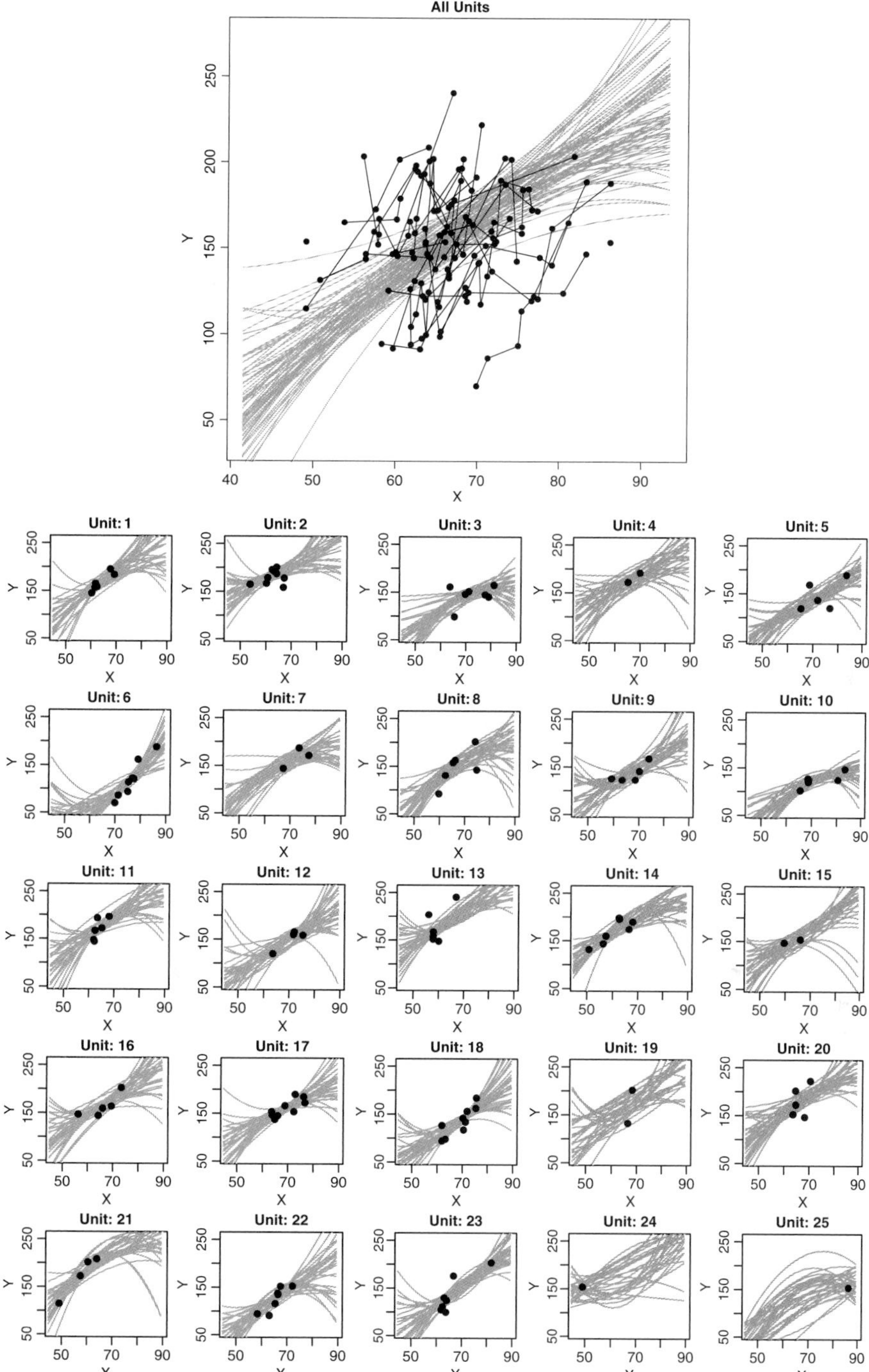

Figure 17.8 Fitting the data of Figure 17.5 with the quadratic-trend model. The flexibility of the quadratic trend yields more *un*certainty in the linear trend, despite the fact that the modal estimate of the quadratic trend is nearly zero.

Exercise 17.1. [Purpose: Apply linear model to quadratic data and do a posterior predictive check.]

(A) Change the script `Jags-Ymet-XmetSsubj-MrobustHier-Example.R` so it uses the family income data. (Notice how simple it is to do that.) Are the data well described by the linear model? What exactly is the "systematic discrepancy" between model and data? Don't just say, "the data are curved." What exactly does that mean in terms of how the data deviate from the model and where?

(B) Is there a way of rejecting the linear model without reference to any other model? In other words, how might you compute a "Bayesian p value" for this situation? Should you? *See Kruschke (2013b) for information and references.*

Exercise 17.2. [Purpose: Observe the autocorrelation in JAGS when data are not standardized.] Change the (simple, nonhierarchical) JAGS program `Jags-Ymet-Xmet-Mrobust.R` so that it uses the raw data instead of the standardized data. Be sure to rename it so that you don't destroy the original program. You will have to remove the standardization from the data block and remove the transformation to the original scale. Make sure that the prior is appropriate for the original scale (perhaps use the prior in the Stan version for guidance). Show the diagnostic graphs of the chains and discuss. Finally, run the chains a long time, with thinning if needed to conserve computer memory, and show that the chains eventually converge to the same posterior as when using standardized data.

Exercise 17.3. [Purpose: Examining chain convergence in JAGS and Stan.] As mentioned in Footnote 5 (p. 504), both JAGS and Stan show some difficulty converging when the quadratic-trend model is applied to the fictitious data of Figure 17.5. The JAGS programs are provided in files `Jags-Ymet-XmetSsubj-MrobustHierQuadWt-Example.R` and `Jags-Ymet-XmetSsubj-MrobustHierQuadWt.R`. The corresponding Stan programs are provided in files `Stan-Ymet-XmetSsubj-MrobustHierQuadWt-Example.R` and `Stan-Ymet-XmetSsubj-Mrobust HierQuadWt.R`. The Stan model specification is shown below so that you can study it and compare it to the JAGS version in Section 17.4 without having to be at your computer:

```
data {
  int<lower=1> Nsubj ;
  int<lower=1> Ntotal ;
  real y[Ntotal] ;
  real x[Ntotal] ;
  real<lower=0> w[Ntotal] ;
  int<lower=1> s[Ntotal] ;
}
transformed data {
  // Standardize the data:
  real zx[Ntotal] ;
  real zy[Ntotal] ;
```

```
  real zw[Ntotal] ;
  real wm ;
  real xm ;
  real ym ;
  real xsd ;
  real ysd ;
  xm <- mean(x) ;
  ym <- mean(y) ;
  wm <- mean(w) ;
  xsd <- sd(x) ;
  ysd <- sd(y) ;
  for ( i in 1:Ntotal ) { // could be vectorized...?
    zx[i] <- ( x[i] - xm ) / xsd ;
    zy[i] <- ( y[i] - ym ) / ysd ;
    zw[i] <- w[i] / wm  ;
  }
}
parameters {
  real zbeta0[Nsubj] ;
  real zbeta1[Nsubj] ;
  real zbeta2[Nsubj] ;
  real<lower=0> zsigma ;
  real zbeta0mu ;
  real zbeta1mu ;
  real zbeta2mu ;
  real<lower=0> zbeta0sigma ;
  real<lower=0> zbeta1sigma ;
  real<lower=0> zbeta2sigma ;
  real<lower=0> nuMinusOne ;
}
transformed parameters {
  real<lower=0> nu ;
  real beta0[Nsubj] ;
  real beta1[Nsubj] ;
  real beta2[Nsubj] ;
  real<lower=0> sigma ;
  real beta0mu ;
  real beta1mu ;
  real beta2mu ;
  nu <- nuMinusOne+1 ;
  // Transform to original scale:
  for ( j in 1:Nsubj ) { // could be vectorized...?
    beta2[j] <- zbeta2[j]*ysd/square(xsd) ;
    beta1[j] <- zbeta1[j]*ysd/xsd - 2*zbeta2[j]*xm*ysd/square(xsd) ;
    beta0[j] <- zbeta0[j]*ysd  + ym - zbeta1[j]*xm*ysd/xsd
                        + zbeta2[j]*square(xm)*ysd/square(xsd) ;
  }
  beta2mu <- zbeta2mu*ysd/square(xsd) ;
  beta1mu <- zbeta1mu*ysd/xsd - 2*zbeta2mu*xm*ysd/square(xsd) ;
```

```
  beta0mu <- zbeta0mu*ysd  + ym - zbeta1mu*xm*ysd/xsd
                           + zbeta2mu*square(xm)*ysd/square(xsd) ;
  sigma <- zsigma * ysd ;
}
model {
  zbeta0mu ~ normal( 0 , 10 ) ;
  zbeta1mu ~ normal( 0 , 10 ) ;
  zbeta2mu ~ normal( 0 , 10 ) ;
  zsigma ~ uniform( 1.0E-3 , 1.0E+3 ) ;
  zbeta0sigma ~ uniform( 1.0E-3 , 1.0E+3 ) ;
  zbeta1sigma ~ uniform( 1.0E-3 , 1.0E+3 ) ;
  zbeta2sigma ~ uniform( 1.0E-3 , 1.0E+3 ) ;
  nuMinusOne ~ exponential(1/29.0) ;
  zbeta0 ~ normal( zbeta0mu , zbeta0sigma ) ; // vectorized
  zbeta1 ~ normal( zbeta1mu , zbeta1sigma ) ; // vectorized
  zbeta2 ~ normal( zbeta2mu , zbeta2sigma ) ; // vectorized
  for ( i in 1:Ntotal ) {
    zy[i] ~ student_t(
              nu ,
              zbeta0[s[i]] + zbeta1[s[i]] * zx[i] + zbeta2[s[i]] * square(zx[i]) ,
              zw[i]*zsigma ) ;
  }
}
```

Review Section 14.4 (p. 414) for hints about programming Stan.

(A) Run Stan on the family-income data, so it achieves the same ESS as the JAGS program for the group-level trend coefficients. How long (in real time) do Stan and JAGS take? Does Stan more consistently converge than JAGS? Does Stan produce better chains for the normality and noise parameters?

(B) Now repeat on the fictitious data of Figure 17.8, which typically gives JAGS and Stan troubles of differing sorts as described Footnote 5 (p. 504). Try to produce examples of these troubles and discuss. Which type of trouble is more tolerable, autocorrelation in the normality parameter (in JAGS) or "bumps" in the regression coefficients (in Stan)?

CHAPTER 18

Metric Predicted Variable with Multiple Metric Predictors

Contents

When I was young two plus two equaled four, but
Since I met you things don't add up no more.
My keel was even before I was kissed, but
Now it's an ocean with swells and a twist.[1]

In this chapter, we are concerned with situations in which the value to be predicted is on a metric scale, and there are several predictors, each of which is also on a metric scale. For example, we might predict a person's college grade point average (GPA) from his or her high-school GPA and scholastic aptitude test (SAT) score. Another such situation is predicting a person's blood pressure from his or her height and weight.

We will consider models in which the predicted variable is an additive combination of predictors, all of which have proportional influence on the prediction. This kind of model is called *multiple linear regression*. We will also consider nonadditive combinations of predictors, which are called *interactions*.

[1] This chapter discusses multiple metric predictors. Basic linear regression considers additive combinations of predictors, for which "two plus two equals four." This chapter also smooches multiplicative interactions of predictors, which give the regression surface a twist as shown in Figure 18.8.

Doing Bayesian Data Analysis, 2nd Edition
http://dx.doi.org/10.1016/B978-0-12-405888-0.00018-0

In the context of the generalized linear model (GLM) introduced in Chapter 15, this chapter's situation involves a linear function of multiple metric predictors, as indicated in the fourth column of Table 15.1 (p. 434), with a link function that is the identity along with a normal distribution (or similar) for describing noise in the data, as indicated in the first row of Table 15.2 (p. 443). For a reminder of how this chapter's combination of predicted and predictor variables relates to other combinations, see Table 15.3 (p. 444).

If you seek a compact introduction to Bayesian methods, using multiple linear regression as a guiding example, see the article by Kruschke et al. (2012). Supplementary materials specific to that article are available at the Web site http://www.indiana.edu/kruschke/BMLR/ where BMLR stands for Bayesian multiple linear regression.

18.1. MULTIPLE LINEAR REGRESSION

Figures 18.1 and 18.2 show examples of data generated by a model for multiple linear regression. The model specifies the dependence of y on x_1 and x_2, but the model does not specify the distribution of x_1 and x_2. At any position, $\langle x_1, x_2\rangle$, the values of y are normally distributed in a vertical direction, centered on the height of the plane at that position. The height of the plane is a linear combination of the x_1 and x_2 values. Formally, $y \sim \text{normal}(\mu, \sigma)$ and $\mu = \beta_0 + \beta_1 x_1 + \beta_2 x_2$. For a review of how to interpret the coefficients as the intercept and slopes, see Figure 15.2 (p. 426). The model assumes homogeneity of variance, which means that at all values of x_1 and x_2, the variance σ^2 of y is the same.

18.1.1. The perils of correlated predictors

Figures 18.1 and 18.2 show data generated from the same model. In both figures, $\sigma = 2$, $\beta_0 = 10$, $\beta_1 = 1$, and $\beta_2 = 2$. All that differs between the two figures is the distribution of the $\langle x_1, x_2\rangle$ values, which is not specified by the model. In Figure 18.1, the $\langle x_1, x_2\rangle$ values are distributed independently. In Figure 18.2, the $\langle x_1, x_2\rangle$ values are negatively correlated: When x_1 is small, x_2 tends to be large, and when x_1 is large, x_2 tends to be small. In each figure, the top-left panel shows a 3D-perspective view of the data ($y \sim \text{normal}(\mu, \sigma = 2)$) superimposed on a grid representation of the plane ($\mu = 10 + 1x_1 + 2x_2$). The data points are connected to the plane with vertical dotted lines, to indicate that the noise is a vertical departure from the plane. The other panels of Figures 18.1 and 18.2 show different perspectives on the same data. The top-right panel of each figure shows the y values plotted against x_1 only, collapsed across x_2. The bottom-left panel of each figure shows the y values plotted against x_2 only, collapsed across x_1. Finally, the bottom-right panel of each figure shows the $\langle x_1, x_2\rangle$ values, collapsed across y. By examining these different perspectives, we will see that underlying trends in the data can be misinterpreted when predictors are correlated and not all predictors are included in the analysis.

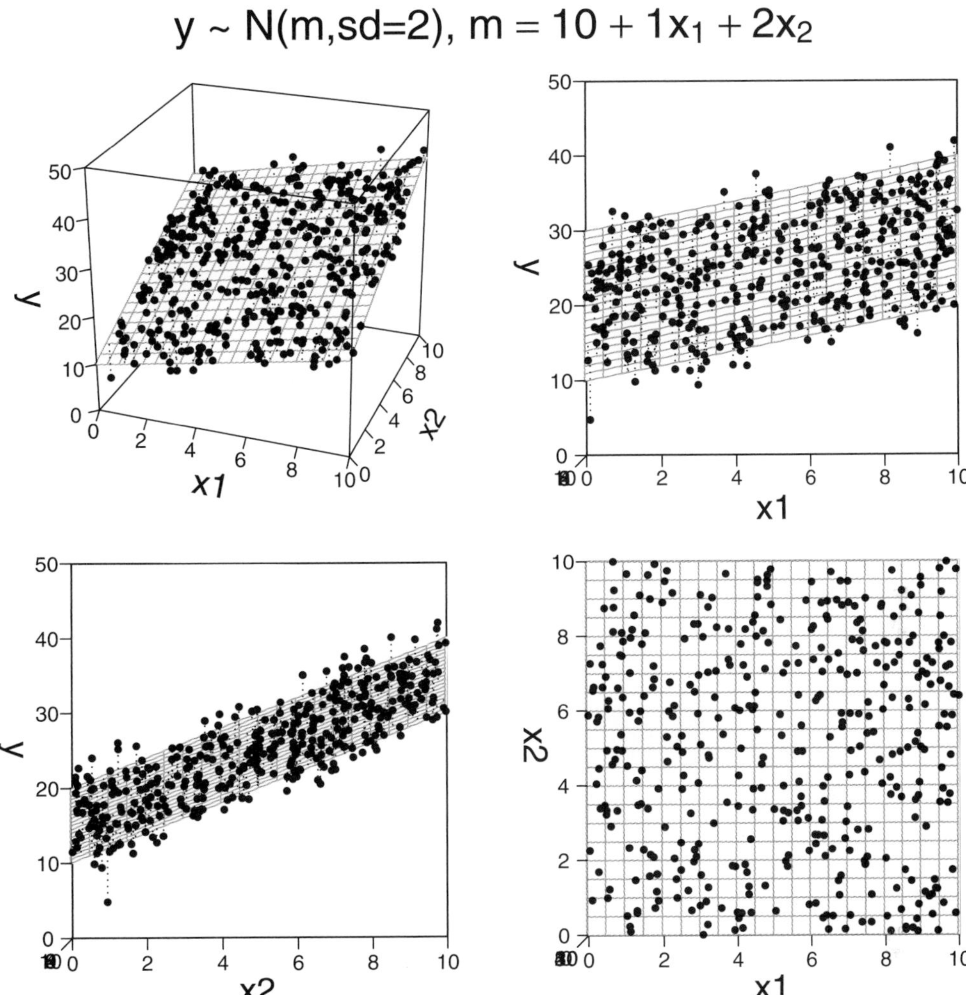

Figure 18.1 Data, y, that are normally distributed around the values in the plane. The $\langle x_1, x_2 \rangle$ values are independent of each other, as shown in the lower-right panel. The panels show different perspectives on the same plane and data. Compare with Figure 18.2.

In Figure 18.1, the $\langle x_1, x_2 \rangle$ values are not correlated, as can be seen in the bottom-right panel. In this case of uncorrelated predictors, the scatter plot of y against x_1 (dots in top-right panel) accurately reflects the true underlying slope, β_1, shown by the grid representation of the plane. And, in the bottom-left panel, the scatter plot of y against x_2 accurately reflects the true underlying slope, β_2, shown by the grid representation of the plane.

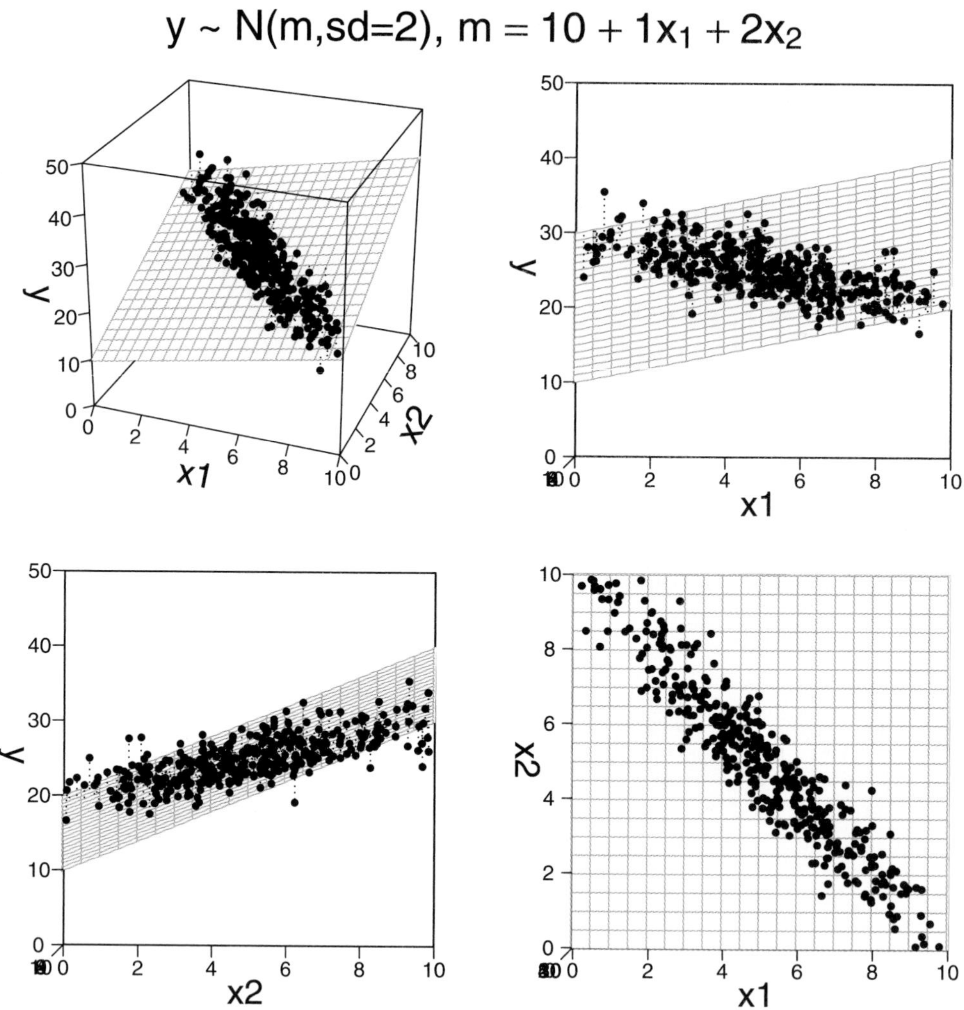

Figure 18.2 Data, y, that are normally distributed around the values in the plane. The $\langle x_1, x_2 \rangle$ values are (anti-)correlated, as shown in the lower-right panel. The panels show different perspectives on the same plane and data. Compare with Figure 18.1.

Interpretive perils arise when predictors are correlated. In Figure 18.2, the $\langle x_1, x_2 \rangle$ values are anticorrelated, as can be seen in the bottom-right panel. In this case of (anti-) correlated predictors, the scatter plot of y against x_1 (dots in top-right panel) does not reflect the true underlying slope, β_1, shown by the grid representation of the plane. The scatter plot of y against x_1 trends downward, even though the true slope is upward ($\beta_1 = +1$). There is no error in the graph; the apparent contradiction is merely an illusion (visual and mathematical) caused by removing the information about x_2. The reason that the y values appear to decline as x_1 increases is that x_2 decreases when

x_1 decreases, and x_2 has a bigger influence on y than the influence of x_1. The analogous problem arises when collapsing across x_1, although less dramatically. The bottom-left panel shows that the scatter plot of y against x_2 does not reflect the true underlying slope, β_2, shown by the grid representation of the plane. The scatter plot of y against x_2 does rise upward, but not steeply enough compared with the true slope β_2. Again, there is no error in the graph; the apparent contradiction is merely an illusion caused by leaving out the information about x_1.

Real data often have correlated predictors. For example, consider trying to predict a state's average high-school SAT score on the basis of the amount of money the state spends per pupil. If you plot only mean SAT against money spent, there is actually a *decreasing* trend, as can be seen in the scatter of data points in the top-right panel of Figure 18.3 (data from Guber, 1999). In other words, SAT scores tend to go down as spending goes up! Guber (1999) explains how some political commentators have used this relationship to argue against funding public education.

The negative influence of spending on SAT scores seems quite counterintuitive. It turns out that the trend is an illusion caused by the influence of another factor which happens to be correlated with spending. The other factor is the proportion of students who take the SAT. Not all students at a high school take the SAT, because the test is used primarily for college entrance applications, and therefore, it is primarily students who intend to apply to college who take the SAT. Most of the top students at a high school will take the SAT, because most of the top students will apply to college. But students who are weaker academically may be less likely to take the SAT, because they are less likely to apply to college. Therefore, the more that a high school encourages mediocre students to take the SAT, the lower will be its average SAT score. It turns out that high schools that spend more money per pupil also have a much higher proportion of students who take the SAT. This correlation can be seen in the lower-right panel of Figure 18.3.

When both predictors are taken into account, the influence of spending on SAT score is seen to be positive, not negative. This positive influence of spending can be seen as the positive slope of the plane along the "Spend" direction in Figure 18.3. The negative influence, of percentage of students taking the SAT, is also clearly shown. To reiterate the main point of this example: It seems that the apparent drop in SAT due to spending is an artifact of spending being correlated with the percentage of students taking the SAT, with the latter having a whoppingly negative influence on SAT scores.

The separate influences of the two predictors could be assessed in this example because the predictors had only mild correlation with each other. There was enough independent variation of the two predictors that their distinct relationships to the outcome variable could be detected. In some situations, however, the predictors are so tightly correlated that their distinct effects are difficult to tease apart. Correlation of predictors causes the estimates of their regression coefficients to trade-off, as we will see when we examine the posterior distribution.

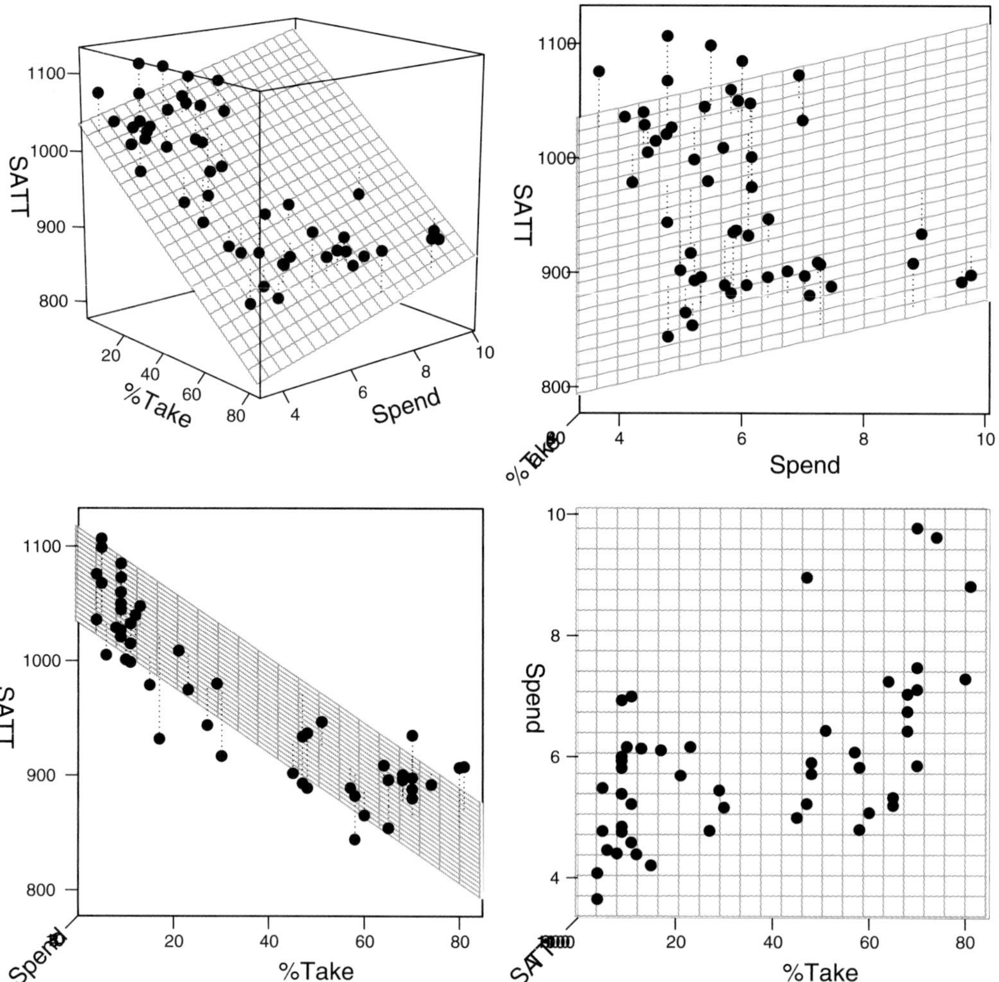

Figure 18.3 The data (Guber, 1999) are plotted as dots, and the grid shows the best fitting plane. "SATT" is the average total SAT score in a state. "%Take" is the percentage of students in the state who took the SAT. "Spend" is the spending per pupil, in thousands of dollars.

18.1.2. The model and implementation

The hierarchical diagram for multiple linear regression is shown in Figure 18.4. It is merely a direct expansion of the diagram for simple linear regression in Figure 17.2 (p. 480). Instead of only one slope coefficient for a single predictor, there are distinct slope coefficients for the multiple predictors. For every coefficient, the prior is normal, just as shown in Figure 17.2. The model also uses a t distribution to describe the noise around

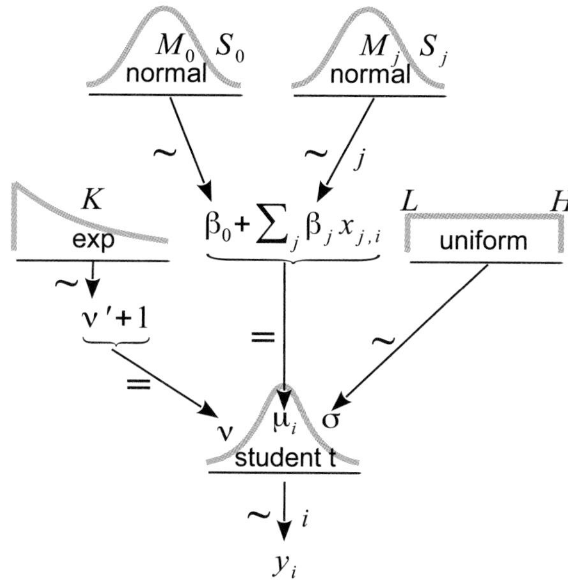

Figure 18.4 Hierarchical diagram for multiple linear regression. Compare with Figure 17.2 (p. 480).

the linear predicted value. This heavy-tailed t distribution accommodates outliers, as was described at length in Section 16.2 and subsequent sections. This model is therefore sometimes referred to as *robust* multiple linear regression.

As with the model for simple linear regression, the Markov Chain Monte Carlo (MCMC) sampling can be more efficient if the data are mean-centered or standardized. Now, however, there are multiple predictors to be standardized. To understand the code for standardizing the data (shown below), note that the predictor values are sent into JAGS as a matrix named x that has a column for each predictor and a row for each data point. The data block of the JAGS code then standardizes the predictors by looping through the columns of the x matrix, as follows:

```
data {
  ym <- mean(y)
  ysd <- sd(y)
  for ( i in 1:Ntotal ) { # Ntotal is the number of data rows
    zy[i] <- ( y[i] - ym ) / ysd
  }
  for ( j in 1:Nx ) {      # Nx is the number of x predictors
    xm[j]  <- mean(x[,j]) # x is a matrix, each column a predictor
    xsd[j] <-   sd(x[,j])
    for ( i in 1:Ntotal ) {
      zx[i,j] <- ( x[i,j] - xm[j] ) / xsd[j]
    }
  }
}
```

The model uses the standardized data, zx and zy, to generate credible values for the standardized parameters. The standardized parameters are then transformed to the original scale by generalizing Equation 17.2 (p. 485) to multiple predictors:

$$
\begin{aligned}
z_{\hat{y}} &= \zeta_0 + \sum_j \zeta_j z_{x_j} \\
\frac{(\hat{y} - M_y)}{\mathrm{SD}_y} &= \zeta_0 + \sum_j \zeta_j \frac{(x_j - M_{x_j})}{\mathrm{SD}_{x_j}} \\
\hat{y} &= \underbrace{\mathrm{SD}_y\, \zeta_0 + M_y - \mathrm{SD}_y \sum_j \zeta_j M_{x_j}/\mathrm{SD}_{x_j}}_{\beta_0} + \sum_j \underbrace{\mathrm{SD}_y\, \zeta_j/\mathrm{SD}_{x_j}}_{\beta_j}\, x_j
\end{aligned}
\tag{18.1}
$$

The estimate of σ_y is merely $\sigma_{z_y}\mathrm{SD}_y$, as was the case for single-predictor linear regression.

As usual, the model specification has a line of code corresponding to every arrow in the hierarchical diagram of Figure 18.4. The JAGS model specification looks like this:

```
model {
  for ( i in 1:Ntotal ) {
    zy[i] ~ dt( zbeta0 + sum( zbeta[1:Nx] * zx[i,1:Nx] ) , 1/zsigma^2 , nu )
  }
  # Priors vague on standardized scale:
  zbeta0 ~ dnorm( 0 , 1/2^2 )
  for ( j in 1:Nx ) {
    zbeta[j] ~ dnorm( 0 , 1/2^2 )
  }
  zsigma ~ dunif( 1.0E-5 , 1.0E+1 )
  nu <- nuMinusOne+1
  nuMinusOne ~ dexp(1/29.0)
  # Transform to original scale:
  beta[1:Nx] <- ( zbeta[1:Nx] / xsd[1:Nx] )*ysd
  beta0 <- zbeta0*ysd  + ym - sum( zbeta[1:Nx] * xm[1:Nx] / xsd[1:Nx] )*ysd
  sigma <- zsigma*ysd
}
```

The prior on the standardized regression coefficients, zbeta[j], uses an arbitrary standard deviation of 2.0. This value was chosen because standardized regression coefficients are algebraically constrained to fall between −1 and +1 in least-squares regression, and therefore, the regression coefficients will not exceed those limits by much. A normal distribution with standard deviation of 2.0 is reasonably flat over the range from −1 to +1. The complete program is in the file Jags-Ymet-XmetMulti-Mrobust.R and a high-level script that calls it is the file Jags-Ymet-XmetMulti-Mrobust-Example.R.

18.1.3. The posterior distribution

Figure 18.5 shows the posterior distribution from the SAT data in Figure 18.3 and model in Figure 18.4. You can see that the slope on spending (Spend) is credibly above zero, even taking into account a modest ROPE and MCMC instability. The slope on spending has a mode of about 13, which suggests that SAT scores rise by about 13 points for every extra \$1000 spent per pupil. The slope on percentage taking the exam (PrcntTake) is also credibly non-zero, with a mode around −2.8, which suggests that SAT scores fall by about 2.8 points for every additional 1% of students who take the test.

The scatter plots in the bottom of Figure 18.5 show correlations among the credible parameter values in the posterior distribution. (These are pairwise scatter plots of credible parameter values from the MCMC chain; these are not scatter plots of data.) In particular, the coefficient for spending (Spend) trades off with the coefficient on percentage taking the exam (PrcntTake). The correlation means that if we believe that the influence of spending is smaller, then we must believe that the influence of percentage taking is larger. This makes sense because those two predictors are correlated in the data.

Figure 18.5 shows that the normality parameter for these data is fairly large, suggesting that there are not many outliers for this particular selection of predictors. It is worth noting that values of y are not inherently outliers or nonoutliers; they are only outliers relative to a spread of predicted values for a particular model. A value of y that seems spurious according to one set of predictors might be nicely linearly predicted by other predictors.

Finally, Figure 18.5 also shows a posterior distribution for a statistic labeled R^2, which is called the *proportion of variance accounted for* in traditional least-squares multiple regression. In least-squares regression, the overall variance in y is algebraically decomposed into the variance of the linearly predicted values and the residual variance: $\sum_i (y_i - \bar{y})^2 = \sum_i (y_i - \hat{y}_i)^2 + \sum_i (\hat{y}_i - \bar{y})^2$, where $\hat{y}_i$ is the linearly predicted value of y_i. The proportion of variance accounted for is $R^2 = \sum_i (\hat{y}_i - \bar{y})^2 / \sum_i (y_i - \bar{y})^2$ when $\hat{y}_i$ is the linear prediction using coefficients that minimize $\sum_i (y_i - \hat{y}_i)^2$. If that makes no sense to you because you have no previous experience with least-squares regression, do not worry, because in Bayesian analysis no such decomposition of variance occurs. But for people familiar with least-squares notions who crave a statistic analogous to R^2, we can compute a surrogate. At each step in the MCMC chain, a credible value of R^2 is computed as $R^2 = \sum_j \zeta_j r_{y,x_j}$, where ζ_j is the standardized regression coefficient for the jth predictor at that step in the MCMC chain, and r_{y,x_j} is the correlation of the predicted values, y, with the jth predictor values, x_j. These correlations are constants, fixed by the data. The equation for expressing R^2 in terms of the regression coefficients is used merely by analogy to least-squares regression, in which the equation is exactly true (e.g., Hays, 1994, Equation 15.14.2, p. 697). The mean value in the distribution of R^2, when using vague priors, is essentially the least-squares estimate and the maximum-likelihood estimate when using a normal likelihood function. The posterior distribution reveals

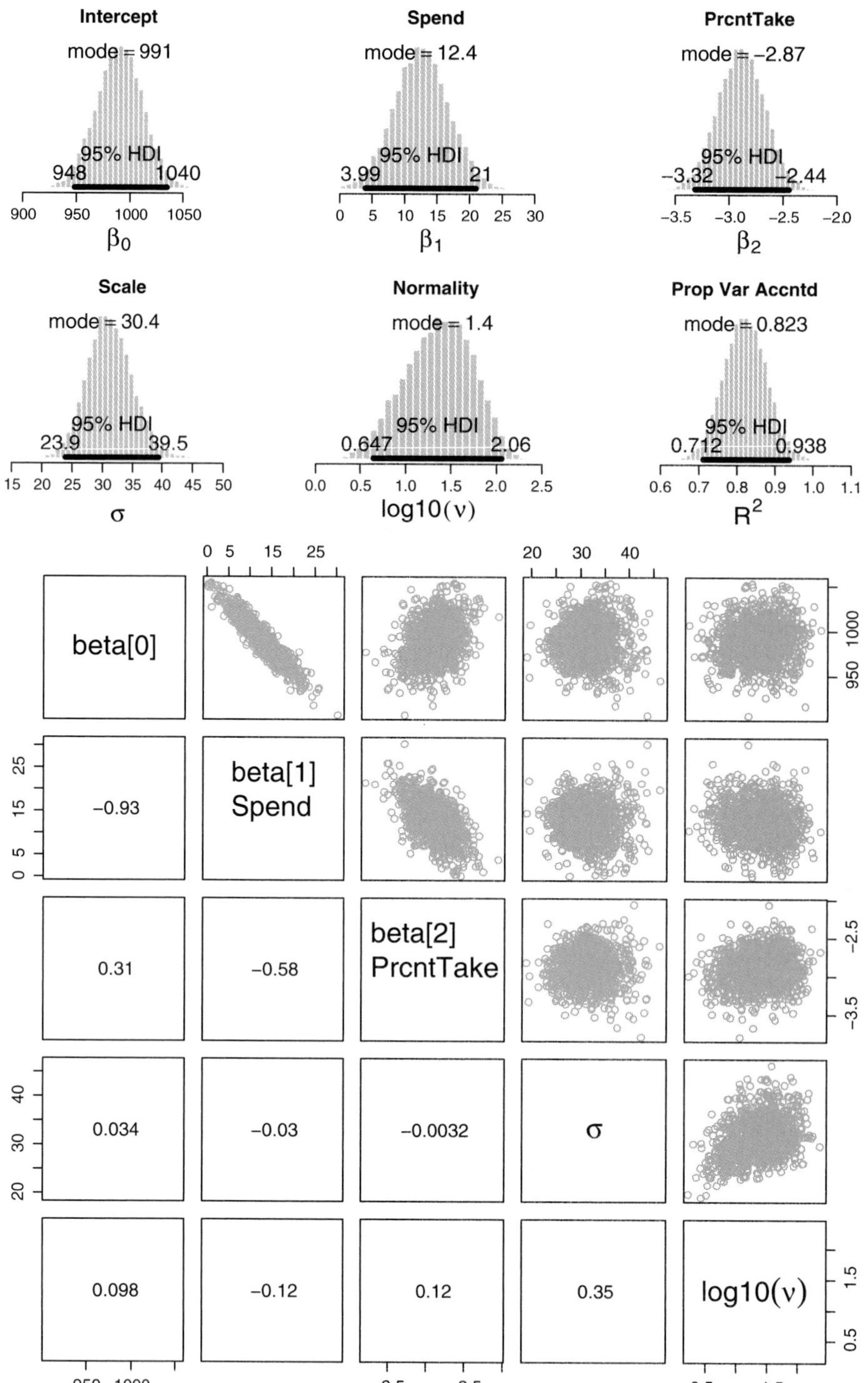

Figure 18.5 Posterior distribution for data in Figure 18.3 and model in Figure 18.4. Scatter plots reveal correlations among credible parameter values; in particular, the coefficient on Spending ("Spend") trades off with the coefficient on Percentage taking the exam ("PrcntTake"), because those predictors are correlated in the data.

the entire distribution of credible R^2 values. The posterior distribution of R^2, defined this way, can exceed 1.0 or fall below 0.0, because R^2 here is a linear combination of credible regression coefficients, not the singular value that minimizes the squared deviations between predictions and data.

Sometimes we are interested in using the linear model to predict y values for x values of interest. It is straight forward to generate a large sample of credible y values for specified x values. At each step in the MCMC chain, the combination of credible parameter values is inserted into the model and random y values are generated. From the distribution of y values, we can compute the mean and highest density interval (HDI) to summarize the centrally predicted y value and the uncertainty of the prediction. As was the case for simple linear regression, illustrated back in Figure 17.3 (p. 481), the uncertainty in predicted y is greater for x values outside the bulk of the data. In other words, extrapolation is more uncertain than interpolation.

18.1.4. Redundant predictors

As a simplified example of correlated predictors, think of just two data points: Suppose $y = 1$ for $\langle x_1, x_2 \rangle = \langle 1, 1 \rangle$ and $y = 2$ for $\langle x_1, x_2 \rangle = \langle 2, 2 \rangle$. The linear model, $y = \beta_1 x_1 + \beta_2 x_2$, is supposed to satisfy both data points, and in this case both are satisfied by $1 = \beta_1 + \beta_2$. Therefore, many different combinations of β_1 and β_2 satisfy the data. For example, it could be that $\beta_1 = 2$ and $\beta_2 = -1$, or $\beta_1 = 0.5$ and $\beta_2 = 0.5$, or $\beta_1 = 0$ and $\beta_2 = 1$. In other words, the credible values of β_1 and β_2 are anticorrelated and trade-off to fit the data.

One of the benefits of Bayesian analysis is that correlations of credible parameter values are explicit in the posterior distribution. Another benefit of Bayesian analysis is that the estimation doesn't "explode" when predictors are strongly correlated. If predictors are correlated, the joint uncertainty in the regression coefficients is evident in the posterior, but the analysis happily generates a posterior distribution regardless of correlations in the predictors. In extreme cases, when the predictors are very strongly correlated, the marginal posteriors will simply reflect the prior distributions on the regression coefficients, with a strong trade-off in their joint posterior distribution.

For illustration, we will use a completely redundant predictor, namely the proportion of students *not* taking the exam. Thus, if PrcntTake is the percentage of students taking the exam, then PropNotTake $= (100 - \text{PrcntTake})/100$ is the proportion of students not taking the exam. For example, if PrcntTake $= 37$, then PropNotTake $= 0.63$. These sorts of redundant predictors can show up in real analyses. Sometimes, the redundant predictors are included because the analyst does not realize (initially) that they are redundant, perhaps because the predictors are labeled differently and come from different sources and are on seemingly different scales. Other times, the predictors are not inherently redundant, but happen to be extremely strongly correlated in the data. For example, suppose we use temperature as a predictor, and we measure the temperature with two thermometers sitting side by side. Their readings should be almost perfectly correlated, even if one has a Celsius scale and the other has a Fahrenheit scale.

Figure 18.6 shows the posterior distribution. One sign of redundant predictors is the (very nearly) perfect correlation between the credible values of the slopes on the predictors, revealed in the pairwise scatter plots. Because the predictors are redundant, the credible regression coefficients trade-off with each other but still fit the data equally well. Consequently, the marginal posterior distribution of either predictor is extremely broad, as can be seen in the upper panels of Figure 18.6. Thus, an extremely broad marginal posterior distribution is another clue that there might be redundancies in the predictors.

Another important clue to redundancy in predictors is autocorrelation in the MCMC chains for the regression coefficients of the predictors. When you run the script, you will see that diagnostic graphs of the chains are generated (not shown here). The chains for the regression coefficients of redundant predictors are highly autocorrelated and very highly correlated with each other.

Unfortunately, the signs of predictor redundancy in the posterior distribution get diffused when there are three of more strongly correlated predictors. In particular, the *pairwise* scatter plots are not sufficient to show a *three-way* trade-off of regression coefficients. Autocorrelation remains high, however.

Of course, the most obvious indicator of redundancy in predictors is not in the posterior distribution of the regression coefficients, but in the predictors themselves. At the beginning of the program, the correlations of the predictors are displayed in the R console, like this:

```
CORRELATION MATRIX OF PREDICTORS:
             Spend PrcntTake PropNotTake
Spend        1.000     0.593      -0.593
PrcntTake    0.593     1.000      -1.000
PropNotTake -0.593    -1.000       1.000
```

If any of the nondiagonal correlations are high (i.e., close to $+1$ or close to -1), be careful when interpreting the posterior distribution. Here, we can see that the correlation of PrcntTake and PropNotTake is -1.0, which is an immediate sign of redundant predictors.

Traditional methods for multiple linear regression can break down when predictors are perfectly (or very strongly) correlated because there is not a unique solution for the best fitting parameter values. Bayesian estimation has no inherent problems with such cases. The posterior distribution merely reveals the trade-off in the parameters and the resulting large uncertainty in their individual values. The extent of the uncertainty is strongly influenced by the prior distribution in this case, because there is potentially an infinite trade-off among equally well-fitting parameter values, and only the prior distribution tempers the infinite range of possibilities. Figure 18.7 shows the prior

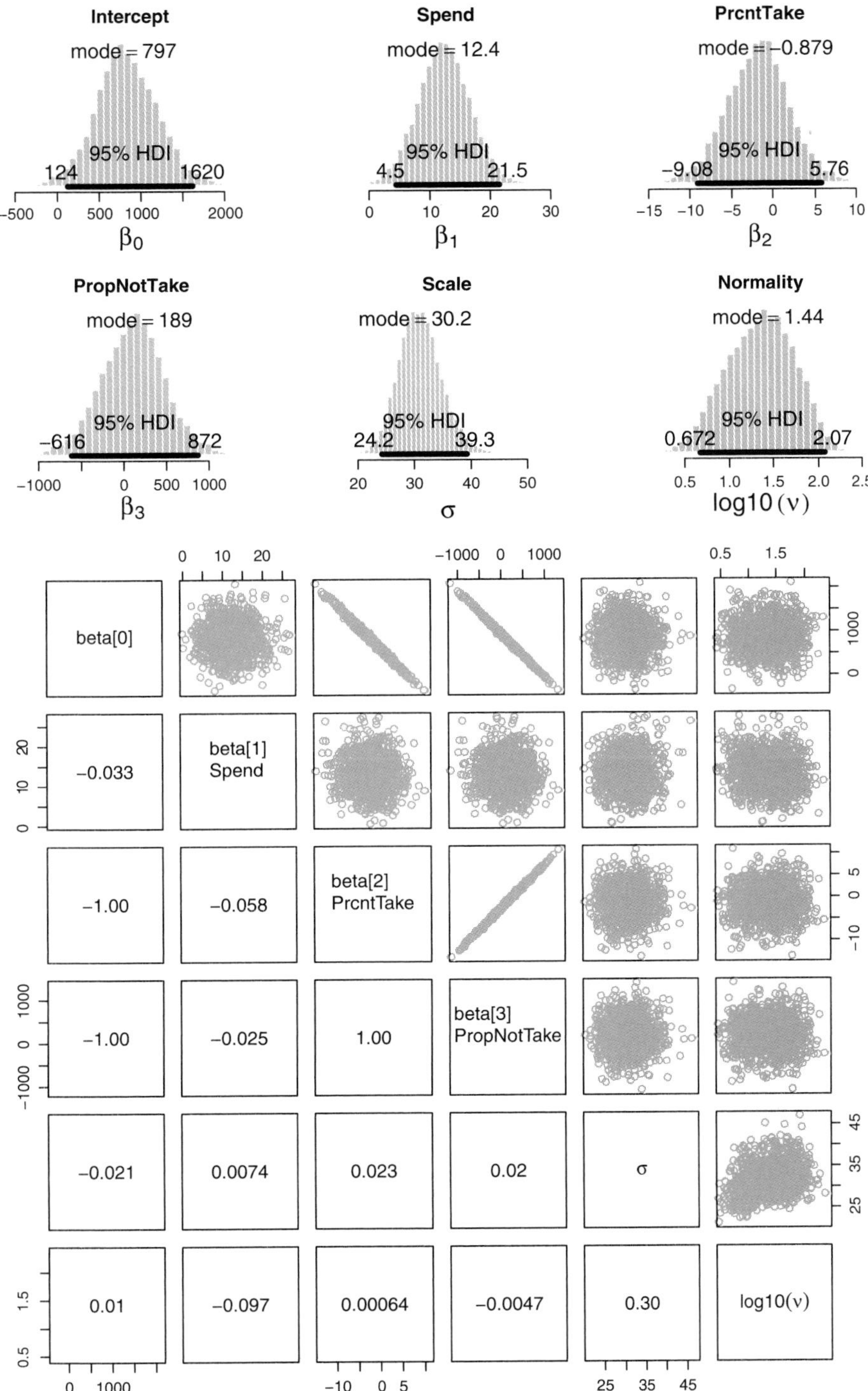

Figure 18.6 Posterior distribution for data in Figure 18.3 with a redundant predictor, the proportion of students not taking the exam. Compare with the result without a redundant predictor in Figure 18.5. Notice the perfect correlation between credible values of the regression coefficients on percentage taking the exam (PrcntTake) and proportion not taking the exam (PropNotTake). The posterior on the redundant predictors is strongly reflective of the prior distribution, which is shown in Figure 18.7.

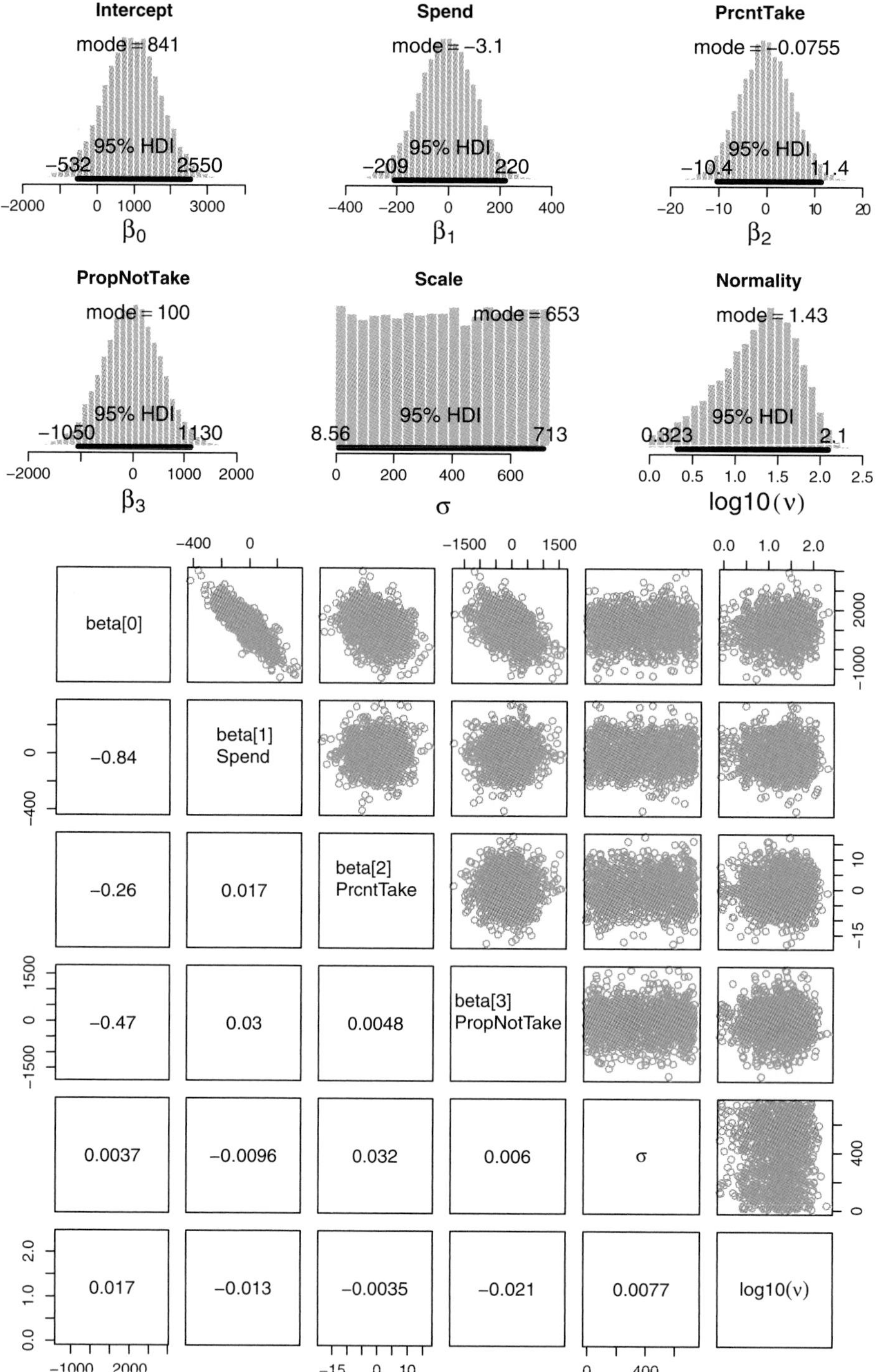

Figure 18.7 The prior distribution for the posterior distribution in Figure 18.6. Notice that the marginal posterior distributions of the redundant predictors (in Figure 18.6) is only a little narrower than the priors shown here.

distribution for this example, transformed to the original scale of the data.[2] Notice that the posterior distribution in Figure 18.6 has ranges for the redundant parameters that are only a little smaller than their priors. If the priors were wider, the posteriors on the redundant parameters would also be wider.

What should you do if you discover redundant predictors? If the predictors are perfectly correlated, then you can simply drop all but one, because the predictors are providing identical information. In this case, retain the predictor that is most relevant for interpreting the results. If the predictors are not perfectly correlated, but very strongly correlated, then there are various options directed at extracting an underlying common factor for the correlated predictors. One option is arbitrarily to create a single predictor that averages the correlated predictors. Essentially, each predictor is standardized, inverted as appropriate so that the standardized values have positive correlation, and then the average of the standardized values is used as the unitary predictor that represents all of the correlated predictors. More elaborate variations of this approach use principal components analysis. Finally, instead of creating a deterministic transform of the predictors, an underlying common factor can be estimated using factor analysis or *structural equation modeling* (SEM). These methods can be implemented in Bayesian software, of course, but go beyond the intended scope of this book. For an introduction to SEM in BUGS (hence easily convertible to JAGS and Stan), see the article by Song and Lee (2012). Another introductory example of Bayesian SEM is presented by Zyphur and Oswald (2013) but using the proprietary software Mplus.

18.1.5. Informative priors, sparse data, and correlated predictors

The examples in this book tend to use mildly informed priors (e.g., using information about the rough magnitude and range of the data). But a benefit of Bayesian analysis is the potential for cumulative scientific progress by using priors that have been informed from previous research.

Informed priors can be especially useful when the amount of data is small compared to the parameter space. A strongly informed prior essentially reduces the scope of the credible parameter space, so that a small amount of new data implies a narrow zone of credible parameter values. For example, suppose we flip a coin once and observe a head. If the prior distribution on the underlying probability of heads is vague, then the single datum leaves us with a broad, uncertain posterior distribution. But suppose we have

[2] The prior distribution was created in JAGS in a different way than explained in Section 8.5 (p. 211). In general, to get JAGS to sample from the prior, we give it empty data. In previous models we accomplished that by commenting out the data (`y`) in the `dataList`. But we cannot do that here because the model needs the `y` values for computing `sd(y)`, which is used for transforming parameters. Instead, we comment out the specification of standardized `zy` in the JAGS data block. The entire loop, `for ( i in 1:Ntotal ) { zy[i] <- ( y[i] - ym ) / ysd }`, is commented out.

prior knowledge that the coin is manufactured by a toy company that creates trick coins that either always come up heads or always come up tails. This knowledge constitutes a strongly informed prior distribution on the underlying probability of heads, with a spike of 50% mass at zero (always tails) and a spike of 50% mass at one (always heads). With this strong prior, the single datum yields a posterior distribution with complete certainty: 100% mass at one (always heads).

As another example of using strong prior information with sparse data, recall the linear regression of weight on height for 30 people in Figure 17.3 (p. 481). The marginal posterior distribution on the slope has a mode of about 4.5 and a fairly broad 95% HDI that extends from about 2.0 to 7.0. Furthermore, the joint posterior distribution on the slope and intercept shows a strong trade-off, illustrated in the scatter plot of the MCMC chain in Figure 17.3. For example, if the slope is about 1.0, then credible intercepts would have to be about +100, but if the slope is about 8.0, then credible intercepts would have to be about −400. Now, suppose that we have strong prior knowledge about the intercept, namely, that a person who has zero height has zero weight. This "knowledge" might seem to be a logical truism, but actually it does not make much sense because the example is referring to adults, none of whom have zero height. But we will ignore reality for this illustration and suppose that we know the intercept must be at zero. From the trade-off in credible intercepts and slopes, an intercept of zero implies that the slope must be very nearly 2.0. Thus, instead of a broad posterior distribution on the slopes that is centered near 4.5, the strong prior on the intercept implies a very narrow posterior distribution on the slopes that is centered near 2.0.

In the context of multiple linear regression, sparse data can lead to usefully precise posteriors on regression coefficients if some of the regression coefficients have informed priors *and* the predictors are strongly correlated. To understand this idea, it is important to remember that when predictors are correlated, their regression coefficients are also (anti-)correlated. For example, recall the SAT data from Figure 18.3 (p. 514) in which spending-per-pupil and percent-taking-the-exam are correlated. Consequently, the posterior estimates of the regression coefficients had a negative correlation, as shown in Figure 18.5 (p. 518). The correlation of credible regression coefficients implies that a strong belief about the value of one regression coefficient constrains the value of the other coefficient. Look carefully at the scatter plot of the two slopes shown in Figure 18.5. It can be seen that if we believe that the slope on percent-taking-the-exam is −3.2, then credible values of the slope on spending-per-pupil must be around 15, with an HDI extending roughly from 10 to 20. Notice that this HDI is smaller than the marginal HDI on spending-per-pupil, which goes from roughly 4 to 21. Thus, constraining the possibilities of one slope also constrains credible values of the other slope, because estimates of the two slopes are correlated.

That influence of one slope estimate on another can be used for inferential advantage when we have prior knowledge about one of the slopes. If some previous or auxiliary research informs the prior of one regression coefficient, that constraint can propagate to the estimates of regression coefficients on other predictors that are correlated with the first. This is especially useful when the sample size is small, and a merely mildly informed prior would not yield a very precise posterior. Of course, the informed prior on the first coefficient must be cogently justified. This might not be easy, especially in the context of multiple linear regression, where the inclusion of additional predictors can greatly change the estimates of the regression coefficients when the predictors are correlated. A robustness check also may be useful, to show how strong the prior must be to draw strong conclusions. If the information used for the prior is compelling, then this technique can be very useful for leveraging novel implications from small samples. An accessible discussion and example from political science is provided by Western and Jackman (1994), and a mathematical discussion is provided by Learner (1978, p. 175+).

18.2. MULTIPLICATIVE INTERACTION OF METRIC PREDICTORS

In some situations, the predicted value might not be an additive combination of the predictors. For example, the effects of drugs are often nonadditive. Consider the effects of two drugs, A and B. The effect of increasing the dose of drug B might be positive when the dose of drug A is small, but the effect of increasing drug B might be negative when the dose of drug A is large. Thus, the effects of the two drugs are not additive, and the effect of a drug depends on the level of the other drug. As another example, consider trying to predict subjective happiness from income and health. If health is low, then an increase in income probably has only a small effect. But if health is high, then an increase from low income to high income probably has a large effect. Thus, the effects of the two factors are not additive, and the effect of one factor depends on the level of the other factor.

Formally, interactions can have many different specific functional forms. We will consider *multiplicative* interaction. This means that the nonadditive interaction is expressed by multiplying the predictors. The predicted value is a weighted combination of the individual predictors and, additionally, the multiplicative product of the predictors. For two metric predictors, regression with multiplicative interaction has these algebraically equivalent expressions:

$$\mu = \beta_0 + \beta_1 x_1 + \beta_2 x_2 + \beta_{1\times 2} x_1 x_2 \tag{18.2}$$

$$= \beta_0 + \underbrace{(\beta_1 + \beta_{1\times 2} x_2)}_{\text{slope of } x_1} x_1 + \beta_2 x_2 \tag{18.3}$$

$$= \beta_0 + \beta_1 x_1 + \underbrace{(\beta_2 + \beta_{1\times 2} x_1)}_{\text{slope of } x_2} x_2 \tag{18.4}$$

These three expressions emphasize different interpretations of interaction, as illustrated in Figure 18.8. The form of Equation 18.2 is illustrated in the left panel of Figure 18.8. The vertical arrows show that the curved-surface interaction is created by adding the product, $\beta_{1\times 2}x_1x_2$, to the planar linear combination.

The form of Equation 18.3 is illustrated in the middle panel of Figure 18.8. Its dark lines show that the slope in the x_1 direction depends on the value of x_2. In particular, when $x_2 = 0$, then the slope along x_1 is $\beta_1 + \beta_{1\times 2}x_2 = -1 + 0.2 \cdot 0 = -1$. But when $x_2 = 10$, then the slope along x_1 is $\beta_1 + \beta_{1\times 2}x_2 = -1 + 0.2 \cdot 10 = +1$. Again, the slope in the x_1 direction changes when x_2 changes, and β_1 only indicates the slope along x_1 when $x_2 = 0$.

The form of Equation 18.4 is illustrated in the right panel of Figure 18.8. It shows that the interaction can be expressed as the slope in the x_2 direction changing when x_1 changes. (Exercise 18.1 has you compute the numerical slopes.) This illustration is exactly analogous to the middle panel of Figure 18.8, but with the roles of x_1 and x_2 exchanged. It is important to realize, and visualize, that the interaction can be expressed in terms of the slopes on either predictor.

Great care must be taken when interpreting the coefficients of a model that includes interaction terms (Braumoeller, 2004). In particular, low-order terms are especially difficult to interpret when higher-order interactions are present. In the simple two-predictor case, the coefficient β_1 describes the influence of predictor x_1 *only* at $x_2 = 0$, because the slope on x_1 is $\beta_1 + \beta_{1\times 2}x_2$, as was shown in Equation 18.3 and graphed in the middle panel of Figure 18.8. In other words, it is not appropriate to say that β_1 indicates the *overall* influence of x_1 on y. Indeed, in many applications, the value of x_2

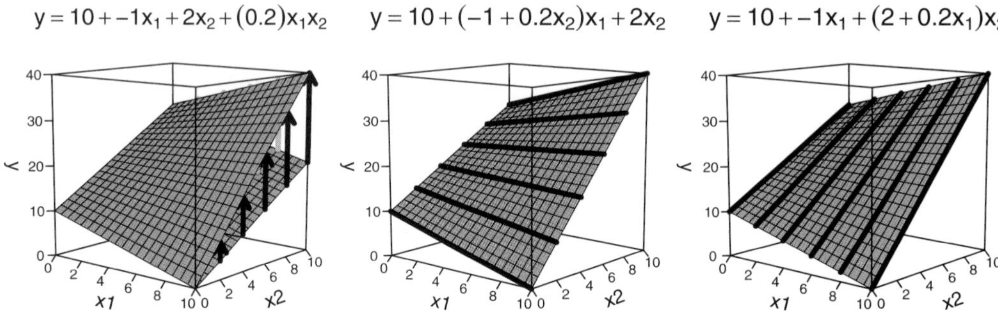

Figure 18.8 A multiplicative interaction of x_1 and x_2 parsed three ways. The left panel emphasizes that the interaction involves a multiplicative component that adds a vertical amount to the planar additive model, as indicated by the arrows that mark $\beta_{1\times 2}x_1x_2$. The middle panel shows the same function, but with the terms algebraically re-grouped to emphasize that the slope in the x_1 direction depends on the value of x_2, as shown by the darkened lines that mark $\beta_1 + \beta_{1\times 2}x_2$. The right panel again shows the same function, but with the terms algebraically re-grouped to emphasize that the slope in the x_2 direction depends on the value of x_1, as shown by the darkened lines that mark $\beta_2 + \beta_{1\times 2}x_1$. Compare with Figure 15.3 (p. 428).

never realistically gets close to zero, and therefore, β_1 has no realistic interpretation at all. For example, suppose we are predicting college GPA (y) from parental income (x_1) and high-school GPA (x_2). If there is interaction, then the regression coefficient, β_1, on parental income, only indicates the slope on x_1 *when* x_2 *(GPA) is zero*. Of course, there are no GPAs of zero, and therefore, β_1 by itself is not very informative.

18.2.1. An example

To estimate the parameters of a model with multiplicative interaction, we could create a new program in JAGS or Stan that takes the unique predictors as input and then multiplies them internally for the desired interactions. This approach would be conceptually faithful because it maintains the idea that there are two predictors and the model combines them nonadditively. But instead of creating a new program, we will use the previously applied additive (noninteraction) model by inventing another predictor that expresses the product of the individual predictors. To do this, we conceptualize the interaction term of Equation 18.2 as an additional additive predictor, like this:

$$\mu = \beta_0 + \beta_1 x_1 + \beta_2 x_2 + \underbrace{\beta_{1\times 2}}_{\beta_3} \underbrace{x_1 x_2}_{x_3} \tag{18.5}$$

We create the new variable $x_3 = x_1 x_2$ outside the model and then submit the new variable as if it were another additive predictor. One benefit of this approach is that we do not have to create a new model, and it is easy, in cases of many predictors, to set up interaction variables for many different combinations of variables. Another key benefit is that we can examine the correlations of the single predictors with the interaction variables. Often the single variables will be correlated with the interaction variables, and therefore, we can anticipate trade-offs in the estimated parameter values that widen the marginal posterior distributions on single parameters.

To illustrate some of the issues involved in interpreting the parameters of a model with interaction, consider again the SAT data from Figure 18.3. Recall that the mean SAT score in a state was predicted from the spending per pupil (Spend) and the percentage of students who took the test (PrcntTake). When no interaction term was included in the model, the posterior distribution looked like Figure 18.5, which indicated a positive influence of Spend and a negative influence of PrcntTake.

We will include a multiplicative interaction of Spend and PrcntTake. Does it make sense that the effect of spending might depend on the percentage of students taking the test? Perhaps yes, because if very few students are taking the test, they are probably already at the top of the class and therefore might not have as much head-room for increasing their scores if more money is spent on them. In other words, it is plausible that the effect of spending is larger when the percentage of students taking the test is larger, and we would not be surprised if there were a positive interaction between those predictors. Therefore, it is theoretically meaningful to include an interaction term in the model.

The computer code for this example is in one of the sections of the file Jags-Ymet-XmetMulti-Mrobust-Example.R. The commands read the data and then create a new variable that is appended as another column on the data frame, after which the relevant column names are specified for the analysis:

```
# Read in data:
myData = read.csv( file="Guber1999data.csv" )
# Append the new interaction variable:
myData = cbind( myData , SpendXPrcnt = myData[,"Spend"]  * myData[,"PrcntTake"] )
# Specify names of data columns to use in the analysis:
yName = "SATT" ; xName = c("Spend","PrcntTake","SpendXPrcnt")
```

When the analysis is run, the first thing it does is display the correlations of the predictors:

```
CORRELATION MATRIX OF PREDICTORS:
            Spend PrcntTake SpendXPrcnt
Spend       1.000     0.593       0.775
PrcntTake   0.593     1.000       0.951
SpendXPrcnt 0.775     0.951       1.000
```

We can see that the interaction variable is strongly correlated with both predictors. Therefore, we know that there will be strong trade-offs among the regression coefficients, and the marginal distributions of single regression coefficients might be much wider than when there was no interaction included.

When we incorporate a multiplicative interaction into the model, the posterior looks like Figure 18.9. The marginal distribution for β_3, also labeled as SpendXPrcnt, indicates that the modal value of the interaction coefficients is indeed positive, as we anticipated it could be. However, the 95% HDI includes zero, which indicates that we do not have very strong precision in the estimate of the magnitude of the interaction.

Notice that the inclusion of the interaction term has changed the apparent marginal distributions for the regression coefficients on Spend and PrcntTake. In particular, the regression coefficient on Spend now clearly includes zero. This might lead a person, inappropriately, to conclude that there is not a credible influence of spending on SAT scores, because zero is among the credible values of β_1. This conclusion is not appropriate because β_1 only indicates the slope on spending *when the percentage of students taking the test is zero*. The slope on Spend depends on the value of PrcntTake because of the interaction.

To properly understand the credible slopes on the two predictors, we must consider the credible slopes on each predictor as a function of the value of the other predictor.

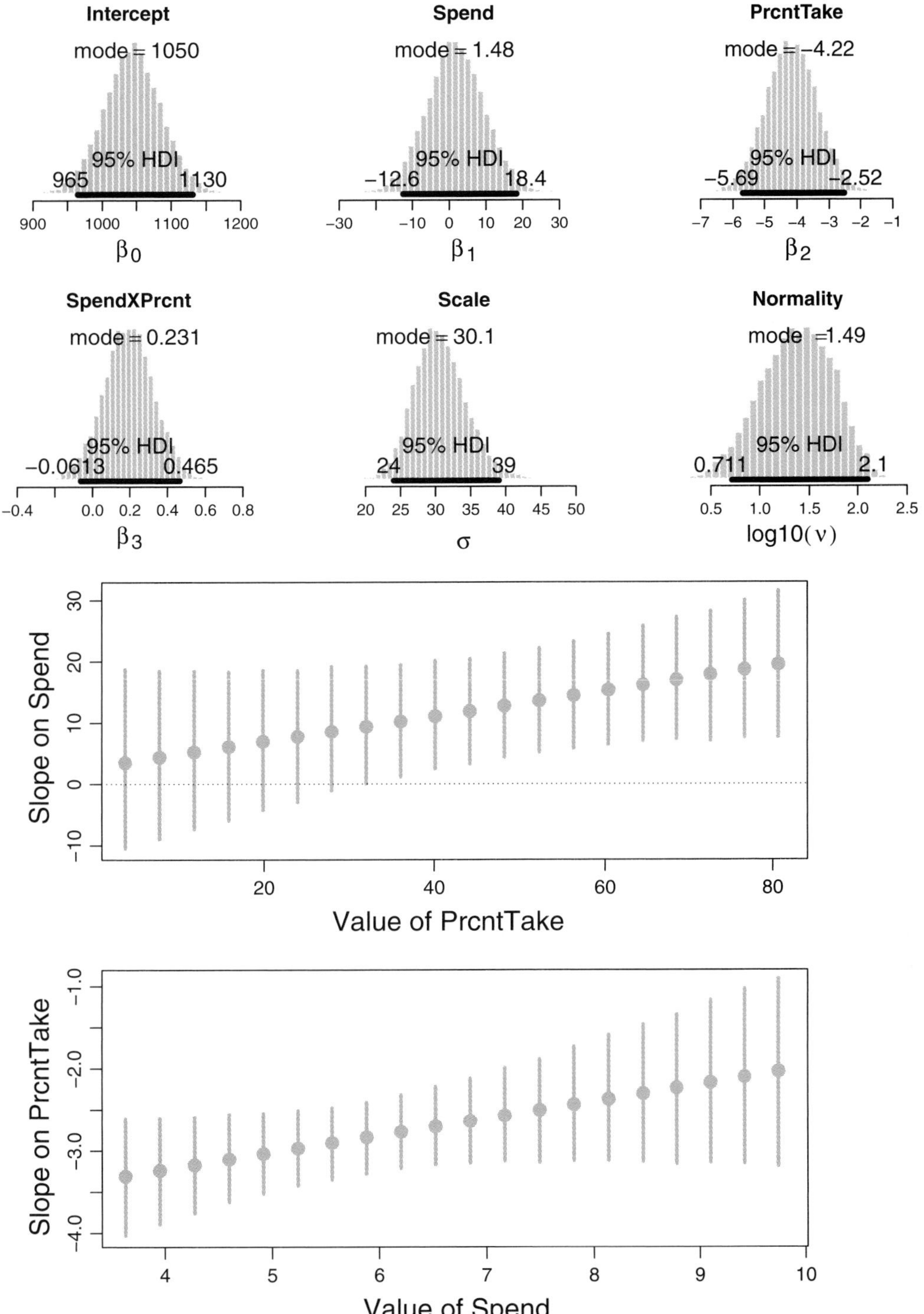

Figure 18.9 Posterior distribution when including a multiplicative interaction of Spend and PrcntTake. The marginal distribution of β_1 is the slope on Spend when PrcntTake=0, and the marginal distribution of β_2 is slope on PrcntTake when Spend=0. Lower panels show 95% HDIs and median values of slopes for other values of predictors. Slope on Spend is $\beta_1 + \beta_3 \cdot$ PrcntTake and slope on PrcntTake is $\beta_2 + \beta_3 \cdot$ Spend.

Recall from Equations 18.3 that the slope on x_1 is $\beta_1 + \beta_{1\times 2}x_2$. Thus, for the present application, the slope on Spend is $\beta_1 + \beta_3 \cdot \text{PrcntTake}$ because $\beta_{1\times 2}$ is β_3 and x_2 is PrcntTake. Thus, for any particular value of PrcntTake, we get the distribution of credible slopes on Spend by stepping through the MCMC chain and computing $\beta_1 + \beta_3 \cdot \text{PrcntTake}$ at each step. We can summarize the distribution of slopes by its median and 95% HDI. We do that for many candidate values of PrcntTake, and the result is plotted in the middle panel of Figure 18.9. You can see that when PrcntTake is large, the credible slopes on Spend clearly exceed zero. You can also mentally extrapolate that when PrcntTake is zero, the median and HDI will match the marginal distribution of β_1 shown in the top of Figure 18.9.

The bottom panel of Figure 18.9 shows the credible slopes on PrcntTake for particular values of Spend. At each step in the MCMC chain, a credible slope was computed as $\beta_2 + \beta_3 \cdot \text{Spend}$. You can see that the median slope on PrcntTake is not constant but depends on the value of Spend. This dependency of the effect of one predictor on the level of the other predictor is the meaning of interaction.

In summary, when there is interaction, then the influence of the individual predictors can *not* be summarized by their individual regression coefficients alone, because those coefficients only describe the influence when the other variables are at zero. A careful analyst considers credible slopes across a variety of values for the other predictors, as in Figure 18.9. Notice that this is true even though the interaction coefficient did not exclude zero from its 95% HDI. In other words, if you include an interaction term, you cannot ignore it even if its marginal posterior distribution includes zero.

18.3. SHRINKAGE OF REGRESSION COEFFICIENTS

In some research, there are many candidate predictors which we suspect could possibly be informative about the predicted variable. For example, when predicting college GPA, we might include high-school GPA, high-school SAT score, income of student, income of parents, years of education of the parents, spending per pupil at the student's high school, student IQ, student height, weight, shoe size, hours of sleep per night, distance from home to school, amount of caffeine consumed, hours spent studying, hours spent earning a wage, blood pressure, etc. We can include all the candidate predictors in the model, with a regression coefficient for every predictor. And this is not even considering interactions, which we will ignore for now.

With so many candidate predictors of noisy data, there may be some regression coefficients that are spuriously estimated to be non-zero. We would like some protection against accidentally nonzero regression coefficients. Moreover, if we are interested in *explaining* variation in the predicted variable, we would like the description of the data to emphasize the predictors that are most clearly related to variation in the predicted variable. In other words, we would like the description to de-emphasize weak or spurious predictors.

One way to implement such a description is by using a *t* distribution for the prior on the regression coefficients. By setting its mean to zero, its normality parameter to a small value, and its scale parameter to a moderately small value, the *t*-distributed prior dictates that regression coefficients should probably be near zero, where the narrow peak of the *t* distribution is. But if a regression coefficient is clearly nonzero, then it could be large, as is allowed by the heavy tail of the *t*-distributed prior.

Figure 18.10 shows a diagram of a multiple linear regression model that has a *t*-distributed prior on the regression coefficients. Compare the diagram with the one in Figure 18.4 (p. 515), and you will see that the only difference from before is the prior on the regression coefficients. The empty braces in the top of Figure 18.10 refer to optional aspects of the model. As was mentioned in the previous paragraph, the normality parameter ν_β and the scale parameter σ_β could be set to constants, in which

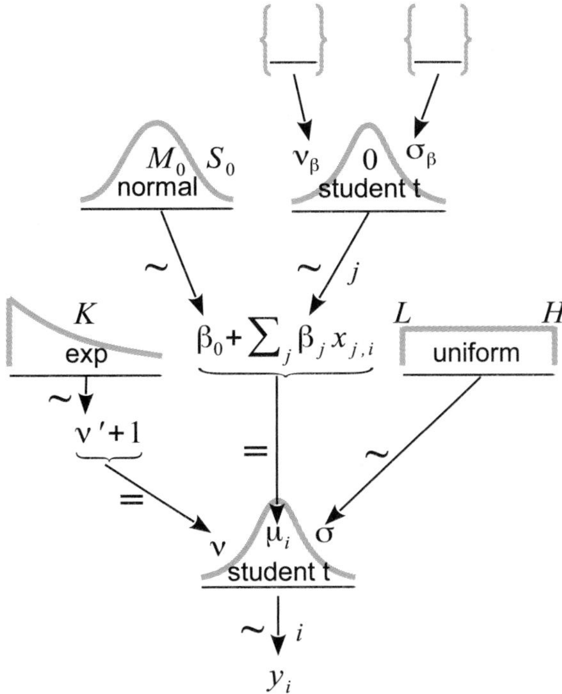

Figure 18.10 Hierarchical diagram for multiple linear regression, with a shrinkage prior across the slope coefficients. Compare with Figure 18.4 (p. 515). The empty braces at the top of the diagram indicate aspects that are optional. Typically the normality parameter ν_β is fixed at a small value, but could be estimated instead. The scale parameter σ_β could be fixed at a small value but could be estimated, in which case the standard deviation across regression coefficients is mutually informed by all the predictors.

case the braces as the top of the diagram would enclose constants and the arrows would be labeled with an equal sign. When the prior has constants, it is sometimes called a *regularizer* for the estimation.

The *t*-distributed prior is just one way to express the notion that regression coefficients near zero should be preferred but with larger coefficients allowed. Alternatively, a *double exponential* distribution could be used. A double exponential is simply an exponential distribution on both $+\beta$ and $-\beta$, equally spread. The double exponential has a single-scale parameter (and no shape parameter). The double exponential is built into JAGS and Stan. A well-known regularization method called *lasso regression* uses a double exponential weighting on the regression coefficients. For a nice explanation of lasso regression in a Bayesian setting, see Lykou and Ntzoufras (2011).

Should the scale parameter (i.e., σ_β in Figure 18.10) be fixed at a constant value or should it be estimated from the data? If it is fixed, then every regression coefficient experiences the same fixed regularization, independently from all the other regression coefficients. If the scale parameter is estimated, then the variability of the estimated regression coefficients across predictors influences the estimate of the scale parameter, which in turn influences all the regression coefficients. In particular, if most of the regression coefficients are estimated to be near zero, then the scale parameter is estimated to be small, which further shrinks the estimates of the regression coefficients.

Neither one of these approaches (using fixed σ_β or estimated σ_β) is inherently "correct." The approaches express different prior assumptions. If the model estimates σ_β (instead of fixing it), the model is assuming that all the regression coefficients are mutually representative of the variability across regression coefficients. In applications where there are many predictors of comparable status, this assumption may be quite realistic. A minimal requirement, for putting all the regression coefficients under a shared overarching distribution with an estimated scale, is that the predictors were selected from some implicit set of reasonably likely predictors, and therefore, we can think of the overarching distribution as reflecting that set. In applications where there are relatively few predictors of different types, this assumption might not be appropriate. Beware of convenience priors that are used in routinized ways.

To illustrate these ideas with a concrete example, consider again the SAT data of Figure 18.3, but now supplemented with 12 more randomly generated predictors. The x values were randomly and independently drawn from a normal distribution centered at zero, and therefore, any correlation between the predictors and any nonzero regression coefficient is an accident of random sampling. We will first apply the simple model of Figure 18.4, for which the regression coefficients have fixed, independent, vague normal priors. The resulting posterior distribution is shown in Figure 18.11. To save space, the results for random predictors 4-9 (xRand4–xRand9) have not been displayed. Attend specifically to the distributions for the regression coefficients on spending (Spend) and random predictor 10 (xRand10). The coefficient on Spend is still

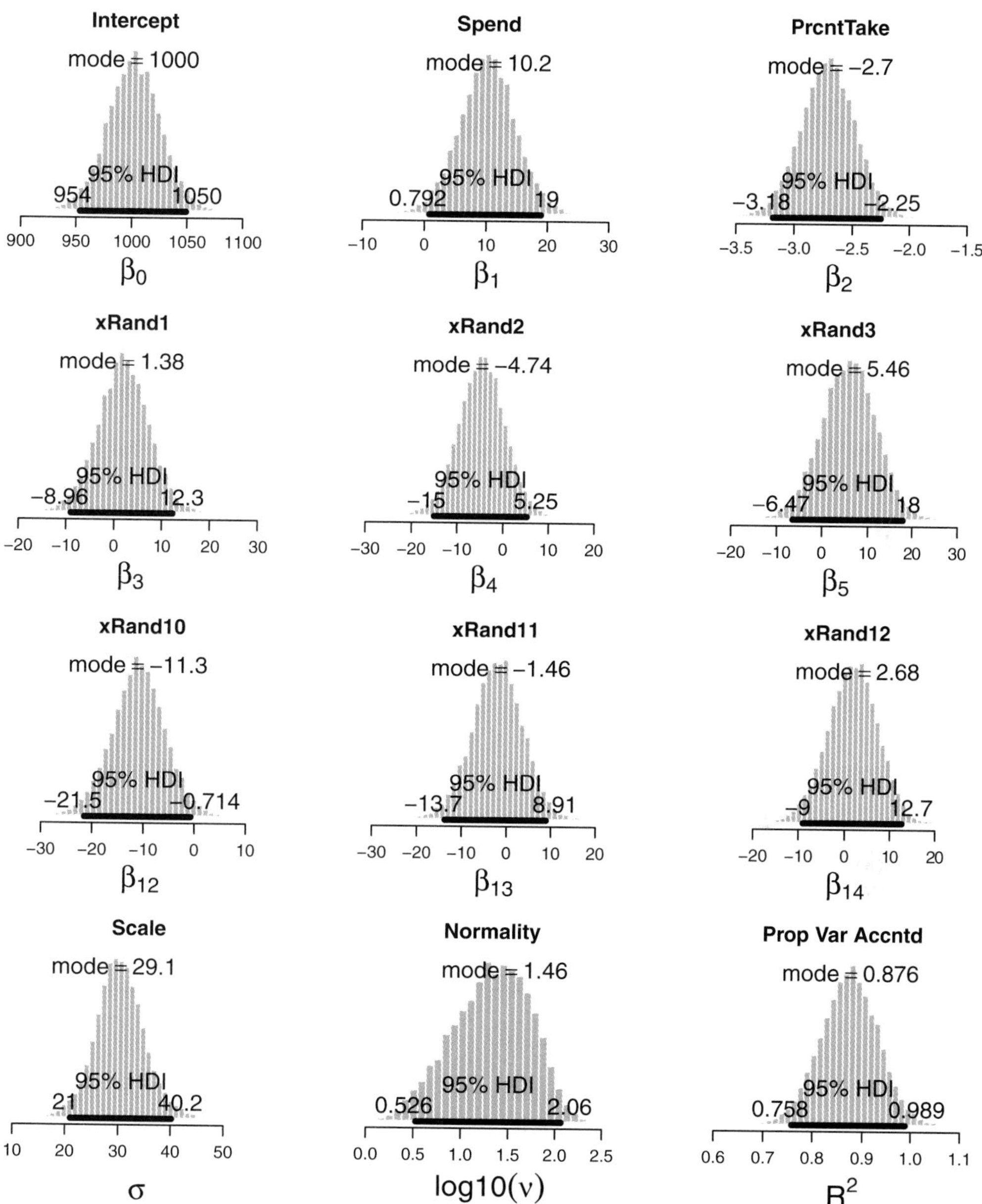

Figure 18.11 Posterior with*out* hierarchical shrinkage, using prior of Figure 18.4. Compare with results when using shrinkage prior in Figure 18.12, especially the coefficients on Spend and xRand10.

estimated to be positive and its 95% HDI falls above zero, but only barely. The inclusion of additional predictors and their parameters has reduced the certainty of the estimate. The coefficient on xRand10 is negative, with its 95% HDI falling below zero, to about the same extent as Spend fell above zero. This apparent relation of xRand10 with SAT score is spurious, an accident of random sampling. We know that the apparently nonzero relation of xRand10 with SAT score is spurious only because we generated the data. For data collected from nature, such as spending and SAT scores, we cannot know which estimates are spurious and which are genuine. Bayesian analysis tells us the best inference we can make, given the data and our choice of descriptive model.

Now we repeat the analysis using the hierarchical model of Figure 18.10, with ν_β fixed at one (i.e., heavy tails), and with σ_β given a gamma prior that has mode 1.0 and standard deviation 1.0 (i.e., broad for the standardized data). The programs for this analysis are in files `Jags-Ymet-XmetMulti-MrobustShrink.R` and `Jags-Ymet-XmetMulti-MrobustShrink-Example.R`. The resulting posterior distribution is shown in Figure 18.12. Notice that the marginal distribution for the regression coefficient on xRand10 is now shifted so that its 95% HDI covers zero. The estimate is shrunken toward zero because many predictors are telling the higher-level *t* distribution that their regression coefficients are near zero. Indeed, the estimate of σ_β (not displayed) has its posterior mode around 0.05, even though its prior mode was at 1.0. The shrinkage also causes the estimate of the regression coefficient on Spend to shift toward zero. Thus, the shrinkage has suppressed a spurious regression coefficient on xRand10, but it has also suppressed what might be a real but small regression coefficient on Spend. Notice, however, that the marginal distribution for the coefficient on PrcntTake has *not* been much affected by shrinkage, presumably because it is big enough that it falls in the tail of the *t* distribution where the prior is relatively flat.

The shrinkage is desirable not only because it shares information across predictors (as expressed by the hierarchical prior) but also because it rationally helps control for false alarms in declaring that a predictor has a nonzero regression coefficient. As the example in Figure 18.11 showed, when there are many candidate predictors, some of them may spuriously appear to have credibly nonzero regression coefficients even when their true coefficients are zero. This sort of false alarm is unavoidable because data are randomly sampled, and there will be occasional coincidences of data that are unrepresentative. By letting each regression coefficient be informed by the other predictors, the coefficients are less likely to be spuriously distorted by a rogue sample.

Finally, notice in Figure 18.12 that the marginal posterior distributions on many of the regression coefficients are (upside-down) funnel shaped, each with a pointy peak near zero and long concave tails, like this: . You can imagine that the posterior distribution from the nonshrinkage model, which is gently rounded at its peak, was pinched at a point on its top edge and lifted over toward zero, like a mommy cat carrying a kitten by the scruff of its neck back to its bed. A funnel shape is characteristic of a posterior

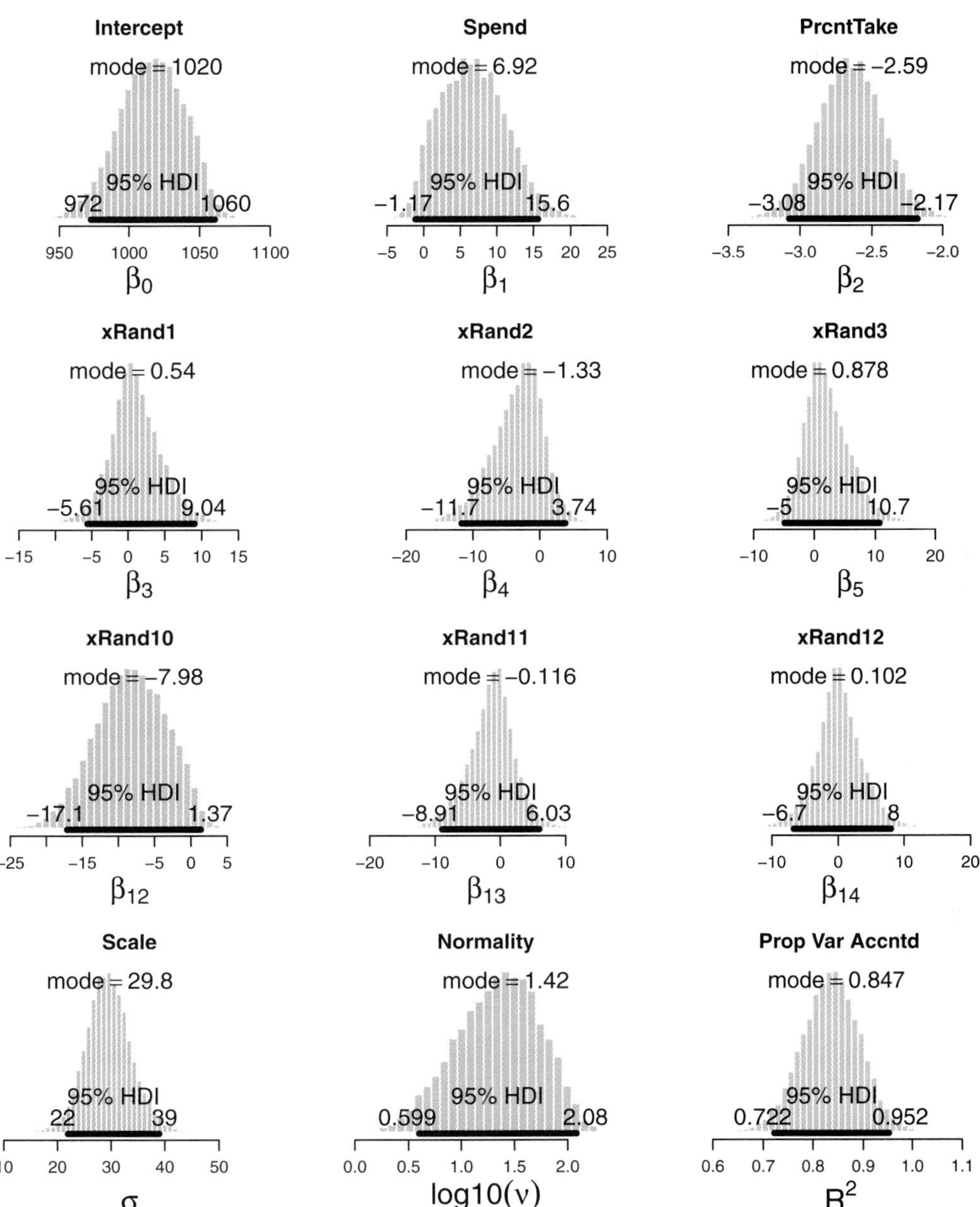

Figure 18.12 Posterior *with* hierarchical shrinkage, using the hierarchical prior of Figure 18.10 with a gamma distribution (mode=1.0, sd=1.0) on standardized σ_β and $\nu_\beta = 1$. Compare with the results when not using hierarchical shrinkage in Figure 18.11, especially the coefficients on Spend and xRand10.

distribution experiencing strong shrinkage. (We have previously seen examples of this, for example way back in Figure 9.10, p. 243. Another example is shown in Exercise 19.1.) If a marginal posterior distribution is displayed only by a dot at its central tendency and a segment for its HDI, without a profile for its shape, then this signature of shrinkage is missed.

18.4. VARIABLE SELECTION

The motivation of the previous section was an assumption that many predictors might have weak predictive value relative to the noise in the data, and therefore, shrinkage would be appropriate to stabilize the estimates. In some applications, it may be theoretically meaningful to suppose that a predictor has literally zero predictive value. In this case, the issue is not estimating a presumed weak predictiveness relative to noise; instead, the issue is deciding whether there is any predictiveness at all. This is almost antithetical to including the predictor in the first place, because including it means that we had some prior belief that the predictor was relevant. Nevertheless, we might want to estimate the credibility that the predictor should be included, in combination with various subsets of other predictors. Deciding which predictors to include is often called *variable selection*.

Some prominent authors eschew the variable-selection approach for typical applications in their fields. For example, Gelman et al. (2013, p. 369) said, "For the regressions we typically see, we do not believe any coefficients to be truly zero and we do not generally consider it a *conceptual* (as opposed to computational) advantage to get point estimates of zero—but regularized estimates such as obtained by lasso can be much better than those resulting from simple least squares and flat prior distributions ...we are not comfortable with an underlying model in which the coefficients can be exactly zero." Other researchers take it for granted, however, that some form of variable selection must be used to make sense of their data. For example, O'Hara and Sillanpää (2009, p. 86) said, "One clear example where this is a sensible way to proceed is in gene mapping, where it is assumed that there are only a small number of genes that have a large effect on a trait, while most of the genes have little or no effect. The underlying biology is therefore sparse: only a few factors (i.e. genes) are expected to influence the trait." They later said, however, "In any real data set, it is unlikely that the 'true' regression coefficients are either zero or large; the sizes are more likely to be tapered towards zero. Hence, the problem is not one of finding the zero coefficients, but of finding those that are small enough to be insignificant, and shrinking them towards zero "(O'Hara & Sillanpää, 2009, p. 95)." Thus, we are entertaining a situation in which there are many candidate predictors that may genuinely have zero real relation to the predicted value or have relations small enough to be counted as insignificant. In this situation, a reasonable question is, which predictors can be credibly included in the descriptive model?

This section introduces some basic ideas and methods of Bayesian variable selection, using some simple illustrative examples. The topic is extensively studied and undergoing rapid development in the literature. The examples presented here are intended to reveal some of the foundational concepts and methods, not to serve as a comprehensive reference for the latest and greatest methods. After studying this section, please see the various cited references and other literature for more details.

There are various methods for Bayesian variable selection (see, e.g., O'Hara & Sillanpää, 2009; Ntzoufras, 2009). The key to models of variable selection (as opposed to shrinkage) is that each predictor has both a regression coefficient and an inclusion indicator, which can be thought of as simply another coefficient that takes on the values 0 or 1. When the inclusion indicator is 1, then the regression coefficient has its usual role. When the inclusion indicator is 0, the predictor has no role in the model and the regression coefficient is superfluous.

To formalize this idea, we modify the basic linear regression equation with a new parameter $\delta_j \in \{0, 1\}$, which is the inclusion indicator for the jth predictor. The predicted mean value of y is then given by

$$\mu_i = \beta_0 + \sum_j \delta_j \beta_j x_{j,i} \tag{18.6}$$

Every combination of δ_j values, across the predictors, constitutes a distinct model of the data. For example, if there are four predictors, then $\langle \delta_1, \delta_2, \delta_3, \delta_4 \rangle = \langle 1, 1, 1, 1 \rangle$ is a model that includes all four predictors, and $\langle \delta_1, \delta_2, \delta_3, \delta_4 \rangle = \langle 0, 1, 0, 1 \rangle$ is a model that includes only the second and fourth predictors, and so on. With four predictors, there are $2^4 = 16$ possible models.

A simple way to put a prior on the inclusion indicator is to have each indicator come from an independent Bernoulli prior, such as $\delta_j \sim \text{dbern}(0.5)$. The constant in the prior affects the prior probability of models with more or fewer predictors included. With a prior inclusion bias of 0.5, all models are equally credible, *a priori*. With a prior inclusion bias less than 0.5, models with less than half the predictors included are *a priori* more credible than models with more than half the predictors.

It is trivial to incorporate the inclusion parameter in a JAGS model specification, as will shown presently.[3] As in all the regression models we have used, the data are standardized in the initial `data` block of the JAGS code, which is not repeated here. Recall that the standardized data are denoted as `zx[i,1:Nx]`, where the index `i` denotes the ith individual and where `Nx` indicates the number of predictors. The standardized regression coefficients are denoted as `zbeta[1:Nx]`, and the new inclusion indicators

[3] The method for variable selection demonstrated in this section, using discrete inclusion indicators, cannot be directly implemented in Stan because Stan does not allow discrete parameters (in its present version). But Stan can be used for hierarchical shrinkage models, which involve continuous parameters.

are denoted as delta[1:Nx]. The model specification (which you may compare with the model specification in Section 18.1.2, p. 514) is as follows:

```
model {
  for ( i in 1:Ntotal ) {
    zy[i] ~ dt( zbeta0 + sum( delta[1:Nx] * zbeta[1:Nx] * zx[i,1:Nx] ) ,
                1/zsigma^2 , nu )
  }
  # Priors vague on standardized scale:
  zbeta0 ~ dnorm( 0 , 1/2^2 )
  for ( j in 1:Nx ) {
    zbeta[j] ~ dnorm( 0 , 1/2^2 )
    delta[j] ~ dbern( 0.5 )
  }
  zsigma ~ dunif( 1.0E-5 , 1.0E+1 )
  nu <- nuMinusOne+1
  nuMinusOne ~ dexp(1/29.0)
  # Transform to original scale:
  beta[1:Nx] <- ( delta[1:Nx] * zbeta[1:Nx] / xsd[1:Nx] )*ysd
  beta0 <- zbeta0*ysd  + ym - sum( delta[1:Nx] * zbeta[1:Nx] * xm[1:Nx]
                                   / xsd[1:Nx] )*ysd
  sigma <- zsigma*ysd
}
```

There are only three lines above that use the new inclusion parameter. The first usage is in the specification of the likelihood for zy[i]. Instead of the predicted mean involving sum(zbeta[1:Nx] * zx[i,1:Nx]), it uses sum(delta[1:Nx] * zbeta[1:Nx] * zx[i,1:Nx]). The second appearance of the inclusion parameter is the specification of its Bernoulli prior. Finally, near the end of the specification above, the transformation from standardized to original scale also incorporates the inclusion indicators. This is necessary because zbeta[j] is irrelevant when it is not used to model the data.

As a first example of applying the variable-selection method, recall the SAT data of Figure 18.3 and the posterior distribution shown in Figure 18.5. For each of 50 states, the average SAT score was regressed on two predictors: average spending per pupil (Spend) and percentage of students who took the test (PrcntTake). With two predictors, there are four possible models involving different subsets of predictors. Because the prior inclusion probability was set at 0.5, each model had a prior probability of $0.5^2 = 0.25$.

Figure 18.13 shows the results. Of the four possible models, only two had a non-negligible posterior probability, namely the model that included both predictors and the model that included only PrcntTake. As shown in the upper panel of Figure 18.13, the model with both predictors has a posterior probability of about 70%. This value is simply the number of times that the MCMC chain had $\langle \delta_1, \delta_2 \rangle = \langle 1, 1 \rangle$ divided by the total number of steps in the chain. Like any MCMC estimate, it is based on

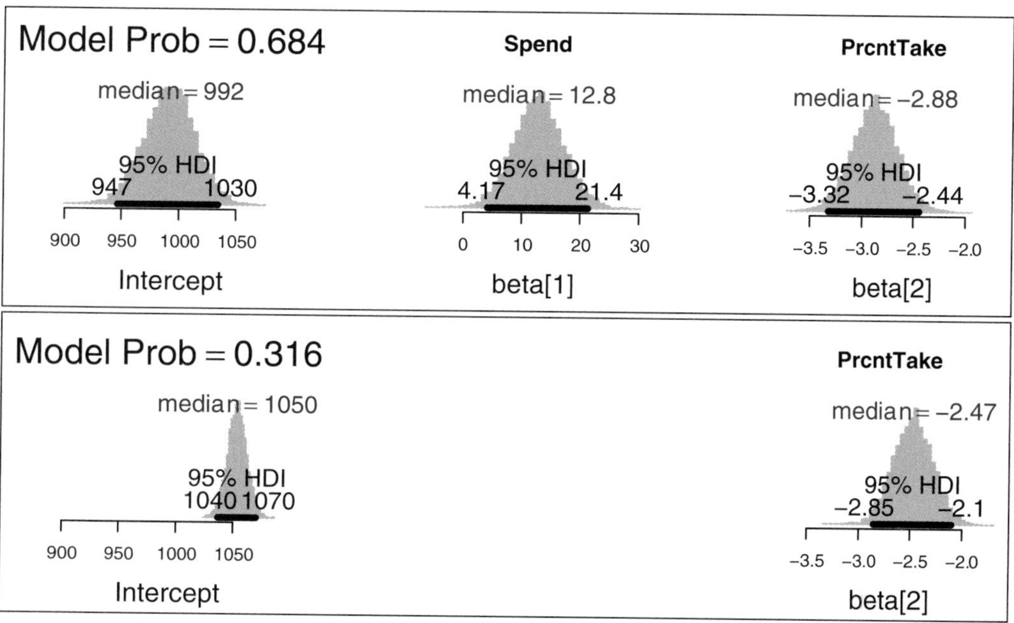

Figure 18.13 Posterior probabilities of different subsets of predictors along with the marginal posterior distributions of the included regression coefficients. The two other possible models, involving only Spend or only the intercept, had essentially zero probability. The prior probability of each model was $0.5^2 = 0.25$.

a random sample and will be somewhat different on different runs. As shown in the lower panel of Figure 18.13, the model with only PrcntTake has a posterior probability of about 30%. This value is simply the number of times that the MCMC chain had $\langle\delta_1, \delta_2\rangle = \langle 0, 1\rangle$ divided by the total number of steps in the chain. Thus, the model involving both predictors is more than twice as credible as the model involving only one predictor.

Figure 18.13 also shows the marginal posterior distributions of the included regression coefficients. These are credible values taken only from the corresponding steps in the chain. Thus, the histograms in the upper panel involve only ≈70% of the chain for which $\langle\delta_1, \delta_2\rangle = \langle 1, 1\rangle$, and the histograms in the lower panel involve only the ≈30% of the chain for which $\langle\delta_1, \delta_2\rangle = \langle 0, 1\rangle$. Notice that the parameter estimates are different for different models. For example, the estimate of the intercept is quite different for different included predictors.

18.4.1. Inclusion probability is strongly affected by vagueness of prior

We will now see that the degree of vagueness of the prior on the regression coefficient can have an enormous influence on the inclusion probability, even though the degree of vagueness has little influence on the estimate of the regression coefficient itself.

Recall that the prior in the model was specified as a generic broad distribution on the standardized regression coefficients, like this:

```
model {
  ...
  # Priors vague on standardized scale:
  zbeta0 ~ dnorm( 0 , 1/2^2 )     # SD=2
  for ( j in 1:Nx ) {
    zbeta[j] ~ dnorm( 0 , 1/2^2 ) # SD=2
    delta[j] ~ dbern( 0.5 )
  }
  ...
}
```

The choice of SD=2 was arbitrary but reasonable because standardized regression coefficients cannot exceed ±1 in least-squares regression. When running the model on the SAT data with two candidate predictors, the result was shown in Figure 18.13 (p. 539).

We now re-run the analysis using different arbitrary degrees of vagueness on the priors for the standardized regression coefficients. We will illustrate with SD=1, like this:

```
model {
  ...
  # Priors vague on standardized scale:
  zbeta0 ~ dnorm( 0 , 1/1^2 )     # SD=1
  for ( j in 1:Nx ) {
    zbeta[j] ~ dnorm( 0 , 1/1^2 ) # SD=1
    delta[j] ~ dbern( 0.5 )
  }
  ...
}
```

and with SD=10, like this:

```
model {
  ...
  # Priors vague on standardized scale:
  zbeta0 ~ dnorm( 0 , 1/10^2 )     # SD=10
  for ( j in 1:Nx ) {
    zbeta[j] ~ dnorm( 0 , 1/10^2 ) # SD=10
    delta[j] ~ dbern( 0.5 )
  }
  ...
}
```

Notice that the prior probability on the inclusion parameters has not changed. The prior inclusion probability is always 0.5 in these examples.

Figure 18.14 shows the results. The upper pair of panels shows the posterior probabilities when the prior on the standardized regression coefficients has SD=1.

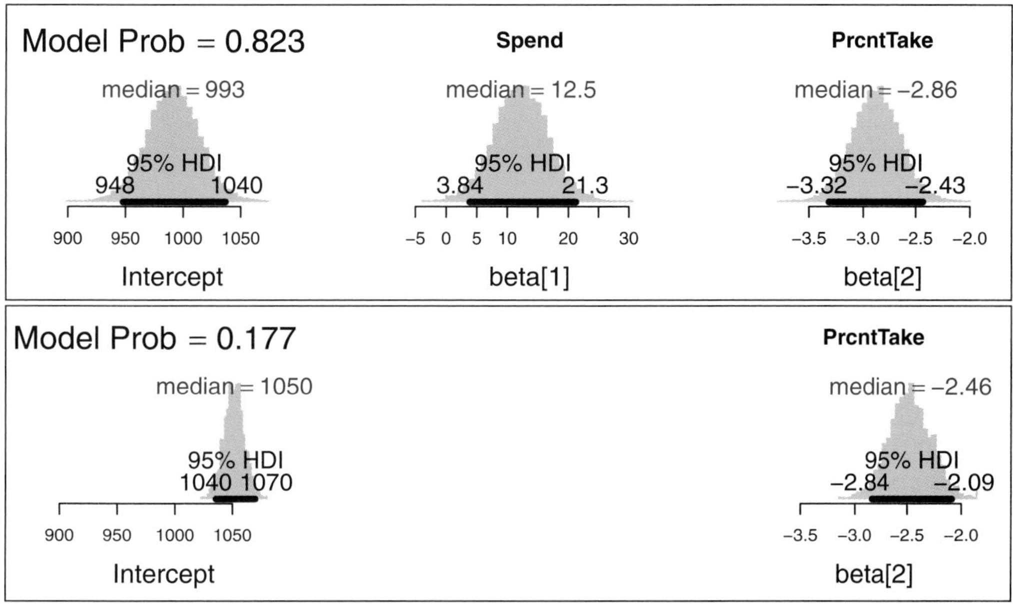

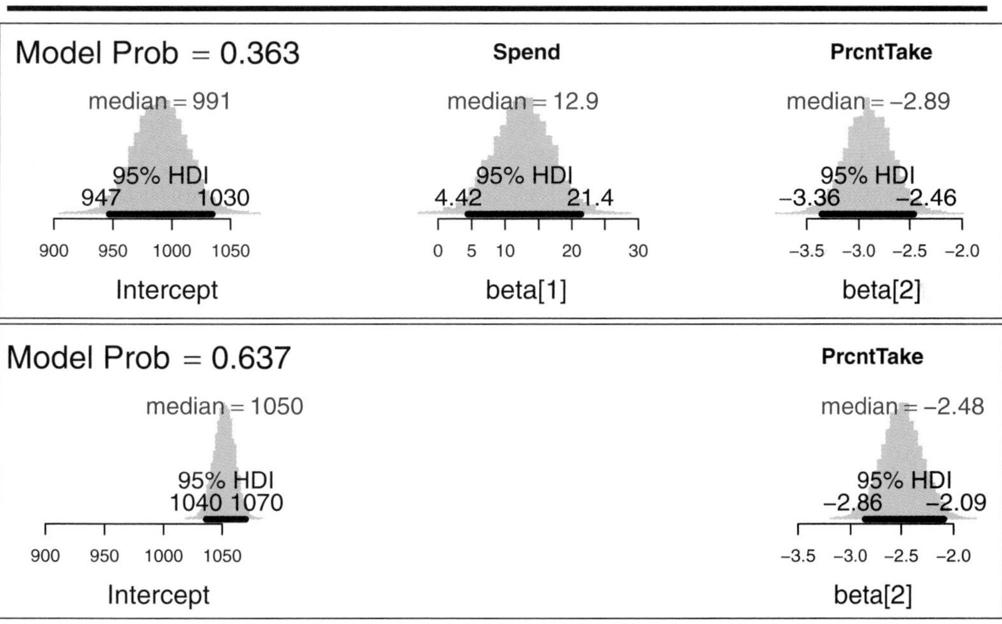

Figure 18.14 Posterior probabilities of different subsets of predictors along with the marginal posterior distributions of the included regression coefficients. Upper two panels show results when the prior on the standardized regression coefficients has SD=1; lower two panels are for SD=10. In both cases, the prior probability of each model was $0.5^2 = 0.25$.

You can see that there is a strong advantage for the two-predictor model, with the posterior inclusion probability of Spend being about 0.82. The lower pair of panels in Figure 18.14 shows the posterior probabilities when the prior on the standardized regression coefficients has `SD=10`. You can see that *there is now an advantage for the one-predictor model*, with the posterior inclusion probability of Spend being only about 0.36. How could that be? After all, a model with all predictors included must be able to fit the data at least as well as a model with only a subset of predictors excluded.

The reason for the lower probability of the more complex model is that each extra parameter dilutes the prior density on the pre-existing parameters. This idea was discussed in Section 10.5 (p. 289). Consider the PrcntTake-only model. In this model, one of the most credible values for the regression coefficient on PrcntTake is $\beta_2 = -2.5$. The likelihood at that point involves $p(D|\beta_2 = -2.5)$ and the prior density at the point involves $p(\beta_2 = -2.5)$, with the posterior probability proportional to their product. The very same likelihood can be achieved by the model that also includes Spend, merely by setting the regression coefficient on Spend to zero: $p(D|\beta_2 = -2.5) = p(D|\beta_2 = -2.5, \beta_1 = 0)$. But the prior density at that point, $p(\beta_2 = -2.5, \beta_1 = 0) = p(\beta_2 = -2.5)\, p(\beta_1 = 0)$, will typically be less than $p(\beta_2 = -2.5)$, because the prior is $p(\beta_2 = -2.5)$ multiplied by a probability density that is almost certainly less than one. Thus, models that include more predictors will pay the cost of lower prior probability. Models with additional predictors will be favored only to the extent that their benefit in higher likelihood outweighs their cost in lower prior probability. When the prior on the regression coefficient is broader, the prior density at any particular value tends to get smaller.

On the other hand, the change in vagueness of the prior distribution has hardly affected the estimates of the regression coefficients at all. Figure 18.14 shows that the estimate of the regression coefficient on Spend has a 95% HDI from about 4 to 21 for both prior distributions, regardless of whether its inclusion probability is low or high. From these results, do we conclude that Spend should be included or not? For me, the robustness of the explicit estimate of the regression coefficient, showing that it is non-zero, trumps the model comparison. As has been emphasized previously in Section 10.6 (p. 292) and in Section 12.4 (p. 354), Bayesian model comparison can be strongly affected by the degree of vagueness in the priors, even though explicit estimates of the parameter values may be minimally affected. Therefore, be very cautious when interpreting the results of Bayesian variable selection. The next section discusses a way to inform the prior by using concurrent data instead of previous data.

18.4.2. Variable selection with hierarchical shrinkage

The previous section emphasized the importance of using appropriate priors on the regression coefficients when doing variable selection, because the vagueness of the priors can have surprisingly large influence on the posterior inclusion probabilities. If you have strong previous research that can inform the prior, then it should be used. But if previous knowledge is weak, then the uncertainty should be expressed in the prior. This is an underlying mantra of the Bayesian approach: Any uncertainty should be expressed in the prior. Thus, if you are not sure what the value of σ_β should be, you can estimate it and include a higher-level distribution to express your prior uncertainty. In other words, in Figure 18.10 (p. 531), we estimate σ_β and give it a prior distribution in place of the open braces. An additional benefit of this approach is that all the predictors simultaneously inform the estimate of σ_β, whereby concurrent data from all the predictors provide an informed prior for each individual predictor.

In the present application, we have uncertainty, and therefore, we place a broad prior on σ_β. The code below shows a few different options in commented lines. In the code, σ_β is denoted `sigmaBeta`. One option sets `sigmaBeta` to a constant, which produces the results reported in the previous section. Another option puts a broad uniform prior on `sigmaBeta`. A uniform prior is intuitively straight forward, but a uniform prior must always have some arbitrary bounds. Therefore, the next option, not commented out of the code below, is a gamma distribution that has a mode at 1.0 but is very broad with a standard deviation of 10.0:

```
model {
  ...
  # Priors vague on standardized scale:
  zbeta0 ~ dnorm( 0 , 1/2^2 )
  for ( j in 1:Nx ) {
    zbeta[j] ~ dt( 0 , 1/sigmaBeta^2 , 1 ) # Notice sigmaBeta
    delta[j] ~ dbern( 0.5 )
  }
  zsigma ~ dunif( 1.0E-5 , 1.0E+1 )
  ## Uncomment one of the following specifications for sigmaBeta:
  # sigmaBeta <- 2.0
  # sigmaBeta ~ dunif( 1.0E-5 , 1.0E+2 )
  sigmaBeta ~ dgamma(1.1051,0.1051) # mode 1.0, sd 10.0
  # sigmaBeta <- 1/sqrt(tauBeta) ; tauBeta ~ dgamma(0.001,0.001)
  ...
}
```

The code is in the files named `Jags-Ymet-XmetMulti-MrobustVarSelect.R` and `Jags-Ymet-XmetMulti-MrobustVarSelect-Example.R`. When it is run, it produces results very similar to those in Figure 18.13. This similarity suggests that the earlier choice of 2.0 for `sigmaBeta` was a lucky proxy for explicitly expressing higher-level uncertainty.

The example used in these sections has involved only two candidate predictors, merely for simplicity in explanation. In most applications of variable selection, there are numerous candidate predictors. As a small extension of the example, it turns out that the

SAT data from Guber (1999) had two additional predictors, namely the average student-teacher ratio (StuTeaRat) in each state and the average salary of the teachers (Salary). These variables are also plausible predictors of SAT score. Should they be included?

First we consider the correlations of the candidate predictors:

```
CORRELATION MATRIX OF PREDICTORS:
           Spend PrcntTake StuTeaRat Salary
Spend      1.000     0.593    -0.371  0.870
PrcntTake  0.593     1.000    -0.213  0.617
StuTeaRat -0.371    -0.213     1.000 -0.001
Salary     0.870     0.617    -0.001  1.000
```

Notice above that Salary is strongly correlated with Spend, and therefore, a model that includes both Salary and Spend will show a strong trade-off between those predictors and consequently will show inflated uncertainty in the regression coefficients for either one. Should only one or the other be included, or both, or neither?

The prior inclusion bias was 0.5 for each predictor, and therefore, each of $2^4 = 16$ models had a prior probability of $0.5^4 = 0.0625$. The prior on the regression coefficients was hierarchical with σ_β having a gamma prior with mode 1.0 and standard deviation of 10.0, as explained in the previous paragraphs.

Figure 18.15 shows the results. The most probable model, shown at the top of Figure 18.15, includes only two predictors, namely Spend and PrcntTake, and has a posterior probability of roughly 50%. The runner up (in the second row of Figure 18.15) has a posterior probability of about half that much and includes only PrcntTake. The Salary predictor is included only in the third most probable model, with a posterior probability of only about 8%.

Notice that in any model that includes both Spend and Salary, the marginal posterior distributions on the two regression coefficients are considerably wider than when including only one or the other. This widening is due to the correlation of the two predictors, and the consequent trade-off in the regression coefficients: A relatively small regression coefficient on one predictor can be compensated by a relatively large regression coefficient on the other predictor.

Notice that the predictors that are most likely to be included tend to be the ones with the largest magnitude *standardized* regression coefficients. For example, PrcntTake is included in every credible model, and its regression coefficient is estimated to be far from zero. The next most probably included predictor is Spend, and its regression coefficient is clearly nonzero but not by as much. The next most probably included predictor is Salary, and its regression coefficient barely excludes zero.

18.4.3. What to report and what to conclude

From the results of variable-selection analysis, such as in Figure 18.15, what should be reported and what should be concluded? Which candidate predictors should be

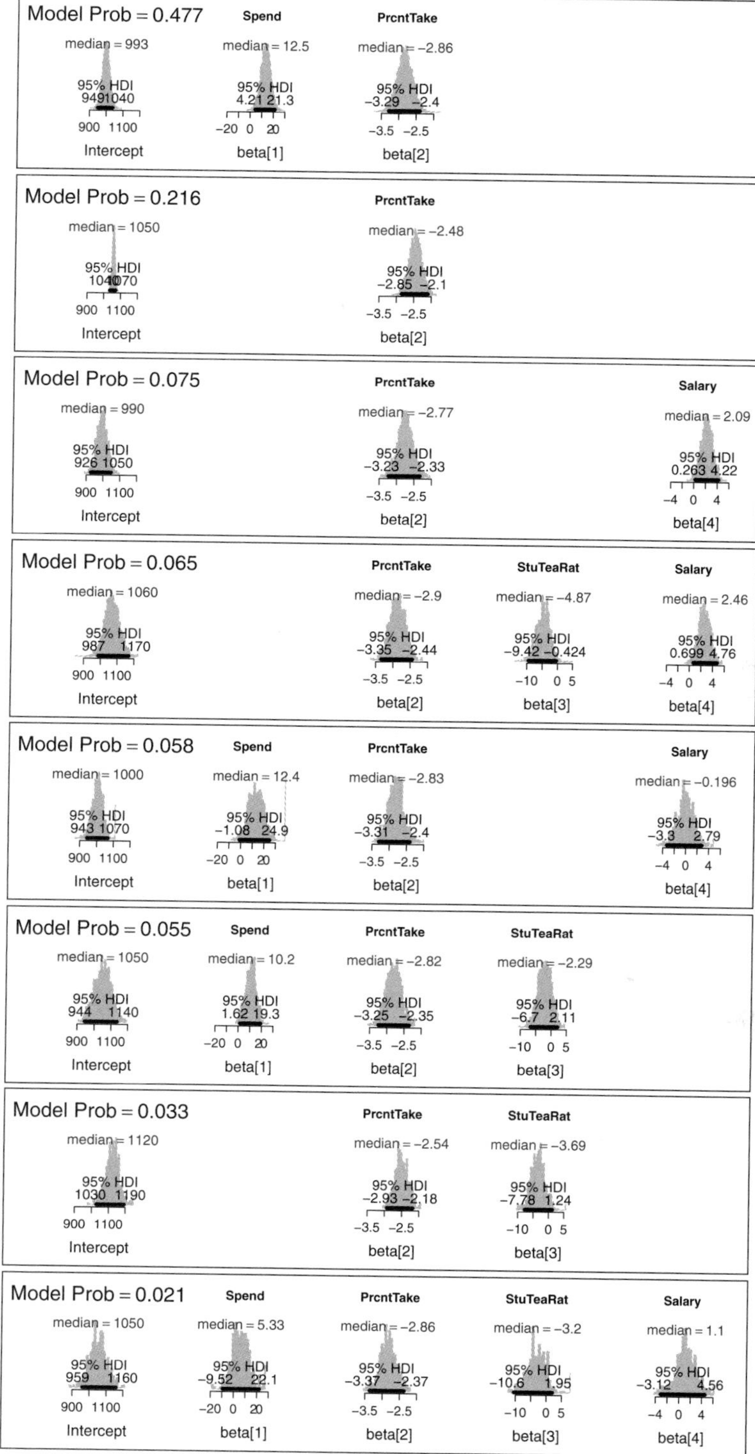

Figure 18.15 Posterior probabilities of different subsets of predictors along with the marginal posterior distributions of the included regression coefficients. The remaining eight possible models had essentially zero probability. The prior probability of each model was $0.5^4 = 0.0625$. The histograms of improbable models are jagged because the MCMC chain visited those models only rarely.

included in an explanatory model? How should predictions of future data be made? Unfortunately, there is no singular "correct" answer. The analysis tells us the relative posterior credibilities of the models for our particular choice of prior. It might make sense to use the single most credible model, especially if it is notably more credible than the runner up, and if the goal is to have a parsimonious explanatory description of the data. But it is important to recognize that using the single best model, when it excludes some predictors, is concluding that the regression coefficients on the excluded predictors are exactly zero. For example, in Figure 18.15, if we used only the best model, then we would be concluding that both student-teacher ratio and teacher salary have *zero* relation to SAT scores. When you exclude variables you are deciding that the regression coefficient is zero. This might be alright for the purpose of parsimonious explanation, but the report should inform the reader about competing models.

A forthright report should state the posterior probabilities of the several top models. Additionally it can be useful to report, for each model, the ratio of its posterior probability relative to that of the best model. For example, from Figure 18.15 we can compute that second-best model has posterior probability that is only $0.21/0.48 \approx 0.45$ of the best model, and the third-best model has posterior probability that is only $0.08/0.48 \approx 0.16$ of the best model. One arbitrary convention is to report all models that have a posterior probability that is at least $1/3$ of the posterior probability of the best model, which would be only the top two models in this example. But any model of theoretical interest should be reported.

Another useful perspective on the posterior distribution is the overall posterior inclusion probability of each predictor. The posterior inclusion probability of a predictor is simply the sum of the posterior probabilities of the models that include it: $p(\delta_j = 1|D) = \sum_{m:\delta_j=1} p(m|D)$. Even more simply, it is the proportion of steps in the overall MCMC chain that include the predictor. In the present example, the marginal inclusion probabilities are approximately 1.0 for PrcntTake, 0.61 for Spend, 0.22 for Salary, and 0.17 for StuTeaRat. While the overall inclusion probabilities provide a different perspective on the predictors than individual models, be careful not to think that the marginal inclusion probabilities can be multiplied to derive the model probabilities. For example, the probability of the model that includes Spend and PrcntTake (i.e., about 0.48) is *not* equal to the product of the probabilities of including Spend (0.61), including PrcntTake (1.0), excluding StuTeaRat $(1 - 0.17)$, and excluding Salary $(1 - 0.22)$.

The report should also indicate how robust are the model probabilities and inclusion probabilities when the prior is changed. As was emphasized in Section 18.4.1, the model probabilities and inclusion probabilities can be strongly affected by the vagueness of the prior on the regression coefficients. If the prior is changed from a gamma on σ_β to a uniform, what happens to the model probabilities? And, of course, the model probabilities and inclusion probabilities are directly affected by the prior on the inclusion indicators themselves.

For each of the reported models, it can be useful to report the marginal posterior distribution of each regression coefficient and other parameters. This can be done graphically, as in Figure 18.15, but typically a more compressed summary will be needed for research reports, in which case the central tendency and 95% HDI limits might suffice for each parameter.

When the goal is prediction of y for interesting values of the predictors, as opposed to parsimonious explanation, then it is usually not appropriate to use only the single most probable model. Instead, predictions should be based on as much information as possible, using all models to the extent that they are credible. This approach is called *Bayesian model averaging* (BMA) and was discussed in Section 10.4 (p. 289). To generate predictions, we merely step through the MCMC chain, and at every step, use the parameters to randomly simulate data from the model. This procedure is exactly what we have done for generating posterior predictions for any application. The only difference from before is that here we call different values of the inclusion coefficients different models. In reality, there is just one overarching model, and some of its parameters are inclusion coefficients. Thus, BMA is really no different than posterior predictions derived from any other application.[4]

18.4.4. Caution: Computational methods

The computer code that created the examples of this section is in the files `Jags-Ymet-XmetMulti-MrobustVarSelect-Example.R` and `Jags-Ymet-XmetMulti-MrobustVarSelect.R`. The code is meant primarily for pedagogical purposes, and it does not scale well for larger applications, for reasons explained presently.

There are a variety of approaches to Bayesian variable selection, and MCMC is just one. MCMC is useful only when there are a modest number of predictors. Consider that when there are p predictors, there are 2^p models. For example, with 10 predictors, there are 1024 models, and with 20 predictors, there are 1,048,576 models. A useful MCMC chain will need ample opportunity to sample from all the models, which would require an impractically long chain even for moderately large numbers of predictors.

Even for a modest number of predictors, the MCMC chain can be badly autocorrelated in the model indices or inclusion indicators. A variety of sampling algorithms and models have been proposed to make MCMC variable selection more efficient (for reviews, see Ntzoufras, 2009; O'Hara & Sillanpää, 2009). The method presented

[4] Alternatively, in applications for which using all (or many) submodels is computationally difficult, Barbieri and Berger (2004) suggest using the single model that includes all variables for which the marginal inclusion probability is greater than 0.5, $p(\delta_j = 1|D) > 0.5$, which is called the *median probability model*.

earlier (sometimes attributed to Kuo & Mallick, 1998) is straightforward but may suffer inefficiency because when an indicator parameter is set to zero during the MCMC walk, its regression coefficient is sampled from the broad prior (unconstrained by the data) and may get a value far from anything that mimics the data. The chain will have low probability of subsequently setting the indicator parameter back to 1. In other words, the chains for the indicator variables might be badly autocorrelated. One consequence is that the estimated model probabilities and inclusion probabilities can be unstable, so very long chains are needed. A possible solution is to use the pseudoprior method (Carlin & Chib, 1995), as was discussed in Section 10.3.2.1 (p. 279). This application of pseudopriors to variable selection was discussed by Dellaportas, Forster, and Ntzoufras (2002), who called it Gibbs variable selection. I will not further discuss it here because, while it is straight forward conceptually, it involves many implementation details (see Ntzoufras, 2009). For additional examples of estimating inclusion coefficients for multiple regression, using various approaches and programmed in BUGS (hence easily adapted to JAGS), see Lykou and Ntzoufras (2011), Ntzoufras (2002), Ntzoufras (2009, Section 11.7), and O'Hara and Sillanpää (2009).

To conclude this section regarding variable selection, it is appropriate to recapitulate the considerations at the beginning of the section. Variable selection is a reasonable approach only if it is genuinely plausible and meaningful that candidate predictors have zero relation to the predicted variable. The results can be surprisingly sensitive to the seemingly innocuous choice of prior for the regression coefficients, and, of course, the prior for the inclusion probability. Because of these limitations, hierarchical shrinkage priors may be a more meaningful approach.

18.4.5. Caution: Interaction variables

The preceding sections, regarding shrinkage of regression coefficients and variable selection, did not mention interactions. When considering whether to include interaction terms, there are the usual considerations with respect to inclusion of any predictor, and additional considerations specific to interaction variables.

When considering the inclusion of *interaction* terms, and the goal of the analysis is explanation, then the main criterion is whether it is theoretically meaningful that the effect of one predictor should depend on the level of another predictor. Inclusion of an interaction term can cause loss of precision in the estimates of the lower-order terms, especially when the interaction variable is correlated with component variables. Moreover, interpretation of interactions and their lower-order terms can be subtle, as we saw, for example, in Figure 18.9.

When interaction terms are included in a model that also has hierarchical shrinkage on regression coefficients, the interaction coefficients should not be put under the same higher-level prior distribution as the individual component coefficients, because interaction coefficients are conceptually from a different class of variables than individual components. For example, when the individual variables are truly additive, then there will be very small magnitude interaction coefficients, even with large magnitude individual regression coefficients. Thus, it could be misleading to use the method of Equation 18.5 with the hierarchical-shrinkage program of Section 18.3, because that program puts all variables' coefficients under the same higher-level distribution. Instead, the program should be modified so that two-way interaction coefficients are under a higher-level prior that is distinct from the higher-level prior for the single-component coefficients. And, of course, different two-way interaction coefficients should be made mutually informative under a higher-level distribution only if it is meaningful to do so.

Whenever an interaction term is included in a model, it is important also to include all lower-order terms. For example, if an $x_i \cdot x_j$ interaction is included, then both of x_i and x_j should also be included in the model. It is also possible to include three-way interactions such as $x_i \cdot x_j \cdot x_k$, if it is theoretically meaningful to do so. A three-way interaction means that the magnitude of a two-way interaction depends on the level of a third variable. When a three-way interaction is included, it is important to include all the lower-order interactions and single predictors, including $x_i \cdot x_j$, $x_i \cdot x_k$, $x_j \cdot x_k$, x_i, x_j, and x_k. When the lower-order terms are omitted, this is artificially setting their regression coefficients to zero, and thereby distorting the posterior estimates on the other terms. For clear discussion and examples of this issue, see Braumoeller (2004) and Brambor, Clark, and Golder (2006). Thus, it would be misleading to use the method of Equation 18.5 with the variable-selection program of Section 18.4 because that program would explore models that include interactions without including the individual components. Instead, the program should be modified so that only the meaningful models are available for comparison. One way to do this is by multiplying each interaction by its own inclusion parameter and all the component inclusion parameters. For example, the interaction $x_j x_k$ is multiplied by the product of inclusion parameters, $\delta_{j \times k} \delta_j \delta_k$. The product of these inclusion parameters can be 1 only if all three are 1. Keep in mind, however, that this also reduces the prior probability of including the interaction.

18.5. EXERCISES

Look for more exercises at https://sites.google.com/site/doingbayesiandataanalysis/

Exercise 18.1. [Purpose: Understanding multiplicative interaction.] Consider the right panel of Figure 18.8. Use Equation 18.4 to compute the slope of the dark lines when $x_1 = 0$ and $x_1 = 10$. Show your work (algebraically). Confirm your answer by computing the slopes from the figure: Visually inspect how much each line rises as x_2 goes from 0 to 10, and compute the ratio of the rise over the run for each line.

Exercise 18.2. [Purpose: Understand the effect of including/excluding predictors, even when they are not correlated. This is also a prelude to analysis of covariance.] The fictitious data in Figure 18.1, p. 511, involves two predictors that are uncorrelated. The data are in the file MultLinRegrPlotUnif.csv.

(A) Run a multiple regression on the two predictors. What is the correlation of the two predictors? Are the estimates of the intercept, slopes, and standard deviation close to the values indicated in Figure 18.1?

(B) Run the regression of y on the single predictor x_1. What parameter estimates have noticeably changed? In particular, why is the estimate of σ so much bigger? Discuss with respect to the upper right panel of Figure 18.1.

(C) Repeat the previous two parts, but this time using only lines 101–150 of the data file (i.e., fewer data points). How does the interpretation of the regression coefficient on x_1 change when x_2 is included?

Exercise 18.3. [Purpose: View the prior distribution.] Figure 18.7 (p. 522) showed the prior distribution for a multiple linear regression. Your job for this exercise is to produce the graph. To do so, read Footnote 2 (p. 523), about commenting out the specification of `zy[i]` in the JAGS data block.

Exercise 18.4. [Purpose: Hands-on experience with variable selection and its sensitivity to priors.] Your goal in this exercise is to produce Figure 18.15 (p. 545) and explore some variations. The relevant programs are `Jags-Ymet-XmetMulti-MrobustVarSelect.R` and `Jags-Ymet-XmetMulti-MrobustVarSelect-Example.R`. For all the parts of this exercise, read the SAT data file with all four candidate predictors. At the top of `Jags-Ymet-XmetMulti-MrobustVarSelect-Example.R` uncomment and comment out lines so that you have the equivalent of

```
myData = read.csv( file="Guber1999data.csv" )
yName = "SATT"
xName = c("Spend","PrcntTake","StuTeaRat","Salary")
fileNameRoot = "Guber1999data-Jags-4X-VarSelect-" # change for distinct saved files
numSavedSteps=15000 ; thinSteps=20
```

(A) In the program `Jags-Ymet-XmetMulti-MrobustVarSelect.R`, be sure that the line

```
sigmaBeta ~ dgamma(1.1051,0.1051) # mode 1.0, sd 10.0
```

is being used (i.e., is the only line of its section not commented out). Run the high-level script. Does its output resemble Figure 18.15 (p. 545)? (It should.)

(B) In the program `Jags-Ymet-XmetMulti-MrobustVarSelect.R`, comment out the line for `sigmaBeta` that gives it a gamma prior, and instead use

```
sigmaBeta <- 10.0
```

Run the high-level script. In what ways is the posterior different than the previous part? Discuss model probabilities, inclusion probabilities, and HDIs of regression coefficients.

(C) Set the prior on `sigmaBeta` back to the gamma distribution of the first part. Now change the prior on the inclusion indices so that

```
delta[j] ~ dbern( 0.2 )
```

Run the high-level script. In what ways is the posterior different than the first part? Discuss model probabilities, inclusion probabilities, and HDIs of regression coefficients.

CHAPTER 19

Metric Predicted Variable with One Nominal Predictor

Contents

Put umpteen people in two groups at random.
Social dynamics make changes in tandem:
Members within groups will quickly conform;
Difference between groups will soon be the norm.[1]

This chapter considers data structures that consist of a metric predicted variable and a nominal predictor. This sort of structure is often encountered in real research. For example, we might want to predict monetary income from political party affiliation, or we might want to predict galvanic skin response to different categories of visual stimulus, or, as we will investigate later in the chapter, we might want to predict life span from categories of sexual activity. This type of data structure can arise from experiments or from observational studies. In experiments, the researcher assigns the categories (at random) to the experimental subjects. In observational studies, both the nominal predictor value and the metric predicted value are generated by processes outside

[1] The models in this chapter are analogous to traditional analysis of variance (ANOVA), which partitions variance into within-group variance and between-group variance. The poem suggests that for groups of people, within-group variance tends to decrease while between-group variance tends to increase.

Doing Bayesian Data Analysis, 2nd Edition
http://dx.doi.org/10.1016/B978-0-12-405888-0.00019-2

the direct control of the researcher. In either case, the same mathematical description can be applied to the data (although causality is best inferred from experimental intervention).

The traditional treatment of this sort of data structure is called single-factor analysis of variance (ANOVA), or sometimes one-way ANOVA. Our Bayesian approach will be a hierarchical generalization of the traditional ANOVA model. The chapter will also consider the situation in which there is also a metric predictor that accompanies the primary nominal predictor. The metric predictor is sometimes called a covariate, and the traditional treatment of this data structure is called analysis of covariance (ANCOVA). The chapter also considers generalizations of the traditional models, because it is straight forward in Bayesian software to implement heavy-tailed distributions to accommodate outliers, along with hierarchical structure to accommodate heterogeneous variances in the different groups, etc.

In the context of the generalized linear model (GLM) introduced in Chapter 15, this chapter's situation involves a linear function of a single nominal predictor, as indicated in the fifth column of Table 15.1 (p. 434), with a link function that is the identity, along with a normal distribution for describing noise in the data, as indicated in the first row of Table 15.2 (p. 443). For a reminder of how this chapter's combination of predicted and predictor variables relates to other combinations, see Table 15.3, p. 444.

19.1. DESCRIBING MULTIPLE GROUPS OF METRIC DATA

As emphasized in Section 2.3 (p. 25), after identifying the relevant data, the next step of Bayesian data analysis is formulating a meaningful mathematical description of the data. For our present application, each group's data are described as random variation around a central tendency. The central tendencies of the groups are conceptualized as deflections from an overall baseline. Details of this model were introduced back in Section 15.2.4.1 (p. 429), and illustrated in Figure 15.4 (p. 431). The ideas are briefly recapitulated in the following.

Figure 19.1 illustrates the conventional description of grouped metric data. Each group is represented as a position on the horizontal axis. The vertical axis represents the variable to be predicted by group membership. The data are assumed to be normally distributed within groups, with equal standard deviation in all groups. The group means are deflections from overall baseline, such that the deflections sum to zero. Figure 19.1 provides a specific numerical example, with data that were randomly generated from the model. For real data, we do not know what process generated them, but we infer credible parameter values of a meaningful mathematical description.

As you may recall from Section 15.2.4.1, we represent the nominal predictor by a vector $\vec{x} = \langle x_{[1]}, \ldots, x_{[J]} \rangle$, where J is the number of categories that the predictor

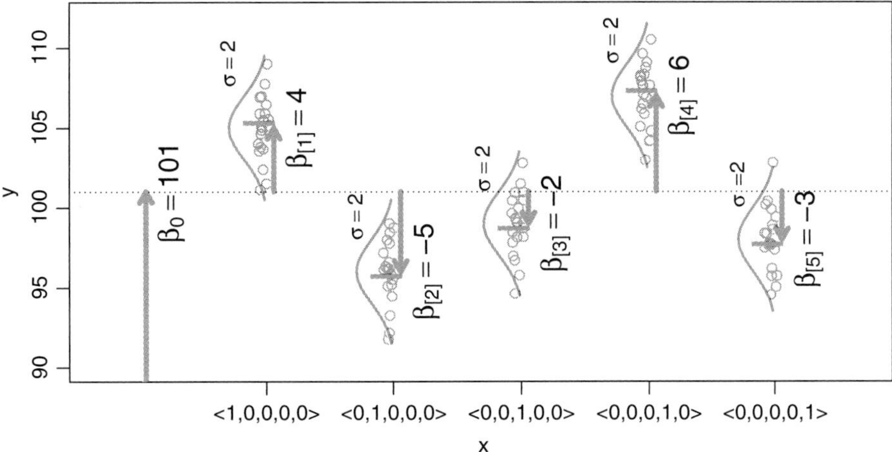

Figure 19.1 Description of data as normally distributed around group means that are conceptualized as deflections from an overall baseline. Data are indicated by circular dots (jittered left-right for visibility). The standard deviation of the data within groups is assumed to be the same for all groups and is indicated as σ. Baseline and deflections are indicated by arrows and β values. Notice that the deflections from baseline sum to zero.

has. When an individual falls in group j of the nominal predictor, this is represented by setting $x_{[j]} = 1$ and $x_{[i \neq j]} = 0$. The x axis of Figure 19.1 marks the levels of $\vec{x}$ using this notation. The predicted value, denoted μ, is the overall baseline plus the deflection of a group:

$$\begin{aligned} \mu &= \beta_0 + \sum_j \beta_{[j]} x_{[j]} \\ &= \beta_0 + \vec{\beta} \cdot \vec{x} \end{aligned} \tag{19.1}$$

where the notation $\vec{\beta} \cdot \vec{x}$ is called the "dot product" of the vectors. In Equation 19.1, the coefficient $\beta_{[j]}$ indicates how much the predicted value of y changes when x changes from neutral to category j. The overall baseline is constrained so that the deflections sum to zero across the categories:

$$\sum_j \beta_{[j]} = 0 \tag{19.2}$$

The expression of the model in Equation 19.1 is not complete without the constraint in Equation 19.2.

The sum-to-zero constraint is implemented in the JAGS (or Stan) program in two steps. First, at any point in the MCMC chain, JAGS finds jointly credible values of a baseline and deflections without directly respecting the sum-to-zero constraint.

Second, the sum-to-zero constraint is imposed by simply subtracting out the mean of the deflections from the deflections and adding it to the baseline. The algebra is now described formally. Let's denote the unconstrained values of the parameters by α, and denote the sum-to-zero versions by β. At any point in the MCMC chain, the predicted value is

$$\mu = \alpha_0 + \sum_j \alpha_{[j]} x_{[j]} \tag{19.3}$$

$$= \underbrace{(\alpha_0 + \overline{\alpha})}_{\beta_0} + \sum_j \underbrace{(\alpha_{[j]} - \overline{\alpha})}_{\beta_{[j]}} x_{[j]} \tag{19.4}$$

$$\text{where } \overline{\alpha} = \frac{1}{J} \sum_{j=1}^{J} \alpha_{[j]}$$

It's easy to show that the $\beta_{[j]}$ values in Equation 19.4 really do sum to zero: $\sum_{j=1}^{J} \beta_{[j]} = \sum_{j=1}^{J} (\alpha_{[j]} - \overline{\alpha}) = \sum_{j=1}^{J} \alpha_{[j]} - \sum_{j=1}^{J} \overline{\alpha} = J\overline{\alpha} - J\overline{\alpha} = 0$. A later section shows how Equations 19.3 and 19.4 are implemented in JAGS.

The descriptive model presented in Figure 19.1 is the traditional one used by classical ANOVA (which is described a bit more in the next section). More general models are straight forward to implement in Bayesian software. For example, outliers could be accommodated by using heavy-tailed noise distributions (such as a t distribution) instead of a normal distribution, and different groups could be given different standard deviations. A later section of this chapter explores these generalizations.

19.2. TRADITIONAL ANALYSIS OF VARIANCE

The terminology, "analysis of variance," comes from a decomposition of overall data variance into within-group variance and between-group variance (Fisher, 1925). Algebraically, the sum of squared deviations of the scores from their overall mean equals the sum of squared deviations of the scores from their respective group means plus the sum of squared deviations of the group means from the overall mean. In other words, the total variance can be partitioned into within-group variance plus between-group variance. Because one definition of the word "analysis" is separation into constituent parts, the term ANOVA accurately describes the underlying algebra in the traditional methods. That algebraic relation is not used in the hierarchical Bayesian approach presented here. The Bayesian method can estimate component variances, however. Therefore, the Bayesian approach is not ANOVA, but is analogous to ANOVA.

Traditional ANOVA makes decisions about equality of groups (i.e., null hypotheses) on the basis of p values. As was discussed at length in Chapter 11, and illustrated in Figure 11.1 on p. 299, p values are computed by imaginary sampling from a null hypothesis. In traditional ANOVA, the null hypothesis assumes (i) the data are normally distributed within groups, and (ii) the standard deviation of the data within each group is the same for all groups. The second assumption is sometimes called "homogeneity of variance." These assumptions are important for mathematical derivation of the sampling distribution. The sample statistic is the ratio of between-group variance to within-group variance, called the F ratio after Ronald Fisher, and therefore the sampling distribution is called the F distribution. For the p value to be accurate, the assumptions of normality and homogeneity of variance should be respected by the data. (Of course, the p value also assumes that the stopping intention is fixed sample size, but that is a separate issue.) These assumptions of normally distributed data with homogeneous variances are entrenched in the traditional approach. That entrenched precedent is why basic models of grouped data make those assumptions, and why the basic models presented in this chapter will also make those assumptions. Fortunately, it is straight forward to relax those assumptions in Bayesian software, where we can use different variance parameters for each group, and use non-normal distributions to describe data within groups, as will be shown later in the chapter.

19.3. HIERARCHICAL BAYESIAN APPROACH

We start with the basic descriptive model that was illustrated in Figure 19.1. Our goal is to estimate its parameters in a Bayesian framework. Therefore, all the parameters need to be given a meaningfully structured prior distribution, which is shown in Figure 19.2. As usual, the hierarchical diagram is scanned from the bottom to the top. At the bottom of Figure 19.2, we see that the data, y_i, are distributed normally around the predicted value, μ_i. The predicted value is specified by Equation 19.1, which is shown in the center of the hierarchical diagram. All the parameters have generic noncommittal prior distributions. Thus, the within-group standard deviation, σ_y, is given a broad uniform prior distribution (as recommended by Gelman, 2006). The baseline parameter, β_0, is given a normal prior distribution, made broad on the scale of the data. The group deflection parameters, β_j, are given a normal prior distribution that has a mean of zero, because the deflection parameters are supposed to sum to zero. (The sum-to-zero constraint is algebraically imposed later, as described below.)

A key novelty of the Bayesian approach is the treatment of σ_β, which is the standard deviation of the distribution of deflection parameters. The diagram in Figure 19.2 shows the prior on σ_β as a set of empty braces, to suggest that there are options for the prior. One option is that we could just set σ_β to a constant. This setting would cause each group deflection to be estimated separately from the other groups, insofar as no group has any influence on the value of σ_β. Setting σ_β to a large constant results in estimates most analogous to traditional ANOVA.

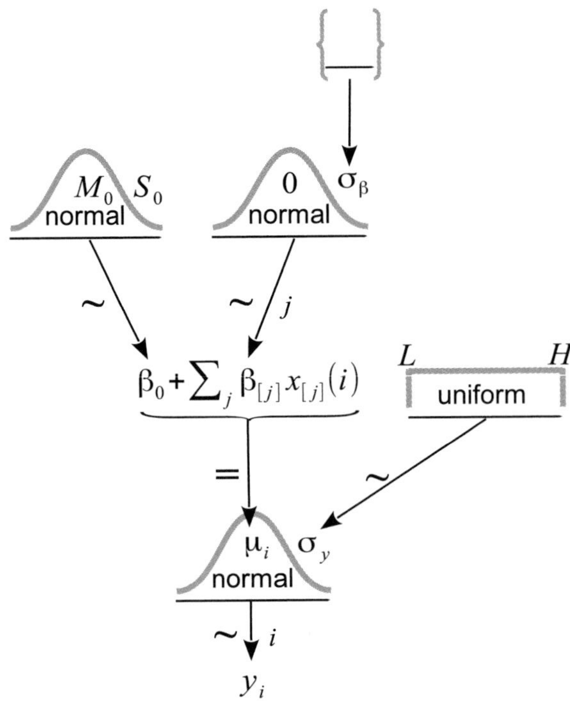

Figure 19.2 Hierarchical diagram for model that describes data from several groups of a single factor. At the top of the diagram, the empty braces indicate the prior distribution on the between-group standard deviation, σ_β, which could be a folded-*t* as recommended by Gelman (2006), a gamma distribution with non-zero mode, or a constant if no sharing across groups is desired.

Instead of fixing σ_β at a constant, it can be estimated from the data (Gelman, 2005). If the data suggest that many of the groups have small deflections from baseline, then σ_β is estimated to be small. Notice that a small value of σ_β imposes strong shrinkage of the estimates of the group deflections. (Shrinkage was explained in Section 9.3 and many examples have been given throughout the book.) When σ_β is estimated, the data dictate how much shrinkage to apply to the estimates of the deflection parameters. If many groups are near baseline, then σ_β is estimated to be small and the there is strong shrinkage of the group deflections, but if many groups are far from baseline, then σ_β is estimated to be large and there is only modest shrinkage of the group deflections.

The form of prior distribution on σ_β should reflect genuine prior beliefs and also produce sensible posterior distributions in real applications. Gelman (2006) recommended using a distribution that does not put too much emphasis on values near zero nor on values far from zero, because too much emphasis near zero causes extreme shrinkage, and too much emphasis far from zero causes insufficient shrinkage. He recommended a half-Cauchy or folded-*t* distribution, which is the positive side of a *t* distribution.

(See Figure 16.4, p. 458, for an illustration of a t distribution.) This distribution puts a finite probability density at zero (unlike a generic gamma (0.001,0.001) distribution on precision), and puts small probability density on large values but with heavy tails that allow for the possibility of large deflections.

In practice, when data sets are small, even a folded-t prior on σ_β can yield implosive shrinkage. This can happen because the prior on σ_β puts moderately large prior credibility on zero deflections between groups, and the data can be accommodated by setting the group deflections to zero and using a larger value for the within-group noise, σ_y. (In general, when data sets are small, the prior can have a large influence on the posterior.) There are various reactions we might have to this situation. We could simply accept the implosive shrinkage as the logically correct implication of the assumptions about the prior, if we are committed to the distribution as a firm expression of our prior beliefs. In this case, we need more data to overcome the firm prior beliefs. But if the prior was selected as a default, then the implosive shrinkage could instead be a signal that we should reconsider our assumptions about the prior. For example, the prior distribution on σ_β could instead express a belief that deflections of zero are implausible. One way to do that is to use a gamma distribution that has a nonzero mode. This option is implemented in the programs described below.[2]

A crucial pre-requisite for estimating σ_β from all the groups is an assumption that all the groups are representative and informative for the estimate. It only makes sense to influence the estimate of one group with data from the other groups if the groups can be meaningfully described as representative of a shared higher-level distribution. Another way of conceptualizing the hierarchy is that the value of σ_β constitutes an informed prior on each group deflection, such that each group deflection has a prior informed by all the other groups. Again, this only makes sense to the extent that the groups can act as prior information for each other. Often this assumption is plausible. For example, in an experiment investigating blood-pressure drugs, all the groups involve blood pressures measured from the same species, and therefore all the groups can reasonably inform an overarching distribution. But if the groups are dominated by a particular subtype, then it might not be appropriate to put them all under a single higher-level distribution. For example, if the experiment involves many different control groups (e.g., various placebos, sham treatments, and no treatments) and only one treatment group, then the presumably small variance between the control groups will make the estimate of σ_β small, causing excessive shrinkage of the estimated deflection of the treatment group. One way to address this situation is to use a heavy-tailed distribution to describe the

[2] In the 1st edition of this book, the problem of implosive shrinkage was addressed by arbitrarily shifting the folded t distribution away from zero by a small amount (Kruschke, 2011b, p. 496). While that solution is not inherently wrong, it expresses a belief that values of σ_β near zero have zero prior credibility. That solution is superseded by the smooth gamma distribution used here.

group deflections. The heavy tails allow some deflections to be large. The shape of the higher-level distribution does not change the underlying assumption, however, that all the groups mutually inform each other. If this assumption is not satisfied, it might be best to set σ_β to a constant.

19.3.1. Implementation in JAGS

Implementing the model of Figure 19.2 in JAGS is straight forward. Every arrow in the diagram has a corresponding line of code in JAGS. The only truly novel part is implementing the sum-to-zero constraint on the coefficients. We will use the algebraic forms that were presented in Equations 19.3 and 19.4.

In the JAGS model specification below, the noise standard deviation σ_y is called `ySigma`, the unconstrained baseline α_0 from Equation 19.3 is called `a0`, and the unconstrained deflection $\alpha_{[j]}$ is called `a[j]`. The baseline and deflections are subsequently converted to sum-to-zero values called `b0` and `b[j]`, respectively. The $\vec{x}$ value of the ith individual is not coded in JAGS as a vector. Instead, group membership of the ith individual is coded by a simple scalar index called `x[i]`, such that `x[i]` has value j when the score comes from the jth group. The model specification begins by stating that the individual y_i values come from a normal distribution centered at a baseline `a0` plus a deflection `a[x[i]]` for the group, with standard deviation `ySigma`:

```
model {
  for ( i in 1:Ntotal ) { y[i] ~ dnorm( a0 + a[x[i]], 1/ySigma^2 ) }
```

The code above does not bother to specify `mu[i]` on a separate line, although it could be written that way and produce the same results.

Next, the priors on `ySigma` and the baseline `a0` are specified. From Figure 19.2, you can see that the prior on `ySigma` is assumed to be uniform over a range that is wide relative to the scale of the data. We could just ask the user, "What is a typical variance for the type of measurement you are predicting?" and then set the prior wide relative to the answer. Instead, we let the data serve as a proxy and we set the prior wide relative to the variance in the data, called `ySD`. Analogously, the normal prior for the baseline `a0` is centered on the data mean and made very wide relative to the variance of the data. The goal is merely to achieve scale invariance, so that whatever is the measurement scale of the data, the prior will be broad and noncommittal on that scale. Thus,

```
ySigma ~ dunif( ySD/100 , ySD*10 )
a0 ~ dnorm( yMean , 1/(ySD*5)^2 )
```

The model specification continues with the prior on the deflections, `a[j]`. The standard deviation of the deflections, σ_β, is coded as `aSigma`. The prior on `aSigma` is a gamma distribution that is broad on the scale of the data, and that has a nonzero mode so that its probability density drops to zero as `aSigma` approaches zero. Specifically, the shape and rate parameters of the gamma distribution are set so its mode is `sd(y)/2` and its standard deviation is `2*sd(y)`, using the function `gammaShRaFromModeSD` explained in Section 9.2.2. The resulting shape and rate values are stored in the two-component vector `agammaShRa`. Thus,

```
for ( j in 1:NxLvl ) { a[j] ~ dnorm( 0.0 , 1/aSigma^2 ) }
aSigma ~ dgamma( agammaShRa[1] , agammaShRa[2] )
```

Finally, the sum-to-zero constraint is satisfied by recentering the baseline and deflections according to Equation 19.4. At each step in the MCMC chain, the predicted group means are computed as `m[j] <- a0 + a[j]`. The baseline is computed as the mean of those group means: `b0 <- mean( m[1:NxLvl] )` where `NxLvl` is the number of groups (the number of levels of the predictor x). Then the sum-to-zero deflections are computed as the group means minus the new baseline: `b[j] <- m[j] - b0`. This results in adding `mean(a[1:NxLvl])` to `a0` and subtracting `mean(a[1:NxLvl])` from all the `a[j]`, just as in Equation 19.4, but the more elaborate process used here can be generalized to multiple factors (as will be done in the next chapter). Thus, the final section of the model specification is as follows:

```
  # Convert a0,a[] to sum-to-zero b0,b[] :
  for ( j in 1:NxLvl ) { m[j] <- a0 + a[j] }
  b0 <- mean( m[1:NxLvl] )
  for ( j in 1:NxLvl ) { b[j] <- m[j] - b0 }
}
```

The full model is specified in the file `Jags-Ymet-Xnom1fac-MnormalHom.R`. As explained in Section 8.3 (p. 206), the filename begins with `Jags-` because it uses JAGS, continues with `Ymet-` because the predicted variable is metric, then has `Xnom1fac-` because the predictor is nominal involving a single factor, and finishes with `MnormalHom` because the model assumes normally distributed data and homogeneous variances. The model is called from the script `Jags-Ymet-Xnom1fac-MnormalHom-Example.R`.

19.3.2. Example: Sex and death

To illustrate use of the model, we consider the life span of males as a function of the amount of their sexual activity. The data (from Hanley & Shapiro, 1994) are derived from an experiment wherein "sexual activity was manipulated by supplying individual males with receptive virgin females at a rate of one or eight virgins per day. The longevity of these males was recorded and compared with that of two control types. The first control consisted of two sets of individual males kept with newly inseminated females equal in number to the virgin females supplied to the experimental males. The second control was a set of individual males kept with no females. (Partridge & Farquhar, 1981, p. 580)" The researchers were interested in whether male sexual activity reduced life span, as this was already established for females. A deleterious effect of sexual activity would be relatively surprising if found in males because the physiological costs of sexual activity are presumably much lower in males than in females. Oh, I almost forget to mention that the species in question is the fruit fly, *Drosophila melanogaster*. It is known for this species that newly inseminated females will not remate (for at least two days), and males will not actively court pregnant females. There were 25 male fruit flies in each of the five groups.

The data are plotted as dots in Figure 19.3. The vertical axis indicates longevity in days. The horizontal axis, labeled "CompanionNumber," indicates the group. Group None0 indicates the group of males with zero companions. Group Pregnant1 indicates the group of males with one pregnant female. Pregnant8 refers to males accompanied by eight pregnant females. Virgin1 refers to males accompanied by one virgin female, and Virgin8 indicates the males accompanied by eight virgin females. The scatter plot of data suggests that the groups with the receptive females, and hence more sexual activity, had lesser life spans. Our goal is estimate the life spans and the magnitude of differences between groups. (You may recall from Exercise 16.1, p. 473, that sexually deprived male fruit flies are driven to alcohol consumption.)

The five groups in this experiment are all the same type of subjects (male *Drosophila*) housed in similar apparatus and handled in similar ways except for the specific experimental treatment. Moreover, there was not an overwhelming preponderance of any one type of treatment. Therefore, it is not unreasonable to let the five groups mutually inform each other via the hierarchical model of Figure 19.2, with all groups influencing the estimation of σ_β (the standard deviation of the deflections).

It is easy to run the model using the high-level script `Jags-Ymet-Xnom1fac-MnormalHom-Example.R`. The first step is loading the data file and specifying the column names of the predicted and predictor variables:

```
myDataFrame = read.csv( file="FruitflyDataReduced.csv" )
# Specify the column names in the data file relevant to the analysis:
yName="Longevity"
xName="CompanionNumber"
```

After that, the function that calls JAGS is invoked in the usual way:

```
# Load the relevant model into R's working memory:
source("Jags-Ymet-Xnom1fac-MnormalHom.R")
# Generate the MCMC chain:
mcmcCoda = genMCMC( datFrm=myDataFrame , yName=yName , xName=xName ,
                    numSavedSteps=11000 , thinSteps=10 , saveName=fileNameRoot )
```

The argument `thinSteps=10` in the function call, above, specifies thinning the chain. As usual, this is merely to keep the saved file size small, but at the cost of somewhat less

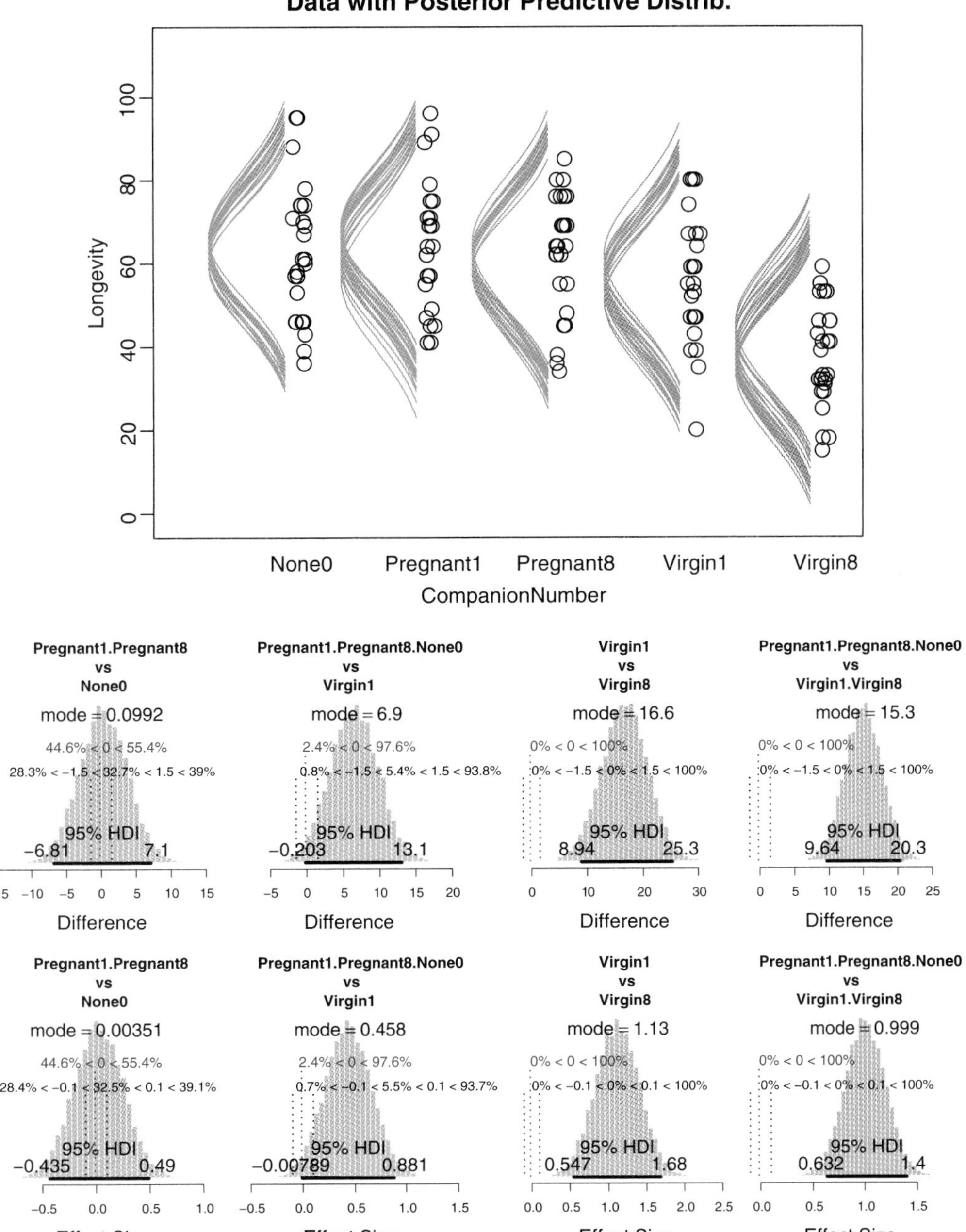

Figure 19.3 Data and posterior distribution for fruit fly longevity. Model assumes normal distributions with homogeneous variances. (Data are plotted with random left-right jitter for visibility.)

stability in the posterior estimates. Thinning is not necessary if you don't mind larger files (in which case the number of saved steps should be increased so that the effective sample size is at least 10,000).

The next part of the script is the usual display of MCMC chain diagnostics:

```
# Display diagnostics of chain, for specified parameters:
parameterNames = varnames(mcmcCoda)
show( parameterNames ) # show all parameter names, for reference
for ( parName in parameterNames ) {
  diagMCMC( codaObject=mcmcCoda , parName=parName ,
            saveName=fileNameRoot , saveType=graphFileType )
}
```

The diagnostic results, not shown here, indicate that the chains are very well behaved, but with some autocorrelation in the standard deviation of the group deflections (i.e., `aSigma` in the code and σ_β in the dependency diagram). Despite the autocorrelation in the high-level variance parameter, the chains are well mixed and the ESS of all the other parameters is at least 10,000.

Figure 19.3 shows posterior predictive distributions superimposed over the data. The curves were created by examining a step in the MCMC chain and plotting normal curves that have the group means and standard deviation at that point in the chain, and repeating for many widely spaced steps in the chain. The smattering of curves looks like a sideways mustache on every group's data. It is important to understand that the bushiness of the mustache represents the uncertainty of the estimate, whereas the width of the mustache represents the value of the standard deviation. For example, a wide mustache that is pencil-thin indicates a large standard deviation that is estimated with high certainty (i.e., small uncertainty). But a narrow mustache that is bushy indicates a small standard deviation that is estimated with poor certainty (i.e., large uncertainty). The bushiness is affected by uncertainty in both the mean and the standard deviation. To get a visual impression of the uncertainty of the mean alone, focus on the spread of the peaks of the curves, not the tails.

Figure 19.3 suggests that the normal distributions with homogeneous variances appear to be reasonable descriptions of the data. There are no dramatic outliers relative to the posterior predicted curves, and the spread of the data within each group appears to be reasonably matched by the width of the posterior normal curves. (Be careful when making visual assessments of homogeneity of variance because the visual spread of the data depends on the sample size; for a reminder see the right panel of Figure 15.9, p. 442.) The range of credible group means, indicated by the peaks of the normal curves, suggests that the group Virgin8 is clearly lower than the others, and the group Virgin1 might be lower than the controls. To find out for sure, we need to examine the differences of group means, which we do in the next section.

19.3.3. Contrasts

For data structured by groups, the researcher is virtually always interested in comparisons between groups or between subsets of groups. In the present application, we might be interested in a large number of paired comparisons. For example, we might be interested in the difference between each of the four treatment groups and the None0 control group. We might be interested in the difference between the Virgin1 group and the Virgin8 group, and the difference between the Pregnant1 group and the Pregnant8 group. We might also be interested in a variety of complex comparisons, which involve differences between averages of sets of groups. For example, we might be interested in the difference between the average of the three control groups (None0, Pregnant1, and Pregnant8) and the average of the two sexually active groups (Virgin1 and Virgin8). Or, we might be interested in the difference between the average of the two unreceptive female groups (Pregnant1 and Pregnant8) and the average of the two sexually active groups. And there are many other comparisons we could make.

It is straight forward to examine the posterior distribution of credible differences. Every step in the MCMC chain provides a combination of group means that are jointly credible, given the data. Therefore, every step in the MCMC chain provides a credible difference between groups, for whatever difference we care to consider. This is the same logic that has been used extensively in this book, for example in Section 9.5.1 regarding baseball batting abilities. Don't forget that the posterior distribution of credible differences cannot be directly discerned from the marginal distributions of individual parameters because the parameters might be correlated, as was explained in Section 12.1.2.1 (p. 340).

To construct the credible differences of group 1 and group 2, at every step in the MCMC chain we compute

$$\begin{aligned}\mu_1 - \mu_2 &= (\beta_0 + \beta_1) - (\beta_0 + \beta_2)\\ &= (+1)\cdot\beta_1 + (-1)\cdot\beta_2\end{aligned}$$

In other words, the baseline cancels out of the calculation, and the difference is a sum of weighted group deflections. Notice that the weights sum to zero. To construct the credible differences of the average of groups 1-3 and the average of groups 4-5, at every step in the MCMC chain we compute

$$\begin{aligned}&(\mu_1 + \mu_2 + \mu_3)/3 - (\mu_4 + \mu_5)/2\\ &\quad= ((\beta_0 + \beta_1) + (\beta_0 + \beta_2) + (\beta_0 + \beta_3))\,/3 - ((\beta_0 + \beta_4) + (\beta_0 + \beta_5))\,/2\\ &\quad= (\beta_1 + \beta_2 + \beta_3)/3 - (\beta_4 + \beta_5)/2\\ &\quad= (+1/3)\cdot\beta_1 + (+1/3)\cdot\beta_2 + (+1/3)\cdot\beta_3 + (-1/2)\cdot\beta_4 + (-1/2)\cdot\beta_5\end{aligned}$$

Again, the difference is a sum of weighted group deflections. The coefficients on the group deflections have the properties that they sum to zero, with the positive coefficients summing to +1 and the negative coefficients summing to −1. Such a combination is called a contrast.[3] The differences can also be expressed in terms of effect size, by dividing the difference by σ_y at each step in the chain.

Contrasts can be specified in the high-level script using meaningful group names instead of arbitrary numerical indices. For example, the *Drosophila* data file codes the group membership of each subject with a meaningful term. Here are a few lines from the *Drosophila* data file:

```
Longevity,CompanionNumber,Thorax
35,Pregnant8,0.64
40,None0,0.64
46,Pregnant1,0.64
```

The first line of the data file above specifies the column names. (The "Thorax" column has not yet been used, but will make an appearance later in the chapter.) The important point here is that the CompanionNumber for each subject is indicated by meaningful labels such as Pregnant8, None0, and Pregnant1. We will take advantage of these meaningful labels when specifying contrasts.

In the programs I wrote, a contrast is specified as a list with four components. The first component is a vector of group names whose average constitutes the first element of the comparison. The second component of the list is a vector of group names whose average constitutes the second element of the comparison. The third component of the list is the comparison value, which is typically zero, but could be a nonzero value. The fourth component of the list is a vector specifying the limits of the ROPE, and which could be `NULL`. Here is an example for the *Drosophila* data:

```
contrasts = list(
  list( c("Virgin1") , c("Virgin8") , compVal=0.0 , ROPE=c(-1.5,1.5) ) ,
  list( c("Pregnant1","Pregnant8","None0") , c("Virgin1","Virgin8") ,
        compVal=0.0 , ROPE=c(-1.5,1.5) )
)
```

The code above defines the variable `contrasts` as a list of lists. Each component list is a specific contrast that we want to have performed. In the `contrasts` list above, there are two specific contrasts. The first is a simple paired comparison between Virgin1 and Virgin8. The second is a complex contrast between the average of the three control groups and the average of the two sexually active groups. In both cases, the comparison value is specified as zero (i.e., `compVal=0.0`) and the ROPE is specified as plus or minus 1.5 days (i.e., `ROPE=c(-1.5,1.5)`). The list of contrasts can be expanded to include all desired contrasts.

[3] Traditionally, the definition of contrast coefficients only requires summing to zero, without the positive coefficients summing to +1 and the negative coefficients summing to −1.

The `contrasts` list is supplied as an argument to the functions that compute summary statistics and plots of the posterior distribution (`plotMCMC` and `smryMCMC`). Some examples of the output of `plotMCMC` are shown in Figure 19.3. The lower panels show the posterior distribution of the contrasts. The contrasts are displayed on the original scale (days) and on the scale of effect size. Notice that there is a whopping difference of about 15 days (effect size of about 1.0) between the average of the control groups and the average of the sexually active groups, with the 95% HDI falling far outside any reasonable ROPE. On the other hand, the difference between the average of the control groups and the Virgin1 group, while having a mode of about 7 days (effect size just under 0.5), has a posterior distribution that notably overlaps zero and the ROPE.

In traditional ANOVA, analysts often perform a so-called omnibus test that asks whether it is plausible that all the groups are simultaneously exactly equal. I find that the omnibus test is rarely meaningful, however. When the null hypothesis of all-equal groups is rejected, then the analyst is virtually always interested in specific contrasts. Importantly, when the null hypothesis of all-equal groups is not rejected, there can still be specific contrasts that exhibit clear differences (as was shown in Section 12.2.2, p. 348), so again the analyst should examine the specific contrasts of interest. In the hierarchical Bayesian estimation used here, there is no direct equivalent to an omnibus test in ANOVA, and the emphasis is on examining all the meaningful contrasts. An omnibus test can be done with Bayesian model comparison, which is briefly described in Section 20.6.

19.3.4. Multiple comparisons and shrinkage

The previous section suggested that an analyst should investigate all contrasts of interest. This recommendation can be thought to conflict with traditional advice in the context on null hypothesis significance testing, which instead recommends that a minimal number of comparisons should be conducted in order to maximize the power of each test while keeping the overall false alarm rate capped at 5% (or whatever maximum is desired). The issue of multiple comparisons was discussed extensively in Section 11.4 (p. 325), and by specific example in Section 11.1.5 (p. 310). One theme of those sections was that no analysis is immune to false alarms, but a Bayesian analysis eschews the use of p values to control false alarms because p values are based on stopping and testing intentions. Instead, a Bayesian analysis can mitigate false alarms by incorporating prior knowledge into the model. In particular, hierarchical structure (which is an expression of prior knowledge) produces shrinkage of estimates, and shrinkage can help rein in estimates of spurious outlying data. For example, in the posterior distribution from the fruit fly data, the modal values of the posterior group means have a range of 23.2. The sample means of the groups have a range of 26.1. Thus, there is some shrinkage in the estimated means. The amount of shrinkage is dictated only by the data and by the prior structure, not by the intended tests.

Caution: The model can produce implosive shrinkage of group deflections when there are few data points in each group and σ_β is estimated (instead of set to a constant). This strong shrinkage occurs because the data can be accommodated by setting all the deflections closer to zero (with small σ_β) and with larger noise standard deviation σ_y. In other words, the model prefers to attribute the overall variance to within-group variability rather than to between-group variability. This preference is not wrong; it is the correct implication of the assumed model structure. If your data involves small sample sizes in each group, and the estimated group means shrink more than seems to be a reasonable description of the data, then it may be a sign that the hierarchical prior is too strong an assumption for the data. Instead, you could set the prior on σ_β to a (large) constant. You can try this yourself in Exercise 19.1.

19.3.5. The two-group case

A special case of our current scenario is when there are only two groups. The model of the present section could, in principle, be applied to the two-group case, but the hierarchical structure would do little good because there is virtually no shrinkage when there are so few groups (and the top-level prior on σ_β is broad as assumed here). That is why the two-group model in Section 16.3 did not use hierarchical structure, as illustrated in Figure 16.11 (p. 468). That model also used a *t* distribution to accommodate outliers in the data, and that model allowed for heterogeneous variances across groups. Thus, for two groups, it is more appropriate to use the model of Section 16.3. The hierarchical multi-group model is generalized to accommodate outliers and heterogeneous variances in Section 19.5.

19.4. INCLUDING A METRIC PREDICTOR

In Figure 19.3, the data within each group have a large standard deviation. For example, longevities in the Virgin8 group range from 20 to 60 days. With such huge variance of longevities within a group, it is impressive that differences across experimental treatments could be detected at all. For example, the difference between the Virgin1 group and the control groups has an effect size of approximately 0.45 (see Figure 19.3), but the uncertainty in its estimate is large. To improve detectability of differences between groups, it would be useful if some of the within-group variance could be attributed to another measurable influence, so that the effect of the experimental treatment would better stand out.

For instance, suppose that there were some separate measure of how robust a subject is, such that more robust subjects tend to live longer than less robust subjects. We could try to experimentally control the robustness, putting only high-robustness subjects in the experiment. In that way, the variance in longevities, within groups, might be much smaller. Of course, we would also want to run the experiment with medium-robustness subjects and low-robustness subjects. On the other hand, instead of arbitrarily binning the robustness as a nominal predictor, we could simply measure the robustness of randomly included subjects, and use the robustness as a separate metric predictor of longevity. This would allow us to model the distinct influences of robustness and experiment treatment.

An analogous idea was explored in the context of multiple regression in Exercise 18.2 (p. 550). The influence of a primary metric predictor on a predicted variable may appear to be weak because of high residual variance. But a second metric predictor, even if it is uncorrelated with the first predictor, might account for some of the variance in the predicted variable, thereby making the estimate of the slope on the first predictor much more certain.

The additional metric predictor is sometimes called a covariate. In the experimental setting, the focus of interest is usually on the nominal predictor (i.e., the experimental treatments), and the covariate is typically thought of as an ancillary predictor to help isolate the effect of the nominal predictor. But mathematically the nominal and metric predictors have equal status in the model. Let's denote the value of the metric covariate for subject i as $x_{\text{cov}}(i)$. Then the expected value of the predicted variable for subject i is

$$\mu(i) = \beta_0 + \sum_j \beta_{[j]} x_{[j]}(i) + \beta_{\text{cov}} x_{\text{cov}}(i) \tag{19.5}$$

with the usual sum-to-zero constraint on the deflections of the nominal predictor stated in Equation 19.2. In words, Equation 19.5 says that the predicted value for subject i is a baseline plus a deflection due to the group of i plus a shift due to the value of i on the covariate.

It is easy to express the model in a hierarchical diagram, as shown in Figure 19.4. The model has only one change from the diagram in Figure 19.2, namely the inclusion of the covariate in the central formula for μ_i and a prior distribution on the parameter β_{cov}. The model substructure for the covariate is just like what was used for linear regression in Figures 17.2 (p. 480) and 18.4 (p. 515).

Notice that the baseline in Equation 19.5 is playing double duty. On the one hand, the baseline is supposed to make the nominal deflections sum to zero, so that the baseline represents the overall mean of the predicted data. On the other hand, the baseline is simultaneously acting as the intercept for the linear regression on x_{cov}. It makes sense to set the intercept as the mean of predicted values if the covariate is re-centered at its mean value, which is denoted $\overline{x}_{\text{cov}}$. Therefore Equation 19.5 is algebraically reformulated to make the baseline respect those constraints. We start by rewriting Equation 19.5 using unconstrained coefficients denoted as α instead of as β, because we will convert the α expressions into corresponding β values. The first equation below is simply Equation 19.5 with x_{cov} recentered on its mean, $\overline{x}_{\text{cov}}$. The second line below merely

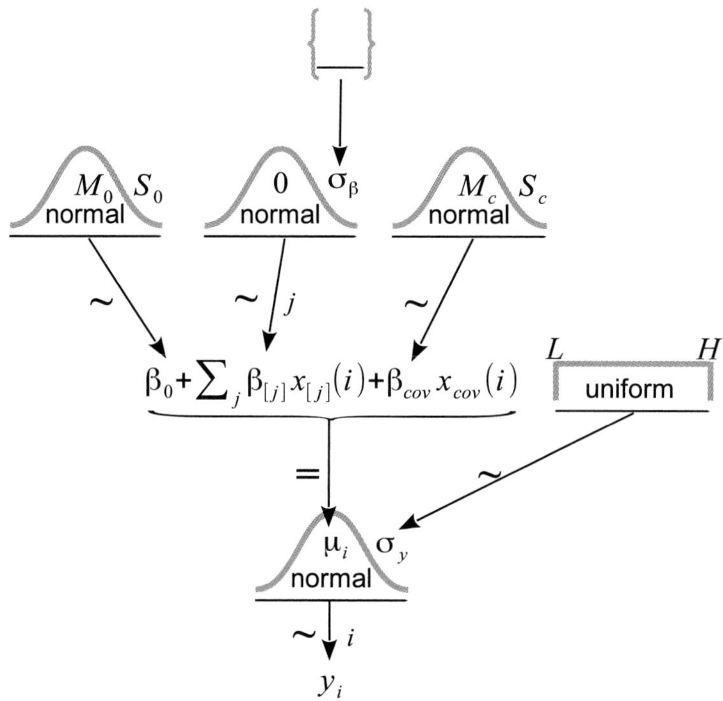

Figure 19.4 Hierarchical diagram for model that describes data from several groups of a single factor, along with a single metric covariate. Compare with Figure 19.2.

algebraically rearranges the terms so that the nominal deflections sum to zero and the constants are combined into the overall baseline:

$$\mu = \alpha_0 + \sum_j \alpha_{[j]} x_{[j]} + \alpha_{\text{cov}} \left(x_{\text{cov}} - \overline{x}_{\text{cov}} \right) \tag{19.6}$$

$$= \underbrace{\alpha_0 + \overline{\alpha} - \alpha_{\text{cov}} \overline{x}_{\text{cov}}}_{\beta_0} + \sum_j \underbrace{\left(\alpha_{[j]} - \overline{\alpha} \right)}_{\beta_{[j]}} x_{[j]} + \underbrace{\alpha_{\text{cov}}}_{\beta_{\text{cov}}} x_{\text{cov}} \tag{19.7}$$

$$\text{where } \overline{\alpha} = \frac{1}{J} \sum_{j=1}^{J} \alpha_{[j]}$$

In the JAGS (or Stan) program for this model, each step in the MCMC chain generates jointly credible values of the α parameters in Equation 19.6, which are then converted to β values that respect the sum-to-zero constraint as indicated by the underbraces in Equation 19.7. The JAGS implementation is in the program `Jags-Ymet-Xnom1met1-MnormalHom.R`.

19.4.1. Example: Sex, death, and size

We continue with the example from Section 19.3.2, regarding the effect of sexual activity on longevity of male fruit flies. The data were displayed in Figure 19.3 (p. 563) along with posterior predictive distributions. The standard deviation of the noise, σ_y, which corresponds to the spread (along the y axis) of the normal distributions plotted in Figure 19.3, had a posterior modal value of approximately 14.8 days. This large within-group variance makes it challenging to estimate small between-group differences. For example, the contrast between the sexually deprived males and the Virgin1 group, shown in the second panel of the lower row, suggests about a 7-day difference in longevity, but the uncertainty of the estimate is large, such that the 95% HDI of the difference extends from about zero to almost 14 days.

It turns out that the life span of a fruit fly is highly correlated with its overall size (which asymptotes at maturity). Larger fruit flies live longer. Because fruit flies were randomly assigned to the five treatments, each treatment had a range of fruit flies of different sizes, and therefore much of the within-group variation in longevity could be due to variation is size alone. The researchers measured the size of each fruit fly's thorax, and used the thorax as a covariate in the analysis (Hanley & Shapiro, 1994; Partridge & Farquhar, 1981).

The high-level script `Jags-Ymet-Xnom1met1-MnormalHom-Example.R` shows how to load the data and run the analysis. The only difference from the previous analysis is that the name of the covariate must be specified, which in this case is "Thorax." The results of the analysis are shown in Figure 19.5. The panels in the upper row show the data within each group, plotted as a function of thorax size on the abscissa. Credible posterior descriptions are superimposed on the data. Within the jth group, the superimposed lines show $\beta_0+\beta_{[j]}+\beta_{\text{cov}}x_{\text{cov}}$ for jointly credible values of the parameters at various steps in the MCMC chain. The sideways-plotted normal distributions illustrate the corresponding values of σ_y for each line.

A key feature to notice is that the within-group noise standard deviation is smaller compared to the previous analysis without the covariate. Specifically, the modal σ_y is about 10.5 days. The contrast between the sexually deprived males and the Virgin1 group, shown in the second panel of the lower row, now has a 95% HDI that extends from about 4 days to almost 14 days, which clearly excludes a difference of zero and is a more certain estimate than without the covariate. The HDI widths of all the contrasts have gotten smaller by virtue of including the covariate in the analysis.

19.4.2. Analogous to traditional ANCOVA

In traditional frequentist methods, use of the model described above (without hierarchical structure) is called ANCOVA. The best fitting parameter values are derived

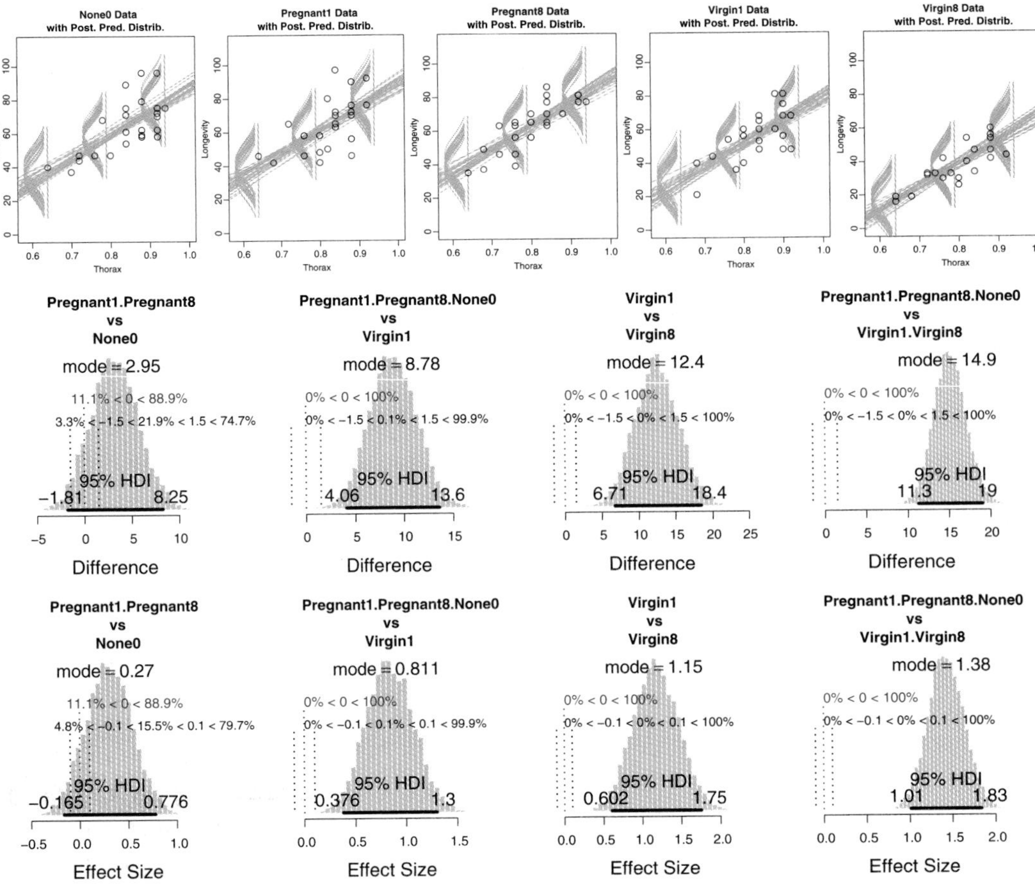

Figure 19.5 Posterior distribution for fruit fly longevity, described by normal distributions with homogeneous variances and a linear function of a covariate. Upper row shows that the within-group variance is smaller than in Figure 19.3. Lower rows show that contrasts are more precise than in Figure 19.3. In particular, here the contrast of Pregnant1 and Pregnant8 and None0 vs Virgin1 is clearly nonzero.

as least-squares estimates, and p values are computed from a null hypothesis and fixed-N sampling intention. As mentioned earlier in the context of ANOVA (Section 19.2), Bayesian methods do not partition the least-squares variance to make estimates, and therefore the Bayesian method is analogous to ANCOVA but is not ANCOVA. Frequentist practitioners are urged to test (with p values) whether the assumptions of (a) equal slope in all groups, (b) equal standard deviation in all groups, and (c) normally distributed noise can be rejected. In a Bayesian approach, the descriptive model is generalized to address these concerns, as will be discussed in Section 19.5.

19.4.3. Relation to hierarchical linear regression

The model in this section has similarities to the hierarchical linear regression of Section 17.3. In particular, Figure 17.5 (p. 492) bears a resemblance to Figure 19.5, in that different subsets of data are being fit with lines, and there is hierarchical structure across the subsets of data.

In Figure 17.5, a line was fit to data from each individual, while in Figure 19.5, a line was fit to data from each treatment. Thus, the nominal predictor in Figure 17.5 is individuals, while the nominal predictor in Figure 19.5 is groups. In either case, the nominal predictor affects the intercept term of the predicted value.

The main structural difference between the models is in the slope coefficients on the metric predictor. In the hierarchical linear regression of Section 17.3, each individual is provided with its own distinct slope, but the slopes of different individuals mutually informed each other via a higher-level distribution. In the model of Figure 19.5, all the groups are described using the same slope on the metric predictor. For a more detailed comparison of the model structures, compare the hierarchical diagrams in Figures 17.6 (p. 493) and 19.4 (p. 570).

Conceptually, the main difference between the models is merely the focus of attention. In the hierarchical linear regression model, the focus was on the slope coefficient. In that case, we were trying to estimate the magnitude of the slope, simultaneously for individuals and overall. The intercepts, which describe the levels of the nominal predictor, were of ancillary interest. In the present section, on the other hand, the focus of attention is reversed. We are most interested in the intercepts and their differences between groups, with the slopes on the covariate being of ancillary interest.

19.5. HETEROGENEOUS VARIANCES AND ROBUSTNESS AGAINST OUTLIERS

As was mentioned earlier in the chapter, we have assumed normally distributed data within groups, and equal variances across the groups, merely for simplicity and for consistency with traditional ANOVA. We can relax those assumptions in Bayesian software. In this section, we use *t* distributed noise instead of normal distributions, and we provide every group with its own standard-deviation parameter. Moreover, we put a hierarchical prior on the standard-deviation parameters, so that each group mutually informs the standard deviations of the other groups via the higher-level distribution.

Figure 19.6 shows a hierarchical diagram of the new model. It is merely an extension of the model in Figure 19.2. At the bottom of the diagram, the data y_i are described by a *t* distribution instead of a normal distribution. The *t* distribution has a normality parameter ν annotated on its left, with its prior being the usual prior we have seen previously (e.g., robust estimation of two groups in Figure 16.11, and robust regression in Figure 17.2). The main novelty of Figure 19.6 involves the right-hand side, which

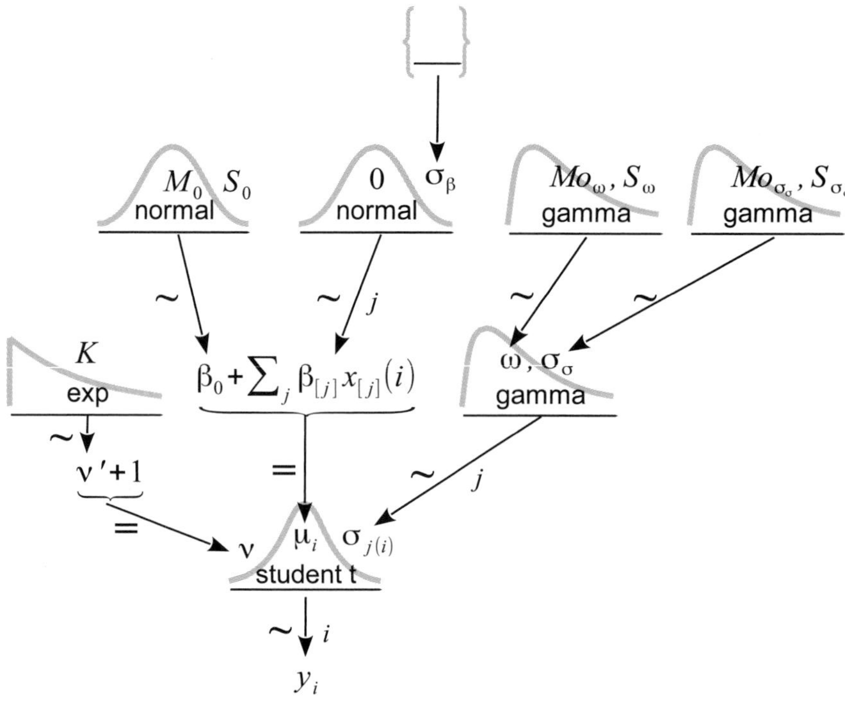

Figure 19.6 Hierarchical diagram for model that describes data from several groups of a single factor, using a heavy-tailed noise distribution and different standard deviations for each group. Compare with Figure 19.2. (The gamma distributions are parameterized by mode and standard deviation.)

shows the hierarchical prior on the scale parameter of the t distribution. Instead of there being a single σ_y parameter that applies to all groups, each group has its own scale parameter, σ_j. The scale parameters of the groups all come from a gamma distribution that has mode ω and standard deviation σ_σ. The mode and standard deviation of the gamma distribution could be set to constants, so that each group's scale is estimated separately. Instead, we will estimate the modal scale value, and estimate the standard deviation of the scale values. The hierarchical diagram shows that both ω and σ_σ are given gamma priors that have modes and standard deviations that make them vague on the scale of the data.

The JAGS implementation of the model is a straight-forward extension of the previous model. The only novel part is deciding what should be the constants for the top-level gamma distributions. I have found it convenient to use the same vague prior for all the top-level gamma distributions, but you may, of course, adjust this as appropriate for your application. Recall that the shape and rate values were chosen so that the mode would be `sd(y)/2` and the standard deviation would be `2*sd(y)`:

```
aGammaShRa = unlist( gammaShRaFromModeSD( mode=sd(y)/2 , sd=2*sd(y) ) )
```

This choice makes the prior broad on the scale of the data, whatever the scale might happen to be.

In the JAGS model statement, below, you should be able to find corresponding lines of code for each arrow in Figure 19.6. One bit of code that is not in the diagram is the conversion from a gamma distribution's mode and standard deviation to its shape and rate, from Equation 9.8 (p. 238). Hopefully you can recognize that conversion in the following JAGS model specification:

```
model {
  for ( i in 1:Ntotal ) {
    y[i] ~ dt(  a0 + a[x[i]] , 1/ySigma[x[i]]^2 , nu )
  }
  nu <- nuMinusOne+1
  nuMinusOne ~  dexp(1/29)
  for ( j in 1:NxLvl ) { ySigma[j] ~ dgamma( ySigmaSh , ySigmaRa ) }
  ySigmaSh <- 1 + ySigmaMode * ySigmaRa
  ySigmaRa <- ( ( ySigmaMode + sqrt( ySigmaMode^2 + 4*ySigmaSD^2 ) )
                / ( 2*ySigmaSD^2 ) )
  ySigmaMode ~ dgamma( aGammaShRa[1] , aGammaShRa[2] )
  ySigmaSD ~ dgamma( aGammaShRa[1] , aGammaShRa[2] )
  a0 ~ dnorm(yMean,1/(ySD*10)^2)
  for ( j in 1:NxLvl ) { a[j] ~ dnorm( 0.0 , 1/aSigma^2 ) }
  aSigma ~ dgamma( aGammaShRa[1] , aGammaShRa[2] )
  # Convert a0,a[] to sum-to-zero b0,b[] :
  for ( j in 1:NxLvl ) { m[j] <- a0 + a[j] }
  b0 <- mean( m[1:NxLvl] )
  for ( j in 1:NxLvl ) { b[j] <- m[j] - b0 }
}
```

The full program is in the file named `Jags-Ymet-Xnom1fac-MrobustHet.R`, and the high-level script that calls it is the file named `Jags-Ymet-Xnom1fac-MrobustHet-Example.R`.

19.5.1. Example: Contrast of means with different variances

To illustrate the potential usefulness of a model with unequal variances across groups, I have contrived an artificial data set, shown as the dots in Figures 19.7 and 19.8. In this set of data, there are four groups, labeled A, B, C, and D, with means (respectively) of 97, 99, 102, and 104. The standard deviations of the groups are dramatically different, with groups A and D having very large standard deviations relative to groups B and C. The data are randomly sampled from normal distributions.

We first examine the results of applying that model that assumes homogeneous variances (diagrammed in Figure 19.2). The results are shown in Figure 19.7. You can see that the estimated within-group standard deviation seems to be much too big for groups

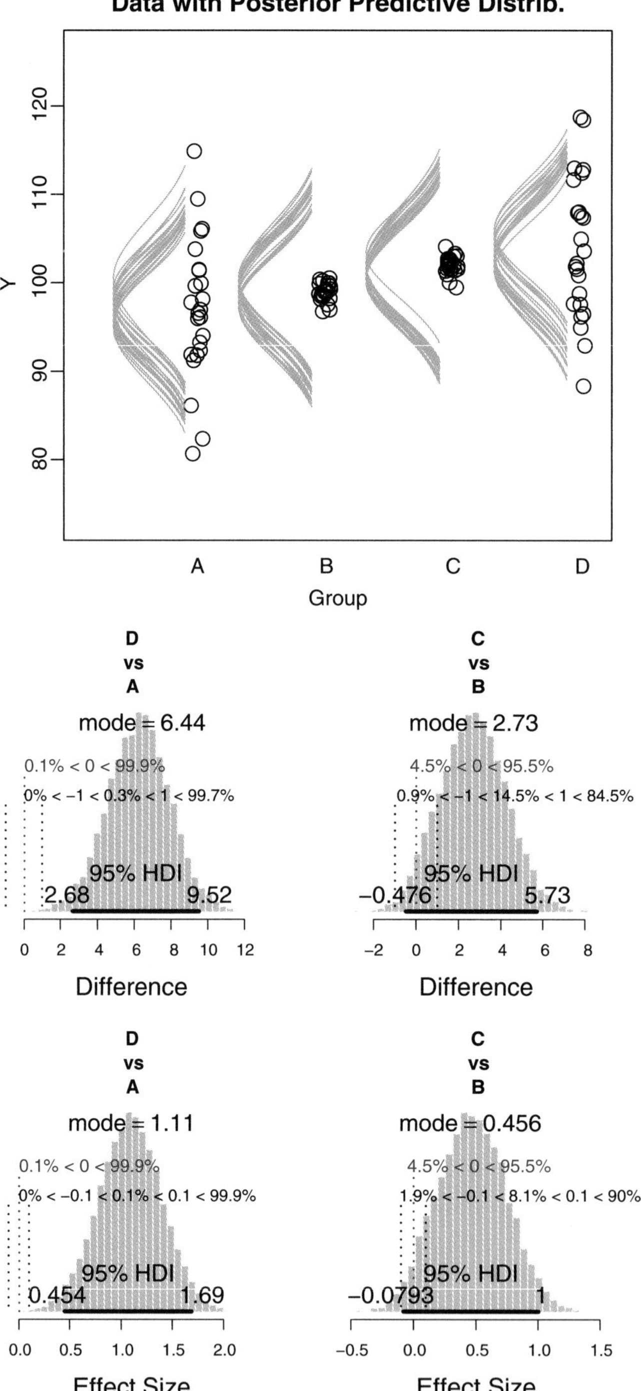

Figure 19.7 Fictitious data to illustrate groups with different variances. Here, the model assumes equal variances across groups. Compare with Figure 19.8.

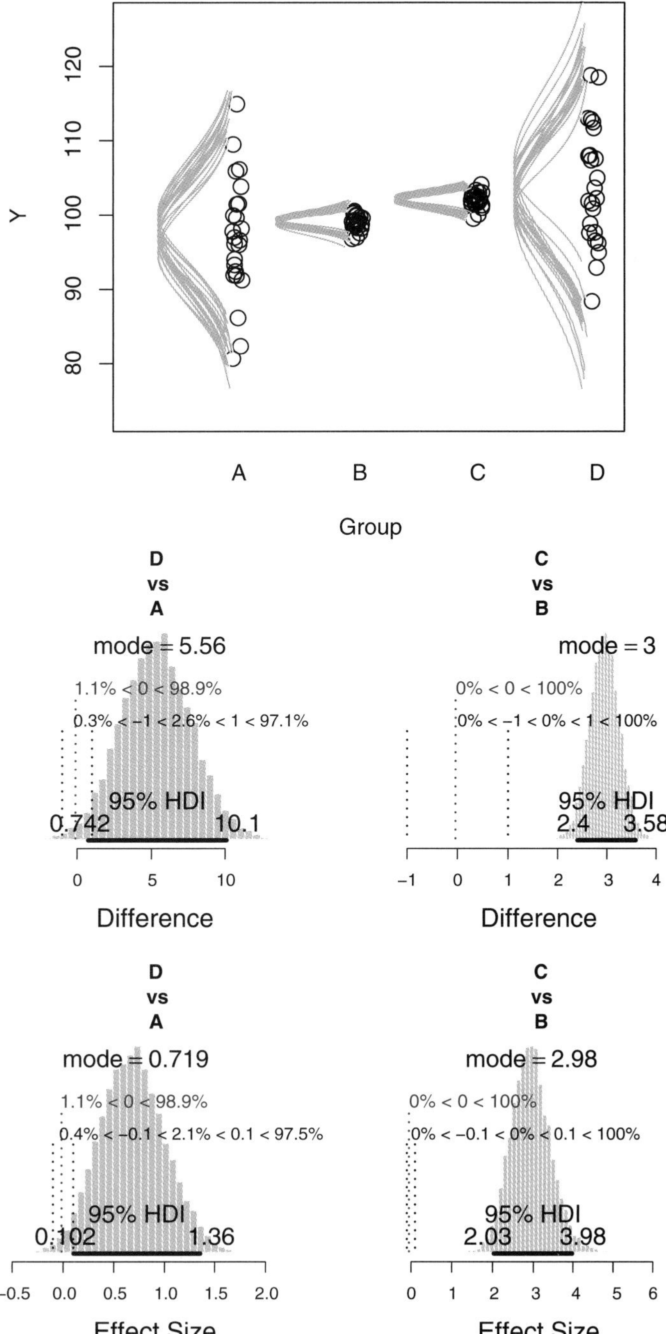

Figure 19.8 Fictitious data to illustrate groups with different variances. Here, the model assumes different variances across groups. Compare with Figure 19.7.

B and C, but too small for groups A and D. In other words, the estimate of the within-group standard deviation ends up being a compromise between the small-variance and large-variance groups.

An important ramification of the estimates of the within-group variance appears in the contrasts of the means, shown in the bottom of Figure 19.7. The posterior distribution on $\mu_D - \mu_A$ indicates that the difference is clearly greater than zero. On the other hand, the posterior distribution on $\mu_C - \mu_B$ indicates that the difference, while positive, is uncertain enough that zero is within the 95% HDI. But these conclusions seem to conflict with what intuition might suggest from the data themselves. The scatter of data in groups B and C barely overlap, and therefore it seems that the difference should be credibly nonzero. But the scatter of data in groups A and D overlap quite substantially, so it would take a lot of data to convince us that the central tendencies are credibly different.

The problem is that the comparison of μ_C and μ_B is using a standard deviation that is too big. The large value of σ widens the normal distribution for describing the data, which lets the μ value slide up and down quite a ways while still keeping the data under the tall part of the normal distribution. Thus, the estimates of μ_C and μ_B are "sloppy" and the posterior distribution of their difference is artificially wide. On the other hand, the comparison of μ_D and μ_A is using a standard deviation that is too small. The small value of σ keeps the estimate of μ close to the center of the data. Thus, the estimate of $\mu_D - \mu_A$ is artificially constrained.

When we analyze the data using the model that allows different variances in each group, we get the results shown in Figure 19.8. Notice that the estimated standard deviations of groups B and C are now much smaller than for groups A and D. The description seems far more appropriate than when using equal variances for all the groups.

Importantly, the contrasts of the means in the lower portion of Figure 19.8 are quite different than in Figure 19.7 and make more sense. In Figure 19.8, the posterior estimate of the difference $\mu_C - \mu_B$ is clearly greater than zero, which makes sense because the data of those groups barely overlap. But the posterior estimate of $\mu_D - \mu_A$, while positive, has a 95% HDI that overlaps a modest ROPE around zero, which also makes sense because the data from the two groups overlap a lot and have only modest sample sizes.

Finally, because each group has its own estimated scale (i.e., σ_j), we can investigate differences in scales across groups. The difference of scales between groups was implemented by default for the two-group case back in Figure 16.12 (p. 471). The multigroup program does not have contrasts of scales built in, but you can easily create them. For example, at any point after running the `mcmcCoda = genMCMC(...)` line of the script, you can run the following commands to display the posterior distribution of $\sigma_1 - \sigma_2$:

```
mcmcMat = as.matrix( mcmcCoda ) # convert coda object to a matrix
openGraph()                     # open a new graphics window
plotPost( mcmcMat[,"ySigma[1]"] - mcmcMat[,"ySigma[2]"] ,
          main=expression(sigma[1]-sigma[2]) , xlab="Difference" ,
          cex.main=2 )
```

In fact, that would make a good exercise; let's call it Exercise 19.3.

19.6. EXERCISES

Look for more exercises at https://sites.google.com/site/doingbayesiandataanalysis/

Exercise 19.1. [Purpose: Notice implosive shrinkage on means when sample size is small, and lack of shrinkage when setting σ_β to a large constant.]

(A) Consider the data file named AnovaShrinkageData.csv, which has columns named Group and Y. How many groups are there? What are the group labels? How many data points per group are there? What are the means of the groups? (*Hints:* Consider using `myDataFrame = read.csv(file="AnovaShrinkageData. csv")` then `aggregate`.)

(B) Adapt the high-level script `Jags-Ymet-Xnom1fac-MnormalHom-Example.R` for reading in AnovaShrinkageData.csv. Set up three contrasts between groups: U vs A, M vs A, and G vs A. Do any of the contrasts suggest a credible non-zero difference between groups? For each of the contrasts, what is the estimated difference between the groups, and what is the actual difference between the sample means of the groups? Include the resulting graphs in your report; see Figure 19.9 for an example. Do any of the graphs look like an inverted funnel, ⋏ as was mentioned in Section 18.3?

(C) In the program file `Jags-Ymet-Xnom1fac-MnormalHom.R`, change the model specification so that σ_β is not estimated but is instead fixed at a large value. Specifically, find `aSigma` and make this change:

```
# aSigma ~ dgamma( agammaShRa[1] , agammaShRa[2] )
aSigma <- ySD*10
```

Save the program, and then rerun the high-level script with the three contrasts from the previous part. Answer the questions of the previous part, applied to this output. See Figure 19.9 for an example. When you are done with this part, be sure to change the program back to the way it was.

(D) Why are the results of the previous two parts so different? *Hint:* What is the estimate of `aSigma` (i.e., σ_β) when it is estimated instead of fixed? Discuss why there is so much shrinkage of the group means, for this particular data set, when σ_β is estimated.

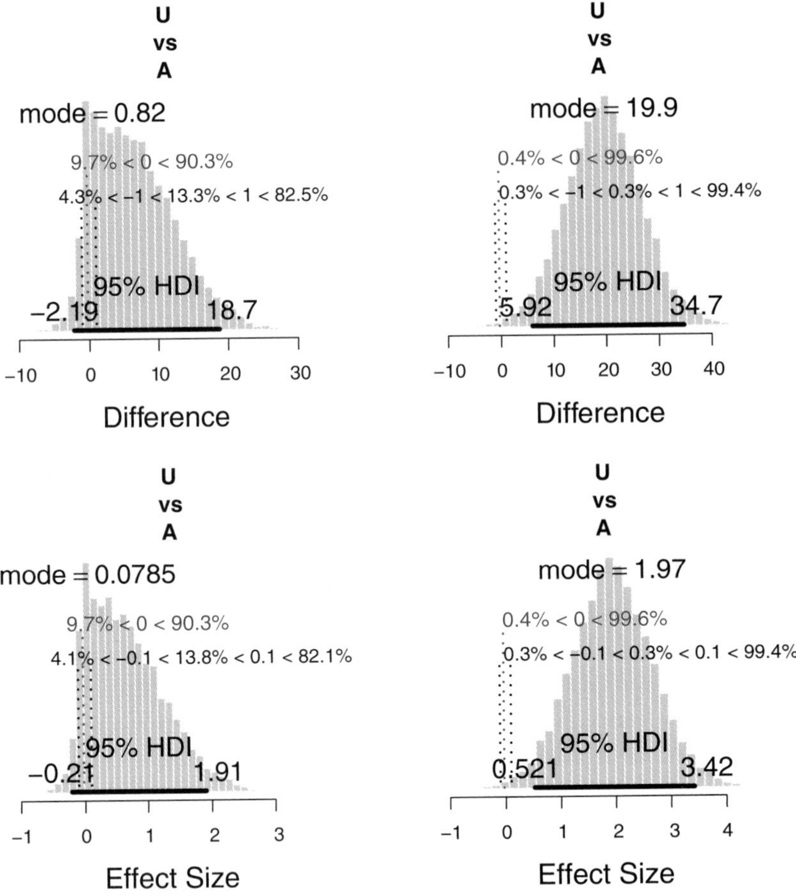

Figure 19.9 For Exercise 19.1. Left: Implosive shrinkage resulting from estimated σ_β. Right: Results for fixed σ_β.

Exercise 19.2. [Purpose: Explicit understanding of the prior distribution for the ANOVA-like model.]

(A) Run the script `Jags-Ymet-Xnom1fac-MnormalHom-Example.R` with the *Drosophila* longevity data, such that the prior distribution is produced. See Section 8.5 for a reminder about how to generate the prior from JAGS. *Hint:* You only need to comment out a single line in the `dataList`. Change the `fileNameRoot` in the script so that you don't overwrite the posterior distribution output.

(B) Continuing with the previous part: Make a fine-bin histogram of the lower part of the prior on σ_β (`aSigma`). Superimpose a curve of the exact gamma distribution using the shape and rate values from the prior specification. Superimpose a vertical line at the mode of the exact gamma distribution. *Hints:* Make the histogram by using `hist( as.matrix(mcmcCoda)[,"aSigma"] ... )` with the `breaks` and `xlim` set so to reveal the low end of the distributions. Get the shape and rate values of the gamma by using `gammaShRaFromModeSD` as in the model specification, and then plot the gamma using the `lines` command. Superimpose a vertical line by using the `abline(v=...)` command. If you've done it all correctly, the curve for the gamma will closely hug the histogram, and the vertical line will intersect the mode of the curve.

(C) For every parameter, state whether or not the prior is reasonably broad in the vicinity of where the posterior ended up.

Exercise 19.3. [Purpose: Use the heterogeneous-variance model to examine differences of scales across groups.] Using R code like that explained at the end of Section 19.5.1, with the data in Figure 19.8 (NonhomogVarData.csv), create graphs of the posterior distributions of $\sigma_1 - \sigma_2$, $\sigma_2 - \sigma_3$, and $(\sigma_1 + \sigma_4)/2 - (\sigma_2 + \sigma_3)/2$.

Exercise 19.4. [Purpose: Working with Stan.] This exercise is a bit of a programing project and therefore might take some time.

(A) Convert the JAGS model in file `Jags-Ymet-Xnom1fac-MnormalHom.R` to Stan (see Chapter 14). Run it on the *Drosophila* longevity data and confirm that it produces the same posterior distribution (except for randomness in the MCMC sample).

(B) The JAGS version of the program had fairly low autocorrelation for all parameters except σ_β (`aSigma`). Does Stan show lower autocorrelation for this parameter? To achieve equivalent ESS on the parameters, which of Stan or JAGS takes more real time? (In answering these questions, set thinning to 1 to get a clear comparison.)

CHAPTER 20

Metric Predicted Variable with Multiple Nominal Predictors

Contents

Sometimes I wonder just how it could be, that
Factors aligned so you'd end up with me.
All of the priors made everyone think, that
Our interaction was destined to shrink.[1]

This chapter considers data structures that consist of a metric predicted variable and two (or more) nominal predictors. This chapter extends ideas introduced in the previous chapter, so please read the previous chapter if you have not already. Data structures of the type considered in this chapter are often encountered in real research. For example, we might want to predict monetary income from political party affiliation and religious affiliation, or we might want to predict galvanic skin response to different combinations

[1] One of the topics of this chapter is interaction of nominal predictors. The interaction deflections can experience a lot of shrinkage in a hierarchical model.

Doing Bayesian Data Analysis, 2nd Edition
http://dx.doi.org/10.1016/B978-0-12-405888-0.00020-9

of categories of visual stimulus and categories of auditory stimulus. As mentioned in the previous chapter, this type of data structure can arise from experiments or from observational studies. In experiments, the researcher assigns the categories (at random) to the experimental subjects. In observational studies, both the nominal predictor values and the metric predicted value are generated by processes outside the direct control of the researcher. In either case, the same mathematical description can be applied to the data (although causality is best inferred from experimental intervention).

The traditional treatment of this sort of data structure is called multifactor analysis of variance (ANOVA). Our Bayesian approach will be a hierarchical generalization of the traditional ANOVA model. The chapter also considers generalizations of the traditional models, because it is straight forward in Bayesian software to implement heavy-tailed distributions to accommodate outliers, along with hierarchical structure to accommodate heterogeneous variances in the different groups.

In the context of the generalized linear model (GLM) introduced in Chapter 15, this chapter's situation involves a linear function of multiple nominal predictors, as indicated in the final column of Table 15.1 (p. 434), with a link function that is the identity along with a normal distribution for describing noise in the data, as indicated in the first row of Table 15.2 (p. 443). For a reminder of how this chapter's combination of predicted and predictor variables relates to other combinations, see Table 15.3 (p. 444).

20.1. DESCRIBING GROUPS OF METRIC DATA WITH MULTIPLE NOMINAL PREDICTORS

The ideas of describing metric data as a function of nominal predictors were explained back in Sections 15.2.4.1–15.2.4.3 (p. 429) and in Section 19.1 (p. 554). Please review those sections now. The material of those sections will be briefly reprised here.

Suppose we have two nominal predictors, denoted $\overrightarrow{x}_1$ and $\overrightarrow{x}_2$. A datum from the jth level of $\overrightarrow{x}_1$ is denoted $x_{1[j]}$, and analogously for the second factor. The predicted value is a baseline plus a deflection due to the level of factor 1 plus a deflection due to the level of factor 2 plus a residual deflection due to the interaction of factors:

$$\begin{aligned} \mu &= \beta_0 + \overrightarrow{\beta}_1 \cdot \overrightarrow{x}_1 + \overrightarrow{\beta}_2 \cdot \overrightarrow{x}_2 + \overrightarrow{\beta}_{1\times2} \cdot \overrightarrow{x}_{1\times2} \\ &= \beta_0 + \sum_j \beta_{1[j]} x_{1[j]} + \sum_k \beta_{2[k]} x_{2[k]} + \sum_{j,k} \beta_{1\times2[j,k]} x_{1\times2[j,k]} \end{aligned} \tag{20.1}$$

The deflections within factors and within the interaction are constrained to sum to zero:

$$\sum_j \beta_{1[j]} = 0 \quad \text{and} \quad \sum_k \beta_{2[k]} = 0 \quad \text{and}$$

$$\sum_j \beta_{1\times2[j,k]} = 0 \text{ for all } k \quad \text{and} \quad \sum_k \beta_{1\times2[j,k]} = 0 \text{ for all } j \tag{20.2}$$

(Equations 20.1 and 20.2 are repetitions of Equations 15.9 and 15.10, p. 434). The actual data are assumed to be randomly distributed around the predicted value.

20.1.1. Interaction

An important concept of models with multiple predictors is interaction. Interaction means that the effect of a predictor depends on the level of another predictor. A little more technically, interaction is what is left over after the main effects of the factors are added: interaction is the nonadditive influence of the factors.

Figure 20.1 shows a simple example of interaction. Both factors have only two levels, so there are four groups altogether. The means of the four groups are plotted as vertical bars; you should imagine that the actual data points are scattered vertically near the tops the bars, as in Figure 19.1 (p. 555). The means are repeated three times in Figure 20.1, with different superimposed lines for different emphases. Within each panel, the left pair of bars indicates level 1 of factor x_1, and the right pair of bars indicates level 2 of factor x_1. Within each pair are the two levels of factor x_2.

The baseline, β_0, is not marked in Figure 20.1, but it is easy to see that the baseline must be five because that is the mean of the four bars. The deflection for level 1 of x_1 is -1.8 and the deflection for level 2 of x_1 is $+1.8$. In other words, to get from the average of the two left bars to the average of the two right bars, we have to go up 3.6 (i.e., two times 1.8). This average influence of factor x_1 is indicated in the left panel by the pair of dashed lines that slope upward from the left pair of bars to the right pair of bars. For the other factor, x_2, the deflection is $+0.2$ for level 1 and -0.2 for level 2. In other words, within each pair of bars, on average to get from level 1 of x_2 to level 2 of x_2 we have to go down 0.4 (i.e., two times 0.2). This average influence of factor x_2 is indicated in

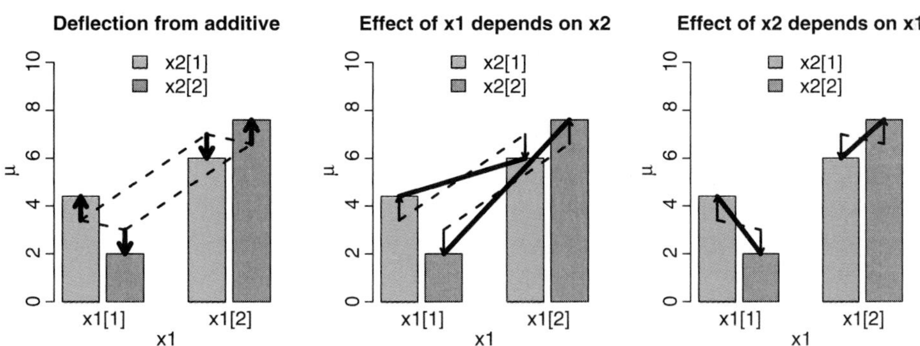

Figure 20.1 An example to illustrate the notion of interaction. Each panel plots the same four means but with different superimposed lines for different emphases expressed in the title of each panel. The dashed lines indicate the average (i.e., main) effects of the factors. Subscripts are elevated to regular size for readability; for example $x_{2[1]}$ is displayed as $x2[1]$. (Compare with Figure 15.5, p. 432.)

the left panel by the dashed lines that slope downward within each pair of bars. These average effects of the factors are often called the main effects.

If the effects of the two factors were purely additive, then the heights of the bars would be at the ends of the dashed lines. For example, the far right bar should be at the baseline plus the deflection due to level 2 of x_1 plus the deflection due to level 2 of x_2, which is $5 + 1.8 + -0.2 = 6.6$. But you can see that the actual height of the far right bar is 7.6. The remaining, nonadditive component is marked by a vertical arrow. Across all four bars, the non-additive interaction components are marked by vertical arrows. Notice that within each level of a factor, the interaction components sum to zero. Thus, within the left pair of pairs (i.e., level 1 of x_1), the two vertical arrows sum to zero. And, for the left bars across the pairs (i.e., level 1 of x_2), the two vertical arrows sum to zero. These sum-to-zero properties were expressed algebraically in Equation 20.2.

Table 20.1 shows the general algebraic method for computing the sum-to-zero deflections from the cell means. We start with the cell means, $m_{1\times 2[j,k]}$, where j refers to the row and k refers to the column. (The cell means correspond to the heights of the bars in Figure 20.1.) We then compute the marginal means, $m_{1[j]}$ and $m_{2[k]}$, and the overall mean, m (in the lower right corner). Then we work "backwards" from the marginal means to the deflections. First, we set the baseline β_0 equal to the overall mean m. Then we determine the main effect deflections as shown in the margins of Table 20.1; for example $\beta_{1[j]} = m_{1[j]} - \beta_0$. Finally, the interaction deflections are set to the cell means minus the sum of the main effects; for example $\beta_{1\times 2[1,1]} = m_{1\times 2[1,1]} - (\beta_{1[1]} + \beta_{2[1]} + \beta_0)$. This method generalizes to any number of levels within factors and any number of factors.

It is straightforward to verify that the deflections computed in Table 20.1 satisfy the sum-to-zero constraints. For example, you can start with $\beta_{1[1]} + \beta_{1[2]}$ and substitute

Table 20.1 How to compute sum-to-zero deflections

$m_{1\times 2[1,1]}$ $\beta_{1\times 2[1,1]}$ $= m_{1\times 2[1,1]}$ $- (\beta_{1[1]} + \beta_{2[1]} + \beta_0)$	$m_{1\times 2[1,2]}$ $\beta_{1\times 2[1,2]}$ $= m_{1\times 2[1,2]}$ $- (\beta_{1[1]} + \beta_{2[2]} + \beta_0)$	$m_{1[1]} = \frac{1}{K}\sum_k^K m_{1\times 2[1,k]}$ $\beta_{1[1]} = m_{1[1]} - \beta_0$
$m_{1\times 2[2,1]}$ $\beta_{1\times 2[2,1]}$ $= m_{1\times 2[2,1]}$ $- (\beta_{1[2]} + \beta_{2[1]} + \beta_0)$	$m_{1\times 2[2,2]}$ $\beta_{1\times 2[2,2]}$ $= m_{1\times 2[2,2]}$ $- (\beta_{1[2]} + \beta_{2[2]} + \beta_0)$	$m_{1[2]} = \frac{1}{K}\sum_k^K m_{1\times 2[2,k]}$ $\beta_{1[2]} = m_{1[2]} - \beta_0$
$m_{2[1]} = \frac{1}{J}\sum_j^J m_{1\times 2[j,1]}$ $\beta_{2[1]} = m_{2[1]} - \beta_0$	$m_{2[2]} = \frac{1}{J}\sum_j^J m_{1\times 2[j,2]}$ $\beta_{2[2]} = m_{2[2]} - \beta_0$	$m = \frac{1}{J\cdot K}\sum_{j,k}^{J,K} m_{1\times 2[j,k]}$ $\beta_0 = m$

Start with the cell means, $m_{1\times 2[j,k]}$, where j refers to the row and k refers to the column. Then compute the marginal means, $m_{1[j]}$, $m_{2[k]}$, and m. Then compute the baseline β_0, the main effect deflections $\beta_{1[j]}$ and $\beta_{2[k]}$, and the interaction deflections $\beta_{1\times 2[j,k]}$

the appropriate means to find that the terms all cancel to yield zero. A less tedious approach is to apply recursively the well-known lemma that deflections from the mean sum to zero. To wit (as the mathematicians say): Consider numbers y_1 through y_N. By definition, their mean is $M = \frac{1}{N}\sum_n^N y_n$. The sum of the deflections from the mean is $\sum_n^N (y_n - M) = \sum_n^N y_n - \sum_n^N M = NM - NM = 0$. The lemma is first applied to the main-effect deflections, then to the interaction deflections.

The average deflection from baseline due to a predictor, in the margins of Table 20.1, is called the main effect of the predictor. The main effects of the predictors correspond to the dashed lines in the left panel of Figure 20.1. When there is nonadditive interaction between predictors, the effect of one predictor depends on the level of the other predictor. The deflection from baseline for a predictor, at a fixed level of the other predictor, is called the simple effect of the predictor at the level of the other predictor. When there is interaction, the simple effects do not equal the main effect.

Now, finally, I can get to the main point of Figure 20.1. The left panel of Figure 20.1 highlights the interaction as the nonadditive component, emphasized by the heavy vertical arrows that mark the departure from additivity. The middle panel of Figure 20.1 highlights the interaction by emphasizing that the effect of x_1 depends on the level of x_2. The heavy lines mark the effect of x_1, that is, the changes from level 1 of x_1 to level 2 of x_1. Notice that the heavy lines have different slopes: The heavy line for level 1 of x_2 has a shallower slope than the heavy line for level 2 of x_2. The right panel of Figure 20.1 highlights the interaction by emphasizing that the effect of x_2 depends on the level of x_1. The heavy lines mark the effect of x_2, that is, the changes from level 1 of x_2 to level 2 of x_2. Notice that the heavy lines have different slopes across levels of x_1, showing that the effect of x_2 depends on the level of x_1.

It may be edifying to compare Figure 20.1, which shows interaction of nominal predictors, with Figure 18.8 (p. 526), which shows interaction of metric predictors. The essential notion of interaction is the same in both cases: interaction is the nonadditive portion of the prediction, and interaction means that the effect of one predictor depends on the level of the other predictor.

20.1.2. Traditional ANOVA

As was explained in Section 19.2 (p. 556), the terminology, "analysis of variance," comes from a decomposition of overall data variance into within-group variance and between-group variance. That algebraic relation is not used in the hierarchical Bayesian approach presented here. The Bayesian method can estimate component variances, however. Therefore the Bayesian approach is not ANOVA, but is analogous to ANOVA. Traditional ANOVA makes decisions about equality of groups (i.e., null hypotheses) on the basis of p values using a null hypothesis that assumes (i) the data are normally distributed within groups, and (ii) the standard deviation of the data within each group is the same for all groups. The second assumption is sometimes called "homogeneity of variance." The entrenched precedent of ANOVA is why basic models of grouped data make those assumptions, and why the basic models presented in this chapter will also make those assumptions. Later in the chapter, those constraints will be relaxed.

20.2. HIERARCHICAL BAYESIAN APPROACH

Our goal is to estimate the main and interaction deflections, and other parameters, based on the observed data. The hierarchical diagram for the model is shown in Figure 20.2. Although the diagram may appear a bit unwieldy, it is simply an expansion of the diagram for single-factor "ANOVA" in Figure 19.2 (p. 558). At the bottom of Figure 20.2, the datum y_i is assumed to be normally distributed around the predicted value μ_i. Moving up the diagram, we see that the predicted value is the baseline plus deflections expressed in Equation 20.1. Each of the parameters is given a prior distribution exactly analogous

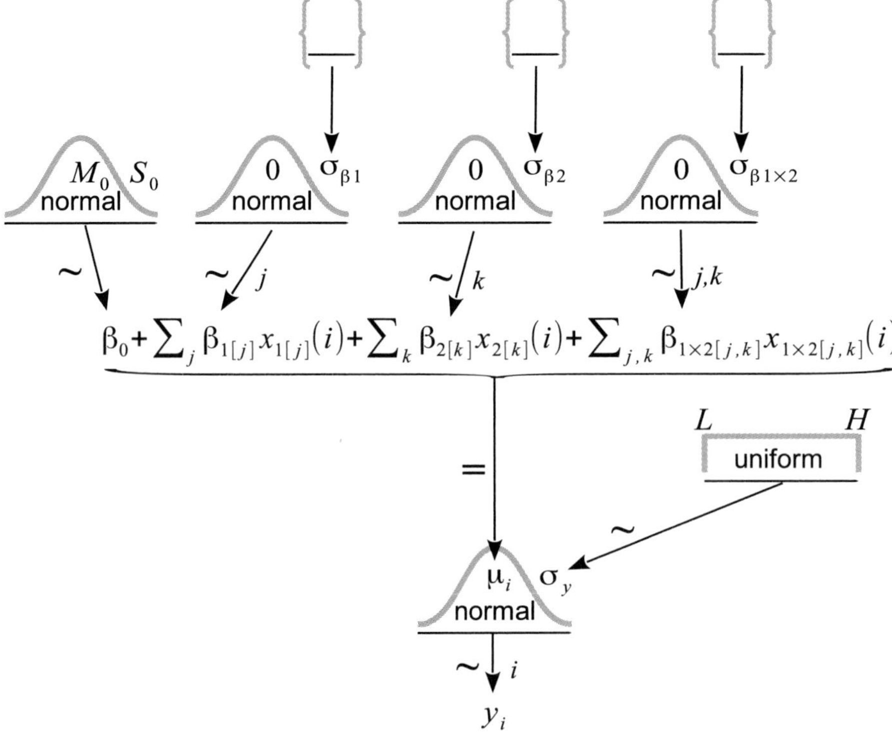

Figure 20.2 Hierarchical diagram for model that describes data from two nominal predictors. At the top of the diagram, the empty braces indicate the prior distribution on the standard deviations of the deflections, which could be a folded-*t* as recommended by Gelman (2006), a gamma distribution with nonzero mode, or a constant if no sharing across levels is desired. Compare with Figure 19.2 (p. 558).

to the single-factor model of Figure 19.2 (p. 558). In particular, at the lower-right of the diagram, there is only one within-group standard deviation that is used for all groups, which is to say that the model assumes homogeneity of variance.

A key conceptual aspect of the model structure is that top-level distributions apply separately to the different predictors and interactions. In other words, there is not just one top-level distribution that describes all deflections for all predictors and interactions together. Instead, the separation reflects a prior assumption that the magnitude of the effect of one predictor might not be very informative regarding the magnitude of the effect of a different predictor. But, within a predictor, the magnitude of deflection produced by one level may inform the magnitude of deflection produced by other levels of that same predictor.[2] The interaction deflections have their own prior distribution, as indicated in the diagram. This separation of variances is not only conceptual, but also respects the fact that main effects and interactions are often of very different magnitudes. The diagram in Figure 20.2 does not show the sum-to-zero constraints of Equation 20.2. These constraints are applied in the computer implementation.

20.2.1. Implementation in JAGS

The model is implemented in JAGS in the usual way, with every arrow in Figure 20.2 having a corresponding expression in the JAGS model specification. The model is specified in file `Jags-Ymet-Xnom2fac-MnormalHom.R`, and it is called by the high-level script `Jags-Ymet-Xnom2fac-MnormalHom-Example.R`. Like the one-factor model of the previous chapter, the baseline and deflections are initially denoted by `a0`, `a1[]`, `a2[]`, and `a1a2[,]`, and then transformed to sum-to-zero versions that are denoted by `b0`, `b1[]`, `b2[]`, and `b1b2[,]`.

Before the model specification itself, the program establishes some constants that will be used for scaling the prior distribution. Specifically, the program computes the mean of the data, the standard deviation of the data (`sd(y)`), and the shape and rate constants for a gamma distribution that has a mode at half `sd(y)` and standard deviation of twice `sd(y)`:

```
yMean = mean(y)
ySD = sd(y)
agammaShRa = unlist( gammaShRaFromModeSD( mode=sd(y)/2 , sd=2*sd(y) ) )
```

[2] By analogy to multiple regression with a shrinkage prior in Figure 18.10 (p. 531), if there are many predictors included in a model, it might be reasonable in principle to include a higher-level distribution across predictors such that the estimated variance of one predictor informs the estimated variance of another predictor. This would be especially useful if the application includes many nominal predictors, each with many levels, but only if the predictors can be thought to be mutually informative. Such applications are rare.

The gamma distribution with those shape and rate constants is broad on the scale of the data and will be used as the prior for the standard deviation parameters. These constants are used merely as a proxy for querying the researcher about the typical magnitude and range of the sort of data in the study.

The model specification then proceeds as follows: note that `Nx1Lvl` is the number of levels in factor 1 and `Nx2Lvl` is the number of levels in factor 2. While reading the specification and comparing it with the hierarchical diagram in Figure 20.2, it can help to scan the arrows in the diagram from the bottom up.

```
model {
  for ( i in 1:Ntotal ) {
    y[i] ~ dnorm( mu[i] , 1/ySigma^2 )
    mu[i] <- a0 + a1[x1[i]] + a2[x2[i]] + a1a2[x1[i],x2[i]]
  }
  ySigma ~ dunif( ySD/100 , ySD*10 )
  a0 ~ dnorm( yMean , 1/(ySD*5)^2 )
  for ( j1 in 1:Nx1Lvl ) { a1[j1] ~ dnorm( 0.0 , 1/a1SD^2 ) }
  a1SD ~ dgamma(agammaShRa[1],agammaShRa[2])
  for ( j2 in 1:Nx2Lvl ) { a2[j2] ~ dnorm( 0.0 , 1/a2SD^2 ) }
  a2SD ~ dgamma(agammaShRa[1],agammaShRa[2])
  for ( j1 in 1:Nx1Lvl ) { for ( j2 in 1:Nx2Lvl ) {
    a1a2[j1,j2] ~ dnorm( 0.0 , 1/a1a2SD^2 )
  } }
  a1a2SD ~ dgamma(agammaShRa[1],agammaShRa[2])
```

The model specification continues with the conversion to deflections that satisfy the sum-to-zero constraints. First the predicted cell means are computed, and then they are converted to sum-to-zero deflections using the method described with Table 20.1:

```
  # Convert a0,a1[],a2[],a1a2[,] to sum-to-zero b0,b1[],b2[],b1b2[,] :
  for ( j1 in 1:Nx1Lvl ) { for ( j2 in 1:Nx2Lvl ) {
    m[j1,j2] <- a0 + a1[j1] + a2[j2] + a1a2[j1,j2] # cell means
  } }
  b0 <- mean( m[1:Nx1Lvl,1:Nx2Lvl] )
  for ( j1 in 1:Nx1Lvl ) { b1[j1] <- mean( m[j1,1:Nx2Lvl] ) - b0 }
  for ( j2 in 1:Nx2Lvl ) { b2[j2] <- mean( m[1:Nx1Lvl,j2] ) - b0 }
  for ( j1 in 1:Nx1Lvl ) { for ( j2 in 1:Nx2Lvl ) {
    b1b2[j1,j2] <- m[j1,j2] - ( b0 + b1[j1] + b2[j2] )
  } }
}
```

20.2.2. Example: It's only money

Although we all know the adage that "money can't buy happiness," there is also a kernel of truth to the saying that "happiness is expensive." One of the most effective ways to become unhappy is to compare your personal income with the incomes of other people, especially people who have a higher income than you. In this section, we will be looking at some real-world salaries, so prepare yourself for the possibility that the data described in this section might make you unhappy. If you find yourself becoming glum, just look at the puppies on the cover of the book.[3]

People work hard to make a living. I am continually impressed by how many hours people work, in all walks of life. Whether it's 50 to 60 h per week (or more) on a farm, in a factory, in service, or in an office, many people work very hard for long hours. Despite equal long hours at work, people are paid vastly different amounts of money, depending on what line of work they are in. A field hand working 60 h a week makes a small fraction of the salary of a corporate executive working 60 h a week. Not only the type of work but also the type of payer affects pay. For example, within academia, a person doing research on consumer decision making would be paid much more for that work if she were in a business school than if she were in a department of psychology. Another influence on salary is experience or seniority. People with more experience tend to be paid more. In this section, we consider these factors in the microcosm of academia.

The data are annual salaries of 1,080 tenure-track professors at a large-enrollment, small-city, midwestern-American, research-oriented, state university. (Salaries at big-city and private universities tend to be higher, while salaries at liberal-arts colleges and teaching-oriented state universities tend to be lower.) The data span 60 academic departments that had at least seven members. The data also include the professor's seniority. In American academia, usually faculty are initially hired as assistant professors after they have spent up to 10 years in graduate school and postdoctoral positions. Assistant professors work in a probationary period, typically for six or seven years, at which point they are promoted to associate professor if they have sufficient research and teaching achievements (otherwise they are "let go"). Professors usually rise through the ranks from assistant to full over the course of perhaps 10–15 years, and then remain at the rank of full for the remainder of their career. Only a small subset of professors acquire an endowed salary or distinguished rank. It should be noted, however, that there are exceptions to this typical rise through the ranks. For example, a person might previously have a prominent career in government or business and then be hired directly at full or higher rank. Moreover, the data do not specify actual years in rank, so it is possible that all the full professors in one department happen to have only a few years in rank while all the full professors in another department happen to have many years in rank.

[3] Or, take solace in Max Ehrmann's 1927 *Desiderata*, which says, in part, "…If you compare yourself with others, you may become vain and bitter; for always there will be greater and lesser persons than yourself. Enjoy your achievements as well as your plans. Keep interested in your own career, however humble; it is a real possession in the changing fortunes of time.…You are a child of the universe, no less than the trees and the stars; you have a right to be here.…With all its sham, drudgery, and broken dreams, it is still a beautiful world. Be cheerful. Strive to be happy." (Ehrmann, 1995)

In summary, there are five ranks: assistant professor (Assis), associate professor (Assoc), full professor (Full), full professor with endowed salary (Endow), and full professor of distinguished rank (Disting). Professors can also have administrative ranks but these were excluded from the analysis; administrative salaries tend to be higher, all else being equal.

Data from four of the 60 academic departments are shown in Figure 20.3. You can see within panels that salaries tend to be higher for higher ranks, and you can see across panels that salaries differ across departments. There might also be interactions, in the sense that the effect of rank might be of different magnitudes in different departments, or the effect of department might be of different magnitudes at different ranks. The

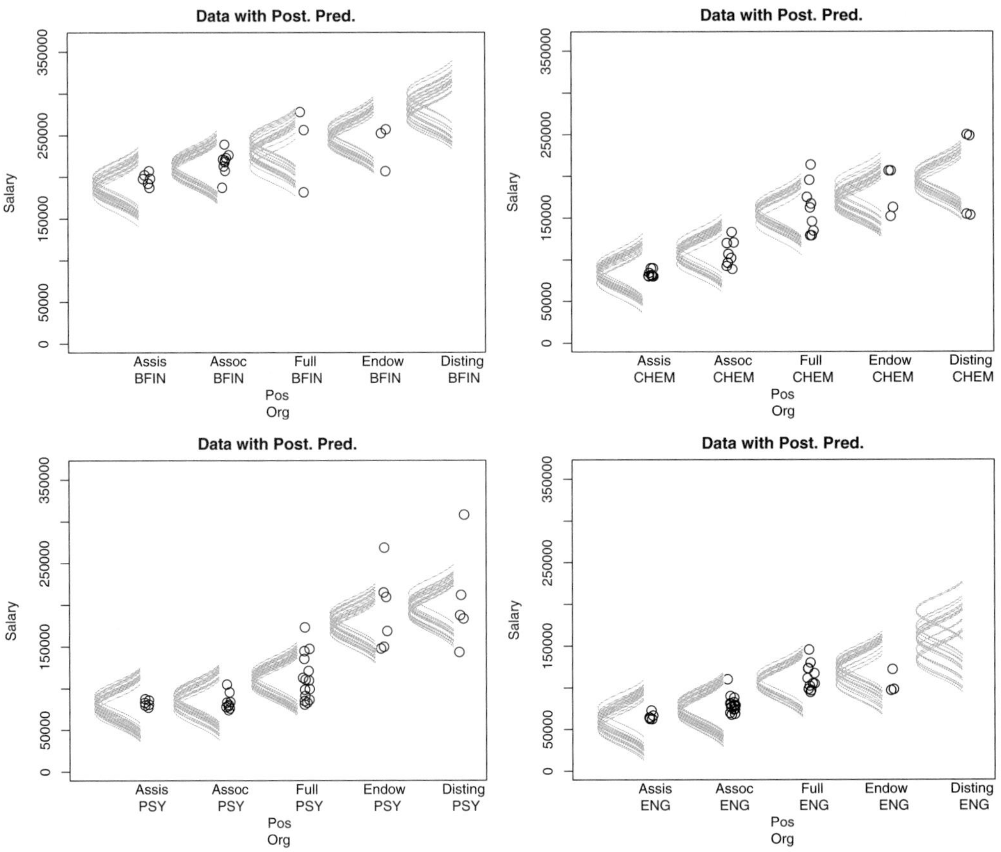

Figure 20.3 Salary data for four departments and five seniorities. The full set of data includes 60 departments. Posterior predictive distributions are from a model that assumes homogeneous variances and normally distributed data within cells. (BFIN, business finance; PSY, psychology; CHEM, chemistry; ENG, English; Pos, position or rank; Org, organization or department.)

goal of our analysis is to describe salaries as a function of two nominal predictors: the academic department and the rank of the professor.[4] Our analysis will estimate the salary deflections due to rank and department, along with interactions. The parameter estimates will provide meaningful information about the trends in the data and the uncertainty in those trends.

The high-level script that loads the data and calls the model is named `Jags-Ymet-Xnom2fac-MnormalHom-Example.R`. The first task accomplished by the script is reading in the data file:

```
myDataFrame = read.csv( file="Salary.csv" )
```

The column that specifies the rank of each professor is called "Pos" for position. As is the case for many real-life data files, its levels are coded obscurely, so the next task is renaming the levels mnemonically and ordering them. This is done with the `factor` function that was explained back in Section 3.4.2 (p. 46):

```
myDataFrame$Pos = factor( myDataFrame$Pos ,
                          levels=c("FT3","FT2","FT1","NDW","DST") ,
                          ordered=TRUE ,
                          labels=c("Assis","Assoc","Full","Endow","Disting") )
```

The script then specifies which columns of the data frame hold the predicted and predictor variables. The first predictor, x_1, is plotted on the horizontal axis of the graphs, so it should not have so many levels that they would exceed the maximal width of a graphing window. Therefore, we set position (rank) as the x_1 factor, as follows:

```
yName = "Salary"
x1Name = "Pos" # column name for rank (position)
x2Name = "Org" # column name for department (organization)
```

The basic results of the analysis are shown as the posterior predictive distributions superimposed on the data in Figure 20.3. The bushiness of the moustache represents the uncertainty of the estimate, whereas the width of the moustache represents the value of the standard deviation. The distributions also show that the model assumes homogeneity of variance: Whether the data within a group are tightly clustered or broadly spread out, the predictive distribution has the same width for every group. The predictive distributions also show that the model happily makes predictions for cells that have no data. The uncertainty, visually represented by bushiness of the moustache, tends to be higher in the cells with no data or small numbers of data points than in cells that have lots of data.

[4] Although rank (seniority) could be treated as an ordinal variable, we will treat it as a nominal predictor.

In this application, we know in advance that the different levels of both factors almost certainly have nonzero effects, and therefore null hypothesis testing is not the main focus. Instead, the emphasis is on estimation of the magnitudes of effects and their uncertainties. Table 20.2 shows a few rows of the summary table produced by the `smryMCMC` function in the script `Jags-Ymet-Xnom2fac-MnormalHom-Example.R`. When you run the script, your results will differ somewhat because of randomness in the MCMC process. The summary table has a row for each parameter, and the full table also has rows for the cell means and for the contrasts of means that will be discussed later. For each parameter, the table shows the estimated mean, median, mode, and 95% HDI limits. The column labeled ESS reports the effective MCMC sample size, which was defined in Equation 7.11 (p. 184). Although the table show many digits, only the first few digits are stable because of randomness in the MCMC process.

Table 20.2 indicates that the baseline salary across all departments and all ranks is about $127,000 (shown as parameter b0), but there is large variation across departments and ranks. For example, on average, assistant professors earn about $46,000 less than the baseline (shown as parameter b1[1]), and even regular full professors earn about $3000 less than the baseline (shown as parameter b1[3]). To those deflections from baseline due to rank, we also add deflections due to department. For example, on average, professors of English earn about $19,000 less than baseline (shown as parameter b2[21]), while professors in business finance earn about $109,000 more than baseline (shown as parameter b2[8]).

The predicted salary from the main effects alone is the sum of their deflections. For example, the additive prediction for the salary of a regular full professor in psychology is the baseline plus the main-effect deflection for full rank plus the main-effect deflection for psychology, b0 + b1[3] + b2[49]. But the actual salaries in that cell may differ from that additive prediction, and the estimated interaction deflection is also shown in Table 20.2 as parameter b1b2[3,49] (which has a value of about −$15,000). Thus, the predicted salary for regular full professors in psychology is b0 + b1[3] + b2[49] + b1b2[3,49]. This sum is reported in the full summary table as the parameter m[3,49] (which is not shown in the excerpts in Table 20.2).

Individual salaries vary tremendously around the predicted cell mean. The estimated standard deviation within a cell is shown in the final row of Table 20.2 as parameter ySigma, which has mean value of about $17,000. It is important to remember that this estimate assumes there is equal standard deviation in every cell, as shown graphically by the posterior predictive distributions plotted in Figure 20.3. Visual inspection of the plot suggests that the assumption of homogeneous variances is not a good description of the data, because some cells have data tightly clustered (much narrower than the predicted distribution) while other cells have data extensively spread out (much wider than the predicted distribution). Later in the chapter we will use a model that has different standard deviation parameters for every cell.

Table 20.2 Excerpt from summary table produced by function `smryMCMC` in script `Jags-Ymet-Xnom2fac-MnormalHom-Example.R`

Parameter	Mean	Median	Mode	ESS	HDI low	HDI high
b0	127,124	127,131	127,108	12,299	124,785	129,396
b1[1] Assis	−46,394	−46,415	−46,483	13,341	−49,467	−43,310
b1[2] Assoc	−33,108	−33,096	−33,052	12,987	−35,987	−30,378
b1[3] Full	−3,156	−3,159	−3,031	12,097	−6,106	−229
b1[4] Endow	26,966	26,980	27,285	13,405	22,424	31,583
b1[5] Disting	55,692	55,738	56,531	12,229	48,404	62,670
b2[21] ENG	−19,412	−19,380	−19,041	12,280	−27,416	−11,812
b2[49] PSY	6,636	6,653	6,686	12,604	353	12,494
b2[13] CHEM	19,159	19,152	19,221	14,597	12,698	25,582
b2[8] BFIN	109,184	109,200	109,156	14,287	100,185	118,579
b1b2[1,49] Assis PSY	−3,249	−3,136	−2,060	15,000	−13,588	6,682
b1b2[3,49] Full PSY	−14,993	−14,997	−15,474	11,963	−23,360	−6,463
b1b2[1,13] Assis CHEM	−12,741	−12,692	−13,110	13,224	−22,151	−3,457
b1b2[3,13] Full CHEM	12,931	12,971	13,087	12,772	3,471	22,240
ySigma	17,997	17,985	17,953	11,968	17,144	18,852

ESS is effective sample size, defined in Equation 7.11, p. 184. In all cases, the HDI mass is 95%. All values are in units of dollars except for the ESS. Although these numbers show many digits, only the first few digits are stable because of randomness in the MCMC process.

20.2.3. Main effect contrasts

In applications with multiple levels of the factors, it is virtually always the case that we are interested in comparing particular levels with each other. For example, we might be interested in comparing two science departments such as chemistry and psychology, or we might be interested in comparing business departments against other departments. Of course we could also compare ranks, such as associate versus assistant professors. These sorts of comparisons, which involve levels of a single factor and collapse across the other factor(s), are called main effect comparisons or contrasts.

Main effect contrasts are specified in the scripts that accompany this book by using math and syntax that was explained in Section 19.3.3 (p. 565). Each main effect contrast is a list that includes two vectors that specify the names of the levels to be compared. We can assemble as many contrasts as we want into a list of contrasts. For example, to compare full professors against associate, and to compare associate against assistant, we would create this list of lists:

```
x1contrasts = list(
  list( c("Full") , c("Assoc") , compVal=0.0 , ROPE=c(-1000,1000) ) ,
  list( c("Assoc") , c("Assis") , compVal=0.0 , ROPE=c(-1000,1000) )
)
```

The list of contrasts is called "x1contrasts" because it specifies the main-effect contrasts for factor x_1, which in this case is rank.[5] Both of the contrasts also specify a comparison value of 0, and an arbitrary ROPE from −$1,000 to +$1,000. The comparison value and ROPE could be omitted or specified as NULL. Main effect contrasts for the other factor are specified analogously. For example,

```
x2contrasts = list(
  list( c("CHEM") , c("ENG") , compVal=0.0 , ROPE=c(-1000,1000) ) ,
  list( c("CHEM") , c("PSY") , compVal=0.0 , ROPE=c(-1000,1000) ) ,
  list( c("BFIN") , c("PSY","CHEM","ENG") , compVal=0.0 , ROPE=c(-1000,1000) )
)
```

The final contrast in the list above compares business finance (BFIN) against the average of the departments of psychology, chemistry, and English.

The results of some of the contrasts specified above are displayed in Figure 20.4. The histograms show the credible values of the differences, given the data. For example, the left panel shows that associate professors make about $13,400 more than assistant professors, on average. The distribution also shows the uncertainty of that estimate, given these data. The 95% HDI goes from roughly $10,000 to $17,000. The middle and right panels show differences between departments. For example, the right panel shows that faculty in business finance make about $108,000 more than the average of faculty in psychology, chemistry, and English (with a 95% HDI from roughly $97,000 to $117,000).

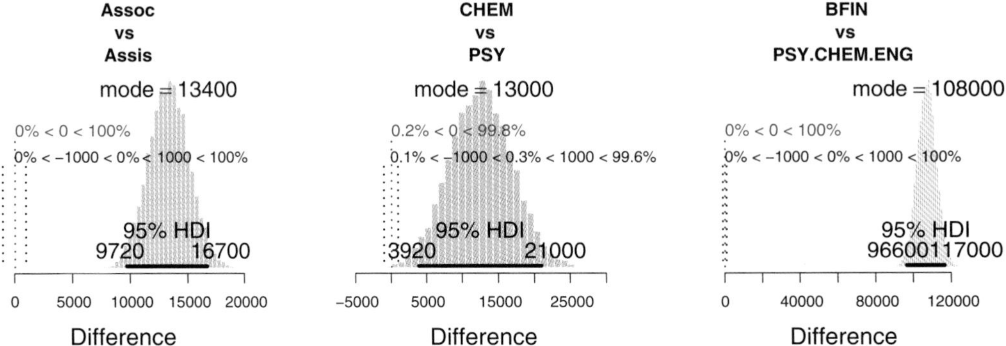

Figure 20.4 Three main effect contrasts. The left panel shows a contrast of two ranks. The right panel shows a "complex" comparison of business finance against the average of three other departments.

[5] A future version of the programs might streamline the specification by including the factor name at the beginning of each contrast instead relying on the arbitrary and non-mnemonic x1 and x2 notation.

20.2.4. Interaction contrasts and simple effects

Just as we can ask about the magnitude of a difference among particular levels of a predictor, we can ask how much that difference depends on the levels of the other predictor. Consider the data back in Figure 20.3, and the posterior summary in Table 20.2. It appears that in the chemistry department the difference between full and assistant professors is larger than the difference in the psychology department. Is the increase in pay for rising to full professor greater in the chemistry department than in the psychology department? How big is the difference of differences? And what is the uncertainty of the difference of differences? The answers are displayed in Figure 20.5. The left panel shows the so-called "simple" comparisons, which are differences between levels of one factor within a single level of another factor. The right panel shows the difference of the differences, where it can be seen that the difference between full and assistant professors is about \$38,000 more in chemistry than in psychology. (There are many possible causes of this interaction other than department per se, such as coincidental differences of years in rank between full professors in the two departments, or coincidental differences in the proportion of full professors who are shifted into endowed or distinguished ranks.)

Interaction contrasts are specified in the script analogously to main-effect contrasts. Each interaction contrast is a list of two lists, in which each sublist specifies vectors of level names. For example, to specify an interaction contrast of full versus assistant professors in chemistry and psychology, the syntax is

```
list( list( c("Full") , c("Assis") ) ,
      list( c("CHEM") , c("PSY") ) ,
      compVal=0.0 , ROPE=c(-1000,1000) )
```

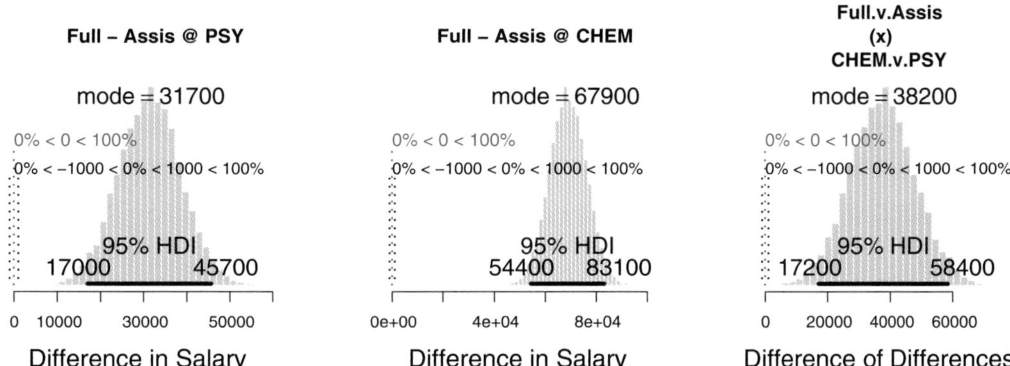

Figure 20.5 The left and middle panel show two "simple" comparisons: Each is a contrast of ranks within a level of department. The right panel shows an interaction contrast, namely, the difference of differences in the simple comparisons.

We can specify as many interactions contrasts as we like, all assembled into a list. The interaction contrasts can also involve averages of combined levels within a factor. For example, the second interaction contrast below considers the difference between full and assistant professors in business finance versus the average of three other departments:

```
x1x2contrasts = list(
  list( list( c("Full") , c("Assis") ) ,
        list( c("CHEM") , c("PSY") ) ,
        compVal=0.0 , ROPE=c(-1000,1000) ) ,
  list( list( c("Full") , c("Assis") ) ,
        list( c("BFIN") , c("PSY","CHEM","ENG") ) ,
        compVal=0.0 , ROPE=c(-1000,1000) )
)
```

20.2.4.1 *Interaction effects: High uncertainty and shrinkage*

It is important to realize that the estimates of interaction contrasts are typically much more uncertain than the estimates of simple effects or main effects. For example, in Figure 20.5, the widths of the 95% HDIs for the two simple effects are just over 28,000, but the width of the 95% HDI for the interaction contrast is more than 40,000. This large uncertainty of an interaction contrast is caused by the fact that it involves at least four sources of uncertainty (i.e., at least four groups of data), unlike its component simple effects which each involve only half of those sources of uncertainty. In general, interaction contrasts require a lot of data to estimate accurately.

The interaction contrasts also can experience notable shrinkage from the hierarchical model. In the present application, for example, there are 300 interaction deflections (5 levels of seniority times 60 departments) that are assumed to come from a higher-level distribution that has an estimated standard deviation, denoted $\sigma_{\beta 1\times 2}$ in Figure 20.2. Chances are that most of the 300 interaction deflections will be small, and therefore the estimated standard deviation of the interaction deflections will be small, and therefore the estimated deflections themselves will be shrunken toward zero. This shrinkage is inherently neither good nor bad; it is simply the correct consequence of the model assumptions. The shrinkage can be good insofar as it mitigates false alarms about interactions, but the shrinkage can be bad if it inappropriately obscures meaningful interactions. Shrinkage can be especially extreme when there are many groups with relatively few data points within groups (as was illustrated, for example, in Exercise 19.1, with Figure 19.9, p. 580). If the shrinkage seems too severe to be meaningful in a particular application, it may be a sign that the hierarchical model structure is an inappropriate description of the data. If this is the case, the model could be changed so that $\sigma_{\beta 1\times 2}$ is set to a constant, which implies that the interaction deflections are not mutually informative.

20.3. RESCALING CAN CHANGE INTERACTIONS, HOMOGENEITY, AND NORMALITY

When interpreting interactions, it can be important to consider the scale on which the data are measured. This is because an interaction means non-additive effects when measured on the current scale. If the data are nonlinearly transformed to a different scale, then the non-additivity can also change.

Consider an example, using utterly fictional numbers merely for illustration. Suppose the average salary of Democratic women is 10 monetary units, for Democratic men it's 12 units, for Republican women it's 15 units, and for Republican men it's 18 units. These data indicate that there is a nonadditive interaction of political party and gender, because the change in pay from women to men is 2 units for Democrats, but 3 units for Republicans. Another way of describing the interaction is to notice that the change in pay from Democrats to Republicans is 5 units for women but 6 units for men. A researcher might be tempted to interpret the interaction as indicating some extra advantage attained by Republican men or Democratic women. But such an interpretation may be inappropriate, because a mere rescaling of the data makes the interaction disappear, as will be described next.

Increases in salary are often measured by percentages and ratios, not be additive or subtractive differences. Consider the salary data of the previous paragraph in terms of percentages. Among Democrats, men make 20% more than women (12 vs. 10). Among Republicans, the men again make 20% more than the women (18 vs. 15). Among women, Republicans make 50% more than Democrats (15 vs. 10). Among men, Republicans again make 50% more than Democrats (18 vs. 12). In these ratio terms, there is no interaction of gender and political party: Change from female to male predicts a 20% increase in salary regardless of party, and change from Democrat to Republican predicts a 50% increase in salary regardless of gender.

Equal ratios are transformed to equal distances by a logarithmic transformation. If we measure salary in terms of the logarithm of monetary units, then the salary of Democratic women is $\log_{10}(10) = 1.000$, the salary of Democratic men is $\log_{10}(12) = 1.079$, the salary of Republican women is $\log_{10}(15) = 1.176$, and the salary of Republican men is $\log_{10}(18) = 1.255$. With this logarithmic scaling, the increase in salary from women to men is 0.079 for both parties, and the increase from Democrat to Republican in 0.176 for both genders. In other words, when salary is measured on a logarithmic scale, there is no interaction of gender and political party.

It may seem strange to measure salary on a logarithmic scale, but there are many situations for which the scale is arbitrary. The pitch of a sound can be measured in terms of frequency (i.e., cycles per second), or in terms of perceived pitch, which is essentially the logarithm of the frequency. The magnitude of an earthquake can be measured by its energy, or by its value on the Richter scale, which is the logarithm of energy. The pace of a dragster on a race track can be measured by the average speed during the run, or by the duration from start to finish (which is the reciprocal of average speed). Thus, measurement scales are not unique, and are instead determined by convention.

The general issue is illustrated in Figure 20.6. Suppose that predictor x_1 has two levels, as does predictor x_2. Suppose we have three data points at each combination of levels, yielding twelve data points altogether. The means at each combination of levels are shown in the top-left graph of Figure 20.6. You can see that there is an interaction, with the effect of x_1 being bigger when $x_2 = 2$ than when $x_2 = 1$. But this interaction goes away when the data are transformed by taking the logarithm, as shown in the lower left graph. Each individual data point was transformed, and then the means in each cell were computed. Of course, the transformation can produce the opposite change: Data with no interaction, as in the lower-left plot, can be made to have an interaction when they are rescaled as in the upper-left plot, via an exponential transformation.

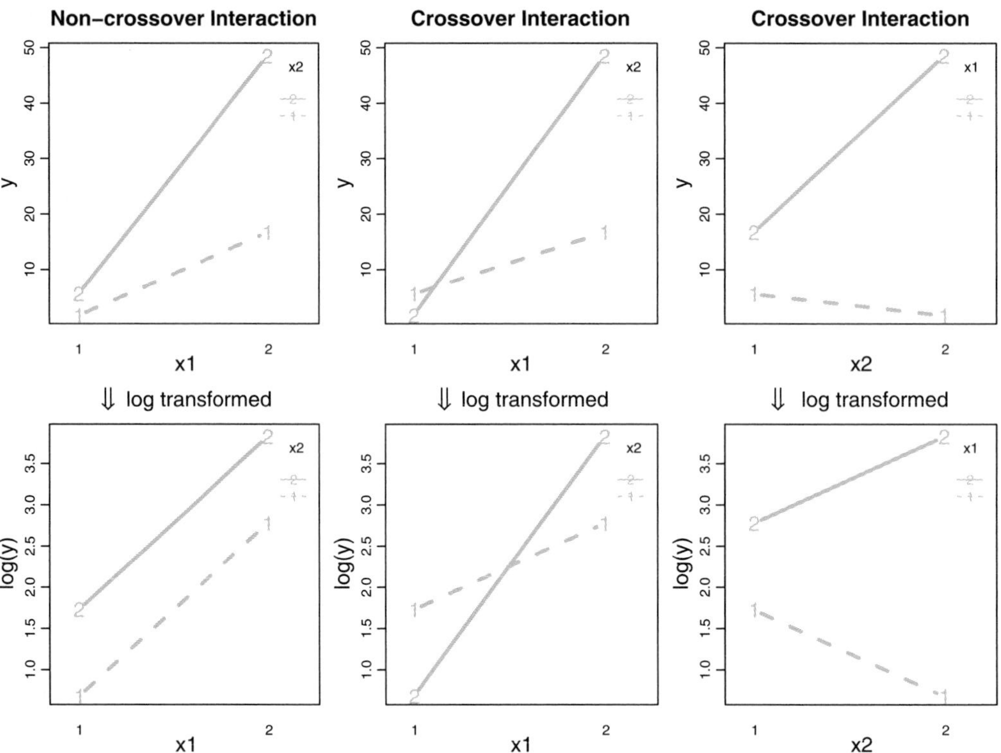

Figure 20.6 Top row shows means of original data; bottom row shows means of logarithmically transformed data. Left column shows a non-crossover interaction. Middle and right columns show a crossover interaction, the same in both columns, but plotted against $x1$ or $x2$ on the abscissa.

The transformability from interaction to non-interaction is only possible for noncrossover interactions. This terminology, "noncrossover," is merely a description of the graph: The lines do not cross over each other and they have the same sign slope. In this situation, the y-axis can have different portions stretched or shrunken so that the lines become parallel. If, however, the lines cross, as in the middle column of Figure 20.6, then there is no way to uncross the lines merely by stretching or shrinking intervals of the y-axis. The right column of Figure 20.6 shows the same data as the middle column, but plotted with the roles of x_1 and x_2 exchanged. When plotted this way, the lines do not cross, but they do have opposite-sign slopes (i.e., one slope is positive and the other slope is negative). There is no way that stretching or shrinking the y-axis can change the signs of the slopes, hence the interaction cannot be removed merely by transforming the data. Because these data have crossing lines when plotted as in the middle column, they are said to have a crossover interaction even when they are plotted as in the right column. (Test your understanding: Is the interaction in Figure 20.1 a crossover interaction?)

It is important to note that the transformation applies to individual raw data values, not to the means of the conditions. A consequence of transforming the data, therefore, is alteration of the variances of the data within each condition. For example, suppose one condition has data values of 100, 110, and 120, while a second condition has data values of 1100, 1110, and 1120. For both conditions, the variance is 66.7, so there is homogeneity of variance. When the data are logarithmically transformed, the variance of the first group becomes 1.05e−3, but the variance of the second group becomes two orders of magnitude smaller, namely 1.02e−5. In the transformed data there is not homogeneity of variance.

Therefore, when applying the hierarchical model of Figure 20.2, we must be aware that it assumes homogeneity of variance. If we transform the data, we are changing the variances within the levels of the predictors. The transformed variances might or might not be reasonably homogeneous. If they are not, then either the data should be transformed in such a way as to respect homogeneity of variance, or the model should be changed to allow unequal variances.

The models we have been using also assume a normal likelihood function, which means that the data in every cell should be normally distributed. When the data are transformed to a different scale, the shape of their distribution also changes. If the distributions become radically non-normal, it may be misleading to use a model with a normal likelihood function.

In summary, this section has made two main points. First, if you have a noncrossover interaction, be careful what you claim about it. A noncrossover interaction merely means nonadditivity on the scale you are using. If this scale is the only meaningful scale, or if this scale is the overwhelmingly dominant scale used in that field of research, then you can cautiously interpret the nonadditive interaction with respect to that scale. But if transformed scales are reasonable, then keep in mind that nonadditivity is scale-specific, and there might be no interaction in a different scale. With a crossover interaction, however, no rescaling can undo the interaction. Second, nonlinear transformations change the within-cell variances and the shapes of the within-cell distributions. Be sure that the model you are using is appropriate to the homogeneity or nonhomogeneity of variances in the data, and to the shapes of the distributions, on whatever scale you are using.

20.4. HETEROGENEOUS VARIANCES AND ROBUSTNESS AGAINST OUTLIERS

As has been mentioned several times, the traditional model for ANOVA assumes equal standard deviations in all cells. This assumption was evident in the posterior predictive distributions shown in Figure 20.3 (p. 592). Unfortunately, this assumption appears to be a poor description of the data. For example, the posterior predictive distribution seems to be too wide for the data from assistant professors, but the posterior predictive distribution seems to be too narrow for some of the data from endowed or distinguished professors. The data seem also to contain outliers, in the sense of being beyond what normally distributed data would typically produce.

Fortunately, it is reasonably straight forward to relax the constraints in Bayesian software such as JAGS and Stan. We did this in the previous chapter for a single-factor ANOVA-style model, and we take the same approach here for the two-factor model. The hierarchical model of Figure 20.2 (p. 588) can be enhanced to provide each cell with its own standard deviation parameter. Instead of a single σ_y parameter that is used simultaneously for all cells, each cell has its own $\sigma_{[j,k]}$ parameter, and those parameters are described as being gamma-distributed across cells analogously to what was shown in Figure 19.6 (p. 574). The result is the hierarchical model in Figure 20.7. It might seem daunting, but its components are all familiar from previous applications.

The model is implemented in the program `Jags-Ymet-Xnom2fac-MrobustHet.R` (where "`MrobustHet`" indicates that the model is robust to outliers and has heterogeneous variances) and is called from the high-level script `Jags-Ymet-Xnom2fac-MrobustHet-Example.R`. The high-level script is essentially the same as before, merely `source`-ing `Jags-Ymet-Xnom2fac-MrobustHet.R` instead of `Jags-Ymet-Xnom2fac-MnormalHom.R`. The JAGS model specification implements the prior on $\sigma_{[j,k]}$ as shown in Figure 20.7, using the reparameterization of the shape and rate parameters into mode and standard deviation that was explained in Equation 9.8 (p. 238).

Results of applying the model are shown in Figure 20.8. It is quite clear from the figure that the posterior predictive distribution has different standard deviations in different cells. The "moustache" over assistant professors is much narrower than the moustaches over higher seniorities. The bottom row of Figure 20.8 shows the marginal

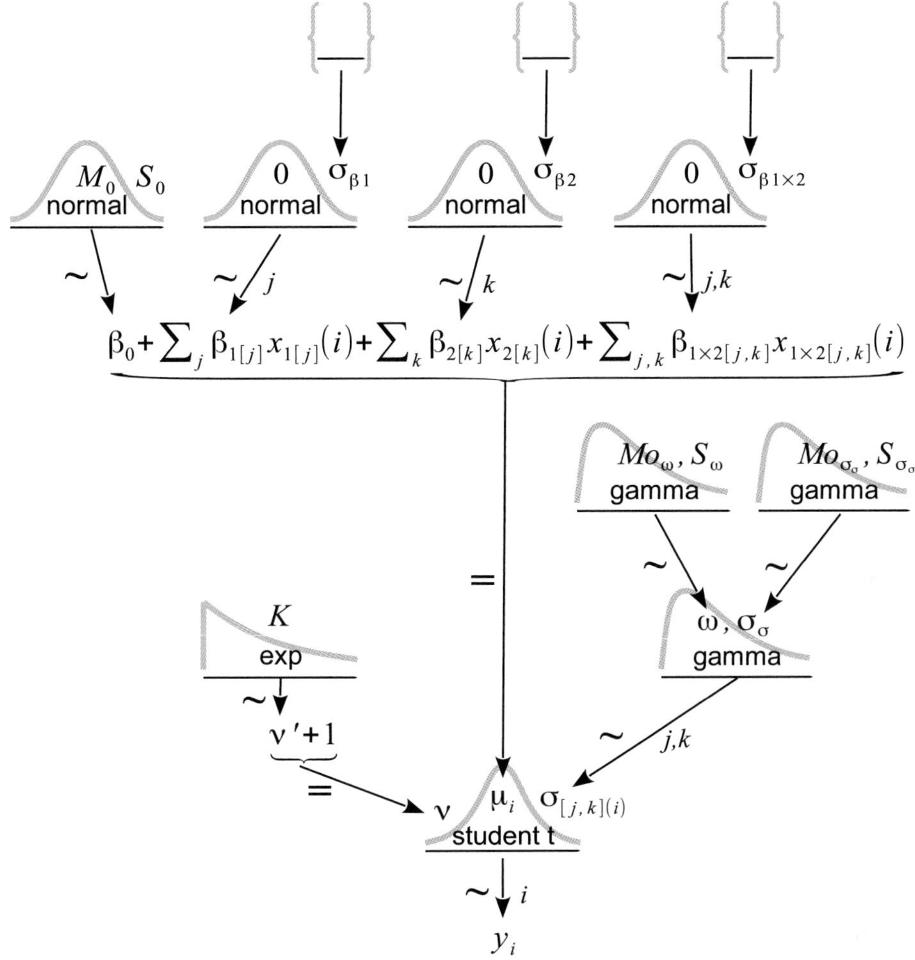

Figure 20.7 Hierarchical diagram for a model that describes data from two nominal predictors, wherein the noise distribution (at bottom of diagram) is robust to outliers and has a different standard deviation parameter, $\sigma_{[j,k]}$, for every cell. Compare with Figure 20.2 (p. 588), which assumes normally distributed noise and homogeneous variances across cells.

posterior distribution on the normality parameter (ν), which has most of its mass on small values. Small values of ν indicate heavy tails and suggest that there are outliers in the data (relative to a normal distribution). The bottom row of Figure 20.8 also shows the modal cell standard deviation (ω in Figure 20.7), and the standard deviation of the estimated cell standard deviations (σ_σ in Figure 20.7). The standard deviation of the cell standard deviations is very large, which means that there is strong heterogeneity in the standard deviations across the 300 cells of the data.

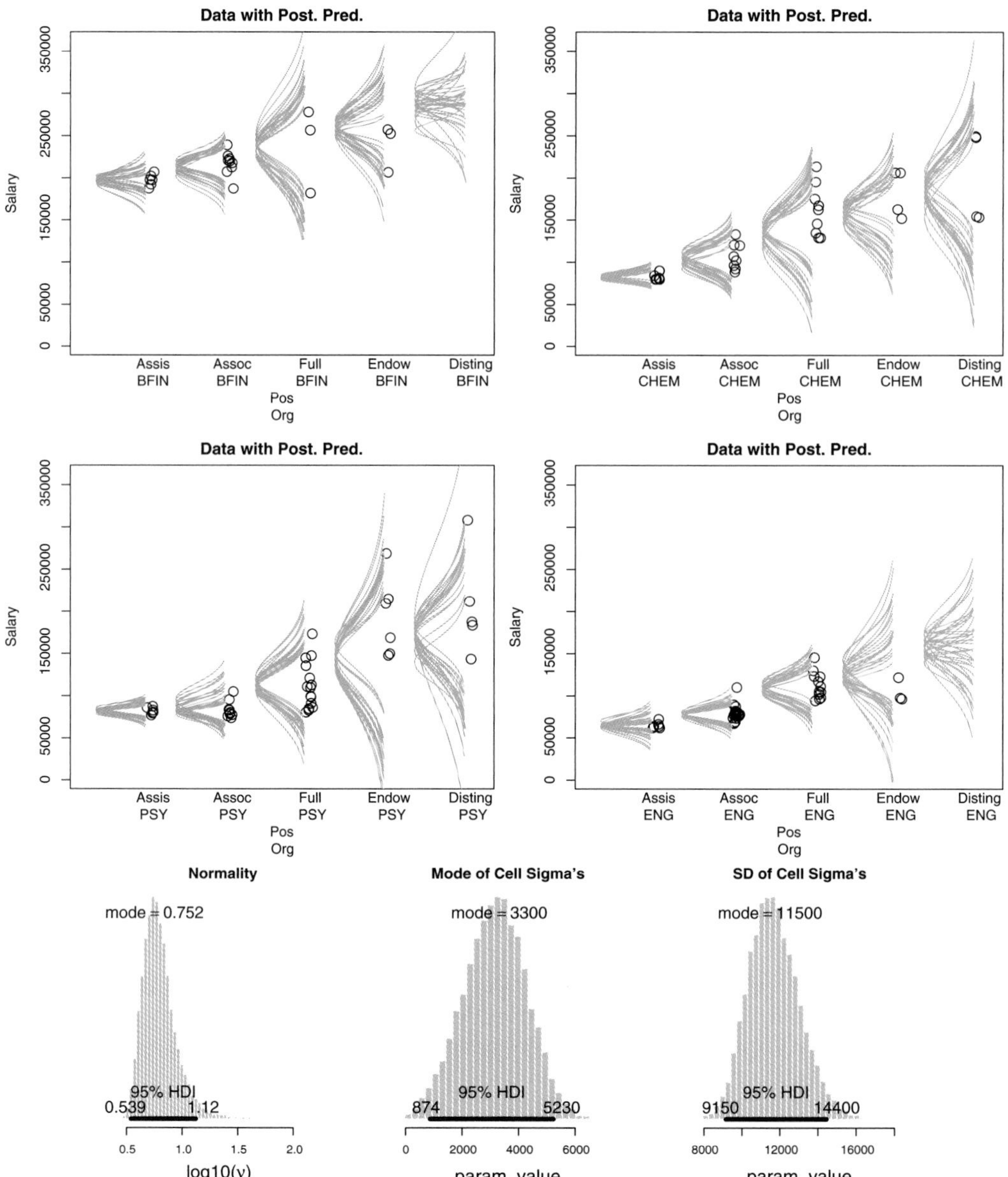

Figure 20.8 Salary data for four (of 60) departments and five seniorities. The model assumes heterogeneous variances and t-distributed data within cells. The bottom row shows marginal posterior distribution of the normality parameter, the modal cell standard deviation (ω in Figure 20.7), and the standard deviation of the estimated cell standard deviations (σ_σ in Figure 20.7). Compare with the results from a model that assumes homogeneous variances and normally distributed data within cells, shown in Figure 20.3. (BFIN, business finance; PSY, psychology; CHEM, chemistry; ENG, english; Pos, position or rank; Org, organization or department.)

Visual inspection of the posterior predictive distributions in Figure 20.8 suggests that the model trades off interaction against within-cell variance. Consider, for example, the data for endowed professors. For all four departments, the means of the posterior predictive distributions do not align with the means of the data in those cells. But the posterior predictive distributions accommodate the off-center data by having large standard deviations. Thus, the model prefers to shrink the interaction deflections and accommodate data by expanding the variances in affected cells.

As further evidence that the model with heterogeneous variances tends to reduce the magnitude of the interaction deflections for these data, consider Figure 20.9, which shows the same interaction contrast as Figure 20.5 (p. 597). For the heterogeneous-variance model, the interaction contrast has magnitude of about \$20,800, but for the homogeneous-variance model, the interaction contrast had magnitude of about \$38,200. More generally, the estimated value of the standard deviation of the interaction deflections is smaller in the heterogeneous-variance model. The parameter being referred to is labeled as $\sigma_{\beta 1\times 2}$ in Figures 20.2 and 20.7, and is called `a1a2SD` in the JAGS model specification. For the homogeneous-variance model, the modal value of $\sigma_{\beta 1\times 2}$ is \$9,700, but for the heterogeneous-variance model, the modal value of $\sigma_{\beta 1\times 2}$ is \$5,300.

Which model is a better description of the data? The homogeneous-variance model fails to represent the obviously different variances in the different cells of the data, and therefore that model might be overcertain about differences between high-variance groups and undercertain about differences between low-variance groups (as was explained with Figures 19.7 and 19.8). On the other hand, the heterogeneous-variance model seems too eager to forego interactions in favor of increased within-cell variances.

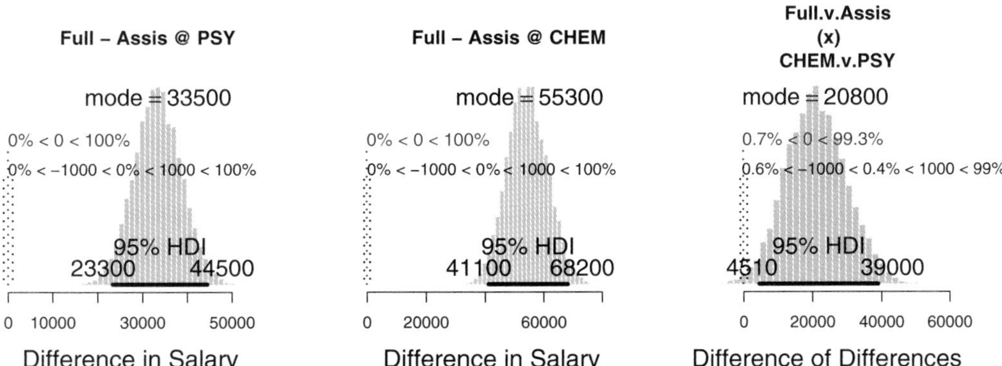

Figure 20.9 The left and middle panel show two "simple" comparisons: Each is a contrast of ranks within a level of department. The right panel shows an interaction contrast, namely, the difference of differences in the simple comparisons. The model assumes heterogeneous variances and t-distributed data within cells. Compare with Figure 20.5 (p. 597).

In principle, an intrepid programmer could do a Bayesian model comparison, putting both models under a higher-level indexical parameter. But the posterior probabilities of the model indices might be overly sensitive to the arbitrary vagueness of the priors on the parameters within each model (as was discussed in Section 10.6). Moreover, both models assume that the data within cells are distributed symmetrically above and below their central tendency, either as a normal distribution or a t-distribution. The data instead seem to be skewed toward larger values, especially for advanced seniorities. Therefore, we might want to create a model that describes the data within each cell as a skewed distribution such as a Weibull (which will not be defined further here because it would lead us too far afield). And, instead of allowing every cell to have a distinct variance, we could allow the model to have distinct variances only for the different seniorities, and not for the different departments. It is straight forward to create and analyze such a model in JAGS or Stan. This flexibility to explore and analyze models that are descriptively appropriate is a strength of the Bayesian approach.

20.5. WITHIN-SUBJECT DESIGNS

In many situations, a single subject (e.g., person, animal, plant, and device) contributes data to multiple levels of the predictors. For example, suppose we are studying how quickly people can respond to stimuli while driving a car and talking on a mobile phone. We want to establish reference response times, so we are interested in how quickly people can press a button in response to a stimulus onset. The stimulus could appear in the visual modality as a light, or in the auditory modality as a tone. The subject could respond with his/her dominant hand, or with his/her nondominant hand. Thus, there are two nominal predictors, namely modality and hand. The novel aspect is that a single subject contributes data to all combinations of the predictors. On many successive trials, the subject gets either a tone or light, and is instructed to respond with either the dominant or nondominant hand. Because the levels of the predictors change within subjects, this situation is called a within-subject design. If there are different people in every cell of the design (as there were, for example, in the professor salary data) then the levels of the predictors change across or between subjects, and the design is called between-subjects.

In the scenario just described, regarding response times with different hands for stimuli in different modalities, each subject contributes multiple response times for repeated trials within a single combination of levels (such as visual stimulus with nondominant hand). Therefore, the design is also referred to as having repeated measures. Sometimes the term "repeated measures" is used to refer to measures from the same subject in different conditions, not only within a single condition, and therefore the terms "repeated measures" and "within-subject" are sometimes used synonymously. I prefer to use "within-subject" because it refers explicitly to the fact that there are multiple conditions applied within a subject. In a within-subject design, it could be that each subject provides only a single measurement in each condition. In this case there would not be repeated measures within cells, only across cells.

When every subject contributes many measurements to every cell, then the model of the situation is a straight-forward extension of the models we have already considered. We merely add "subject" as another nominal predictor in the model, with each individual subject being a level of the predictor. If there is one predictor other than subject, the model becomes

$$y = \beta_0 + \vec{\beta}_1 \vec{x}_1 + \vec{\beta}_S \vec{x}_S + \vec{\beta}_{1\times S} \vec{x}_{1\times S}$$

This is exactly the two-predictor model we have already considered, with the second predictor being subject. When there are two predictors other than subject, the model becomes

$$\begin{aligned} y = \beta_0 & \quad && \text{baseline} \\ + \vec{\beta}_1 \vec{x}_1 + \vec{\beta}_2 \vec{x}_2 + \vec{\beta}_S \vec{x}_S & && \text{main effects} \\ + \vec{\beta}_{1\times 2} \vec{x}_{1\times 2} + \vec{\beta}_{1\times S} \vec{x}_{1\times S} + \vec{\beta}_{2\times S} \vec{x}_{2\times S} & && \text{two-way interactions} \\ + \vec{\beta}_{1\times 2\times S} \vec{x}_{1\times 2\times S} & && \text{three-way interaction} \end{aligned}$$

This model includes all the two-way interactions of the factors, plus the three-way interaction. We will not discuss three-way interactions in depth, but the idea is analogous to two-way interactions. Just as a two-way interaction means that the simple effect of a factor depends on the level of another (second) factor, a three-way interaction means that the two-way interaction depends on the level of another (third) factor. For example, recall the two-way interaction contrast in Figure 20.5 (p. 597), which showed that difference between full professors and assistant professors was larger in chemistry than in psychology. Suppose we had a third nominal predictor, such as different universities. It could be that the two-way interaction has different magnitudes at different universities. The main point here, however, is that subject merely plays the role of a third nominal predictor in a standard ANOVA-style model when there are multiple measurements for each subject in each condition.

There are other situations, however, in which each subject contributes only one datum to a cell. For example, suppose the value to be predicted is IQ, as measured by a lengthy exam, with one predictor being type of noise during the exam (e.g., vocal noises, ocean noises, and quiet) and the other predictor being format (e.g., on paper, computer only, computer with paper scratch pad). Although it is conceivable that subjects could be repeatedly tested in each condition, it would be challenging enough to get people to sit through all combinations even once. Thus, each subject would contribute one value to each condition.

In the situation when each subject contributes only one datum per condition, the models described above, with all the interaction terms, break down. Another way of thinking about the problem is with reference to Table 20.1 (p. 586). In that table, the cell means are perfectly redescribed in terms of a baseline plus main-effect deflections plus interaction deflections. The data are randomly distributed around the cell means within the cells, and the standard deviation parameter, σ_y, describes that variability. But if there is only a single datum in each cell, then the mean of the cell *is* the single datum, and the parameters perfectly fit the data with zero noise variance. In other words, there are more parameters than data, and we have gained nothing by the analysis. Therefore, instead of attempting to estimate all the interactions of subjects with other predictors, we assume a simpler model in which the only influence of subjects is a main effect:

$$\begin{aligned} y = \beta_0 &+ \vec{\beta}_1 \vec{x}_1 + \vec{\beta}_2 \vec{x}_2 + \vec{\beta}_{1\times 2} \vec{x}_{1\times 2} \\ &+ \vec{\beta}_S \vec{x}_S \end{aligned}$$

In other words, we assume a main effect of subject, but no interaction of subject with other predictors. In this model, the subject effect (deflection) is constant across treatments, and the treatment effects (deflections) are constant across subjects. Notice that the model makes no requirement that every subject contributes a datum to every condition. Indeed, the model allows zero or multiple data per subject per condition. Bayesian estimation makes no assumptions or requirements that the design is balanced (i.e., has equal numbers of measurement in each cell). If there are many observations per subject in every cell, then one of the previously described models may be considered.

20.5.1. Why use a within-subject design? And why not?

The primary reason to use a within-subject design is that you can achieve greater precision in the estimates of the effects than in a between-subject design. For example, suppose you are interested in measuring the effect on response time of using the dominant versus non-dominant hand. Suppose there is a population of four subjects from whom we could measure data. Suppose that if we could measure every subject in every condition, we would know that for the first subject, his or her response times for dominant and nondominant hands are 300 and 320 ms. For the second subject, the response times are 350 and 370. For the third subject, the response times are 400 and 420, and for the fourth subject, the response times are 450 and 470. Thus, for every subject, the difference between dominant and nondominant hands is exactly 20 ms, but there are big differences across subjects in overall response times. Suppose we have the resources to measure only two data points in each condition. Suppose we measure response times from the dominant hands of two of the subjects. Should we measure response times from the nondominant hands of the same two subjects, or the nondominant hands of the two other subjects? If we measure from the same two subjects, then the estimated effect for each subject is 20 ms, and we have high certainty in the magnitude of the effect. But if we measure from the two other subjects, then the estimated effect of dominant versus nondominant hand is the average of the first two subjects versus the average of the second two subjects, and the difference is badly affected by the big differences between subjects. The between-subject design yields lower precision in the estimate of the effect.

Because of the gain in precision, it is desirable to use within-subject designs. But there are many dangers of within-subject designs that need to be considered before they are applied in any particular situation. The key problem is that, in most situations, when you measure the subject you change the subject, and therefore subsequent measurements are not measuring the same subject. The simplest examples of this are mere fatigue or generic practice effects. In measures of response time, if you measure repeatedly from the same subject, you will find improvement over the first several trials because of the subject gaining practice with the task, but after a while, as the subject tires, there will be a decline in performance. The problem is that if you measure the dominant hand in the early trials, and the nondominant hand in the later trials, then the effect of practice or fatigue will contaminate the effect of handedness. The repeated measurement process affects and contaminates the measure that is supposed to be a signature of the predictor.

Practice and fatigue effects can be overcome by randomly distributing and repeating the conditions throughout the repeated measures, if the practice and fatigue effects influence all conditions equally. Thus, if practice improves both the dominant and nondominant hand by 50 ms, then the difference between dominant and nondominant hands is unaffected by practice. But practice might affect the nondominant hand much more than the dominant hand. You can imagine that in complex designs with many predictors, each with many levels, in can become difficult to justify an assumption that repeated measures have comparable effects on all conditions.

Worse yet, in some situations there can be differential carryover effects from one condition to the next. For example, having just experienced practice in the visual modality with the nondominant hand might improve subsequent performance in the auditory modality with the nondominant hand, but might not improve subsequent performance in the visual modality with the dominant hand. Thus, the carryover effect is different for different subsequent conditions.

When you suspect strong differential carryover effects, you may be able to explicitly manipulate the ordering of the conditions and measure the carryover effects, but this might be impossible mathematically and impractical, depending on the specifics of your situation. In this case, you must revert to a between subjects design, and simply include many subjects to average out the between-subject noise.

In general, all the models we have been using assume independence of observations: the probability of the set of data is the product of the probabilities of the individual data points. When we use repeated measures, this assumption is much less easy to justify. On the one hand, when we repeatedly flip a coin, we might be safe to assume that its underlying bias does not change much from one flip to the next. But, on the other hand, when we repeatedly test the response time of a human subject, it is less easy to justify an assumption that the underlying response time remains unaffected by the previous trial. Researchers will often make the assumption of independence merely as a convenient approximation, hoping that by arranging conditions randomly across many repeated measures, the differential carryover effects will be minimized.

20.5.2. Split-plot design

"'All industrial experiments are split-plot experiments.' This provocative remark has been attributed to the famous industrial statistician, Cuthbert Daniel, by Box, Hunter, and Hunter (2005) in their well-known text on the design of experiments. Split-plot experiments were invented by Fisher (1925) and their importance in industrial experimentation has been long recognized (Yates, 1935)." (B. Jones & Nachtsheim, 2009, p. 340) Split-plot designs are also common in psychology and in agriculture, where they originated and got their name.

Consider an agricultural experiment investigating the productivity of different soil tilling methods and different fertilizers. It is relatively easy to provide all the farmers with the several different fertilizers. But it might be relatively difficult to provide all farmers with all the machinery for several different tilling methods. Therefore, any particular farmer will use a single (randomly assigned) tilling method on his whole plot, and tilling methods will differ between whole plots. Each farmer will split his field into subplots and apply all the fertilizers to different (randomly assigned) split plots, and fertilizers will differ across split plots within whole plots. This type of experiment inspires the name, split-plot design. The generic experiment-design term for the farmer's field is "block." Then, the factor that varies within every field is called the within-block factor and the factor that varies between fields is called the between-block factor. Notice also that each split plot yields a single measurement (in this case the productivity measured in bushels per acre), not multiple measurements.

Split plot designs are also common in psychology experiments. Some factors are difficult or impossible to manipulate in human subjects, such as political or religious affiliation, handedness, age, and sex. Other factors are relatively easy to manipulate, such as whether a text is presented to the subject visually or auditorily. Each human subject is analogous to a block or whole plot. Each subject experiences a single level of the factor that varies between blocks, that is, between subjects. But each subject experiences all levels of the factor that varies within blocks, that is, within subjects. In psychology, therefore, researchers refer to between-subject and within-subject factors in split-plot designs (e.g., Maxwell & Delaney, 2004, chap. 12). As with the agricultural scenario described in the previous paragraph, the basic split-plot design assumes there is a single datum per cell.

20.5.2.1 Example: Knee high by the fourth of July

For purposes of illustration, consider an experiment in agronomy that measured corn production (in bushels per acre) as a function of how the fields were tilled and how fertilizer was applied. There were three tilling methods, namely moldboard plow, chisel plow, and ridge tilling. A moldboard plow turns the soil over, whereas a chisel plow stirs the soil without inverting it. Ridge tilling scrapes the tops of the planting ridges that rise between furrows. The tilling methods differ in how they mix organic materials into the soil and how they resist erosion and protect against weeds. Each field used a single tilling method, hence tilling is the between-block or between-subject factor. There were three fertilization methods, namely broadcast, deep banding, and surface banding. In broadcast fertilization, the fertilizer is spread across all the soil. In banding, the fertilizer is concentrated in bands near the corn. In surface banding, the fertilizer is applied near the surface of soil, whereas in deep banding, the fertilizer is applied several inches below the seed. The location of the fertilizer is especially important for fertilizers such as phosphorus which are essentially immobile and do not disperse through the soil. Each field used all three fertilization methods on different subplots, hence fertilizer is the within-block or within-subject factor.

Figure 20.10 shows data from the experiment. There is a separate panel for each tilling method. Within panels, each dot shows the corn production of a subplot that was fertilized as indicated by the abscissa. Subplots from the same field are connected by

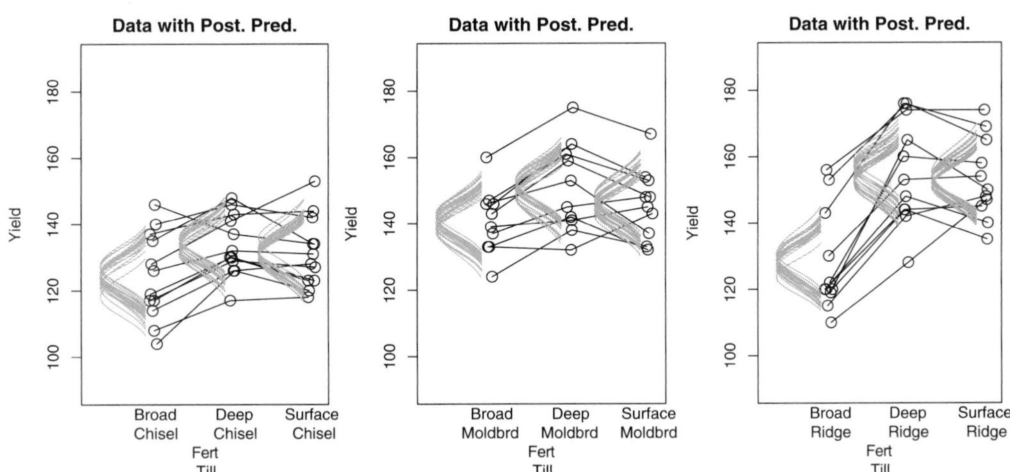

Figure 20.10 Corn production yields (bushels per acre) for different tilling (Till) and phosphorus fertilization (Fert) methods. Panels show different tilling methods (Chisel, chisel plow; Moldbrd, moldboard plow; Ridge, ridge tilling), which varied between fields. Within panels, abscissa shows different phosphorus fertilizer placements (Broad, broadcast; Deep, deep banding; Surface, surface banding), which varied within fields. Dots connected by lines indicate the same field. (Please note that these data are completely fictitious! I made them up one afternoon after skimming some information online.)

lines. There are 12 fields for chisel tilling, 10 fields for moldboard tilling, and 11 fields for ridge tilling. You can see by the relative heights of the lines that some fields are generally more productive than other fields, even when treated the same way. This variation between fields treated the same could be caused by any number of other factors, such as differences in weather, soil, insects, or plant disease. Our goal is to describe the trends in the data, and to estimate the differences between different tilling and fertilization methods.

20.5.2.2 The descriptive model

In the classical ANOVA-style model for a split-plot design, the overall variance is conceptually decomposed into five components: the main effect of the between-subjects factor, the main effect of the within-subjects factor, the interaction of the two factors, the effect of subject within levels of the between-subject factor, and the interaction of subject with the within-subject factor. Unfortunately, because there is only a single datum per cell, the five components exactly match the data, which is to say that there are as many parameters as there are data points. (If every subject contributed multiple data points to every cell then the five-component model could be used.) Because there is no residual noise within cells, the classical approach is to treat the final component as noise, that is, treat the interaction of subject with the within-subject factor as noise. That component is not included in the model (at least, not distinct from noise). We will do the same for the descriptive model in our Bayesian analysis. The next few paragraphs provide mathematical details to justify and explain the points just made. To do this, we need to define some notation for the five effects. Then we will convert the effects to sum-to-zero deflections. The sum-to-zero calculations will also be used in the JAGS implementation of the model.

Notation and terminology. Because we can anthropomorphize each field as a subject in the experiment, I will use the terminology of within-subject and between-subject factors (instead of within/between-field or within/between-block). Then we need some mathematical notation for the factors and levels. Define $B[i]$ as the ith level of the between-subject factor, which has I levels total. $W[j]$ is the jth level of the within-subject factor, which has J levels total. $S|B[k|i]$ is subject k in level i of the between-subject factor, which has $K|i$ subjects. There can be different numbers of subjects, $K|i$ in different levels i, as indeed there are in the data of Figure 20.10. (This notation, $S|B[k|i]$, might be confusing at first, but it is correct: Subject $S|B[k|i]$ is in all levels of factor W and in one level, namely level i, of factor B.) The single datum in a cell is denoted $y_{B\times W\times S|B[i,j,k|i]}$.

Marginal means and sum-to-zero deflections. The goal of the next few paragraphs is to motivate the model by considering the traditional description of the data in terms of main effects and interactions. We consider the various marginal means, and then how to express them as sum-to-zero deflections. The mean for subject $S|B[k|i]$ (across levels of W, within the level of B) is

$$m_{S|B[k|i]} = \frac{1}{J}\sum_{j}^{J} y_{B\times W\times S|B[i,j,k|i]}$$

The mean for treatment combination $B \times W[i,j]$ (across subjects) is

$$m_{B\times W[i,j]} = \frac{1}{K|i}\sum_{k|i}^{K|i} y_{B\times W\times S|B[i,j,k|i]}$$

The mean for level $B[i]$ (across W and S) is

$$m_{B[i]} = \frac{1}{J}\sum_{j}^{J} m_{B\times W[i,j]}$$

The mean for level $W[j]$ (across B and S) is

$$m_{W[j]} = \frac{1}{I}\sum_{i}^{I} m_{B\times W[i,j]}$$

The overall mean is

$$m = \frac{1}{I\cdot J}\sum_{i,j}^{I,J} m_{B\times W[i,j]}$$

We now convert the means to sum-to-zero deflections (analogous to Table 20.1). We set the baseline to the overall mean, and then define the main effect deflections as differences from the baseline:

$$\beta_0 = m \tag{20.3}$$

$$\begin{aligned}\beta_{B[i]} &= m_{B[i]} - \beta_0 \\ &= m_{B[i]} - m\end{aligned} \tag{20.4}$$

$$\begin{aligned}\beta_{W[j]} &= m_{W[j]} - \beta_0 \\ &= m_{W[j]} - m\end{aligned} \tag{20.5}$$

For the interaction of the factors, the deflections are

$$\begin{aligned}\beta_{B\times W[i,j]} &= m_{B\times W[i,j]} - (\beta_0 + \beta_{B[i]} + \beta_{W[j]}) \\ &= m_{B\times W[i,j]} - m_{B[i]} - m_{W[j]} + m\end{aligned} \tag{20.6}$$

The deflection of subject $k|i$ is

$$\begin{aligned}\beta_{S|B[k|i]} &= m_{S|B[k|i]} - \left(\beta_0 + \beta_{B[i]}\right) \\ &= m_{S|B[k|i]} - m_{B[i]}\end{aligned} \tag{20.7}$$

Finally, the deflections for the interaction of subjects with the within-subject factor are

$$\begin{aligned}\beta_{W\times S|B[j,k|i]} &= y_{B\times W\times S|B[i,j,k|i]} - \left(\beta_0 + \beta_{B[i]} + \beta_{W[j]} + \beta_{B\times W[i,j]} + \beta_{S|B[k|i]}\right) \\ &= y_{B\times W\times S|B[i,j,k|i]} - m_{B\times W[i,j]} - m_{S|B[k|i]} + m_{B[i]}\end{aligned} \tag{20.8}$$

It is straight forward to verify that the total of the sum-to-zero effects (in Equations 20.3–20.8) exactly equals the data: $y_{B\times W\times S|B[i,j,k|i]} = \beta_0 + \beta_{B[i]} + \beta_{W[j]} + \beta_{B\times W[i,j]} + \beta_{S|B[k|i]} + \beta_{W\times S|B[j,k|i]}$. This leaves no residual variance for noise. Therefore, we treat $\beta_{W\times S|B[j,k|i]}$ as noise, that is, as random variation that we can not identify separately from noise. Consequently, individual data values are modeled as randomly distributed around the sum of the other effects, as follows:

$$y_{B\times W\times S|B[i,j,k|i]} \sim \text{normal}(\mu_{[i,j,k|i]}, \sigma) \tag{20.9}$$

$$\mu_{[i,j,k|i]} = \beta_0 + \beta_{B[i]} + \beta_{W[j]} + \beta_{B\times W[i,j]} + \beta_{S|B[k|i]} \tag{20.10}$$

where the deflections all respect the sum-to-zero contraints that fall out of Equations. 20.3–20.7.

20.5.2.3 Implementation in JAGS

The file named `Jags-Ymet-XnomSplitPlot-MnormalHom.R` implements the model, which is called from the high-level script named `Jags-Ymet-XnomSplitPlot-MnormalHom-Example.R`. The model specification is simply Equations 20.9 and 20.10 with the usual priors put on the deflection parameters. Notice from Equation 20.9 that a single noise parameter, σ, is used for all levels of the factors, which is to say that the model assumes homogeneity of variance.

The only novel part of the JAGS specification, relative to previous models, is computing the sum-to-zero deflections. The logical procedure is the same as before: first let the MCMC process find credible values of the baseline and deflections without the sum-to-zero constraint, and then recenter them to respect the sum-to-zero constraint. But doing this turns out to require some creative use of arrays because JAGS has more limited array-indexing abilities than R. I will not take the space here to explain its details, and instead rely on the intrepid reader to inspect the program if interested.

20.5.2.4 Results

Figure 20.10 shows the basic results in the form of posterior predictive distributions superimposed on the data. The predictive normal distributions are plotted with means at $\beta_0 + \beta_B + \beta_W + \beta_{B \times W}$ (collapsed across β_S) and with standard deviation σ from Equation 20.9. Thus, the normal distributions have a width that represents variation from the predicted value within a single subject's curve, not across subjects.

The results suggest that, overall, the moldboard plow and ridge tilling produced about equal yields. A comparison of the yields (specified in the high-level script named `Jags-Ymet-XnomSplitPlot-MnormalHom-Example.R`) is shown in the left panel of Figure 20.11. Not only is the difference very nearly zero, but the estimate of the difference is fairly precise, and we might even want to say that the difference is equivalent to zero for practical purposes, depending on how we specified the limits of the ROPE. On the other hand, chisel tilling produces lower yields than the average of moldboard plow and ridge tilling, as shown in the contrast in the second panel of Figure 20.11. These two contrasts are on the between-subject factor.

The Bayesian approach yields more powerful estimates of between-subject effects than traditional NHST. This is because traditional NHST uses different and larger denominator-error terms when computing F ratios for between-subject effects than for within-subject effects. Bayesian estimation, on the other hand, simply finds parameter values that are jointly credible, given the data.

The results also suggest that broadcast fertilization is less productive than banding fertilization. The middle-right panel of Figure 20.11 shows the contrast of the average of the two banding fertilization methods against the broadcast fertilization. The difference is large and the uncertainty is narrow.

Finally, the right panel of Figure 20.11 shows an interaction contrast. The difference between banding and broadcast fertilization seems to be especially pronounced for ridge tilling. So, we compare that difference of fertilization methods for ridge tilling versus the average of the other two tilling techniques, and the difference of differences is seen to be large.

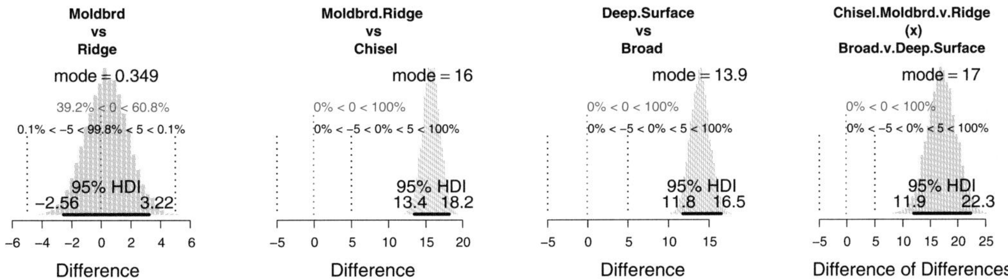

Figure 20.11 Main-effect contrasts and an interaction contrast for the corn-production data in Figure 20.10. (It is worth reiterating that these data are fictitious and might not reflect reality. The tilling and fertilization methods may also differ in effects other than current-year yield, such as cost or future-year soil quality.)

Analyzed as a two-factor between-subject design. To understand better the power of within-subject designs (and split-plot designs as a special case), it can be useful to analyze the data without taking into account the fact that the same field (i.e., subject) was used for all levels of fertilizer. We would not want to do this analysis for real research; I am presenting the analysis here merely to educate your intuition.

Consider the data for chisel tilling with broadcast fertilization, plotted at the top left of Figure 20.10 (p. 611). There is a lot of variation in yield across the identically treated fields, because some fields are more productive than others due to influences other than tilling and fertilizing. In the split-plot design, each field is measured with all types of fertilization, and therefore we can estimate each field's background level of productivity separately from the effects of the manipulated factors. The background level is suggested by the overall level of the lines connecting data from the same field, shown in Figure 20.10 (p. 611).

Suppose those same 99 data points came from all different fields instead of from 33 fields each with three fertilization methods. Then we would have a two-factor between-subject design, which was the primary focus at the beginning of this chapter in Figure 20.2 (p. 588). The data are replotted in Figure 20.12 with field/subject coding suppressed, hence no lines connecting data from the same field/subject. Because the between-subject variation can only be attributed to noise, the posterior predictive distribution has a much larger standard deviation than in Figure 20.10. The contrasts (shown in the bottom row) are much less certain than in Figure 20.11. In particular, the interaction contrast has experienced more shrinkage and its 95% HDI overlaps zero. Thus, if you can design research using within-subject factors, heeding the caveats of Section 20.5.1, you may be able to estimate effects with greater precision.

20.6. MODEL COMPARISON APPROACH

At the end of Section 19.3.3, I briefly discussed the omnibus test in ANOVA, and why I find it to be of only limited usefulness. The omnibus test of a factor or interaction asks whether there is some nonzero deviation anywhere across levels of a factor or the interaction. I argued that the omnibus test is not very useful in most applications because we want to know which groups differ and by how much, not merely whether there is some nonzero difference somewhere among the groups.

If, however, you would really like to compute the posterior probability that there is a nonzero deflection somewhere among the groups, an easy way to do it is with a factor-inclusion parameter, analogous to predictor-inclusion parameters for variable selection in multiple linear regression (Section 18.4, p. 536). In the JAGS model specification (in this case, for a between-subject design as in `Jags-Ymet-Xnom2fac-MnormalHom.R`), the factor-deflection parameters are multiplied by factor-inclusion parameters which can have value 0 or 1. Thus, the line in the JAGS model that was like this:

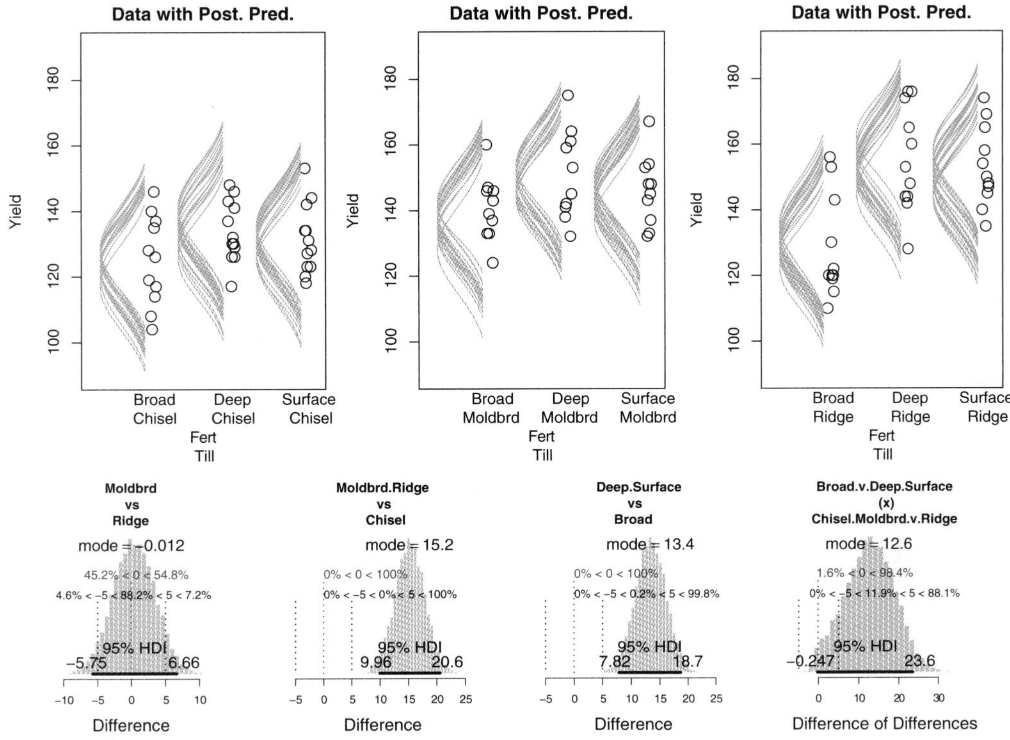

Figure 20.12 Data from Figure 20.10 (p. 611), with field/subject coding suppressed (hence no lines connecting data from the same field/subject). Because between-subject variation is modeled as noise, the posterior predictive distribution has a much larger standard deviation, and the contrasts (shown in the bottom row) are much less certain than in Figure 20.11 (p. 615).

```
mu[i] <- a0 + a1[x1[i]] + a2[x2[i]] + a1a2[x1[i],x2[i]]
```

is modified to incorporate factor inclusion parameters like this:

```
mu[i] <- ( a0  +  delta1 * a1[x1[i]]  +  delta2 * a2[x2[i]]
           +  delta1x2 * delta1 * delta2 * a1a2[x1[i],x2[i]] )
```

The factor-inclusion parameters are given Bernoulli priors that express the prior probability of including the factors:

```
delta1 ~ dbern( 0.5 )
delta2 ~ dbern( 0.5 )
delta1x2 ~ dbern( 0.5 )
```

For a reminder of further details involved in modifying a JAGS model, review Section 17.5.2 (p. 502).

When a factor-inclusion parameter is 1, the factor's deflections are used to describe the data. When the factor-inclusion parameter is 0, the factor is not used to describe the data, and the factor-deflection values are irrelevant (and randomly sampled by JAGS from the prior distribution of the deflection parameter). The posterior probability that the factor-inclusion parameter is 1 indicates the credibility of the factor-deflection model relative to a model in which the factor has zero influence, which is analogous to an omnibus test of the factor.

You may have noticed in the expression for `mu[i]`, above, that the interaction deflection was multiplied by the product of all three inclusion parameters, `deltalx2 * delta1 * delta2`, instead of only by `deltalx2`. This was done so that the interaction deflections have an influence only if both component factors are also included. It rarely makes sense to include an interaction without the component factors, as was discussed in Section 18.4.5 (p. 548). This use of the product of all three inclusion parameters implies that the prior probability of incorporating the interaction is lower than the component factors.

Section 18.4.1 (p. 539) discussed important caveats regarding the vagueness of the prior distribution on the included parameters, and Section 18.4.4 (p. 547) cautioned about autocorrelation in the MCMC chains. Those caveats apply here as well! In particular, it is easy to (inappropriately) exclude factors by setting the prior on their deflections to be extremely broad.

Bayes' factor approaches to hypothesis tests in ANOVA were presented by Rouder, Morey, Speckman, and Province (2012) and Wetzels, Grasman, and Wagenmakers (2012). Morey and Rouder's BayesFactor package for R is available at the Web site http://bayesfactorpcl.r-forge.r-project.org/.

20.7. EXERCISES

Look for more exercises at https://sites.google.com/site/doingbayesiandataanalysis/

Exercise 20.1. [Purpose: Using a novel data file and specifying meaningful contrasts.] The data file SeaweedData.csv (adapted from Qian & Shen, 2007) records how quickly seaweed regenerates when in the presence of different types of grazers. Data were collected from eight different tidal areas of the Oregon coast. We want to predict the amount of seaweed from the two predictors: grazer type and tidal zone. The tidal zones are simply labeled A–H. The grazer type was more involved, with six levels: No grazers (None), small fish only (f), small and large fish (fF), limpets only (L), limpets and small fish (Lf), limpets and small fish and large fish (LfF). We would like to know the effects of the different types of grazers, and we would also like to know about the different zones.

(A) Modify the script `Jags-Ymet-Xnom2fac-MnormalHom-Example.R` so it reads in the data from SeaweedData.csv. The column name for seaweed amount is SeaweedAmt, and the column names for the predictors are Grazer and Zone. In the script, specify

Grazer as the predictor that will appear as the abscissa in the plots of the data. For now, set all the contrasts to NULL. Run the script to check that it works.

(B) What is the average effect of small fish across all the zones? Answer this question by setting up the following three contrasts: none versus small fish only; limpets only versus limpets and small fish; the average of none and limpets versus the average of small fish only and limpets with small fish. Discuss the results.

(C) What is the average effect of limpets? There are several contrasts that can address this question, but be sure to include the average of none, small fish only, small and large fish, versus the average of limpets only, limpets and small fish, and limpets with small fish and large fish.

(D) Are there noticeable differences between zones? In particular, set up a contrast between zone A and zone D. Briefly discuss the result.

(E) Does the effect of limpets depend on whether it's in zone A or D? Set up appropriate interaction contrasts.

(F) Would it make any sense to run the heterogeneous-variance model on these data? (The answer is no; explain why. *Hint:* Notice how few data points there are in each group.)

Try to do the above without looking at the R commands below. But, to reduce any frustration, here is a hint (you might try different contrasts):

```
myDataFrame = read.csv( file="SeaweedData.csv" )
yName="SeaweedAmt"
x1Name="Grazer"
x2Name="Zone"
x1contrasts = list(
  # effect of small fish, f:
  list( c("None") , c("f") , compVal=0.0 , ROPE=c(-5,5) ) ,
  list( c("L") , c("Lf") , compVal=0.0 , ROPE=c(-5,5) ) ,
  list( c("None","L") , c("f","Lf") , compVal=0.0 , ROPE=c(-5,5) ) ,
  # effect of large fish, F:
  list( c("f","Lf") , c("fF","LfF") , compVal=0.0 , ROPE=c(-5,5) ) ,
  # effect of limpets, L:
  list( c("None","f","fF") , c("L","Lf","LfF") , compVal=0.0 , ROPE=c(-5,5) )
)
x2contrasts=list(
  list( c("D") , c("A") , compVal=0.0 , ROPE=c(-5,5) )
)
x1x2contrasts = list(
  # interaction of limpets in zones
  list( list( c("None","f","fF") , c("L","Lf","LfF") ) ,
        list( c("D") , c("A") ) ,
        compVal=0.0 , ROPE=c(-5,5) )
)
fileNameRoot = "SeaweedData-" # or whatever you prefer
graphFileType = "eps"         # or whatever you prefer
```

Exercise 20.2. [Purpose: Examine effect of transforming data on interaction, heterogeneity of variance, and skew.]

(A) Use the script `Jags-Ymet-Xnom2fac-MnormalHom-Example.R` with the data in Salary.csv. Run it once, as is, and verify that you get results similar to those shown in Figures 20.3-20.5. Answer this: Is the interaction contrast in Figure 20.5 a cross-over interaction or not? Explain.

(B) Transform the salary data by taking the logarithm (base 10). For a review of what this might do, see Section 20.3 (p. 599). To accomplish the transformation, try the following additions to the script right after reading in the data file:

```
myDataFrame = cbind( myDataFrame , LogSalary = log10(myDataFrame$ Salary) )
yName="LogSalary"
```

You will also want to change all the contrast `ROPE`s to `NULL` because the scale has changed. Run the analysis on these transformed data. Report the graphs analogous to Figure 20.3. Is there (by visual inspection) heterogeneity of variance in the transformed data? Is the interaction contrast analogous to Figure 20.5 still credibly nonzero? Is there any noticeable change in the skew of data distributions within groups?

(C) Use the script `Jags-Ymet-Xnom2fac-MrobustHet-Example.R` with the data in Salary.csv. Run it once, as is, and verify that you get results similar to those shown in Figures 20.8 and 20.9. Is the interaction contrast analogous to Figure 20.5 still credibly nonzero?

(D) As you did in a previous part of this exercise, transform the data by taking the base-10 logarithm (and change the `ROPE`s). Run the analysis on the transformed data. Is the interaction contrast analogous to Figure 20.5 still credibly nonzero? Does the resulting description seem any better than the previous part? For example, are the upper and lower bounds of the posterior predictive distributions more sensible?

CHAPTER 21

Dichotomous Predicted Variable

Contents

Fortune and Favor make fickle decrees, it's
Heads or it's tails with no middle degrees.
Flippant commandments decreed by law gods, have
Reasons so rare they have minus log odds.[1]

This chapter considers data structures that consist of a dichotomous predicted variable. The early chapters of the book were focused on this type of data, but now we reframe the analyses in terms of the generalized linear model. Data structures of the type considered in this chapter are often encountered in real research. For example, we might want to predict whether a person in a demographic study would be male or female based on their height and weight. Or we might want to predict whether or not a person will vote, based on their annual income. Or we might want to predict whether a baseball batter will get a hit, based on what their primary fielding position is.

The traditional treatment of these sorts of data structure is called "logistic regression." In Bayesian software it is easy to generalize the traditional models so they are robust

[1] This chapter is about logistic regression, and one of the concepts is called "log odds," explained in Section 21.2.1. I was fortunate to rhyme "log odds" with "law gods" and then work backwards to their names, Fortune and Favor.

Doing Bayesian Data Analysis, 2nd Edition
http://dx.doi.org/10.1016/B978-0-12-405888-0.00021-0

to outliers, allow different variances within levels of a nominal predictor, and have hierarchical structure to share information across levels or factors as appropriate.

In the context of the generalized linear model (GLM) introduced in Chapter 15, this chapter's situation involves an inverse-link function that is logistic along with a Bernoulli distribution for describing noise in the data, as indicated in the second row of Table 15.2 (p. 443). For a reminder of how this chapter's combination of predicted and predictor variables relates to other combinations, see Table 15.3 (p. 444).

21.1. MULTIPLE METRIC PREDICTORS

We begin by considering a situation with multiple metric predictors, because this case makes it easiest to visualize the concepts of logistic regression. Suppose we measure the height, weight, and gender (male or female) of a sample of full-grown adults. From everyday experience, it seems plausible that a tall, heavy person is more likely to be male than a short, light person. But exactly how predictive of gender is height or weight?

Some representative data are shown in Figure 21.1. The data are fictitious but realistic, generated by an accurate model of a large population survey (Brainard & Burmaster, 1992). The data are plotted as 1's or 0's, with gender arbitrarily coded as male = 1 and female = 0. All 0's are located on the bottom plane and all 1's are located on the top plane. You can see that the 1's tend to be at larger values of height and weight, while the 0's tend to be at smaller values of height and weight. (A "top down" view is shown in Figure 21.4, p. 628.)

21.1.1. The model and implementation in JAGS

How can we describe the rise in the probability of 1's as height and weight increase in Figure 21.1? We will use a logistic function of a linear combination of the predictors. This type of model was introduced in Section 15.3.1.1 (p. 436), and in Figure 15.10 (p. 443). The idea is that a linear combination of metric predictors is mapped to a probability value via the logistic function, and the predicted 0's and 1's are Bernoulli distributed around the probability. Restated formally:

$$\mu = \text{logistic}\,(\beta_0 + \beta_1 x_1 + \beta_2 x_2)$$
$$y \sim \text{Bernoulli}(\mu)$$

where

$$\text{logistic}(x) = \frac{1}{(1 + \exp(-x))}$$

The intercept and slope parameters (i.e., β_0, β_1, and β_2) are given prior distributions in the usual ways, although the interpretation of logistic regression coefficients requires careful thought, as will be explained in Section 21.2.1.

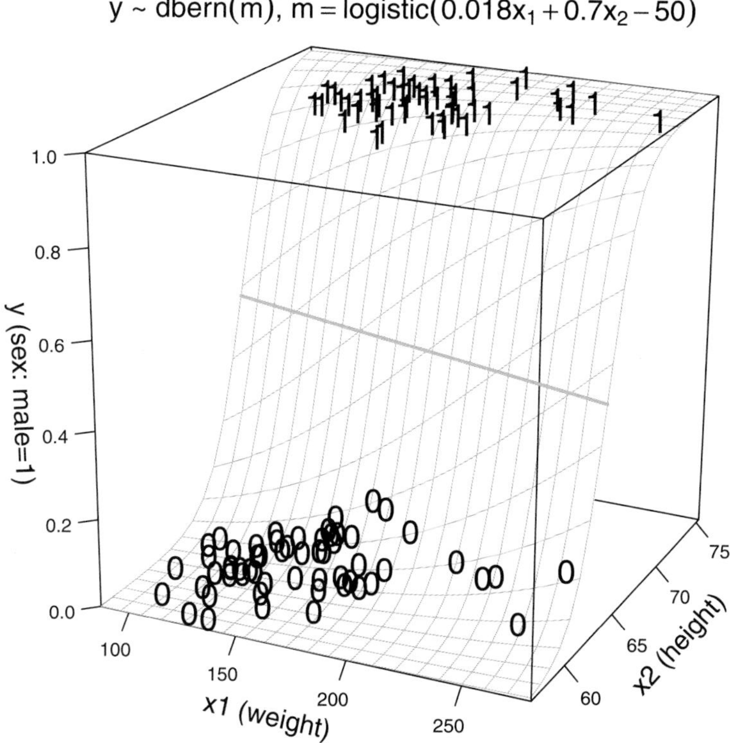

Figure 21.1 Data show gender (arbitrarily coded as male = 1, female = 0) as a function of weight (in pounds) and height (in inches). All 0's are located on the bottom plane of the cube, and all 1's are located on the top plane of the cube. Logistic surface shows maximum-likelihood estimate. Heavy line shows 50% level contour.

A diagram of the model is presented in Figure 21.2. At the bottom of the diagram, each dichotomous y_i value comes from a Bernoulli distribution with a "bias" of μ_i. (Recall that I refer to the parameter in the Bernoulli distribution as the bias, regardless of its value. Thus, a coin with a bias value of 0.5 is fair.) The μ value is determined as the logistic function of the linear combination of predictors. Finally, the intercept and slope parameters are given conventional normal priors at the top of the diagram.

It can be useful to compare the diagram for multiple logistic regression (in Figure 21.2) with the diagram for robust multiple linear regression in Figure 18.4 (p. 515). The linear core of the model is the same in both diagrams. What differs across diagrams is the bottom of the diagram that describes y, because y is on different types of scales in the two models. In the present application (Figure 21.2), y is dichotomous and therefore is described as coming from a Bernoulli distribution. In the earlier application

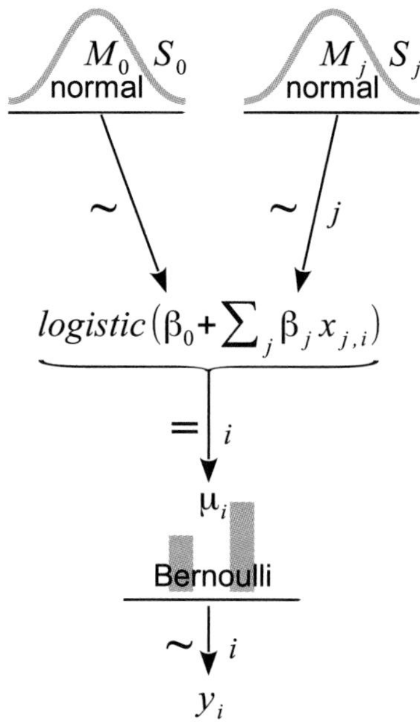

Figure 21.2 Dependency diagram for multiple logistic regression. Compare with the diagram for robust multiple linear regression in Figure 18.4 (p. 515).

(Figure 18.4, p. 515), y was metric and therefore was described as coming from a t distribution (for robustness against outliers).

It is straight forward to compose JAGS (or Stan) code for the model in Figure 21.2. Before getting to the JAGS code itself, there are a couple more preliminary details to explain. First, as we did for linear regression, we will try to reduce autocorrelation in the MCMC chains by standardizing the data. This is done merely to improve the efficiency of the MCMC process, not out of logical necessity. The y values must be 0's and 1's and therefore are not standardized, but the predictor values are metric and are standardized. Recall from Equation 17.1 (p. 485) that we denote the standardized value of x as $z = (x - \overline{x}) / s_x$, where $\overline{x}$ is the mean of x and s_x is the standard deviation of x. And, recall from Equation 15.15 (p. 439) that the inverse logistic function is called the logit function. We standardize the data and let JAGS find credible parameter values, denoted ζ_0 and ζ_j, for the standardized data. We then transform the standardized parameter values to the original scale, as follows:

$$
\begin{aligned}
\text{logit}(\mu) &= \zeta_0 + \sum_j \zeta_j z_j \\
&= \zeta_0 + \sum_j \zeta_j \, \frac{x_j - \overline{x}_j}{s_{x_j}}
\end{aligned}
$$

$$= \underbrace{\zeta_0 - \sum_j \frac{\zeta_j}{s_{x_j}} \bar{x}_j}_{\beta_0} + \sum_j \underbrace{\frac{\zeta_j}{s_{x_j}}}_{\beta_j} x_j \tag{21.1}$$

The transformation indicated by the underbraces in Equation 21.1 is applied at every step in the MCMC chain.

One more detail before getting to the JAGS code. In JAGS, the logistic function is called `ilogit`, for inverse logit. Recall from Section 15.3.1.1 (p. 436) that the inverse link function goes from the linear combination of predictors to the predicted tendency of the data, whereas the link function goes from the predicted tendency of the data to the linear combination of predictors. In this application, the link function is the logit. The inverse link function is, therefore, the inverse logit, abbreviated in JAGS as `ilogit`. The inverse link function is also, of course, the logistic. The `ilogit` notation helps distinguish the logistic function from the logistic distribution. We will not be using the logistic distribution, but just for your information, the logistic distribution is the probability density function which has a cumulative probability given by the logistic function. In other words, the logistic distribution is to the logistic function as the normal distribution is to the cumulative normal function in Figure 15.8 (p. 440). The logistic distribution is similar to the normal distribution but with heavier tails. Again, we will not be using the logistic distribution, and the point of this paragraph has been to explain JAGS' use of the term `ilogit` for the logistic function.

The JAGS code for the logistic regression model in Figure 21.2 is presented below. It begins with a `data` block that standardizes the data. Then the `model` block expresses the dependency arrows in Figure 21.2, and finally the parameters are transformed to the original scale using Equation 21.1:

```
# Standardize the data:
data {
  for ( j in 1:Nx ) {
    xm[j]  <- mean(x[,j])
    xsd[j] <-   sd(x[,j])
    for ( i in 1:Ntotal ) {
      zx[i,j] <- ( x[i,j] - xm[j] ) / xsd[j]
    }
  }
}
# Specify the model for standardized data:
model {
  for ( i in 1:Ntotal ) {
    # In JAGS, ilogit is logistic:
    y[i] ~ dbern( ilogit( zbeta0 + sum( zbeta[1:Nx] * zx[i,1:Nx] ) ) )
  }
```

```
  # Priors vague on standardized scale:
  zbeta0 ~ dnorm( 0 , 1/2^2 )
  for ( j in 1:Nx ) {
    zbeta[j] ~ dnorm( 0 , 1/2^2 )
  }
  # Transform to original scale:
  beta[1:Nx] <- zbeta[1:Nx] / xsd[1:Nx]
  beta0 <- zbeta0 - sum( zbeta[1:Nx] * xm[1:Nx] / xsd[1:Nx] )
}
```

The line in the JAGS code that specified the Bernoulli distribution did not use a separate line for specifying μ_i because we did not want to record the μ_i values. If we wanted to, we would instead explicitly create μ_i as follows:

```
y[i] ~ dbern( mu[i] )
mu[i] <- ilogit( zbeta0 + sum( zbeta[1:Nx] * zx[i,1:Nx] ) )
```

The complete program is in the file named `Jags-Ydich-XmetMulti-Mlogistic.R`, and the high-level script is the file named `Jags-Ydich-XmetMulti-Mlogistic-Example.R`.

21.1.2. Example: Height, weight, and gender

Consider again the data in Figure 21.4, which show the gender (male = 1, female = 0), height (in inches), and weight (in pounds) for 110 adults. We would like to predict gender from height and weight. We will first consider the results of using a single predictor, weight, and then consider the results of using both predictors.

Figure 21.3 shows the results of predicting gender from weight alone. The data are plotted as points that fall only at 0 or 1 on the y-axis. Superimposed on the data are logistic curves that have parameter values from various steps in the MCMC chain. The spread of the logistic curves indicates the uncertainty of the estimate; the steepness of the logistic curves indicates the magnitude of the regression coefficient. The 50% probability threshold is marked by arrows that drop down from the logistic curve to the x-axis, near a weight of approximately 160 pounds. The threshold is the x value at which $\mu = 0.5$, which is $x = -\beta_0/\beta_1$. You can see that the logistic appears to have a clear positive (nonzero) rise as weight increases, suggesting that weight is indeed informative for predicting gender. But the rise is gentle, not steep: There is no sharp threshold weight below which most people are female and above which most people are male.

The lower panels of Figure 21.3 show the marginal posterior distribution on the parameters. In particular, the slope coefficient, β_1, has a mode larger than 0.03 and a 95% HDI that is well above zero (presumably by enough to exclude at least some non-zero ROPE). A later section will discuss how to interpret the numerical value of the regression coefficient.

Now we consider the results when using two predictors, namely both height and weight, as shown in Figure 21.4. The data are plotted as 1's and 0's with x_1 (weight)

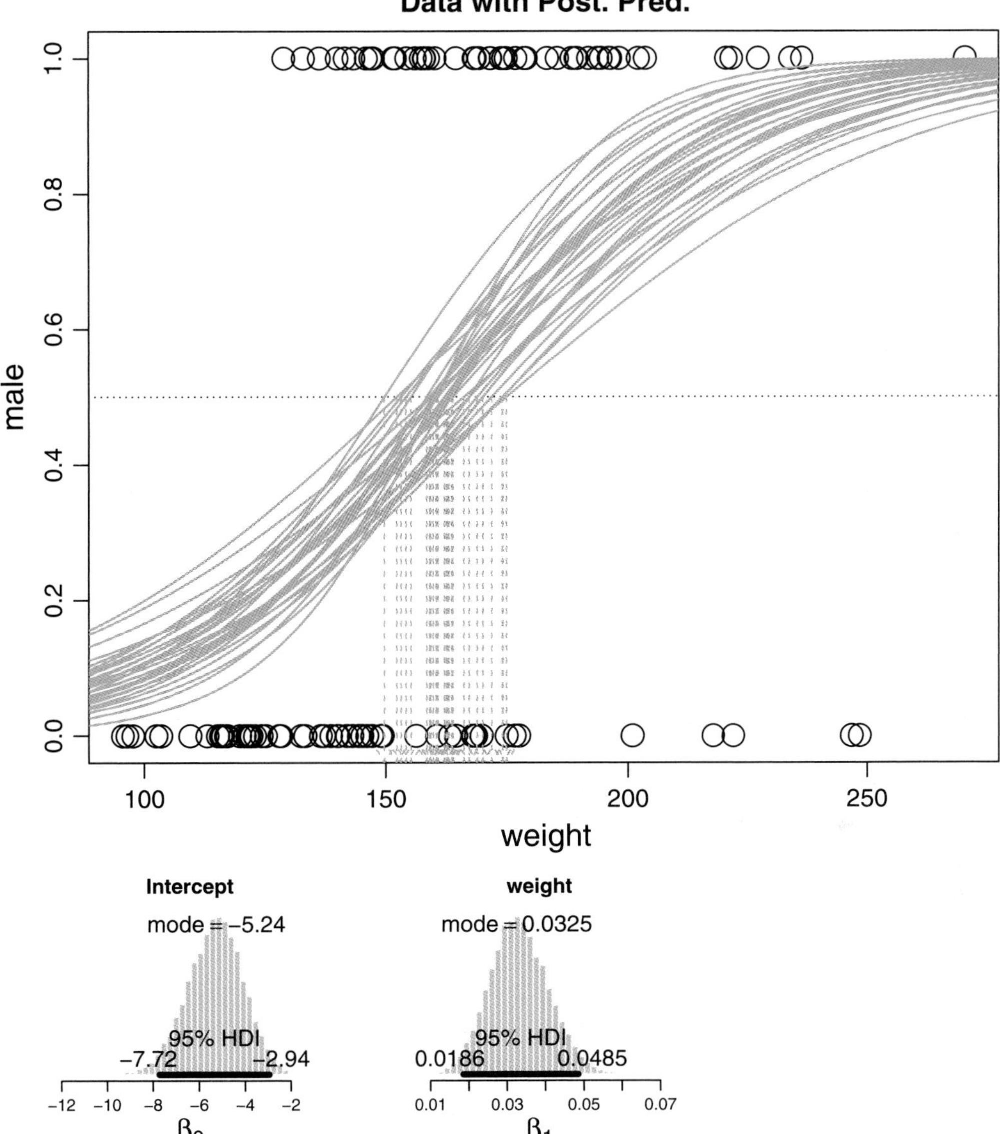

Figure 21.3 Predicting gender (arbitrarily coded as male = 1, female = 0) as a function of weight (in pounds), using logistic regression. Upper panel: Data are indicated by dots. Logistic curves are a random sample from the MCMC posterior. Descending arrows point to threshold weights at which the probability of male is 50%. Lower panels: Marginal posterior distribution on intercept and slope.

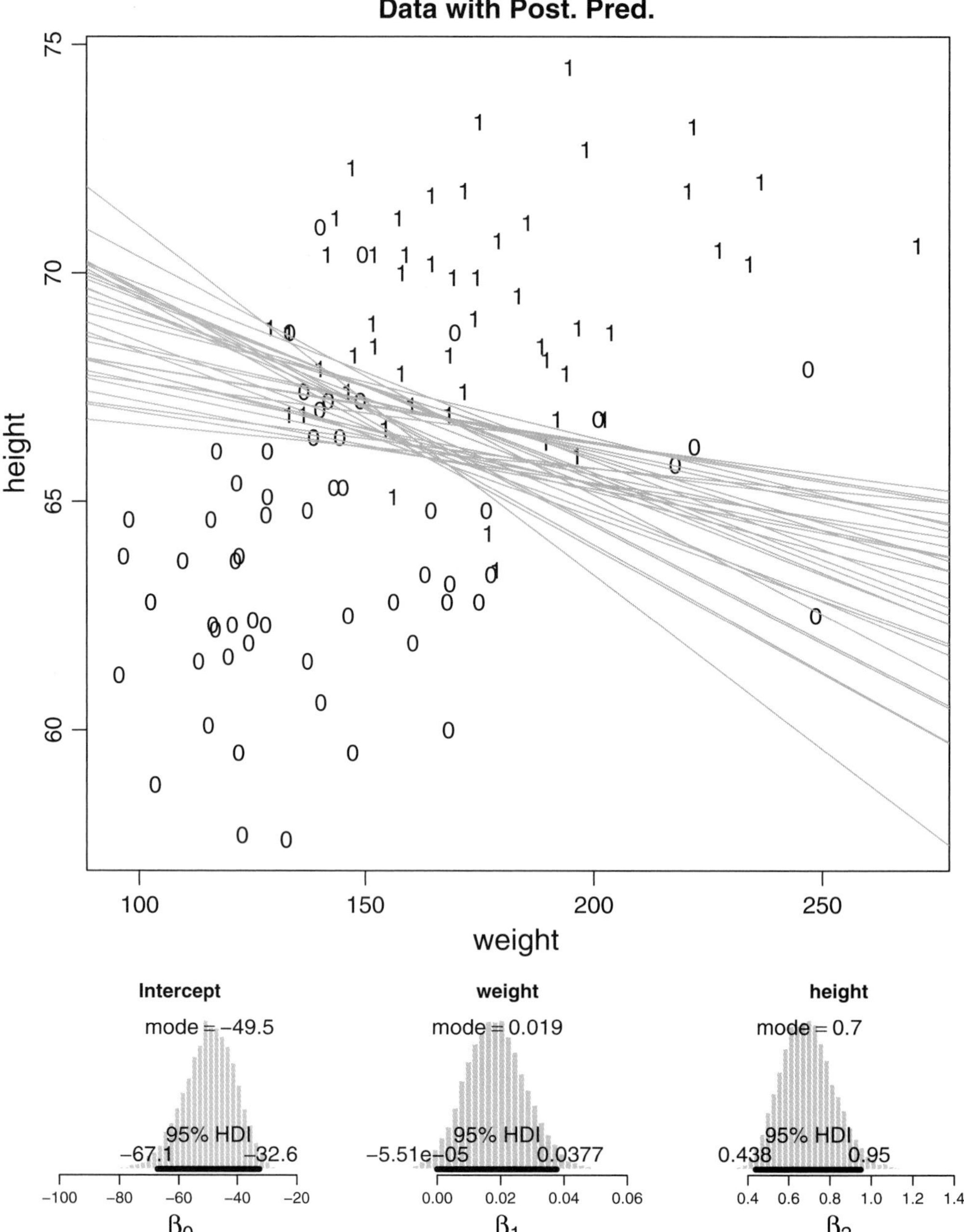

Figure 21.4 Predicting gender (arbitrarily coded as male = 1, female = 0) as a function of weight (in pounds) and height (in inches), using logistic regression. Upper panel: Data are indicated by 0's and 1's. Lines show a random sample from the MCMC posterior of thresholds at which the probability of male is 50%. Lower panels: Marginal posterior distribution on intercept and slopes.

on the horizontal axis and x_2 (height) on the vertical axis. You can see from the scatter plot of points that weight and height are positively correlated. Superimposed on the data are credible level contours at which $p(\text{male}) = 50\%$. Look back at Figure 21.1 (p. 623) for a picture of how a 50% level contour falls on a logistic surface. The 50% level contour is the set of x_1, x_2 values for which $\mu = 0.5$, which is satisfied by $x_2 = (-\beta_0/\beta_2) + (-\beta_1/\beta_2)x_1$. On one side of the level contour there is less than 50% chance of being male, and on the other side of the level contour there is greater than 50% chance of being male, according to the model (not necessarily in reality). The spread of the credible level contours indicates the uncertainty in the parameter estimates. The perpendicular to the level contour indicates the direction in which probability changes the fastest. The angle of the level contours in Figure 21.4 suggests that the probability of being male increases rapidly as height goes up, but the probability of being male increases only a little as weight goes up. This interpretation is confirmed by the lower panels of Figure 21.4, which show the marginal posterior on the regression coefficients. In particular, the regression coefficient on weight has a modal value less than 0.02 and its 95% HDI essentially touches zero.

The regression coefficient on weight is smaller when height is included in the regression than when height is not included (compare Figure 21.4 with Figure 21.3). Why would this be? Perhaps you anticipated the answer: Because the predictors are correlated, and it's really height that's doing most of the predictive work in this case. When weight is the only predictor included, it has some predictive power because weight correlates with height, and height predicts gender. But when height is included as a second predictor, we find that the independent predictiveness of weight is relatively small. This issue of interpreting correlated predictors is the same here, in the context of logistic regression, as was discussed at length in Section 18.1.1 (p. 510) in the context of linear regression.

21.2. INTERPRETING THE REGRESSION COEFFICIENTS

In this section, I'll discuss how to interpret the parameters in logistic regression. The first subsection explains how to interpret the numerical magnitude of the slope coefficients in terms of "log odds." The next subsection shows how data with relatively few 1's or 0's can yield ambiguity in the parameter estimates. Then an example with strongly correlated predictors reveals tradeoffs in slope coefficients. Finally, I briefly describe the meaning of multiplicative interaction for logistic regression.

21.2.1. Log odds

When the logistic regression formula is written using the logit function, we have $\text{logit}(\mu) = \beta_0 + \beta_1 x_1 + \beta_2 x_2$. The formula implies that whenever x_1 goes up by 1 unit (on the x_1 scale), then $\text{logit}(\mu)$ goes up by an amount β_1. And whenever x_2 goes up by 1 unit (on the x_2 scale), then $\text{logit}(\mu)$ goes up by an amount β_2. Thus, the regression coefficients are telling us about increases in logit(μ). To understand the regression coefficients, we need to understand logit(μ).

The logit function is the inverse of the logistic function. Formally, for $0 < \mu < 1$, $\text{logit}(\mu) = \log(\mu/(1-\mu))$, where the logarithm is the natural logarithm, that is, the inverse of the exponential function. It's easy to verify through algebra that this expression for the logit really does make it the inverse of the logistic: If $x = \text{logit}(\mu) = \log(\mu/(1-\mu))$, then $\mu = \text{logistic}(x) = 1/(1+\exp(-x))$ and vice versa. Now, in applications to logistic regression, μ is the probability that $y = 1$, and therefore we can write $\text{logit}(\mu) = \text{logit}\left(p(y{=}1)\right) = \log\left(p(y{=}1)/(1-p(y{=}1))\right) = \log\left(p(y{=}1)/p(y{=}0)\right)$. The ratio, $p(y{=}1)/p(y{=}0)$, is called the odds of outcome 1 to outcome 0, and therefore $\text{logit}(\mu)$ is the log odds of outcome 1 to outcome 0.

Combining the previous two paragraphs, we can say that the regression coefficients are telling us about increases in the log odds. Let's consider a numerical example. Rounding the modal values in Figure 21.4, suppose that $\beta_0 = -50.0$, $\beta_1 = 0.02$, and $\beta_2 = 0.70$.

- Consider a hypothetical person who weighs 160 pounds, that is, $x_1 = 160$. If that person were 63 inches tall, then the predicted probability of being male is $\text{logistic}(\beta_0 + \beta_1 x_1 + \beta_2 x_2) = \text{logistic}(-50.0 + 0.02 \cdot 160 + 0.70 \cdot 63) = 0.063$. That probability is a log odds of $\log(0.063/(1-0.063)) = -2.70$. Notice the negative value of the log odds, which indicates the probability is less than 50%. If that person were 1 inch taller, at 64 inches, then the predicted probability of being male is 0.119, which has a log odds of -2.00. Thus, when x_2 was increased by 1 unit, the probability went up 0.056 (from 0.063 to 0.119), and the log odds increased 0.70 (from -2.70 to -2.00) which is exactly β_2.
- Again consider a hypothetical person who weighs 160 pounds, that is, $x_1 = 160$. Now consider an increase in height from 67 to 68 inches. If the person were 67 inches tall, then the predicted probability of being male is $\text{logistic}(\beta_0 + \beta_1 x_1 + \beta_2 x_2) = \text{logistic}(-50.0 + 0.02 \cdot 160 + 0.70 \cdot 67) = 0.525$. That probability is a log odds of $\log(0.525/(1-0.525)) = 0.10$. Notice the positive value of the log odds, which indicates the probability is greater than 50%. If that person were 1 inch taller, at 68 inches, then the predicted probability of being male is 0.690, which has a log odds of 0.80. Thus, when x_2 was increased by 1 unit, the probability went up 0.165 (from 0.525 to 0.690), and the log odds increased 0.70 (from 0.10 to 0.80) which is exactly β_2.

The two numerical examples show that an increase of 1 unit of x_j increases the log odds by β_j. But a constant increase in log odds does not mean a constant increase in probability. In the first numerical example, the increase in probability was 0.056, but in the second numerical example, the increase in probability was 0.165.

Thus, a regression coefficient in logistic regression indicates how much a 1 unit change of the predictor increases the log odds of outcome 1. A regression coefficient of 0.5 corresponds to a rate of probability change of about 12.5 percentage points per x-unit at the threshold x value. A regression coefficient of 1.0 corresponds to a rate of probability change of about 24.4 percentage points per x-unit at the threshold x value. When x is much larger or smaller than the threshold x value, the rate of change in probability is smaller, even though the rate of change in log odds is constant.

21.2.2. When there are few 1's or 0's in the data

In logistic regression, you can think of the parameters as describing the boundary between the 0's and the 1's. If there are many 0's and 1's, then the estimate of the boundary parameters can be fairly accurate. But if there are few 0's or few 1's, the boundary can be difficult to identify very accurately, even if there are many data points overall.

In the data of Figure 21.1, involving gender (male, female) as a function of weight and height, there were approximately 50% 1's (males) and 50% 0's (females). Because there were lots of 0's and 1's, the boundary between them could be estimated relatively accurately. But many realistic data sets have only a small percentage of 0's or 1's. For example, suppose we are studying the incidence of heart attacks predicted by blood pressure. We measure the systolic blood pressure at the annual check-ups of a random sample of people, and then record whether or not they had a heart attack during the following year. We code not having a heart attack as 0 and having a heart attack as 1. Presumably the incidents of heart attack will be very few, and the data will have very few 1's. This dearth of 1's in the data will make it difficult to estimate the intercept and slope of the logistic function with much accuracy.

Figure 21.5 shows an example. For both panels, the x values of the data are identical, random values from a standardized normal distribution. For both panels, the y values were generated randomly from a Bernoulli distribution with bias given by a logistic function with the same slope, $\beta_1 = 1$. What differs between panels is where the threshold of the logistic falls relative to the mean of the x values. For the left panel, the threshold is at 3 (i.e., $\beta_0 = -3$), whereas for the right panel, the threshold is at 0. You can see that the data in the left panel have relatively few 1's (in fact, only about 7% 1's), whereas the data in the right panel have about 50% 1's. The right panel is analogous to the case of gender as a function of weight. The left panel is analogous to the case of heart attack as a function of blood pressure.

You can see in Figure 21.5 that the estimate of the slope (and of the intercept) is more certain in the right panel than in the left panel. The 95% HDI on the slope, β_1, is much wider in the left panel than in the right panel, and you can see that the logistic curves in the left panel have greater variation in steepness than the logistic curves in the right panel. The analogous statements hold true for the intercept parameter.

Thus, if you are doing an experimental study and you can manipulate the x values, you will want to select x values that yield about equal numbers of 0's and 1's for the y values overall. If you are doing an observational study, such that you cannot control

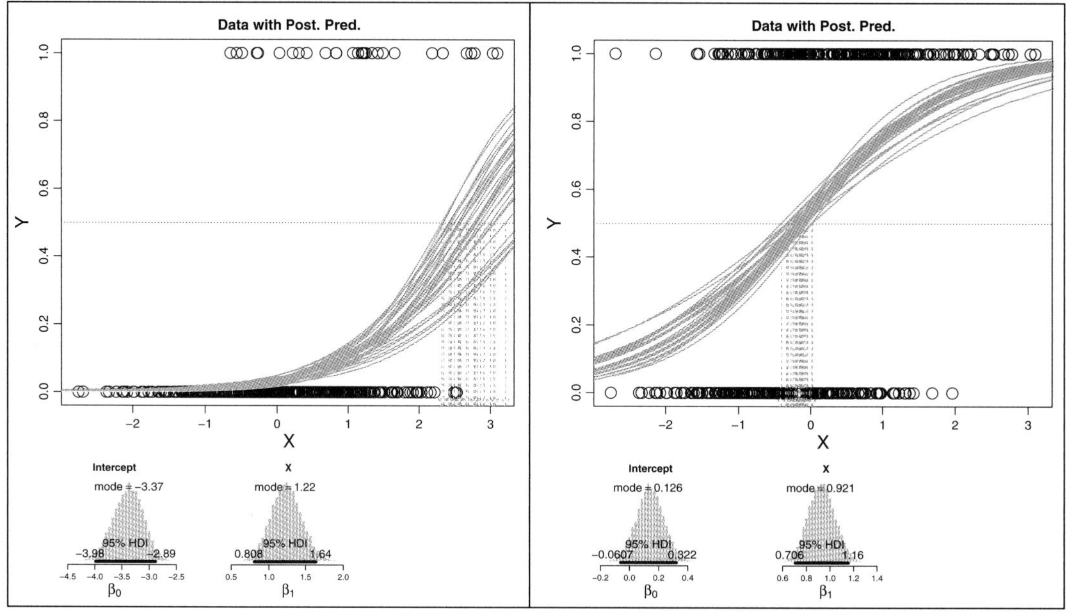

Figure 21.5 Parameter estimates are more uncertain when there are few 0's or 1's in the data. The main panels have data with the same *x* values, and *y* values randomly generated by logistic functions with the same slope ($\beta_1 = 1$) but different intercepts ($\beta_0 = -3$ and $\beta_0 = 0$).

any independent variables, then you should be aware that the parameter estimates may be surprisingly ambiguous if your data have only a small proportion of 0's or 1's. Conversely, when interpreting the parameter estimates, it helps to recognize when the threshold falls at the fringe of the data, which is one possible cause of relatively few 0's or 1's.

21.2.3. Correlated predictors

Another important cause of parameter uncertainty is correlated predictors. This issue was previously discussed at length, but the context of logistic regression provides novel illustration in terms of level contours.

Figure 21.6 shows a case of data with strongly correlated predictors. Both predictors have values randomly drawn from a standardized normal distribution. The y values are generated randomly from a Bernoulli distribution with bias given by slope parameters $\beta_1 = 1$ and $\beta_2 = 1$. The threshold falls in their midst, with $\beta_0 = 0$, so there are about half 0's and half 1's.

The posterior estimates of the parameter values are shown in Figure 21.6. In particular, notice that the 50% level contours (the threshold lines) are extremely ambiguous, with many different possible angles. The right panel shows strong anticorrelation of credible β_1 and β_2 values.

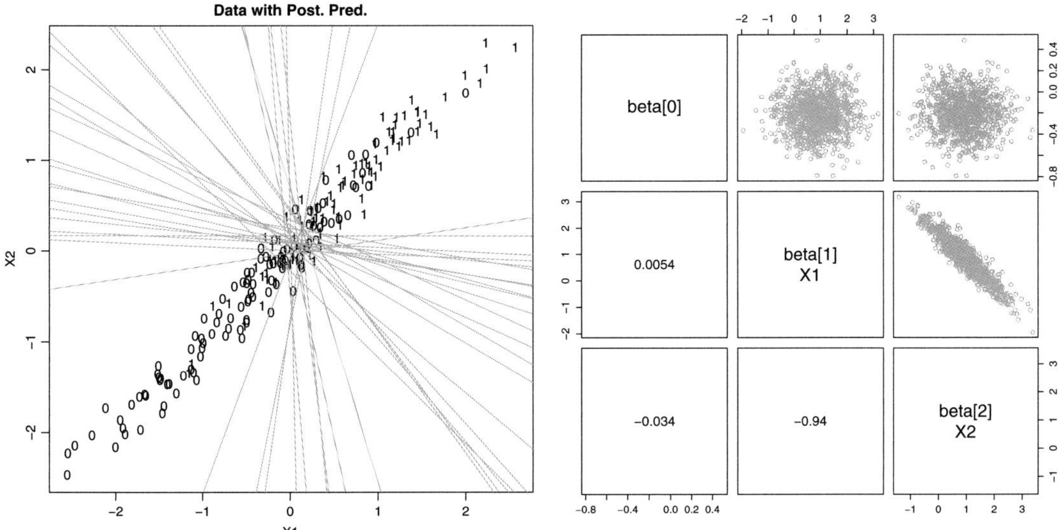

Figure 21.6 Estimates of slope parameters trade off when the predictors are correlated. Left panel shows credible 50% level contours superimposed on data. Right panel shows strong anticorrelation of credible β_1 and β_2 values.

The same ambiguity can arise for correlations of multiple predictors, but higher-dimensional correlations are difficult to graph. As in the case of linear regression, it is important to consider the correlations of the predictors when interpreting the parameters of logistic regression. The high-level scripts display the predictor correlations on the R console.

21.2.4. Interaction of metric predictors

There may be applications in which it is meaningful to consider a multiplicative interaction of metric predictors. For example, in the case of predicting gender (male vs female) from weight and height, it could be that an additive combination of the predictors is not very accurate. For example, it might be that only the combination of tall and heavy is a good indicator of being male, while being tall but not heavy, or heavy but not tall, both indicate being female. This sort of conjunctive combination of predictors can be expressed by their multiplication (or other ways).

Figure 21.7 shows examples of logistic surfaces with multiplicative interaction of predictors. Within Figure 21.7, the title of each panel displays the formula being plotted. The left column of Figure 21.7 shows examples without interaction, for reference. The right column shows the same examples from the left column but with a

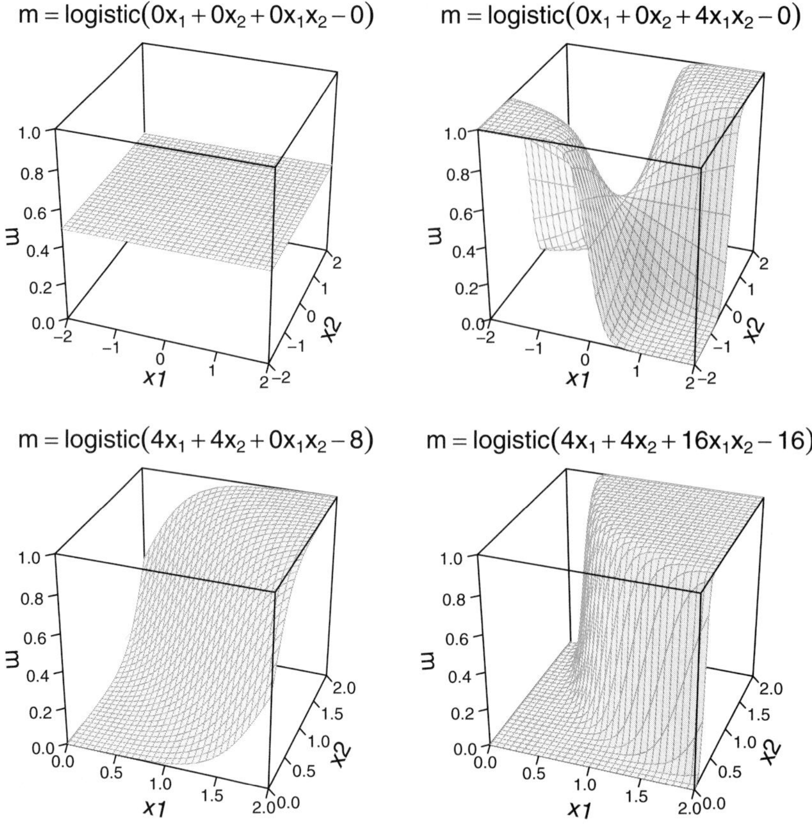

Figure 21.7 Multiplicative interaction of metric predictors in logistic regression. Left column shows examples of no interaction. Right column shows corresponding logistic surfaces with interaction. Title of each plot shows the coefficients on the predictors.

multiplicative interaction included. In the top row, you can see that including the term $+4x_1x_2$ makes the logistic surface rise whenever x_1 and x_2 are both positive or both negative. In the bottom row, you can see that including a multiplicative interaction (with an adjustment of the intercept) produces the sort of conjunctive combination described in the previous paragraph. Thus, when either x_1 or x_2 is zero, the logistic surface is nearly zero, but when both predictors are positive, the logistic surface is high. Importantly, the 50% level contour is straight when there is no interaction, but is curved when there is interaction.

A key thing to remember when using an interaction term is that regression coefficients on individual predictors only indicate the slope when all the other predictors are set to zero. This issue was illustrated extensively in the context of linear regression in Section 18.2 (p. 525). Analogous points apply to logistic regression.

21.3. ROBUST LOGISTIC REGRESSION

Look again at Figure 21.3 (p. 627), which showed gender as a function of weight alone. Notice that there are several unusual data points in the lower right that represent heavy females. For these data points to be accommodated by a logistic function, its slope must not be too extreme. If the slope is extreme, then the logistic function gets close to its asymptote at $y = 1$ for heavy weights, which then makes the probability of data points with $y = 0$ essentially nil. In other words, the outliers relative to the logistic function can only be accommodated by reducing the magnitude of the estimated slope.

One way to address the problem of outliers is by considering additional predictors, such as height in addition to weight. It could be that the heavy females also happen to be short, and therefore a logistic function with a large slope coefficient on height would account for the data, without having to artificially reduce its slope coefficient on weight.

Usually we do not have other predictors immediately at hand to account for the outliers, and instead we use a model that incorporates a description of outliers as such. Because this sort of model provides parameter estimates that are relatively stable in the presence of outliers, it is called robust against outliers. We have routinely considered robust models in previous chapters involving metric variables (e.g., Section 16.2, p. 458). In the present context of a dichotomous predicted variable, we need a new mathematical formulation.

We will describe the data as being a mixture of two different sources. One source is the logistic function of the predictor(s). The other source is sheer randomness or "guessing," whereby the y value comes from the flip of a fair coin: $y \sim \text{Bernoulli}(\mu = 1/2)$. We suppose that every data point has a small chance, α, of being generated by the guessing process, but usually, with probability $1 - \alpha$, the y value comes from the logistic function of the predictor. With the two sources combined, the predicted probability that $y = 1$ is

$$\mu = \alpha \cdot \frac{1}{2} + (1 - \alpha) \cdot \text{logistic}\left(\beta_0 + \sum_j \beta_j x_j\right) \tag{21.2}$$

Notice that when the guessing coefficient is zero, then the conventional logistic model is completely recovered. When the guessing coefficient is one, then the y values are completely random. (This could also be achieved by setting all the slope coefficients in the logistic function to zero.)

Our goal is to estimate the guessing coefficient along with the logistic parameters. For this, we need to establish a prior on the guessing coefficient, α. In most applications we would expect the proportion of random outliers in the data to be small, and therefore the prior should emphasize small values of α. For the example presented below, I have set the prior as $\alpha \sim \text{dbeta}(1, 9)$. This prior distribution gives values of α greater than 0.5 very low but non-zero probability.

The JAGS model specification for robust logistic regression is a modest extension of the model specification for ordinary logistic regression. As before, it starts with standardizing the data and ends with transforming the parameters back to the original scale. The mixture of Equation 21.2 is directly expressed in JAGS, with α coded as `guess`:

```
for ( i in 1:Ntotal ) {
  y[i] ~ dbern( mu[i] )
  mu[i] <- ( guess * (1/2)
             + (1.0-guess) * ilogit(zbeta0+sum(zbeta[1:Nx] * zx[i,1:Nx])) )
}
guess ~ dbeta(1,9)
```

The model definitions are in the file named `Jags-Ydich-XmetMulti-MlogisticRobust.R` and the high-level script that calls the model is named `Jags-Ydich-XmetMulti-MlogisticRobust-Example.R`.

Figure 21.8 shows the fit of the robust logistic regression model for predicting gender as a function of weight alone. The superimposed curves show μ from Equation 21.2. Notice that the curves asymptote at levels away from 0 and 1, unlike the ordinary logistic curves in Figure 21.3. The modal estimate of the guessing parameter is almost 0.2. This implies that the asymptotes are around $y = 0.1$ and $y = 0.9$. Especially for heavy weights, the non-1 asymptote lets the model accommodate outlying data while still having a steep slope at the threshold. Notice that the slope of the logistic is larger than in Figure 21.3.

Figure 21.9 shows pairwise posterior parameters. In particular, consider the panel that plots the value of β_1 (i.e., the slope on weight) against the value of the guessing parameter. Notice the strong positive correlation between those two parameters. This means that as the guessing parameter gets larger, credible values of the slope get larger also.

For some data sets with extreme outliers (relative to the predictions of ordinary logistic regression), the inclusion of a guessing parameter for robust logistic regression can make the difference between slope estimates that are unrealistically small and slope estimates that are large and certain enough to be useful. But if the data do not have extreme outliers, they can be difficult to detect, as exemplified in Exercise 21.1. There are other ways to model outliers, as we will see in Exercise 23.2 (p. 701) in the chapter on ordinal predicted variables.

21.4. NOMINAL PREDICTORS

We now turn our attention from metric predictors to nominal predictors. For example, we might be interested in predicting whether a person voted for candidate A or candidate B (the dichotomous predicted variable), based on the person's political party and religious affiliations (nominal predictors). Or we might want to predict the probability that a

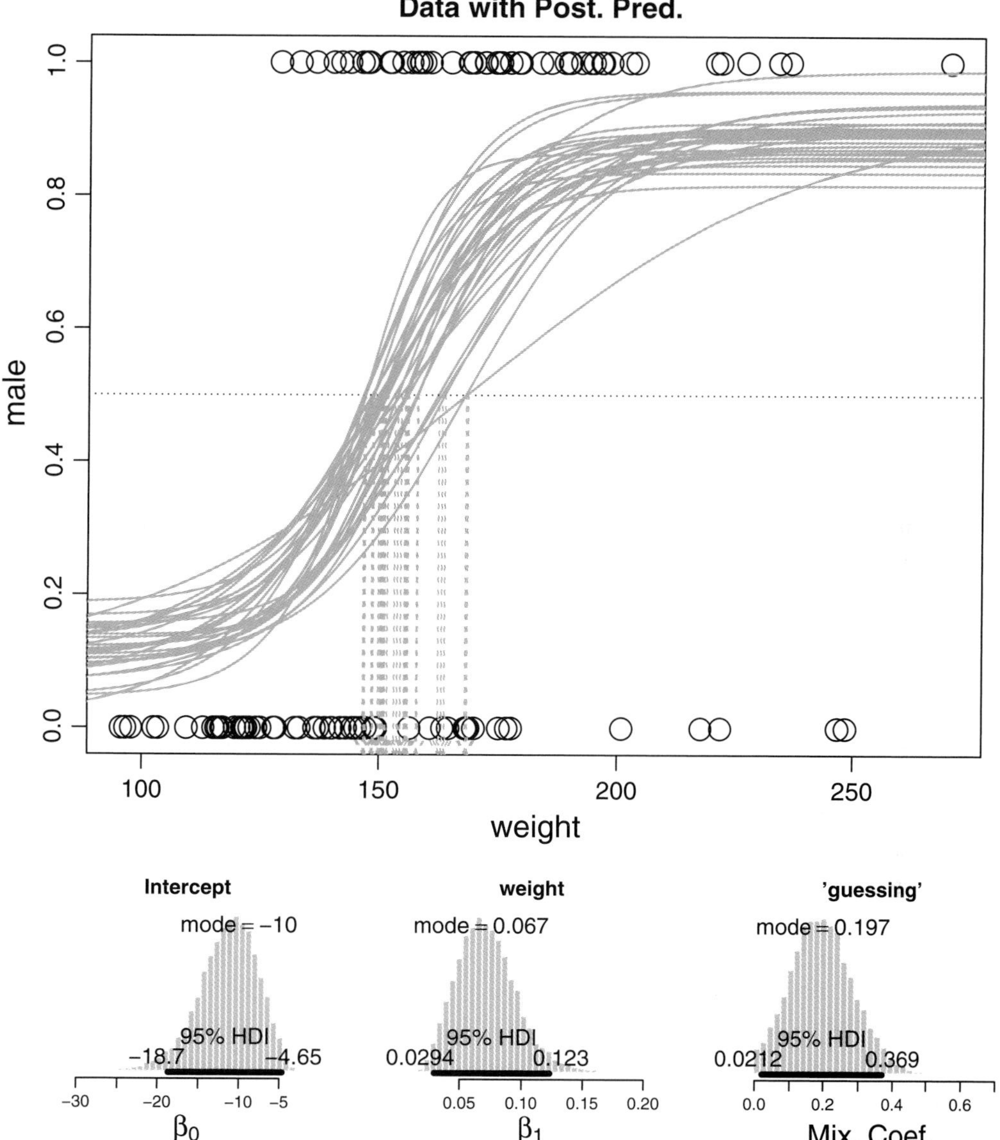

Figure 21.8 Predicting gender (arbitrarily coded as male = 1, female = 0) as a function of weight (in pounds), using robust logistic regression. Upper panel: Data are indicated by dots, the same as in Figure 21.3. Curves are a random sample from the MCMC posterior; notice asymptotes away from 0,1 limits. Descending arrows point to threshold weights at which the probability of male is 50%. Lower panels: Marginal posterior distribution on baseline, slope, and guessing coefficient. Pairwise plots are shown in Figure 21.9.

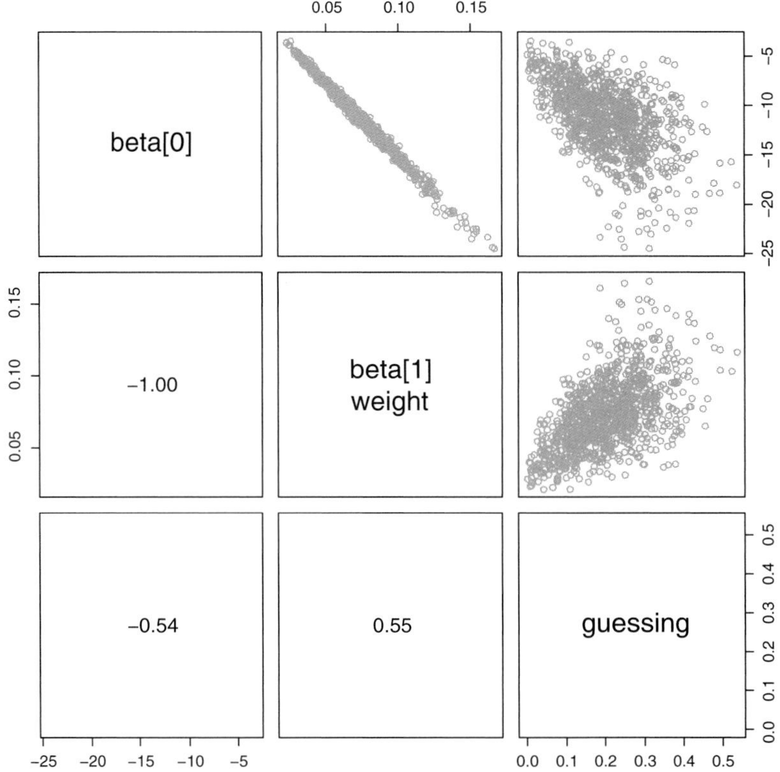

Figure 21.9 Pairwise plots for posterior distribution in Figure 21.8. Notice here that the guessing coefficient is correlated with the slope, beta[1].

baseball player will get a hit when at bat (dichotomous predicted), based on the player's fielding position (nominal predictor).

21.4.1. Single group

We start with the simplest, trivial case: A single group. In this case, we have several observed 0's and 1's, and our goal is to estimate the underlying probability of 1's. This is simply the case of estimating the underlying bias of a coin by observing several flips. The novelty here is that we treat it as a case of the generalized linear model, using a logistic function of a baseline.

Back in Chapter 6, we formalized this situation by denoting the bias of the coin as θ, and giving θ a beta prior. Thus, the model was $y \sim \text{Bernoulli}(\theta)$ with $\theta \sim \text{dbeta}(a, b)$. Now, the bias of the coin is denoted μ (instead of θ). The value of μ is given by a logistic function of a baseline, β_0. Thus, in the new model:

$$y \sim \text{Bernoulli}(\mu) \quad \text{with} \quad \mu = \text{logistic}(\beta_0)$$

The model expresses the bias in the coin in terms of β_0. When $\beta_0 = 0$, the coin is fair because $\mu = 0.5$. When $\beta_0 > 0$, then $\mu > 0.5$ and the coin is head biased. When $\beta_0 < 0$, then $\mu < 0.5$ and the coin is tail biased. Notice that β_0 ranges from $-\infty$ to $+\infty$, while μ ranges from 0 to 1. Therefore the prior on β_0 must also support $-\infty$ to $+\infty$. The conventional prior is a normal distribution:

$$\beta_0 \sim \text{normal}(M_0, S_0)$$

When $M_0 = 0$, the prior distribution is centered at $\mu = 0.5$. Figure 21.10 presents a diagram of the model, which should be compared with the beta prior in Figure 8.2 (p. 196).

Because μ is a logistic function of a normally distributed value β_0, it is not immediately obvious what the prior distribution of μ actually looks like. Figure 21.11 gives some examples. In each example, random values of β_0 were generated from its normal prior, and then those values of β_0 were converted to values of μ via the logistic function. The resulting values of μ were then plotted in a histogram. The histograms in Figure 21.11 are outlined by the exact mathematical form of the implied prior on μ, derived from the reparameterization formula explained in Section 25.3 (p. 729).[2]

For some values of prior mean and standard deviation on β_0, the prior distribution of μ looks somewhat similar to a beta distribution. But there are many choices of prior mean and standard deviation that make the distribution of μ look nothing like a beta distribution. In particular, there is no choice of prior mean and standard deviation that give μ a uniform distribution. Also, while the beta-prior constants have a natural interpretation in terms of previously observed data (recall Section 6.2.1, p. 127), the normal-prior constants for the logistic mapping to μ are not as intuitive.

It is straight forward to convert the beta-prior program, Jags-Ydich-Xnom1subj-MbernBeta.R, to a version that uses the logistic of a baseline. (The new program could be called Jags-Ydich-Xnom1subj-Mlogistic.R.) The model specification from Jags-Ydich-Xnom1subj-MbernBeta.R is as follows:

```
model {
  for ( i in 1:Ntotal ) {
    y[i] ~ dbern( theta )
  }
  theta ~ dbeta(1,1)
}
```

[2] The reparameterization formula, Equation 25.1 (p. 729) in Section 25.3, requires the inverse and derivative of the logistic function. We already know that the inverse of the logistic is the logit, which is $\text{logistic}^{-1}(\mu) = \log(\mu/(1-\mu))$. It turns out that the derivative of the logistic function is conveniently expressed in terms of the logistic function itself: $\text{logistic}'(\beta) = \text{logistic}(\beta) \cdot (1 - \text{logistic}(\beta))$. We substitute $\text{logistic}^{-1}(\mu)$ for β in the derivative. Then Equation 25.1 (p. 729) is used to plot the probability density of μ.

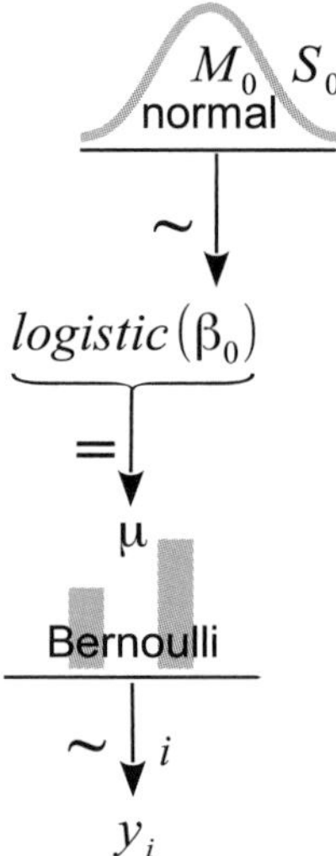

Figure 21.10 Estimating the underlying bias of a coin using a logistic function of a baseline. When $\beta_0 = 0$, then $\mu = 0.5$, and the coin is fair. Compare with the beta prior in Figure 8.2 (p. 196).

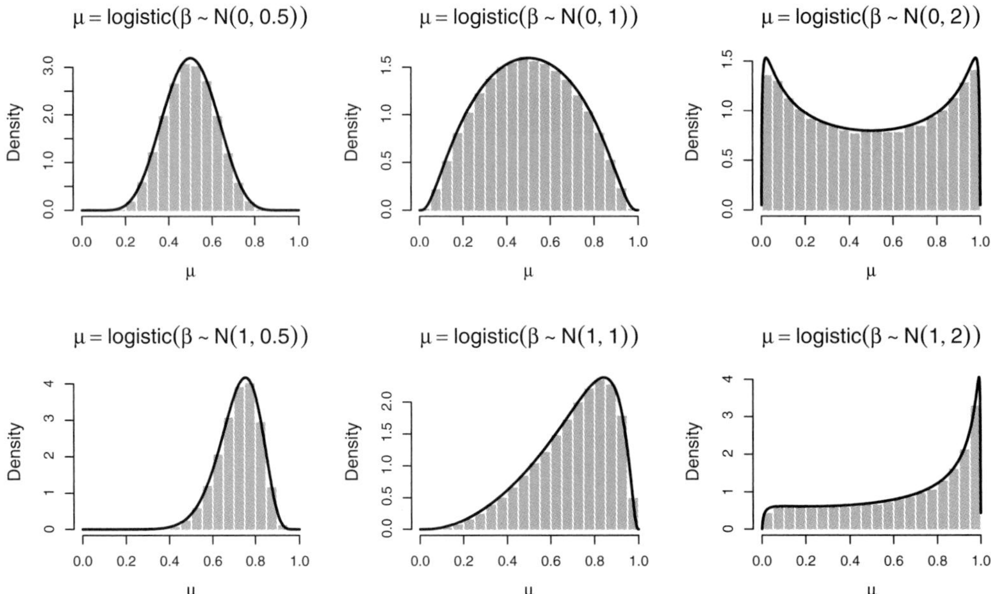

Figure 21.11 Prior distributions on μ for different choices of M_0 and S_0 in Figure 21.10.

The logistic version could instead use the following:

```
model {
  for ( i in 1:Ntotal ) {
    y[i] ~ dbern( ilogit( b0 ) )
  }
  b0 ~ dnorm( 0 , 1/2^2)
}
```

The prior on β_0 was given a mean of $M_0 = 0$ and a standard deviation of $S_0 = 2$. Notice that the model specification, above, did not use a separate line to explicitly name μ. If you want to keep track of μ, you could do it after the fact by creating $\mu = \text{logistic}(\beta_0)$ at each step in the chain, or you could do it this way:

```
model {
  for ( i in 1:Ntotal ) {
    y[i] ~ dbern( mu )
  }
  mu <- ilogit( b0 )
  b0 ~ dnorm( 0 , 1/2^2 )
}
```

Review Section 17.5.2 (p. 502) for tips about what other aspects of the program would need to be changed. For any given data set, the posterior distribution created by the logistic version of the program would not exactly match the posterior distribution created by the beta version of the program, because the prior distributions are not identical. You have the chance to try all this, hands on, in Exercise 21.2.

21.4.2. Multiple groups

The previous section discussed the trivial case of a single group, primarily for the purpose of introducing concepts. In the present section, we consider a more realistic case in which there are several groups. Moreover, instead of each "subject" within a group contributing a single dichotomous score, each subject contributes many dichotomous measures.

21.4.2.1 Example: Baseball again

We will focus on the baseball data introduced in Section 9.5.1 (p. 253), in which each baseball player has some opportunities at bat and gets a hit on some of those occasions. Thus, each opportunity at bat has a dichotomous outcome. We are interested in estimating the probability of getting a hit based on the primary fielding position of the player. Position is a nominal predictor.

In this situation, each player contributes many dichotomous outcomes, not just one. Moreover, each player within a position is thought to have his own batting ability distinct from other players, although all players are thought to have abilities that are representative of their position. Thus, each position is described as having an underlying ability, and each player is a random variant of that position's ability, and the data from a player are generated by the individual player's ability.

21.4.2.2 The model

The model we will use is diagrammed in Figure 21.12. The top part of the structure is based on the ANOVA-like model of Figure 19.2 (p. 558). It shows that the modal ability of position j, denoted ω_j, is a logistic function of a baseline ability β_0 plus a deflection for the position, $\beta_{[j]}$. The baseline and deflection parameters are given the usual priors for an ANOVA-style model.

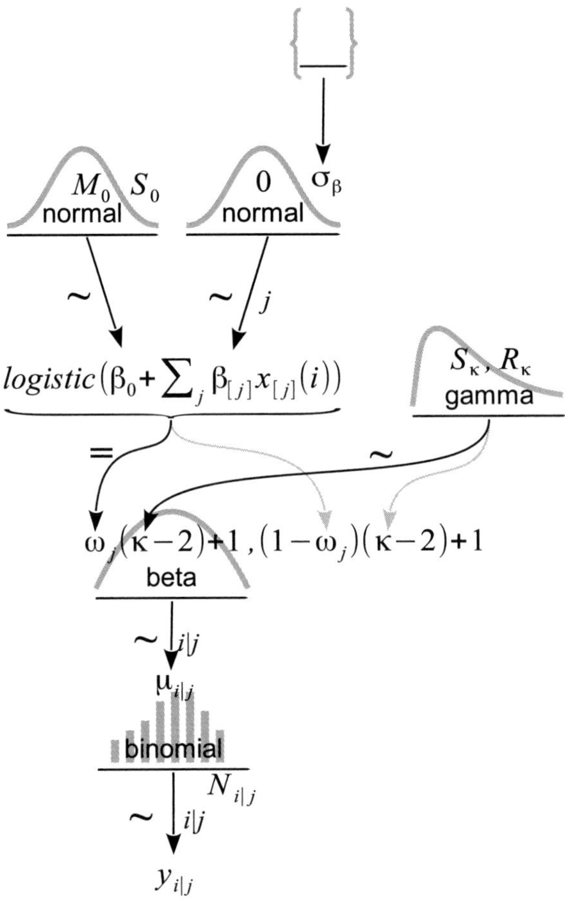

Figure 21.12 Hierarchical diagram for logistic ANOVA-like model. The top part of the structure is based on the ANOVA-like model of Figure 19.2 (p. 558). The lower part of the structure is based on the models of Figure 9.7 (p. 236) and Figure 9.13 (p. 252).

The lower part of Figure 21.12 is based on the models of Figure 9.7 (p. 236) and Figure 9.13 (p. 252). The ability of position j, denoted ω_j, is the mode of a beta distribution that describes the spread of individual abilities within the position. The concentration of the beta distribution, denoted κ, is estimated and given a gamma prior. I've chosen to use a single concentration parameter to describe all groups simultaneously, by analogy to an assumption of homogeneity of variance in ANOVA. (Group-specific concentration parameters could be used, as shown in Figure 9.13, p. 252.) The ability of player i in position j is denoted $\mu_{i|j}$. The number of hits by player i, out of $N_{i|j}$ opportunities at bat, is denoted $y_{i|j}$, and is assumed to be a random draw from a binomial distribution with mean $\mu_{i|j}$. We use the binomial distribution as a computational shortcut for multiple independent dichotomous events, not as a claim that the measuring event was conditional on fixed N, as was discussed is Section 9.4 (p. 249; see also Footnote 5, p. 249).

The model functions are in the file name `Jags-Ybinom-Xnom1fac-Mlogistic.R`, and the high-level script is the file named `Jags-Ybinom-Xnom1fac-Mlogistic-Example.R`. The JAGS model specification implements all the arrows in Figure 21.12. I hope that by this point in the book you can scan the JAGS model specification, below, and understand how each line of code corresponds with the arrows in Figure 21.12:

```
model {
  for ( i in 1:Ntotal ) {
    y[i] ~ dbin( mu[i] , N[i] )
    mu[i] ~ dbeta( omega[x[i]]*(kappa-2)+1, (1-omega[x[i]])*(kappa-2)+1 )
  }
  for ( j in 1:NxLvl ) {
    omega[j] <- ilogit( a0 + a[j] ) # In JAGS, ilogit is logistic
    a[j] ~ dnorm( 0.0 , 1/aSigma^2 )
  }
  a0 ~ dnorm( 0.0 , 1/2^2 )
  aSigma ~ dgamma( 1.64 , 0.32 )  # mode=2, sd=4
  kappa <- kappaMinusTwo + 2
  kappaMinusTwo ~ dgamma( 0.01 , 0.01 )  # mean=1 , sd=10 (generic vague)
  # Convert a0,a[] to sum-to-zero b0,b[] :
  for ( j in 1:NxLvl ) { m[j] <- a0 + a[j] } # cell means
  b0 <- mean( m[1:NxLvl] )
  for ( j in 1:NxLvl ) { b[j] <- m[j] - b0 }
}
```

The only novelty in the code, above, is the choice of constants in the prior for `aSigma`. Recall that `aSigma` represents σ_β in Figure 21.12, which is the standard deviation of the deflection parameters across groups. The gamma prior on `aSigma` was made broad on its scale. The reasonable values of σ_β are not very intuitive because they refer to standard deviations of deflections that are the inverse-logistic of probabilities. Consider two extreme probabilities near 0.001 and 0.999. The corresponding inverse-logistic values are about -6.9 and $+6.9$, which have a standard deviation of almost 10. This represents the high end of the prior distribution on σ_β, and more typical values will be around, say, 2. Therefore the gamma prior was given a mode of 2 and a standard deviation of 4. This is broad enough to have minimal influence on moderate amounts of data. You should change the prior constants if appropriate for your application.

21.4.2.3 Results

Each player's ratio of hits to at-bats is shown as a dot in Figure 21.13, with the dot size indicating the number of at-bats. Larger dots correspond to more data, and therefore should be more influential to the parameter estimation. The posterior predictive distribution is superimposed along the side of each group's data. The profiles are beta distributions from the model of Figure 21.12 with credible values of ω_j and κ. The beta distributions have their tails clipped so they span 95%. Each "moustache" is very thin, meaning that the estimate of the group modes is fairly certain. This high degree of certainty makes intuitive sense, because the data set is fairly large. The model assumed equal concentration in every group, which might be a poor description of the data. In particular, visual inspection of Figure 21.13 suggests that there might be greater variation within pitchers than within other positions.

Contrasts between positions are shown in the lower panels of Figure 21.13. The contrasts are shown for both the log odds deflection parameters (β_j or `b`) and the group modes (ω or `omega`). Compare the group-mode contrasts with the previously reported contrasts shown in Figure 9.14 (p. 256). They are very similar, despite the difference in model structure.

The present model focuses on descriptions at the group level, not at the individual-player level. Therefore the program does not have contrasts for individual players built in. But the JAGS model does record the individual-player estimates as `mu[i]`, so contrasts of individual players can be conducted.

Which model is better, the ANOVA-style model of Figure 21.12 in this section, or the tower of beta distributions used in Figures 9.7 (p. 236) and 9.13 (p. 252)? The answer is: It depends. It depends on meaningfulness, descriptive adequacy, and generalizability. The tower of beta distributions might be more intuitively meaningful because the mode parameters of the beta distributions can be directly interpreted on the scale of the data, unlike the deflection parameters inside the logistic function. Descriptive adequacy for these data might demand that each group has its own concentration parameter, regardless of upper-level structure of the model. The ANOVA-style model may be more generalizable for applications that have multiple predictors, including covariates. Indeed, the top structure of Figure 21.12 can be changed to any combination of metric and nominal predictors that are of interest.

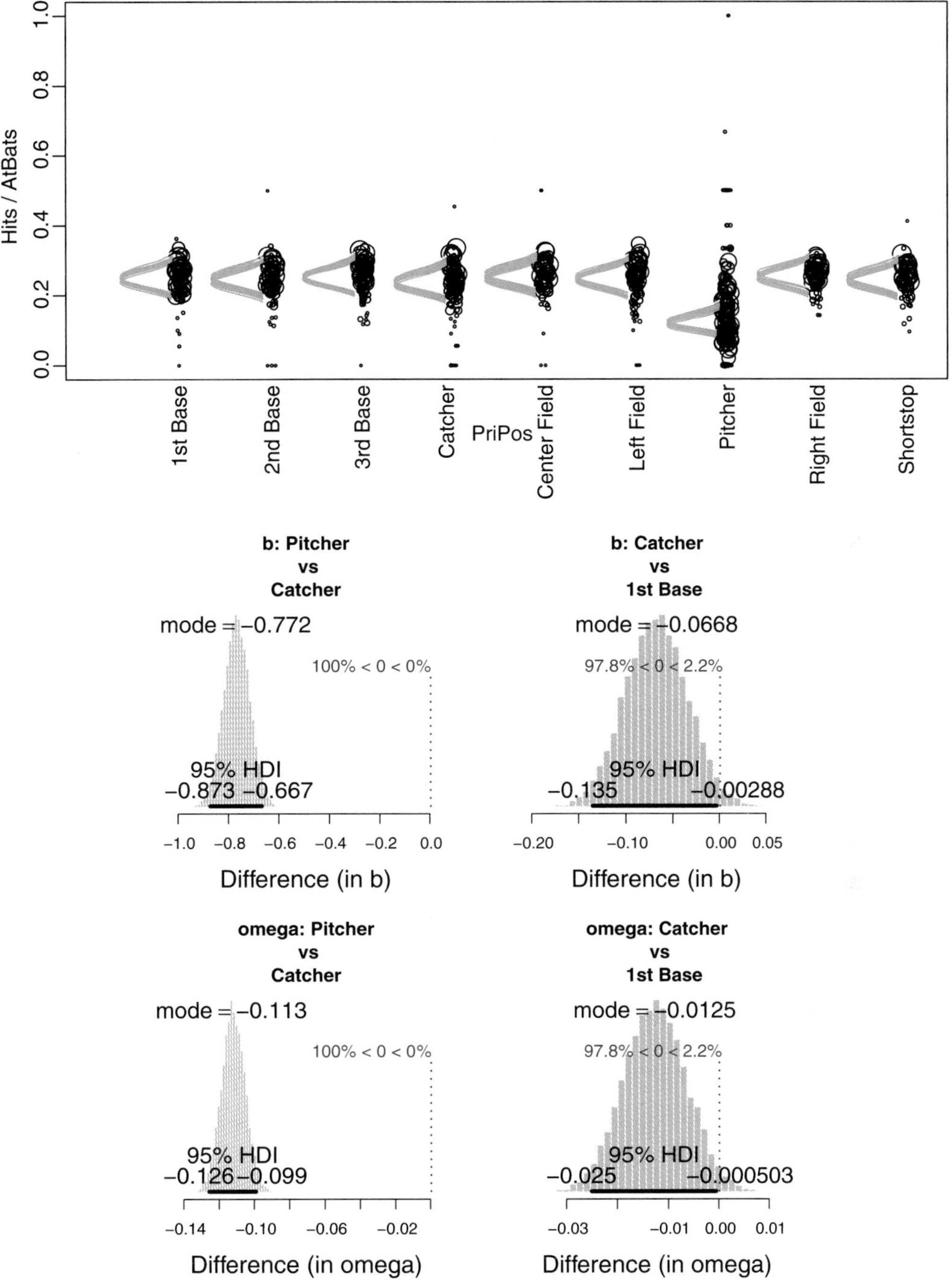

Figure 21.13 Baseball batting data are shown in the upper panel with dot size proportional to number of at-bats. The posterior predictive distributions are credible beta distributions assuming homogeneous concentration across positions. Lower panels show selected contrasts, which can be compared with the contrasts shown in Figure 9.14 (p. 256).

21.5. EXERCISES

Look for more exercises at https://sites.google.com/site/doingbayesiandataanalysis/

Exercise 21.1. [Purpose: Robust logistic regression: Building understanding and seeing its usefulness and difficulty.] When there are extreme outliers, robust logistic regression makes a big difference in the parameter estimates. But when the outliers are only modest in extremity or numerosity, they can be difficult to detect.

(A) Create some data with outliers. Type the following into R:

```
N=500                   # number of data points to generate.
x = runif(N)            # random values from a uniform distribution.
x = (x-mean(x))/sd(x)   # standardize the x values.
b0 = 0                  # the true intercept.
b1 = 4                  # the true slope.
guess = 0.1             # the true guessing amount.
# Probability of y=1 is mixture of guessing and logistic:
mu = guess*(1/2) + (1-guess)*1/(1+exp(-(b0+b1*x)))
# Generate random y values:
y = rep(NA,N)  # set up place holder for y values
for ( i in 1:length(x) ) {
  y[i] = sample( c(1,0) , size=1 , prob=c(mu[i],1-mu[i]) )
}
# Write the data into a file:
write.csv( cbind( Y=y , X=x ) ,
           file=paste0("RobustLogisticExercise-",b0,"-",b1,"-", guess,".csv") ,
           row.names=FALSE )
```

Make sure that the current working directory is set appropriately so that the saved data file is where you want it.

(B) Run the nonrobust logistic regression on the data. In the top part of the high-level script, `Jags-Ydich-XmetMulti-Mlogistic-Example.R`, use the following to read in the data file:

```
myData = read.csv("RobustLogisticExercise-0-4-0.1.csv")
yName = "Y" ; xName = "X"
fileNameRoot = "RobustLogisticExercise-0-4-0.1-REGULAR-"
numSavedSteps=15000 ; thinSteps=2
```

In your report, include the graph of the data with superimposed logistic curves, and the marginal posterior distribution on β_1. Is β_1 accurately estimated?

(C) Run the robust logistic regression on the data. In the top part of the high-level script, `Jags-Ydich-XmetMulti-MlogisticRobust-Example.R`, use the following to read in the data file:

```
myData = read.csv("RobustLogisticExercise-0-4-0.1.csv")
yName = "Y" ; xName = "X"
fileNameRoot = "RobustLogisticExercise-0-4-0.1-ROBUST-"
numSavedSteps=15000 ; thinSteps=2
```

In your report, include the graph of the data with superimposed logistic curves, and the marginal posterior distribution on β_1. Is β_1 accurately estimated? Is the estimate of β_1 different than for the nonrobust model of the previous part?

(D) Repeat the above three parts, but this time generate random x values from a normal distribution instead of from a uniform distribution. That is, change a single line, from `x = runif(N)` to `x = rnorm(N)`. You should find that the robust logistic regression does not recover the true slope (or guessing rate) very well, although a little better than nonrobust logistic regression. Discuss why. The moral of the examples was stated at the beginning of the exercises. Is there any disadvantage to using robust logistic regression?

Exercise 21.2. [Purpose: Practice modifying JAGS programs, and comparing the beta prior with the logistic-normal prior for the bias of a coin.]

(A) Create a program for estimating the bias of a single coin, using a logistic-normal prior, as outlined in Section 21.4.1 (p. 638). Hint: Make a copy of `Jags-Ydich-Xnom1subj-MbernBeta.R` and call it `Jags-Ydich-Xnom1subj-Mlogistic.R`. Make the required changes in the copied program. Do the same for the high-level script, `Jags-Ydich-Xnom1subj-MbernBeta-Example.R`.

(B) Create graphs of the prior on `mu` (review Section 8.5, p. 211, for how). See Figure 21.11 for reference.

(C) Consider a data set with 1 head in 1 flip. Show the posterior distribution for the new model, and for the beta-prior model. Are the posterior distributions noticeably different?

(D) Consider a data set with 30 heads in 40 flips. Show the posterior distribution for the new model, and for the beta-prior model. And the posterior distributions noticeably different?

Exercise 21.3. [Purpose: Agony and frustration for little gain while extending the logistic "ANOVA" model to have different concentration parameters for different groups. Sort of like climbing Everest, but it's less fun to tell people about this afterwards.] The baseball data in Figure 21.13 suggested that some positions (especially pitchers) might have different ranges of individual abilities than other positions. In other words, the model's assumption of equal concentration (κ) in all groups might be inadequate. For this exercise, your task is to extend the model of Figure 21.12

so it has distinct concentration parameters for each group. Use Figure 9.13 (p. 252) as a guide for distinct κ_j without estimating their variability. Or, for overachievers, use Figure 19.6 (p. 574) as a guide for distinct κ_j *with* estimating their variability. Copy the files `Jags-Ybinom-Xnom1fac-Mlogistic.R` and `Jags-Ybinom-Xnom1fac-Mlogistic-Example.R` to new files, renamed appropriately, for your starting point.

CHAPTER 22

Nominal Predicted Variable

Contents

Just when dichotomous, metric, were sure,
Multiple nominal outcomes occur.
Just when you thought you might rest and relax,
Here come logistics of linking softmax.[1]

This chapter considers data structures that have a nominal predicted variable. When the nominal predicted variable has only two possible values, this reduces to the case of the dichotomous predicted variable considered in the previous chapter. In the present chapter, we generalize to cases in which the predicted variable has three or more categorical values. For example, we might want to predict a person's political party affiliation (a nominal variable) based on their annual income and years of education. Or we might want to predict the species of fish (a nominal variable) we are likely to observe based on the salinity and temperature of the water.

The traditional treatment of this sort of data structure is called multinomial logistic regression or conditional logistic regression. We will consider Bayesian approaches to these methods. As usual, in Bayesian software it is easy to generalize the traditional models so they are robust to outliers, allow different variances within levels of a nominal predictor, and have hierarchical structure to share information across levels or factors as appropriate.

[1] This bit of doggerel is almost self explanatory, but notice that "logistics" simultaneously means the coordination of a complicated operation and the plural of "logistic."

Doing Bayesian Data Analysis, 2nd Edition
http://dx.doi.org/10.1016/B978-0-12-405888-0.00022-2

In the context of the generalized linear model (GLM) introduced in Chapter 15, this chapter's situation involves a link function that is the so-called softmax along with a categorical distribution for describing noise in the data, as indicated in the third row of Table 15.2 (p. 443). For a reminder of how this chapter's combination of predicted and predictor variables relates to other combinations, see Table 15.3 (p. 444).

22.1. SOFTMAX REGRESSION

The usual generalizations of logistic regression are of two types: conditional logistic regression, which will be explored later in the chapter, and so-called multinomial logistic regression, which we explore in this section. I am not fond of the traditional name, however, because the model does not use the logistic function per se, so the name "logistic" is a misnomer, and all the models in this chapter describe multinomial data, so the name "multinomial" is not informative. Instead, the key descriptor of the model is its inverse-link function, which is the softmax function (which will be defined below). Therefore, I refer to the method as softmax regression instead of multinomial logistic regression.

In the previous chapter, we used the logistic function to go from a linear combination of predictors to the probability of outcome 1 relative to outcome 0. In this chapter, we want to generalize to multiple categorical outcomes. The generalization of the logistic function requires a bit of mathematical development, but it's really just repeated algebraic manipulation of exponential functions, so don't be deterred. To keep the notation reasonably simple, we will suppose that we have a single metric predictor, x. All the derivations below easily generalize to multiple predictors. The underlying linear propensity of outcome k is denoted

$$\lambda_k = \beta_{0,k} + \beta_{1,k}x \tag{22.1}$$

The subscripts k indicate that there is an equation like Equation 22.1 for every outcome category. We call the set of possible outcomes S. Now a novelty: The probability of outcome k is given by the softmax function:

$$\phi_k = \text{softmax}_S(\{\lambda_k\}) = \frac{\exp(\lambda_k)}{\sum_{c \in S} \exp(\lambda_c)} \tag{22.2}$$

In words, Equation 22.2 says that the probability of outcome k is the exponentiated linear propensity of outcome k relative to the sum of exponentiated linear propensities across all outcomes in the set S. You may be wondering, Why exponentiate? Intuitively, we have to go from propensities that can have negative values to probabilities that can only have non-negative values, and we have to preserve order. The exponential function satisfies that need.

The softmax function is used in many applications as a way of mapping several real-valued variables to order-preserving outcome probabilities (e.g., Bishop, 2006). It is called the softmax function because when it is given another parameter, called the gain γ, that amplifies all the inputs, then the softmax assigns the maximum input with nearly 100% probability when the gain is large:

$$\text{As } \gamma \to \infty, \qquad \frac{\exp(\gamma\lambda_k)}{\sum_{c\in S}\exp(\gamma\lambda_c)} \to \begin{cases} 1 \text{ if } \lambda_k = \max(\{\lambda_c\}) \\ 0 \text{ otherwise} \end{cases}$$

The softmax formulation can be useful for applications that use the derivative (i.e., gradient) to find optimal input values, because gradient ascent requires smooth differentiable functions (cf., Kruschke & Movellan, 1991).

It turns out that there are indeterminacies in the system of Equations 22.1 and 22.2. We can add a constant C_0 to every $\beta_{0,k}$, and add a constant C_1 to every $\beta_{1,k}$, and get exactly the same probabilities of responding with each category:

$$\begin{aligned}
&\frac{\exp\left((\beta_{0,k}+C_0)+(\beta_{1,k}+C_1)x\right)}{\sum_{c\in S}\exp\left((\beta_{0,c}+C_0)+(\beta_{1,c}+C_1)x\right)} \\
&\quad= \frac{\exp\left((C_0+C_1x)+\beta_{0,k}+\beta_{1,k}x\right)}{\sum_{c\in S}\exp\left((C_0+C_1x)+\beta_{0,c}+\beta_{1,c}x\right)} \\
&\quad= \frac{\exp(C_0+C_1x)\exp\left(\beta_{0,k}+\beta_{1,k}x\right)}{\sum_{c\in S}\exp(C_0+C_1x)\exp\left(\beta_{0,c}+\beta_{1,c}x\right)} \\
&\quad= \frac{\exp\left(\beta_{0,k}+\beta_{1,k}x\right)}{\sum_{c\in S}\exp\left(\beta_{0,c}+\beta_{1,c}x\right)} \\
&\quad= \phi_k
\end{aligned} \tag{22.3}$$

Therefore, we can set the baseline and slope for one of the response categories to arbitrary convenient constants. We will set the constants of one response category, called the reference category r, to zero: $\beta_{0,r} = 0$ and $\beta_{1,r} = 0$.

Because of the indeterminacy in the regression coefficients, we can interpret the regression coefficients only relative to the reference category. Recall from Section 21.2.1 (p. 629) that the regression coefficients in logistic regression can be conceived in terms of log odds of outcome 1 relative to outcome 0. In the present application, the regression coefficients can be conceived in terms of the log odds of each outcome relative to the reference outcome:

$$\begin{aligned}
\log\left(\frac{\phi_k}{\phi_r}\right) &= \log\left(\frac{\exp\left(\beta_{0,k}+\beta_{1,k}x\right)}{\exp\left(\beta_{0,r}+\beta_{1,r}x\right)}\right) \\
&= \log\left(\frac{\exp\left(\beta_{0,k}+\beta_{1,k}x\right)}{\exp(0+0x)}\right) \\
&= \beta_{0,k}+\beta_{1,k}x
\end{aligned} \tag{22.4}$$

In other words, the regression coefficient $\beta_{1,k}$ is the increase in log odds of outcome k relative to the reference outcome r for a one-unit increase in x.

Figure 22.1 shows two examples of data generated from the multinomial logistic model. The examples use two predictors (x_1 and x_2) to better illustrate the variety of patterns that can be produced by the model. The regression coefficients are displayed in the titles of the panels within the figure. The regression coefficients are large on the scale of x so that the transitions across outcomes are visually crisp. In realistic data, the regression coefficients are usually of much smaller magnitude and the outcomes of all types extensively overlap.

For the examples in Figure 22.1, the reference outcome was chosen to be 1, so all the regression coefficients describe the log odds of the other outcomes relative to outcome 1. Consider outcome 2, which has log odds specified by the regression coefficients for λ_2. If, for λ_2, the regression coefficient on x_1 is positive, then the probability of outcome 2, relative to outcome 1, goes up as x_1 increases. And, still for λ_2, if the regression coefficient on x_2 is positive, then the probability of outcome 2, relative to outcome 1, goes up as x_2 increases. The baseline, β_0, determines how large x_1 and x_2 need to become for the probability of outcome 2 to exceed the probability of outcome 1: The larger the baseline, the smaller x_1 and x_2 need to be.

You can see these trends occur in the two panels of Figure 22.1. In each panel, locate the region in which the reference outcome (i.e., 1) occurs. If λ_2's coefficient

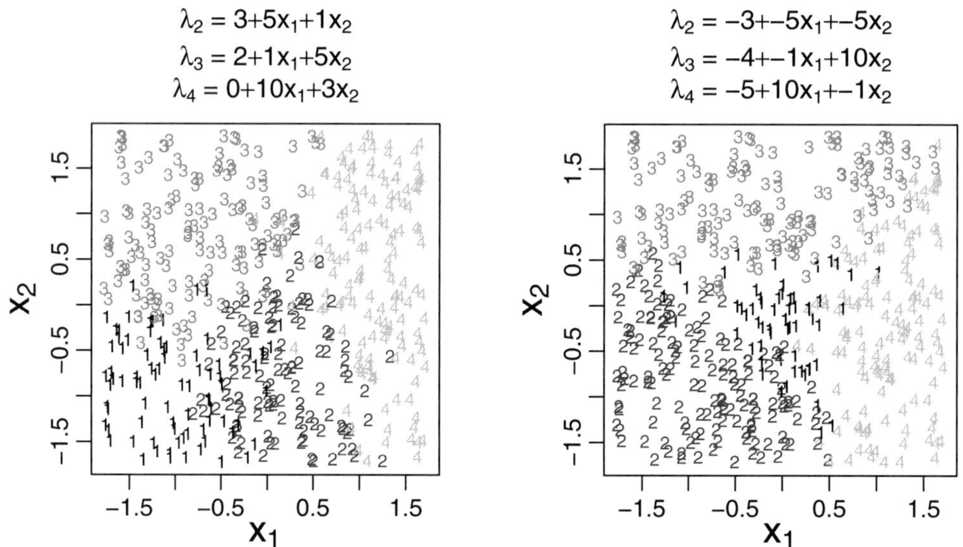

Figure 22.1 Examples of data generated from the softmax regression model. Above each panel are the specific instantiations of Equation 22.1 (for two predictors instead of only one predictor), with the reference outcome chosen to be 1, so $\lambda_1 \equiv 0$. The outcomes were sampled according to probabilities computed from the softmax function of Equation 22.2.

on x_1 is positive, then the region in which outcome 2 occurs will tend to be on the right of the region where the reference outcome occurs. If λ_2's coefficient on x_1 is negative, then the region in which outcome 2 occurs will tend to be on the left of the region where the reference outcome occurs. Analogously for x_2: If λ_2's coefficient on x_2 is positive, then the region in which outcome 2 occurs will tend to be above the region where the reference outcome occurs. If λ_2's coefficient on x_2 is negative, then the region in which outcome 2 occurs will tend to be below the region where the reference outcome occurs. These trends apply to the other outcomes as well.

Another general conclusion to take away from Figure 22.1 is that the regions of the different outcomes do not necessarily have piecewise linear boundaries. Later in the chapter we will consider a different sort of model that always produces outcome regions that have piecewise linear boundaries.

22.1.1. Softmax reduces to logistic for two outcomes

When there are only two outcomes, the softmax formulation reduces to the logistic regression of Chapter 21. The reference outcome is declared to be outcome $y = 0$, and the regression coefficients describe the log odds of outcome $y = 1$ relative to outcome $y = 0$. To make this explicit, let's see how the softmax function of exponentiated linear propensities becomes the logistic function when there are just two outcome categories. We start with the definition of the softmax function in Equation 22.2 and algebraically re-arrange it for the case of two outcome categories:

$$
\begin{aligned}
\phi_1 &= \frac{\exp(\lambda_1)}{\sum_{c\in\{1,0\}} \exp(\lambda_c)} \\
&= \frac{\exp(\lambda_1)}{\exp(\lambda_1) + \exp(\lambda_0)} \\
&= \frac{\exp(\lambda_1)}{\exp(\lambda_1) + 1} && \text{because } \lambda_0 \equiv 0 + 0x \\
&= \frac{\exp(\lambda_1)/\exp(\lambda_1)}{\exp(\lambda_1)/\exp(\lambda_1) + 1/\exp(\lambda_1)} \\
&= \frac{1}{1 + \exp(-\lambda_1)} \\
&= \text{logistic}(\lambda_1) && (22.5)
\end{aligned}
$$

Thus, softmax regression is one natural generalization of logistic regression. We will see a different generalization later in the chapter.

22.1.2. Independence from irrelevant attributes

An important property of the softmax function of Equation 22.2 is known as independence from irrelevant attributes (Luce, 1959, 2008). The model implies that the ratio of probabilities of two outcomes is the same regardless of what other possible outcomes are included in the set. Let S denote the set of possible outcomes. Then, from the definition of the softmax function, the ratio of outcomes j and k is

$$\frac{\phi_j}{\phi_k} = \frac{\exp(\lambda_j) / \sum_{c \in S} \exp(\lambda_c)}{\exp(\lambda_k) / \sum_{c \in S} \exp(\lambda_c)} \tag{22.6}$$

The summation in the denominators cancels and has no effect on the ratio of probabilities. Obviously if we changed the set of outcomes S to any other set S^* that still contains outcomes j and k, the summation $\sum_{c \in S^*}$ would still cancel and have no effect on the ratio of probabilities.

An intuitive example that obeys independence from irrelevant attributes is as follows. Suppose there are three ways to get from home to work, namely walking, bicycling, or bussing. Suppose that a person prefers walking the best, followed by bicycling, followed by bussing, and suppose that choosing a method is probabilistic with ratios of 3:2:1, which is to say that walking is chosen 3 to 2 over bicycling, which is chosen 2 to 1 over bussing. In other words, there is a 50% chance of walking, a 33.3% chance of bicycling, and a 16.7% chance of bussing. The ratio of the probability of walking to the probability of bussing is 3:1. Now, suppose one day the bicycle has a flat tire, so the outcome set is reduced. It makes intuitive sense that there should be the same ratio of probabilities among the remaining options, which is to say that walking should still be preferred to bussing in the ratio 3:1.

But not all situations will be accurately described by independence from irrelevant attributes. Debreu (1960) pointed out an example that violates independence from irrelevant attributes. Suppose there are three ways to get from home to work, namely walking, taking the red bus, or taking the blue bus. Suppose that a person prefers walking to bussing in the ratio 3:1, but is indifferent about red or blue buses. When all three options are available, there is a 75% probability of walking and a 25% probability of taking a bus, which means a 12.5% probability of taking a red bus and a 12.5% probability of taking a blue bus. The ratio of walking to taking a red bus is therefore 6:1. Now, suppose one day the blue bus company breaks down, so the outcome set is reduced. It makes intuitive sense that there should still be a 75% preference for walking and a 25% preference for taking a bus, but now the ratio of walking to taking a red bus is 3:1, not 6:1.

Thus, when applying the descriptive model of Equation 22.2, we are implicitly assuming independence from irrelevant attributes. This might or might not be a reasonable assumption in any given application.

22.2. CONDITIONAL LOGISTIC REGRESSION

Softmax regression conceives of each outcome as an independent change in log odds from the reference outcome, and a special case of that is dichotomous logistic regression. But we can generalize logistic regression another way, which may better capture some patterns of data. The idea of this generalization is that we divide the set of outcomes into a hierarchy of two-set divisions, and use a logistic to describe the probability of each branch of the two-set divisions. The underlying propensity to respond with any outcome in the subset of outcomes S^* relative to (i.e., conditional on) a larger set S is denoted

$$\lambda_{S^*|S} = \beta_{0,S^*|S} + \beta_{1,S^*|S}\, x \tag{22.7}$$

and the conditional response probability is

$$\phi_{S^*|S} = \text{logistic}(\lambda_{S^*|S}) \tag{22.8}$$

The thrust of Equations 22.7 and 22.8 is that the regression coefficients refer to the conditional probability of outcomes for the designated subsets, not necessarily to a single outcome among the full set of outcomes.

Figure 22.2 shows two examples of hierarchies of binary divisions of four outcomes. In each example, the full set of (four) outcomes appears at the top of the hierarchy, and the binary divisions progress downward. Each branch is labeled with its conditional probability. For example, the top left branch indicates the occurrence of outcome 1 given the full set, and the conditional probability of this occurrence is labeled $\phi_{\{1\}|\{1,2,3,4\}}$. This conditional probability will be modeled by a logistic

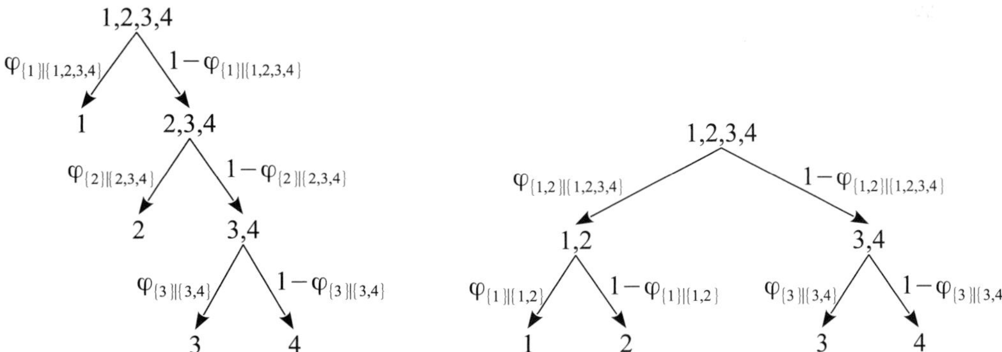

Figure 22.2 Two hierarchies of binary divisions of outcomes 1, 2, 3, and 4. Each branch is labeled with its conditional probability. In conditional logistic regression, each binary conditional probability is modeled by a logistic function. An example of data generated from the left hierarchy is shown in the left side of Figure 22.3, and an example of data generated from the right hierarchy is shown in the right side of Figure 22.3.

function of the predictors. We will now explore detailed numerical examples of the two hierarchies.

In the left hierarchy of Figure 22.2, we opt to split the outcomes into a hierarchy of binary divisions as follows:

- 1 versus 2, 3, or 4
- 2 versus 3 or 4
- 3 versus 4

At each level in the hierarchy, the conditional probability of the options is described by a logistic function of a linear combination of the predictors. For our example, we assume there are two metric predictors. The linear combinations of predictors are denoted

$$\begin{aligned}
\lambda_{\{1\}|\{1,2,3,4\}} &= \beta_{0,\{1\}|\{1,2,3,4\}} + \beta_{1,\{1\}|\{1,2,3,4\}}x_1 + \beta_{2,\{1\}|\{1,2,3,4\}}x_2 \\
\lambda_{\{2\}|\{2,3,4\}} &= \beta_{0,\{2\}|\{2,3,4\}} + \beta_{1,\{2\}|\{2,3,4\}}x_1 + \beta_{2,\{2\}|\{2,3,4\}}x_2 \\
\lambda_{\{3\}|\{3,4\}} &= \beta_{0,\{3\}|\{3,4\}} + \beta_{1,\{3\}|\{3,4\}}x_1 + \beta_{2,\{3\}|\{3,4\}}x_2
\end{aligned} \tag{22.9}$$

and the conditional probabilities of the outcome sets are simply the logistic function applied to each of the λ values:

$$\begin{aligned}
\phi_{\{1\}|\{1,2,3,4\}} &= \text{logistic}(\lambda_{\{1\}|\{1,2,3,4\}}) \\
\phi_{\{2\}|\{2,3,4\}} &= \text{logistic}(\lambda_{\{2\}|\{2,3,4\}}) \\
\phi_{\{3\}|\{3,4\}} &= \text{logistic}(\lambda_{\{3\}|\{3,4\}})
\end{aligned} \tag{22.10}$$

The ϕ values, above, should be thought of as conditional probabilities, as marked on the arrows in Figure 22.2. For example, $\phi_{\{2\}|\{2,3,4\}}$ is the probability of outcome 2 given that outcomes 2, 3, or 4 occurred. And the complement of that probability, $1 - \phi_{\{2\}|\{2,3,4\}}$ is the probability of outcomes 3 or 4 given that outcomes 2, 3, or 4 occurred.

Finally, the most important equations that actually determine the relations between the expressions above, are the equations that specify the probabilities of the individual outcomes as the appropriate combinations of the conditional probabilities. These probabilities of the individual outcomes are determined simply by multiplying the conditional probabilities on the branches of the hierarchy in Figure 22.2 that lead to the individual outcomes. Thus,

$$\begin{aligned}
\phi_1 &= \phi_{\{1\}|\{1,2,3,4\}} \\
\phi_2 &= \phi_{\{2\}|\{2,3,4\}} \cdot (1 - \phi_{\{1\}|\{1,2,3,4\}}) \\
\phi_3 &= \phi_{\{3\}|\{3,4\}} \cdot (1 - \phi_{\{2\}|\{2,3,4\}}) \cdot (1 - \phi_{\{1\}|\{1,2,3,4\}}) \\
\phi_4 &= (1 - \phi_{\{3\}|\{3,4\}}) \cdot (1 - \phi_{\{2\}|\{2,3,4\}}) \cdot (1 - \phi_{\{1\}|\{1,2,3,4\}})
\end{aligned} \tag{22.11}$$

Notice that the sum of the outcome probabilities is indeed 1 as it should be; that is, $\phi_1 + \phi_2 + \phi_3 + \phi_4 = 1$. This result is easiest to verify in this case by summing in the order $((\phi_4 + \phi_3) + \phi_2) + \phi_1$.

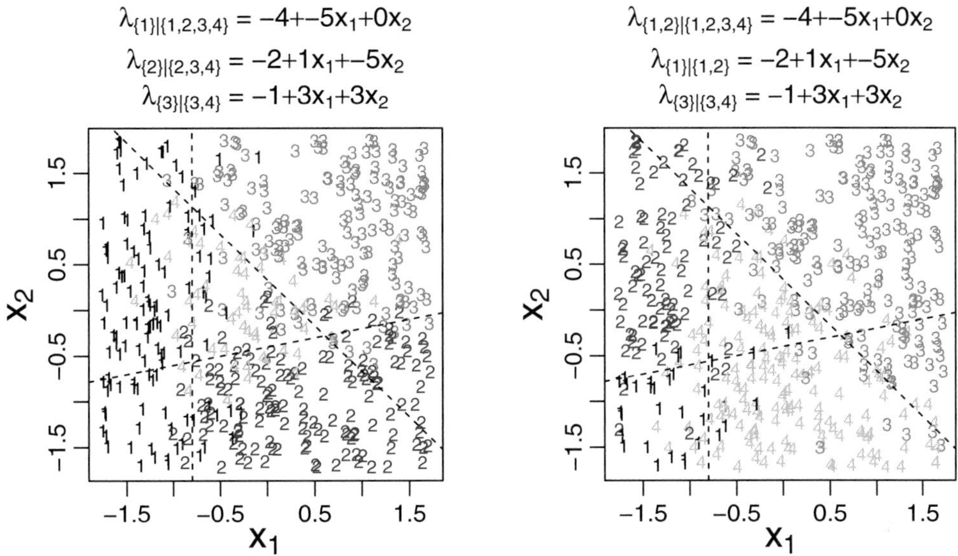

Figure 22.3 *Left panel:* Example of data generated from the conditional logistic model of Equations 22.9–22.11, which express the hierarchy in the left side of Figure 22.2. *Right panel:* Example of data generated from the conditional logistic model of Equations 22.12–22.14, which express the hierarchy in the right side of Figure 22.2. Dashed lines indicate 50% level contours of the conditional logistic functions.

The left panel of Figure 22.3 shows an example of data generated from the conditional logistic model of Equations 22.9–22.11. The x_1 and x_2 values were randomly generated from uniform distributions. The regression coefficients are displayed in the title of the panel within the figure. The regression coefficients are large compared to the scale of x so that the transitions across outcomes are visually crisp. In realistic data, the regression coefficients are usually of much smaller magnitude and outcomes of all types extensively overlap. You can see in the left panel of Figure 22.3 that there is a linear separation between the region of 1 outcomes and the region of 2, 3, or 4 outcomes. Then, within the region of 2, 3, or 4 outcomes, there is a linear separation between the region of 2 outcomes and the region of 3 or 4 outcomes. Finally, within the region of 3 or 4 outcomes, there is a linear separation between the 3's and 4's. Of course, there is probabilistic noise around the linear separations; the underlying linear separations are plotted where the logistic probability is 50%.

We now consider another example of conditional logistic regression, again involving four outcomes and two predictors, but with a different parsing of the outcomes. The hierarchy of binary partitions of the outcomes is illustrated in the right side of Figure 22.2. In this case, we opt to split the outcomes into the following hierarchy of binary choices:

- 1 or 2 versus 3 or 4
- 1 versus 2
- 3 versus 4

At each level in the hierarchy, the conditional probability of the options is described by a logistic function of a linear combination of the predictors. For the choices above, the linear combinations of predictors are denoted

$$\begin{aligned}
\lambda_{\{1,2\}|\{1,2,3,4\}} &= \beta_{0,\{1,2\}|\{1,2,3,4\}} + \beta_{1,\{1,2\}|\{1,2,3,4\}}x_1 + \beta_{2,\{1,2\}|\{1,2,3,4\}}x_2 \\
\lambda_{\{1\}|\{1,2\}} &= \beta_{0,\{1\}|\{1,2\}} + \beta_{1,\{1\}|\{1,2\}}x_1 + \beta_{2,\{1\}|\{1,2\}}x_2 \\
\lambda_{\{3\}|\{3,4\}} &= \beta_{0,\{3\}|\{3,4\}} + \beta_{1,\{3\}|\{3,4\}}x_1 + \beta_{2,\{3\}|\{3,4\}}x_2
\end{aligned} \tag{22.12}$$

and the conditional probabilities of the outcome sets are simply the logistic function applied to each of the λ values:

$$\begin{aligned}
\phi_{\{1,2\}|\{1,2,3,4\}} &= \text{logistic}(\lambda_{\{1,2\}|\{1,2,3,4\}}) \\
\phi_{\{1\}|\{1,2\}} &= \text{logistic}(\lambda_{\{1\}|\{1,2\}}) \\
\phi_{\{3\}|\{3,4\}} &= \text{logistic}(\lambda_{\{3\}|\{3,4\}})
\end{aligned} \tag{22.13}$$

Finally, the most important equations, that actually determine the relations between the expressions above, are the equations that specify the probabilities of the individual outcomes as the appropriate combinations of the conditional probabilities, which can be gleaned from the conditional probabilities on the branches of the hierarchy in the right side of Figure 22.2. Thus:

$$\begin{aligned}
\phi_1 &= \phi_{\{1\}|\{1,2\}} \cdot \phi_{\{1,2\}|\{1,2,3,4\}} \\
\phi_2 &= (1 - \phi_{\{1\}|\{1,2\}}) \cdot \phi_{\{1,2\}|\{1,2,3,4\}} \\
\phi_3 &= \phi_{\{3\}|\{3,4\}} \cdot (1 - \phi_{\{1,2\}|\{1,2,3,4\}}) \\
\phi_4 &= (1 - \phi_{\{3\}|\{3,4\}}) \cdot (1 - \phi_{\{1,2\}|\{1,2,3,4\}})
\end{aligned} \tag{22.14}$$

Notice that the sum of the outcome probabilities is indeed 1 as it should be; that is, $\phi_1 + \phi_2 + \phi_3 + \phi_4 = 1$. This result is easiest to verify in this case by summing in the order $(\phi_1 + \phi_2) + (\phi_3 + \phi_4)$. The only structural difference between this example and the previous example is the difference between the forms of Equations 22.11 and 22.14. Aside from that, both examples involve coefficients on predictors where the meaning of the coefficients is determined by how their values are combined in Equations 22.11 and 22.14. The coefficients were given mnemonic subscripts in anticipation of the particular combinations Equations 22.11 and 22.14.

The right panel of Figure 22.3 shows an example of data generated from the conditional logistic model of Equations 22.12–22.14. The regression coefficients are displayed in the title of the panel within the figure. You can see in the right panel of Figure 22.3 that there is a linear separation between the region of 1 or 2 outcomes and the region of 3 or 4 outcomes. Then, within the region of 1 or 2 outcomes, there is a linear separation between the region of 1 outcomes and the region of 2 outcomes. Finally, within the region of 3 or 4 outcomes, there is a linear separation between the 3's and 4's. Of course, there is probabilistic noise around the linear separations; the linear separations are plotted as dashed lines where the logistic probability is 50%.

In general, conditional logistic regression requires that there is a linear division between two subsets of the outcomes, and then within each of those subsets there is a linear division of smaller subsets, and so on. This sort of linear division is not required of the softmax regression model, examples of which we saw in Figure 22.1. You can see that the outcome regions of the softmax regression in Figure 22.1 do not appear to have the hierarchical linear divisions required of conditional logistic regression. Real data can be extremely noisy, and there can be multiple predictors, so it can be challenging or impossible to visually ascertain which sort of model is most appropriate. The choice of model is driven primarily by theoretical meaningfulness.

22.3. IMPLEMENTATION IN JAGS

22.3.1. Softmax model

Figure 22.4 shows a hierarchical diagram for softmax regression. It is directly analogous to the hierarchical diagram for logistic regression, shown here again for convenience (repeated from Figure 21.2, p. 624). The juxtaposition of the diagrams presents another perspective on the fact that logistic regression is a special case of softmax regression.

There are only a few novelties in the diagram for softmax regression in Figure 22.4. Most obviously, at the bottom of the diagram is a distribution named "categorical." It is just like the Bernoulli distribution but with several outcomes instead of only two. The outcomes are typically named as consecutive integers starting at 1, but this naming scheme does not connote ordering or distance between values. Thus, y_i at the bottom of the diagram indicates an integer outcome label for the ith data point. Outcome k has probability denoted $\mu_{[k]}$, and the outcome probabilities are visually suggested by the heights of the bars in the icon for the categorical distribution, just as the outcomes of the Bernoulli distribution have probabilities indicated by the heights of its two bars. Moving up the diagram, each outcome's probability is given by the softmax function. The subscript $[k]$ in the softmax function merely suggests that the function is applied to every outcome k.

The model is expressed in JAGS much like logistic regression on metric predictors. The predictor values are standardized to make MCMC sampling more efficient, and

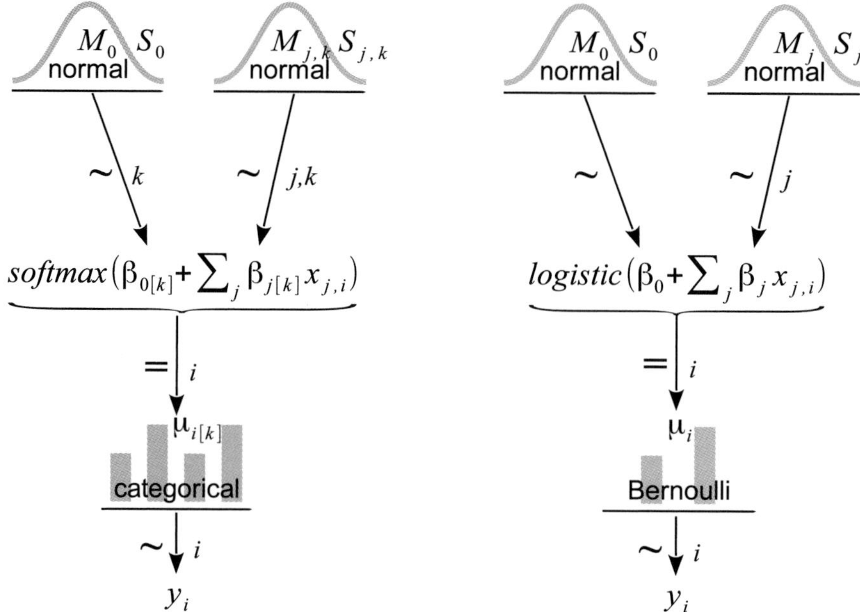

Figure 22.4 Hierarchical diagrams for softmax regression (on left) and for logistic regression (on right), both for metric predictors. Right diagram is repeated from Figure 21.2 (p. 624).

the parameters are transformed to the original scale, just as in logistic regression. A novelty comes in computing the softmax function, because JAGS does not have softmax built in (but Stan does). The JAGS code uses a `for` loop to go through the outcomes and compute the exponentiated λ_k values from Equation 22.1. The variable `explambda[k,i]` is the exponentiated λ_k for the ith data point. Those values are then normalized and used as the probabilities in the categorical distribution, which is denoted in JAGS as `dcat`. In the JAGS code below, `Nout` is the number of outcome categories, `Nx` is the number of predictors, and `Ntotal` is the number data points. Please read the code below and see if you can make sense of each line. As usual, the lines of JAGS code start at the bottom of the dependency diagram.

```
model {
  for ( i in 1:Ntotal ) {
    y[i] ~ dcat( mu[1:Nout,i] )
    mu[1:Nout,i] <- explambda[1:Nout,i] / sum(explambda[1:Nout,i])
    for ( k in 1:Nout ) {
      explambda[k,i] <- exp( zbeta0[k] + sum( zbeta[k,1:Nx] * zx[i,1:Nx] ) )
    }
  }
  # The reference outcome 1 is given coefficients of zero:
```

```
  zbeta0[1] <- 0
  for ( j in 1:Nx ) { zbeta[1,j] <- 0 }
  # Priors vague on standardized scale:
  for ( k in 2:Nout ) { # notice this starts at outcome 2
    zbeta0[k] ~ dnorm( 0 , 1/20^2 )
    for ( j in 1:Nx ) {
      zbeta[k,j] ~ dnorm( 0 , 1/20^2 )
    }
  }
  # Transform to original scale ...
}
```

The `dcat` distribution in JAGS automatically normalizes its argument vector, so we do not need to prenormalize its argument. Thus, the explicit normalizing we did above, like this:

```
y[i] ~ dcat( mu[1:Nout,i] )
mu[1:Nout,i] <- explambda[1:Nout,i] / sum(explambda[1:Nout,i])
```

could instead be simplified and stated like this:

```
y[i] ~ dcat( explambda[1:Nout,i] )
```

The simplified form is used in the program `Jags-Ynom-XmetMulti-Msoftmax.R`, which is called from the high-level script `Jags-Ynom-XmetMulti-Msoftmax-Example.R`.

22.3.2. Conditional logistic model

The conditional logistic model has all its primary "action" in the outcome-partition hierarchies of Figure 22.2. You can imagine combining an outcome-partition hierarchy of Figure 22.2 with the logistic regression diagram on the right side of Figure 22.4 to create a diagram of a conditional logistic model. Each $\phi_{S^*|S}$ in the outcome-partition hierarchy would have its own logistic function. Note that every different outcome-partition hierarchy yields a different conditional logistic model.

Although making a diagram of the model might be challenging, implementing it in JAGS is easy. Consider the outcome-partition hierarchy on the *left* side of Figure 22.2, with corresponding formal expression in Equations 22.9–22.11. The outcome probabilities, ϕ_k, for the *i*th data point are coded in JAGS as `mu[k,i]`. Take a look at the JAGS code, below, to see the direct implementation of Equations 22.9–22.11. Remember that the logistic function in JAGS is `ilogit`.

```
model {
  for ( i in 1:Ntotal ) {
    y[i] ~ dcat( mu[1:Nout,i] )
    mu[1,i] <- phi[1,i]
    mu[2,i] <- phi[2,i] * (1-phi[1,i])
    mu[3,i] <- phi[3,i] * (1-phi[2,i]) * (1-phi[1,i])
```

```
    mu[4,i] <- (1-phi[3,i]) * (1-phi[2,i]) * (1-phi[1,i])
    for ( r in 1:(Nout-1) ) {
      phi[r,i] <- ilogit( zbeta0[r] + sum( zbeta[r,1:Nx] * zx[i,1:Nx] ) )
    }
  }
  # Priors vague on standardized scale:
  for ( r in 1:(Nout-1) ) {
    zbeta0[r] ~ dnorm( 0 , 1/20^2 )
    for ( j in 1:Nx ) {
      zbeta[r,j] ~ dnorm( 0 , 1/20^2 )
    }
  }
  # Transform to original scale ...
}
```

The outcome-partition hierarchy on the right-hand side of Figure 22.2, with corresponding formal expression in Equations 22.12–22.14, is virtually identical. The only difference is the specification of the outcome probabilities:

```
model {
  for ( i in 1:Ntotal ) {
    y[i] ~ dcat( mu[1:Nout,i] )
    mu[1,i] <- phi[2,i] * phi[1,i]
    mu[2,i] <- (1-phi[2,i]) * phi[1,i]
    mu[3,i] <- phi[3,i] * (1-phi[1,i])
    mu[4,i] <- (1-phi[3,i]) * (1-phi[1,i])
    for ( r in 1:(Nout-1) ) {
      phi[r,i] <- ilogit( zbeta0[r] + sum( zbeta[r,1:Nx] * zx[i,1:Nx] ) )
    }
  }
  # Priors same as above ...
  # Transform to original scale ...
}
```

The models are defined in the files named `Jags-Ynom-XmetMulti-McondLogistic1.R` and `Jags-Ynom-XmetMulti-McondLogistic2.R`, and are called from the high-level scripts named `Jags-Ynom-XmetMulti-McondLogistic1-Example.R` and `Jags-Ynom-XmetMulti-McondLogistic2-Example.R`.

22.3.3. Results: Interpreting the regression coefficients

22.3.3.1 Softmax model

We start by applying the softmax model to some data generated by a softmax model. Figure 22.5 shows the results. The data are reproduced in the left of the figure, and the posterior distribution is shown on the right. The data graph includes, in its title, the values of the regression coefficients that actually generated the data. The estimated

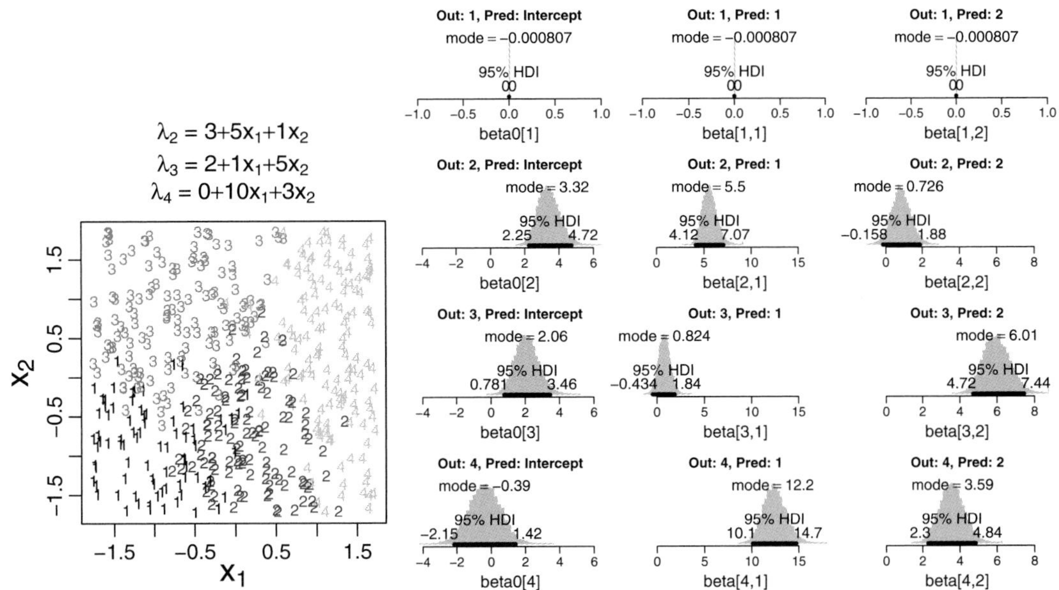

Figure 22.5 Posterior parameter estimates of the softmax model applied to data generated from a softmax model. Data are shown on the left with the true parameter values (reproduced from Figure 22.1). The four rows of marginal posterior distributions correspond to the four outcomes, and the columns of distributions correspond to β_0, β_1, and β_2.

parameter values should be near the generating values, but not exactly the same because the data are merely a finite random sample. Each row of marginal posterior distributions corresponds to an outcome value. The top row corresponds to the reference outcome 1, which has regression coefficients set to zero, and the posterior distributions show spikes at zero that confirm this choice of reference outcome. The second row corresponds to outcome 2. Its true regression coefficients are shown in the equation for λ_2 above the data graph. You can see that the estimated values are close to the true values. This correspondence of true and estimated parameter values obtains for all the outcome values.

For real data, we usually do not know what process truly generated the data, much less its true parameter values. All we have is a smattering of very noisy data and a posterior distribution on the parameter values of the model that we chose to describe the data. Interpreting the parameters is always contextualized relative to the model. For the softmax model, the regression coefficient for outcome k on predictor x_j indicates that rate at which the log odds of that outcome increase relative to the reference outcome for a one unit increase in x_j, assuming that a softmax model is a reasonable description of the data.

It is easy to transform the estimated parameter values to a different reference category. Recall from Equation 22.3 (p. 651) that arbitrary constants can be added to all the regression coefficients without changing the model prediction. Therefore, to change the parameters estimates so they are relative to outcome R, we simply subtract $\beta_{j,R}$ from $\beta_{j,k}$ for all predictors j and all outcomes k. We do this at every step in the MCMC chain. For example, in Figure 22.5, consider the regression coefficient on x_1 for outcome 2. Relative to reference outcome 1, this coefficient is positive, meaning that the probability of outcome 2 increases relative to outcome 1 when x_1 increases. You can see this in the data graph, as the region of 2's falls to right side (positive x_1 direction) of the region of 1's. But if the reference outcome is changed to outcome 4, then the coefficient on x_1 for outcome 2 changes to a negative value. Algebraically this happens because the coefficient on x_1 for outcome 4 is larger than for outcome 2, so when the coefficient for outcome 4 is subtracted, the result is a negative value for the coefficient on outcome 2. Visually, you can see this in the data graph, as the region of 2's falls to the left side (negative x_1 direction) of the region of 4's. Thus, interpreting regression coefficients in a softmax model is rather different than in linear regression. In linear regression, a positive regression coefficient implies that y increases when the predictor increases. But not in softmax regression, where a positive regression coefficient is only positive with respect to a particular reference outcome.

22.3.3.2 Conditional logistic model

We now apply the conditional logistic models to the data generated by those models. The upper halves of Figures 22.6 and 22.7 show the data and the posterior distributions of the parameters. (The lower halves of the figures will be discussed later.) Superimposed on the data are a smattering of credible 50% threshold lines for each of the conditional logistic functions. You can see that the true parameter values a recovered reasonably well.

In the upper half of Figure 22.6, consider the estimates of regression coefficients for "Lambda 1." This corresponds to $\lambda_{\{1\}|\{1,2,3,4\}}$, which indicates the probability of outcome 1 versus all other outcomes. The estimated coefficient on x_1 is negative, indicating that the probability of outcome 1 increases as x_1 decreases. The estimated coefficient on x_2 is essentially zero, indicating that the probability of outcome 1 is unaffected by changes in x_2. This interpretation can also be seen in the threshold lines superimposed on data: The lines separating 1's from other outcomes are essentially vertical. Now consider the regression coefficients for "Lambda 2," which corresponds to $\lambda_{\{2\}|\{2,3,4\}}$ and indicates the probability of outcome 2 within the zone where outcome 1 does not occur. The estimated coefficient on x_2 is negative, which means that the probability of outcome 2 increases as x_2 decreases, again within the zone where outcome 1 does not occur. In general, regression coefficients in conditional logistic regression need to be interpreted for the zone in which they apply.

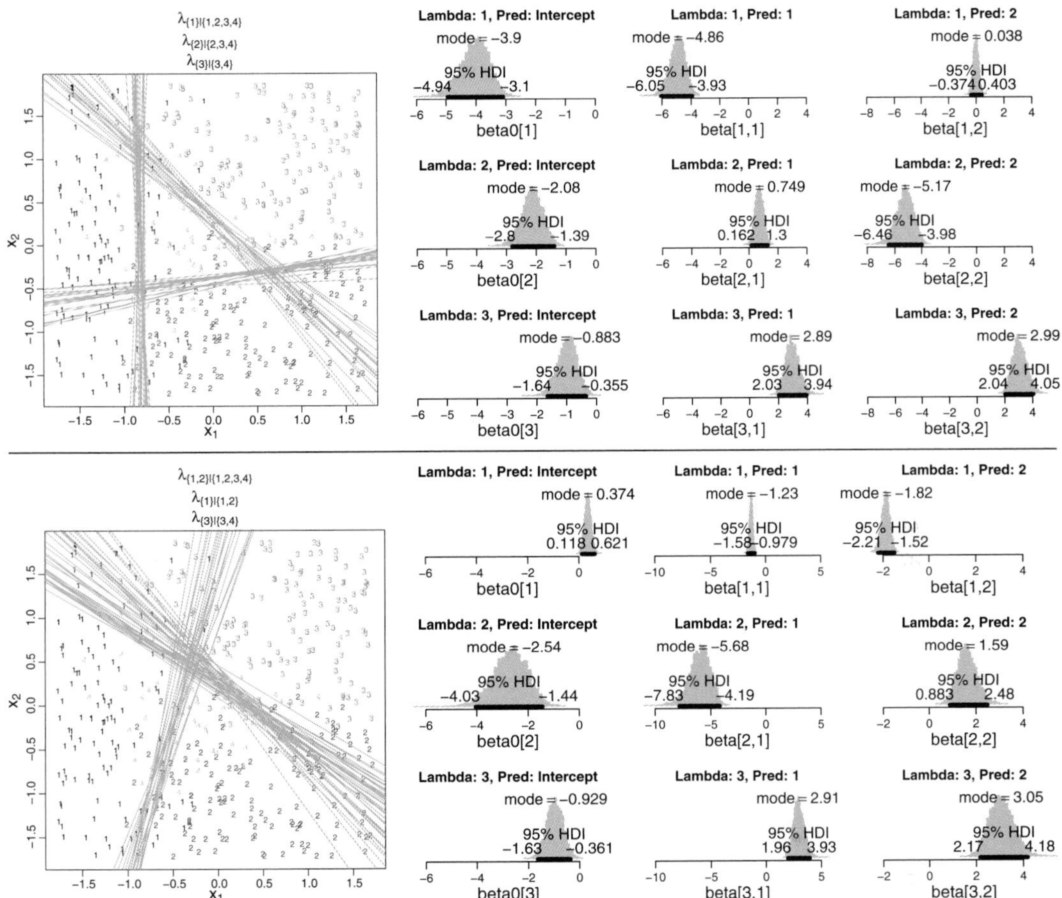

Figure 22.6 *Upper half:* Posterior parameter estimates of the conditional logistic model of Equation 22.11 for data in left panel of Figure 22.3 (i.e., the model structure matches the data generator). The three rows of distributions correspond to the three lambda functions. The columns of distributions correspond to β_0, β_1, and β_2 of the lambda function. *Lower half:* Posterior parameter estimates of the conditional logistic model of Equation 22.14 (i.e., model structure does not match data generator).

In the upper half of Figure 22.7, notice that the estimates for λ_2 are more uncertain, with wider HDI's, than the other coefficients. This uncertainty is also shown in the threshold lines on the data: The lines separating the 1's from the 2's have a much wider spread than the other boundaries. Inspection of the scatter plot explains why: There is only a small zone of data that informs the separation of 1's from 2's, and therefore the estimate must be relatively ambiguous. Compare that with the relatively large zone of data that informs the separation of 3's from 4's (described by λ_3) and which is relatively certain.

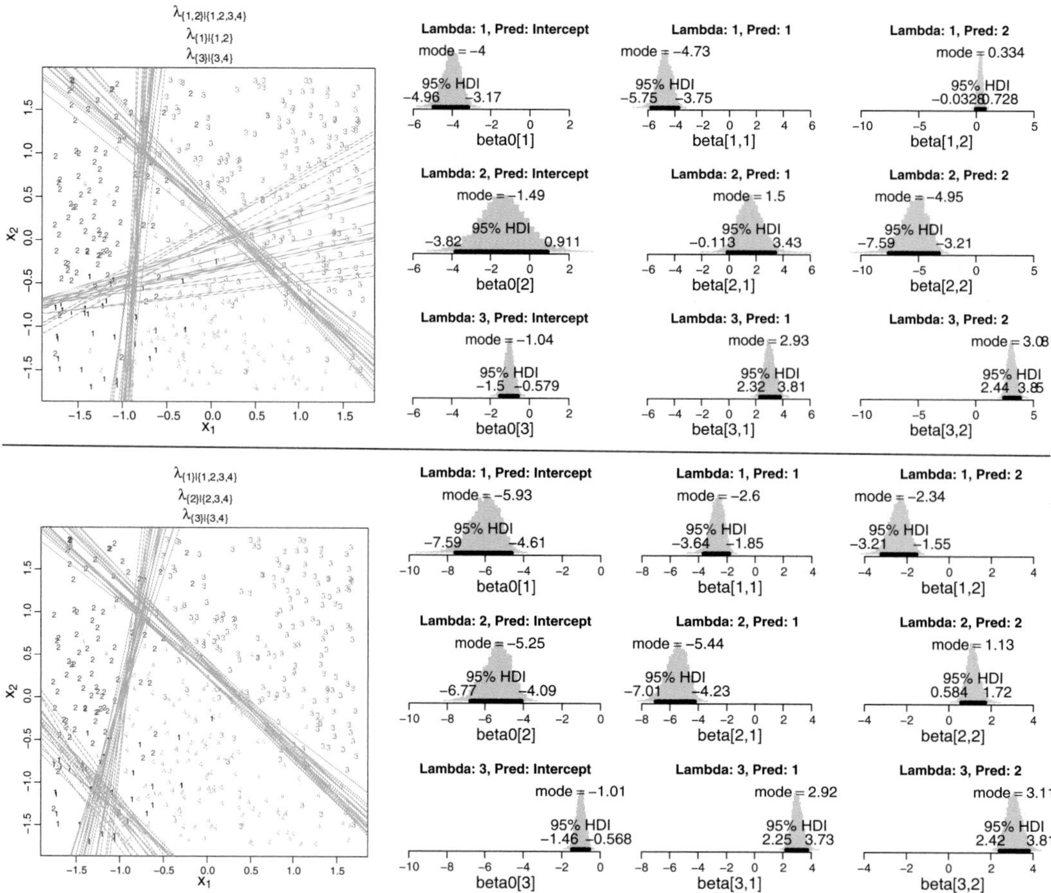

Figure 22.7 *Upper half:* Posterior parameter estimates of the conditional logistic model of Equation 22.14 for data in right panel of Figure 22.3 (i.e., the model structure matches the data generator). The three rows of distributions correspond to the three lambda functions. The columns of distributions correspond to β_0, β_1, and β_2 of the lambda function. *Lower half:* Posterior parameter estimates of the conditional logistic model of Equation 22.11 (i.e., model structure does not match data structure).

The lower halves of Figures 22.6 and 22.7 show the results of applying the wrong descriptive model to the data. Consider the lower half of Figure 22.6, which applies the conditional logistic model that splits the outcomes first into zones of 1 and 2 versus 3 and 4, whereas the data were generated by the conditional logistic model that splits outcomes first by 1 versus all other outcomes. You can see that the estimate for "Lambda 1" (which corresponds to $\lambda_{\{1,2\}|\{1,2,3,4\}}$) has negative coefficients for both x_1 and x_2. The corresponding diagonal boundary lines on the data do not do a very good job of cleanly separating the 1's and 2's from the 3's and 4's. In particular, notice that a lot of 4's fall on the wrong side of the boundary. Curiously in this example, the estimated boundary between the 3's and 4's, within their zone, falls at almost the same place as the boundary between for 1's and 2's versus 3's and 4's.

The lower half of Figure 22.7 applies the conditional logistic model that first splits 1's versus the other outcomes to the data generated by the conditional logistic model that first splits 1's and 2's versus the other outcomes. The results show that the lower-left zone of 1's is split off by a diagonal boundary. Then, within the complementary non-1's zone, the 2's are separated by a nearly vertical boundary from the 3's and 4's. If you compare upper and lower halves of Figure 22.7, your can see that the fits of the two models are not that different, and if the data were a bit noisier (as realistic data usually are), then it could be difficult to decide which model is a better description.

In principle, the different conditional logistic models could be put into an overarching hierarchical model comparison. If you have only a few specific candidate models to compare, this could be a feasible approach. But it is not an easily pursued approach to selecting a partition of outcomes from all possible partitions of outcomes when there are many outcomes. For example, with four outcomes, there are two types of partition structures, as shown in Figure 22.2 (p. 655), and each type has 12 structurally distinct assignments of outcomes to its branches, yielding 24 possible models. With 5 outcomes, there are 180 possible models. And, for any number of outcomes, add one more model to the mix, namely the softmax model. For realistically noisy data, it is unlikely that any single model will stand head and shoulders about the others. Therefore, it is typical to consider a single model, or small set of models, that are motivated by being meaningful in the context of the application, and interpreting the parameter estimates in that meaningful context. Exercise 22.1 provides an example of meaningful interpretation of parameter estimates. Exercise 22.4 considers applying the softmax model to data generated by a conditional logistic model and vice versa.

Finally, when you run the models in JAGS, you may find that there is high autocorrelation in the MCMC chains (even with standardized data), which requires a very long chain for adequate ESS. This suggests that Stan might be a more efficient approach. See Exercise 22.5. Examples of softmax regression programmed in BUGS are given by Ntzoufras (2009, pp. 298–300) and by Lunn, Jackson, Best, Thomas, and Spiegelhalter (2013, pp. 130–131).

22.4. GENERALIZATIONS AND VARIATIONS OF THE MODELS

The goal of this chapter is to introduce the concepts and methods of softmax and conditional logistic regression, not to provide an exhaustive suite of programs for all applications. Fortunately, it is usually straight forward in principle to program in JAGS or Stan whatever model you may need. In particular, from the examples given in this chapter, it should be easy to implement any softmax or conditional logistic model on metric predictors.

Extreme outliers can affect the parameter estimates of softmax and conditional logistic regression. Fortunately it is easy to generalize and implement the robust modeling approach that was used for dichotomous logistic regression in Section 21.3 (p. 635). The predicted probabilities of the softmax or conditional logistic model are mixed with a "guessing" probability as in Equation 21.2 (p. 635), with the guessing probability being 1 over the number of outcomes.

Variable selection can be easily implemented. Just as predictors in linear regression or logistic regression can be given inclusions parameters, so can predictors in softmax or conditional logistic regression. The method is implemented just as was demonstrated in Section 18.4 (p. 536), and the same caveats and cautions still apply, as were explained throughout that section including subsection 18.4.1 regarding the influence of the priors on the regression coefficients.

The model can have nominal predictors instead of or in addition to metric predictors. For inspiration, consult the model diagram in Figure 21.12 (p. 642). A thorough development of this application would involve discussion of multinomial and Dirichlet distributions, which are generalizations of binomial and beta distributions. But that would take me beyond the intended scope of this chapter.

22.5. EXERCISES

Look for more exercises at https://sites.google.com/site/doingbayesiandataanalysis/

Exercise 22.1. [Purpose: Interpreting regression coefficients in softmax regression.]

(A) Consider a situation in which there is a nominal predicted variable that has three values, and there is a single metric predictor. Suppose the outcome 1's are mostly on the left end of the predictor, the outcome 2's are mostly in the mid-range of the predictor, and the outcome 3's are mostly on the right end of the predictor. We use the softmax model. If outcome 2 is the reference outcome, what will be the signs (i.e., positive or negative) of the baseline and slope for outcome 1, and what will the signs of the baseline and slope for outcome 3? If outcome 1 is the reference outcome, will the slope for outcome 2 be greater or less than the slope for outcome 3? Explain. (To check your intuition, you could create a simple data set and run the model.)

(B) Run the softmax model on the data from the right side of Figure 22.1. (The top parts of the example files already have the relevant code for loading the data.) The estimated parameter values should be close to the generating values. Discuss the meaning of the parameter values, with special emphasis on the reference outcome.

Exercise 22.2. [Purpose: Thinking about the applicability of the assumption of independence from irrelevant attributes, and the applicability of hierarchical partition of outcomes.] Suppose ecological researchers are monitoring the fish species in an estuary. The researchers check various locations around the estuary. At each location, they measure the temperature of the water, the salinity of the water, and the counts of various species of fish (using sonar or nets from which the fish are quickly freed unharmed). Each fish is a data point: Species is the predicted nominal value, while salinity and temperature are the metric predictors.

(A) Suppose that the species of fish are known to be highly sensitive to salinity, such that some are essentially fresh water fish and others are salt water fish. Moreover, suppose it is known that the species are also fairly sensitive to temperature. With this background knowledge, discuss whether it might be more appropriate to use a softmax model or a conditional logistic model.

(B) Suppose that the species of fish are known to be very adaptable, and can survive a range of salinities and temperatures. With this background knowledge, discuss whether it might be more appropriate to use a softmax model or a conditional logistic model.

(C) Discuss whether or not independence from irrelevant attributes might apply to this scenario.

Exercise 22.3. [Purpose: More thinking about the applicability of the assumption of independence from irrelevant attributes, and the applicability of hierarchical partition of outcomes.] Suppose marketing researchers want to predict the brand of car owned by people as a function of their annual income and years of education.

(A) Discuss whether or not independence from irrelevant attributes might apply to this scenario.

(B) Discuss whether a softmax model or a conditional logistic model could be meaningful for this scenario.

Exercise 22.4. [Purpose: Explore softmax versus conditional logistic estimates of the same data.]

(A) Apply the softmax model to the data generated by the conditional logistic model that has 1 versus 2, 3, or 4 as its first division. (The top parts of the example files already have the relevant code for loading the data.) Interpret the estimated regression coefficients. In particular, why is the coefficient on x_2 negative for λ_2, positive for λ_3, and around zero for λ_4? And, why does it make sense for the coefficient on x_1 to be of smaller magnitude for λ_4 than for λ_2 and λ_3?

(B) Apply the conditional logistic model that has 1 or 2 versus 3 or 4 as its first division to the softmax data in the left of Figure 22.1. (The top parts of the example files already have the relevant code for loading the data.) Interpret the estimated regression coefficients. Does the model describe the data reasonably well, despite being the wrong model? (The answer to that last question is no, not really, because, while the model divides the 1's and 2's nicely, and divides the 3's and 4's nicely, it gets a lot of outcomes on the wrong side of the first division of 1's or 2's versus 3's or 4's. However, if the data were realistically noisier, we might not notice.)

Exercise 22.5. [Purpose: Practice with Stan, and (hopefully) speeding up the softmax model.] Program the softmax model of `Jags-Ynom-XmetMulti-Msoftmax.R` in Stan. (Ever notice how the exercises that take the fewest words to state take the most hours to do?) Note that Stan has a softmax function built in; see the Stan reference manual. Run it on the data of Figure 22.1 (and check that it gets the same results as the JAGS model). Compare the real time it takes to generate a result with the same ESS as JAGS.

CHAPTER 23

Ordinal Predicted Variable

Contents

The winner is first, and that's all that he knows, whether
Won by a mile or won by a nose. But
Second recalls every inch of that distance, in
Vivid detail and with haunting persistence.[1]

This chapter considers data that have an ordinal predicted variable. For example, we might want to predict people's happiness ratings on a 1-to-7 scale as a function of their total financial assets. Or we might want to predict ratings of movies as a function of the year they were made.

One traditional treatment of this sort of data structure is called *ordinal or ordered probit regression*. We will consider a Bayesian approach to this model. As usual, in Bayesian software, it is easy to generalize the traditional model so it is robust to outliers, allows different variances within levels of a nominal predictor, or has hierarchical structure to share information across levels or factors as appropriate.

[1] This chapter is about modeling ordinal data. The poem emphasizes the emotional difference between ordinal and metric measurement.

Doing Bayesian Data Analysis, 2nd Edition
http://dx.doi.org/10.1016/B978-0-12-405888-0.00023-4

In the context of the generalized linear model (GLM) introduced in Chapter 15, this chapter's situation involves an inverse-link function that is a thresholded cumulative normal with a categorical distribution for describing noise in the data, as indicated in the fourth row of Table 15.2 (p. 443). For a reminder of how this chapter's combination of predicted and predictor variables relates to other combinations, see Table 15.3 (p. 444).

23.1. MODELING ORDINAL DATA WITH AN UNDERLYING METRIC VARIABLE

Suppose we ask a bunch of people how happy they are, and we make them respond on a discrete rating scale with integer values from 1 to 7, with 1 labeled "extremely unhappy" and 7 labeled "extremely happy." How do people generate a discrete ordinal response? It is intuitively plausible that people have some internal feeling of happiness that varies on a continuous metric scale, and they have some sense of thresholds for each response category. People respond with the discrete ordinal value that has thresholds that bracket their underlying continuous metric feeling of happiness. A key aspect of this idea is that the underlying metric value is randomly distributed across people (or across moments within a person). We will assume that the underlying metric value is normally distributed.

As another example, suppose we ask a teacher to assign letter grades to essays from a class of students, and we ask her to respond on a discrete rating scale from 1 to 5, with those integers labeled "F," "D," "C," "B," and "A." How does the teacher generate a discrete ordinal grade? It is intuitively plausible that she carefully reads each essay and gets a sense of its quality on an underlying continuous metric scale. She also has some thresholds for each outcome category, and she assigns the discrete ordinal value that has thresholds that bracket the underlying metric value. We will assume that the underlying metric value is normally distributed across students (or across time within the teacher).

You can imagine that the distribution of ordinal values might not resemble a normal distribution, even though the underlying metric values are normally distributed. Figure 23.1 shows some examples of ordinal outcome probabilities generated from an underlying normal distribution. The horizontal axis is the underlying continuous metric value. Thresholds are plotted as vertical dashed lines, labeled θ. In all examples, the ordinal scale has 7 levels, and hence, there are 6 thresholds. The lowest threshold is set at $\theta_1 = 1.5$ (to separate outcomes 1 and 2), and the highest threshold is set at $\theta_6 = 6.5$ (to separate outcomes 6 and 7). The normal curve in each panel shows the distribution of underlying continuous values. What differs across panels are the settings of means, standard deviations, and remaining thresholds.

The crucial concept in Figure 23.1 is that the probability of a particular ordinal outcome is the area under the normal curve between the thresholds of that outcome. For example, the probability

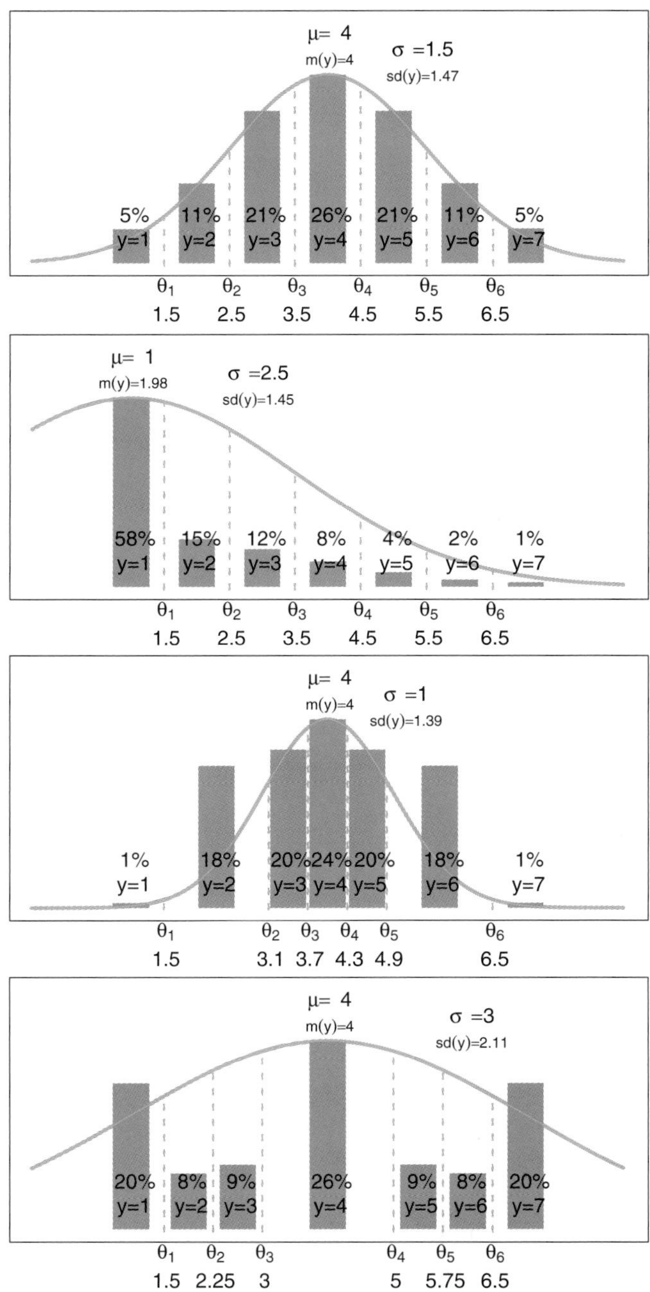

Figure 23.1 Examples of ordinal data generated by the thresholded cumulative-normal model. Ordinal outcome values (y) are indicated by bars. The horizontal axis is the underlying continuous value, which has a normal distribution with mean μ and standard deviation σ. The mean and standard deviation of the ordinal data values (treated as if metric) are annotated as m(y) and sd(y).

of outcome 2 is the area under the normal curve between thresholds θ_1 and θ_2. The vertical bars in Figure 23.1 indicate the probabilities of the outcomes by the heights of the bars. Each bar is annotated with its corresponding outcome value and probability (rounded to the nearest 1%).

How is the area of an interval computed? The idea is that we consider the cumulative area under the normal up the high-side threshold, and subtract away the cumulative area under the normal up to the low-side threshold. Recall that the cumulative area under the standardized normal is denoted $\Phi(z)$, as was illustrated in Figure 15.8 (p. 440). Thus, the area under the normal to the left of θ_k is $\Phi((\theta_k - \mu)/\sigma)$, and the area under the normal to the left of θ_{k-1} is $\Phi((\theta_{k-1} - \mu)/\sigma)$. Therefore, the area under the normal curve between the two thresholds, which is the probability of outcome k, is

$$p(y = k|\mu, \sigma, \{\theta_j\}) = \Phi((\theta_k - \mu)/\sigma) - \Phi((\theta_{k-1} - \mu)/\sigma) \tag{23.1}$$

Equation 23.1 applies even to the least and greatest ordinal values if we append two "virtual" thresholds at $-\infty$ and $+\infty$. Thus, for the least ordinal value, namely $y = 1$, the probability is

$$\begin{aligned} p(y = 1|\mu, \sigma, \{\theta_j\}) &= \Phi((\theta_1 - \mu)/\sigma) - \Phi((\theta_0 - \mu)/\sigma) \\ &= \Phi((\theta_1 - \mu)/\sigma) - \Phi((-\infty - \mu)/\sigma) \\ &= \Phi((\theta_1 - \mu)/\sigma) - 0 \end{aligned} \tag{23.2}$$

And, for the greatest ordinal value, denoted $y = K$, the probability is

$$\begin{aligned} p(y = K|\mu, \sigma, \{\theta_j\}) &= \Phi((\theta_K - \mu)/\sigma) - \Phi((\theta_{K-1} - \mu)/\sigma) \\ &= \Phi((\infty - \mu)/\sigma) - \Phi((\theta_{K-1} - \mu)/\sigma) \\ &= 1 - \Phi((\theta_{K-1} - \mu)/\sigma) \end{aligned} \tag{23.3}$$

The top panel of Figure 23.1 shows a case in which the thresholds are equally spaced and the normal distribution is centered over the middle of the scale, with a moderate standard deviation. In this case, the bar heights mimic the shape of the normal curve fairly well. Exercise 23.1 shows you how to verify this in R.

The second panel of Figure 23.1 shows a case in which the mean of the normal distribution falls below the lowest threshold, and the standard deviation is fairly wide. In this case, the lowest ordinal response has a high probability because much of the normal distribution falls below the lowest threshold. But higher ordinal responses also occur because the standard deviation is wide. Thus, in this case, the distribution of ordinal values does not look very normal, even though it was generated by an underlying normal distribution. An example of real data with this type of distribution will be discussed later.

The third and fourth panels of Figure 23.1 show cases in which the thresholds are not equally spaced. In the third panel, the middle values have relatively narrowly spaced thresholds compared to the penultimate thresholds. The normal distribution in this case is not very wide, and the result is a "broad shouldered" distribution of ordinal values. This type of distribution can arise, for example, when the end categories are labeled too extremely for them to occur very often (e.g., How often do you tell the truth? 1 = I have never told the truth; 7 = I have always told the truth, never even telling a "white lie"). The fourth panel shows a case in which the middle interval is wide compared to the others, and the normal distribution is wide, which yields a three-peaked distribution of ordinal values.

Thus, a normally distributed underlying metric value can yield a clearly non-normal distribution of discrete ordinal values. This result does not imply that the ordinal values can be treated as if they were themselves metric and normally distributed; in fact it implies the opposite: We might be able to model a distribution of ordinal values as consecutive intervals of a normal distribution on an underlying metric scale with appropriately positioned thresholds.

23.2. THE CASE OF A SINGLE GROUP

Suppose we have a set of ordinal scores from a single group. Perhaps the scores are letter grades from a class and we would like to know to what extent the mean grade falls above or below a "C." Or perhaps the scores are agree-disagree ratings and we would like to know the extent to which the mean ratings fall above or below the neutral midpoint. Our goal is to describe the set of ordinal scores according to the model illustrated in Figure 23.1. We will use Bayesian inference to estimate the parameters.

If there are K ordinal values, the model has $K + 1$ parameters: $\theta_1, \ldots, \theta_{K-1}$, μ, and σ. If you think about it a moment, you'll realize that the parameter values trade-off and are undetermined. For instance, we could add a constant to all the thresholds, but have the same probabilities if we added the same constant to the mean. In terms of the graph of Figure 23.1, this is like sliding the x-axis underneath the data an arbitrary distance left or right. Moreover, we could expand or contract the axis to an arbitrary extent, centered at the mean, but have the same probabilities by making a compensatory adjustment to the standard deviation (and thresholds). Therefore, we have to "pin down" the axis by setting two of the parameters to arbitrary constants. There is no uniquely correct choice of which parameters to fix, but we will fix the two extreme thresholds to meaningful values on the outcome scale. Specifically, we will set

$$\theta_1 \equiv 1 + 0.5 \quad \text{and} \quad \theta_{K-1} \equiv K - 0.5 \tag{23.4}$$

This setting implies that our estimates of the other parameters are all with respect to these meaningful anchors. For example, in Figure 23.1, the thresholds were set according to Equation 23.4, and therefore, all the other parameter values make sense with respect to these anchors.

23.2.1. Implementation in JAGS

To estimate the parameters, we need to express Equations 23.1–23.4 in JAGS (or Stan), and establish prior distributions. The first part of the JAGS model specification says that each ordinal datum comes from a categorical distribution that has probabilities computed as in Equations 23.1–23.3. In the JAGS code below, `Ntotal` is the total number of data points, `nYlevels` is the number of outcome levels, and the matrix `pr[i,k]` holds the predicted probability that datum `y[i]` has value `k`. In other words, `pr[i,k]` is $p(y_i = k|\mu, \sigma, \{\theta_j\})$. Recall that the cumulative normal function in JAGS (and R) is called `pnorm`. Now, see if you can find Equations 23.1–23.3 in the first part of the model specification:

```
model {
  for ( i in 1:Ntotal ) {
    y[i] ~ dcat( pr[i,1:nYlevels] )
    pr[i,1] <- pnorm( thresh[1] , mu , 1/sigma^2 )
    for ( k in 2:(nYlevels-1) ) {
      pr[i,k] <- max( 0 ,  pnorm( thresh[ k ] , mu , 1/sigma^2 )
                         - pnorm( thresh[k-1] , mu , 1/sigma^2 ) )
    }
    pr[i,nYlevels] <- 1 - pnorm( thresh[nYlevels-1] , mu , 1/sigma^2 )
  }
```

In the JAGS model specification above, the cumulative normal probabilities that are written as

```
pnorm( thresh[k] , mu , 1/sigma^2 )
```

could instead be coded as

```
pnorm( (thresh[k]-mu)/sigma , 0 , 1 )
```

This alternative form (above) better matches the form in Equation 23.1 and may be used instead if it suits your aesthetic preferences. I avoid the form with 0 and 1 because it can be dangerous, from a programming perspective, to use unlabeled constants.

You may be wondering why the expression for `pr[i,k]`, above, uses the maximum of zero and the difference between cumulative normal probabilities. The reason is that the threshold values are randomly generated by the MCMC chain, and it is remotely possible that the value of `thresh[k]` would be less than the value of `thresh[k-1]`, which would produce a negative difference of cumulative normal probabilities, which makes no sense. By setting the difference to zero, the likelihood of $y = k$ is zero, and the candidate threshold values are not accepted.

The model specification continues with the prior distribution on the parameters. As usual, in the absence of specific prior knowledge, we use a prior that is broad on the scale of data. The mean and standard deviation should be somewhere in the vicinity of the data, which can only range from 1 to `nYlevels` on the ordinal outcome scale. Thus, the prior is declared as:

```
mu ~ dnorm( (1+nYlevels)/2 , 1/(nYlevels)^2 )
sigma ~ dunif( nYlevels/1000 , nYlevels*10 )
```

The prior on the thresholds centers θ_k at $k + 0.5$ but allows a considerable range:

```
for ( k in 2:(nYlevels-2) ) { # 1 and nYlevels-1 are fixed, not stochastic
  thresh[k] ~ dnorm( k+0.5 , 1/2^2 )
}
} # end of model
```

Notice above that only the non-end thresholds are given priors, because only they are estimated. The end thresholds are fixed as specified in Equation 23.4. But how is Equation 23.4 implemented in JAGS? This requires specifying the fixed components of the threshold vector in the data list that is delivered to JAGS separately. The other components of the threshold vector are estimated (i.e., stochastically set by the MCMC process). Estimated components are given the value `NA`, which JAGS interprets as meaning that JAGS should fill them in. Fixed components are given whatever constant values are desired. Thus, in R, we state:

```
thresh = rep(NA,nYlevels-1)           # start by putting NA in all components
thresh[1] = 1 + 0.5                   # overwrite 1st component with 1+0.5
thresh[nYlevels-1] = nYlevels - 0.5 # overwrite last component with nYlevels-0.5
dataList = list(                      # package thresh into the dataList
  thresh = thresh ,
  # ... other data and constants
)
```

The full specification is in the program `Jags-Yord-Xnom1grp-Mnormal.R`, which is called from the high-level script `Jags-Yord-Xnom1grp-Mnormal-Example.R`.

23.2.2. Examples: Bayesian estimation recovers true parameter values

Figures 23.2 and 23.3 show results of two examples. In both cases, the data were generated from the model with particular known parameter values, and the goal of the

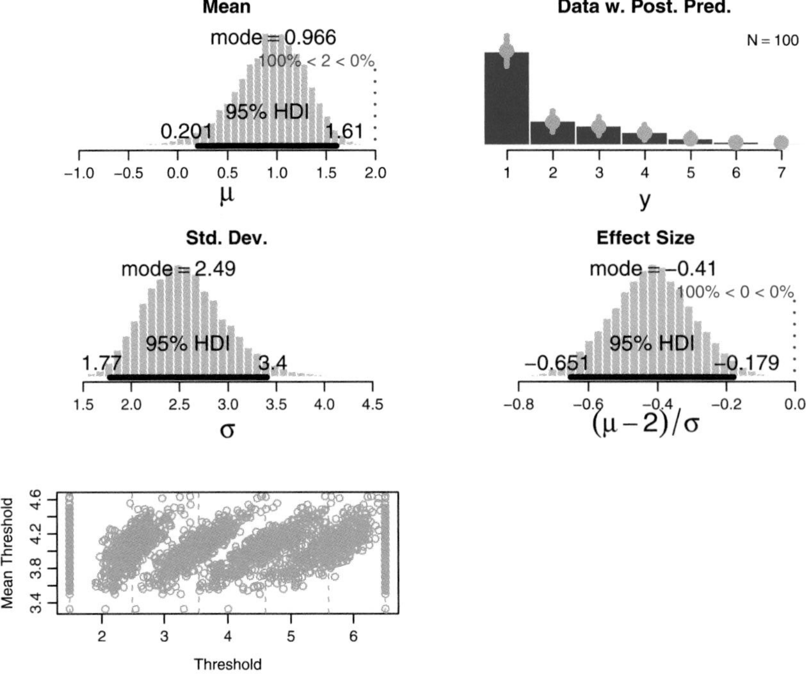

Figure 23.2 Bayesian posterior distribution for one group of ordinal data. True generating parameters are $\mu = 1.0$, $\sigma = 2.5$, $\theta_1 = 1.5$, $\theta_2 = 2.5$, $\theta_3 = 3.5$, $\theta_4 = 4.5$, $\theta_5 = 5.5$, and $\theta_6 = 6.5$. The Bayesian estimation accurately recovers the generating parameters. Posterior distribution clearly excludes a comparison value of $\mu = 2.0$. The posterior predictive distribution accurately describes the data distribution. *NHST treating data as metric:* The mean is *not* significantly different from a comparison value of $\mu = 2.0$: $M = 1.95$, $t = 0.36$, $p = 0.722$, with 95% CI of 1.67–2.23, and with effect size $d = 0.036$. The sample standard deviation is $S = 1.40$. The t test describes the data as normally distributed, which clearly is not the case here.

examples is to demonstrate that Bayesian estimation accurately recovers the parameter values even when μ is extreme and when the thresholds are not evenly spaced.

Figure 23.2 shows an example in which the ordinal data happen to be piled up on one end of the scale. The data are shown as the histogram in the top-right subpanel of the figure. (A real case of this sort of distribution will be provided later.) The generating parameter values are $\mu = 1.0$ and $\sigma = 2.5$, with the thresholds equally spaced at $\theta_1 = 1.5$, $\theta_2 = 2.5$, $\theta_3 = 3.5$, $\theta_4 = 4.5$, $\theta_5 = 5.5$, and $\theta_6 = 6.5$. As you can see in the marginal posterior distributions of μ and σ, the Bayesian estimation very accurately recovers the generating parameters.

The top-right subpanel of Figure 23.2 superimposes posterior predictive probabilities of the outcomes. At each outcome value, a dot plots the median posterior predictive probability and a vertical segment indicates the 95% HDI of posterior predictive probabilities. A key aspect of this thresholded cumulative-normal model is that it accurately describes the ordinal data distribution.

The posterior distribution of an effect size is shown in the second row, right column of Figure 23.2. Effect size for a single group must be defined with respect to a comparison value C; for purposes of illustration, the comparison value here is chosen to be $C = 2.0$. Effect size is defined as the difference between the mean and the comparison value, relative to the standard deviation: $(\mu - C)/\sigma$. (This type of effect size was introduced in the case of a single group of metric values, as displayed in Figure 16.3, p. 457.) The posterior distribution leaves no doubt that the effect size is different from zero. A similar conclusion is reached by examining the posterior distribution of μ relative to the comparison value $C = 2$: We see that credible values of μ are clearly different from 2. Later, we will see that a conventional t test, treating the ordinal data as if they were metric, comes to a very different conclusion.

The bottom panel of Figure 23.2 shows the posterior distribution on the thresholds. The plot does not show the marginal distributions of individual thresholds because the threshold values are strongly correlated with each other, and therefore, the marginal distributions can be misleading. Remember that, at each step in the MCMC chain, the parameter values are *jointly* credible. Consider what happens at a particular step in the chain. If θ_2 is randomly chosen to be a bit higher than usual, the upward shift of the threshold not only increases the likelihood of outcome 2 but also decreases the likelihood of outcome 3. To compensate and maintain the likelihood of outcome 3, threshold θ_3 also is chosen to be a bit higher than usual. This domino effect continues through all the thresholds. Consequently, if one threshold is higher than usual, all thresholds tend to be higher than usual, and if one threshold is lower than usual, then all thresholds tend to be lower than usual. In other words, the thresholds are strongly correlated across steps in the chain. To display this linkage of the thresholds, I have chosen to plot the threshold values against the mean of the thresholds. At a particular step s in the MCMC chain, the jointly credible threshold values are $\theta_1(s), \ldots, \theta_{K-1}(s)$, and the mean threshold *at that step in the chain* is $\overline{\theta}(s) = \sum_k^{K-1} \theta_k(s)/(K-1)$. The threshold values are then plotted as the points $\langle\theta_1(s), \overline{\theta}(s)\rangle, \ldots, \langle\theta_{K-1}(s), \overline{\theta}(s)\rangle$. Notice that these points are all at the same height on the vertical axis. At different steps in the chain, the mean threshold will be higher or lower. You can see in the bottom panel of Figure 23.2 that this plotting method produces distinct clusters of points for each threshold value. You can see that the clusters for θ_1 and θ_{K-1} have fixed values at 1.5 and $K-0.5$, as they should. Each cluster has a vertical dashed line that indicates its mean. The mean values for the estimated thresholds are very near the values that generated the data. You can also see that the higher thresholds have greater uncertainty than the lower thresholds, which makes sense because there are fewer high data points than low data points.

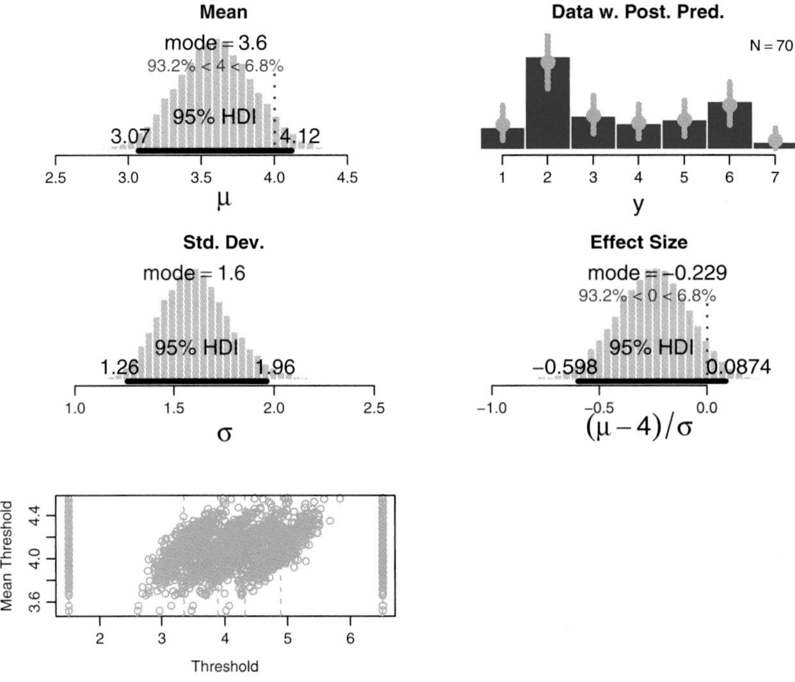

Figure 23.3 Bayesian posterior distribution for one group of ordinal data. True generating parameters are $\mu = 3.5$ and $\sigma = 1.5$, with $\theta_1 = 1.5$, $\theta_2 = 3.3$, $\theta_3 = 3.8$, $\theta_4 = 4.2$, $\theta_5 = 4.7$, and $\theta_6 = 6.5$. The Bayesian estimation accurately recovers the generating parameters. Posterior distribution includes (does not reject) a comparison value of $\mu = 4.0$. Importantly, the posterior predictive distribution nicely describes the data distribution. *NHST treating data as metric:* The mean is significantly different from $\mu = 4.0$: $M = 3.47$, $t = 2.47$, $p = 0.016$, with 95% CI of 3.04 to 3.90, and effect size $d = 0.295$. The sample standard deviation is $S = 1.79$. The t test describes the data as normally distributed which clearly is not the case here.

A second example is shown in Figure 23.3. This example illustrates a case in which the thresholds are not equally spaced, such that the penultimate ordinal values are emphasized and the overall data distribution is bimodal. The caption of the figure reports the true values of the parameters. You can see from the posterior distribution that the true parameter values are accurately estimated, including the thresholds. Importantly, the posterior predictive distribution in the top-right subpanel accurately describes the bimodal distribution of the outcomes.

23.2.2.1 Not the same results as pretending the data are metric

In some conventional approaches to ordinal data, the data are treated as if they were metric and normally distributed. When the data of Figure 23.2 are treated this way, the result is that they are described as normally distributed with a mean of 1.95 and a standard deviation of 1.40, which badly mispredicts the probabilities of the ordinal values in this case, putting highest predicted probability on an outcome of 2 and nearly equal predicted probabilities on outcomes 1 and 3. When a NHST t test is applied to the data, the mean is *not* significantly different from a comparison value of $\mu = 2.0$: $M = 1.95$, $t = 0.36$, $p = 0.722$, with 95% CI of 1.67–2.23, and with effect size $d = 0.036$. The sample standard deviation is $S = 1.40$. The t test assumes that the data are normally distributed, which clearly is not the case here.

When the data of Figure 23.3 are treated as metric and normally distributed, the estimated mean is not far from the Bayesian estimate, but the predictions of individual outcome probabilities are terrible (because the actual probabilities are bimodal, not normal), and the t test comes to a different conclusion than the Bayesian analysis: The mean is significantly different from $\mu = 4.0$: $M = 3.47$, $t = 2.47$, $p = 0.016$, with 95% CI of 3.04–3.90, and effect size $d = 0.295$. The sample standard deviation is $S = 1.79$.

In the two examples, the conclusions of the two approaches differed. For the data of Figure 23.2, the cumulative-normal Bayesian estimation showed that the underlying mean differed from a comparison value of 2.0, but the ordinal-as-metric t test concluded that the mean was not significantly different from 2.0. For the data of Figure 23.3, the cumulative-normal Bayesian estimation showed that the underlying mean had a 95% HDI that included a comparison value of 4.0, but the ordinal-as-metric t test concluded that the mean was significantly different from 4.0. Which of the analyses yields the more trustworthy conclusion? The one that describes the data better. In these cases, there is no doubt that the cumulative-normal model is the better description of the data. Bayesian estimation is an excellent way to estimate the parameters of the model and examine the uncertainty of the estimates.

23.2.2.2 Ordinal outcomes versus Likert scales

In the social sciences, the most common source of ordinal data is questionnaire items that have an ordinal-response interface. For example, rate how much you agree with the statement, "Bayesian estimation is more informative than null-hypothesis significance testing," by selecting one option from the following: 1 = strongly disagree; 2 = disagree; 3 = undecided; 4 = agree; 5 = strongly agree. This sort of ordinal response interface is often called a Likert-type response (Likert, 1932, pronounced LICK-ert not LIKE-ert). Sometimes, it is called a Likert "scale" but the term "scale" in this context is more properly reserved for referring to an underlying metric variable that is indicated by the arithmetic mean of several meaningfully related Likert-type responses (e.g., Carifio & Perla, 2007, 2008; Norman, 2010).

A metric Likert scale derives from several ordinal-response items. The idea is that all the ordinal responses are generated randomly from the same underlying metric variable. If it is assumed that the ordinal responses are linearly related to the underlying metric scale, then the ordinal values are averaged, and the average is described as normally distributed on a metric scale. More sophisticated approaches can treat the underlying metric scale as a latent factor and generate the ordinal responses with a thresholded cumulative-normal model.

The main point for our purposes is that this chapter assumes that we are interested in describing the ordinal outcomes themselves, not (necessarily) the arithmetic average of several ordinal responses. If there are many related items on a multiitem questionnaire, then the model of the ordinal data can use latent factors to express relations among the items. We will not be exploring such models in this book, however.

23.3. THE CASE OF TWO GROUPS

There are many situations in which we would like to compare two groups of ordinal outcomes. Consider a questionnaire that has people indicate how much they agree with various statements about social issues. The ordinal response interface allows people to select one level of 1 = strongly disagree; 2 = disagree; 3 = undecided; 4 = agree; and 5 = strongly agree. One statement on the questionnaire is "Left-handed people should be given equal rights under the law." Another statement on the questionnaire is "Homosexual people should be given equal rights under the law." The two statements could be given to separate groups of respondents if within-subject contrast effects are a concern. We may be interested in how different the responses are to the two questions.

As another example, suppose we ask people to rate their happiness on a 5-point ordinal scale, from very unhappy to very happy. One group of people just sat through 10 min of video advertisements for luxury items such as jewelry and sports cars. Another group of people just sat through 10 min of video advertisements for local charities. We may be interested in how different the responses are of the two groups.

In both examples in the preceding text, the two groups of outcomes were on the same ordinal scale. In the first example, both questionnaire statements were answered on the same disagree–agree scale. In the second example, both groups responded on the same very unhappy–very happy scale. Therefore, we assume that both groups have the same underlying metric variable with the same thresholds. What differs across groups is the mean and standard deviation. Thus, in the case of the questionnaire, ratings of all statements access the same underlying metric scale for disagreement or agreement, with the same thresholds for ordinal responses, and what differs across statements is the mean and variance of the feeling of agreement. In the case of the happiness ratings, we assume that all respondents access a similar underlying metric scale with thresholds for happiness ratings, and what differs across groups is the mean and variance of the feeling of happiness.

23.3.1. Implementation in JAGS

The model specification in JAGS is a simple extension of the specification for one group that was explained in the previous sections. The extension merely uses two means and two standard deviations (one for each group). It uses the same thresholds for both groups. As in the one-group case, the number of outcome levels is denoted `nYlevels` and the total number data points across both groups is denoted `Ntotal`. As was the case for two groups of metric data explored in Section 16.3 (p. 468), the group index of respondent `i` is denoted `x[i]` (and can have values 1 or 2). The JAGS model specification is virtually identical to the case of one group, except what was `mu` is now `mu[x[i]]` and what was `sigma` is now `sigma[x[i]]`:

```
model {
  for ( i in 1:Ntotal ) {
    y[i] ~ dcat( pr[i,1:nYlevels] )
    pr[i,1] <- pnorm( thresh[1] , mu[x[i]] , 1/sigma[x[i]]^2 )
    for ( k in 2:(nYlevels-1) ) {
      pr[i,k] <- max( 0 ,  pnorm( thresh[ k ] , mu[x[i]] , 1/sigma[x[i]]^2 )
                         - pnorm( thresh[k-1] , mu[x[i]] , 1/sigma[x[i]]^2 ) )
    }
    pr[i,nYlevels] <- 1 - pnorm( thresh[nYlevels-1] , mu[x[i]] , 1/sigma[x[i]]^2 )
  }
  for ( j in 1:2 ) { # 2 groups
    mu[j] ~ dnorm( (1+nYlevels)/2 , 1/(nYlevels)^2 )
    sigma[j] ~ dunif( nYlevels/1000 , nYlevels*10 )
  }
  for ( k in 2:(nYlevels-2) ) {  # 1 and nYlevels-1 are fixed, not stochastic
    thresh[k] ~ dnorm( k+0.5 , 1/2^2 )
  }
}
```

The complete model specification and other functions are defined in `Jags-Yord-Xnom2grp-MnormalHet.R`, and the high-level script that calls the functions is named `Jags-Yord-Xnom2grp-MnormalHet-Example.R`.

23.3.2. Examples: Not funny

Figure 23.4 shows a case with artificial data to demonstrate accurate recovery of known parameters. In this case, the ordinal data emphasize the low end of the scale. The true parameters have different means but equal variances, and the thresholds are equally spaced. The figure caption provides exact values. The posterior distribution shows that the parameter values are accurately recovered, and the posterior predictive distribution accurately describes the distribution of data. However, the uncertainty is large enough that 95% HDI on the effect size includes zero.

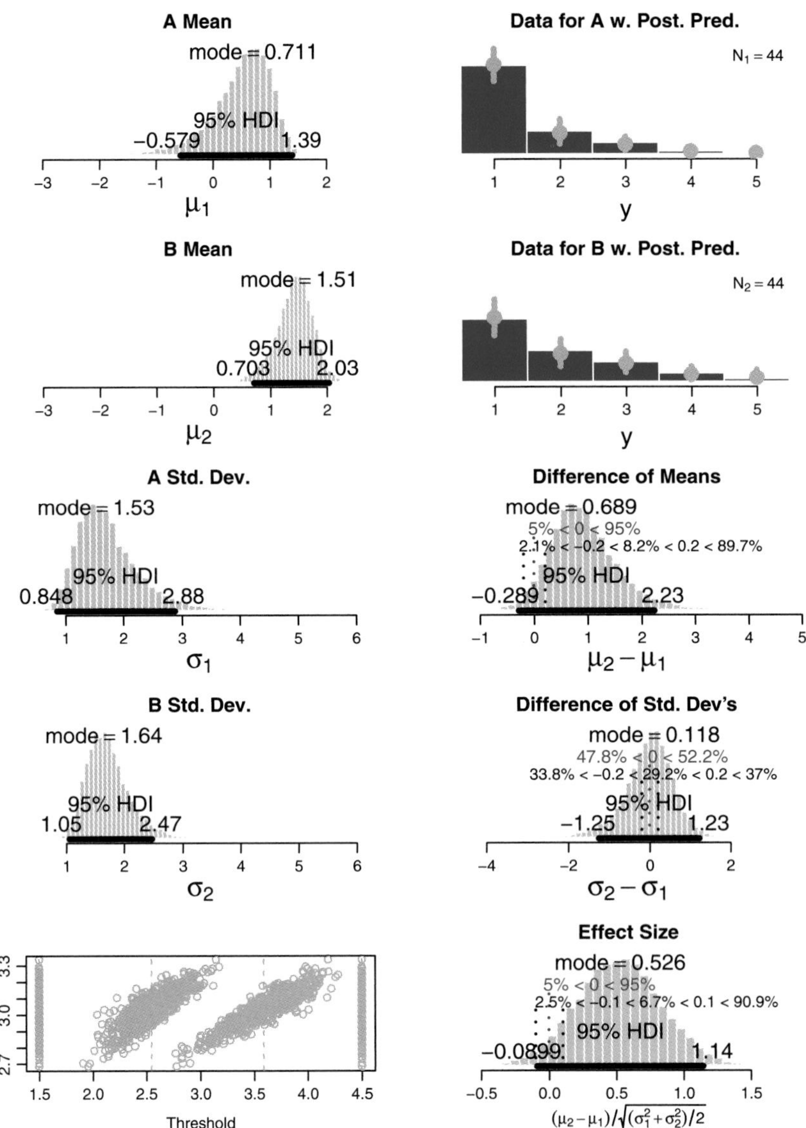

Figure 23.4 Bayesian posterior distribution for two groups of ordinal data. The true generating parameters are $\mu_1 = 0.7, \mu_2 = 1.5, \sigma_1 = 1.6, \sigma_2 = 1.6$ (notice equal variances), with $\theta_1 = 1.5, \theta_2 = 2.5, \theta_3 = 3.5$, and $\theta_4 = 4.5$. The Bayesian estimation accurately recovers the generating parameters. The 95% HDIs include zero effect size and zero difference of standard deviations. *NHST treating data as metric:* The means are significantly different: $M_1 = 1.43, M_2 = 1.86, t = 2.18, p = 0.032$, with effect size $d = 0.466$ with 95% CI of 0.036–0.895. An *F* test of the variances concludes that the standard deviations are significantly different: $S_1 = 0.76, S_2 = 1.07, p = 0.027$. Notice in this case that treating the values as metric greatly underestimates their variances, as well as erroneously concludes the variances are different.

When the data of Figure 23.4 are treated as if they were metric and submitted to NHST, the conclusions are different. In particular, a t test concludes that the means are significantly different *and* an F test concludes that the variances are also significantly different (see the figure caption for details). These conclusions are not to be trusted because their p values assume that the data are normally distributed, which is clearly not the case here.

Figure 23.5 shows a case with real ordinal data that hug one end of the response scale. The data are funniness ratings of two jokes.[2] For each of 25 jokes, people were asked "How funny did you find the joke?" on a rating scale from 1 to 7 with 1 marked "Not at all" and 7 marked "Very." Figure 23.5 shows results from two of the jokes that were rated as relatively less funny than others.[3] I won't report the jokes here because, after all, people did not think these particular jokes were very funny! As can be seen in the posterior distribution of Figure 23.5, Bayesian estimation shows clear differences in the underlying means and variances of the two jokes, while simultaneously describing the data distributions very well. However, if the data are treated as metric and submitted to NHST, no significant differences emerge, as detailed in the caption. Again, the conclusions from NHST are not to be trusted because the p values assume that the data are normally distributed, which is clearly not the case here.

23.4. THE CASE OF METRIC PREDICTORS

We now consider the case of an ordinal predicted value with metric predictors. For example, we could predict people's subjective ratings of happiness as a function of their monetary income and years of education. Or we could predict people's ratings of the funniness of a joke as a function of how much alcohol they have consumed and their age. You can glance ahead at Figures 23.7 and 23.8 for examples of a single metric predictor, and at Figures 23.9 and 23.10 for examples with two metric predictors.

We will use a model that combines the basic ideas of linear regression with the thresholded cumulative-normal model of ordinal outcome probabilities. The model is illustrated in Figure 23.6. The right side of Figure 23.6 shows the familiar linear regression model, with the metric predictor on the horizontal axis, and the predicted value μ on the vertical axis. This portion of the diagram is analogous to the linear regression diagram in Figure 17.1 (p. 478). The predicted value of the linear regression is of an underlying metric variable, however, not the observed ordinal variable. To

[2] Data in Figure 23.5 are from an as-yet unpublished study I conducted with the collaboration of Allison Vollmer as part of her undergraduate honors project.

[3] The ratings for the two jokes in Figure 23.5 came from the same subjects, and therefore, an analyst might instead want to compute a difference in ratings for every subject, and then create a model of the difference scores. Indeed, for the full set of 25 jokes, it is possible to model subject effects in addition to joke effects. This would be overkill for present purposes.

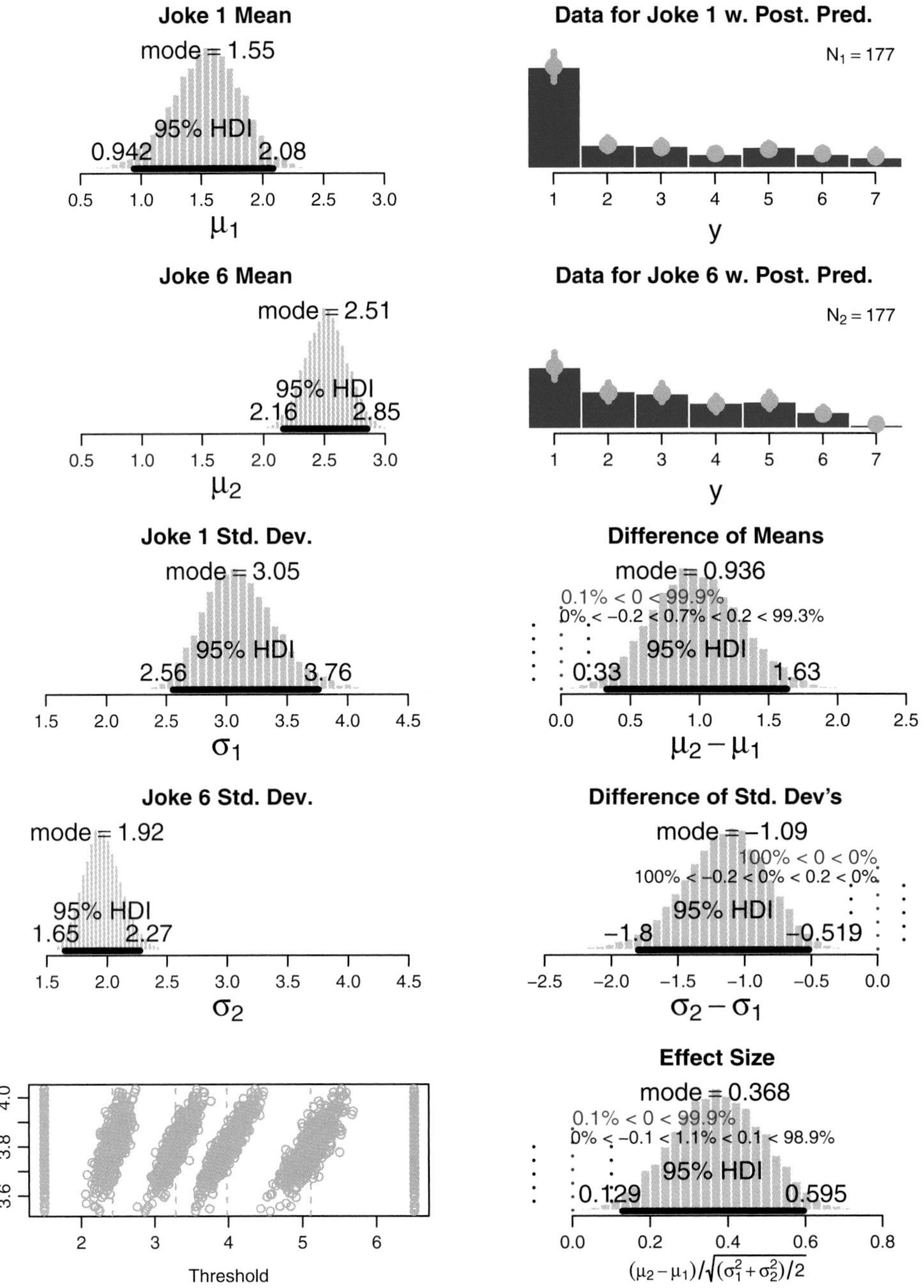

Figure 23.5 Bayesian posterior distribution for two groups of ordinal data from funniness ratings of jokes. Notice that the mean funniness ratings and standard deviations are clearly different. *NHST treating data as metric:* The means are not significantly different: $M_1 = 2.59$, $M_2 = 2.91$, $t = 1.67$, $p = 0.096$, with effect size $d = 0.178$ with 95% CI of -0.032–0.387. An F test of the variances concludes that the standard deviations are not significantly different: $S_1 = 1.96$, $S_2 = 1.73$, $p = 0.116$.

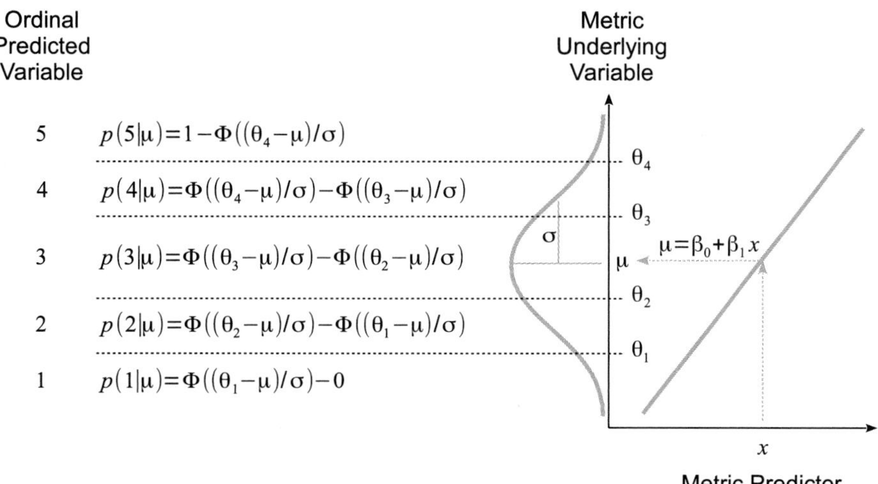

Figure 23.6 Thresholded cumulative-normal regression. Right side shows metric predictor variable mapped to metric underlying variable, as in simple linear regression of Figure 17.1 (p. 478). Left side shows mapping from metric underlying to observed ordinal variable, displaying Equations 23.1–23.3 at the corresponding intervals between thresholds.

get from the underlying metric variable to the observed ordinal variable, we use the thresholded cumulative-normal model. The left side of Figure 23.6 shows the mapping from underlying metric to observed ordinal variables, displaying Equations 23.1–23.3 at the corresponding intervals between thresholds. As before, we must pin down the metric underlying variable by setting anchors as in Equation 23.4: $\theta_1 \equiv 1.5$ and $\theta_{K-1} \equiv K - 0.5$.

It is important to understand what happens in Figure 23.6 for different values of the predictor. The figure is displayed with x at a middling value. Notice that if x were set at a larger value, then μ would be larger (because of the positive slope in this example). When μ is larger, the normal distribution is pushed upward on the vertical axis (because μ is the mean of normal distribution). Consequently, there is larger area under the normal in higher ordinal intervals, and smaller area under the normal in lower ordinal intervals, which is to say that the probability of higher ordinal values increases and the probability of lower ordinal values decreases. The analogous logic implies that lower values of the predictor x produce increased probability of lower ordinal values (when the slope is positive).

This type of model is often referred to as *ordinal probit regression* or *ordered probit regression* because the probit function is the link function corresponding to the cumulative-normal inverse-link function (as you will fondly recall from Section 15.3.1.2, p. 439). But, as you know by now, I find it more meaningful to name the models by their inverse-link function, which in this case is the thresholded cumulative normal.

23.4.1. Implementation in JAGS

The model of Figure 23.6 is easy to implement in JAGS (or Stan). In the program, we generalize to multiple metric predictors (instead of only one metric predictor). There are only two changes from the model specification of the previous applications. First, μ is defined as a linear function of the predictors, just as in the previous programs for multiple regression (linear, logistic, softmax, or conditional logistic). Second, the predictors are standardized to improve MCMC efficiency, again as in the programs for multiple regression. Here is the JAGS model specification:

```
# Standardize the data:
data {
  for ( j in 1:Nx ) {
    xm[j]  <- mean(x[,j])
    xsd[j] <-   sd(x[,j])
    for ( i in 1:Ntotal ) {
      zx[i,j] <- ( x[i,j] - xm[j] ) / xsd[j]
    }
  }
}
# Specify the model for standardized data:
model {
  for ( i in 1:Ntotal ) {
    y[i] ~ dcat( pr[i,1:nYlevels] )
    pr[i,1] <- pnorm( thresh[1] , mu[i] , 1/sigma^2 )
    for ( k in 2:(nYlevels-1) ) {
      pr[i,k] <- max( 0 ,  pnorm( thresh[ k ] , mu[i] , 1/sigma^2 )
                         - pnorm( thresh[k-1] , mu[i] , 1/sigma^2 ) )
    }
    pr[i,nYlevels] <- 1 - pnorm( thresh[nYlevels-1] , mu[i] , 1/sigma^2 )
    mu[i] <- zbeta0 + sum( zbeta[1:Nx] * zx[i,1:Nx] )
  }
  # Priors vague on anchored ordinal scale:
  zbeta0 ~ dnorm( (1+nYlevels)/2 , 1/(nYlevels)^2 )
  for ( j in 1:Nx ) {
    zbeta[j] ~ dnorm( 0 , 1/(nYlevels)^2 )
  }
  zsigma ~ dunif( nYlevels/1000 , nYlevels*10 )
  for ( k in 2:(nYlevels-2) ) {  # 1 and nYlevels-1 are fixed
    thresh[k] ~ dnorm( k+0.5 , 1/2^2 )
  }
  # Transform to original scale:
  beta[1:Nx] <- ( zbeta[1:Nx] / xsd[1:Nx] )
  beta0 <- zbeta0  - sum( zbeta[1:Nx] * xm[1:Nx] / xsd[1:Nx] )
  sigma <- zsigma
}
```

You will have noticed in the code, above, that the predicted values were not standardized. Why? Because the predicted values are ordinal and do not have meaningful metric intervals between them. The ordinal values are just ordered categories, with probabilities described by the categorical distribution `dcat` in JAGS, which is a distribution over integer indices. Because the predicted values are not standardized, the transformation from standardized parameters to original-scale parameters is exactly like what was used for logistic regression, back in Equation 21.1 (p. 625). Full details are in the files `Jags-Yord-XmetMulti-Mnormal.R` and `Jags-Yord-XmetMulti-Mnormal-Example.R`.

23.4.2. Example: Happiness and money

We start with examples that have a single metric predictor. First, a case with fictitious data generated by known parameter values, which demonstrates that Bayesian estimation accurately recovers the parameter values. Figure 23.7 shows the data in its upper panel. The ordinal predicted values range from 1 to 7 in this case. You can see that when the predictor value (x) is small, there are a preponderance of small y values, and when the predictor value is large, there are mostly large y values. The parameter values that were used to generated the data are specified in the figure caption.

Figure 23.7 displays aspects of the posterior predictive distribution superimposed on the data. A smattering of credible regression lines is displayed. It must be remembered, however, that the regression lines refer to the underlying metric predicted variable, not to the ordinal predicted variable. Thus, the regression lines are merely suggestive and should be used to get a visual impression of the uncertainty in the slope and intercept. More directly pertinent are the posterior predicted probabilities of each outcome at selected x values marked by vertical lines. At the selected x values, the posterior predicted probabilities of the outcomes are plotted as horizontal bars "rising" leftward away from the vertical baseline. To see the probability distribution, it is easiest to tilt your head to the left, and imagine the bars as if marking the probabilities of intervals under the normal curve. In this example, when x is small, the bars on small values of y are tall relative to the bars on large values of y. And when x is large, the bars on small values of y are short relative to the bars on large values of y.

The main point of the example in Figure 23.7, other than familiarizing you with the graphical conventions, is that Bayesian estimation accurately recovers the true parameter values that generated the data. This is in contrast with estimates derived by treating the ordinal data as if they were metric. The least-squares estimates of the slope, intercept, and residual standard deviation are reported in the caption of the figure, and they are much too small in this case.

Our second example involving a single metric predictor uses real data from the Chinese household income project of 2002, which surveyed personal income and other aspects of people in urban and rural areas of the People's Republic of China (Shi,

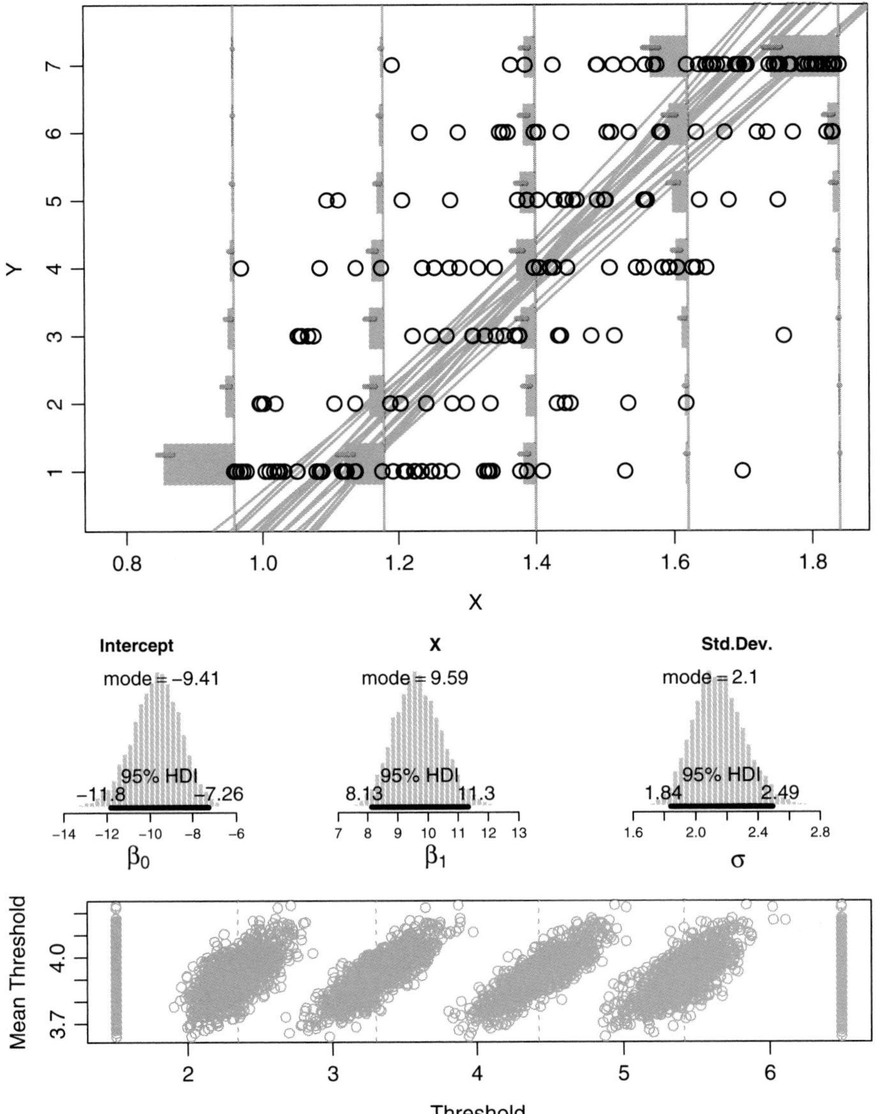

Figure 23.7 Upper panel: Data are shown as dots. Horizontal bars show mean posterior predicted probability at selected values of the predictor as marked by the vertical lines. Grey segments at tops (i.e., left end) of bars show the 95% HDI of posterior predicted probability. A smattering of credible regression lines is superimposed. The true generating parameters are $\beta_0 = -10.0$, $\beta_1 = 10.0$, $\sigma = 2.0$, $\theta_1 = 1.5$, $\theta_2 = 2.5$, $\theta_3 = 3.5$, $\theta_4 = 4.5$, $\theta_5 = 5.5$, $\theta_6 = 6.5$. Lower panels: The Bayesian estimation accurately recovers the generating parameters, as indicated by the marginal posterior distributions. *Least-squares estimate treating data as metric:* $\beta_0 = -5.42$ (SE=0.61), $\beta_1 = 6.71$ (SE=0.43), $\sigma = 1.52$. Notice that the slope is badly underestimated by least-squares estimation in this case.

2009). In particular, one survey item was a person's total household assets, measured in Chinese yuan (which had an exchange rate in 2002 of about 8.27 yuan per 1.00 US dollar). Another survey item asked, "Generally speaking, do you feel happy?" with response options of 1 = not happy at all, 2 = not very happy, 3 = so-so, 4 = happy, and 5 = very happy (and "don't know," which was given just 1% of the time and is excluded from this analysis).[4] There were 6,835 data points, as plotted in Figure 23.8.

One thing to notice from the data in Figure 23.8 is that most happiness ratings are 4, with the next most prominent being 3. One way that this preponderance of "4" ratings is handled by the parameter estimates is by setting the thresholds for categories 2 and 3 to relatively low values, as shown in the bottom panel. This makes the interval for 4 relatively wide.

The posterior distribution in Figure 23.8 indicates that the mean happiness rating clearly increases as assets increase. But it's not a huge increase: If assets go up by 82,770 yuan (the equivalent of 10,000 US dollars in 2002), then mean happiness increases 0.34 points (on the underlying metric scale). Moreover, at any given level of assets, there is large variability in happiness ratings: The marginal posterior distribution of σ has a mode of about 0.85 (on a response scale that ranges from 1 to 5). Note that the value of σ is estimated with high certainty, in that its 95% HDI extends only from about 0.83 to 0.88. This is worth reiterating to be sure the meaning of σ is clear: σ indicates variance in the data, and it is big, but the posterior estimate of the variance is also precise because the sample size is large.

If we were to treat the ordinal ratings as if they were metric values, the least-squares estimates of the regression parameters are a bit less than the Bayesian estimates, as reported in the caption of Figure 23.8, despite this being a case in which the data fall mostly in the middle of the response scale. For an example of treating these happiness data as metric, see Jiang, Lu, and Sato (2009), who defended the practice by citing comparisons reported by Ferrer-i-Carbonell and Frijters (2004).

The general trends indicated by this analysis are typical findings in many studies. But the analysis given here is meant only for pedagogical purposes, not for drawing strong conclusions about the relation of happiness to money. In particular, this example used linear trend for simplicity, and more sophisticated analyses (perhaps with other data) could examine nonlinear trends. Of course, there are many influences on happiness other than money, and the link between money and happiness can be mediated by other factors (e.g., Blanchflower & Oswald, 2004; Johnson & Krueger, 2006, and references cited therein).

[4] As reported by Shi (2009), the actual survey item was reverse scaled, with 1 = very happy and 5 = not happy at all. The original ordering was changed in the analysis presented here so that very happy is the high end of the scale.

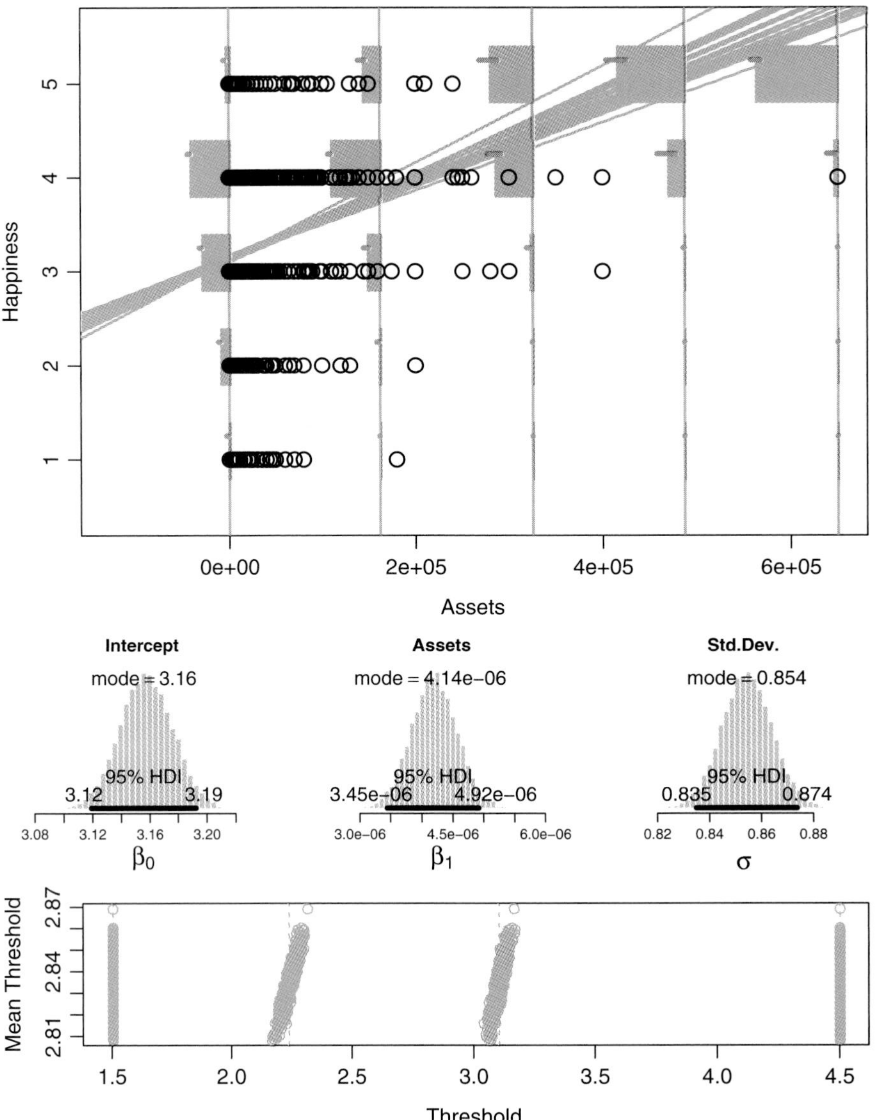

Figure 23.8 Upper panel: Happiness as a function of total household assets ($N = 6,835$, data from Shi, 2009). Assets are in 2002 Chinese yaun (2e+05 yuan was equivalent to about 24,200 US dollars in 2002). Horizontal bars show mean posterior predicted probability at selected values of the predictor as marked by the vertical lines. Grey segments at tops (i.e., left end) of bars show the 95% HDI of posterior predicted probability. A smattering of credible regression lines is superimposed. Lower panels show marginal posterior distribution on parameters. *Least-squares estimate treating data as metric:* $\beta_0 = 3.425$ (SE=0.012), $\beta_1 = 3.82e - 6$ (SE=3.39e-7), $\sigma = 0.847$. Notice that the slope is estimated to be smaller by least-squares estimation in this case, and predictive probabilities are different than Bayesian.

23.4.3. Example: Movies—They don't make 'em like they used to

We now consider a case with two predictors. A movie critic rated many movies on a 1–7 scale[5] and the data about the movies included their length (i.e., duration in minutes) and year of release. These data were assembled by Moore (2006) and come from reviews by Maltin (1996). It's not necessary that there should be any relation between length or year of release and the reviewer's ratings, linear or otherwise. Nevertheless, we can apply thresholded cumulative-normal linear regression to explore possible trends.

Figure 23.9 shows the data and results of the analysis. The upper panel of Figure 23.9 shows a scatter plot in which each point is a movie, with each datum plotted as a numeral 1–7 that indicates the movie's rating, positioned on axes that indicate the length and year of the movie. You can see that there is considerable variation in the ratings that is not accounted for by the predictors, because movies of many different ratings are intermingled among each other.

Despite the "noise" in the data, the analysis reveals credibly nonzero relationships between the predictors and the reviewer's ratings. The marginal posterior distribution on the regression coefficients indicates that ratings tend to rise as length increases, but ratings tend to decline as year of release increases. In other words, the reviewer tends to like old, long movies more than recent, short movies.

To get a visual impression of the linear trend in the data, Figure 23.9 plots a few credible threshold lines. At a step in the MCMC chain, consider that jointly credible thresholds and regression coefficients. We can solve for the $\langle x_1, x_2 \rangle$ loci that fall exactly at threshold θ_k by starting with the definition of the predicted value μ and setting it equal to θ_k:

$$\begin{aligned} \mu &= \theta_k \\ &= \beta_0 + \beta_1 x_1 + \beta_2 x_2 \end{aligned}$$

hence,

$$x_2 = \left(\frac{\theta_k - \beta_0}{\beta_2}\right) + \left(\frac{-\beta_1}{\beta_2}\right) x_1 \tag{23.5}$$

The lines determined by Equation 23.5 are plotted in Figure 23.9 at a few different steps in the MCMC chain, with thresholds from the same step plotted with lines of the same type (solid, dashed, or dotted). Threshold lines from the same step in the chain must be parallel because the regression coefficients are constant at that step, but are different at another step. The threshold lines in Figure 23.9 are level contours on the underlying metric planar surface, and the lines reveal that the ratings increase toward the top left, that is, as x_1 decreases and x_2 increases.

[5] The original rating scale reported by Moore (2006) and Maltin (1996) was ordinal 1–4 "stars" in half steps. Why is it okay to convert this to a 1–7 ordinal scale? By the way, Moore (2006) treated the ordinal data as if they were metric.

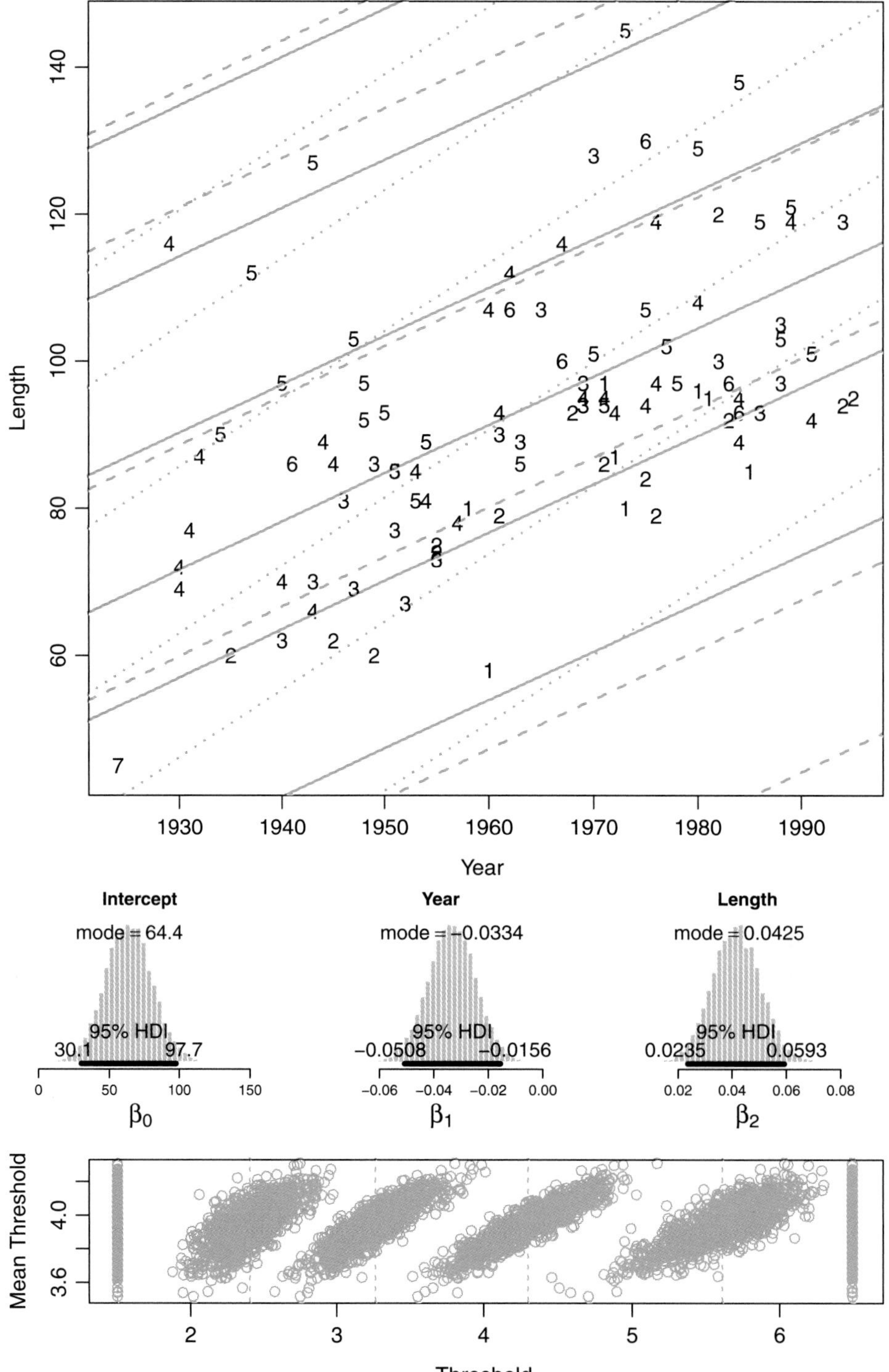

Figure 23.9 Analysis of movie-rating data from Moore (2006). Not shown is the marginal posterior on σ, which has a modal value of about 1.25 and an 95% HDI from about 1.0 to 1.5.

Curiously, however, the extreme thresholds fall beyond where any data occur. For instance, the threshold line for transitioning from 6 to 7 appears at the extreme upper left of the plotting region and there are no data points at all in the region corresponding to 7 above the 6-to-7 threshold. The analogous remarks apply to the lower-right corner: There are no data points in the region corresponding to 1, below the 1-to-2 threshold. How can this be? The next section explains.

23.4.4. Why are some thresholds outside the data?

To motivate an understanding of the thresholds, consider the artificial data presented in Figure 23.10. These data imitate the movie ratings, but with two key differences. First and foremost, the artificial data have much smaller noise, with $\sigma = 0.20$ as opposed to $\sigma \approx 1.25$ in the real data. Second, the artificial data have points that span the entire range of both predictors, unlike the real data which had points mostly in the central region.

Bayesian estimation was applied to the artificial movie data in the same way as the real movie ratings. Notice that the posterior predictive thresholds plotted in Figure 23.10 very clearly cleave parallel regions of distinct ordinal values. The example demonstrates that the threshold lines *do* make intuitive sense when there is low noise and a broad range of data.

This leaves us needing an intuitive explanation for why the thresholds do not seem so clear for the real data. Figure 23.11 is designed to drive your intuition. The figure shows a single predictor on the horizontal axis of the panels, with the probability of outcome on the vertical axis. Each outcome is marked by a different line style (solid, dashed, dotted, and so on). All that changes across panels is the amount of noise. In the top panel, the noise is small, with $\sigma = 0.1$. Notice in this case that each outcome clearly dominates its corresponding interval. For example, outcome 2 occurs with nearly 100% probability between $\theta_1 = 1.5$ and $\theta_2 = 2.5$, and outcome 2 rarely occurs outside that interval. The next panel down has a bit more noise, with $\sigma = 0.5$. You can see that each outcome has maximum probability within its corresponding interval, but there is considerable smearing of outcomes into adjacent intervals. This smearing is worse in the next panel where $\sigma = 1.0$. When the noise is large, as in the bottom panel where $\sigma = 2.0$, there is so much "smearing" of outcomes across thresholds that the most probable outcome within intervals is not necessarily the outcome centered on that interval. For example, between $\theta_1 = 1.5$ and $\theta_2 = 2.5$ the most probable outcome is 1, not 2. To understand why this happens, imagine the normal distribution (from Figure 23.6) centered on $\theta_1 = 1.5$. A full 50% of the normal falls below $\theta_1 = 1.5$, so the probability of outcome 1 is 50%, but the probability of outcome 2 is much less because only a small portion of the normal distribution falls within the interval from $\theta_1 = 1.5$ to $\theta_2 = 2.5$.

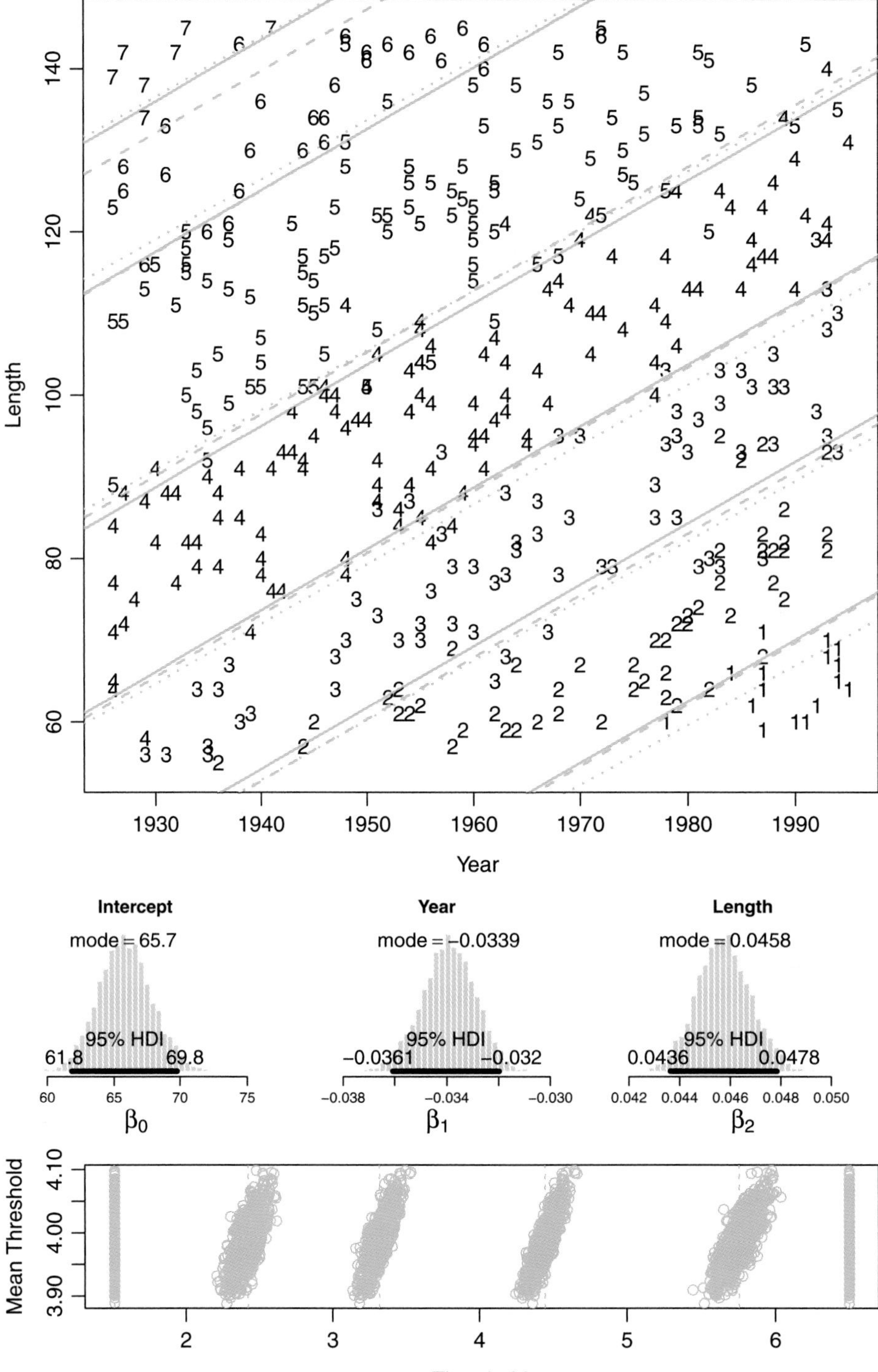

Figure 23.10 Analysis of simulated movie-rating data. These artificial data have very small noise, with generating $\sigma = 0.20$, compared with $\sigma \approx 1.25$ in the actual data of Figure 23.9. (The intercept and slope parameters are set a bit differently than the actual data so that all outcome values are present in the range of the predictors.)

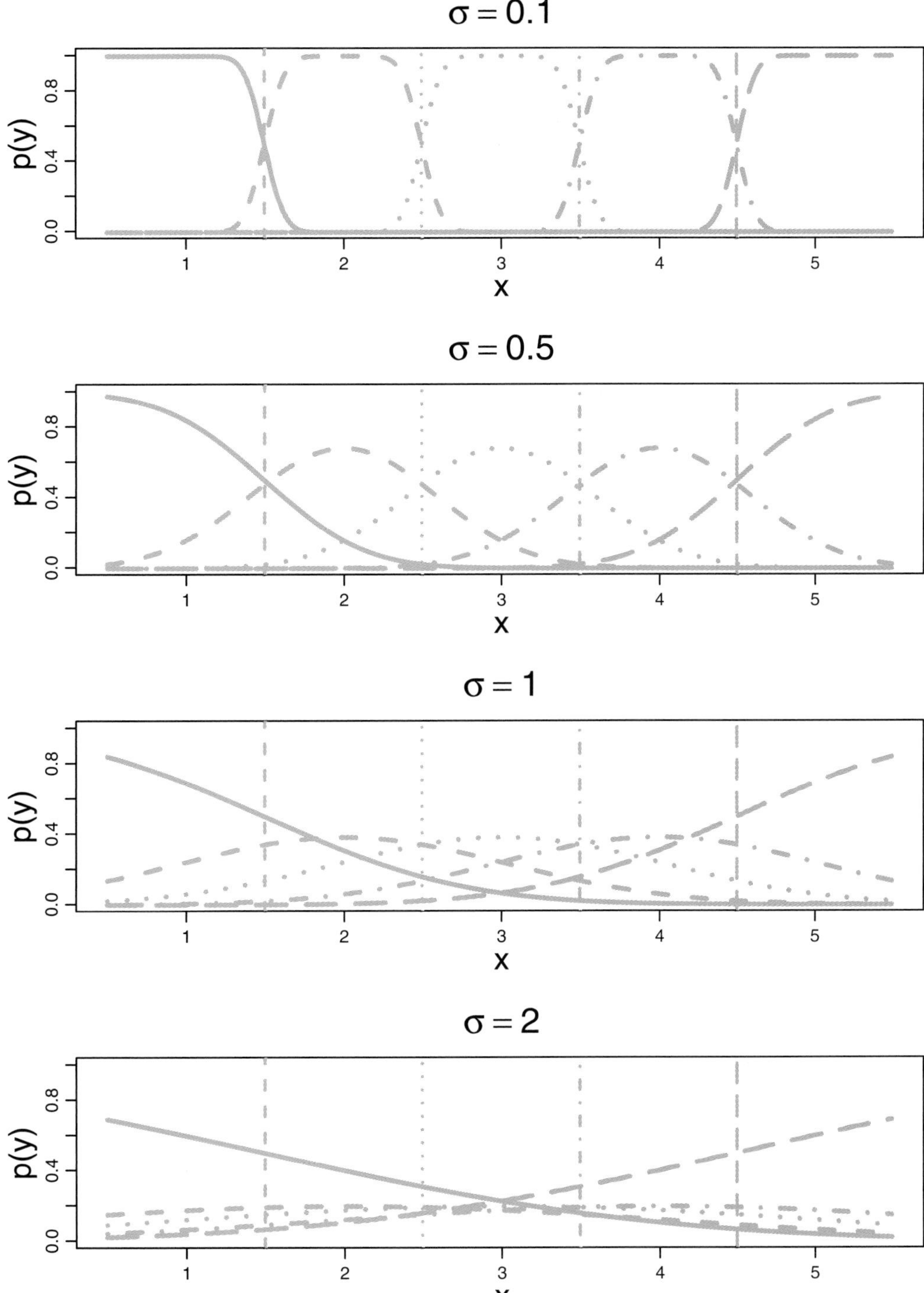

Figure 23.11 The probability of an ordinal response as a function of the predictor x. Thresholds are arbitrarily set at $\theta_1 = 1.5, \theta_2 = 2.5, \theta_3 = 3.5$, and $\theta_4 = 4.5$. Upper panel is for small noise ($\sigma = 0.1$), and lower panel is for large noise ($\sigma = 2.0$). Note that x values in the data might span only a small range on the x axis.

Suppose that the noise is large and that the range of the predictor values in the data is narrow relative to the thresholds. For example, suppose that the predictor values in the data go only from 2 to 4 in the bottom panel of Figure 23.11. Because of the smearing of outcome probabilities, the data would probably have many outcomes of 1 and 5 despite the fact that the predictor values never are less than θ_1 or greater than θ_4. It is this situation that occurs for the real movie data: The most credible parameter values are a large value for σ with the extreme thresholds set a bit outside the predictor values that actually occur in the data.

The preceding discussion has referred to σ as "noise" merely for linguistic ease. Calling the outcomes "noisy" does not mean the underlying generator of the outcomes is inherently wildly random. *The "noise" is merely variation in the outcome that cannot be accounted for by the particular model we have chosen with the particular predictors we have chosen.* A different model and/or different predictors might account for the outcomes well with little residual noise. In this sense, the noise is in the model, not in the data.

23.5. POSTERIOR PREDICTION

Throughout the examples of this chapter, it has been shown that the thresholded cumulative-normal model makes posterior predictions of the probabilities of each outcome. For example, when dealing with a single group of data, the top right panel of Figure 23.3 (p. 680) shows predicted probabilities with 95% HDIs on each outcome. As another example, when dealing with a single predictor, the top panel of Figure 23.8 (p. 692) again shows predicted probabilities with 95% HDIs on each outcome at selected values of the predictor. How is this accomplished and how is it generalized to multiple predictors?

The answer is this: At each step in the MCMC chain, we compute the predicted outcome probabilities $p(y|\mu(x), \sigma, \{\theta_k\})$ using Equations 23.1–23.4 with $\mu(x) = \beta_0 + \sum_j \beta_j x_j$. Then, from the full chain of credible posterior predicted outcome probabilities, we compute the central tendency (e.g., mean, median, or mode) and 95% HDI.

To make this concrete, consider the case of a single metric predictor. The value of x at which we want to know the predicted outcome probabilities is denoted `xProbe`, and we assume it has already been given a value earlier in the program. The following code is processed by R (not by JAGS) and is executed after JAGS has returned the MCMC chain in a coda object named `codaSamples`. Please read through the following code; the comments explain each successive line.

```
# Convert JAGS coda object to a matrix. The matrix has named columns
# for each parameter, with one row per step in the MCMC chain:
mcmcMat = as.matrix( codaSamples )
```

```
# Get the length of the chain:
chainLength = nrow( mcmcMat )
# Merely to simplify reference later, copy the MCMC parameter values
# into new variables:
#   Vector of intercepts:
beta0 = mcmcMat[ , "beta0"]
#   Vector of slopes:
beta1 = mcmcMat[ , "beta"]
#   Matrix of thresholds, row = step in chain, col = threshold:
#   (grep is a pattern matcher. Here, grep gets all columns with
#    "thresh" beginning the name. In R, get help: ?grep)
thresh  = mcmcMat[ , grep("^thresh",colnames(mcmcMat))]
#   Vector of standard deviations:
sigma = mcmcMat[ , "sigma"]
# Compute predicted outcome probabilities at each step in chain:
#   First, declare a matrix to hold the values:
outProb = matrix( NA , nrow=chainLength , ncol=max(y) )
#   For each step in the chain...
for ( stepIdx in 1:chainLength ) {
  # Compute mu at xProbe:
  mu = beta0[stepIdx] + beta1[stepIdx] * xProbe
  # Compute the cumulative normal probability up to each threshold:
  # (remember, thresh is a matrix with a column for each theshold,
  #  so the result of next statement is a vector)
  threshCumProb = pnorm( ( thresh[stepIdx , ] - mu ) / sigma[stepIdx] )
  # Compute difference of cumulative normal probabilities:
  outProb[stepIdx , ] = c(threshCumProb,1) - c(0,threshCumProb)
}
# Compute central tendency of each outcome's predicted prob:
outMed = apply( outProb , 2 , median , na.rm=TRUE )
# Compute HDI of each outcome's predicted prob:
outHdi = apply( outProb , 2 , HDIofMCMC )
```

The last few lines above used the function `apply`, which was explained in Section 3.6 (p. 56). The function `HDIofMCMC` is defined in the utility programs for this book and is described in Section 25.2 (p. 725). The result in `outMed` is a vector of median predicted probabilities for the outcomes, with the corresponding limits of the HDI's in the columns of the matrix `outHdi`.

23.6. GENERALIZATIONS AND EXTENSIONS

The goal of this chapter is to introduce the concepts and methods of thresholded cumulative-normal regression (a.k.a. ordinal or ordered probit regressed), not to provide an exhaustive suite of programs for all applications. Fortunately, it is usually straight forward in principle to program in JAGS or Stan whatever model you may need. In particular, from the programs provided in this chapter, it should be easy to implement any case with one or two groups or metric predictors.

If there are extreme outliers in the data, it is straight forward to modify the programs. There are at least two reasonable approaches to outliers. One approach is to use a heavy-tailed distribution instead of a normal distribution to describe noise. Thus, in Figure 23.6, instead of a normal distribution imagine a *t* distribution. And, in Equations 23.1–23.3, instead of using a cumulative normal function, use a cumulative *t* function. Fortunately, this is easy to do in JAGS (and R and Stan) because the cumulative *t* function is built in as the function `pt` (analogous to `pnorm`). Everywhere in the program that `pnorm` is used, substitute `pt`, making sure to include the normality parameter (which could be estimated or fixed at a small value). A second approach to describing outliers is to treat them not as mean-centered noise but instead as unrelated to the predictor. In particular, the outcome probabilities from the thresholded cumulative-normal model can be mixed with a random-outcome model that assigns equal probabilities to all the outcomes. This is the method that was used for dichotomous logistic regression in Section 21.3 (p. 635). The predicted probabilities of the thresholded cumulative-normal model are mixed with a "guessing" probability as in Equation 21.2 (p. 635), with the guessing probability being 1 over the number of outcomes. Exercise 23.2 provides details.

Variable selection can be easily implemented. Just as predictors in linear regression or logistic regression can be given inclusions parameters, so can predictors in thresholded cumulative-normal regression. The method is implemented just as was demonstrated in Section 18.4 (p. 536), and the same caveats and cautions still apply, as were explained throughout that section including subsection 18.4.1 regarding the influence of the priors on the regression coefficients.

The model can have nominal predictors instead of or in addition to metric predictors. For inspiration, consult the model diagram in Figure 21.12 (p. 642). The only change is putting thresholds on the normal noise distribution to create probabilities for the ordinal outcomes.

23.7. EXERCISES

Look for more exercises at https://sites.google.com/site/doingbayesiandataanalysis/

Exercise 23.1. [Purpose: Hands on experience with computing the cumulative normal, to build intuitions for the predicted probabilities in the figures and for using `pnorm` in R and JAGS.]

(A) Confirm the interval probabilities displayed in Figure 23.1 by using the `pnorm` function in R. Hint: For each panel, do something like the following:

```
mu = 3.5                                    # change as appropriate
sigma = 2.0                                 # change as appropriate
thresh = c( 1.5, 2.5, 3.5, 4.5, 5.5, 6.5 )  # change as appropriate
```

```
pToThresh = pnorm( thresh , mu , sigma )        # What does this compute?!?
pToThresh = pnorm( (thresh-mu)/sigma , 0 , 1 ) # or use this instead
c( pToThresh , 1 ) - c( 0 , pToThresh )         # Explain this!
```

(B) Confirm, approximately, the posterior predicted probabilities shown in Figure 23.3 (p. 680) by using the modal posterior estimates of μ, σ, and thresholds displayed in the figure. Just "eyeball" the modal thresholds from the dashed vertical lines. Use R code from the previous part of the exercise.

(C) Confirm, approximately, the posterior predicted probabilities shown in Figure 23.8 for assets= 1.6e5 yuan and for assets= 4.9e5 yuan. Use the modal posterior estimates of the parameters shown in Figure 23.8, and "eyeball" the modal estimates of the thresholds. Extend the R code from the first part; you'll have to compute `mu` from the intercept and slope at the probed values of x.

Exercise 23.2. [Purpose: Modifying the program to handle outliers.] Suppose we suspect that the data have a lot of outlying values, relative to an underlying normal distribution.

(A) In this part, we extend the thresholded cumulative-normal model so it includes a random guessing mixture. For a review of the idea, see Section 21.3 (p. 635). Copy `Jags-Yord-XmetMulti-Mnormal.R` to a new file with a new name for the guessing model. In the new program, make the following modifications. In the `data` section of the JAGS model, add

```
for ( k in 1:nYlevels ) {
  guessVec[k] <- 1/nYlevels
}
```

In the `model` section, change the `y[i]` line from

```
y[i] ~ dcat( pr[i,1:nYlevels] )
```

to

```
y[i] ~ dcat( (1-alpha)*pr[i,1:nYlevels] + alpha*guessVec )
```

At the end of the model, give the mixture coefficient a prior like this:

```
alpha ~ dbeta(1,9)
```

Run the model on the movie data. Be sure you `source` the new program, not the original program; and, change the `fileNameRoot` for saving the output files so that the nonrobust files are not overwritten. Report the results and discuss any differences from the results of the nonguessing model in Figure 23.9. In particular: (i) Show the posterior distribution of the guessing parameter (`alpha`). (ii) Are the regression coefficients a little more extreme (yes), and why? (iii) Is there anything unusual about the posterior distribution on the thresholds (see Figure 23.12), and why? Hint: Even if the

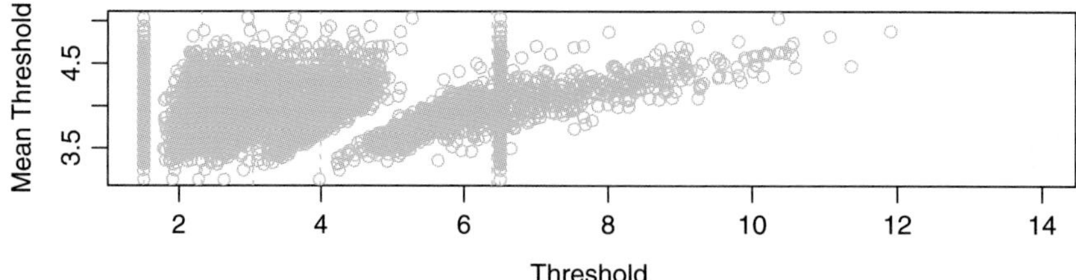

Figure 23.12 For Exercise 23.2 Posterior distribution on thresholds for movie-rating data when the model is a mixture with random guessing.

thresholds are randomly inverted in the MCMC chain, and the outcome probabilities from that component are zero, the random component still gives the outcomes a nonzero probability.

(B) In this part, change the thresholded cumulative-normal model to a thresholded cumulative-*t*-distribution model (with no random guessing mixture). For a review of using the *t* distribution to handle outliers in linear regression, see Section 17.2.1 (p. 483). Copy `Jags-Yord-XmetMulti-Mnormal.R` to a new file with a new name and work on the new copy. You will only need to change the lines of the JAGS model specification that involve `pr[i,...]` from `pnorm` to `pt` (yes, `pt` not `dt`, because we want the cumulative probability function), being sure to include the normality parameter. Run the model on the movie data. Answer the same questions as the previous part. In particular, are the thresholds ever inverted? Hint: The thresholds will not be inverted in this model because that would yield zero probability for the outcomes in the affected intervals.

CHAPTER 24

Count Predicted Variable

Contents

Count me the hours that we've been together, I'll
Count you the hours I'm light as a feather, but
'Cause every hour you're all that I see, there's
No telling if there's a contingency.[1]

Consider a situation in which we observe two nominal values for every item measured. For example, suppose there is an election, and we poll randomly selected people regarding their political party affiliation (which is a nominal variable) and their candidate preference (which is also a nominal variable). Every person increases the count of one particular combination of party affiliation and candidate preference. Across the whole sample, the result is a table of counts for each combination of values of the nominal variables. The counts are what we are trying to predict and the nominal variables are the predictors. This is the type of situation addressed in this chapter.

Traditionally, this sort of data structure could be addressed with a chi-square test of independence. But we will take a Bayesian approach, and consequently have no need to compute chi-square values or p values (except for comparison with the Bayesian conclusions, as in Exercise 24.3). Bayesian analysis provides rich information, including credible intervals on the joint proportions and on any desired comparison of conditions.

In the context of the generalized linear model (GLM) introduced in Chapter 15, this chapter's situation involves a predicted value that is a count, for which we will use an

[1] This chapter is about predicted data that are on a count-valued measurement scale. Such data are often arranged in so-called contingency tables. The poem plays with the terms "count" and "contingency."

Doing Bayesian Data Analysis, 2nd Edition
http://dx.doi.org/10.1016/B978-0-12-405888-0.00024-6

Table 24.1 Counts of combinations of hair color and eye color

	Hair color				
Eye color	**Black**	**Brunette**	**Red**	**Blond**	**Marginal (eye color)**
Blue	20	94	84	17	215
Brown	68	7	119	26	220
Green	5	16	29	14	64
Hazel	15	10	54	14	93
Marginal (hair color)	108	127	286	71	592

Data adapted from Snee (1974).

inverse-link function that is exponential along with a Poisson distribution for describing noise in the data, as indicated in the final row of Table 15.2 (p. 443). For a reminder of how this chapter's combination of predicted and predictor variables relates to other combinations, see Table 15.3 (p. 444).

24.1. POISSON EXPONENTIAL MODEL

I refer to the model that will be explained in this section as Poisson exponential because, as we will see, the noise distribution is a Poisson distribution and the inverse-link function is exponential. Because our models are meant to describe data—as was explained in the first two steps of Bayesian data analysis in Section 2.3 (p. 25)—the best way to motivate the model is with a concrete data set that we want to describe.

24.1.1. Data structure

Table 24.1 shows counts of hair color and eye color from a classroom poll of students at the University of Delaware (Snee, 1974). These are the same data as in Table 4.1 (p. 90) but with the original counts instead of proportions, and with levels re-ordered alphabetically. Respondents self-reported their hair color and eye color, using the labels shown in Table 24.1. The cells of the table indicate the frequency with which each combination occurred in the sample. Each respondent fell in one and only one cell of the table. The data to be predicted are the cell counts. The predictors are the nominal variables. This structure is analogous to two-way analysis of variance (ANOVA), which also had two nominal predictors, but had several metric values in each cell instead of a single count.

For data like these, we can ask a number of questions. We could wonder about one predictor at a time, and ask questions such as, "How much more frequent are brown-eyed people than hazel-eyed people?" or "How much more frequent are black-haired people than blond-haired people?" Those questions are analogous to main effects in ANOVA. But usually we display the joint counts specifically because we're interested in the relationship between the variables. We would like to know if the distribution of counts across one predictor is contingent upon the level of the other predictor. For example, does the distribution of hair colors depend on eye color, and, specifically, is the proportion of blond-haired people the same for brown-eyed people as for blue-eyed people? These questions are analogous to interaction contrasts in ANOVA.

24.1.2. Exponential link function

The model can be motivated two different ways. One way is simply to start with the two-way ANOVA model for combining nominal predictors, and find a way to map the predicted value to count data. The predicted μ from the ANOVA model (recall Equations 20.1 and 20.2, p. 584) could be any value from negative to positive infinity, but frequencies are non-negative. Therefore, we must transform the ANOVA predictions to non-negative values, while preserving order. The natural way to do this, mathematically, is with the exponential transformation. But this transformation only gets us to an underlying continuous predicted value, not to the probability of discrete counts. A natural candidate for the needed likelihood distribution is the Poisson (described later), which takes a non-negative value λ and gives a probability for each integer from zero to infinity. But this motivation may seem a bit arbitrary, even if there's nothing wrong with it in principle.

A different motivation starts by treating the cell counts as representative of underlying cell probabilities, and then asking whether the two nominal variables contribute independent influences to the cell probabilities. For example, in Table 24.1, there's a particular marginal probability that hair color is black, and a particular marginal probability that eye color is brown. If hair color and eye color are independent, then the joint probability of black hair and brown eyes is the product of the marginal probabilities. The attributes of hair color and eye color are independent if that relationship holds for every cell in the table. (Recall the definition of independence in Section 4.4.2, p. 92.) Independence of hair color and eye color means that the proportion of black hair among brown-eyed people is the same as the proportion of black hair among blue-eyed people, and so on for all hair and eye colors.

To check for independence of attributes, we need to estimate the marginal probabilities of the attributes. Denote the marginal (i.e., total) count of the rth row as y_r, and denote the marginal count of the cth column as y_c. Then the marginal proportions are y_r/N and y_c/N, where N is the total of the entire table. If the attributes are independent, then the *predicted* joint probability, $\hat{p}(r,c)$, should equal the product of the marginal probabilities, which means $\hat{p}(r,c) = p(r) \cdot p(c)$, hence $\hat{y}_{r,c}/N = y_r/N \cdot y_c/N$. Because the models we've been dealing with involve additive combinations, not multiplicative combinations, we convert the multiplicative expression of independence into an additive expression by using the facts that $\log(a \cdot b) = \log(a) + \log(b)$ and $\exp(\log(x)) = x$, as follows:

$$\begin{aligned}\hat{y}_{r,c}/N &= y_r/N \cdot y_c/N \\ \hat{y}_{r,c} &= 1/N \cdot y_r \cdot y_c \\ \underbrace{\hat{y}_{r,c}}_{\lambda_{r,c}} &= \exp\Bigg(\underbrace{\log(1/N)}_{\beta_0} + \underbrace{\log(y_r)}_{\beta_r} + \underbrace{\log(y_c)}_{\beta_c}\Bigg)\end{aligned} \tag{24.1}$$

If we abstract the form of Equation 24.1 away from the specific counts, we get the equation $\lambda_{r,c} = \exp(\beta_0 + \beta_r + \beta_c)$. The idea is that whatever are the values of the β's, the resulting λ's will obey multiplicative independence. An example is shown in Table 24.2. The choice of β's is shown in the margins of the table, and the resulting λ's are shown in the cells of the table. Notice that every row has the same relative probabilities, namely, 10, 100, and 1. In other words, the row and column attributes are independent.

We have dealt before with additive combinations of row and column influences, in the context of ANOVA. In ANOVA, β_0 is a baseline representing the overall central tendency of the outcomes, and β_r is a deflection away from baseline due to being in the rth row, and β_c is a deflection away from baseline due to being in the cth column. The deflections are constrained to sum to zero, as required by Equation 20.2 (p. 584), and the example in Table 24.2 respects this constraint.

In ANOVA, when the cell mean is not captured by an additive combination of row and column effects, we include interaction terms, denoted here as $\beta_{r,c}$. The interaction terms are constrained so that every row and every column sums to zero: $\sum_r \beta_{r,c} = 0$ for all c and $\sum_c \beta_{r,c} = 0$ for all r. The key idea is that the interaction terms in the model, which indicate violations of additivity in standard ANOVA, indicate violations of multiplicative independence in exponentiated ANOVA. To summarize, the model of the cell tendencies is Equation 24.1 extended with interaction terms:

$$\lambda_{r,c} = \exp\left(\beta_0 + \beta_r + \beta_c + \beta_{r,c}\right) \tag{24.2}$$

with the constraints

$$\sum_r \beta_r = 0, \quad \sum_c \beta_c = 0, \quad \sum_r \beta_{r,c} = 0 \text{ for all } c, \quad \text{and} \quad \sum_c \beta_{r,c} = 0 \text{ for all } r$$

Table 24.2 Example of exponentiated linear model with zero interaction

$\beta_0 = 4.60517$	$\beta_{c=1} = 0$	$\beta_{c=2} = 2.30259$	$\beta_{c=3} = -2.30259$
$\beta_{r=1} = 0$	100	1000	10
$\beta_{r=2} = 2.30259$	1000	10000	100
$\beta_{r=3} = -2.30259$	10	100	1

Margins show the values of the β's, and cells show $\lambda_{r,c} = \exp(\beta_0 + \beta_r + \beta_c)$ as in Equation 24.1. Notice that every row has the same relative probabilities, namely, 10, 100, 1. In other words, the row and column attributes are independent. Notice also that the row and column deflections sum to zero, as required by Equation 20.2 (p. 584).

If the researcher is interested in violations of independence, then the interest is on the magnitudes of the $\beta_{r,c}$ interaction terms, and specifically on meaningful interaction contrasts. The model is especially convenient for this purpose, because specific interaction contrasts can be investigated to determine in more detail where the nonindependence is arising.

24.1.3. Poisson noise distribution

The value of $\lambda_{r,c}$ in Equation 24.2 is a cell tendency, not a predicted count *per se*. In particular, the value of $\lambda_{r,c}$ can be any non-negative real value, but counts can only be integers. What we need is a likelihood function that maps the parameter value $\lambda_{r,c}$ to the probabilities of possible counts. The Poisson distribution is a natural choice. The Poisson distribution is named after the French mathematician Simon-Denis Poisson (1781–1840) and is defined as

$$p(y|\lambda) = \lambda^y \exp(-\lambda)/y! \tag{24.3}$$

where y is a non-negative integer and λ is a non-negative real number. The mean of the Poisson distribution is λ. Importantly, the variance of the Poisson distribution is also λ (i.e., the standard deviation is $\sqrt{\lambda}$). Thus, in the Poisson distribution, the variance is completely yoked to the mean. Examples of the Poisson distribution are shown in Figure 24.1. Notice that the distribution is discrete, having masses only at non-negative integer values. Notice that the visual central tendency of the distribution does indeed correspond with λ. And notice that as λ increases, the width of the distribution also increases. The examples in Figure 24.1 use noninteger values of λ to emphasize that λ is not necessarily an integer, even though y is an integer.

The Poisson distribution is often used to model discrete occurrences in time (or across space) when the probability of occurrence is the same at any moment in time (e.g., Sadiku & Tofighi, 1999). For example, suppose that customers arrive at a retail store at an *average* rate of 35 people per hour. Then the Poisson distribution, with $\lambda = 35$, is a model of the probability that any particular number of people will arrive in an hour. As another example, suppose that 11.2% of the students at the University of Delaware in the early 1970s had black hair and brown eyes, and suppose that an average of 600 students per term will fill out a survey. That means an average of 67.2 (= 11.2% · 600) students per term will indicate they have black hair and brown eyes. The Poisson distribution, $p(y|\lambda{=}67.2)$, gives the probability that any particular number of people will give that response in a term.

We will use the Poisson distribution as the likelihood function for modeling the probability of the observed count, $y_{r,c}$, given the mean, $\lambda_{r,c}$, from Equation 24.2. The idea is that each particular r, c combination has an underlying average rate of occurrence, $\lambda_{r,c}$. We collect data for a certain period of time, during which we happen to observe

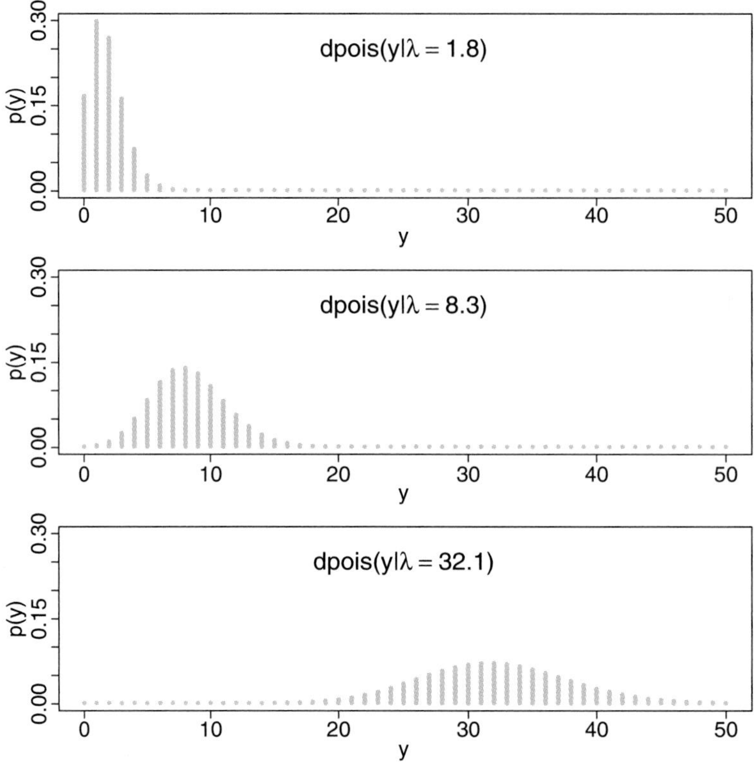

Figure 24.1 Examples of the Poisson distribution. The range of y includes integers from zero to positive infinity. The value of λ is displayed within each panel. The notation "dpois" means the Poisson probability distribution of Equation 24.3.

particular frequencies, $y_{r,c}$, of each combination. From the observed frequencies, we infer the underlying average rates.

24.1.4. The complete model and implementation in JAGS

In summary, the preceding discussion says that the predicted tendency of the cell counts is given by Equation 24.2 (including the sum-to-zero constraints), and the probability of the observed cell count is given by the Poisson distribution in Equation 24.3. All we have yet to do is provide prior distributions for the parameters. Fortunately, we have already done this (at least structurally) for two-factor "ANOVA." The resulting model is displayed in Figure 24.2. The top part shows the prior from two-factor "ANOVA," reproduced from Figure 20.2 (p. 588). The mathematical expression in the middle of Figure 24.2 is the same as in Figure 20.2 except for the new exponential inverse-link function. The lower part of Figure 24.2 simply illustrates the Poisson noise distribution.

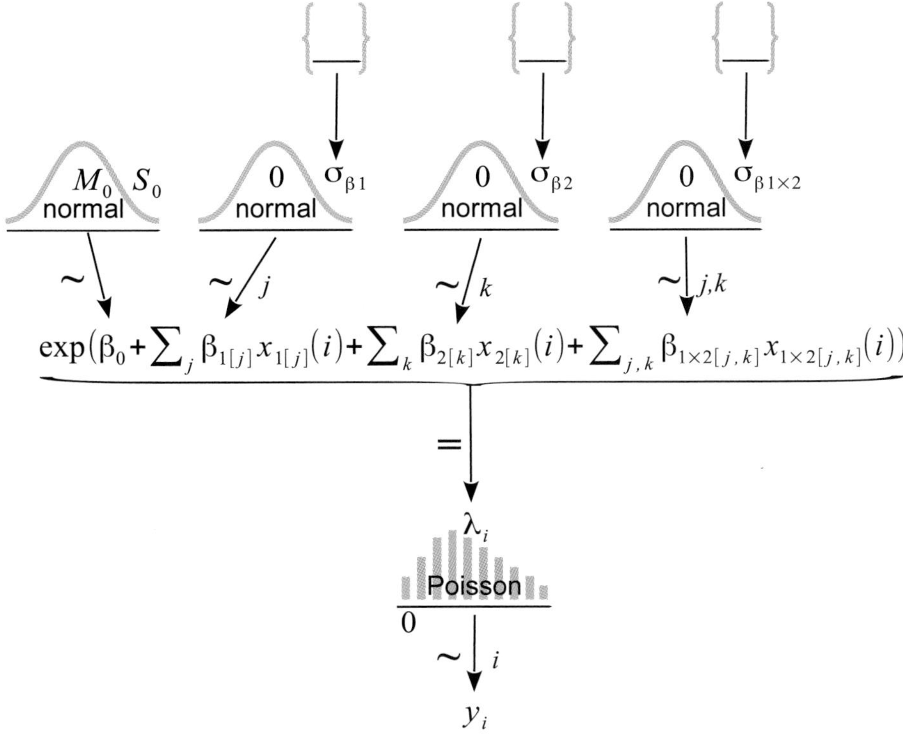

Figure 24.2 Hierarchical dependency diagram of the Poisson exponential model for two nominal predictors. Compare with the diagram for two-factor "ANOVA" in Figure 20.2 (p. 588).

The model is implemented in JAGS as a variation on the model for two-factor "ANOVA." As in that model, the sum-to-zero constraint is imposed by re-centering the deflection parameters that are found by MCMC without the constraint. The deflection parameters that are directly sampled by the MCMC process are denoted `a1`, `a2`, and `a1a2`, and the sum-to-zero versions are denoted `b1`, `b2`, and `b1b2`. To understand the following JAGS code, it is helpful to understand that the data file has one row per cell, such that the *i*th line of the data file contains the count `y[i]` from table row `x1[i]` and column `x2[i]`. The first part of the JAGS model specification directly implements the Poisson noise distribution and exponential inverse-link function:

```
model {
  for ( i in 1:Ncell ) {
    y[i] ~ dpois( lambda[i] )
    lambda[i] <- exp( a0 + a1[x1[i]] + a2[x2[i]] + a1a2[x1[i],x2[i]] )
  }
```

The next part of the JAGS model specifies the prior distribution on the baseline and deflection parameters. The prior is supposed to be broad on the scale of the data, but we must be careful about what scale is being modeled by the baseline and deflections. The counts are being directly described by λ, but it is $\log(\lambda)$ being described by the baseline and deflections. Thus, the prior on the baseline and deflections should be broad on the scale of the logarithm of the data. To establish a generic baseline, consider that if the data points were distributed equally among the cells, the mean count would be the total count divided by the number of cells. The biggest possible standard deviation across cells would occur when all the counts were loaded into a single cell and all the other cells were zero. We take the logarithms of those values to define the following constants:

```
yLogMean = log( sum(y) / (Nx1Lvl*Nx2Lvl) )
yLogSD = log( sd( c( rep(0,Ncell-1) , sum(y) ) ) )
agammaShRa = unlist( gammaShRaFromModeSD( mode=yLogSD , sd=2*yLogSD ) )
```

Those constants are used to set broad priors, as follows:

```
a0 ~ dnorm( yLogMean , 1/(yLogSD*2)^2 )
for ( j1 in 1:Nx1Lvl ) { a1[j1] ~ dnorm( 0.0 , 1/a1SD^2 ) }
a1SD ~ dgamma(agammaShRa[1],agammaShRa[2])
for ( j2 in 1:Nx2Lvl ) { a2[j2] ~ dnorm( 0.0 , 1/a2SD^2 ) }
a2SD ~ dgamma(agammaShRa[1],agammaShRa[2])
for ( j1 in 1:Nx1Lvl ) { for ( j2 in 1:Nx2Lvl ) {
  a1a2[j1,j2] ~ dnorm( 0.0 , 1/a1a2SD^2 )
} }
a1a2SD ~ dgamma(agammaShRa[1],agammaShRa[2])
```

Near the end of the JAGS model specification, the baseline and deflections are recentered to respect the sum-to-zero constraints, exactly as was done for two-factor "ANOVA":

```
# Convert a0,a1[],a2[],a1a2[,] to sum-to-zero b0,b1[],b2[],b1b2[,] :
for ( j1 in 1:Nx1Lvl ) { for ( j2 in 1:Nx2Lvl ) {
  m[j1,j2] <- a0 + a1[j1] + a2[j2] + a1a2[j1,j2] # cell means
} }
b0 <- mean( m[1:Nx1Lvl,1:Nx2Lvl] )
for ( j1 in 1:Nx1Lvl ) { b1[j1] <- mean( m[j1,1:Nx2Lvl] ) - b0 }
for ( j2 in 1:Nx2Lvl ) { b2[j2] <- mean( m[1:Nx1Lvl,j2] ) - b0 }
for ( j1 in 1:Nx1Lvl ) { for ( j2 in 1:Nx2Lvl ) {
  b1b2[j1,j2] <- m[j1,j2] - ( b0 + b1[j1] + b2[j2] )
} }
```

Finally, a novel section of the model specification computes posterior predicted cell probabilities. Each cell mean (which was computed when re-centering for sum to zero) is exponentiated to compute the predicted count for each cell, and then the counts are normalized to compute the predicted proportion in each cell, denoted `ppx1x2p`. The predicted marginal proportions are also computed, as follows:

```
  # Compute predicted proportions:
  for ( j1 in 1:Nx1Lvl ) { for ( j2 in 1:Nx2Lvl ) {
    expm[j1,j2] <- exp(m[j1,j2])
    ppx1x2p[j1,j2] <- expm[j1,j2]/sum(expm[1:Nx1Lvl,1:Nx2Lvl])
  } }
  for ( j1 in 1:Nx1Lvl ) { ppx1p[j1] <- sum(ppx1x2p[j1,1:Nx2Lvl]) }
  for ( j2 in 1:Nx2Lvl ) { ppx2p[j2] <- sum(ppx1x2p[1:Nx1Lvl,j2]) }
}
```

The model specification and other functions are defined in the file `Jags-Ycount-Xnom2fac-MpoissonExp.R`, and a high-level script that calls the functions is the file `Jags-Ycount-Xnom2fac-MpoissonExp-Example.R`.

24.2. EXAMPLE: HAIR EYE GO AGAIN[2]

The analysis was run on the counts of eye and hair colors in Table 24.1. The posterior predicted cell proportions are shown in Figure 24.3. The top of Figure 24.3 is arrayed like the cells of Table 24.1, with the original counts in the titles of each panel. The data proportions (i.e., cell count divided by total count) are displayed as small triangles on the horizontal axis. In this case, the data proportions fall near the middles of the 95% HDIs in each cell. A nice feature of Bayesian analysis is the explicit posterior predictive uncertainty displayed in Figure 24.3.

Often we are interested in whether the attributes are independent, in the sense that the relative proportions of levels of one attribute do not depend on the level of the other attribute. As was explained earlier in the formal motivation of the model, lack of independence is captured by the model's interaction deflections. We consider any interaction contrasts that may be of interest. It is important to notice that independence refers to *ratios* of proportions, which corresponds to *differences* of underlying deflection parameters. In other words, it is not very meaningful to consider interaction contrasts of proportions; instead we consider interaction contrasts of deflection parameters. Suppose we are interested in the interaction contrast of blue eyes versus brown eyes for black hair versus blond hair. The lower panels of Figure 24.3 shows the corresponding differences of the underlying interaction deflection parameters. On the lower left is the difference of blue and brown eyes for black hair, which you can see has a negative value. This means that for black hair, the interaction deflection for brown eyes is larger than the interaction deflection for blue eyes, which makes intuitive sense insofar as we known brown eyes and black hair is a prominent combination. In the lower-middle panel of Figure 24.3 is the difference of blue and brown eyes for blond hair, which you can see has a positive value. Again this makes intuitive sense, as blue eyes and blond hair tend to go together. The lower-right panel shows the interaction contrast (i.e., the difference of differences),

[2] The section header is a pun making fun of the protracted use of this example: Here I go again.

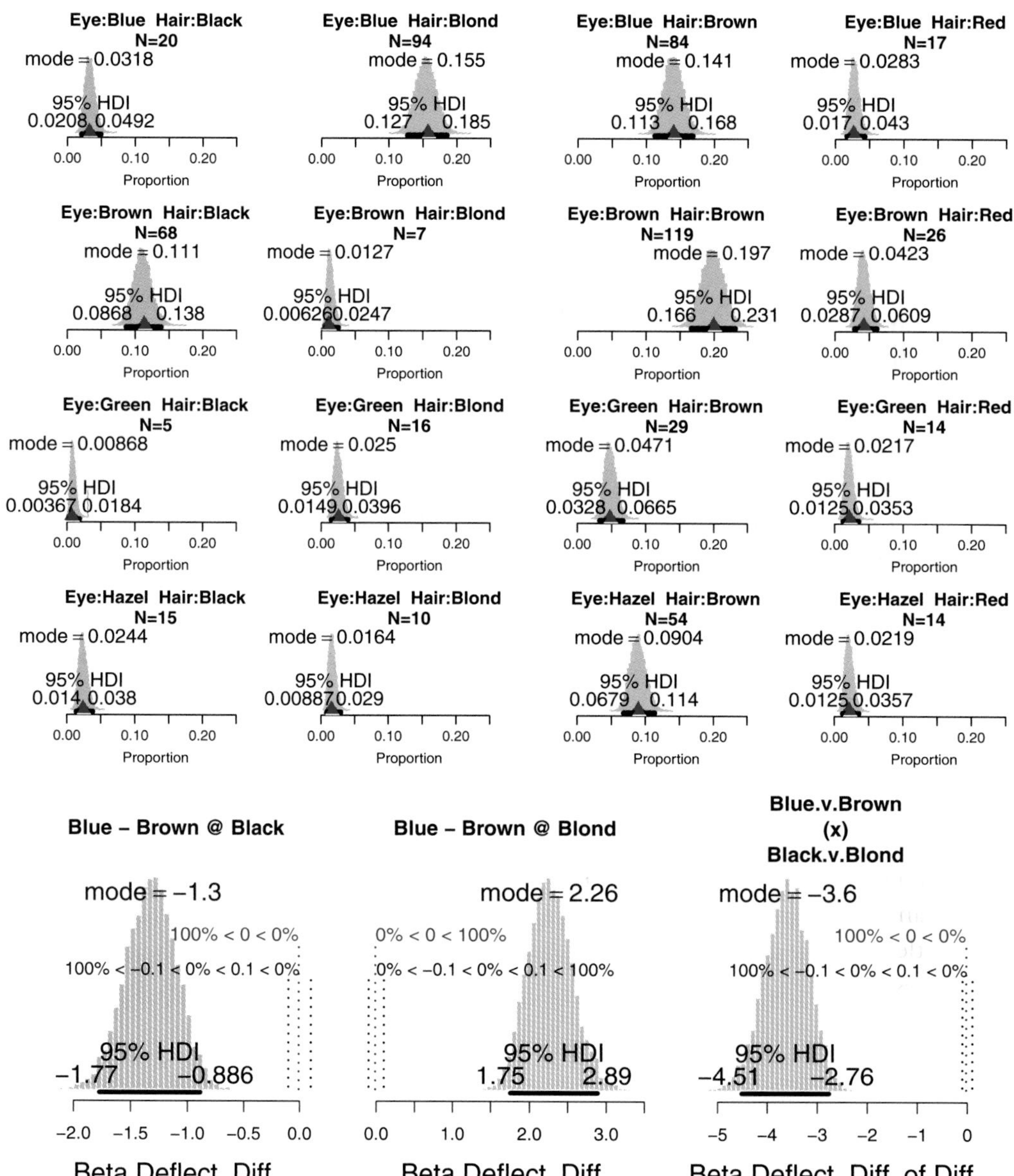

Figure 24.3 *Upper panels:* Posterior distribution of estimated cell proportions for data in Table 24.1. Triangles indicate actual data proportions. *Lower row:* Interaction contrast.

which is strongly nonzero. The nonzero interaction contrast means that the attributes are not independent.

The posterior distributions of differences in the lower row of Figure 24.3 are displayed with ROPEs. When specifying a ROPE for the interaction deflection parameters, it is important to remember that the parameters are on the scale of the logarithm of the count. Thus, a difference of 0.1 corresponds to about a 10% change in the counts. In this case, the ROPE was chosen arbitrarily merely for purposes of illustration.

24.3. EXAMPLE: INTERACTION CONTRASTS, SHRINKAGE, AND OMNIBUS TEST

In this section, we consider some contrived data to illustrate aspects of interaction contrasts. Like the eye and hair data, the fictitious data have two attributes with four levels each. Figure 24.4 shows the cell labels and counts in the titles of the panels. The rows are the levels of attribute A (A1–A4) and the columns are the levels of attribute B (B1–B4). You can see that in the upper-left and lower-right quadrants, all the cells have counts of 22. In the upper-right and lower-left quadrants, all the cells have counts of 11. This pattern of data suggests that there may be an interaction of the attributes, which is to say that the attributes are not independent.

When the JAGS program of the previous section is run on these data, it produces the posterior predictive distribution shown in Figure 24.4. You can see that cells with identical counts have nearly identical posterior predictive distributions, with variation between them caused only by randomness in the MCMC chain (a longer chain would yield less variation between cells with the same count). You can also see shrinkage in the predicted cell proportions: The modal predicted proportion for high-count cells is a little less than the data proportion, and the modal predicted proportion for the low-count cells is a little more than the data proportion. Exercise 24.2 has you examine shrinkage and its causes. In particular, you will find that if you change the structural assumptions about the prior, shrinkage is greatly reduced.

To assess the apparent lack of independence, we conduct interaction contrasts. One plausible interaction contrast involves the central four cells: $\langle A2, B2\rangle$, $\langle A2, B3\rangle$, $\langle A3, B2\rangle$, and $\langle A3, B3\rangle$. The interaction contrast asks whether the difference between A2 and A3 is the same at B2 as at B3. The result is in the bottom left of Figure 24.4, where it can be seen that the magnitude of the interaction contrast is non-zero but the 95% HDI nearly touches zero. Does this imply that there is *not* a sufficiently strong interaction and the attributes could be considered to be independent? No, because there are other interaction contrasts to consider. In particular, we can consider the interaction contrast that involves comparing the quadrants, averaging over the four cells in each quadrant. The result is in the bottom right of Figure 24.4, where it can be seen that the interaction contrast is clearly nonzero.

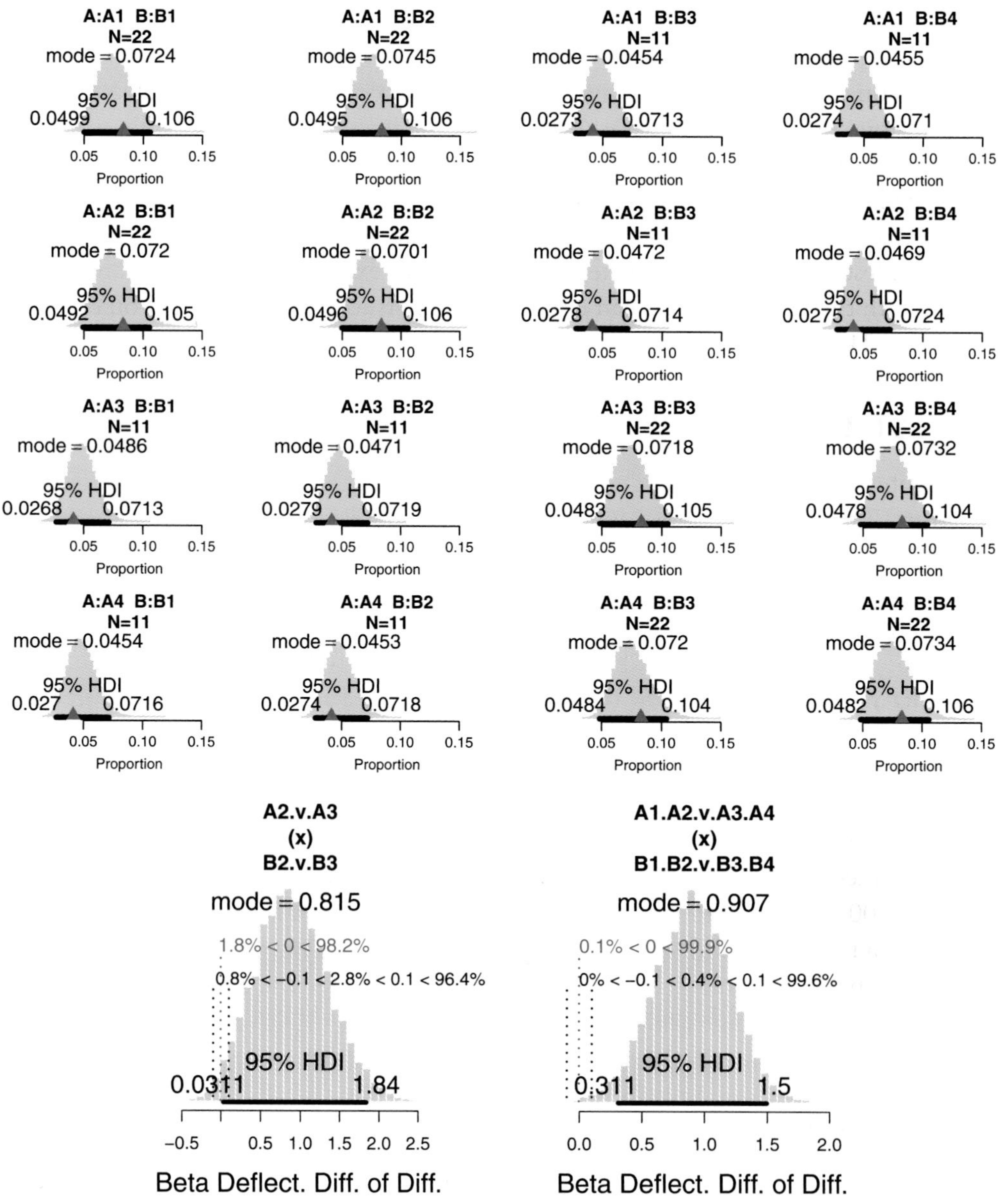

Figure 24.4 *Upper panels:* Posterior distribution of estimated cell proportions. Triangles indicate actual data proportions. *Lower row:* Interaction contrasts.

The model presented here has no way to conduct an "ominbus" test of interaction. However, like the ANOVA-style models presented in Chapters 19 and 20, it is easy to extend the model so it has an inclusion coefficient on the interaction deflections. The inclusion coefficient can have values of 0 or 1, and is given a Bernoulli prior. This extended model is fine in principle, but in practice the MCMC chain can be highly autocorrelated, and pseudo-priors may be helpful (as described in Section 10.3.2.1, p. 279). Also, the results must be carefully filtered so that steps in the chain when the inclusion coefficient is 1 are examined separately from steps in the chain when the inclusion coefficient is 0. Even when there is a high posterior probability that the inclusion coefficient is 1, we still must consider specific interaction contrasts to determine where the interaction "lives" in the data.

The traditional NHST omnibus test of independence for these type of data is the chi-square test, which I will not take the space here to explain. Readers who are already familiar with the chi-square test can compare its results to Bayesian interaction contrasts in Exercise 24.3.

24.4. LOG-LINEAR MODELS FOR CONTINGENCY TABLES

This chapter only scratches the surface of methods for analyzing count data from nominal predictors. This type of data is often displayed as a table, and the counts in each cell are thought of as contingent upon the level of the nominal predictor. Therefore, the data are referred to as "contingency tables." There can be more than two predictors, and models are generalized in the same way as ANOVA is generalized to more than two predictors. The formulation presented here has emphasized the inverse-link expressed as $\lambda = \exp(\beta_0 + \beta_r + \beta_c + \beta_{r,c})$ in Equation 24.2, but this equation can also be written as $\log(\lambda) = \beta_0 + \beta_r + \beta_c + \beta_{r,c}$. This latter form motivates the usual name for these models: *log-linear models for contingency tables*. This is the terminology to use when you want to explore these models more deeply. Agresti and Hitchcock (2005) provide a brief review of Bayesian log-linear models for contingency tables, but the method used in this chapter is not included, because the hierarchical ANOVA model was only popularized later (Gelman, 2005, 2006). For a description of Bayesian inference regarding contingency tables with a model like the one presented in this chapter but without the Gelman-style hierarchical prior, see Marin and Robert (2007, pp. 109–118) and Congdon (2005, p. 134 and p. 202).

24.5. EXERCISES

Look for more exercises at https://sites.google.com/site/doingbayesiandataanalysis/

Exercise 24.1. [Purpose: Trying the analysis on another data set.]

(A) A set of data from Snee (1974) reports counts of criminals on two attributes: the type of crime they committed and whether or not they regularly drink alcohol. Figure 24.5 shows the data in the titles of the panels. The data are in the file CrimeDrink.csv. Run the analysis on the data using the script, `Jags-Ycount-Xnom2fac-MpoissonExp-Example.R`.

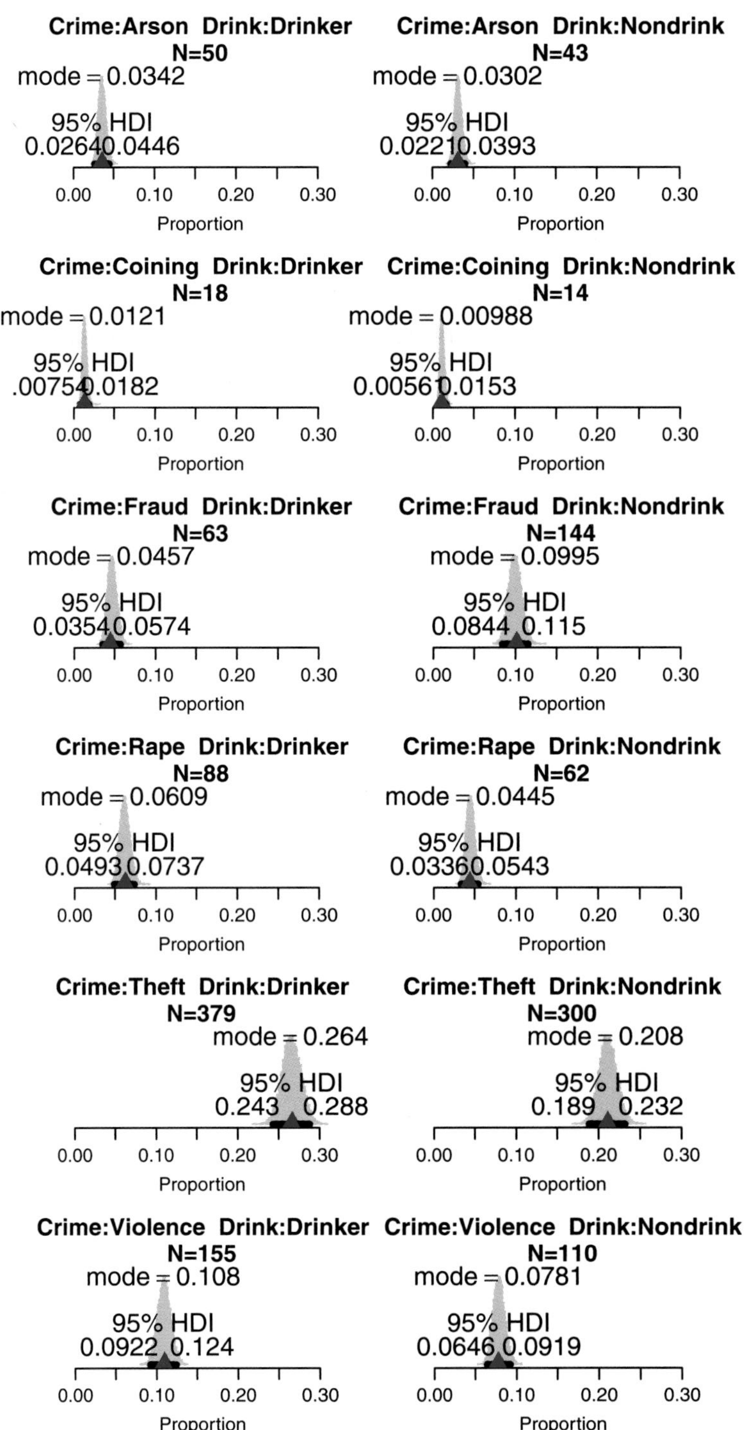

Figure 24.5 For Exercise 24.1. Posterior distribution of estimated cell proportions for crime and drinking data. Triangles indicate actual data proportions.

(B) What is the posterior estimate of the proportion of crimes that is committed by drinkers? Is the precision good enough to say that credibly more crimes are committed by drinkers than by nondrinkers? (This question is asking about a main-effect contrast.)

(C) What is the posterior estimate of the proportion of crimes that are fraud and the proportion that are violent (other than rape, which is a separate category in this data set)? Overall, is the precision good enough to say that those proportions are credibly different?

(D) Perform an interaction contrast of Fraud and Violence versus Drinker and Nondrink. What does this interaction contrast mean, in plain language?

Exercise 24.2. [Purpose: Exploring shrinkage of interaction contrasts.]

(A) Run the script `Jags-Ycount-Xnom2fac-MpoissonExp-Example.R` using the FourByFourCount.csv data with `countMult=3`. Figure 24.6 shows results, for your reference. Notice the shrinkage of cell-proportion estimates, and the funnel-shaped posteriors (like this: ⋏) on interaction contrasts.

(B) Modify the model specification in `Jags-Ycount-Xnom2fac-Mpoisson Exp.R` as follows. Change

```
a1a2SD ~ dgamma(agammaShRa[1],agammaShRa[2])
```

to

```
a1a2SD <- yLogSD
```

and run the analysis again. (Be sure to save the new version of `Jags-Ycount-Xnom2fac-MpoissonExp.R` so that the changes are `source`d when you run the script.) Notice that the shrinkage disappears from the interaction contrast. Discuss why there is strong shrinkage of the interaction contrasts in the previous part but not in the version of this part.

(C) Continuing with the results of the previous part, notice that the modal cell proportions are smaller than the sample proportions. Despite the modes being less than the data proportions, verify that the posterior cell proportions really do sum to 1.0 on every step in the chain. Why are the modes of the marginals less than the sample proportions? (Hints: To extract the posterior cell proportions at every step, try something like the following.

```
mcmcMat = as.matrix(mcmcCoda)
round(rowSums( mcmcMat[ , grep("ppx1x2p",colnames(mcmcMat)) ] ) ,5)
```

Explain what that R code does! The modes are less than the sample proportion because of marginalizing over the skew in the posterior.)

Exercise 24.3. [Purpose: Compare results of NHST chi-square test to results of Bayesian approach.] For this exercise, you must already be familiar with NHST chi-square tests. Also, you must have installed the `reshape2` package, as described back

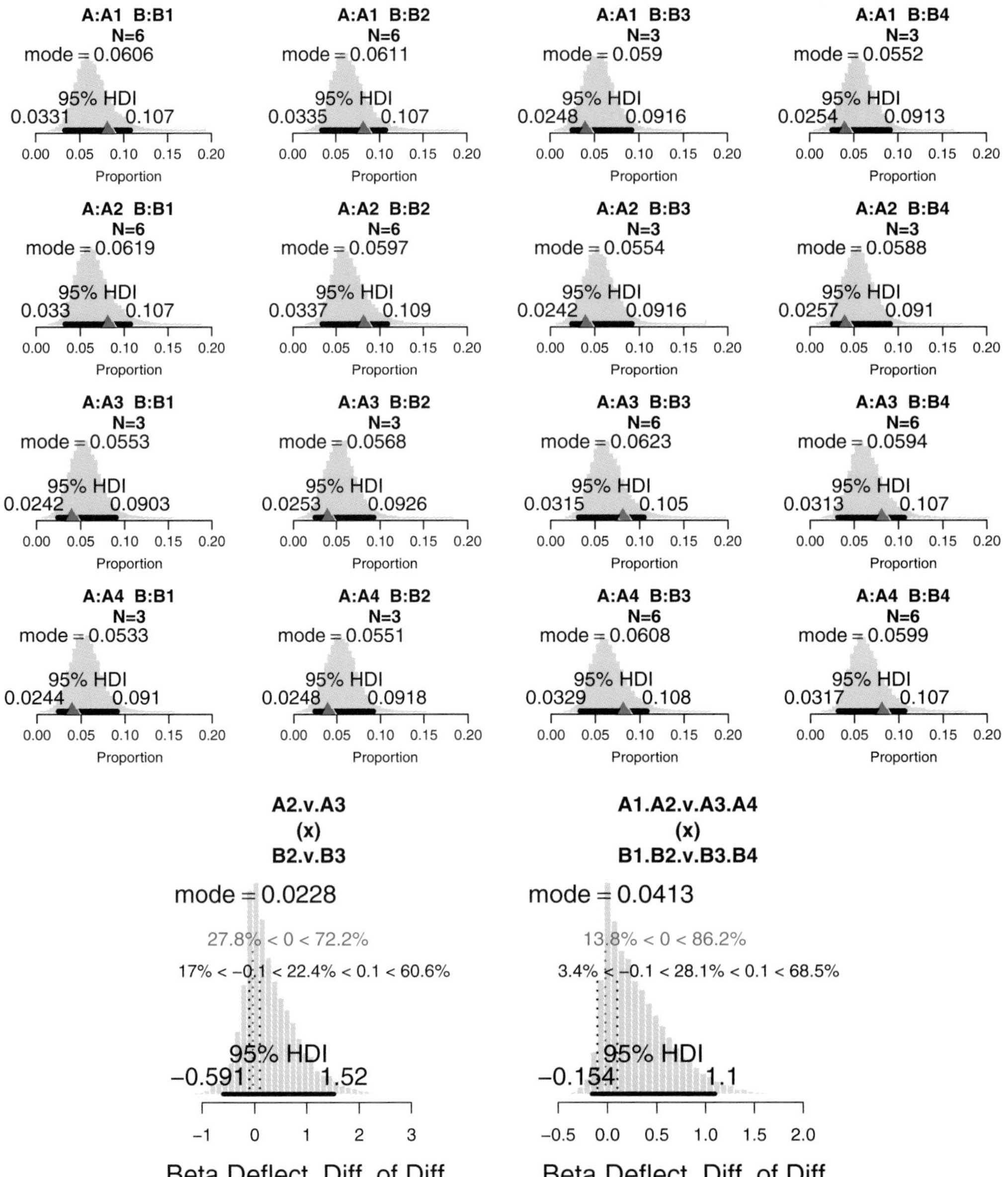

Figure 24.6 For Exercise 24.2 regarding shrinkage. *Upper panels:* Posterior distribution of estimated cell proportions. Triangles indicate actual data proportions. *Lower row:* Interaction contrast.

in Section 3.6 (p. 56). Before doing the remainder of the exercise, load the `reshape2` functions by typing the following at R's command line:

```
library(reshape2)
```

(A) Let's do a traditional chi-square test of the hair-color eye-color data of Table 24.1. Type the following commands into R:

```
myDF = read.csv( file="HairEyeColor.csv" )
myMat = acast( myDF , Eye ~ Hair , value.var="Count" )
Xsq = chisq.test(myMat)
Xsq$observed
round(Xsq$expected,2)
round(Xsq$residuals,2)
Xsq
```

Explain each line. What exactly do we conclude from this (omnibus) test? Do we know which subset of cells is responsible for the "significant" result? How might we find out? Does the omnibus test really tell us much?

(B) Let's do a traditional chi-square test of the data in Figures 24.4 and 24.6. Run the following lines three times, using `countMult=11`, `countMult=7`, and `countMult=3`.

```
myDF = read.csv( file="FourByFourCount.csv" )
countMult = 11 # try 3, 7, and 11
myDF$Count = round( myDF$Count * countMult )
myMat = acast( myDF , A ~ B , value.var="Count" )
Xsq = chisq.test(myMat)
Xsq$observed
round(Xsq$expected,2)
round(Xsq$residuals,2)
Xsq
```

Do the conclusions of the chi-square test match the conclusions of the Bayesian estimation? (You'll have to run the Bayesian estimation for `countMult=7`.) Can you compare an omnibus chi-square with the Bayesian interaction contrasts? Extra credit for overachievers: Run the Bayesian estimations with `a1a2SD` set to a constant instead of estimated, as in Exercise 24.2.

CHAPTER 25

Tools in the Trunk

Contents

She changes her hair, and he changes his style,
She paints on her face, and he wears a fake smile,
She shrink wraps her head, and he stretches the truth,
But they'll always be stuck with their done wasted youth.[1]

This chapter includes some important topics that apply to many different models throughout the book. Please see the list of contents, above. The sections can be read independently of each other and at any time.

25.1. REPORTING A BAYESIAN ANALYSIS

Bayesian data analyses are not yet standard procedure in many fields of research, and no conventional format for reporting them has been established. Therefore, the researcher who reports a Bayesian analysis must be sensitive to the background knowledge of his or her specific audience, and must frame the description accordingly.

[1] One of the topics in this chapter is reparameterization, in which parameters of a model are transformed into new parameters. For example, a rectangle can be described in terms of its length and width, or in terms of its area and aspect ratio, where area = length × width and aspect ratio = length/width. Either way, it's the same rectangle. The poem personifies reparameterization.

Doing Bayesian Data Analysis, 2nd Edition
http://dx.doi.org/10.1016/B978-0-12-405888-0.00025-8

I once assigned an exercise to students in which they had to write up the results of a Bayesian analysis as it would appear in a research journal. Because I am a psychologist, a student then earnestly asked me, "Do we have to write the report as if it were for a "psychology" journal or for a "science" journal?" After swallowing my feeling of injury at the implication that psychology is not science, I politely asked the student to clarify the question. The student said, "For a psychology journal you have to explain what you did, but for a science journal the reader has to puzzle it out." I hope that all science journals will allow you to explain what you did.

Thinking about how to clearly report a Bayesian analysis also gets you to think clearly about the analysis itself. One recent article in a statistics education journal described a classroom exercise in which students had to generate topics to include in a report (Pullenayegum, Guo, & Hopkins, 2012, with discussion on my blog at http://doingbayesiandataanalysis.blogspot.com/2012/05/how-to-report-bayesian-analysis.html). The primary focus of the exercise was the conceptual clarification it encouraged in the students, not the resulting list. Nevertheless, it is interesting that the list generated by the students missed some points that I had indicated were essential in the 1st edition of this book, and which appear again below. I hope that the motivation for the points I offer below helps you to produce clear reports and to better understand Bayesian analysis.

25.1.1. Essential points

Recall the basic steps of a Bayesian analysis from Section 2.3 (p. 25): Identify the data, define a descriptive model, specify a prior, compute the posterior distribution, interpret the posterior distribution, and, check that the model is a reasonable description of the data. Those steps are in logical order, with each step building on the previous step. That logical order should be preserved in the report of the analysis. The essential points listed below mirror the basic mechanical steps of Bayesian analysis, preceded by an important preliminary step that might recede in importance as Bayesian methods become standard procedure.

- Motivate the use of Bayesian (non-NHST) analysis: Many audiences, including journal editors and reviewers, are comfortable with NHST and are not familiar with Bayesian methods. I have found that most editors and reviewers are eager to publish research that uses the best possible methods, including Bayesian analyses, but the editors and reviewers appreciate an explanation of why you have used Bayesian analysis instead of NHST. You may motivate your use of Bayesian data analysis on several grounds, depending in part on the particular application and audience. For example, Bayesian models are designed to be appropriate to the data structure, without having to make approximation assumptions typical in NHST (e.g., homogeneity of variances across groups, and normally distributed noise). The inferences from a Bayesian analysis are richer and more informative than NHST because the posterior distribution reveals joint probabilities of combinations of parameter values. And, of course, there is no reliance on sampling distributions and p values to interpret the parameter estimates.

- Clearly describe the data structure, the model, and the model's parameters: Ultimately you want to report the meaningful parameter values, but you can't do that until you explain the model, and you can't do that until you explain the data being modeled. Therefore, recapitulate the data structure, reminding your reader of the predicted and predictor variables. Then describe the model, emphasizing the meaning of the parameters. This task of describing the model can be arduous for complex hierarchical models, but it is necessary and crucial if your analysis is to mean anything to your audience.
- Clearly describe and justify the prior: It is important to convince your audience that your prior is appropriate and does not predetermine the outcome. The prior should be amenable to a skeptical audience. The prior should be at least mildly informed to match the scale of the data being modeled. If there is copious previous research using very similar methods, it should not be ignored. Optionally, as mentioned again below, it may be helpful to try different priors and report the robustness of the posterior. When the goal is continuous parameter estimation, the posterior distribution is usually robust against reasonable changes in vague priors, but when the goal is model comparison (such as with inclusion coefficients) then the posterior probabilities of the models can be strongly affected by the choice priors.
- Report the MCMC details, especially evidence that the chains were converged and of sufficient length. Section 7.5 (p. 178) explained in detail how to examine the chains for convergence, and every program in this book produces diagnostic graphics of chains for at least some of the parameters. The programs all produce summaries of the parameters that include the ESS. Your report should indicate that the chains were checked for convergence, and your report should indicate the ESS of the relevant parameters. Usually this section of the report can be brief.
- Interpret the posterior: Many models have dozens or even hundreds of parameters, and therefore it is impossible to summarize all of them. The choice of which parameters or contrasts to report is driven by domain-specific theory and by the results themselves. You want to report the parameters and contrasts that are theoretically meaningful. You can report the posterior central tendency of a parameter and its HDI in text alone; histograms of posteriors are useful for the analyst to understand the posterior and for explanation in a textbook, but may be unnecessary in a concise report. Describe effects of shrinkage if appropriate. If your model includes interactions of predictors, be careful how you interpret lower-order effects. Finally, if you are using a ROPE for decisions, justify its limits.

25.1.2. Optional points

The following points are not necessarily crucial to address in every report, but should be considered. Whether or not to include these points depends on the particulars of the application domain, the points you want to make, and the audience to which the report is being addressed.

- Robustness of the posterior for different priors: When there is contention about the prior, it can be most convincing simply to conduct the analysis with different priors and demonstrate that the essential conclusions from the posterior do not change. This may be especially important when doing model comparisons, such as when using inclusion coefficients (or using Bayes factors to assess null values). Which priors should be used? Those that are meaningful and amenable to your audience, such as reviewers and editors of the journal to which the report is submitted.
- Posterior predictive check: By generating simulated data from the credible parameter values of the model, and examining the qualities of the simulated data, the veracity of the model can be further bolstered, if the simulated data do resemble the actual data. On the other hand, if the simulated data are discrepant from the actual data in systematic and interpretable ways, then the posterior predictive check can inspire new research and new models. For a perspective on posterior predictive checks, see the article by Kruschke (2013b) and Section 17.5.1 (among others) of this book.
- Power analysis: If there is only a weak effect in your results, what sample size would be needed to achieve some desired precision in the estimate? If you found a credibly nonzero difference, what was the retrospective power of your experiment and what is its replication power? This sort of information can be useful not only for the researcher's own planning, but it can also be useful to the audience of the report to anticipate potential follow-up research and to assess the robustness of the currently reported results. Chapter 13 described power analysis in depth.

25.1.3. Helpful points

Finally, to help science be cumulative, make your results available on the web:

- Post the raw data: There are at least two benefits of posting the original data. One benefit is that subsequent researchers can analyze the data with different models. New insights can be gained by alternative modeling interpretations. The longevity of the original research is enhanced. A second benefit is that if an exact or near-exact replication is conducted, the original data set can be concatenated with the new data set, to enhance sensitivity of the new data. Either way, your work gets cited!
- Post the MCMC sample of the posterior: There are at least two benefits of making the posterior publicly available. One is that other researchers can explore the posterior for effects and comparisons that were not in the report. Complex models have many parameters, and no single report can cover every possible perspective on the posterior

distribution. The longevity and impact of the research is thereby enhanced. A second benefit is that if subsequent researchers do follow-up research with a similar design and model, then the posted posterior can inform the prior of the subsequent analysis. Because the full posterior automatically incorporates all the covariations between all the parameters, the full posterior can be more useful than summaries of marginal distributions in a report. Either way, your work gets cited!

25.2. FUNCTIONS FOR COMPUTING HIGHEST DENSITY INTERVALS

HDIs have been used routinely throughout the book to describe the widths of distributions. Recall Figure 4.5 (p. 88) for examples, and review Figure 12.2 (p. 342) for an explanation of how HDIs differ from equal-tailed intervals. The present section provides details regarding how the HDIs are computed. The algorithm for computing an HDI on a grid approximation applies to any dimensionality and any shape distribution. The algorithms for computing an HDI of an MCMC sample or for a mathematical function apply only to single parameters with unimodal distributions. The R functions are defined in the file `DBDA2E-utilities.R`.

25.2.1. R code for computing HDI of a grid approximation

We can imagine the grid approximation of a distribution as a landscape of poles sticking up from each point on the parameter grid, with the height of each pole indicating the probability mass at that discrete point. We can imagine the highest density region by visualizing a rising tide: We gradually flood the landscape, monitoring the total mass of the poles that protrude above water, stopping the flood when 95% (say) of the mass remains protruding. The waterline at that moment defines the highest density region (e.g., Hyndman, 1996).

The function, listed below, finds the approximate highest density region in a somewhat analogous way. It uses one extra trick at the beginning, however. It first sorts all the poles in order of height, from tallest to shortest. The idea is to move down the sorted queue of poles until the cumulative probability has just barely exceeded 95% (or whatever mass is desired). The resulting height is the "waterline" that defines all points inside the highest density. See the comments in the top of the code for details of how to use the function.

```
HDIofGrid = function( probMassVec , credMass=0.95 ) {
  # Arguments:
  #   probMassVec is a vector of probability masses at each grid point.
  #   credMass is the desired mass of the HDI region.
  # Return value:
  #   A list with components:
  #   indices is a vector of indices that are in the HDI
  #   mass is the total mass of the included indices
```

```
  #   height is the smallest component probability mass in the HDI
  # Example of use: For determining HDI of a beta(30,12) distribution
  #   approximated on a grid:
  #   > probDensityVec = dbeta( seq(0,1,length=201) , 30 , 12 )
  #   > probMassVec = probDensityVec / sum( probDensityVec )
  #   > HDIinfo = HDIofGrid( probMassVec )
  #   > show( HDIinfo )
  sortedProbMass = sort( probMassVec , decreasing=TRUE )
  HDIheightIdx = min( which( cumsum( sortedProbMass ) >= credMass ) )
  HDIheight = sortedProbMass[ HDIheightIdx ]
  HDImass = sum( probMassVec[ probMassVec >= HDIheight ] )
  return( list( indices = which( probMassVec >= HDIheight ) ,
                mass = HDImass , height = HDIheight ) )
}
```

25.2.2. HDI of unimodal distribution is shortest interval

The algorithms for computing the HDI of an MCMC sample or of a mathematical function rely on a crucial property: For a unimodal probability distribution on a single variable, the HDI of mass M is the narrowest possible interval of that mass. Figure 25.1 illustrates why this is true. Consider the 90% HDI as shown. We construct another interval of 90% mass by moving the limits of the HDI to right, such that each limit is

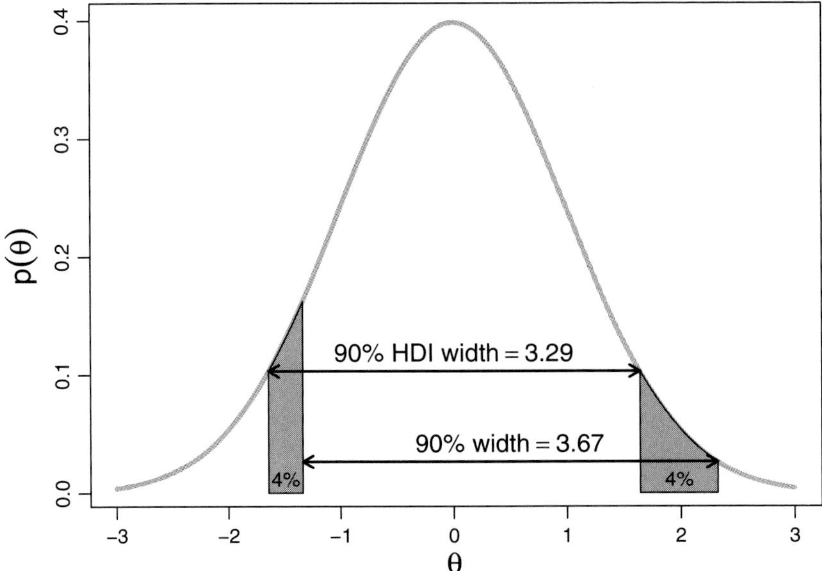

Figure 25.1 For a unimodal distribution, the HDI is the narrowest interval of that mass. This figure shows the 90% HDI and another interval that has 90% mass.

moved to a point that covers 4%, as marked in grey in Figure 25.1. The new interval must also cover 90%, because the 4% lost on the left is replaced by the 4% gained on the right.

Consider the grey regions in Figure 25.1. Their left edges have the same height, because the left edges are defined by the HDI. Their areas are the same, because, by definition, the areas are both 4%. Notice, however, that the left grey area is narrower than the right grey area, because the left area falls at a point where the distribution is increasing, but the right area falls at a point where the distribution is decreasing. Consequently, the distance between right edges of the two grey zones must be greater than the HDI width. (The exact widths are marked in Figure 25.1.) This argument applies for any size of grey zone, going right or left of the HDI, and for any mass HDI. The argument relies on unimodality, however.

Given the argument and diagram in Figure 25.1, it is not too hard to believe the converse: For a unimodal distribution on one variable, for any mass M, the interval containing mass M that has the narrowest width is the HDI for that mass. The algorithms described below are based on this property of HDIs. The algorithms find the HDI by searching among candidate intervals of mass M. The shortest one found is declared to be the HDI. It is an approximation, of course. See Chen and Shao (1999) for more details, and Chen, He, Shao, and Xu (2003) for dealing with the unusual situation of multimodal distributions.

25.2.3. R code for computing HDI of a MCMC sample

It is important to remember that an MCMC sample is only a random and noisy representation of the underlying distribution, and therefore the HDI found from the MCMC sample is also noisy. For details, review the discussion of HDI approximation error that accompanies Figures 7.13 (p. 185) and 7.14 (p. 186). Below is the code for finding the HDI of an MCMC sample. It is brief and hopefully self-explanatory after the discussion of the previous section.

```
HDIofMCMC = function( sampleVec , credMass=0.95 ) {
  # Computes highest density interval from a sample of representative values,
  #   estimated as shortest credible interval.
  # Arguments:
  #   sampleVec
  #     is a vector of representative values from a probability distribution.
  #   credMass
  #     is a scalar between 0 and 1, indicating the mass within the credible
  #     interval that is to be estimated.
  # Value:
  #   HDIlim is a vector containing the limits of the HDI
  sortedPts = sort( sampleVec )
  ciIdxInc = ceiling( credMass * length( sortedPts ) )
```

```
  nCIs = length( sortedPts ) - ciIdxInc
  ciWidth = rep( 0 , nCIs )
  for ( i in 1:nCIs ) {
    ciWidth[ i ] = sortedPts[ i + ciIdxInc ] - sortedPts[ i ]
  }
  HDImin = sortedPts[ which.min( ciWidth ) ]
  HDImax = sortedPts[ which.min( ciWidth ) + ciIdxInc ]
  HDIlim = c( HDImin , HDImax )
  return( HDIlim )
}
```

25.2.4. R code for computing HDI of a function

The function described in this section finds the HDI of a unimodal probability density function that is specified mathematically in R. For example, the function can find HDI's of normal densities or of beta densities or of gamma densities, because those densities are specified as functions in R.

What the program accomplishes is just a search of HDI's, converging to the shortest one, but it does this by using some commands and R functions that may not have been used elsewhere in the book. One function that the program uses is the inverse cumulative density function (ICDF) for whatever probability distribution is being targeted. We have seen one case of an ICDF previously, namely the probit function, which is the inverse of the cumulative-density function for the normal distribution. In R, the ICDF of the normal is the `qnorm(x)` function, where the argument `x` is a value between zero and one. The program for finding the HDI takes, as one of its arguments, R's name for the ICDF of the function. For example, if we want to find an HDI of a normal density, we pass in `ICDFname="qnorm"`.

The crucial function called by the program is R's `optimize` routine. The `optimize` routine searches and finds the minimum of a specified function over a specified domain. In the program below, we define a function called `intervalWidth` that returns the width of the interval that starts at `lowTailPr` and has 95% mass. This `intervalWidth` function is repeatedly called from the `optimize` routine until it converges to a minimum.

```
HDIofICDF = function( ICDFname , credMass=0.95 , tol=1e-8 , ... ) {
  # Arguments:
  #   ICDFname is R's name for the inverse cumulative density function
  #     of the distribution.
  #   credMass is the desired mass of the HDI region.
  #   tol is passed to R's optimize function.
  # Return value:
  #   Highest density interval (HDI) limits in a vector.
  # Example of use: For determining HDI of a beta(30,12) distribution, type
  #   > HDIofICDF( qbeta , shape1 = 30 , shape2 = 12 )
```

```
  #   Notice that the parameters of the ICDFname must be explicitly named;
  #   e.g., HDIofICDF( qbeta , 30 , 12 ) does not work.
  # Adapted and corrected from Greg Snow's TeachingDemos package.
  incredMass =  1.0 - credMass
  intervalWidth = function( lowTailPr , ICDFname , credMass , ... ) {
    ICDFname( credMass + lowTailPr , ... ) - ICDFname( lowTailPr , ... )
  }
  optInfo = optimize( intervalWidth , c( 0 , incredMass ) , ICDFname=ICDFname ,
                      credMass=credMass , tol=tol , ... )
  HDIlowTailPr = optInfo$minimum
  return( c( ICDFname( HDIlowTailPr , ... ) ,
             ICDFname( credMass + HDIlowTailPr , ... ) ) )
}
```

25.3. REPARAMETERIZATION

There are situations in which one parameterization is intuitive to express a distribution, but a different parameterization is required for mathematical convenience. For example, we may think intuitively of the standard deviation of a normal distribution, but have to parameterize the distribution in terms of the precision (i.e., reciprocal of the variance). The question is, if we express a probability distribution on one scale, what is the equivalent distribution on a transformed scale? The answer is not difficult to figure out, especially for single parameters.

Let the "destination" parameter be denoted θ, and suppose that $\theta = f(\phi)$ for the "source" parameter ϕ, with a monotonic and differentiable function f. Let the probability distribution on ϕ be denoted $p(\phi)$. Then the corresponding probability distribution on θ is

$$p(\theta) = \frac{p(f^{-1}(\theta))}{|f'(f^{-1}(\theta))|} \tag{25.1}$$

where $f'(\phi)$ is the derivative of f with respect to ϕ.

Here's why. Consider a small (actually infinitesimal) interval under the distribution $p(\phi)$, at a particular value ϕ. Call the width of the interval $d\phi$. The probability mass in that interval is the product of the density and the width: $p(\phi) \cdot d\phi$. We want to construct a probability density on θ, which we denote $p(\theta) = p\left(f(\phi)\right)$, that has the same probability mass in the corresponding interval at $\theta = f(\phi)$. The width of the corresponding interval on θ is $d\theta = d\phi \cdot |f'(\phi)|$ because, by definition of the derivative, $f'(\phi) = d\theta/d\phi$. So, the probability mass in that interval is $p(\theta)\cdot d\theta = p\left(f(\phi)\right)\cdot d\phi\cdot|f'(\phi)|$. Therefore, to equate the probability masses in the corresponding intervals, we require that $p\left(f(\phi)\right) \cdot d\phi \cdot |f'(\phi)| = p(\phi) \cdot d\phi$, which, when rearranged, yields $p\left(f(\phi)\right) = p(\phi) \big/ |f'(\phi)|$, which is Equation 25.1.

25.3.1. Examples

An example was given in Figure 21.11 (p. 640), which had a normal distribution on parameter β_0 transformed to a distribution over parameter μ via a logistic function. The question was, What is the implied distribution on μ? Footnote 2 on p. 639 explained the application of Equation 25.1.

As another example, we can apply Equation 25.1 to the case in which $\theta = f(\phi) = \text{logistic}(\phi) = 1/[1 + \exp(-\phi)]$, and the distribution on ϕ is given by $p(\phi|a, b) = \left(f(\phi)\right)^a \left(1 - f(\phi)\right)^b / B(a, b)$. An example of this distribution, for $a = 1$ and $b = 1$, is shown in the top-left panel of Figure 25.2. Notice that the derivative of the logistic function f is $f'(\phi) = \exp(-\phi)/[1 + \exp(-\phi)]^2 = f(\phi)\,(1 - f(\phi)) = \theta(1 - \theta)$. Therefore, according to Equation 25.1, the equivalent probability density at $\theta = f(\phi)$ is $p(\theta) = p(\phi)/f'(\phi) = \theta^a\,(1 - \theta)^b/[\theta(1 - \theta)B(\theta|a, b)] = \theta^{(a-1)}\,(1 - \theta)^{(b-1)}/B(a, b) = \text{beta}(\theta|a, b)$. The upper row of Figure 25.2 shows this situation when $a = b = 1$. An intuitive way to think of this situation is that the probability on ϕ is dense near $\phi = 0$, but that is exactly where the logistic transformation stretches the distribution. On the other hand, the probability on ϕ is sparse at large positive or large negative values, but that is exactly where the logistic transformation compresses the distribution.

As another example, consider a case in which $\theta = f(\psi) = 1 - \exp(-\psi)$, with the probability density $p(\psi) = \exp(-\psi)$. Notice that the derivative of the transformation is $f'(\psi) = \exp(-\psi)$, and therefore the equivalent density at $\theta = f(\phi)$ is $p\left(f(\psi)\right) = p(\psi)/f'(\psi) = 1$. In other words, the equivalent density on θ is the uniform density, as shown in the lower row of Figure 25.2.

As a final example, Figure 25.3 shows a uniform distribution on standard deviation transformed to the corresponding distribution on precision. By definition, precision is the reciprocal of squared standard deviation.

25.3.2. Reparameterization of two parameters

When there is more than one parameter being transformed, the calculus becomes a little more involved. Suppose we have a probability density on a two-parameter space, $p(\alpha_1, \alpha_2)$. Let $\beta_1 = f_1(\alpha_1, \alpha_2)$ and $\beta_2 = f_2(\alpha_1, \alpha_2)$. Our goal is to find the probability density $p(\beta_1, \beta_2)$ that corresponds to $p(\alpha_1, \alpha_2)$. We do this by considering infinitesimal corresponding regions. Consider a point α_1, α_2. The probability mass of a small region near that point is the density at that point times the area of the small region: $p(\alpha_1, \alpha_2)\, d\alpha_1\, d\alpha_2$. The corresponding region in the transformed parameters should have the same mass. That mass is the density at the transformed point times the area of the region mapped to from the originating region. In vector calculus textbooks, you can find discussions demonstrating that the area of the mapped-to region

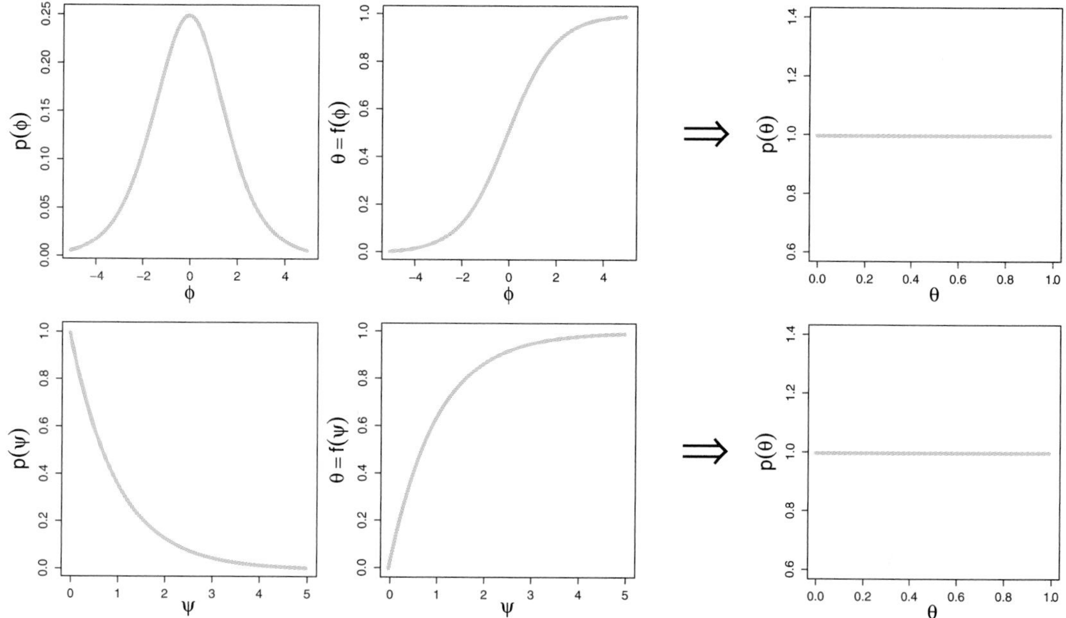

Figure 25.2 *Top row:* A reparameterization that maps a peaked distribution over $\phi \in [-\infty, +\infty]$ to a uniform distribution over $\theta \in [0, 1]$. *Bottom row:* A reparameterization that maps a descending exponential distribution over $\psi \in [0, +\infty]$ to a uniform distribution over $\theta \in [0, 1]$.

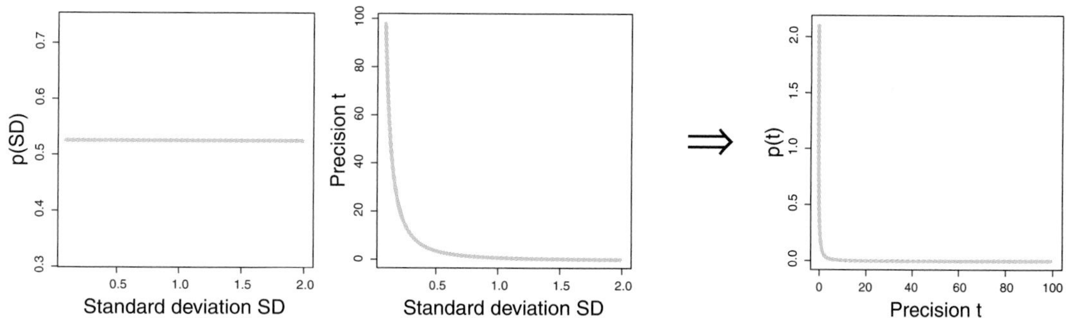

Figure 25.3 A uniform distribution on standard deviation, mapped to the corresponding distribution on precision.

is $|\det(J)|\, d\alpha_1\, d\alpha_2$ where J is the Jacobian matrix: $J_{r,c} = df_r(\alpha_1, \alpha_2)/d\alpha_c$ and $\det(J)$ is the determinant of the Jacobian matrix. Setting the two masses equal and rearranging yields $p(\beta_1, \beta_2) = p(\alpha_1, \alpha_2)/|\det(J)|$. You can see that Equation 25.1 is a special case. As we have had no occasions to apply this transformation, no examples will be provided here.

But the method has been mentioned for those intrepid souls who may wish to venture into the wilderness of multivariate probability distributions with nothing more than pen and paper.

25.4. CENSORED DATA IN JAGS

In many situations some data are censored, which means that their values are known only within a certain range. For example, in response-time experiments, the experimenter is usually willing to wait for a response for only a limited duration such as 5 s, beyond which the measurement trial is terminated. If the responder was "on task" but did not respond before the 5 s limit, the datum is still informative, because it means the task took longer than 5 s. Censored data should not be discarded because the remaining data are biased to be below the censoring cut-off in this case. An analogous situation occurs when measuring life span after a treatment: Often the researcher cannot wait for all subjects to expire, and the subjects still surviving after the limited measurement duration contribute censored data. Data can also be censored on the low side, for example when a measuring device has a lower limit. Sometimes data can be interval censored, which means that there are some intervals in which we do not know the exact value, only the bounds on the interval. A contrived example of interval censoring is when we measure the distance that a putted miniature-golf ball rolls across the carpet, with its path partially covered by a tunnel. If the ball stops within the tunnel, all we know is that its distance is somewhere between the entrance and exit of the tunnel.

In all the examples of this section, we are assuming that the censoring is independent of the variable being measured. For example, we assume that the experimenter's 5 s time limit does not affect the responder's response time, and the responder's time does not affect the experimenter's limit. Similarly, we assume that tunnel does not affect the distance travelled by the ball, and we assume that the distance traveled by the ball does not affect the position of the tunnel.

To illustrate why it is important to include censored data in the analysis, consider a case in which $N = 500$ values are generated randomly from a normal distribution with $\mu = 100$ and $\sigma = 15$. Suppose that values above 106 are censored, as are values in the interval between 94 and 100. For the censored values, all we know is the interval in which they occurred, but not their exact value. Figure 25.4 shows results from two analyses of the data. The upper quartet of panels in Figure 25.4 shows results when the censored data are excluded from the analysis. There are only $N = 255$ uncensored values, and they have a mean far less than 100. The posterior estimate of μ and σ are both less than the underlying true values. The lower quartet of Figure 25.4 shows results when the censored data are retained in the analysis. The posterior estimate of μ and σ is very accurate in this case, despite the uncertainty in the censored data values.

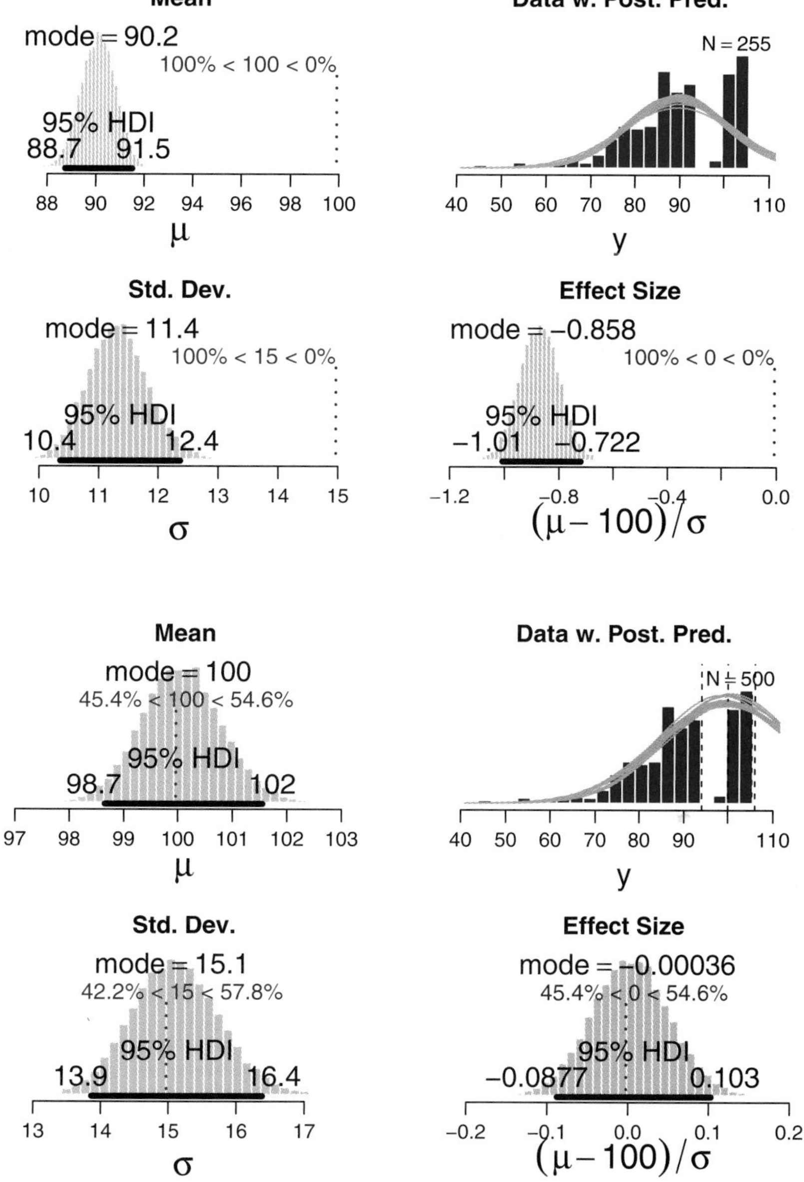

Figure 25.4 Data randomly generated from a normal distribution with $\mu = 100$ and $\sigma = 15$, then censored between 94 and 100 and above 106. *Upper quartet:* Censored data omitted from analysis; parameter estimates are too small. *Lower quartet:* Censored data imputed in known bins; parameter estimates are accurate.

The remainder of this section explains how the analysis of censored data is implemented in JAGS. The first crucial feature is the coding of the data. The censored values need to indicate not merely that they were censored, but also which censoring interval they came from, and the thresholds of the intervals. To save space, the intervals are here called "bins." Some lines from the data file for the example in Figure 25.4 follow:

```
     y   ybin  thresh1  thresh2  thresh3   #   yOrig
 79.36    0       94.0    100.0    106.0   #   79.36
 82.97    0       94.0    100.0    106.0   #   82.97
    NA    1       94.0    100.0    106.0   #   95.61
    NA    1       94.0    100.0    106.0   #   99.87
103.28    2       94.0    100.0    106.0   #  103.28
104.40    2       94.0    100.0    106.0   #  104.40
    NA    3       94.0    100.0    106.0   #  125.74
    NA    3       94.0    100.0    106.0   #  130.60
```

The column named y contains the main data. Notice that censored values are coded as NA. Each row of the data file also specifies the bin in which the y value falls, in a column titled ybin. Notice that the bin numbering for JAGS begins at 0, not at 1. The thresholds that define the bin limits are specified in the next columns. In this example there are three thresholds, at 94.0, 100.0, and 106.0. Values of y less than the first threshold have ybin=0. Values of y greater than the first threshold but less than the second threshold have ybin=1. And so on. In principle, the thresholds could be different for each datum (as long as they are independent of the data), and therefore the threshold values are repeated on every row of the data file. For conceptual clarity, the original random values from the normal distribution are included in the column labeled yOrig, but no real data file would have such a column, of course.

There are two aspects to understanding how JAGS models censored data. First, JAGS simultaneously describes both y and ybin. In this sense JAGS uses a multivariate model, predicting two variables from its underlying parameters. The key lines in the model specification look like this:

```
y[i] ~ dnorm( mu , 1/sigma^2 )
ybin[i] ~ dinterval( y[i] , threshMat[i, ] )
```

The first line above says that the y[i] value comes from a normal distribution with mean mu and standard deviation sigma. The second line above says that the ybin[i] value comes from an "interval" distribution, for which the first argument is the value y[i] and the second argument is a vector of the threshold values, here denoted threshMat[i,]. If you think of the interval distribution as a generator of ybin values, then it produces ybin=0 when y is less than the lowest threshold in threshMat[i,], it produces ybin=1 when y is between the lowest threshold and next lowest threshold in threshMat[i,], and so on. If you think of the interval distribution as a probability distribution, then it has probability 1.0 whenever the value of ybin is appropriate for the value of y relative to the values of the thresholds in threshMat[i,], and the interval distribution has probability 0.0 whenever the value of ybin is not appropriate for the value of y relative to the values of the thresholds in threshMat[i,]. If every y[i] value were present (i.e., uncensored) then the line that specifies the interval distribution would have no effect on the model because its probability would always be 1.0.

The second aspect to understanding how JAGS models censored data is the fact that when JAGS encounters a missing data value, JAGS automatically imputes a random value generated from the model and the credible parameter values at that step in the MCMC chain. In other words, JAGS treats missing data values as if they were parameters, and the whole model (informed by the other data) acts as a prior distribution for generating the missing parameter. If JAGS knew only that a missing data value y[i] came from a normal distribution, then JAGS would impute a random value sampled from the normal distribution. But JAGS knows also that the missing data value y[i] must satisfy the interval distribution, and therefore the imputed y[i] value must also come from the appropriate bin.

The full model specification is otherwise familiar:

```
model {
  for ( i in 1:Ntotal ) {
    y[i] ~ dnorm( mu , 1/sigma^2 )
    ybin[i] ~ dinterval( y[i] , threshMat[i, ] )
  }
  mu ~ dnorm( meanY , 1/(100*sdY)^2 )
  sigma ~ dunif( sdY/1000 , sdY*1000 )
}
```

The only other essential for implementing the model in JAGS is initializing the chains for the missing values of y. While JAGS can initialize many parameters automatically, the imputed data in this model structure need to be initialized explicitly. The data vector, y, has both known and imputed values, and we only need to initialize the imputed components, leaving the known values at the values supplied in the data file. There is a complementarity between the data vector y, which has NA at imputed components, and the initial value for the chain y, which has NA at known components. A similar approach was taken to initializing thresholds in the thresholded cumulative-normal model of Section 23.2.1 (p. 676). The following R commands set up initial values for the y chain, denoted yInit:

```
yInit = rep( NA , length(y) )                   # Define placeholder.
for ( i in 1:length(y) ) {                      # For each datum,
  if ( is.na(y[i]) ) {                          # if y is censored, then:
```

```
    if ( ybin[i]==0 ) {                          # if it's from the lowest bin
      yInit[i] = threshMat[i,1]-1                # initialize at below low thresh;
    } else if ( ybin[i]==ncol(threshMat) ) {     # if it's from the highest bin
      yInit[i] = threshMat[i,ncol(threshMat)]+1  # initialize at above high thresh;
    } else {                                     # else initialize at middle of bin.
      yInit[i] = (threshMat[i,ybin[i]]+threshMat[i,ybin[i]+1])/2
    }
  }
}
initsList = list( y=yInit )  # Assemble into the list later sent to JAGS.
```

The model and its supporting functions can be found in the file named `Jags-YmetBinned-Xnom1grp-MnormalInterval.R`, and the high-level script that calls the functions is `Jags-YmetBinned-Xnom1grp-Mnormal Interval-Example.R`.

25.5. WHAT NEXT?

If you have made it this far and you are looking for more, you might peruse posts at my blog, http://doingbayesiandataanalysis.blogspot.com/, and search there for topics that interest you. When you are ready for more applications of frequently-used statistical models, consider the book by Ntzoufras (2009), which provides many examples in R and WinBUGS. The WinBUGS examples can be easily converted to JAGS. The book by Gelman et al. (2013) uses Stan and covers many advanced topics including nonparametric models such as Gaussian process and Dirichlet process models. Finally, the book by Lee and Wagenmakers (2014) examines applications of WinBUGS to models of cognitive processes such as memory, categorization, and decision making.

BIBLIOGRAPHY

Abelson, R. P. (1997). On the surprising longevity of flogged horses: Why there is a case for the significance test. *Psychological Science*, *8*(1), 12-15.

Adcock, C. J. (1997). Sample size determination: A review. *The Statistician*, *46*, 261-283.

Agresti, A., & Hitchcock, D. B. (2005). Bayesian inference for categorical data analysis. *Statistical Methods & Applications*, *14*(3), 297-330.

Albert, J. H., & Rossman, A. J. (2001). *Workshop statistics: Discovery with data, a Bayesian approach*. Emeryville, CA: Key College Publishing.

Andrews, M. (2011). Doing Bayesian data analysis: A tutorial using R and BUGS. *British Journal of Mathematical and Statistical Psychology*, *64*, 538-540. http://dx.doi.org/10.1111/j.2044-8317.2011.02027.x (Book review).

Anscombe, F. J. (1954). Fixed sample size analysis of sequential observations. *Biometrics*, *10*, 89-100.

Barbieri, M. M., & Berger, J. O. (2004). Optimal predictive model selection. *The Annals of Statistics*, *32*(3), 870-897.

Barry, J. (2011). Doing Bayesian data analysis: A tutorial with R and BUGS. *Europe's Journal of Psychology*, *7*, 778-779 (Book review).

Bayes, T., & Price, R. (1763). An essay towards solving a problem in the doctrine of chances. By the late Rev. Mr. Bayes, F.R.S. Communicated by Mr. Price, in a letter to John Canton, A.M.F.R.S. *Philosophical Transactions*, *53*, 370-418. http://dx.doi.org/10.1098/rstl.1763.0053

Berger, J. O. (1985). *Statistical decision theory and Bayesian analysis* (2nd ed.). New York: Springer.

Berger, J. O., & Berry, D. A. (1988). Statistical analysis and the illusion of objectivity. *American Scientist*, *76*(2), 159-165.

Berger, R. L., Boos, D. D., & Guess, F. M. (1988). Tests and confidence sets for comparing two mean residual life functions. *Biometrics*, *44*(1), 103-115.

Berger, J. O., & Pericchi, L. R. (1996). The intrinsic Bayes factor for model selection and prediction. *Journal of the American Statistical Association*, *91*(433), 109-122.

Berger, J. O., & Pericchi, L. R. (2001). Objective Bayesian methods for model selection: Introduction and comparison. In *IMS Lecture notes—Monograph series, Model selection*, (Vol. 38, pp. 135-207). Beachwood, Ohio, USA: Institute of Mathematical Statistics.

Berry, D. A. (2006). Bayesian clinical trials. *Nature Reviews Drug Discovery*, *5*, 27-36.

Berry, D. A. (2011). Adaptive clinical trials: The promise and the caution. *Journal of Clinical Oncology*, *29*(6), 606-609.

Berry, D. A., & Hochberg, Y. (1999). Bayesian perspectives on multiple comparisons. *Journal of Statistical Planning and Inference*, *82*(1-2), 215-227.

Berry, S. M., Carlin, B. P., Lee, J. J., & Müller, P. (2011). *Bayesian adaptive methods for clinical trials*. Boca Raton, FL: CRC Press.

Bishop, C. M. (2006). *Pattern recognition and machine learning*. New York: Springer.

Blanchflower, D. G., & Oswald, A. J. (2004). Money, sex and happiness: An empirical study. *The Scandinavian Journal of Economics*, *106*(3), 393-415.

Bliss, C. I. (1934). The method of probits. *Science*, *79*(2037), 38-39.

Bolstad, W. M. (2009). *Understanding computational Bayesian statistics*. Hoboken, NJ: Wiley.

Box, G., Hunter, W., & Hunter, S. (2005). *Statistics for experimenters: Design, innovation, and discovery* (2nd ed.). New York: Wiley-Interscience.

Brainard, J., & Burmaster, D. E. (1992). Bivariate distributions for height and weight of men and women in the United States. *Risk Analysis*, *12*(2), 267-275.

Brambor, T., Clark, W. R., & Golder, M. (2006). Understanding interaction models: Improving empirical analyses. *Political Analysis*, *14*, 63-82.

Braumoeller, B. F. (2004). Hypothesis testing and multiplicative interaction terms. *International Organization*, *58*(04), 807-820.

Brehmer, B. (1974). Hypotheses about relations between scaled variables in the learning of probabilistic inference tasks. *Organizational Behavior and Human Performance, 11*, 1-27.

Brooks, S. P., & Gelman, A. (1998). Alternative methods for monitoring convergence of iterative simulations. *Journal of Computational and Graphical Statistics*, 7, 434-455.

Carifio, J., & Perla, R. (2007). Ten common misunderstandings, misconceptions, persistent myths and urban legends about Likert scales and Likert response formats and their antidotes. *Journal of Social Sciences, 3*, 106-116.

Carifio, J., & Perla, R. (2008). Resolving the 50-year debate around using and misusing Likert scales. *Medical Education, 42*, 1150-1152.

Carlin, B. P., & Chib, S. (1995). Bayesian model choice via Markov chain Monte Carlo methods. *Journal of the Royal Statistical Society: Series B, 57*(3), 473-484.

Carlin, B. P., & Louis, T. A. (2009). *Bayesian methods for data analysis* (3rd ed.). Boca Raton, FL: CRC Press.

Cavagnaro, D. R., Myung, J. I., Pitt, M. A., & Kujala, J. V. (2010). Adaptive design optimization: A mutual information-based approach to model discrimination in cognitive science. *Neural computation, 22*(4), 887-905.

Chaloner, K., & Verdinelli, I. (1995). Bayesian experimental design: A review. *Statistical Science, 10*(3), 273-304.

Chen, M.-H., He, X., Shao, Q.-M., & Xu, H. (2003). A Monte Carlo gap test in computing HPD regions. In H. Zhang & J. Huang (Eds.), *Development of modern statistics and related topics* (pp. 38-52). Singapore: World Scientific.

Chen, M.-H., & Shao, Q.-M. (1999). Monte Carlo estimation of Bayesian credible and HPD intervals. *Journal of Computational and Graphical Statistics, 8*, 69-92.

Chib, S., & Greenberg, E. (1995). Understanding the Metropolis-Hastings algorithm. *The American Statistician, 49*(4), 327-335.

Cohen, J. (1988). *Statistical power analysis for the behavioral sciences* (2nd ed.). Hillsdale, NJ: Erlbaum.

Cohen, J. (1994). The world is round ($p < .05$). *American Psychologist, 49*, 997-1003.

Colvin, K. F. (2013). Doing Bayesian data analysis: A tutorial with R and BUGS. *Journal of Educational Measurement, 50*, 469-471. http://dx.doi.org/10.1111/jedm.12029 (Book review).

Congdon, P. (2005). *Bayesian models for categorical data.* West Sussex, England: Wiley.

Cumming, G. (2012). *Understanding the new statistics: Effect sizes, confidence intervals, and meta-analysis.* New York: Routledge.

Cumming, G., & Fidler, F. (2009). Confidence intervals: Better answers to better questions. *Zeitschrift für Psychologie/Journal of Psychology, 217*(1), 15-26.

Dale, A. I. (1999). *A history of inverse probability: From Thomas Bayes to Karl Pearson* (2nd ed.). New York: Springer.

Damgaard, L. H. (2007). Technical note: How to use WinBUGS to draw inferences in animal models. *Journal of Animal Science, 85*, 1363-1368.

De Santis, F. (2004). Statistical evidence and sample size determination for Bayesian hypothesis testing. *Journal of Statistical Planning and Inference, 124*, 121-144.

De Santis, F. (2007). Using historical data for Bayesian sample size determination. *Journal of the Royal Statistical Society: Series A, 170*, 95-113.

Debreu, G. (1960). Review of R. D. Luce, individual choice behavior: A theoretical analysis. *American Economic Review, 50*, 186-188.

DeGroot, M. H. (2004). *Optimal statistical decisions.* New York: Wiley-Interscience.

Dellaportas, P., Forster, J. J., & Ntzoufras, I. (2002). On Bayesian model and variable selection using MCMC. *Statistics and Computing, 12*(1), 27-36.

Deng, L.-Y., & Lin, D. K. (2000). Random number generation for the new century. *The American Statistician, 54*(2), 145-150.

Denwood, M. J. (2013). runjags: An R package providing interface utilities, parallel computing methods and additional distributions for MCMC models in JAGS. *Journal of Statistical Software.* Retrieved from http://cran.r-project.org/web/packages/runjags/ (in review)

Diaconis, P., Holmes, S., & Montgomery, R. (2007). Dynamical bias in the coin toss. *SIAM Review, 49*(2), 211-235.

Dienes, Z. (2008). *Understanding psychology as a science: An introduction to scientific and statistical inference.* Hampshire, UK: Palgrave Macmillan.

Dienes, Z. (2011). Bayesian versus orthodox statistics: Which side are you on? *Perspectives on Psychological Science, 6*(3), 274-290.

Ding, C. (2011). Incorporating our own knowledge in data analysis: Is this the right time? *PsycCRITIQUES, 56* (Book review). http://dx.doi.org/10.1037/a0024579.

Doyle, A. C. (1890). *The sign of four.* London: Spencer Blackett.

Duane, S., Kennedy, A. D., Pendleton, B. J., & Roweth, D. (1987). Hybrid Monte Carlo. *Physics Letters B, 195*(2), 216-222.

Dunnett, C. W., & Gent, M. (1977). Significance testing to establish equivalence between treatments, with special reference to data in the form of 2 × 2 tables. *Biometrics, 33*, 593-602.

Eckhardt, R. (1987). Stan Ulam, John von Neumann, and the Monte Carlo method. *Los Alamos Science, 15*, 131-137.

Edwards, W., Lindman, H., & Savage, L. J. (1963). Bayesian statistical inference for psychological research. *Psychological Review, 70*, 193-242.

Ehrmann, M. (1995). *The desiderata of happiness: A collection of philosophical poems.* New York: Crown.

Fellingham, G. W. (2012). Doing Bayesian data analysis: A tutorial with R and BUGS. *The American Statistician, 66*(2), 139-140 (Book review).

Ferrer-i-Carbonell, A., & Frijters, P. (2004). How important is methodology for the estimates of the determinants of happiness? *The Economic Journal, 114*(497), 641-659.

Fisher, R. A. (1925). *Statistical methods for research workers.* Edinburgh: Oliver and Boyd.

Flegal, J. M., Haran, M., & Jones, G. L. (2008). Markov chain Monte Carlo: Can we trust the third significant figure? *Statistical Science, 23*(2), 250-260.

Freedman, L. S., Lowe, D., & Macaskill, P. (1984). Stopping rules for clinical trials incorporating clinical opinion. *Biometrics, 40*, 575-586.

Friel, N., & Wyse, J. (2012). Estimating the evidence—A review. *Statistica Neerlandica, 66*(3), 288-308.

Gallistel, C. R. (2009). The importance of proving the null. *Psychological Review, 116*(2), 439-453.

Gardner, M. J., & Altman, D. G. (1986). Confidence intervals rather than *p* values: Estimation rather than hypothesis testing. *British Medical Journal, 292*(6522), 746-750.

Gelfand, A. E. (2000). Gibbs sampling. *Journal of the American Statistical Association, 95*(452), 1300-1304.

Gelfand, A. E., & Dey, D. K. (1994). Bayesian model choice: Asymptotics and exact calculations. *Journal of the Royal Statistical Society: Series B, 56*, 501-514.

Gelfand, A. E., & Smith, A. F. M. (1990). Sampling-based approaches to calculating marginal densities. *Journal of the American Statistical Association, 85*(410), 398-409.

Gelman, A. (2005). Analysis of variance—Why it is more important than ever. *The Annals of Statistics, 33*(1), 1-53.

Gelman, A. (2006). Prior distributions for variance parameters in hierarchical models. *Bayesian Analysis, 1*(3), 515-533.

Gelman, A., Carlin, J. B., Stern, H. S., Dunson, D. B., Vehtari, A., & Rubin, D. B. (2013). *Bayesian data analysis* (3rd ed.). Boca Raton, FL: CRC Press.

Gelman, A., Hill, J., & Yajima, M. (2009). *Why we (usually) don't have to worry about multiple comparisons.* Retrieved from http://www.stat.columbia.edu/~gelman/research/unpublished/multiple2.pdf.

Gelman, A., & Nolan, D. (2002). You can load a die, but you can't bias a coin. *The American Statistician, 56*(4), 308-311.

Gelman, A., & Rubin, D. B. (1992). Inference from iterative simulation using multiple sequences. *Statistical Science, 7*, 457-472.

Geman, S., & Geman, D. (1984). Stochastic relaxation, Gibbs distributions, and the Bayesian restoration of images. *IEEE Transactions on Pattern Analysis and Machine Intelligence, 6*, 721-741.

Gentle, J. E. (2003). *Random number generation and Monte Carlo methods* (2nd ed.). New York: Springer.

Gerard, P. D., Smith, D. R., & Weerakkody, G. (1998). Limits of retrospective power analysis. *The Journal of Wildlife Management, 62*(2), 801-807.

Gershman, S. J., & Blei, D. M. (2012). A tutorial on Bayesian nonparametric models. *Journal of Mathematical Psychology*, *56*, 1-12.

Gigerenzer, G., & Hoffrage, U. (1995). How to improve Bayesian reasoning without instruction: Frequency formats. *Psychological Review*, *102*, 684-704.

Gigerenzer, G., Krauss, S., & Vitouch, O. (2004). The null ritual: What you always wanted to know about significance testing but were afraid to ask. In D. Kaplan (Ed.), *The Sage handbook of quantitative methodology for the social sciences* (pp. 391-408). Thousand Oaks, CA: Sage.

Gill, J. (2009). *Bayesian methods: A social and behavioral sciences approach* (2nd ed.). Boca Raton, FL: CRC Press.

Goldstein, D. G. (2011). Doing Bayesian data analysis: A tutorial with R and BUGS. *Journal of Economic Psychology*, *32*(5), 724-725. http://dx.doi.org/10.1016/j.joep.2011.05.010 (Book review).

Gopalan, R., & Berry, D. (1998). Bayesian multiple comparisons using Dirichlet process priors. *Journal of the American Statistical Association*, *93*, 1130-1139.

Gosset, W. S. (1908). The probable error of a mean. *Biometrika*, *6*, 1-25.

Grimmer, J. (2011). An introduction to Bayesian inference via variational approximations. *Political Analysis*, *19*(1), 32-47.

Guber, D. L. (1999). Getting what you pay for: The debate over equity in public school expenditures. *Journal of Statistics Education*, 7(2). Retrieved from http://www.amstat.org/publications/JSE/secure/v7n2/datasets.guber.cfm.

Gubernatis, J. E. (2005). Marshall Rosenbluth and the Metropolis algorithm. *Physics of Plasmas*, *12*, 057303 1-5.

Guenther, W. C. (1965). *Concepts of statistical inference*. New York: McGraw-Hill.

Hahn, U., Chater, N., & Richardson, L. B. (2003). Similarity as transformation. *Cognition*, *87*(1), 1-32.

Han, C., & Carlin, B. P. (2001). Markov chain Monte Carlo methods for computing Bayes factors: A comparative review. *Journal of the American Statistical Association*, *96*(455), 1122-1132.

Hand, D. J., Daly, F., Lunn, A. D., McConway, K. J., & Ostrowski, E. (1994). *A handbook of small data sets*. London: Chapman & Hall.

Hanley, J. A., & Shapiro, S. H. (1994). Sexual activity and the lifespan of male fruitflies: A dataset that gets attention. *Journal of Statistics Education*, *2*(1). http://www.amstat.org/publications/jse/v2n1/datasets.hanley.html.

Hastings, W. K. (1970). Monte Carlo sampling methods using Markov chains and their applications. *Biometrika*, *57*(1), 97-109.

Hays, W. L. (1994). *Statistics* (5th ed.). Fort Worth, TX: Harcourt Brace College Publishers.

Hobbs, B. P., & Carlin, B. P. (2008). Practical Bayesian design and analysis for drug and device clinical trials. *Journal of Biopharmaceutical Statistics*, *18*(1), 54-80.

Hoenig, J. M., & Heisey, D. M. (2001). The abuse of power: The pervasive fallacy of power calculations for data analysis. *The American Statistician*, *55*(1), 19-24.

Hoffman, P. J., Earle, T. C., & Slovic, P. (1981). Multidimensional functional learning (MFL) and some new conceptions of feedback. *Organizational Behavior and Human Performance*, *27*(1), 75-102.

Hoffman, M. D. & Gelman, A. (2014). The no-U-turn sampler: Adaptively setting path lengths in Hamiltonian Monte Carlo. *Journal of Machine Learning Research*, 15, 1593–1623.

Holcomb, J., & Spalsbury, A. (2005). Teaching students to use summary statistics and graphics to clean and analyze data. *Journal of Statistics Education*, *13*(3). Retrieved from http://www.amstat.org/publications/jse/v13n3/datasets.holcomb.html.

Howson, C., & Urbach, P. (2006). *Scientific reasoning: The Bayesian approach* (3rd ed.). Chicago: Open Court.

Hyndman, R. J. (1996). Computing and graphing highest density regions. *The American Statistician*, *50*(2), 120-126.

Jefferys, W. H., & Berger, J. O. (1992). Ockham's razor and Bayesian analysis. *American Scientist*, *80*, 64-72.

Jeffreys, H. (1961). *Theory of probability*. Oxford, UK: Oxford University Press.

Jiang, S., Lu, M., & Sato, H. (2009). *Happiness in the dual society of urban China: Hukou identity, horizontal inequality and heterogeneous reference*. Global COE Hi-Stat Discussion Paper Series. No. 20, Tokyo, Japan: Hitotsubashi University. http://gcoe.ier.hit-u.ac.jp/research/discussion/2008/pdf/gd08-020.pdf

Johnson, W., & Krueger, R. F. (2006). How money buys happiness: Genetic and environmental processes linking finances and life satisfaction. *Journal of Personality and Social Psychology, 90*(4), 680-691.

Jones, M. C., & Faddy, M. J. (2003). A skew extension of the *t*-distribution, with applications. *Journal of the Royal Statistical Society: Series B (Statistical Methodology), 65*(1), 159-174.

Jones, B., & Nachtsheim, C. J. (2009). Split-plot designs: What, why, and how. *Journal of Quality Technology, 41*(4), 340-361.

Joseph, L., Wolfson, D. B., & du Berger, R. (1995a). Sample size calculations for binomial proportions via highest posterior density intervals. *The Statistician, 44*, 143-154.

Joseph, L., Wolfson, D. B., & du Berger, R. (1995b). Some comments on Bayesian sample size determination. *The Statistician, 44*, 167-171.

Kalish, M. L., Griffiths, T. L., & Lewandowsky, S. (2007). Iterated learning: Intergenerational knowledge transmission reveals inductive biases. *Psychonomic Bulletin & Review, 14*(2), 288.

Kass, R. E., Carlin, B. P., Gelman, A., & Neal, R. M. (1998). Markov chain Monte Carlo in practice: A roundtable discussion. *The American Statistician, 52*, 93-100.

Kass, R. E., & Raftery, A. E. (1995). Bayes factors. *Journal of the American Statistical Association, 90*, 773-795.

Kelley, K. (2013). Effect size and sample size planning. In T. D. Little (Ed.), *Oxford handbook of quantitative methods* (Vol. 1: Foundations, pp. 206-222). New York: Oxford University Press.

Kolmogorov, A. N. (1956). *Foundations of the theory of probability*. New York: Chelsea.

Krauss, S., Martignon, L., & Hoffrage, U. (1999). Simplifying Bayesian inference: The general case. In L. Magnani, N. J. Nersessian, & P. Thagard (Eds.), *Model-based reasoning in scientific discovery* (pp. 165-180). New York: Springer.

Kruschke, J. K. (2008). Bayesian approaches to associative learning: From passive to active learning. *Learning & Behavior, 36*(3), 210-226.

Kruschke, J. K. (2009). Highlighting: A canonical experiment. In B. Ross (Ed.), *The psychology of learning and motivation* (Vol. 51, pp. 153-185). Burlington, MA: Academic Press/Elsevier.

Kruschke, J. K. (2010). Bayesian data analysis. *Wiley Interdisciplinary Reviews: Cognitive Science, 1*(5), 658-676. http://dx.doi.org/10.1002/wcs.72

Kruschke, J. K. (2011a). Bayesian assessment of null values via parameter estimation and model comparison. *Perspectives on Psychological Science, 6*(3), 299-312.

Kruschke, J. K. (2011b). *Doing Bayesian data analysis: A tutorial with R and BUGS* (1st ed.). Burlington, MA: Academic Press/Elsevier.

Kruschke, J. K. (2013a). Bayesian estimation supersedes the *t* test. *Journal of Experimental Psychology: General, 142*(2), 573-603. http://dx.doi.org/10.1037/a0029146

Kruschke, J. K. (2013b). Posterior predictive checks can and should be Bayesian: Comment on Gelman and Shalizi, philosophy and the practice of Bayesian statistics. *British Journal of Mathematical and Statistical Psychology, 66*, 45-56. http://dx.doi.org/10.1111/j.2044-8317.2012.02063.x

Kruschke, J. K., Aguinis, H., & Joo, H. (2012). The time has come: Bayesian methods for data analysis in the organizational sciences. *Organizational Research Methods, 15*, 722-752. http://dx.doi.org/10.1177/1094428112457829

Kruschke, J. K., & Movellan, J. R. (1991). Benefits of gain: Speeded learning and minimal hidden layers in back-propagation networks. *IEEE Transactions on Systems, Man and Cybernetics, 21*, 273-280.

Kruschke, J. K., & Vanpaemel, W. (in press). Bayesian estimation in hierarchical models. In J. R. Busemeyer, J. T. Townsend, Z. J. Wang, & A. Eidels (Eds.), *Oxford handbook of computational and mathematical psychology*. Oxford: Oxford University Press.

Kuo, L., & Mallick, B. (1998). Variable selection for regression models. *Sankhyā: The Indian Journal of Statistics: Series B, 60*, 65-81.

Lange, K. L., Little, R. J. A., & Taylor, J. M. G. (1989). Robust statistical modeling using the *t* distribution. *Journal of the American Statistical Association, 84*(408), 881-896.

Lazarus, R. S., & Eriksen, C. W. (1952). Effects of failure stress upon skilled performance. *Journal of Experimental Psychology, 43*(2), 100-105. http://dx.doi.org/10.1037/h0056614

Learner, E. E. (1978). *Specification searches*. New York: Wiley.

Lee, M. D., & Wagenmakers, E.-J. (2014). *Bayesian cognitive modeling: A practical course*. Cambridge, England: Cambridge University Press.

Lee, M. D., & Webb, M. R. (2005). Modeling individual differences in cognition. *Psychonomic Bulletin & Review, 12*(4), 605-621.

Likert, R. (1932). A technique for the measurement of attitudes. *Archives of Psychology, 140*, 1-55.

Lindley, D. V. (1975). The future of statistics: A Bayesian 21st century. *Advances in Applied Probability, 7*, 106-115.

Lindley, D. V. (1997). The choice of sample size. *The Statistician, 46*, 129-138.

Lindley, D. V., & Phillips, L. D. (1976). Inference for a Bernoulli process (a Bayesian view). *The American Statistician, 30*(3), 112-119.

Lindquist, M. A., & Gelman, A. (2009). Correlations and multiple comparisons in functional imaging—A statistical perspective. *Perspectives in Psychological Science, 4*(3), 310-313.

Link, W. A., & Eaton, M. J. (2012). On thinning of chains in MCMC. *Methods in Ecology and Evolution, 3*(1), 112-115.

Liu, C. C., & Aitkin, M. (2008). Bayes factors: Prior sensitivity and model generalizability. *Journal of Mathematical Psychology, 52*, 362-375.

Lodewyckx, T., Kim, W., Lee, M. D., Tuerlinckx, F., Kuppens, P., & Wagenmakers, E.-J. (2011). A tutorial on Bayes factor estimation with the product space method. *Journal of Mathematical Psychology, 55*(5), 331-347.

Luce, R. D. (1959). *Individual choice behavior.* New York: Wiley.

Luce, R. D. (2008). Luce's choice axiom. *Scholarpedia, 3*(12), 8077 (Revision #121550).

Lunn, D., Jackson, C., Best, N., Thomas, A., & Spiegelhalter, D. (2013). *The BUGS book: A practical introduction to Bayesian analysis.* Boca Raton, FL: CRC Press.

Lunn, D., Thomas, A., Best, N., & Spiegelhalter, D. (2000). WinBUGS—A Bayesian modelling framework: Concepts, structure, and extensibility. *Statistics and Computing, 10*(4), 325-337.

Lykou, A., & Ntzoufras, I. (2011). WinBUGS: A tutorial. *WIREs Computational Statistics, 3*, 385-396.

Mace, A. E. (1964). *Sample size determination.* New York: Reinhold.

MacKay, D. J. C. (2003). *Information theory, inference & learning algorithms.* Cambridge, UK: Cambridge University Press.

Maltin, L. (1996). *Leonard Maltin's 1996 movie and video guide.* New York: Penguin Books.

Marin, J.-M., & Robert, C. P. (2007). *Bayesian core: A practical approach to computational Bayesian statistics.* New York: Springer.

Maxwell, S. E., & Delaney, H. D. (2004). *Designing experiments and analyzing data: A model comparison perspective* (2nd ed.). Mahwah, NJ: Erlbaum.

Maxwell, S. E., Kelley, K., & Rausch, J. R. (2008). Sample size planning for statistical power and accuracy in parameter estimation. *Annual Review of Psychology, 59*, 537-563.

McCullagh, P., & Nelder, J. (1989). *Generalized linear models* (2nd ed.). Boca Raton, FL: Chapman and Hall/CRC.

McGrayne, S. B. (2011). *The theory that would not die.* New Haven: Yale University Press.

McIntyre, L. (1994). Using cigarette data for an introduction to multiple regression. *Journal of Statistics Education, 2*(1). Retrieved from http://www.amstat.org/publications/jse/v2n1/datasets.mcintyre.html.

Meehl, P. E. (1967). Theory-testing in psychology and physics: A methodological paradox. *Philosophy of Science, 34*, 103-115.

Meehl, P. E. (1978). Theoretical risks and tabular asterisks: Sir Karl, Sir Ronald, and the slow progress of soft psychology. *Journal of Consulting and Clinical Psychology, 46*(4), 806.

Meehl, P. E. (1997). The problem is epistemology, not statistics: Replace significance tests by confidence intervals and quantify accuracy of risky numerical predictions. In L. L. Harlow, S. A. Mulaik, & J. H. Steiger (Eds.), *What if there were no significance tests* (pp. 395-425). Mahwah, NJ: Erlbaum.

Meng, C. Y. K., & Dempster, A. P. (1987). A Bayesian approach to the multiplicity problem for significance testing with binomial data. *Biometrics, 43*(2), 301-311.

Metropolis, N., Rosenbluth, A. W., Rosenbluth, M. N., Teller, A. H., & Teller, E. (1953). Equations of state calculations by fast computing machines. *Journal of Chemical Physics, 21*, 1087-1091.

Meyer, R., & Yu, J. (2000). BUGS for a Bayesian analysis of stochastic volatility models. *Econometrics Journal, 3*(2), 198-215.

Miller, J. (2009). What is the probability of replicating a statistically significant effect? *Psychonomic Bulletin & Review, 16*(4), 617-640.

Moore, T. L. (2006). Paradoxes in film ratings. *Journal of Statistics Education, 14*(1). Retrieved from www.amstat.org/publications/jse/v14n1/datasets.moore.html

Muller, P., Parmigiani, G., & Rice, K. (2007). FDR and Bayesian multiple comparisons rules. In J. M. Bernardo et al. (Eds.), *Bayesian statistics* (Vol. 8). Oxford, UK: Oxford University Press.

Müller, P., & Mitra, R. (2013). Bayesian nonparametric inference—Why and how. *Bayesian Analysis, 8*(2), 269-302.

Myung, I. J., & Pitt, M. A. (1997). Applying Occam's razor in modeling cognition: A Bayesian approach. *Psychonomic Bulletin and Review, 4*, 79-95.

Nakagawa, S., & Foster, T. M. (2004). The case against retrospective statistical power analyses with an introduction to power analysis. *Acta Ethologica*, 7(2), 103-108.

Neal, R. M. (1994). An improved acceptance procedure for the hybrid Monte Carlo algorithm. *Journal of Computational Physics, 111*(1), 194-203.

Neal, R. M. (2011). Mcmc using Hamiltonian dynamics. In S. Brooks, A. Gelman, G. L. Jones, & X. L. Meng (Eds.), *Handbook of Markov chain Monte Carlo* (pp. 113-162). New York: Chapman & Hall.

Nelder, J. A., & Wedderburn, R. W. M. (1972). Generalized linear models. *Journal of the Royal Statistical Society: Series A (General), 135*(3), 370-384.

Nietzsche, F. (1967). *The will to power* (W. Kaufmann & R. J. Hollingdale, Trans.). New York: Random House.

Nittono, H., Fukushima, M., Yano, A., & Moriya, H. (2012). The power of Kawaii: Viewing cute images promotes a careful behavior and narrows attentional focus. *PLoS One*, 7(9), e46362.

Norman, G. (2010). Likert scales, levels of measurement and the "laws" of statistics. *Advances in Health Science Education, 15*, 625-632.

Ntzoufras, I. (2002). Gibbs variable selection using BUGS. *Journal of Statistical Software*, 7(7), 1-19. Retrieved from http://www.jstatsoft.org/v07/i07.

Ntzoufras, I. (2009). *Bayesian modeling using WinBUGS*. Hoboken, NJ: Wiley.

O'Hagan, A. (1995). Fractional Bayes factors for model comparison. *Journal of the Royal Statistical Society: Series B (Methodological), 57*, 99-138.

O'Hagan, A. (1997). Properties of intrinsic and fractional Bayes factors. *Test, 6*(1), 101-118.

O'Hara, R. B., & Sillanpää, M. J. (2009). A review of Bayesian variable selection methods: What, how and which. *Bayesian Analysis, 4*(1), 85-117.

O'Keefe, D. J. (2007). Post hoc power, observed power, a priori power, retrospective power, prospective power, achieved power: Sorting out appropriate uses of statistical power analyses. *Communication Methods and Measures, 1*(4), 291-299.

Pan, Z., & Kupper, L. L. (1999). Sample size determination for multiple comparison studies treating confidence interval width as random. *Statistics in Medicine, 18*, 1475-1488.

Partridge, L., & Farquhar, M. (1981). Sexual activity reduces lifespan of male fruitflies. *Nature, 294*, 580-582.

Pham-Gia, T., & Turkkan, N. (1992). Sample size determination in Bayesian analysis. *The Statistician, 41*, 389-392.

Plummer, M. (2003). JAGS: A program for analysis of Bayesian graphical models using Gibbs sampling. In *Proceedings of the 3rd international workshop on distributed statistical computing (dsc 2003), Vienna, Austria.* ISSN 1609-395X.

Plummer, M. (2012). *JAGS version 3.3.0 user manual* [Computer software manual]. (October 1, 2012).

Poole, C. (1987). Beyond the confidence interval. *American Journal of Public Health*, 77(2), 195-199.

Pullenayegum, E. M., Guo, Q., & Hopkins, R. B. (2012). Developing critical thinking about reporting of Bayesian analyses. *Journal of Statistics Education, 20*(1). www.amstat.org/publications/jse/v20n1/pullenayegum.pdf.

Qian, S. S., & Shen, Z. (2007). Ecological applications of multilevel analysis of variance. *Ecology, 88*(10), 2489-2495.

R Core Team. (2013). *R: A language and environment for statistical computing* [Computer software manual]. Vienna, Austria: R Foundation for Statistical Computing. Retrieved from http://www.R-project.org/.

Robert, C. P., & Casella, G. (2004). *Monte Carlo statistical methods* (2nd ed.). New York: Springer.

Rogers, J. L., Howard, K. I., & Vessey, J. T. (1993). Using significance tests to evaluate equivalence between two experimental groups. *Psychological Bulletin, 113*(3), 553-565.

Rosa, L., Rosa, E., Sarner, L., & Barrett, S. (1998). A close look at therapeutic touch. *Journal of the American Medical Association, 279*(13), 1005-1010.

Rota, G.-C. (1964). The number of partitions of a set. *The American Mathematical Monthly, 71*(5), 498-504.

Rouder, J. N., Lu, J., Speckman, P., Sun, D., & Jiang, Y. (2005). A hierarchical model for estimating response time distributions. *Psychonomic Bulletin & Review, 12*(2), 195-223.

Rouder, J. N., Morey, R. D., Speckman, P. L., & Province, J. M. (2012). Default Bayes factors for ANOVA designs. *Journal of Mathematical Psychology, 56*, 356-374.

Rouder, J. N., Speckman, P. L., Sun, D., Morey, R. D., & Iverson, G. (2009). Bayesian *t*-tests for accepting and rejecting the null hypothesis. *Psychonomic Bulletin & Review, 16*, 225-237.

Roy, A., Ghosal, S., & Rosenberger, W. F. (2009). Convergence properties of sequential Bayesian D-optimal designs. *Journal of Statistical Planning and Inference, 139*, 425-440.

RStudio. (2013). *RStudio: Integrated development environment for R (version 0.97.551).* Boston, MA. Available from http://www.rstudio.org/.

Sadiku, M. N. O., & Tofighi, M. R. (1999). A tutorial on simulation of queueing models. *International Journal of Electrical Engineering Education, 36*, 102-120.

Schmidt, F. L. (1996). Statistical significance testing and cumulative knowledge in psychology: Implications for training of researchers. *Psychological methods, 1*(2), 115-129.

Schweder, T., & Hjort, N. L. (2002). Confidence and likelihood. *Scandinavian Journal of Statistics, 29*, 309-332.

Scott, J. G., & Berger, J. O. (2006). An exploration of aspects of Bayesian multiple testing. *Journal of Statistical Planning and Inference, 136*(7), 2144-2162.

Serlin, R. C., & Lapsley, D. K. (1985). Rationality in psychological research: The good-enough principle. *American Psychologist, 40*(1), 73-83.

Serlin, R. C., & Lapsley, D. K. (1993). Rational appraisal of psychological research and the good-enough principle. In G. Keren & C. Lewis (Eds.), *Methodological and quantitative issues in the analysis of psychological data* (pp. 199-228). Hillsdale, NJ: Erlbaum.

Shi, L. (2009). *Chinese household income project, 2002.* ICPSR21741-v1. Ann Arbor, MI: Inter-university Consortium for Political and Social Research [distributor], 2009-08-14. http://dx.doi.org/10.3886/ICPSR21741.v1.

Shohat-Ophir, G., Kaun, K. R., Azanchi, R., Mohammed, H., & Heberlein, U. (2012). Sexual deprivation increases ethanol intake in drosophila. *Science, 335*, 1351-1355.

Singh, K., Xie, M., & Strawderman, W. E. (2007). Confidence distribution (CD)—Distribution estimator of a parameter. In R. Liu et al. (Ed.), *Complex datasets and inverse problems* (Vol. 54, pp. 132-150). Beachwood, OH: Institute of Mathematical Statistics.

Smit, H. J., & Rogers, P. J. (2000). Effects of low doses of caffeine on cognitive performance, mood and thirst in low and higher caffeine consumers. *Psychopharmacology, 152*(2), 167-173.

Smithson, M. (2011). Doing Bayesian data analysis: A tutorial with R and BUGS. *Journal of Mathematical Psychology, 55*(5), 397-398. http://dx.doi.org/10.1016/j.jmp.2011.05.002 (Book review).

Snee, R. D. (1974). Graphical display of two-way contingency tables. *The American Statistician, 28*(1), 9-12.

Song, X.-Y., & Lee, S.-Y. (2012). A tutorial on the Bayesian approach for analyzing structural equation models. *Journal of Mathematical Psychology, 56*(3), 135-148.

Spiegelhalter, D. J., Freedman, L. S., & Parmar, M. K. B. (1994). Bayesian approaches to randomized trials. *Journal of the Royal Statistical Society: Series A, 157*, 357-416.

Stan Development Team. (2012). *Stan: A C++ library for probability and sampling, version 1.1.* Retrieved from http://mc-stan.org/citations.html.

Steidl, R. J., Hayes, J. P., & Schauber, E. (1997). Statistical power analysis in wildlife research. *The Journal of Wildlife Management, 61*(2), 270-279.

Stevens, S. S. (1946). On the theory of scales of measurement. *Science, 103*(2684), 677-680.

Sullivan, K. M., & Foster, D. A. (1990). Use of the confidence interval function. *Epidemiology, 1*(1), 39-42.

Sun, S., Pan, W., & Wang, L. L. (2011). Rethinking observed power: Concept, practice and implications. *Methodology*, 7(3), 81-87.

Thomas, L. (1997). Retrospective power analysis. *Conservation Biology, 11*(1), 276-280.

Thomas, A., O'Hara, B., Ligges, U., & Sturtz, S. (2006). Making BUGS open. *R News, 6*(1), 12-17.

Tsionas, E. G. (2002). Bayesian inference in the noncentral Student-*t* model. *Journal of Computational and Graphical Statistics, 11*(1), 208-221.

Vanpaemel, W. (2010). Prior sensitivity in theory testing: An apologia for the Bayes factor. *Journal of Mathematical Psychology, 54*, 491-498.

Vanpaemel, W., & Tuerlinckx, F. (2012). Doing Bayesian data analysis in the classroom: An experience based review of John K. Kruschke's (2011) 'Doing Bayesian data analysis: A tutorial with R and BUGS'. *Journal of Mathematical Psychology, 56*, 64-66.

Wabersich, D., & Vandekerckhove, J. (2014). Extending JAGS: A tutorial on adding custom distributions to JAGS (with a diffusion model example). *Behavior Research Methods, 46*, 15-28. Retrieved from http://www.cidlab.com/supp.php?o=wv13.

Wagenmakers, E.-J. (2007). A practical solution to the pervasive problems of *p* values. *Psychonomic Bulletin & Review, 14*(5), 779-804.

Wagenmakers, E.-J., Lodewyckx, T., Kuriyal, H., & Grasman, R. (2010). Bayesian hypothesis testing for psychologists: A tutorial on the Savage? Dickey method. *Cognitive Psychology, 60*, 158-189.

Walker, L. J., Gustafson, P., & Frimer, J. A. (2007). The application of Bayesian analysis to issues in developmental research. *International Journal of Behavioral Development, 31*(4), 366.

Wang, F., & Gelfand, A. E. (2002). A simulation-based approach to Bayesian sample size determination for performance under a given model and for separating models. *Statistical Science, 17*, 193-208.

Weiss, R. (1997). Bayesian sample size calculations for hypothesis testing. *The Statistician, 46*, 185-191.

Western, B., & Jackman, S. (1994). Bayesian inference for comparative research. *The American Political Science Review, 88*(2), 412-423.

Westlake, W. J. (1976). Symmetrical confidence intervals for bioequivalence trials. *Biometrics, 32*, 741-744.

Westlake, W. J. (1981). Response to bioequivalence testing—A need to rethink. *Biometrics, 37*, 591-593.

Wetzels, R., Grasman, R. P., & Wagenmakers, E.-J. (2012). A default Bayesian hypothesis test for ANOVA designs. *The American Statistician, 66*(2), 104-111.

Wetzels, R., Matzke, D., Lee, M. D., Rouder, J., Iverson, G., & Wagenmakers, E.-J. (2011). Statistical evidence in experimental psychology: An empirical comparison using 855 *t* tests. *Perspectives on Psychological Science, 6*(3), 291-298.

Wickham, H. (2007). Reshaping data with the reshape package. *Journal of Statistical Software, 21*(12), 1-20. http://www.jstatsoft.org/v21/i12.

Winkler, R. L. (1967). The assessment of prior distributions in Bayesian analysis. *American Statistical Association Journal, 62*(319), 776-800.

Yates, F. (1935). Complex experiments, with discussion. *Journal of the Royal Statistical Society: Series B, 2*, 181-223.

Zhang, Z., Lai, K., Lu, Z., & Tong, X. (2013). Bayesian inference and application of robust growth curve models using Student's *t* distribution. *Structural Equation Modeling: A Multidisciplinary Journal, 20*(1), 47-78.

Zhu, M., & Lu, A. Y. (2004). The counter-intuitive non-informative prior for the Bernoulli family. *Journal of Statistics Education, 12*(2). http://www.amstat.org/publications/jse/v12n2/zhu.pdf.

Zyphur, M. J., & Oswald, F. L. (2013). Bayesian estimation and inference: A user's guide. *Journal of Management*. http://dx.doi.org/10.1177/0149206313501200.

INDEX

Note: Page numbers followed by *f* indicate figures and *t* indicate tables.

L

M

Q

R

S

T